TOWARDS RESILIENT SOCIETIES: THE SYNERGY OF RELIGION, EDUCATION, HEALTH, SCIENCE, AND TECHNOLOGY

Resilience and sustainability are essential in navigating today's global challenges. *Towards Resilient Societies: The Synergy of Religion, Education, Health, Science, and Technology* presents innovative interdisciplinary research that explores how diverse fields contribute to building adaptive and inclusive communities. This book highlights the intersections of governance, education, health, science, technology, social transformation, and ethical perspectives in achieving sustainable development.

This proceedings publication features 164 peer-reviewed papers by scholars all over the world, and delves into seven key themes: education and psychology in resilience-building; governance and political transformation; economic and legal frameworks for sustainability; scientific and technological advancements for societal resilience; religion, ethics, and sustainability; language, communication, and humanities in cultural and social sustainability; and gender equity and inclusive development. By integrating these themes, the book aligns with the United Nations Sustainable Development Goals (SDGs) and provides theoretical and practical insights for shaping a sustainable future.

This is an essential resource for academics, researchers, policymakers, and professionals in sustainability, governance and development studies, science and technology, education and health, and social sciences. It offers evidence-based insights and strategic recommendations for fostering more resilient and equitable societies.

PROCEEDINGS OF THE INTERNATIONAL CONFERENCE ON INNOVATIVE INTERDISCIPLINARY SCIENCE FOR INCLUSIVE AND SUSTAINABLE FUTURE (GREENTHINK 2024), JAKARTA, INDONESIA, SEPTEMBER 12TH–13TH, 2024

Towards Resilient Societies: The Synergy of Religion, Education, Health, Science, and Technology

Volume 1

Edited by

Maila D.H. Rahiem
Universitas Islam Negeri Syarif Hidayatullah, Indonesia

CRC Press
Taylor & Francis Group
Boca Raton London New York Leiden

CRC Press is an imprint of the
Taylor & Francis Group, an **informa** business
A BALKEMA BOOK

First published 2026
by CRC Press/Balkema
4 Park Square, Milton Park, Abingdon, Oxon, OX14 4RN

and by CRC Press/Balkema
2385 NW Executive Center Drive, Suite 320, Boca Raton FL 33431

CRC Press/Balkema is an imprint of the Taylor & Francis Group, an informa business

British Library Cataloguing-in-Publication Data
A catalogue record for this book is available from the British Library

Set

ISBN: 978-1-041-10418-6 (hbk)
ISBN: 978-1-041-10419-3 (pbk)

Volume 1
ISBN: 978-1-041-08471-6 (hbk)
ISBN: 978-1-041-08473-0 (pbk)
ISBN: 978-1-003-64554-2 (ebk)

DOI: 10.1201/9781003645542

Volume 2
ISBN: 978-1-041-10416-2 (hbk)
ISBN: 978-1-041-10417-9 (pbk)
ISBN: 978-1-003-65494-0 (ebk)

DOI: 10.1201/9781003654940

Typeset in Times New Roman
by MPS Limited, Chennai, India

Towards Resilient Societies: The Synergy of Religion, Education, Health, Science, and Technology – Rahiem (Ed.)
© 2026 The Editor(s), ISBN: 978-1-041-08471-6
Open Access: www.taylorfrancis.com, CC BY-NC-ND 4.0 license

Table of Contents

Session 3: Transforming economies and legal frameworks for a sustainable future

Towards Resilient Societies: The Synergy of Religion, Education, Health, Science, and Technology – Rahiem (Ed.)
© 2026 The Editor(s), ISBN: 978-1-041-08471-6
Open Access: www.taylorfrancis.com, CC BY-NC-ND 4.0 license

Preface

Resilience has become an urgent global concern as societies navigate the complexities of environmental, economic, technological, and socio-political change. Achieving sustainability and inclusivity requires a multidisciplinary approach that integrates knowledge across education, governance, health, science, and technology. This proceedings book, *Towards Resilient Societies: The Synergy of Religion, Education, Health, Science, and Technology*, presents a diverse collection of research that contributes to the ongoing discourse on resilience and sustainability.

With contributions from authors across Indonesia, the UK, the US, Australia, Singapore, Cambodia, Japan, Thailand, Afghanistan, Malaysia, Gambia, Pakistan, India, Iran, UEA, Hungary, South Africa, Yemen, China, Bangladesh and the Philippines, this book encapsulates the latest research, theories, and practical solutions that aim to foster societal resilience. The 163 peer-reviewed papers compiled in this volume explore resilience from economic, legal, scientific, educational, and sociocultural perspectives, providing insights that are valuable not only for scholars but also for policymakers and practitioners working on sustainability-driven initiatives. The research addresses key global challenges, ranging from policy adaptation, educational innovation, advancements in health and technology, to the role of religion and ethics in sustainability.

This book originates from GreenThink: The International Conference on Innovative Interdisciplinary Science for Inclusive and Sustainable Future, which was held online on 12–13 September 2024. Despite being conducted virtually, the conference successfully facilitated intellectual exchange among academics and professionals worldwide, fostering discussions on the synergy of religion, education, health, science, and technology in shaping a more sustainable and resilient future. The overarching theme of the conference, *Towards Resilient Societies: The Synergy of Religion, Education, Health, Science, and Technology*, underscores the necessity of multidisciplinary collaboration in addressing contemporary global challenges.

This volume also includes research presented in collaboration with international conferences hosted by different faculties at Universitas Islam Negeri Syarif Hidayatullah Jakarta, reflecting the institution's strong commitment to advancing interdisciplinary scholarship. These conferences include the International Conference on Social and Political Sciences 2024, Faculty of Social and Political Sciences; the International Conference on Islam Nusantara 2024, Faculty of Adab and Humanities; the International Conference on Education in Muslim Society 2024, Faculty of Educational Sciences; the Seminar and International Conference on Contemporary Social Movements, Faculty of Da'wah and Communication; the International Conference on Law and Justice 2024, Faculty of Sharia and Law; the International Conference on Health Sciences 2024, Faculty of Health Sciences; the International Colloquium on Interdisciplinary Islamic Studies 2024, School of Graduate Studies; the International Meeting on Medical Network 2024, Faculty of Medicine; and UIN Annual Conference on Economics and Business (UINACEB) 2024–Faculty of Economics and Business.

The 164 peer-reviewed papers in this book are structured into seven thematic chapters, each highlighting a critical domain of resilience-building. The first chapter, *Education and Psychology in Building Resilient and Inclusive Societies*, consists of 42 papers examining how education and psychological resilience contribute to societal well-being. Topics in this section include the role of the Merdeka curriculum in fostering critical thinking, religiosity and academic resilience in pesantren students, teacher well-being in Islamic education, and integrating Islamic values into mathematics education. Several studies focus on the integration of technology in education, such as research on artificial intelligence literacy among prospective teachers and digital leadership in schools. The discussion extends to environmental education, inclusive education leadership, and curriculum innovations aimed at equipping future generations with the skills necessary for sustainable development.

The second chapter, *Governance, Social Resilience, and Political Transformations in the 21st Century*, comprises 31 papers addressing public policy, governance, and social sustainability. Research in this chapter covers the effectiveness of social policies, water infrastructure development, sustainable urban planning, and the alignment of Sustainable Development Goals with local governance strategies. Other studies explore public diplomacy, security issues, and political economy, offering insights into how governance models can be strengthened to promote resilience in diverse sociopolitical landscapes.

The third chapter, *Transforming Economies and Legal Frameworks for a Sustainable Future*, includes 25 papers that delve into economic policies, legal studies, and financial sustainability. Topics range from Islamic banking spin-offs, zakat-based economic models, and blockchain applications in waqf governance to the role of Islamic philanthropy in poverty alleviation and economic independence in pesantren communities. Several articles focus on regional policy reforms, the legal challenges of cross-border data transfers, and the impact of financial regulations on economic resilience. This chapter emphasises the critical role of legal and financial frameworks in shaping sustainable economic growth.

The fourth chapter, *Innovations in Science, Health, and Technology for Societal Resilience*, comprises 20 papers that address technological advancements and healthcare solutions. Topics include digital transformation in health services, the role of artificial intelligence in medical diagnostics, bacterial explorations for pharmaceutical advancements, and innovative interventions in response to the COVID-19 pandemic. Several studies explore maternal and child health, workplace lactation support, and strategies to prevent stunting in young children. These papers provide valuable insights into scientific innovations that contribute to public health, environmental conservation, and digital transformation in healthcare systems.

The fifth chapter, *Interdisciplinary Perspectives on Religion, Ethics, and Sustainability*, presents 19 papers that investigate the role of religion and ethics in promoting sustainability and resilience. Research in this section includes Islamic perspectives on climate change, fatwa reformulation for sustainable development, conservation in Islamic wills, and the integration of religious values in environmental ethics. Several studies analyse how religious institutions and teachings contribute to social harmony, economic equity, and ecological preservation.

The sixth chapter, *Language, Communication, and Humanities in Shaping Cultural and Social Sustainability*, features 15 papers exploring the power of discourse, media, and cultural narratives in shaping perceptions of sustainability. Topics include Eco linguistics and sustainable development, digital da'wah strategies, the social construction of beauty in K-pop fandoms, and political discourse in Arabic and Persian languages.

The seventh and final chapter, *Advancing Gender Equity and Inclusive Development in a Changing World*, consists of 11 papers addressing issues of gender justice, women's empowerment, and policy transformation. Research topics include female ulama and Islamic judicial leadership, gender justice in divorce cases, and digital activism for eradicating violence against women.

This book would not have been possible without the dedication of numerous contributors. We acknowledge the authors whose research forms the foundation of this volume, as well as the peer reviewers who ensured the quality and integrity of the selected papers. Special appreciation is extended to the rector and vice rectors of Universitas Islam Negeri Syarif Hidayatullah Jakarta, as well as the head and secretary of the Institute of Research and Community Empowerment for their support in fostering a strong research culture. We also recognise the contributions of partner institutions, including the Ministry of National Development Planning (Bappenas) and Universitas Islam Negeri Mataram, for their collaboration in advancing interdisciplinary research.

Additionally, we extend our gratitude to the distinguished speakers whose papers are included in this book, namely Assoc. Prof. Steven Krauss from the University of Missouri, Assoc. Prof. Ritesh Chugh from Central Queensland University, Australia, and Prof. Ayesha Farooq from Aligarh Muslim University. Their perspectives have enriched this volume with global insights into sustainability, education, governance, and social transformation.

The themes explored in this book resonate with global research agendas and policy frameworks, particularly in relation to the United Nations Sustainable Development Goals. The intersection of education, governance, technology, and social systems plays a crucial role in achieving these goals, making the findings of this volume highly relevant to current academic and policy discussions. By integrating perspectives from multiple disciplines, this book enhances ongoing debates on building more resilient and inclusive societies while contributing to evidence-based policy formulation and institutional practices.

It is hoped that this book will serve as a catalyst for deeper exploration, constructive dialogue, and practical solutions that contribute to more resilient, sustainable, and equitable societies.

Prof. Maila Dinia Husni Rahiem, M.A., Ph.D.
Editor

Towards Resilient Societies: The Synergy of Religion, Education, Health, Science, and Technology – Rahiem (Ed.)
© 2026 The Editor(s), ISBN: 978-1-041-08471-6
Open Access: www.taylorfrancis.com, CC BY-NC-ND 4.0 license

Foreword

The pursuit of resilient and sustainable societies has never been more urgent. As we face complex global challenges—ranging from climate change, economic instability, and public health crises to rapid technological shifts and social transformations—interdisciplinary research becomes essential in providing solutions that bridge theory and practice. This proceedings book, *Towards Resilient Societies: The Synergy of Religion, Education, Health, Science, and Technology*, represents a significant academic contribution to this critical discourse. It brings together a diverse range of scholars, researchers, and practitioners to examine the interconnected nature of resilience and sustainability from multiple perspectives, fostering a deeper understanding of how these challenges can be addressed holistically.

The research compiled in this volume originates from *GreenThink: The International Conference on Innovative Interdisciplinary Science for Inclusive and Sustainable Future*, hosted online on 12–13 September 2024. This conference successfully brought together experts from across Indonesia and international institutions in the tthe UK, the US, Australia, Singapore, Cambodia, Japan, Thailand, Afghanistan, Malaysia, Gambia, Pakistan, India, Iran, UEA, Hungary, South Africa, Yemen, China, Bangladesh and the Philippines. The discussions within this book reflect a broad spectrum of interdisciplinary inquiry, covering education, governance, health, science, technology, economics, and social development. By integrating knowledge from various disciplines, this book not only contributes to academic scholarship but also provides actionable insights for policymakers, educators, and professionals dedicated to sustainability-driven initiatives.

The significance of the research presented in this book extends beyond academia. In an era where global development is increasingly framed by the United Nations Sustainable Development Goals (SDGs), the studies in this volume provide evidence-based recommendations and innovative frameworks that can guide policy and practice. Topics such as the role of education in fostering psychological resilience, economic and legal frameworks for sustainable development, advancements in health and technology, and the ethical and religious dimensions of sustainability highlight the depth and breadth of knowledge needed to address today's pressing challenges. The discussions also emphasize the importance of governance, digital transformation, and gender equity in creating inclusive and adaptable societies.

This book is not merely a collection of research papers—it is a call to action. Scholars will find valuable theoretical perspectives and empirical studies that advance academic discourse. Policymakers will gain insights into governance strategies, legal frameworks, and policy interventions that align with global sustainability efforts. Practitioners, including educators, healthcare professionals, and social workers, will discover practical applications that can be implemented in their respective fields. By fostering dialogue between academia and real-world practice, this volume serves as a catalyst for further collaboration and innovation in resilience-building.

As Rector of Universitas Islam Negeri Syarif Hidayatullah Jakarta, I am proud to support and endorse this remarkable academic endeavour. This book is a testament to the power of collaboration, bringing together scholars from multiple disciplines to contribute to a shared vision of a more sustainable future. The success of this publication is a reflection of the dedication and commitment of many individuals and institutions. I extend my deepest appreciation to all the contributors, reviewers, and partner institutions that have played a role in making this book a reality.

I would like to express my highest gratitude and admiration to Professor Maila Dinia Husni Rahiem and her editorial team, who have worked tirelessly, with unwavering enthusiasm and perseverance, to bring this book to completion. Their commitment to academic excellence and their ability to unite scholars from diverse backgrounds have made this volume a valuable resource for anyone interested in resilience and sustainability.

It is my hope that this book will inspire further research, encourage interdisciplinary collaboration, and provide meaningful contributions to the global effort toward building more resilient societies. May it serve as a foundation for new ideas, constructive dialogue, and innovative solutions that will shape the future of sustainability in academia, policy, and practice.

Prof. Asep Saepudin Jahar, M.A., Ph.D.
Rector Universitas Islam Negeri Syarif Hidayatullah Jakarta

Towards Resilient Societies: The Synergy of Religion, Education, Health, Science, and Technology – Rahiem (Ed.)
© 2026 The Editor(s), ISBN: 978-1-041-08471-6
Open Access: www.taylorfrancis.com, CC BY-NC-ND 4.0 license

Acknowledgements

The publication of *Towards Resilient Societies: The Synergy of Religion, Education, Health, Science, and Technology* is the result of a collaborative effort by numerous individuals and institutions. This book embodies a strong commitment to advancing interdisciplinary research and fostering critical discussions on sustainability and resilience.

We express our sincere gratitude to Prof. Asep Saepudin Jahar, M.A., Ph.D., Rector of Universitas Islam Negeri Syarif Hidayatullah Jakarta, along with the Vice Rectors, for their unwavering support in strengthening a research-oriented academic culture. We also acknowledge the Head and Secretary of the Institute of Research and Community Empowerment for their role in facilitating a research environment that made this publication possible.

This book would not have been realised without the dedication of the editorial team and the organising committee of the conference and proceedings publication. Under the leadership of Prof. Maila Dinia Husni Rahiem, M.A., Ph.D., the team has worked tirelessly to ensure the scholarly integrity and coherence of this volume. Their commitment to coordinating the peer-review process, refining manuscripts, and upholding high publication standards has been invaluable.

We also extend our gratitude to the organisers of GreenThink: The International Conference on Innovative Interdisciplinary Science for Inclusive and Sustainable Future, held online on 12–13 September 2024. Despite being conducted virtually, the conference successfully facilitated intellectual exchanges among scholars and professionals, which are reflected in this book.

Furthermore, we acknowledge the contributions of partner conferences at Universitas Islam Negeri Syarif Hidayatullah Jakarta, whose collaborations have enriched this publication:

- International Conference on Social and Political Sciences (ICSPS) 2024 – Faculty of Social and Political Sciences
- International Conference on Islam Nusantara (INSAN) 2024 – Faculty of Adab and Humanities
- International Conference on Education in Muslim Society (ICEMS) 2024 – Faculty of Educational Sciences
- The Seminar and International Conference on Contemporary Social Movements– Faculty of Da'wah and Communication
- International Conference on Law and Justice (ICLJ) 2024 – Faculty of Sharia and Law
- International Conference on Health Sciences (ICHS) 2024 – Faculty of Health Sciences
- International Colloquium on Interdisciplinary Islamic Studies (ICIIS) 2024 – School of Graduate Studies
- International Meeting on Medical Network (IMMAN) 2024 – Faculty of Medicine
- UIN Annual Conference on Economics and Business (UINACEB) 2024–Faculty of Economics and Business

We also appreciate the support of our institutional partners, including the Ministry of National Development Planning (Bappenas) and Universitas Islam Negeri Mataram, in advancing interdisciplinary research and fostering cross-sector collaborations.

A special note of appreciation goes to the peer reviewers, whose expertise and insights have strengthened the academic quality of the papers in this book. Their role in maintaining rigorous standards has been essential in shaping this volume.

We are also grateful to the distinguished speakers, whose research contributions have provided valuable global perspectives on sustainability, education, governance, and social transformation:

1. Assoc. Prof. Ritesh Chugh – Central Queensland University, Australia
2. Prof. Haslinda Abdullah – Universiti Putra Malaysia, Malaysia
3. Benedict Wermter – Veritas Peduli Lingkungan, Indonesia
4. Prof. Ayesha Farooq – Aligarh Muslim University, India
5. Dr. Guido Schnieders – German Academic Exchange Service (DAAD), Indonesia
6. Assoc. Prof. Steven Krauss – University of Missouri, USA

Lastly, we extend our appreciation to all authors, researchers, and conference participants whose work is featured in this volume. Their dedication and intellectual contributions have enriched the discourse on resilience and sustainability. We hope this book serves as a valuable resource for academic scholarship, policy development, and practical solutions in fostering more resilient societies.

Jakarta, February 2025

Editorial Team and Organizing Committee
Towards Resilient Societies: The Synergy of Religion, Education, Health, Science, and Technology

Towards Resilient Societies: The Synergy of Religion, Education, Health, Science, and Technology – Rahiem (Ed.)
© 2026 The Editor(s), ISBN: 978-1-041-08471-6
Open Access: www.taylorfrancis.com, CC BY-NC-ND 4.0 license

Editor Biography

Prof. Maila Dinia Husni Rahiem, M.A., Ph.D., is a professor of Early Childhood Education and Social Welfare at Universitas Islam Negeri Syarif Hidayatullah Jakarta, Indonesia. She currently serves as the Head of the Center for International Cooperation Services (PLKI) and is a Senior Advisor to the Minister of Primary and Secondary Education of the Republic of Indonesia. Additionally, she is the Director of the Center for the Study of Religion, Environment, and Climate Change and an Adjunct Professor at Central Queensland University, Australia.

Her research interests include resilience, disaster management, sustainability, early childhood education, Islamic education, and narrative inquiry. She has published extensively on child development, education for sustainable development, disaster risk reduction, and the role of religion in fostering community resilience. Recognized among the Top 2% of Scientists Worldwide by Stanford University, she has received multiple academic excellence awards for her impactful research and publications.

Prof. Maila plays a pivotal role in advancing international research collaborations, policy development, and sustainability initiatives. As Chief Editor of the GreenThink 2024 Proceedings, she has contributed significantly to interdisciplinary discussions on the synergy of religion, education, health, science, and technology in fostering resilient societies.

Towards Resilient Societies: The Synergy of Religion, Education, Health, Science, and Technology – Rahiem (Ed.)
© 2026 The Editor(s), ISBN: 978-1-041-08471-6
Open Access: www.taylorfrancis.com, CC BY-NC-ND 4.0 license

Scientific Committee

Advisory Board

- **Prof. Asep Saepudin Jahar, M.A., Ph.D.,** *Universitas Islam Negeri Syarif Hidayatullah Jakarta*
- **Prof. Dr. Ahmad Tholabi, S.Ag., S.H., M.H., M.A.,** *Universitas Islam Negeri Syarif Hidayatullah Jakarta*
- **Prof. Dr. Imam Subchi, M.A.,** *Universitas Islam Negeri Syarif Hidayatullah Jakarta*
- **Prof. Ali Munhanif, M.A., Ph.D.,** *Universitas Islam Negeri Syarif Hidayatullah Jakarta*
- **Din Wahid, M.A., Ph.D.,** *Universitas Islam Negeri Syarif Hidayatullah Jakarta*
- **Prof. Amelia Fauzia, Ph.D.,** *Universitas Islam Negeri Syarif Hidayatullah Jakarta*
- **Prof. Dr. Khamami, M.A.,** *Universitas Islam Negeri Syarif Hidayatullah Jakarta*

Chairs

- **Prof. Haslinda Abdullah,** *Universiti Putra Malaysia*
- **Prof. Ayesha Farooq,** *Aligarh Muslim University*

Members

- **Assoc. Prof. Ritesh Chugh, Ph.D.,** *Central Queensland University*
- **Assoc. Prof. Steven Krauss, Ph.D.,** *University of Missouri*
- **Prof. Siti Nurul Azkiyah, M.Sc., Ph.D.,** *Universitas Islam Negeri Syarif Hidayatullah Jakarta*
- **Prof. Dr. Ibnu Qizam, S.E., M.Si., Ak., CA.,** *Universitas Islam Negeri Syarif Hidayatullah Jakarta*
- **Prof. Dr. apt. Zilhadia, M.Si.,** *Universitas Islam Negeri Syarif Hidayatullah Jakarta*
- **Prof. Dr. Dzuriyatun Toyibah, M.Si., M.A.,** *Universitas Islam Negeri Syarif Hidayatullah Jakarta*
- **Prof. Dr. Flori Ratna Sari, Ph.D.,** *Universitas Islam Negeri Syarif Hidayatullah Jakarta*
- **Prof. Dr. Rena Latifa, M.Psi.,** *Universitas Islam Negeri Syarif Hidayatullah Jakarta*
- **Prof. Kusmana, S.Ag., M.A., Ph.D.,** *Universitas Islam Negeri Syarif Hidayatullah Jakarta*
- **Prof. Drs. Ismatu Ropi, M.A., Ph.D.,** *Universitas Islam Negeri Syarif Hidayatullah Jakarta*
- **Dr. Ade Abdul Hak, M.Hum., CIQnR.,** *Universitas Islam Negeri Syarif Hidayatullah Jakarta*
- **Dr. Muhammad Maksum, M.A.,** *Universitas Islam Negeri Syarif Hidayatullah Jakarta*
- **Dr. Gun Gun Heryanto, M.Si.,** *Universitas Islam Negeri Syarif Hidayatullah Jakarta*
- **Dr. Yuli Yasin, M.A.,** *Universitas Islam Negeri Syarif Hidayatullah Jakarta*
- **Dr. Husni Teja Sukmana, S.T., M.Sc., Ph.D.,** *Universitas Islam Negeri Syarif Hidayatullah Jakarta*
- **Dr. Achmad Zaki, M.Epid., Sp.OT., FICS.,** *Universitas Islam Negeri Syarif Hidayatullah Jakarta*
- **Dr. Yunita Faela Nisa, M.Psi.,** *Universitas Islam Negeri Syarif Hidayatullah Jakarta*
- **Dr. Yeni Ratna Yuningsih, S.Ag., M.A., Ph.D.,** *Universitas Islam Negeri Syarif Hidayatullah Jakarta*
- **Dr. Munasprianto Ramli, M.A., Ph.D.,** *Universitas Islam Negeri Syarif Hidayatullah Jakarta*

- **Dr. Iding Rosyidin, S.Ag., M.Si.,** *Universitas Islam Negeri Syarif Hidayatullah Jakarta*
- **Dr. Salamah Agung, M.A., Ph.D.,** *Universitas Islam Negeri Syarif Hidayatullah Jakarta*
- **Dr. Fuad Fachruddin, Ph.D.,** *Universitas Islam Negeri Syarif Hidayatullah Jakarta*
- **Dr. Sujiyo Miranto, M.Pd.,** *Universitas Islam Negeri Syarif Hidayatullah Jakarta*
- **Dr. Yanti Herlanti, M.Pd.,** *Universitas Islam Negeri Syarif Hidayatullah Jakarta*

Towards Resilient Societies: The Synergy of Religion, Education, Health, Science, and Technology – Rahiem (Ed.)

Organizing Committee

Editor-in-Chief:
Prof. Maila Dinia Husni Rahiem, M.A., Ph.D.

Secretary:
Evi Muliyah, M.Si.

Publication Management:
Dzikri Rahmat Romadhon, M.PFis.
Asmui, M.H.
Yudi, S.Sos.

Secretariat:
Ratna Faeruz, M.Pd.
Iman Matin, M.Pd.
Johan Aristya Lesmana, M.Si.
Eny Supriyati Rosyidatun, S.Si., M.A.
Endi Aulia Garadian, M.Hum.
Nurbaini Futuhat Wulansari, S.Si., M.I.Kom.

Session 1: Governance, social resilience, and political transformations in the 21st century

Towards Resilient Societies: The Synergy of Religion, Education, Health, Science, and Technology – Rahiem (Ed.)

Resilience and the 4-H thriving model: A conceptual comparison of youth development approaches

Steven Krauss, Ph.D.
University of Missouri, USA
ORCID ID: 0000-0003-4168-6224

ABSTRACT: The concept of youth resilience—the ability of young people to mobilise internal and external resources to address challenges in their environment—has garnered significant attention from scholars and service providers in the post-COVID era. Global studies highlight trends related to deteriorating mental health, well-being, and educational outcomes among youth and adolescents. Positive Youth Development (PYD), a programmatic approach that supports SDG 4 (quality education) through promoting healthy transitions to adulthood, has demonstrated considerable promise in fostering resilience among youth through developing key protective factors. However, few studies have compared specific PYD program models to resilience frameworks to uncover the former's potential for enhancing youth resilience. This study provides a focused analysis of the 4-H Thriving Model, the program-wide practice model guiding 4-H programs throughout the United States. I present a conceptual comparison and analysis of the Thriving Model alongside Kempfer and colleagues' resilience framework to explore the model's potential to promote resilience among youth program participants. The paper concludes with essential questions for researchers concerning the need to explore further the 4-H Thriving Model and its impact on resilience among youth.

Keywords: resilience, 4-H Thriving Model, positive youth development, out-of-school time youth programs, non-formal education

1 INTRODUCTION

There is a widespread perception that young people today face unprecedented challenges exacerbated by the global COVID-19 pandemic (De France *et al.* 2022; Mansfield *et al.* 2022). These challenges include increasing trends of negative mental health outcomes, such as anxiety, depression, and emotional dysregulation, as well as changes in substance use and service utilisation (Chavira *et al.* 2022; Fortuna *et al.* 2023). In response to these issues, educators, mental health professionals, youth workers, and other stakeholders have cited the critical importance of resilience to help young people effectively cope with adversity (Dvorsky *et al.* 2021).

Positive Youth Development (PYD) programs are increasingly seen as valuable resources for building resilience, especially for young people from trauma-filled backgrounds. A growing body of case study literature on specific programs indicates positive outcomes. However, few conceptual efforts have been made to explicitly link PYD program models to promoting resilience in youth. Consequently, there is limited understanding of how PYD programs practically build the protective factors necessary for resilience.

This brief paper aims to draw parallels between the goals and processes of one of the United States' flagship PYD programs, 4-H, and its recently adopted national program model, the Thriving Model, with the youth resilience literature. By examining the literature on resilience, particularly the framework presented by Kumpfer and Summerhays (2006), alongside research on 4-H, the current paper provides a concise treatment of the Thriving Model's potential for building resilience among 4-H participants. The paper concludes with key questions and directions for future research around understanding how participation in 4-H can build resilience in different sub-groups of youth participants.

2 YOUTH RESILIENCE TAKES CENTRE STAGE

There is a growing belief that young people today are less resilient than previous generations. Popular media frequently claim that young individuals struggle to cope with routine life hardships (Velez *et al.* 2022). However, whether today's youth are less resilient than in the past is complex, involving factors such as mental health trends, societal changes, and evolving understandings of resilience. Resilience is defined as "the process of effectively mobilising internal and external resources in adapting to or managing significant sources of stress or trauma. Thus, cultivating resilience means fostering adolescents' capacity, flexibility, and coping strategies as they face developmental changes and life stresses to 'bounce back' from difficult experiences and achieve positive outcomes" (Lee *et al.* 2012).

Scholars contend that resilience is a multifaceted concept that results from individual capacity to draw on personal resources as well as those found within their surrounding ecologies (Christmas and Khanlou 2019). Resilience is thus viewed systemically, involving the ability of young people to harness both personal and social resources to manage challenges and stresses. How well a young person copes with being bullied at school, for instance, depends not only on their internal coping strategies—such as negotiation skills and emotional hardiness—but also on their ability to access and leverage resources in their environment. These resources might include support and advocacy from peers, teachers, school administrators, counsellors, and family members. The ability to call on these various resources to address and overcome problems contributes significantly to a young person's resilience (Khanlou and Wray 2014).

3 A STRENGTHS-BASED APPROACH TO COPING WITH ADVERSITY: POSITIVE YOUTH DEVELOPMENT

Programmatic attempts to build resilience in young people often rely on structured interventions and comprehensive developmental strategies (Masten and Obradović 2006). One approach that has received much attention over the past two decades is positive youth development (PYD), which focuses on fostering the strengths and potential of young individuals rather than on remedying their weaknesses or preventing specific problems (Benson *et al.* 2007). According to Lerner *et al.*, (2005), the PYD theoretical perspective is considered part of the third phase of the study of adolescent development as a field, emphasising a strengths-based approach to working with you.

As elucidated by Bowers *et al.* (2014), PYD represents adaptive developmental regulation of young people representing bidirectional relations between the attributes of a young person and the features of their family, school, and community environment. Within these points of contact with one's setting, PYD occurs when there is alignment between a young person's personal attributes that constitute strengths of the individual (e.g., intentional self-regulation skills, hopeful future expectations, school engagement; Lerner *et al.* 2013) and features of the young person's environment that can be described as developmental assets (e.g., warmth and monitoring by engaged parents, sustained relations with a caring and competent mentor or youth program practitioner). This alignment-of-strengths approach to youth development provides a conceptual frame for designing PYD programs to achieve optimal bidirectional fit.

Toward this aim, PYD program designers focus on engaging youth in constructive activities, promoting positive relationships, and providing opportunities for voice in decision-making, leadership, and skill-building, attempting to optimise the strengths of the individual youth with growth opportunities (Lerner *et al.* 2005). High-quality PYD programs have been shown to enhance youth resilience by developing their self-efficacy, improving their social skills, and providing a safe and supportive environment where they can learn how to overcome challenges (Catalano *et al.* 2004; Durlak *et al.* 2010). Despite growing evidence that PYD programs are effective in enhancing resilience, especially among adolescents, few studies have explored how specific program models activate critical processes that help youth build resilience.

PYD and resilience share certain conceptual characteristics. Inherent within the definition of resilience is the idea that a young person has effectively adapted to an environment that has within it certain risks to the child's well-being, stemming from the family, school, community, or due to a personal disability (Masten and Obradović 2006). Studies on resilience have shown that the main difference between individuals who adapt very well despite facing risks and individuals who end up in maladaptation is the existence of protective factors (Lee *et al.* 2012). The activation or cultivation of protective factors in and around a young people helps to buffer the effects of risk existing within the young person's life world. According to Resnick (2000), protective factors include a complex interplay of experiences, assets, and resources categorised by the extrafamilial (relationships with non-familial adults), familial (parenting, family cohesion, family communication), self-system (competence, connectedness, social responsibility), and individual characteristics (self-beliefs, health, cognition), that promote positive functioning in the face of stressors and challenges. The factors support the effective functioning of young people and the achievement of developmental tasks.

Through a thriving orientation, PYD programs do not focus solely on young people who are "at risk" of falling behind or engaging in unhealthy behaviours due to various stressors in their environment. Instead, PYD aims to provide structure, opportunity, and support for all young people (Lerner *et al.* 2013). This is done by promoting opportunities to build skills and pursue personal interests, supportive relationships, and services that meet the needs of youth who can then fully participate in their own development (Villarruel *et al.* 2003). As an approach, PYD is implemented in a variety of youth program settings including schools, community-based organisations, after-school programs, summer camps, and even colleges/universities.

PYD programs focus on building intentional processes (programs, activities) that promote positive outcomes or thriving. Thriving has been conceptualised as a pathway or trajectory on the way to a positive future over time, marked by the acquisition of characteristics termed the "Five Cs" – competence, confidence, character, connection, and caring (Lerner *et al.* 2011). In the context of adolescent development, thriving is fueled by mutual, positive interactions between youth and their developmental contexts. At the heart of thriving is that a young person is animated and motivated intrinsically by his or her spark or special sense of who he or she is as a person (Arnold and Gagnon 2019). Thriving as an orientation rather than a status promotes a developmental journey that helps young people achieve positive outcomes such as academic motivation and success, reduction in risk behaviours, healthful choices, and elevated personal standards. Developmental outcomes, in turn, predict a successful transition to adulthood, marked by quality health and well-being, economic stability, and civic engagement (Gambone *et al.* 2002).

4 PYD IN 4-H: THE THRIVING MODEL

4-H is America's flagship youth development program. Starting back in 1902 as a series of agricultural clubs and after-school programs for youth, 4-H became part of the National Cooperative Extension System in 1914 by the United States Department of Agriculture (USDA). Cooperative Extension "combines the expertise and resources of federal, state, and local governments and is designed to meet the need for research, knowledge, and educational programs" (Lee 2018). The 4-H Clover emblem was adopted in 1924, and 4-H clubs were formed nationwide. 4-H is administered in every U.S. state by land-grant universities.

Today, over six million young people ages 8 to 18 across the United States participate in the 4-H program. The contemporary 4-H program is focused on PYD, providing an array of skill-based activities, projects, camps, and programs aimed at providing experiences to enhance participants' likelihood of success in adulthood (Arnold 2018). Despite the longevity and resources behind the 4-H program, stakeholders have recently articulated a system-wide program model guiding the unique PYD approach of 4-H. In 2018, Arnold proposed the Thriving Model, based on a synthesis of extensive research in PYD and 4-H, along with existing PYD frameworks, such as the 5 C's (Lerner *et al.* 2005), the 4-H Essential Elements (Kress 2005); the 4-H Impact Study (Peterson *et al.* 2001), and the National Research Council's features of high-quality PYD programs (Gootman and Eccles 2002).

The Thriving Model is 4-H's first attempt to adopt an evidence-based system-wide program framework. As articulated by Ferrari and Arnold (2024), the model consists of three main elements that reflect 4-H's unique recipe of PYD setting, process, and outcomes: (1) 4-H programs as *developmental settings* characterised by youth sparks, a safe physical environment and positive emotional climate, and developmental relationships; (2) *indicators of youth thriving* that reflect the process of social, emotional, cognitive, and behavioural development that takes place within 4-H programs; and (3) *developmental outcomes*, which are indicators of PYD that include academic motivation and success, social competence, responsibility, connection with others, and contribution to others (Figure 1). Youth engagement, an outcome of the development context, "drives the thriving" along a developmental pathway to outcomes. The developmental outcomes, in turn, lead to long-term outcomes realised in adulthood (Arnold and Gagnon 2020). The three elements of the model form a predictive theory of change for 4-H programs (Arnold 2018).

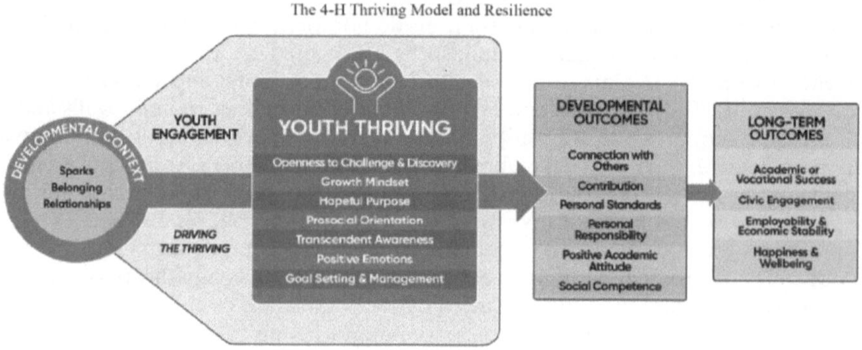

Figure 1. 4-H thriving model (Arnold 2018).

While the Thriving Model is not a model for building resilience, the model shares several key conceptual areas of convergence with theory and research on resilience. Both approaches embrace a PYD orientation, targeting the development and well-being of youth, aiming to equip them with the tools and skills necessary to navigate life's challenges effectively. Both the Thriving Model and resilience foreground the crucial role of supportive relationships in fostering positive outcomes, such as mentoring relationships and other developmental relationships where adults support and guide youth, fostering trust and providing a stable foundation for growth. The resilience literature highlights the importance of strong social connections and supportive relationships with family, peers, and mentors as critical factors in building resilience (Stewart and Sun 2004).

Second, developing self-efficacy and empowerment is a key element in both approaches, as building confidence and a sense of control over one's life is a common theme in both domains. The Thriving Model emphasises setting and achieving personal goals, fostering a growth mindset, and encouraging self-reflection, all enhancing self-efficacy and empowerment (Fields 2020). Research on resilience indicates that a strong sense of self-efficacy is vital for overcoming challenges and returning from adversity (Galli and Vealey 2008).

Both models underscore the importance of skill-building to facilitate resilience and thriving. The Thriving Model focuses on developing life skills, including leadership, communication, and critical thinking, essential for personal and professional success (Ferrari and Arnold 2024). Resilience literature similarly points to developing problem-solving skills, emotional regulation, and adaptive coping strategies to navigate difficulties effectively (Polizzi and Lynn 2021).

An emphasis on adaptability and maintaining an orientation of growth from experienced adversity, or a growth mindset, is evident in both models. The Thriving Model includes a growth mindset as an indicator of thriving, where youth are taught to view challenges as opportunities for growth and to persist in the face of setbacks (Arnold and Gagnon 2020). Resilience is closely linked to adaptability and the ability to adjust one's thoughts, emotions, and behaviours in response to changing circumstances (Wu *et al.* 2013).

Setting goals and having a sense of purpose are integral to both frameworks. In the Thriving Model, goal management is another key element. In this process, youth learn to set, pursue, and reevaluate their goals, helping them align their efforts with their sparks and values (Arnold 2018). In the resilience literature, scholars point to having clear goals and a sense of purpose as helping youth maintain motivation and direction, even when facing adversity (Lewis and Hill 2021).

Finally, both literatures highlight the role of external supports and resources in promoting thriving and resilient outcomes. The 4-H Thriving Model is predicated on participation in 4-H projects, programs, and events, which provide its members access to resources, opportunities, and supportive relationships from adult volunteers and program staff that nurture their development. 4-H is a low-cost program that also encourages participation from parents and family members. 4-H members can participate in the program continuously between the ages of 5 to 18, which ensures time to benefit from the multiple opportunities afforded and build caring, supportive relationships with adults. Its community-based focus also helps to ensure that programs are adequately funded and resourced through donations, and other forms of outside support. The resilience literature emphasises the importance of external supports such as safe communities, effective schools, and accessible services that offer protection against risk factors (Ellis *et al.* 2017).

5 HOW THE 4-H THRIVING MODEL CAN PROMOTE RESILIENCE IN YOUNG PEOPLE

Interpersonal and transactional processes mitigate biological and environmental risks while fostering resilience. As a result, models aimed at promoting resilience in young people emphasise the importance of multiple socialisation processes. These processes typically focus on three key categories of protective factors: personal characteristics, the quality of relationships, and broader environmental or ecological factors (Kumpfer and Summerhays 2006). The 4-H Thriving Model follows a similar framework, known as the developmental context, which includes sparks, developmental relationships, a sense of belonging, and engagement. Several parallels exist between these elements, the core components of high-quality 4-H programming, and the protective factors outlined by Kumpfer and Summerhays (2006).

By focusing on young people's personal interests, known as "sparks," the Thriving Model emphasises the strengths that participants bring to the 4-H program and seeks to develop them through a strengths-based approach. These sparks are nurtured through ongoing developmental relationships with adults, called "sparks champions," who foster positive, supportive connections with youth. By working with participants to cultivate their sparks, these adults build lasting relationships throughout the participants' time in the program and, in some cases, beyond the program. Through this process, 4-H has the potential to build resilience by combining a young person's autonomy to pursue their interests with the support of nurturing adults. These relationships serve as a protective factor against adverse life circumstances.

The Thriving Model promotes healthy social ecologies for its youth participants, thereby strengthening the potential for resilience. One of the ways it does this is through its wide array of project offerings that are carried out in community clubs, along with school and afterschool-based programs and regional, national, and even international programs. 4-H thus helps its members build positive connections in schools, in their communities, and even regionally and nationally with other 4-H youth through such programs as the National 4-H Conference, where members meet other 4-H'ers from other states.

Unlike place-based programs, which limit participation to specific locations, 4-H encourages civic engagement through various service and leadership initiatives that reflect the "heart" and "hands" of the 4-H creed. These activities foster relationships with community organisations, underserved populations, and other entities, fostering civic-mindedness and community connections. Additionally, 4-H maintains a strong family-community orientation by involving parents as volunteers. This helps build relationships among adults in the program, further enhancing community-building, belonging, and healthy social ecologies (Lee 2018).

Through the integration of sparks, developmental relationships, and family and community engagement, 4-H has the potential to build resilience in participants by providing positive socialisation processes that serve as protective factors against risk.

6 CONCLUSION AND FUTURE DIRECTIONS

This paper provides a concise examination of the intersection between the 4-H Thriving Model and research on youth resilience, aiming to understand how 4-H can enhance resilience in young people. While the existing literature on both approaches shows much overlap both in concept and practice, considerably more work is needed, especially from the Thriving Model side, to understand how 4-H participation builds resilience in youth. Specifically, more research is needed to understand 1) how the Thriving model leads to resilience, 2) how it works with young people at different points in the risk spectrum, 3) what "dosage" of 4-H programming (i.e., what types of program experiences and for how long) contributes to resilience, and 4) how 4-H works to build resilience among different sub-groups of young people, particularly those from under-served or historically marginalised backgrounds.

Lastly, little is known about how the Thriving model translates across diverse or international youth development settings. The model was only recently introduced into 4-H within the United States and, therefore, has not yet been evaluated for its efficacy and relevance with diverse groups of young people or outside the United States. Furthermore, scholars have yet to examine whether the model is applicable outside of 4-H program settings. Efforts to translate the model across diverse cultural and ethnic youth populations are, therefore, needed.

REFERENCES

Arnold, M. E. (2018). From context to outcomes: A thriving model for 4-H youth development programs. *Journal of Human Sciences and Extension*, *6*(1), 11. https://doi.org/10.54718/NBNL5438

Arnold, M. E., and Gagnon, R. J. (2019). Illuminating the process of youth development: The mediating effect of thriving on youth development program outcomes. *Journal of Human Sciences and Extension*, *7*(3), 2. https://doi.org/10.54718/GHUP2927

Arnold, M. E., and Gagnon, R. J. (2020). Illuminating the process of youth development: The mediating effect of thriving on youth development program outcomes. *Journal of Human Sciences and Extension*, *8*(3), 105–124. https://doi.org/10.5195/jyd.2020.954

Benson, P. L., Scales, P. C., Hamilton, S. F., and Sesma, A., Jr. (2007). Positive youth development: Theory, research, and applications. *Handbook of Child Psychology*. https://doi.org/10.1002/9780470147658.chpsy0116

Bowers, E. P., Geldhof, G. J., Johnson, S. K., Lerner, J. V., and Lerner, R. M. (2014). Special issue introduction: Thriving across the adolescent years: A view of the issues. *Journal of Youth and Adolescence*, *43*(6), 859–868. https://doi.org/10.1007/s10964-014-0117-8

Catalano, R. F., Berglund, M. L., Ryan, J. A. M., Lonczak, H. S., and Hawkins, J. D. (2004). Positive youth development in the United States: Research findings on evaluations of positive youth development programs. *The Annals of the American Academy of Political and Social Science*, *591*(1), 98–124. https://doi.org/10.1177/0002716203260102

Chavira, D. A., Ponting, C., and Ramos, G. (2022). The impact of COVID-19 on child and adolescent mental health and treatment considerations. *Behaviour Research and Therapy*, *157*, 104169. https://doi.org/10.1016/j.brat.2022.104169

Christmas, C. M., and Khanlou, N. (2019). Defining youth resilience: A scoping review. *International Journal of Mental Health and Addiction*, *17*, 731–742. https://doi.org/10.1007/s11469-018-0002-x

De France, K., Hancock, G. R., Stack, D. M., Serbin, L. A., and Hollenstein, T. (2022). The mental health implications of COVID-19 for adolescents: Follow-up of a four-wave longitudinal study during the pandemic. *American Psychologist*, *77*(1), 85. https://doi.org/10.1037/amp0000838

Durlak, J. A., Weissberg, R. P., and Pachan, M. (2010). A meta-analysis of after-school programs that seek to promote personal and social skills in children and adolescents. *American Journal of Community Psychology*, *45*(3–4), 294–309. https://doi.org/10.1007/s10464-010-9300-6

Dvorsky, M. R., Breaux, R., and Becker, S. P. (2021). Finding ordinary magic in extraordinary times: Child and adolescent resilience during the COVID-19 pandemic. *European Child and Adolescent Psychiatry*, *30*(11), 1829–1831. https://doi.org/10.1007/s00787-020-01583-8

Ellis, B. J., Bianchi, J., Griskevicius, V., and Frankenhuis, W. E. (2017). Beyond risk and protective factors: An adaptation-based approach to resilience. *Perspectives on Psychological Science*, *12*(4), 561–587. https://doi.org/10.1177/1745691617693054

Ferrari, T. M., and Arnold, M. E. (2024). The application of social science in 4-H youth development. *Extension Education and the Social Sciences: Uplifting Children, Youth, Families, and Communities*, 137.

Fields, N. I. (2020). Exploring the 4-H thriving model: A commentary through an equity lens. *Journal of Youth Development*, *15*(6), 171–194.

Fortuna, L. R., Brown, I. C., Woods, G. G. L., and Porche, M. V. (2023). The impact of COVID-19 on anxiety disorders in youth: Coping with stress, worry, and recovering from a pandemic. *Child and Adolescent Psychiatric Clinics*, *32*(3), 531–542. https://doi.org/10.1016/j.chc.2023.02.002

Galli, N., and Vealey, R. S. (2008). "Bouncing back" from adversity: Athletes' experiences of resilience. *The Sport Psychologist*, *22*(3), 316–335. https://doi.org/10.1123/tsp.22.3.316

Gambone, M. A., Klem, A. M., and Connell, J. P. (2002). *Finding out What Matters for Youth*. Philadelphia, PA: Youth Development Strategies.

Gootman, J. A., and Eccles, J. (Eds.). (2002). *Community Programs to Promote Youth Development*. National Academies Press.

Khanlou, N., and Wray, R. (2014). A whole community approach toward child and youth resilience promotion: A review of resilience literature. *International Journal of Mental Health and Addiction*, *12*, 64–79. https://doi.org/10.1007/s11469-013-9470-1

Kress, C. (2005). Essential elements of positive youth development. In *Strengthening Positive Youth Development Environments* (pp. 20–23). Madison: University of Wisconsin Extension 4-H Program.

Kumpfer, K. L., and Summerhays, J. F. (2006). Prevention approaches to enhance resilience among high-risk youth: Comments on the papers of Dishion and Connell and Greenberg. *Annals of the New York Academy of Sciences*, *1094*(1), 151–163. https://doi.org/10.1196/annals.1376.014

Lee, D. (2018). 4-H programs: An overview. *Integrated Studies*, *123*. https://digitalcommons.murraystate.edu/bis437/123

Lee, T. Y., Cheung, C. K., and Kwong, W. M. (2012). Resilience as a positive youth development construct: A conceptual review. *The Scientific World Journal*, *2012*(1), 390450. https://doi.org/10.1100/2012/390450

Lerner, J. V., Bowers, E. P., Minor, K., Boyd, M. J., Mueller, M. K., Schmid, K. L., *et al.* (2013). Positive youth development: Processes, philosophies, and programs. In R. M. Lerner, M. A. Easterbrooks, and J. Mistry (Eds.) (Editor-in-Chief: I. B. Weiner), *Handbook of Psychology, Vol. 6: Developmental Psychology* (2nd ed., pp. 365–392). Hoboken, NJ: Wiley.

Lerner, R. M., Almerigi, J. B., Theokas, C., and Lerner, J. V. (2005). Positive youth development: A view of the issues. *Journal of Early Adolescence*, *25*(1), 10–16. https://doi.org/10.1177/0272431604273211

Lerner, R. M., Lerner, J. V., and Benson, J. B. (2011). Positive youth development: Research and applications for promoting thriving in adolescence. *Advances in Child Development and Behavior*, *41*, 1–17.

Lewis, N. A., and Hill, P. L. (2021). Sense of purpose promotes resilience to cognitive deficits attributable to depressive symptoms. *Frontiers in Psychology*, *12*, 698109. https://doi.org/10.3389/fpsyg.2021.698109

Mansfield, R., Santos, J., Deighton, J., Hayes, D., Velikonja, T., Boehnke, J. R., and Patalay, P. (2022). The impact of the COVID-19 pandemic on adolescent mental health: A natural experiment. *Royal Society Open Science*, *9*(4), 211114. https://doi.org/10.1098/rsos.211114

Masten, A. S., and Obradović, J. (2006). Competence and resilience in development. *Annals of the New York Academy of Sciences*, *1094*(1), 13–27. https://doi.org/10.1196/annals.1376.003

Peterson, B., Gerhard, G., Hunter, L., Marek, L., Phillips, C., and Titcomb, A. (2001). *Prepared and engaged youth serving American communities: The National 4-H Impact Assessment Project*. Washington, DC: National 4-H Headquarters.

Polizzi, C. P., and Lynn, S. J. (2021). Regulating emotionality to manage adversity: A systematic review of the relation between emotion regulation and psychological resilience. *Cognitive Therapy and Research*, 1–21. https://doi.org/10.1007/s10608-020-10186-1

Resnick, M. D. (2000). Protective factors, resiliency, and healthy youth development. *Adolescent Medicine: State of the Art Reviews*, *11*(1), 157–164.

Stewart, D., and Sun, J. (2004). How can we build resilience in primary school-aged children? The importance of social support from adults and peers in family, school, and community settings. *Asia Pacific Journal of Public Health*, *16*(1_suppl), 37–41. https://doi.org/10.1177/101053950401600S10

Velez, G., Hahn, M., and Troyer, B. (2022). Making meaning of COVID-19: An exploratory analysis of US adolescent experiences of the pandemic. *Translational Issues in Psychological Science*, *8*(2), 269. https://doi.org/10.1037/tps0000326

Villarruel, F. A., Perkins, D. F., Borden, L. M., and Keith, J. G. (Eds.). (2003). *Community Youth Development: Programs, Policies, and Practices*. Thousand Oaks, CA: Sage.

Wu, G., Feder, A., Cohen, H., Kim, J. J., Calderon, S., Charney, D. S., and Mathé, A. A. (2013). Understanding resilience. *Frontiers in Behavioral Neuroscience*, *7*, 10. https://doi.org/10.3389/fnbeh.2013.00010

Towards Resilient Societies: The Synergy of Religion, Education, Health, Science, and Technology – Rahiem (Ed.)

The green façade: Exploring questionable green ICT practices

Ritesh Chugh
School of Engineering and Technology, Central Queensland University, Australia
ORCID ID: 0000-0003-0061-7206

Sangeetha Kutty
School of Engineering and Technology, Central Queensland University, Australia
ORCID ID: 0000-0002-1360-0457

ABSTRACT: This study investigates questionable green ICT practices among major technology companies, revealing significant discrepancies between corporate sustainability claims and actual environmental practices. Using a case study approach supported by a review of media and third-party reports, the research identifies key issues such as greenwashing, inadequate e-waste management, and the questionable effectiveness of carbon offsets. The findings highlight that despite companies like Amazon, Google, Apple, and Microsoft promoting robust green initiatives, their practices often fall short of genuine sustainability, with concerns including misleading claims about energy sources, insufficient lifecycle assessments, and ineffective recycling programs. The study emphasises the need for improved transparency, better e-waste management practices, stricter regulatory oversight, and a critical examination of carbon offset practices. These findings align with UN Sustainable Development Goals (SDGs) 12 and 13, which advocate for sustainable consumption, production, and urgent action to combat climate change, urging corporations to adopt transparent and accountable sustainability practices that contribute to environmental preservation and climate resilience. Achieving progress in ICT sustainability will require not only credible environmental claims but also actionable strategies that deliver tangible outcomes for a sustainable future.

Keywords: carbon offsets, e-waste management, green computing, green ICT, greenwashing, ICT, ICT companies, lifecycle assessments, renewable energy, transparency

1 INTRODUCTION

The rapid advancement of Information and Communication Technology (ICT) has significantly transformed modern society, reshaping industries, economies, and daily life. As ICT becomes more pervasive, its environmental footprint has also grown, raising concerns about the sector's sustainability (Belkhir and Elmeligi2018). The concept of "green ICT" or "green computing" has emerged to address these concerns. It advocates for practices that reduce energy consumption, minimise electronic waste (e-waste), and promote renewable resources (Hankel *et al.* 2018; Verdecchia *et al.* 2017). These practices include designing energy-efficient hardware, creating repairable electronic devices, implementing sustainable data centre practices, and encouraging the responsible disposal of e-waste.

However, the push towards greener ICT is fraught with challenges, including the prevalence of questionable practices that undermine genuine sustainability efforts. One such practice is greenwashing, where companies make exaggerated or misleading claims about the environmental benefits of their products or operations (McLean and McLean 2020).

DOI: 10.1201/9781003645542-2

Greenwashing not only misguides consumers but also diverts attention from real environmental issues, weakening collective efforts to tackle global challenges such as climate change and resource depletion (Delmas and Burbano 2011; Lyon and Montgomery 2015). It can create a false sense of progress while perpetuating harmful practices, ultimately slowing the transition to truly sustainable solutions. Additionally, the issue of electronic waste (e-waste) is a critical concern in the ICT industry (Chugh *et al.* 2016). Despite many companies claiming robust recycling programs, instances of e-waste being shipped to countries with inadequate environmental regulations continue to surface (Andeobu *et al.* 2023). This poses environmental risks and raises ethical issues regarding the responsible management of hazardous materials. Another area of concern is the use of carbon offsets by companies to claim carbon neutrality (Guix *et al.* 2022). While offsets can play a role in broader sustainability strategies, they can also be used to mask ongoing emissions if not accompanied by substantial reductions in actual carbon output.

These examples illustrate the complex and often contradictory landscape of green ICT. As the industry faces growing pressure to demonstrate environmental responsibility, it is crucial to critically examine corporate claims and practices. There is a need for greater transparency, accountability, and regulatory oversight to ensure that sustainability efforts are genuine and not merely superficial marketing strategies. The findings of this study could inform policy development and industry standards, urging stricter regulations and more transparent reporting of sustainability practices to prevent greenwashing and encourage authentic environmental responsibility.

Therefore, this study aims to address the following research question: *What are the questionable green ICT practices in the ICT industry?*

The remainder of the paper is organised as follows. Section 2 provides a review of the literature on greenwashing and sustainability practices in the ICT sector, while Section 3 outlines the methodology used to examine the environmental claims of selected ICT companies. Section 4 presents the findings and analysis, highlighting the discrepancies between corporate claims and actual practices. Section 5 outlines a discussion, followed by the implications in Section 6. Finally, Section 7 concludes the paper by summarising the key insights and suggesting directions for future research.

2 LITERATURE REVIEW

Greenwashing involves companies misleading consumers about the environmental benefits of their products or practices. Delmas and Burbano (2011) describe greenwashing as the gap between the positive communication of environmental performance and the actual, less favourable performance. In the ICT sector, greenwashing may include exaggerating energy efficiency claims, misrepresenting data centre impacts, or not disclosing non-renewable energy use. Such practices can erode trust and lead to consumer scepticism, diminishing the impact of authentic green marketing (Lopes *et al.* 2023). The complexity of ICT supply chains often obscures environmental claims, compounded by insufficient regulatory oversight and inconsistent standards, which permit selective disclosure of favourable information (DNI 2022; Hilty and Aebischer 2015; Köhler and Erdmann 2004).

Electronic waste (e-waste) represents a major environmental issue in ICT. The rapid obsolescence of devices and increasing technology demand have escalated e-waste production. Baldé (2017) reports that global e-waste reached 44.7 million metric tons in 2016, with only 20% recycled properly. Improper e-waste disposal poses severe environmental and health risks due to hazardous materials like lead, mercury, and cadmium. The research discusses strategies such as recycling, refurbishment, and extended producer responsibility programs, but effectiveness is often hindered by inadequate infrastructure, poor regulatory enforcement, and global e-waste (Leclerc and Badami 2020; Liu *et al.* 2023; Resmi and Fasila 2017). Much e-waste from developed countries is exported to developing nations, where it is often processed under unsafe conditions (Abalansa *et al.* 2021). The circular economy concept, emphasising material reuse, refurbishment, and recycling, is emerging as a solution (Morseletto 2020).

Carbon offsets are used by companies to mitigate their carbon footprint by investing in emission-reducing projects, like reforestation or renewable energy (Günther *et al.* 2020). However, their efficacy is debated. Bumpus and Liverman (2008) argue that offsets should complement, not replace, direct emissions reductions. In the ICT sector, reliance on offsets may enable companies to maintain status quo practices rather than implement substantial emission reductions (Dowlat 2023). Effective carbon offsets depend on project integrity, as poorly managed projects may not deliver the promised reductions, undermining offset credibility.

Lifecycle assessment (LCA) evaluates a product's environmental impact throughout its lifecycle, from raw material acquisition to disposal (Finnveden *et al.* 2009). For ICT products, LCA is essential for understanding full environmental impacts, including energy use, resource consumption, and waste generation. However, LCA application in ICT is challenging due to rapid technological advancements and diverse products. Arushanyan (2016) emphasises considering both direct impacts (e.g., energy use) and indirect impacts (e.g., manufacturing and disposal). Despite available LCA tools, many companies fail to account for indirect impacts. Researchers have noted that manufacturers often emphasise energy efficiency during use but neglect the environmental costs of production and disposal (Kunkel and Matthess 2020; Lieder and Rashid 2016). Telenko and colleagues (2016) argue that design for environment (DfE) principles can significantly mitigate these impacts by reducing energy consumption and resource use.

Lastly, regulatory frameworks are crucial for promoting sustainability in the ICT sector, though their effectiveness varies by region. The European Commission's Waste from Electrical and Electronic Equipment (WEEE) Directive is a robust framework that sets standards for e-waste collection, recycling, and recovery (European Commission 2023). Nevertheless, enforcement and compliance can be difficult, especially in countries with weaker regulations. Consumer perceptions of environmental, social, and governance (ESG) initiatives are also significant. Research indicates that consumers are increasingly aware of ESG issues and prefer companies with genuine sustainability efforts (Frey *et al.* 2023). However, greenwashing can breed scepticism and undermine trust.

3 METHODOLOGY

The study employs a case study methodology to investigate questionable green ICT practices in the ICT industry through media and third-party reports. This approach highlights the influence of external narratives on public perceptions and critiques of corporate sustainability (Camilleri, 2017). Case study research involves an investigation of one or more instances of observable phenomena within a specific context, utilising multiple sources of evidence to gain a comprehensive understanding of the case (Turnbull *et al.* 2021). In this instance, data is collected from various media outlets and third-party organisations to evaluate the environmental claims and practices of selected ICT companies.

Cases are selected based on a set of criteria: the prominence and influence of the ICT companies, their history of public sustainability commitments, and any related controversies or criticisms of their environmental practices. Selection prioritises large multinational corporations with substantial media coverage, particularly those known for making significant sustainability claims. These criteria ensure that the companies under scrutiny are representative of prominent players in the industry, while controversies surrounding their environmental practices provide insight into potential greenwashing. Hence, this study's purposive shortlist of ICT companies includes Amazon, Google, Apple, and Microsoft, which have been publicly scrutinised for their environmental practices.

Media reports include news articles, feature stories, and opinion pieces from both mainstream and specialised publications such as BBC News, The Guardian, Bloomberg, The Conversation, The Verge, Computer Weekly, Computing (The Channel Company), The Interoopt, Carbon Herald, W.Media, MIT Technology Review, Digital Trends and Drilled.

Third-party reports come from NGOs, environmental watchdogs, and independent research institutions like Greenpeace, Stand.earth, Ethical Consumer, and InfluenceMap. These sources provide insights into corporate claims and potential greenwashing through research and industry benchmarking (Park and Brorson 2005; Yue and Li 2023).

The collected data is analysed using qualitative content analysis, which involves coding and categorising textual data to identify themes and patterns (Schreier 2012). The coding process begins with an open coding phase where key concepts, terms, and indicators of greenwashing (such as unsubstantiated claims or inconsistent environmental practices) are identified in the text. These initial codes are then grouped into broader themes that emerge from the data, such as 'carbon offsets,' 'e-waste management,' and 'greenwashing indicators.' The analysis then refines these themes, allowing for a deeper understanding of how these companies present their environmental claims. Key aspects of the analysis include identifying greenwashing indicators, critiques of lifecycle assessments, e-waste management practices, and the evaluation of carbon offsets. Each of these aspects contributes to understanding the credibility and effectiveness of the companies' environmental claims and practices.

Since the study relies solely on secondary data sources, there are inherent strengths and limitations. A strength of this approach is the ability to analyse a wide range of publicly available data, offering diverse perspectives on the companies' environmental practices. However, a limitation is that secondary data may be subject to biases, as media outlets and third-party organisations may present selective or incomplete information. Additionally, the absence of primary data collection limits the ability to verify the accuracy of claims directly from the companies themselves.

4 FINDINGS AND ANALYSIS

Several questionable green ICT practices across the examined companies have been identified and are outlined in the following subsections.

4.1 *Amazon*

4.1.1 *Greenwashing and energy transparency*

Amazon has frequently promoted its commitment to renewable energy, claiming substantial investments in green projects. However, Amazon's reported energy consumption has been inconsistent with its claims of environmental sustainability, raising questions about the validity of its green credentials (Ethical Consumer 2020). For example, although Amazon asserts that it achieved 100% renewable energy consumption for its operations in 2023, critics argue that this claim may not reflect the actual energy mix powering its data centres (Donnelly 2024). In a report, Greenpeace criticised Amazon for being opaque about its energy mix, suggesting that a significant portion of its power could still be derived from fossil fuels, thereby questioning the authenticity of its environmental claims (Greenpeace 2019).

4.1.2 *E-waste management transparency*

Amazon has also been critiqued for its e-waste management policies (Laville 2020). While the company claims to have robust recycling programs, there is limited public information on the actual processes and the fate of decommissioned equipment (Horwood 2023). This lack of transparency makes it challenging to assess the effectiveness of Amazon's e-waste management strategies.

4.2 *Google*

4.2.1 *Fossil fuel engagement and increased emissions*

Google has positioned itself as a leader in renewable energy investments, often highlighting its efforts to power its global operations with 100% renewable energy (Google 2016, 2021). Despite its claims of a commitment to sustainability, Google continues to provide cloud services to fossil fuel companies, including Saudi Aramco. This partnership has raised concerns about the company's environmental integrity, particularly as it supports operations that contribute to carbon emissions (Biddle 2023). Google's greenhouse gas emissions have reportedly increased by 48% over the past five years, largely due to energy-intensive data centres, contradicting its goal of achieving net-zero carbon emissions by 2030 (Rahman-Jones 2024).

4.2.2 *Carbon offset credibility and impact*

Google uses carbon offsets to balance out its carbon footprint, claiming to be carbon-neutral since 2007 (Rathi 2024). However, the effectiveness of these offsets has been questioned, as critics argue that many offset projects may overestimate their impact and that some projects would have occurred regardless of offset funding (Moises 2024). This has led to concerns about the credibility of the offsets used by Google, particularly after the company purchased nearly 3 million tons of offsets in 2022 (George 2024). Furthermore, Google's recent shift away from purchasing carbon credits in favour of direct emissions reductions highlights a recognition of the limitations of its previous offset strategy, suggesting that its earlier claims of carbon neutrality may not have been fully substantiated (Temple 2024).

4.3 *Apple*

4.3.1 *Planned obsolescence and recycling inefficiencies*

Apple markets its products as environmentally friendly, often highlighting its use of recycled materials and energy-efficient designs (Apple 2024). However, Apple has been criticised for practices that could lead to increased e-waste, such as designing products with limited repairability (Laville 2020). The company's emphasis on recycling, while beneficial, does not fully address the issues associated with planned obsolescence and the limited lifespan of its products. Its design choices limit product reparability, potentially leading to increased e-waste as consumers replace rather than repair their devices (Chugh 2021).

Some media reports question the effectiveness of Apple's take-back and recycling programs. Critics argue that while Apple's initiatives, such as the Apple Reuse and Recycling Program, are well-intentioned, they may not achieve the level of environmental impact the company claims. For instance, some reports highlight that the financial incentives for consumers to participate in these programs are often lower than selling their devices independently (Blake 2023). Additionally, incidents of mismanagement by recycling partners have raised concerns about the actual efficiency of these efforts (Carr 2024).

4.4 *Microsoft*

4.4.1 *Emissions increase and offset reliance*

Microsoft has set ambitious goals to become carbon-negative by 2030, primarily through reducing emissions, using renewable energy, investing in carbon removal technologies, and moving beyond carbon offsets (Smith 2020). However, Microsoft's climate strategy has been criticised for a 46% increase in overall emissions since its 2020 pledge, inadequate renewable energy use, insufficient efforts to decarbonise its supply chain, and reliance on problematic carbon offsets (Zhang 2024). Additionally, there are reported conflicts between Microsoft's sustainability goals and its partnerships with fossil fuel companies focused on automating and accelerating oil extraction (Stone 2024). These issues suggest that Microsoft's direct business operations might not align fully with its environmental commitments, thereby questioning the integrity of its carbon neutrality claims.

4.4.2 *Greenwashing and technology viability*

While Microsoft aims to achieve carbon negativity by 2030, some critics highlight that the technologies required for significant carbon removal, such as direct air capture and advanced carbon sequestration methods, are not yet commercially viable at the scale necessary to meet

Microsoft's ambitious targets (Roberts 2020). The reliance on these technologies raises doubts about whether Microsoft can genuinely deliver on its promises without substantial advancements in the field. Moreover, while Microsoft has set clear goals and invested in various projects, the implementation and actual environmental benefits are still uncertain due to the limitations of current technology and standards in the carbon removal industry. This raises concerns about the potential for greenwashing, where positive communication about environmental efforts overshadows more problematic aspects of their operations. In addition, Microsoft's climate goals appear to be going awry as the company's growing investment in energy-intensive AI operations has led to a significant increase in its greenhouse gas emissions (Calma 2024), raising concerns about greenwashing and the transparency of its commitment to becoming carbon-negative. In addition, evidence of Microsoft's engagement in climate change policy from 2023 to the present is somewhat limited (InfluenceMap 2024).

5 DISCUSSION

The analysis of green ICT practices reveals several discrepancies between corporate claims and actual practices, highlighting the prevalence of greenwashing and the complexity of assessing sustainability. Amazon's claims of renewable energy use are undermined by a lack of transparency about the actual sources used in its data centres. Similarly, despite investing heavily in renewable energy, Google and Microsoft maintain relationships with fossil fuel companies, raising doubts about the sincerity of their environmental commitments (Biddle 2023; Stone 2024). This discrepancy between public claims and actual practices undermines consumer trust and risks normalising superficial sustainability efforts. It reflects a broader trend where marketing strategies often overshadow genuine environmental impact, highlighting the need for stricter regulations and standards that require verifiable data on green claims (de Freitas Netto *et al.* 2020). Such findings resonate strongly with UN Sustainable Development Goal (SDG) 12, which emphasises sustainable consumption and production, urging organisations to adopt transparent and accountable sustainability practices (United Nations 2015). These discrepancies also conflict with SDG 13, which calls for urgent action to combat climate change and its impacts, underscoring the importance of aligning corporate practices with genuine climate goals.

Lifecycle assessments and carbon offsets often fall short of achieving corporate sustainability goals. Despite its carbon neutrality claims, Google's continued support for fossil fuel exploration highlights the limitations of relying on offsets without addressing core emissions issues (Biddle 2023; Moises 2024). Effectiveness concerns are further compounded by additionality, verification standards, and transparency issues in offset reporting (Pan *et al.* 2022). Offsets should serve as a supplementary measure rather than the primary strategy for achieving carbon neutrality.

E-waste management remains a critical challenge, as shown by critiques of Amazon and Apple's practices. Although Apple has improved recycling and the use of recycled materials, its product designs, such as limited repairability, contribute to the e-waste problem (Chugh 2021). Effective e-waste management requires extended producer responsibility through reusability, repairability, and robust take-back schemes, alongside greater transparency in disposal and recycling processes (Leclerc and Badami 2020). Efforts to address e-waste are integral to achieving SDG 12, as it prioritises sustainable management and the reduction of waste generation through prevention, reduction, recycling, and reuse. Addressing e-waste also contributes to SDG 13 by reducing the environmental impact of ICT products, helping mitigate the long-term effects of waste accumulation and resource depletion.

Finally, regulatory and market forces play a crucial role in shaping corporate environmental practices. Media and third-party scrutiny pressure companies to demonstrate genuine sustainability efforts (Park and Brorson 2005). Without stringent regulations, however, companies may persist in greenwashing and rely on superficial solutions like carbon offsets. Consumers sceptical of carbon offsets view reducing emissions as a more sustainable option (Utgård and Gaustad 2024). Stronger regulations, greater transparency, and financial incentives—such as tax breaks for green innovations, subsidies for renewable energy adoption, and penalties for excessive carbon emissions—are essential to rewarding genuine sustainability efforts as stakeholders increasingly demand more substantive environmental strategies (Lu *et al.* 2020). Policymakers must consider the intersection of SDG 13 and these regulatory frameworks to ensure that industry efforts genuinely contribute to combating climate change and not just superficial claims.

6 PRACTICAL IMPLICATIONS

The findings highlight the need for greater transparency in corporate environmental practices. For ICT companies, this means providing detailed, verifiable information about their sustainability efforts, including energy sources, lifecycle assessments, and carbon offset projects. Companies should adopt more rigorous reporting standards and ensure that independent audits substantiate their public claims. This increased transparency can help build consumer trust and reduce the risk of being accused of greenwashing.

The critiques of e-waste management practices indicate a pressing need for more effective strategies. Companies should focus on designing products for longevity, repairability, and recyclability. Implementing robust and economical take-back and recycling programs is crucial, along with ensuring that these programs are managed transparently and ethically. By addressing these issues, companies can mitigate their environmental impact and contribute to a circular economy.

While the findings did not explicitly identify regulatory frameworks, some of the issues discussed are related to or could be influenced by regulatory frameworks. For instance, stricter regulatory requirements could address the lack of transparency in Amazon's energy claims or Apple's recycling inefficiencies. Regulatory frameworks could mandate more detailed and transparent reporting of energy sources and consumption, as well as comprehensive reporting on the effectiveness of take-back and recycling programs, including financial incentives for consumers and the performance of recycling partners, to prevent misleading claims. Policymakers should consider introducing regulations that require ICT companies to disclose their entire energy mix, mandate third-party verification of carbon offset projects, and enforce minimum standards for e-waste management that ensure compliance with environmental and ethical guidelines.

The growing awareness of greenwashing also highlights the importance of education and engagement. Companies should focus on educating their customers about the true environmental impact of their products and services. Providing clear, accessible information about sustainability practices and engaging in open dialogues with stakeholders can enhance credibility and foster a more informed consumer base. Industry leaders can collaborate with educational institutions and environmental NGOs to develop comprehensive sustainability literacy programs aimed at both consumers and employees, enhancing overall awareness of greenwashing and sustainable practices.

Companies should reassess their sustainability investments to ensure that they deliver tangible environmental benefits. This involves evaluating the effectiveness of carbon offset projects, prioritising direct emissions reductions, and investing in innovative technologies that contribute to sustainable practices. It is also recommended that industry leaders advocate for establishing clear guidelines on carbon offsetting practices, including adopting internationally recognised standards to enhance the credibility of claims in the market.

7 CONCLUSION

This study has explored the questionable green ICT practices of leading companies such as Amazon, Google, Apple, and Microsoft, using media and third-party reports as primary data sources. The analysis has revealed significant issues, including greenwashing, challenges in e-waste management, and the complexities surrounding carbon offsetting.

The findings illustrate the gap between corporate sustainability claims and actual practices, highlighting the prevalence of greenwashing and the need for greater transparency and accountability in environmental reporting. Companies often present a façade of environmental responsibility while falling short in areas such as energy sourcing, lifecycle management, and e-waste disposal. The critique of carbon offsets also illustrates the limitations of relying on these mechanisms without addressing direct emissions reductions and systemic sustainability challenges. Practically, the results suggest that ICT companies must enhance their transparency, improve their e-waste management strategies, and engage more proactively with regulatory frameworks.

Despite the contributions, this study has its limitations. First, the reliance on media and third-party reports means that the findings are subject to the biases and limitations inherent in these sources. Media coverage can be selective and may not fully capture the complexity of corporate practices, while third-party reports may vary in their methodologies and focus. Second, the study's scope is limited to a small number of high-profile ICT companies, which may not be representative of the broader industry. Future research could benefit from a more diverse sample of companies and a combination of primary data collection methods to provide a more comprehensive view.

REFERENCES

Abalansa, S., El Mahrad, B., Icely, J., and Newton, A. (2021). Electronic waste, an environmental problem exported to developing countries: The good, the bad and the ugly. *Sustainability*, *13*(9), 1–24.

Andeobu, L., Wibowo, S., and Grandhi, S. (2023). Informal E-waste recycling practices and environmental pollution in Africa: What is the way forward? *International Journal of Hygiene and Environmental Health*, *252*, 114192.

Apple. (2024). *Environmental Progress Report*. https://www.apple.com/au/environment/pdf/Apple_Environmental_Progress_Report_2024.pdf

Arushanyan, Y. (2016). *Environmental impacts of ICT: present and future*, KTH Royal Institute of Technology, 1–60. https://kth.diva-portal.org/smash/get/diva2:933594/FULLTEXT01.pdf

Baldé, C. P., Forti, V., Gray, V., Kuehr, R. and Stegmann, P. (2017). *The Global E-waste Monitor 2017: Quantities, Flows, and Resources*. https://collections.unu.edu/eserv/UNU:6341/Global-Ewaste_Monitor_2017__electronic_single_pages_.pdf

Belkhir, L., and Elmeligi, A. (2018). Assessing ICT global emissions footprint: Trends to 2040 and recommendations. *Journal of Cleaner Production*, *177*, 448–463.

Biddle, S. (2023). *Google greenwashes a dirty partnership with climate-destroying Saudi Aramco*. The Intercept. https://theintercept.com/2023/02/17/google-cloud-saudi-aramco-oil/

Blake, A. (2023). *People are slamming Apple for its 'brutal' trade-in prices*. Digital Trends. https://www.digitaltrends.com/computing/apple-trade-in-prices-brutally-low/

Bumpus, A. G., and Liverman, D. M. (2008). Accumulation by decarbonization and the governance of carbon offsets. *Economic Geography*, *84*(2), 127–155.

Calma, J. (2024). *Microsoft's AI obsession is jeopardizing its climate ambitions*. The Verge. https://www.theverge.com/2024/5/15/24157496/microsoft-ai-carbon-footprint-greenhouse-gas-emissions-grow-climate-pledge

Camilleri, M. A. (2017). *Corporate sustainability, social responsibility and environmental management: An Introduction to Theory and Practice with Case Studies*. Springer International Publishing.

Carr, A. (2024). *What really happens when you trade in an iPhone at the Apple store*. Bloomberg. https://www.bloomberg.com/news/features/2024-04-18/apple-iphone-recycling-program-has-secrets?

Chugh, R. (2021). *Screwed over: how Apple and others are making it impossible to get a cheap and easy phone repair*. The Conversation. https://theconversation.com/screwed-over-how-apple-and-others-are-making-it-impossible-to-get-a-cheap-and-easy-phone-repair-156871

Chugh, R., Wibowo, S., and Grandhi, S. (2016). Environmentally sustainable Information and Communication Technology usage: awareness and practices of Indian Information and Communication Technology professionals. *Journal of Cleaner Production*, *131*, 435–446.

de Freitas Netto, S. V., Sobral, M. F. F., Ribeiro, A. R. B., and Soares, G. R. d. L. (2020). Concepts and forms of greenwashing: A systematic review. *Environmental Sciences Europe*, *32*, 1–12.

Delmas, M. A., and Burbano, V. C. (2011). The drivers of greenwashing. *California Management Review*, *54* (1), 64–87.

DNI. (2022). *Information and communications technology and the supply chain risk.* https://www.dni.gov/files/NCSC/documents/supplychain/ict-supply-chain-risk-2022-5BE169B1-.pdf

Donnelly, C. (2024). *Cloud sustainability: Amazon employees question firm's renewable energy claims.* Computer Weekly. https://www.computerweekly.com/news/366596253/Amazon-employees-question-firms-renewable-energy-claims

Dowlat, C. (2023). *ICT: a sector disconnected from the climate reality?* https://www.carbon4finance.com/files/ICT_a_sector_disconnected_from_the_climate_reality.pdf

Ethical Consumer. (2020). *Amazon and Microsoft: Greenwashing in the technology industry?* Ethical Consumer. https://www.ethicalconsumer.org/technology/amazon-microsoft-greenwashing-technology-industry

European Commission. (2023). *Waste from Electrical and Electronic Equipment (WEEE).* European Commission. https://environment.ec.europa.eu/topics/waste-and-recycling/waste-electrical-and-electronic-equipment-weee_en

Finnveden, G., Hauschild, M. Z., Ekvall, T., Guinée, J., Heijungs, R., Hellweg, S., Koehler, A., Pennington, D., and Suh, S. (2009). Recent developments in life cycle assessment. *Journal of Environmental Management, 91*(1), 1–21.

Frey, S., Am, J., Doshi, V., Malik, A., and Noble, S. (2023). *Consumers care about sustainability—and back it up with their wallets.* https://www.mckinsey.com/industries/consumer-packaged-goods/our-insights/consumers-care-about-sustainability-and-back-it-up-with-their-wallets#/

George, V. (2024). *Google stops its mass purchase of carbon credits.* Carbon Herald. https://carbonherald.com/google-stops-its-mass-purchase-of-carbon-credits/

Google. (2016). *100% renewable is just the beginning.* https://sustainability.google/operating-sustainably/stories/announcement-100/

Google. (2021). *Four consecutive years of 100% renewable energy—and what's next.* https://cloud.google.com/blog/topics/sustainability/google-achieves-four-consecutive-years-of-100-percent-renewable-energy

Greenpeace. (2019). *Clicking clean Virginia: The dirty energy powering data center alley.* https://www.greenpeace.org/usa/reports/click-clean-virginia/

Guix, M., Ollé, C., and Font, X. (2022). Trustworthy or misleading communication of voluntary carbon offsets in the aviation industry. *Tourism Management, 88*, 1–13.

Günther, S. A., Staake, T., Schöb, S., and Tiefenbeck, V. (2020). The behavioral response to a corporate carbon offset program: A field experiment on adverse effects and mitigation strategies. *Global Environmental Change, 64*, 1–14.

Hankel, A., Heimeriks, G., and Lago, P. (2018). A systematic literature review of the factors of influence on the environmental impact of ICT. *Technologies, 6*(3), 1–19.

Hilty, L. M., and Aebischer, B. (2015). ICT for sustainability: An emerging research field. *ICT Innovations for Sustainability*, 3–36.

Horwood. (2023). *AWS V Microsoft Azure: Which is the most sustainable cloud platform?* Computing (The Channel Company). https://www.computing.co.uk/analysis/4113003/aws-microsoft-azure-most-sustainable-cloud-platform/page/4

InfluenceMap. (2024). *Microsoft:InfluenceMap score for climate policy engagement.* InfluenceMap. https://lobbymap.org//company/Microsoft-7d653460f792e07f6b11b6c4fe696330/EN/projectlink/Microsoft-In-Climate-Change

Köhler, A., and Erdmann, L. (2004). Expected environmental impacts of pervasive computing. *Human and Ecological Risk Assessment, 10*(5), 831–852.

Kunkel, S., and Matthess, M. (2020). Digital transformation and environmental sustainability in industry: Putting expectations in Asian and African policies into perspective. *Environmental Science and Policy, 112*, 318–329.

Laville, S. (2020). *Amazon and Apple 'not playing their part' in tackling electronic waste.* The Guardian. https://www.theguardian.com/technology/2020/nov/26/amazon-and-apple-not-playing-their-part-in-tackling-electronic-waste

Leclerc, S. H., and Badami, M. G. (2020). Extended producer responsibility for E-waste management: Policy drivers and challenges. *Journal of Cleaner Production, 251*(2020), 1–10.

Lieder, M., and Rashid, A. (2016). Towards circular economy implementation: a comprehensive review in context of manufacturing industry. *Journal of Cleaner Production, 115*, 36–51.

Liu, K., Tan, Q., Yu, J., and Wang, M. (2023). A global perspective on e-waste recycling. *Circular Economy, 2*, 1–06.

Lopes, J. M., Gomes, S., and Trancoso, T. (2023). The dark side of green marketing: How greenwashing affects circular consumption? *Sustainability, 15*(15), 1–17.

Lu, Y., Khan, Z. A., Alvarez-Alvarado, M. S., Zhang, Y., Huang, Z., and Imran, M. (2020). A critical review of sustainable energy policies for the promotion of renewable energy sources. *Sustainability*, *12*(12), 1–30.

Lyon, T. P., and Montgomery, A. W. (2015). The means and end of greenwash. *Organization and Environment*, 28(2), 223–249.

McLean, J., and McLean, J. (2020). Delivering green digital geographies? More-than-real corporate sustainability and digital technologies. *Changing Digital Geographies: Technologies, Environments and People*, 139–158.

Moises, H. (2024). *Google ends mass purchase of carbon offsets, aims for net-zero emissions by 2030.* https://w.media/google-ends-mass-purchase-of-carbon-offsets-aims-for-net-zero-emissions-by-2030/

Morseletto, P. (2020). Targets for a circular economy. *Resources, Conservation and Recycling*, *153*(2020), 1–12.

Pan, C., Shrestha, A., Innes, J. L., Zhou, G., Li, N., Li, J., He, Y., Sheng, C., Niles, J.-O., and Wang, G. (2022). Key challenges and approaches to addressing barriers in forest carbon offset projects. *Journal of Forestry Research*, *33*(4), 1109–1122.

Park, J., and Brorson, T. (2005). Experiences of and views on third-party assurance of corporate environmental and sustainability reports. *Journal of Cleaner Production*, *13*(10–11), 1095–1106.

Rahman-Jones, I. (2024). *AI drives 48% increase in Google emissions.* BBC. https://www.bbc.com/news/articles/c51yvz51k2xo

Rathi, A. (2024). *Google is no longer claiming to be carbon neutral.* Bloomberg. https://www.bloomberg.com/news/articles/2024-07-08/google-is-no-longer-claiming-to-be-carbon-neutral

Resmi, N., and Fasila, K. (2017). E-waste management and refurbishment prediction (EMARP) model for refurbishment industries. *Journal of Environmental Management*, *201*, 303–308.

Roberts, D. (2020). *Microsoft's astonishing climate change goals, explained.* Vox. https://www.vox.com/energy-and-environment/2020/7/30/21336777/microsoft-climate-change-goals-negative-emissions-technologies

Schreier, M. (2012). *Qualitative content analysis in practice.* Sage Publications.

Smith, B. (2020). *Microsoft will be carbon negative by 2030.* Microsoft. https://blogs.microsoft.com/blog/2020/01/16/microsoft-will-be-carbon-negative-by-2030/

Stone, M. (2024). *Microsoft employees spent years fighting the tech giant's oil ties. Now, they're speaking out.* Drilled. https://drilled.media/news/microsoft-employees

Telenko, C., O'Rourke, J. M., Conner Seepersad, C., and Webber, M. E. (2016). A compilation of design for environment guidelines. *Journal of Mechanical Design*, *138*(3), 1–11.

Temple, J. (2024). *Google, Amazon and the problem with Big Tech's climate claims.* MIT Technology Review. https://www.technologyreview.com/2024/07/17/1095019/google-amazon-and-the-problem-with-big-techs-climate-claims/

Turnbull, D., Chugh, R., and Luck, J. (2021). The use of case study design in learning management system research: A label of convenience? *International Journal of Qualitative Methods*, *20*, 1–11.

United Nations. (2015). Transforming our world: the 2030 agenda for sustainable development. https://sdgs.un.org/2030agenda

Utgård, J., and Gaustad, T. (2024). Carbon Emission Reductions and Offsets: Consumer Perceptions of Firm Environmental Sustainability. *Journal of Cleaner Production*, 470, 1–9.

Verdecchia, R., Ricchiuti, F., Hankel, A., Lago, P., and Procaccianti, G. (2017). Green ICT research and challenges. *Advances and New Trends in Environmental Informatics: Stability, Continuity, Innovation*, 37–48, Springer

Yue, J., and Li, Y. (2023). Media attention and corporate greenwashing behavior: Evidence from China. *Finance Research Letters*, *55*(B), 1–7.

Zhang, X. (2024). *Ctrl-Alt-Incomplete: The gaps in Microsoft's climate leadership.* https://stand.earth/wp-content/uploads/2024/02/Microsoft-An-IT-giant-lacking-leadership-on-climate-and-energy-2.pdf

Towards Resilient Societies: The Synergy of Religion, Education, Health, Science, and Technology – Rahiem (Ed.)

Iranian nuclear technology development and strategies against US sanctions 2018–2020

Adre Alir & M. Adian Firnas
Department of International Relations, Universitas Islam Negeri Syarif Hidayatullah Jakarta, Indonesia

ABSTRACT: This study examines Iran's nuclear development, its response to US sanctions between 2018 and 2020, and the broader implications for global peace, economic resilience, and international cooperation (SDGs 16 and 17). Iran's nuclear programme began in 1957 but was halted in 1979. Negotiations with the P5+1 led to the Joint Comprehensive Plan of Action (JCPOA) in 2015, from which the US withdrew in 2018, reinstating economic sanctions on Iran. In response, Iran strengthened bilateral relations and adopted innovative strategies, including discounted oil sales, to mitigate economic pressures. Using qualitative methods and secondary data sources, the study reveals that while sanctions achieved limited objectives, they failed to coerce Iran into compliance. Instead, Iran demonstrated economic resilience through diversified exports and continued advancement of its nuclear programme. The findings underscore the importance of inclusive diplomacy and sustainable conflict resolution as essential tools for fostering global stability and peaceful international cooperation.

Keywords: economic sanctions, Iran, JCPOA, nuclear, US

1 INTRODUCTION

Iran's nuclear program started in 1957 when under President Dwight Eisenhower's Atoms for Peace program, Iran and the United States signed a nuclear cooperation agreement on March 5th 1957 (Mousavian and Mousavian 2018). In 1974, Shah Reza Pahlavi announced an ambitious plan to develop multiple nuclear reactors immediately. In 1975, construction of the Bushehr nuclear power plant began and was completed in 2011, making it the first reactor in the Middle East (Vaez and Sadjadpour 2013).

In 2009, Iran built the Fordow Fuel Enrichment Plant (FFEP), located near the city of Qom. Iran also enriched uranium to 20% at the Natanz plant, higher than the 3.5% level necessary. Iran operates two enrichment stages at Fordow and plans to double its capacity (International Crisis Group 2012).

In 2015, Iran, European Union, and the P5+1 (US, China, Russia, France, the UK, and Germany) signed the Joint Comprehensive Plan of Action (JCPOA) (Council on Foreign Relations 2023). That agreement required Iran to reduce its uranium stockpile and refrain from building facilities with uranium exceeding 3.67% (European External Action Service 2015). Regrettably, on May 8th 2018, President Donald Trump withdrew the US from the JCPOA and imposed sanctions on Iran (Belal 2019). These sanctions significantly impacted Iran, given its status as one of the world's largest oil reserves. However, in 2019, Iran enriched uranium by more than 3.67% to 4.5% and announced that it would remove its commitment to research and development under the JCPOA (Nuclear Threat Initiative 2020).

DOI: 10.1201/9781003645542-3

2 METHODS

The qualitative research (Creswell and Creswell 2017) approach with descriptive methods was used in the current study. This study utilised secondary data sources that were collected using a literature review. Data sources were selected based on relevance to the research topic, credibility (e.g., peer-reviewed articles and authoritative reports), recency (preferably within the last ten years, unless foundational), and accessibility of reliable databases and repositories. Furthermore, the data analysis involved collecting relevant materials, filtering them based on the criteria, organising them systematically to identify key themes, and synthesising insights to draw meaningful conclusions aligned with the research objectives. This process ensures the study is grounded in reliable and relevant data.

3 FINDINGS & DISCUSSION

3.1 *The US sanctions on Iran and the implications*

A significant decrease in FDI occurred as a result of sanctions. However, during most of the sanctions period, FDI suffered losses in the short term, but these losses decreased slightly in the long term (Figure 1) (Heydarian *et al.* 2022).

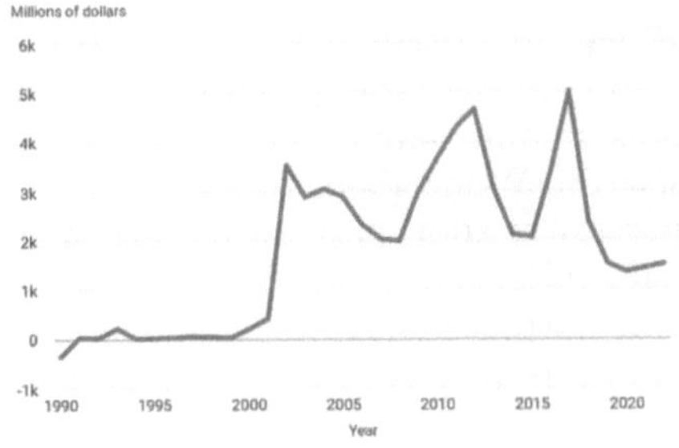

Figure 1. Iran's FDI chart 1990–2020.

In the case of Iran, Executive Order (EO) 13846, 13871, 13902, and 13949 directly resulted in implications for entities and individuals. Many foreign companies were sanctioned and stopped working with Iran. In 2018, FDI in Iran decreased 52.72%, 36.45% and 11.01% in 2019 and 2020 from the previous year (Macrotrends 2023a).

In 2018, Iran's oil sector also experienced declining production and exports (Figure 2). This decrease was primarily due to buyers' efforts to reduce their reliance on Iranian oil. Before the US withdrew from the JCPOA in April 2018, Iran's oil exports reached 2.9 million bpd, but in October 2018, exports shrank to 1.8 million bpd (Ghasseminejad 2023). This downward trend continued in subsequent years, with Iran's oil exports falling to 0.97 million bpd in 2019 and less than 0.4 bpd in 2020 (Jungman and Jerry 2022; US Energy Information Administration 2021).

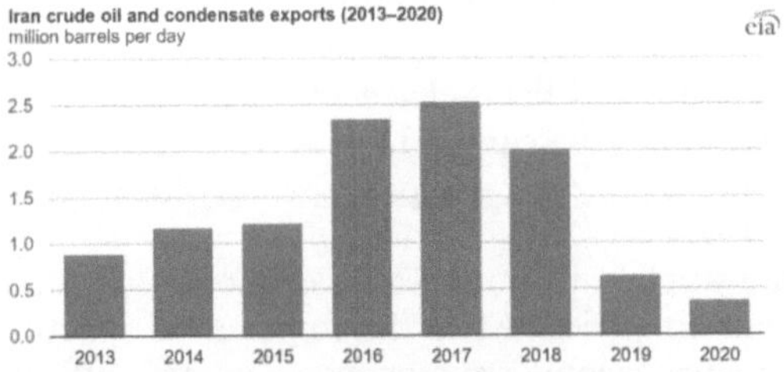

Figure 2. Iran crude oil and condensate export chart 2013–2020.

The inflation in Iran has increased significantly (Figure 3). Inflation increased in the short and long term due to sanctions caused by oil exports and the ban on purchasing US dollars (Iran International 2023; Ture and Khazaei 2022). In 2018, inflation reached 18.01%, which means an increase of 9.97% from the previous year; in 2019 it was 39.91%, an increase of 21.89% from the previous year. Although there was a decrease in inflation in 2020, dropping by 9.31% to 30.59%, it remains significantly higher than in previous years (Macrotrends 2023b; World Bank 2023).

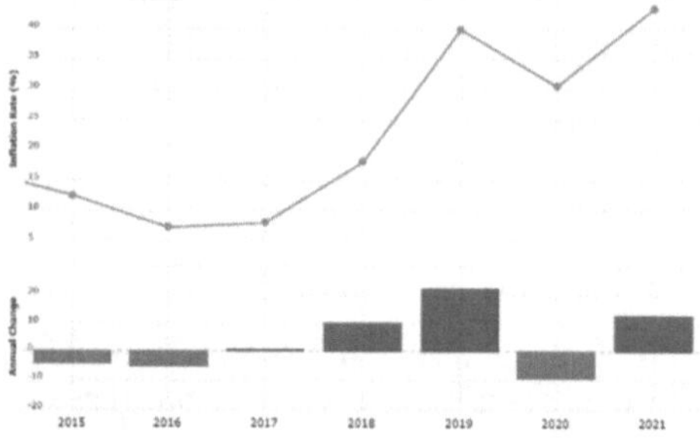

Figure 3. Iran's inflation percentage 2015–2020.

Furthermore, Iran's GDP has decreased. The main factors contributing to this decline were external disruptions to gas and oil production, including new sanctions on industries such as petrochemicals, metals, mining, and maritime, as well as the ending of US waivers for major importers of Iranian oil (World Bank 2019).

The macroeconomic dynamics in Iran, particularly the impact of sanctions, have had significant consequences on the microeconomy (Northam 2020). Unemployment and inflation rates have risen, leading to a worsening economy and a subsequent increase in domestic violence and suicides. Food markets and currency exchanges have also experienced volatility (Morris 2021). Iranian President Hassan Rouhani had accused the US of causing significant economic damage to Iran through sanctions. He stated that Iran had lost $150 billion because of these sanctions and described the US actions as "a barbaric crime" (Reuters 2020).

3.2 *Iran's strategy in response to US sanctions 2018–2020*

Ayatollah Khomeini's disappointment with European countries that did not help Iran made Khomeini try to change direction toward the East. Khomeini said that Iran should 'look' to the East rather than to the West because Europe did not act, which made Iran look underestimated due to its dependence on Europe (Farda 2018).

To avoid US sanctions, Tehran used innovative strategies. In January 2019, Iran offered Asian buyers a $1 per barrel discount. This was the biggest discount Iran had offered against Saudi prices in more than ten years. Iran's total oil exports went to Asia, mostly China, South Korea, India, Turkey, and Japan (The Iran Primer 2019).

Iran had also begun turning off tanker identification systems used by the International Maritime Organization to track vessel movements. In contrast, other vessels changed their names and numbers at sea to avoid being detected. In mid-2019, Iran also stored its oil in bonded storage at Chinese ports. Iran can find buyers in Asia without paying tariffs or other duties because the oil stored there does not pass through local customs or Chinese import data (The Iran Primer 2019).

Another approach Iran applied to improve its economy was cooperating with Turkey, Russia, and Central Asian countries. After the removal of sanctions, economic relations between Turkey and Iran have become a major focus for state officials from both countries.

While Turkey's non-oil exports to Iran remained unaffected in 2018 and 2019, Iran experienced a decline in exports. This has positioned Turkey as an important economic backup channel for Iran. Although Iran will suffer significant losses from the deteriorating economic relations, they are trying to prevent further deterioration, benefiting Turkey greatly. In an attempt to compensate for the lack of hard currency due to sanctions, Iran lowered the price of its gas exports to Turkey (Badawi 2020).

Russia and Iran also have mutual motivations to increase trade relations. Russia's alliance with Iran has expanded exports and strengthened its presence in the Middle East. Iran, currently facing sanctions, sees Russia as a significant market and is eager to benefit from its offerings. An agreement with the Eurasian Economic Union (EAEU) allows Iran to export various goods without tariffs or import duties. The EAEU will export goods like meat, sweets, cosmetics, machinery, and technical equipment to Iran, while Iran will export vegetables, fruits, building materials, and carpets to the EAEU (Islamian 2019).

Despite economic pressure, Iran strives to enhance its economy by engaging with Central Asian countries. One example is its progress in economic relations with Tajikistan and Uzbekistan (Kachmar 2022).

Despite the overall collapse of the Iranian economy, economic sanctions have not achieved the desired goals of the US. Iran's decision to restart its nuclear program is an example of the sanctions' failure. Since the ratification of the JCPOA on October 18th 2015, Tehran has been compliant. However, Iran disrupted the agreement by exceeding its water stockpile and uranium enrichment limits in May 2019. By January 2020, Iran had announced that they were no longer adhering to the operational restrictions of the JCPOA (Drezner 2022).

4 CONCLUSION

The US imposed economic sanctions on Iran from 2018 to 2020, causing disruptions in the oil sector and economic disruption in Iran. Iran experienced a weak economic condition, as indicated by decreased oil exports, FDI figures, GDP, and rising inflation. This can be observed by comparing before and during Iran's sanctions. The changes can be interpreted as the impact of sanctions imposed by the US in the 2018–2020 period through five EOs ordered by President Trump. Iran responded to sanctions in various ways, including export diversification, increased bilateral relations with certain countries—especially in Asia—and innovative strategies such as discounts on oil sales. However, this does not show evidence that the US goal of changing Iran's behaviour or coercing Iran was achieved, but only sanctions' constraining and signalling functions. This is apparent from Iran violating the agreement by exceeding its water supply and uranium enrichment limits and continuing to develop its nuclear technology program.

REFERENCES

Badawi, T. (2020, March 12). *The Economic Turn in Turkish-Iranian Relations*. Carnegie Endowment for International Peace. https://carnegieendowment.org/sada/81273

Belal, K. (2019). Uncertainty over the Joint Comprehensive Plan of Action: Iran, the European Union and the United States. *Policy Perspectives*, *16*(1), 23–39. https://doi.org/10.13169/polipers.16.1.0023

Council on Foreign Relations. (2023, June 21). *What Is the Iran Nuclear Deal?* https://www.cfr.org/backgrounder/what-iran-nuclear-deal

Creswell, J., and Creswell, J. D. (2017). *Research Design: Qualitative, Quantitative, and Mixed Methods Approaches*. SAGE Publications. https://us.sagepub.com/en-us/nam/research-design/book270550

Drezner, D. W. (2022). How not to sanction. *International Affairs*, *98*(5), 1533–1552. https://doi.org/10.1093/ia/iiac065

European External Action Service. (2015, July 14). *Joint Comprehensive Plan of Action*. https://eeas.europa.eu/archives/docs/statements-eeas/docs/iran_agreement/iran_joint-comprehensive-plan-of-action_en.pdf
Farda, R. (2018, October 18). *"Look East!," Says Khamenei, Disappointed With Europe's Efforts To Save JCPOA*. Radio Free Europe / Radio Liberty. https://en.radiofarda.com/a/look-east-says-khamenei-disappointed-with-europe-s-efforts-to-save-jcpoa/29550531.html
Ghasseminejad, S. (2023, September 28). *The Monetary Value of Relaxed Oil Sanctions Enforcement by the Biden Administration*. FDD. https://www.fdd.org/analysis/2023/09/28/the-monetary-value-of-relaxed-oil-sanctions-enforcement-by-the-biden-administration/
Heydarian, S., Pahlavani, M., and Mirjalili, S. H. (2022). The Impact of Financial Sanctions on Capital Inflow and Outflow (case of Iran). *Journal of Money And Economy*, *17*(1), 67–88. https://doi.org/10.52547/jme.17.1.67
International Crisis Group. (2012). *In Heavy Waters: Iran's Nuclear Program, The Risk Of War And Lessons From Turkey* [Dataset]. https://doi.org/10.1163/2210-7975_HRD-9812-0172
Iran International. (2023, October 18). *Iran Claims Currency Reserves Rise Amid Toothless US Sanctions*. Iran International. https://www.iranintl.com/en/202310178229
Islamian, K. (2019, November 20). *The Russia-Iran Relationship in a Sanctions Era*. https://www.ui.se/utrikesmagasinet/analyser/2019/november/the-russia-iran-relationship-in-a-sanction-era/
Jungman, C., and Jerry, C. (2022, October 6). *Analysis of Iranian Oil Sales Under President Trump vs. President Biden*. UANI. https://www.unitedagainstnucleariran.com/blog/analysis-of-iranian-oil-sales-under-president-trump-vs-president-biden
Kachmar, O. (2022, April 14). *Iran Eyes Opportunities in Central Asia*. New Lines Institute. https://newlinesinstitute.org/state-resilience-fragility/power-vacuums/iran-eyes-opportunities-in-central-asia/
Macrotrends. (2023a). *Iran Foreign Direct Investment 1970-2023*. https://www.macrotrends.net/countries/IRN/iran/foreign-direct-investment
Macrotrends. (2023b). *Iran Inflation Rate 1960-2023*. https://www.macrotrends.net/countries/IRN/iran/inflation-rate-cpi
Morris, K. (2021, November 17). *Intro To 2018 U.S. Sanctions On Iran | ACE*. https://ace-usa.org/blog/research/research-foreignpolicy/intro-to-2018-u-s-sanctions-on-iran/
Mousavian, S. H., and Mousavian, M. M. (2018). Building on the Iran Nuclear Deal for International Peace and Security. *Journal for Peace and Nuclear Disarmament*, *1*(1), 169–192. https://doi.org/10.1080/25751654.2017.1420373
Northam, J. (2020, January 16). Why Iran's Economy Has Not Collapsed Amid U.S. Sanctions And "Maximum Pressure." *NPR*. https://www.npr.org/2020/01/16/796781021/why-irans-economy-has-not-collapsed-amid-u-s-sanctions-and-maximum-pressure
Nuclear Threat Initiative. (2020, June 25). *Iran Nuclear Overview. The Nuclear Threat Initiative*. https://www.nti.org/analysis/articles/iran-nuclear/
Reuters. (2020, September 26). *Iranian president accuses U.S. of savagery after new sanctions. Reuters*. https://www.reuters.comarticle/idUSKBN26H0BJ/
The Iran Primer. (2019, September 11). *Iran's Increasing Reliance on China*. https://iranprimer.usip.org/blog/2019/sep/11/irans-increasing-reliance-china
Ture, H. E., and Khazaei, A. R. (2022). Determinants of Inflation in Iran and Policies to Curb It. *IMF Working Papers*, *2022*(181). https://doi.org/10.5089/9798400220555.001.A001
US Energy Information Administration. (2021, August 12). *Iran's crude oil production fell to an almost 40-year low in 2020*. https://www.eia.gov/todayinenergy/detail.php?id=49116
Vaez, A., and Sadjadpour, K. (2013). *Iran's Nuclear Odyssey: Costs and Risks—Carnegie Endowment for International Peace*. https://carnegieendowment.org/2013/04/02/iran-s-nuclear-odyssey-costs-and-risks-pub-51346
World Bank. (2019, October 9). *Iran's Economic Update—October 2019*. World Bank. https://www.worldbank.org/en/country/iran/publication/economic-update-october-2019
World Bank. (2023). *World Development Indicators | DataBank*. https://databank.worldbank.org/source/world-development-indicators/Series/FP.CPI.TOTL.ZG

Towards Resilient Societies: The Synergy of Religion, Education, Health, Science, and Technology – Rahiem (Ed.)

Leveraging SIREKAP for 2024 election accountability: Strengthening Indonesia's democratic resilience

A. Bakir Ihsan
Universitas Islam Negeri Syarif Hidayatullah Jakarta, Indonesia
ORCID ID: 0000-0002-5618-2486

Muhammad Farras Shaka & Davin Julian Hermes
Universitas Islam Negeri Syarif Hidayatullah Jakarta, Indonesia

ABSTRACT: This article examines the role of the Electronic Recapitulation Information System (SIREKAP) in strengthening Indonesia's **democratic resilience** during the 2024 elections. Using a qualitative approach, this study analyses data and documents to evaluate SIREKAP's effectiveness in ensuring transparent and accountable electoral processes. Findings reveal three key insights. First, SIREKAP signifies an enhancement over the preceding Vote Counting Information System (SITUNG), as it reduces state officials' direct involvement in vote counting and mitigates the risk of electoral fraud. Second, operational challenges, such as data reporting inconsistencies by the Election Organising Group (KPPS) and technical limitations in Optical Character Recognition (OCR), have triggered public concerns regarding data accuracy and transparency. Third, the General Election Commission (KPU) did not fully adopt SIREKAP as the primary tool for vote tabulation, limiting its contribution to sustaining electoral integrity and public trust. This study recommends enhancing SIREKAP's operational capabilities and establishing it as the central platform for transparent and accountable vote counting. Effective stakeholder collaboration is essential to fortify democratic resilience, ensuring electoral processes remain credible, transparent, and aligned with the Sustainable Development Goals (SDGs).

Keywords: accountability, election, KPU, KPPS, SITUNG, SIREKAP

1 INTRODUCTION

As stated in the famous Latin phrase "vox populi, vox dei", which translates to the voice of the people is the voice of god, this signifies how precious the people's votes are in elections, making it necessary to be protected at all cost (Turnbull-Dugarte 2019). One of the most essential values of elections is the accountability of the choices voiced by the people, regardless if it is a presidential or legislative election (Kam *et al.* 2020). One of the key challenges to Indonesia's democratic resilience lies in ensuring that electoral systems are efficient, accountable, and capable of withstanding attempts at manipulation, fraud, and public distrust (Aytimur and Bruns 2018; Jung 2017). The adoption of SIREKAP represents not merely a technical upgrade from SITUNG but also a strategic effort to strengthen the resilience of democratic institutions amidst evolving threats to electoral integrity.

A crucial issue observed in Indonesian general elections throughout history is how the people's votes can be safeguarded from any misdemeanour that may occur until an election is done, bigger funding becomes necessary to protect the people's votes and ultimately align the voting's results with the true choices of the people (Halida *et al.* 2022, Sugiharto and

DOI: 10.1201/9781003645542-4

Riyanti 2020). A tendency toward electoral fraud is quite apparent given the cases of violations that occurred in every election, the electoral manipulations involving the political candidates or even the election organisers for the sake of claiming victory, be it during presidential, legislative, mayoral or gubernatorial elections, and they had even been observed in village head elections. The integrity and accountability of elections, subsequently, pose a serious threat to the sustainability of democracy (Bayu 2017). In fact, in certain cases, elections lacking integrity and accountability have led to the emergence of leaders who tend to contradict the core substance of democracy by having restrictions on civil liberties and weak legislative control (Mujani and Liddle 2021).

Numerous efforts have been made to ensure accountable and honest elections through an efficient and effective vote tabulation system, such as the Indonesian government's use of the e-Recap system by means of a virtual private network (VPN) in the 2004 elections (Djuyandi *et al.* 2019). As time goes by from one election to the next, the measures applied to create stronger electoral accountability through the use of information technology have been gradually developed more and more, from using the C1 form scanning method for data entry to using Optical Character Recognition (OCR) and Optical Mark Recognition (OMR) (Lee 2018). Developing the use of information technology in elections is not simply about making the process more effective and efficient, it is also about making the process more secure and accountable, especially in terms of protecting and safeguarding the people's votes.

The link between elections and the various forms of electoral fraud has already been studied by many scholars from political, economic, or legal perspectives (Fortin-Rittberger *et al.* 2017). In theory, electoral fraud occurs for two reasons. First, it is caused by the distorted thinking and behaviour of the actors involved in the process (Aspinall *et al.* 2019). Second, it is caused by a fraudulent system, either the actors in the system or the supporting tools that do not work efficiently (Piccone 2018). The first factor requires a comprehensive and intensive approach to changing people's attitudes to obey the rules, while the second requires improvements and changes to the systems used in elections, including the existence of an accountable media in calculating results.

2 METHODS

The current analysis used qualitative methods to examine the uses of information technology for creating electoral effectiveness and accountability (Creswel 2018), by exploring and understanding the meanings of certain facts and data. What we consider "facts" in this analysis are documented data regarding the 2024 elections, especially those relating to SIREKAP. The data analysis process was done through inductive methods especially referring to SIREKAP usage as a specific case and the tendency of electoral flaws as a general case in electoral politics. Based on the descriptions on the function of information technology, especially SIREKAP in this case, and electoral accountability as the protector of voice of the people, showing the interlinking complexity between information technology as a system, on the one hand, and the level of consciousness or interest of the people in the other. Based on the above statement, a reduction was made to the various data collected in relation to SIREKAP and the 2024 Indonesian elections. The relevant data were then interpreted, and conclusions were drawn from the analysis. The results are expected to provide a new theoretical and practical basis for the organisation of fair, honest, and accountable elections and the sustainable strengthening of democracy.

3 FINDINGS AND DISCUSSION

This article analysed SIREKAP as a programmed technology media with a measurable system based on the targeted main objective, i.e., calculating accountable and transparent vote acquisition results (Kamaly *et al.* 2023). Applying this vote recapitulation information system is an effort to create better elections. Information technology's role in elections is crucial, and it goes hand in hand with the development of information technology that is becoming more easily accessible to the public (Bratt 2022).

Despite SIREKAP not being used as the primary and legitimate tool for counting votes, its presence maximises the information technology used in the 2019 election known as SITUNG. Several weaknesses in SITUNG usage that caused errors in data entry were detected following direct inspections by KPU and public reports. The occurrence of these data entry errors can be attributed to the suboptimal operation of SITUNG, a circumstance that has the potential to erode public confidence in the integrity of election organisers (Pratama 2023; Suriani and Kurniawan 2023).

Based on the SITUNG weaknesses stated above, examining the ability and capacity of SIREKAP in accumulating and processing the voter's data based on the C Plenary form from each polling station by the KPPS could lessen any possibility of flaws occurring. Presented in the following are some of the substances discussed in this article:

3.1 *Weaknesses of SITUNG application in the 2019 Indonesia election*

The analysis revealed several weaknesses in SITUNG usage causing data entry errors detected following the direct inspection by KPU and public reports. The public reported 82 data entry errors and 193 data entry errors based on KPU's inspection report (Zuhri 2019). Furthermore, the involvement of many other parties in using SITUNG also resulted in potential deviations; this occurred because vote counting reports were made after they were finished at the district level. The process, which started from the polling station, would have already gone through two stages, i.e., the residential and district level involving many people, thereby creating possible deviations in the specified vote counting results (Habibi *et al.* 2023).

3.2 *SIREKAP as the new tool and its weaknesses in the 2024 election*

The current case study highlights several weaknesses in the SIREKAP application during the 2024 election, namely SIREKAP's function in facilitating the recapitulation of electoral vote counting as stipulated by KPU Decree Number 66 of 2024. This information technology-based application is divided into two types, which are: the mobile version used by KPPS members in reporting calculations with the main database sourced from the C form at polling stations (TPS); the web version used by the sub-district election committee (PPK). KPU also uses the web version at the city/district and provincial levels to collect and aggregate data uploaded by KPPS via SIREKAP Mobile (Figure 1).

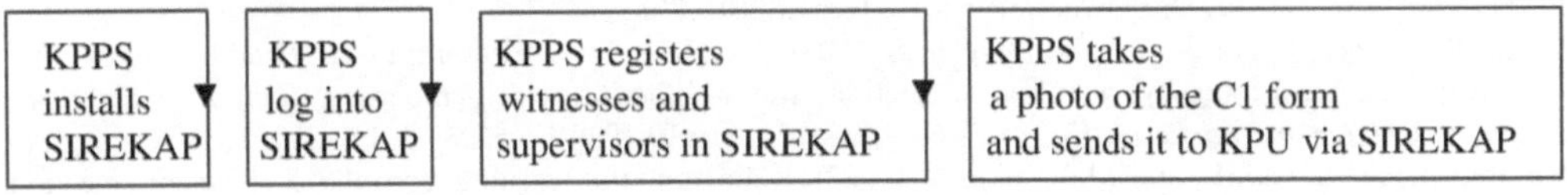

Figure 1. Flowchart of reporting process through SIREKAP.

There were several problems in operating the SIREKAP application. The first problem was data retrieval from the C1 form. The results were handwritten using OCR technology. Another problem was the installation of mobile stamps on each KPPS mobile phone. The different capabilities or qualities of mobile phones in capturing images also led to potential errors or levels of accuracy in reading data. As a result, there were differences in the C1 forms (Setiyawan and Mustofa 2024).

4 CONCLUSION

Based on the analysis, this article concluded that SIREKAP is a form of improvement to the latest SITUNG application, and it limits the possibility of national apparatus playing a role in vote counting in order to minimise potential electoral fraud. SIREKAP counted the voting results by scanning the C1 forms directly in all polling stations (TPS) under the authority of the team of election organisers (KPPS) and were sent directly to KPU's main server. Second, there were problems with using SIREKAP, which led to problems with election accountability. This is apparent in the fluctuation of data reporting by the KPPS, which led the public to believe that there was fraud in recapitulating the vote count results. Third, KPU did not consider SIREKAP the main tool for vote tabulation, leaving its function less than optimal in creating accountability in the electoral process. SIREKAP, despite its limitations, remains a crucial tool for reinforcing democratic resilience by safeguarding the transparency, accountability, and credibility of electoral processes.

The above findings show that SIREKAP has operational weaknesses and functional problems, rendering it lacking in strengthening accountability and transparency in executing elections. This article, thus, suggests that election organisers take proactive measures to improve the functionality of the SIREKAP application, particularly by implementing a policy that strengthens the SIREKAP application to establish it as the primary system for a transparent tabulation of votes. This policy requires the cooperation of all stakeholders to ensure its effective implementation. Consequently, partnership is crucial in conducting elections and is also a key concern stipulated in the SDGs. The implementation of these recommendations will ultimately foster public confidence to actively participate in elections to strengthen sustainable democratic resilience.

REFERENCES

Aspinall, E., Fossati, D., Muhtadi, B., and Warburton, E. (2019). Elites, masses, and democratic decline in Indonesia. *Democratization*, 27(4), 505–526. https://doi.org/10.1080/13510347.2019.1680971

Aytimur, R. E., and Bruns, C. (2018). Accountability with large electorates. *The Economic Journal*, 129(620), 1529–1560. https://doi.org/10.1111/ecoj.12618

Bayu, H. H. (2017). Redesain sertifikat hasil penghitungan suara di tempat pemungutan suara (TPS) dalam meningkatkan integritas hasil pemilu [Master's thesis, Universitas Airlangga].

Bratt, D. T. (2022). Alberta's transition to a two-party system: The 2015 and 2019 elections. *Canadian Political Science Review*, 16(1), 32–41. https://doi.org/10.24124/c677/20221838

Creswell, J. W., and Creswell, J. D. (2018). *Research design: Qualitative, quantitative, and mixed methods approaches*. Sage Publications.

Djuyandi, Y., Herdiansah, A. G., Yulita, I. N., and Sudirman, S. (2019). Using vote e-recapitulation as a means to anticipate public disorders in election security in Indonesia. *Humanities and Social Sciences Reviews*, 7(5), 111–122. https://doi.org/10.18510/hssr.2019.7515

Fortin-Rittberger, J., Harfst, P., and Dingler, S. C. (2017). The costs of electoral fraud: Establishing the link between electoral integrity, winning an election, and satisfaction with democracy. *Journal of Elections, Public Opinion and Parties*, 27(3), 350–368. https://doi.org/10.1080/17457289.2017.1310111

Habibi, M., Destavino, I., and Mahadika, A. (2023). Reflection and learning from the 2019 general election and the 2020 regional election: An analysis of the election implementation experience in Indonesia. *Kybernology: Journal of Government Studies*, 3(1), 1–16. https://doi.org/10.26618/kjgs.v3i1.12007

Halida, R., Susianto, H., Mujani, S., and Pratama, A. J. (2022). Vote-selling as unethical behavior: Effects of voter's inhibitory self-control, decision toward vote-buying money, and candidate's quality in Indonesia election. *Journal of Social and Political Psychology*, 10(2), 570–587. https://doi.org/10.5964/jspp.5643

Jung, D. (2017). Does partisanship hurt electoral accountability? Individual-level and aggregate-level comparisons of Western and Postcommunist democracies. *East European Politics and Societies: And Cultures*, 32(1), 168–201. https://doi.org/10.1177/0888325417728772

Kam, C., Bertelli, A. M., and Held, A. (2020). The electoral system, the party system and accountability in parliamentary government. *American Political Science Review*, 114(3), 744–760. https://doi.org/10.1017/s0003055420000143

Kamaly, N., Zulqiram, M., Faraidiany, M., and Afrijal, A. (2023). Democratic elections: Strategy to protect state civil apparatus from the threat of money politics and intimidation in general elections. *Malikussaleh Social and Political Reviews*, 4(2), 72. https://doi.org/10.29103/mspr.v4i2.13556

Lee, A. (2018). Guarding the elections online: New practices, trust, and empowerment of citizens identities. *Journal of Indonesian Social Sciences and Humanities*, 8(1), 11–28. https://doi.org/10.14203/jissh.v8i1.89

Mujani, S., and Liddle, W. (2021). Indonesia: Jokowi sidelines democracy. *Journal of Democracy*, 32(4), 72–86. https://dx.doi.org/10.1353/jod.2021.0053

Piccone, T. (2018). Democracy and digital technology. *Sur, International Journal on Human Rights*, 15(25). https://sur.conectas.org/en/democracy-and-digital-technology/

Pratama, K.J. (2023). Strengthening the vote counting information system of the general election commission: Regulatory and governance challenges. *Majalah Hukum Nasional*, 53(1). https://doi.org/10.33331/mhn.v53i1.218

Setiyawan, R., and Mustofa, Z. (2024). Comparison of the performance of naive Bayes and support vector machine in SIREKAP sentiment analysis with the lexicon-based approach. *Journal of Soft Computing Exploration*, 5(2), 122–132. https://doi.org/10.52465/joscex.v5i2.367

Sugiharto, I., and Riyanti, R. (2020). The problems with honest and fairness principles in general election in Indonesia. *Proceedings of the International Conference on Agriculture, Social Sciences, Education, Technology and Health (ICASSETH 2019)*. https://doi.org/10.2991/assehr.k.200402.038

Suriani, U., and Kurniawan, T. B. (2023). Comparing the prediction of numeric patterns on Form C1 using the k-nearest neighbors (K-NN) method and a combination of k-nearest neighbors (K-NN) with connected component labeling (CCL). *Journal of Information Systems and Informatics*, 5(4), 1569–1580. https://doi.org/10.51519/journalisi.v5i4.592

Turnbull-Dugarte, S. J. (2019). Explaining the end of Spanish exceptionalism and electoral support for Vox. *Research and Politics*, 6(2). https://doi.org/10.1177/2053168019851680

Zuhri, S. (2019). Urgensi pemanfaatan teknologi informasi dalam penghitungan dan rekapitulasi suara. *Jurnal Komisi Pemilihan Umum*. https://journal.kpu.go.id/index.php/ERE/article/view/135

Towards Resilient Societies: The Synergy of Religion, Education, Health, Science, and Technology – Rahiem (Ed.)

Social media and public opinion: Sweden's policy shift on Syrian refugees

Adelia Kinanti & Khoirun Nisa
Universitas Islam Negeri Syarif Hidayatullah Jakarta, Indonesia

ABSTRACT: This paper examines the role of social media technology in shaping public opinion and influencing Sweden's foreign policy shifts towards Syrian refugees during the 2015 refugee crisis. Using a qualitative approach, the study draws from literature reviews and electronic media sources, framed by the theory of freedom of speech, and the concepts of foreign policy and national interest. The findings reveal that the Swedish government utilised Twitter as a communication tool to share policy updates and engage with the public. At the same time, social media became a platform for Swedish citizens to express their views on refugee-related issues. Public opinion, amplified through Information and Communication Technology (ICT) platforms, significantly influenced the transition in Sweden's refugee policy. This study highlights the reciprocal relationship between public discourse on social media and policy-making processes, emphasising the growing role of digital platforms in shaping modern foreign policy decisions.

Keywords: foreign policy, public opinion, refugee, social media, Sweden, Syrian refugees

1 INTRODUCTION

The end of the cold war era marked an intensification in the use of more advanced technology. Social media has become one of the ICTs that Swedish people use to express their opinions. Especially during the influx of refugees, social media became a platform to voice their public opinion. Sweden has become one of the destination countries for refugees because of the ease of entry given and the concept of the Welfare State that it promotes (Hajighasemi and Oghazi 2022). Syrian refugees are refugees who come because of the instability of the country caused by various factors such as prolonged conflict and the *Arab Spring* wave in the Middle East (Tsourapas 2019). Syrian refugees come in massive numbers to Sweden because of the economic and political stability and easy access provided unlike other European countries.

This is interesting to discuss because it relates to the media transition in the collection of public opinion which is no longer traditional but rather utilises technology. The Swedish policy accommodates welfare state system, which prioritises the well-being of its society and the government in this case also facilitates social media platforms such as Twitter for the public to voice their opinions. Several researches have conducted regarding transition of Swedish refugee's policy. Maha (2021) found that Swedish feminist foreign is influenced by three main actors, namely individuals (leaders), the state and the system level. This foreign policy is the same regulation as the U-Turn policy after the wave of refugees in 2015. This is evidenced by the condition of a country as one of the main actors shaping foreign policy where in the case of the U-Turn policy, there are restrictions on reunification policies, this is due to Sweden upholds feminist values so that the policy change aims to provide protection to minority women who are burdened by early marriage (Nylund *et al.* 2023). Larasati

DOI: 10.1201/9781003645542-5

(2018) also found the existence of gender issues behind changes in Swedish policy towards refugees (Brännström and Gunneflo 2021).

Furthermore, the study of the economic impact of the presence of Syrian refugees in Sweden is also raised in Haodong Qi's article (Qi 2021) which summarizes there was a decrease income of Syrian immigrants in Sweden. This is because of most of the refugees derived from lack competencies skill that they may have a low level of income from labour and tend to rely on public transfers (Abdelhady and Al Ariss 2023). In general, the research has focused some factors and impacts of the change of Syria refugee policy, while the participation of Swedish people in refugee policy changes become an essential point to be discussed. The influx of Syrian refugees in Sweden has led to a clash of values, sparking debates on issues such as extremism and gender issues, which have influenced public opinion, particularly on social media. In response, the Swedish government has leveraged and synthesised these public voices to inform and shape policy changes (Berntzen 2018).

Thus, the issue to be raised in this paper is related to how the participation of Swedish society in political issues related to refugees through social media can influence and lead to changes in refugee policy in Sweden. This paper will be analysed using John Stuart Mill's theory of freedom of speech, which states in Mahon Rathnam that ideas and opinions need to be freely exchanged in order to lead to the possibility of a better society (Mahon Rathnam 2023). Rosenau's concept of foreign policy is defined as the strategies and actions that form the basis of a state in its relations with external actors on the international stage (Adesina 2017). In addition, the concept of national interest is also used as an analytical tool in this research where Felix E. Oppenheim states that national interest is the welfare that a national government seeks to achieve at the international scale in the aspects of security and welfare (Ota and Ecoma 2022). In this case, Syrian refugees as external actors, due to the majority of Swedish public opinion, lead to changes in foreign policy towards refugees.

2 METHODS

This research utilizes secondary data that was obtained through credible data sources from literature studies in the form of journals, books, websites, government records and so on. The research method used in this paper is qualitative method. This research aims at analysing Swedish people's participation in political issues regarding refugees through social media. The research uses Twitter as data that sourced from academic journals and was selected using specific keywords related to Sweden's policy shift on Syrian refugees, analysed through sentiment analysis and its potential influence on policy changes. Through these methods, the author seeks to explain the problem under study by producing descriptive and explanatory data. Therefore, qualitative methods are very compatible to understand the phenomena that occur where Swedish people participate in the refugee problem in their country which then leads to changes in refugee policies in their country.

3 FINDINGS AND DISCUSSION

The change in Sweden's policy towards Syrian refugees was motivated by the issues arising from Syrian refugees that led to public opinion formed among the Swedish people through social media platforms. Some of the policies have changed include the implementation of border checks in Sweden and Denmark. The next policy change is related to changes in residence permit policies and reunification of family members. The rules regarding family reunification were changed under the *Aliens Act* in 2016 which established restrictions on access to family reunification where refugees can apply for reunification within three months of obtaining legal status as a refugee (Hudson *et al.* 2023). The policy that also undergoes transition is labour market policy.

3.1 *Extremism issues implicated in public opinion on social media in Sweden*

Several incidents related to certain issues often lead public and government opinion into an unfavourable perspective towards refugees. Reporting from BBC News, there was a mosque bombing in Uppsala in 2015 (Nilsson and Esholdt 2022). The action was carried out by an anti-immigrant group which invited a variety of responses from the public, both those who support the anti-racism movement and those who have a negative perspective on Muslims (BBC News 2015). Based on survey data from the *Pew Research Center*, countries in Europe namely Hungary, Poland, Germany as the top three in the survey, tend to consider the refugee phenomenon as potentially increasing acts of terrorism in the domestic sphere of each country. Sweden is one of the countries that received more votes in the aspect of refugees potentially increasing terrorism activities, namely 57 votes (Figure 1).

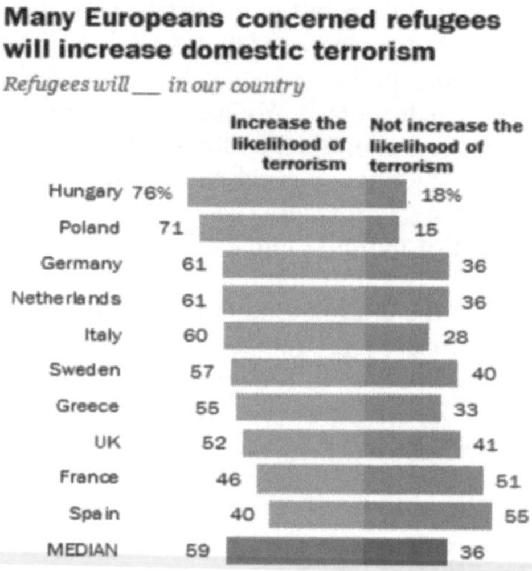

Figure 1. European countries‘ perspectives on refugees in 2015–2016 (BBC News 2015).

Moreover, there are networks of extremism in several Swedish cities. These networks recruit refugees from Syria and Iraq to join the *Islamic State* (IS) group in Syria. According to the BBC, there are more than 300 refugees in Sweden who went to Syria to fight with IS (Rostami *et al.* 2020). The presence of extremist networks in Sweden is in line with Barry Buzan and Lene Hansen's thoughts regarding the concept of security as a goal to be achieved in national interests. Buzan and Hansen mentioned non-traditional security that does not only come from the military aspect and the threat can come from inside and outside the country (Gheciu and Wohlforth 2018). Extremist networks are targeting Syrian refugees who have taken refuge in Sweden. These events have given momentum to the *Sweden Democrats* (SD) party, which is known for its anti-immigrant agenda, especially against Muslims. Sweden's foreign policy of accepting more than 156,000 refugees in 2015 with the majority being Syrian refugees has caused a crisis phenomenon in the country (Lyck-Bowen 2020). The SD party believes that the policy is too open and adding more refugees to the country will only cause problems that can threaten the welfare of the country (PBS 2016).The refugee crisis in 2015 and the Paris attacks in the same year further increased the SD party's ambition to influence policy in Sweden (Oksanen *et al.* 2020). The acts of terrorism were organized by the *Islamic State* (IS) militant group with shootings and bombings that left many people in critical condition.

In addition, the existence of anti-immigrant and anti-Muslim trends and in maintaining security aspects in Sweden, the Riksdag parliament is considering limiting access to incoming refugees through changes in several foreign policies towards refugees. This will reduce the number of refugees entering and settling in Sweden (Zamboni 2019). As Richard Snyder sees it, the government bureaucracy has an important role as decision makers in planning and formulating foreign policy (Kaarbo *et al.* 2024). So that the Riksdag parliament leads to policy regulations that create anti-immigrant trends. Based on this phenomenon, *hashtags* emerged in the Twitter social media application which became the *key hashtags* in the @socialdemokrat account both in the form of *Tweets* and *Retweets* as shown in Figure 2 below.

Table 1. Key Hashtags @socialdemokrat (Ts and RTs), November 01- December 15, 2015

Domestic Themes/Tags		International Themes/Tags	
#svpol	*Swedish politics*	#Paris	—
#MigPol	*Swedish immigration policy*	#Parisattacks	—
#sverigetilsammans	*"Sweden together" refugee program*	#terroristattack	—
#etableringspaket #etablering	*Government refugee settlement program*	#JHA	*EU Justice and Home Affairs Council*
#snabbspår	*Government refugee skills recognition fast-track program*	#FAC	*EU Foreign Affairs Council*
#eupol	*Swedish EU policy*	#security	—
#flykt	*Refuge/refugees*	#youth	—
#mothatbrott #NoHateSE	*Actions against hate attacks in Sweden*	#migration	—
#jobb	*Jobs*	#firearms	—
#skolan	*Schools*	#borders	—
#utbpol	*Swedish Education Policy*	#Ukraine	—
#aftonbladet	*The Aftonbladet*	#ASEM	
#svtnyheter	*SVT—Swedish public TV news*		
#svtagenda	*SVT evening news broadcast*		
#Stockholm	—		

Figure 2. Hashtags on Twitter @socialdemokrat November 01–December 15, 2015 (Krzyżanowski 2018).

Based on this figure, it shows that there are trends in refugee problems that the government needs to play a role in making decisions or policy changes. As the hashtags in the Swedish domestic sphere tend to be related to hashtags about refugee residency programs (*#etableringspaket*) and area-specific programs in Sweden's socioeconomic reality (*#skilan*, *#jobb*). This is how John Stuart Mill expressed his view in the concept of *the marketplace of ideas* through the theory of freedom of speech that in public communication, it is impossible to avoid debate or open discussion (Moorman *et al.* 2022). So based on this, the ideas expressed by the community through social media platforms are inevitable because they are in line with what is called the right to freedom of opinion. The right to freedom of speech in question is related to the rights of Swedish society to have a certain opinion or point of view on a phenomenon.

Meanwhile, looking at the international trends that occurred on the Swedish Twitter social media led to trends after the attacks in Paris in November 2015. This further strengthens the statement in Mill's concept referring to democracy, openness, freedom, competition and government elitism (Mahon Rathnam 2023). In this case, the fear of the Swedish people after the attacks in Paris led to concerns related to public opinion voiced on social media as a form of democracy, openness and freedom. In addition, there was an incident that led to further negative public opinion towards refugees during a festival in Sweden, precisely in the Norrkoping area. In this event, there was sexual harassment, namely the rape of teenage girls by a number of men who were known based on witness testimony, the perpetrators had dark skin and Arab faces (The Daily Beast 2016). This led to hashtags flooding social media such as *#refugeesnotwelcome* and *#rapefugees*. According to John Milton, in his view, the importance of freedom of speech is an effort to reach the truth (Mahon Rathnam 2023). Through people, in this case the Swedish people who freely express their opinions on social media through the use of hashtags, they will judge which truth is really in the common interest and this will lead to majority ideas that influences and shapes policy shifts, particularly in relation to Sweden's stance on Syrian refugees.

The government in this case took part as released through a *press conference on* November 24 regarding political talks about changes in Sweden's policy towards refugees. This can be seen through several *tweets* describing certain situations in Sweden. One of the following tweets describes the refugee shelter and the responsibility Sweden has taken on, announced by Prime Minister Stefan Lofven was, "*Nobody can doubt Sweden's will to help refugees. But right now we cannot handle more. More must find their place in other countries*" (Krzyżanowski 2018). This shows the influence of public opinion on the formation and decision-making by the government in seeing the needs and common interests of the people in the country (Grimmelikhuijsen and Meijer 2022). Seeing the phenomenon that occurs in the realm of social media as a forum for the Swedish government and the public to voice their opinions is in line with the concept of political participation which can be interpreted as the participation of individuals and groups on a media platform can shape and influence policy making and contribute to political outcomes in their country (Theocharis *et al.* 2023).

4 CONCLUSION

This research shows the participation of the Swedish people in the formation of public opinion on social media as an internal factor that drives the transition of the *Open-Door* policy to the *U-Turn Policy* in Sweden. The argumentation is based on the results of research showing the existence of causal factors related to aspects of security and welfare which are in Sweden's national interest. This research shows the role of the government in facilitating information through the Twitter social media page as a forum for communicating policy developments to the public as well as a medium where Swedish people can express their opinions on issues and phenomena that are occurring.

Practical recommendations for policymakers are that it is crucial to actively engage with social media to not only disseminate information but also listen to the concerns and opinions of the populace. By promoting transparency and open communication, policymakers can create a more participatory process, ensuring that diverse viewpoints are considered in shaping policies. Additionally, encouraging respectful and constructive online discussions can help bridge divides, enhance public trust and ultimately guide more effective, inclusive decision-making.

REFERENCES

Abdelhady, D., and Al Ariss, A. (2023). How capital shapes refugees' access to the labour market: the case of Syrians in Sweden. *International Journal of Human Resource Management*, 34(16), 3144–3168. https://doi.org/10.1080/09585192.2022.2110845

Adesina, O. S. (2017). Foreign policy in an era of digital diplomacy. *Cogent Social Sciences*, 3(1). https://doi.org/10.1080/23311886.2017.1297175

BBC News. (2015). *Sweden Protest After Three Mosque Fires in One Week. BBC News*. https://www.bbc.com/news/world-europe-30658482

Berntzen, L. E. (2018). The Anti-Islamic Movement: Far Right and Liberal? May, 1.

Brännström, L., and Gunneflo, M. (2021). Swedish foreign policy feminisms: Women, social democracy and capitalism. *Australian Feminist Law Journal*, 47(2), 207–227. https://doi.org/10.1080/13200968.2022.2088189

Gheciu, A., and Wohlforth, W. C. (2018). The future of security studies. *The Oxford Handbook of International Security*, July, 3–13. https://doi.org/10.1093/oxfordhb/9780198777854.013.1

Grimmelikhuijsen, S., and Meijer, A. (2022). Legitimacy of Algorithmic Decision-Making: Six Threats and the Need for a Calibrated Institutional Response. *Perspectives on Public Management and Governance*, 5 (3), 232–242. https://doi.org/10.1093/ppmgov/gvac008

Hajighasemi, A., and Oghazi, P. (2022). Outcomes of Swedish migration and economics of the welfare system. *Economic Research-Ekonomska Istrazivanja*, 35(1), 986–1010. https://doi.org/10.1080/1331677X.2021.1952089

Hudson, C., Sandberg, L., and Sundström, K. (2023). Integrating the Immigrant the Swedish Way? Understandings of Citizenship and Integration in Swedish Local Civic Integration Projects. *Journal of Intercultural Studies*, 44(4), 553–569. https://doi.org/10.1080/07256868.2022.2107624

Kaarbo, J., G, C., and Thies. (2024). The Oxford Handbook of Foreign Policy Analysis. In *Oxford Handbooks (Online edi)*. Oxford Academic. https://doi.org/10.1093/oxfordhb/9780198843061.001.0001

Krzyżanowski, M. (2018). "We Are A Small Country That Has Done Enormously Lot": The 'Refugee Crisis' and the Hybrid Discourse of Politicizing Immigration in Sweden. *Journal of Immigrant and Refugee Studies*, 16(1–2), 97–117. https://doi.org/10.1080/15562948.2017.1317895

Larasati, P. (2018). *Perspektif Feminisme Terhadap Perubahan Kebijakan Swedia Terkait Krisis Pengungsi Periode 2014-2017*. In Institutional Repository UIN Syarif Hidayatullah. Skripsi Sarjana UIN Syarif Hidayatullah Jakarta.

Lyck-Bowen, M. (2020). Multireligious Cooperation and the Integration of Muslim Migrants in Sweden. *Annals of the American Academy of Political and Social Science*, 690(1), 168–174. https://doi.org/10.1177/0002716220939919

Maha, I. F. (2021). Kebijakan Luar Negeri Feminis Swedia (2014-2018). *Journal of International Relations*, 7 (1), 1–12. https://doi.org/10.14710/jirud.v7i1.29344

Mahon Rathnam, L. (2023). The Marketplace of Ideas and the Agora: Herodotus on the Power of Isegoria. *American Political Science Review*, 117(1), 140–152. https://doi.org/10.1017/S0003055422000661

Moorman, C., van Heerde, H. J., Page Moreau, C., and Palmatier, R. W. (2022). From Vision to Reality: Lessons in Creating a Marketplace of Ideas. *Journal of Marketing*, 86(6), 4–12. https://doi.org/10.1177/00222429221129661

Nilsson, M., and Esholdt, H. F. (2022). After the Caliphate: Changing Mobilization in the Swedish Salafi-Jihadist Environment following the Fall of ISIS. *Studies in Conflict and Terrorism*, 0(0), 1–22. https://doi.org/10.1080/1057610X.2022.2104682

Nylund, M. L., Håkansson, S., and Bjarnegård, E. (2023). The Transformative Potential of Feminist Foreign Policy: The Case of Sweden. *Journal of Women, Politics and Policy*, 44(3), 257–273. https://doi.org/10.1080/1554477X.2022.2113662

Oksanen, A., Kaakinen, M., Minkkinen, J., Räsänen, P., Enjolras, B., and Steen-Johnsen, K. (2020). Perceived Societal Fear and Cyberhate after the November 2015 Paris Terrorist Attacks. *Terrorism and Political Violence*, 32(5), 1047–1066. https://doi.org/10.1080/09546553.2018.1442329

Ota, E. N., and Ecoma, C. S. (2022). Power and National Interest in International Relations. *European Journal of Humanities and Social Sciences*, 2(4), 23–30. https://doi.org/10.24018/ejsocial.2022.2.4.268

PBS. (2016). *Sweden's Welcome to Refugees Disturbed by Violent Backlash. PBS News Hour*. https://www.pbs.org/newshour/show/swedens-welcome-to-refugees-disturbed-by-violent-backlash

Qi, H. (2021). The Economic Impact of Syrian Refugees in Sweden. In Malmo Institute for Studies of Migration, Diversity and Welfare. *Forced Displacement and Refugee-Host Community Solidarity (FOCUS)*.

Rostami, A., Sturup, J., Mondani, H., Thevselius, P., Sarnecki, J., and Edling, C. (2020). The Swedish Mujahideen: An Exploratory Study of 41 Swedish Foreign Fighters Deceased in Iraq and Syria. *Studies in Conflict and Terrorism*, 43(5), 382–395. https://doi.org/10.1080/1057610X.2018.1463615

The Daily Beast. (2016). *Are Migrants Really Raping Swedish Women*. https://www.thedailybeast.com/are-migrants-really-raping-swedish-women

Theocharis, Y., Boulianne, S., Koc-Michalska, K., and Bimber, B. (2023). Platform affordances and political participation: how social media reshape political engagement. *West European Politics*, 46(4), 788–811. https://doi.org/10.1080/01402382.2022.2087410

Tsourapas, G. (2019). The Syrian Refugee Crisis and Foreign Policy Decision-Making in Jordan, Lebanon, and Turkey. *Journal of Global Security Studies*, 4(4), 464–481. https://doi.org/10.1093/jogss/ogz016

Zamboni, M. (2019). Swedish legislation and the migration crisis. *Theory and Practice of Legislation*, 7(2), 101–133. https://doi.org/10.1080/20508840.2020.1729599

Towards Resilient Societies: The Synergy of Religion, Education, Health, Science, and Technology – Rahiem (Ed.)

Water infrastructure development for key commodities in West Nusa Tenggara: Challenges and opportunities

Agus Manshur
Ministry of National Development Planning, Jakarta, Indonesia

Firre An Suprapto
State University of Surabaya, Surabaya, Indonesia

Feita Puspita Murti & Jelita Sari Wiedoko
Ministry of National Development Planning, Jakarta, Indonesia

Editha Praditya
Indonesia Defense University, Bogor, Indonesia

Alya Maharani
IPB University, Bogor, Indonesia

Aziza Firdausy Fauzadha
Padjadjaran University, Sumedang, Indonesia

ABSTRACT: Effective water resource management is essential for regions like West Nusa Tenggara (NTB), where agriculture forms the backbone of the economy. This study examines the development of water infrastructure in NTB, focusing on its role in supporting key agricultural commodities. Using spatial analysis as an innovative approach, the study uncovers the relationship between water resource infrastructure, housing, settlements, and key commodities. This method offers insights into the distribution of clean water, infrastructure efficiency, and demand disparities, forming a foundation for more targeted water management strategies. The findings reveal a water surplus in NTB, accompanied by significant regional disparities. High-demand areas, such as Lombok Barat, Lombok Tengah, and Lombok Timur, face challenges due to population density and intensive agricultural activities, while regions like Sumbawa exhibit lower water demand. These results contribute to SDG 6 by identifying actionable strategies to improve water accessibility and promote sustainable agriculture. Key policy recommendations include modernising irrigation systems, fostering water-based industries, and integrating water resource management into regional development plans. By addressing these challenges with tailored solutions, the study supports sustainable agriculture in NTB and offers a replicable framework for optimising water infrastructure in similar regions, advancing broader goals of water and food security.

Keywords: infrastructure development, key commodities, spatial analysis, sustainable development goal 6, water infrastructure

1 INTRODUCTION

Reliable access to water is essential for improving quality of life, supporting economic activities, and achieving environmental sustainability. As a fundamental human right, clean water also plays a pivotal role in achieving sustainable development goals (Imran *et al.*

DOI: 10.1201/9781003645542-6

2023), particularly SDG 6, which aims for universal access to water and sanitation by 2030. In regions with inadequate infrastructure, water scarcity burdens communities by requiring significant time and energy for water collection, ultimately reducing productivity (Ardiansyah *et al.* 2020; Qiao *et al.* 2020). Investing in reliable water systems can alleviate these challenges, enabling communities to focus on more productive activities (Meeks 2017). For agriculture, which heavily relies on irrigation and livestock water supply, an adequate water source is vital to maintaining food production (Hsu and Shr 2023). Accurate water demand forecasting is thus crucial for sustainable water management and long-term planning (Baseham 2024).

However, focusing solely on the importance of water risks overlooking specific challenges faced by regions like West Nusa Tenggara Barat (NTB). Agriculture in NTB, particularly key commodities such as rice, depends on efficient water management. Rice is a vital crop for NTB, serving as the cornerstone of the local agricultural economy, plays a crucial role in sustaining livelihoods, and ensuring food security for the region. However, the region struggles with unreliable irrigation systems and uneven water distribution, which limit agricultural productivity and economic resilience. These challenges reflect broader issues within Indonesia's agricultural sector which faces mounting pressure to meet the growing demand for staple crops, exacerbated by population growth and shrinking paddy fields (Hatta *et al.* 2023). Futhermore, without sufficient infrastructure, it is difficult for regions to remain competitive in the global market (Estutama and Adhi 2021). NTB and neighbouring East Nusa Tenggara (NTT) encompass nearly half of Indonesia's semi-arid lands, emphasizing the urgent need for improved water management to support food security and economic growth (Nurkhaida *et al.* 2021).

To address these challenges, investing in strategic water infrastructure, such as dams, irrigation systems, and water supply networks is crucial for unlocking the region's agricultural potential (Grigg 2019). NTB has leveraged its position as the Indonesian province with the highest number of operational dams, managing 74 out of 215 dams across 17 provinces as of October 2022 (Databoks 2023). These dams serve various purposes, including irrigation, water supply, and hydroelectric power (Krztoń, *et al.* 2022; Shi *et al.* 2019). Despite this extensive infrastructure, the region continues to face critical challenges in water management. Inadequate irrigation systems and uneven distribution of water remain persistent issues, restricting the productivity of essential crops. Forecasting future water demand for irrigation storage and other uses remains a key challenge for ensuring long-term water sustainability (Schmitt and Rosa 2024).

This study explores the crucial connection between water infrastructure development and the agricultural productivity of key commodities in Nusa Tenggara Barat (NTB). It highlights how improvements in irrigation, distribution networks, and access to clean water can directly enhance agricultural output, particularly for crops like rice. The unique aspect of this study lies in its integrated approach, where water resource management is closely tied to agricultural development. This study merges the two, providing a more holistic view of how effective water management can support long-term agricultural sustainability. By identifying specific challenges and opportunities within NTB's water infrastructure, this study provides tailored policy recommendations applicable not only to NTB but also to other regions facing similar water and agricultural sustainability issues. These insights are key for developing region-specific strategies that contribute to achieving SDG 6, alongside broader goals of food security and economic resilience.

2 METHODS

This study employs spatial analysis using Quantum Geographic Information System (QGIS) software to examine the relationship between water resource infrastructure, housing, settlements, and key commodities in West Nusa Tenggara Barat (NTB). The steps of analysis include integrating 2022 clean water needs data at the district level, overlaying geospatial information on infrastructure and settlements to identify critical areas requiring intervention, and using QGIS to create maps visualising infrastructure gaps and prioritising resource allocation. Data for this study sourced from the Forum Infrastruktur Wilayah (FIW) organised Bappenas which brings together stakeholders to provide insights for infrastructure planning in NTB and East Nusa Tenggara (NTT). The FIW data ensures alignment with current infrastructure priorities, including land, regional resources, and infrastructure needs. The 2022 clean water data was chosen for its relevance to current infrastructure needs, particularly in addressing water scarcity issues. However, limitations such as data gaps and spatial accuracy exist, and cross-referencing and data validation efforts were employed to mitigate these issues. This methodology provides actionable insights for optimising resource allocation and supporting the sustainable development of key commodities in NTB, guiding evidence-based policy decisions and fostering economic resilience in the region.

3 FINDINGS AND DISCUSSION

The analysis below examines the interconnections between water resource infrastructure, housing and settlements, and key commodities in West Nusa Tenggara (NTB). The relationship between water resource infrastructure and the housing, settlement, and agricultural sectors in NTB is depicted in the Figure 1. The image illustrates that dam infrastructure is widely distributed across Lombok and Sumbawa Islands, predominantly near the province's rice fields or agricultural areas. The infrastructure supporting the Drinking Water Supply System (SPAM) is located near residential or urban areas. The largest residential areas are found in Mataram City and Bima City, where the surrounding districts still support clean water needs. Apart from Mataram City, extensive residential areas are located near agricultural regions. The key commodity in NTB is rice, with a total paddy field area of 270,965.01 hectares and agricultural land covering 561,059.90 hectares. The clean water capacity in NTB is supported by 22 dams and 75 lakes/reservoirs. Based on the analysis, considering the water needs for rice commodities and household clean water requirements, NTB experiences an over-supply of water capacity.

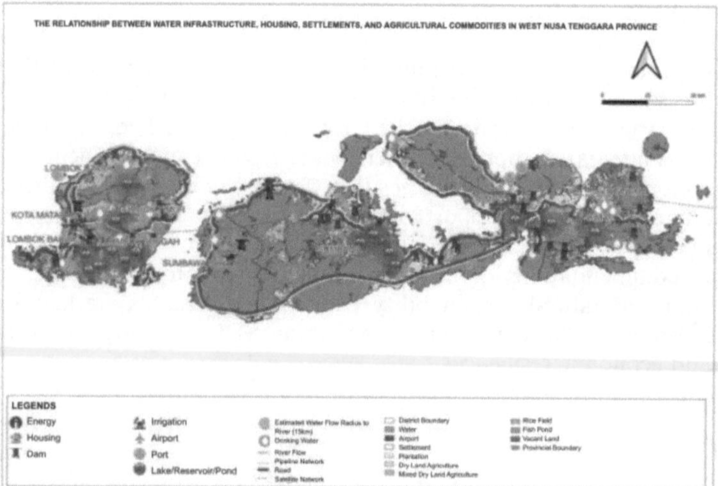

Figure 1. Relationship between water resource infrastructure and housing, settlements, and key commodities in West Nusa Tenggara (NTB).

The construction of dams in West Nusa Tenggara (NTB) has significant potential to transform the region into a major agricultural hub, particularly by supporting essential commodities like rice. Rice is crucial for global food security and any threats to its productivity could have widespread implications (Khairulbahri 2021). Despite having a surplus of water, NTB has not fully utilised this resource to enhance agricultural output or reduce regional disparities. The surplus water could be better harnessed to improve irrigation efficiency, stabilise crop yields, and expand agricultural areas, especially in underdeveloped regions. However, this water surplus does not automatically lead to better agricultural outcomes. Several barriers prevent NTB from fully capitalising on its water resources. Ineffective governance, poor coordination in water management policies, and infrastructure inefficiencies, particularly in irrigation systems, hinder the efficient use of water. These issues are especially noticeable in areas with high agricultural potential, where current systems fail to meet the demands of farmers While efforts to boost water productivity could lead to higher revenue and improved food security, the results are not guaranteed (Wichelns 2015). To address these challenges, the government must prioritise modernising irrigation systems, improving dam infrastructure, and implementing comprehensive water management policies. Additionally, developing post-harvest processing and supporting water-based industries like fish farming could reduce crop losses, add value to agricultural products, and create jobs. Targeted investments in rural areas and better water distribution would promote sustainable agricultural growth. By tackling these issues, NTB can optimise its water use, increase agricultural productivity, and support long-term economic growth.

Furthermore, to meet the growing demand for rice, the government can expand cultivation areas, improve land use, reduce post-harvest losses, and enhance irrigation systems (Rachman *et al.* 2022). Implementing Green Revolution strategies like high-yield varieties and mechanised labour, can help increase self-sufficiency (Sujianto *et al.* 2022). Efforts to increase agricultural productivity have led to disruptions in regional water dynamics, particularly in NTB, where inefficient infrastructure and inadequate water management complicate the effective use of available resources. Unlike East Nusa Tenggara (NTT), which relies on rainwater harvesting and efficient irrigation to address water scarcity. NTB's challenge lies in optimising the surplus water for agricultural use, balancing the need for increased productivity with environmental sustainability. This makes NTB's water management issues distinct from those faced by other semi-arid regions within Indonesia. Additionally, effective waste management is essential for minimising environmental impacts and ensuring long-term sustainability (Muqsit *et al.* 2024).

In NTB, the availability of clean water directly affects both crop productivity and public health. Efficient water management not only enhances agricultural output but also improves the quality of life by ensuring access to clean water for drinking, sanitation, and household needs. The analysis of clean water demand across NTB reveals regional disparities, indicating the need for infrastructure investments that are specifically tailored to the unique requirements of each district.

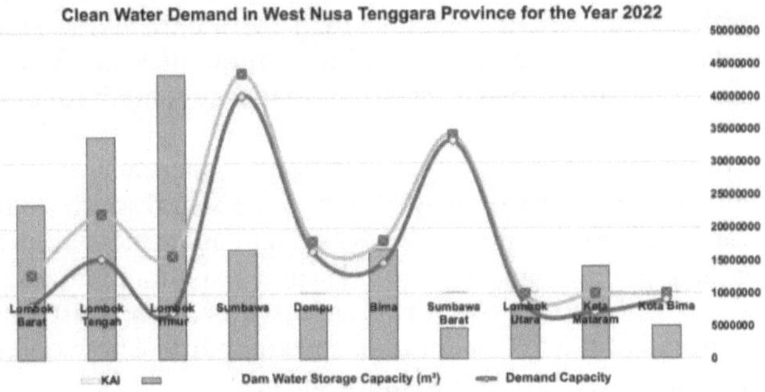

Figure 2. Graph of clean water needs in West Nusa Tenggara province.

The Figure 2 illustrates clean water needs across various regions of West Nusa Tenggara Province, highlighting disparities in water demand among cities. Green bars represent the volume of clean water required for household purposes (KART), while yellow lines show dam storage capacity and red lines indicate actual capacity needs. The data reveals regional variations where Sumbawa, Dompu, and Sumbawa Barat exhibit lower clean water needs than other areas, likely due to more efficient water use practices and lower population density. These regions may have fewer large-scale agricultural activities or more dispersed settlements, which reduce the overall demand for clean water. Conversely, Lombok Barat, Lombok Tengah, and Lombok Timur show higher clean water requirements, which can be attributed to more intensive agricultural activities, such as rice cultivation, which require substantial water resources. Additionally, higher population density in urban areas increases household water demand. Trends in the red and yellow lines suggest that regions with higher dam storage capacity also require more significant volumes of clean water. Peaks in these lines, particularly evident in Sumbawa and Sumbawa Barat, suggest periods of heightened water demand, potentially during dry seasons or peak agricultural periods.

The graph highlights significant variations in clean water needs across NTB districts, emphasising the need for strategic infrastructure planning, particularly when considering the role of dams in meeting these demands. Properly designed and managed dam infrastructure can significantly enhance water distribution and support agriculture. However, challenges in dam management, such as aging infrastructure, inadequate maintenance, and environmental impacts like sedimentation and ecosystem disruption, pose significant obstacles to efficient water use. In high-demand regions, where population density and agricultural activities are concentrated, there is a pressing need for substantial investments in dam infrastructure and irrigation systems to meet growing water demands. Conversely, areas with lower water demand should focus on optimising the use of existing infrastructure to improve efficiency. Incorporating sustainable practices in dam management, such as regular maintenance and the use of eco-friendly technologies, can help mitigate environmental impacts and ensure the long-term reliability of water supply systems. Prioritising dam construction and water management based on these regional needs and environmental considerations will not only support agricultural productivity but also contribute to long-term food security in NTB. Substantial infrastructure investments also crucial for developing NTB's key commodities. Upgrading roads, irrigation systems, storage facilities, and distribution networks will enhance efficiency, reduce post-harvest losses, and lower costs. Improved irrigation not only stabilises yields but also supports rural development (Higgins *et al.* 2021).

Effective water resource management relies on the collaboration of farmers, policymakers, and industry stakeholders. Farmers play a critical role in adopting water-efficient practices and contributing to sustainable agricultural production, while policymakers are tasked with creating and enforcing regulations that ensure equitable water distribution and long-term sustainability. Industry stakeholders, including water-based industries and private-sector partners, can provide technical expertise and investments to improve water infrastructure and practices. Meanwhile, local communities are key in maintaining and managing these infrastructure projects, ensuring that they remain functional and sustainable over time.

Additionally, providing training to farmers and fishermen enhances their skills and improves the quality and quantity of yields. Comprehensive resource management, including water conservation, land use planning, and waste management, will protect NTB's natural resources. Strengthening distribution networks and expanding market access for NTB's agricultural products will lower transportation costs, create export opportunities, and stimulate economic growth, all while promoting sustainability. These collaborative efforts will ensure the region's long-term prosperity and resilience.

4 CONCLUSION

In conclusion, the analysis of water resource infrastructure in NTB underscores the need for strategic planning to optimize water use. The distribution of dam infrastructure across Lombok and Sumbawa Islands, along with the presence of Drinking Water Supply Systems (SPAM) near residential areas and agricultural zones, reveals the complex dynamics of water management in NTB. With 22 dams and 75 lakes/reservoirs providing surplus water capacity, NTB faces both opportunities and challenges in optimising water resources for agricultural sustainability and community welfare. However, limitations such as outdated irrigation techniques, uneven water distribution, and insufficient coordination among water management bodies hinder NTB's ability to realise its potential fully.

Policy interventions are necessary to address these issues. Efforts should focus on modernising irrigation systems to improve water efficiency, ensuring better water use for crop production while minimising waste; investing in sustainable water-based industries like fish farming to diversify income sources and enhance economic resilience in regions where traditional agriculture faces challenges; and improving coordination between regional water management authorities to help align water allocation with local needs, reduce conflicts over resources, and streamline decision-making. In addition, investments in infrastructure, such as constructing dams tailored to district-specific needs, are crucial to boosting agricultural productivity and ensuring that water is stored and distributed efficiently. Aligned with regional development plans, these strategies will foster sustainable growth and strengthen NTB's resilience as an agrarian hub.

The insights gained from NTB's approach hold significant potential for scaling up to other regions in Indonesia that are grappling with similar challenges in water resource management and agricultural sustainability. Furthermore, as these approaches are scaled, they can contribute to achieving broader national objectives, such as those outlined in SDG 6 (Clean Water and Sanitation), by improving water availability and quality for both agricultural and domestic purposes. The potential for scaling these findings beyond NTB offers a valuable opportunity to improve water governance, boost agricultural sustainability, and support economic growth across Indonesia, especially in regions where water scarcity and agricultural challenges are becoming increasingly urgent.

REFERENCES

Ardiansyah, B., Kusmaryo, R. D. H., and Muin, M. F. (2020). Prospects of infrastructure development from the perspective of economic growth. *International Journal of Social Science and Business*, 4(3), 390–399. https://ejournal.undiksha.ac.id/index.php/IJSSB/index

Baseham, A. (2024). Forecasting Urban Water Demand in 2032: A Case Study of Bandung City. *Jurnal Teknik Pengairan: Journal of Water Resources Engineering* Vol. 15 No. 1. DOI:10.21776/ub.pengairan.2024.015.01.1.

Databoks. (2023, May 22). *Ini Provinsi dengan Bendungan Terbanyak pada 2022*. Retrieved from https://databoks.katadata.co.id/datapublish/2023/05/22/ini-provinsi-dengan-bendungan-terbanyak-pada-2022

Estutama P., and Adhi Kurniawan M. (2021). Public Works and Housing Infrastructure Planning using Environmental Carrying Capacity Consideration:Case Study on Planning Dam Development in Kalimantan Island, Indonesia. *The Journal of Indonesia Sustainable Development Planning*, 2(3), 257–271. https://doi.org/10.46456/jisdep.v2i3.97

Grigg, N. S. (2019). Global water infrastructure: State of the art review. *International Journal of Water Resources Development*, 35(2), 181–205. https://doi.org/10.1080/07900627.2017.1401919

Hatta, M., Sulakhudin, Burhansyah, R., Kifli, G. C., Dewi, D. O., Kilmanun, J. C., Permana, D., Supriadi, K., Warman, R., Azis, H., Santari, P. T., and Widiastuti, D. P. (2023). Food self-sufficiency: Managing the newly-opened tidal paddy fields for rice farming in Indonesia (A case study in West Kalimantan, Indonesia). *Heliyon* 9 e13839. https://doi.org/10.1016/j.heliyon.2023.e13839

Higgins, D., Arslan, A., and Winters, P. (2021). What role can small-scale irrigation play in promoting inclusive rural transformation? Evidence from smallholder rice farmers in the Philippines. *Agricultural Water Management*. Volume 243. https://doi.org/10.1016/j.agwat.2020.106437

Hsu, C.-C., and Shr, Y.-H. (2023). The impacts of water storage infrastructure on real property values and crop production. *Water Research*, 247, 120786. https://doi.org/10.1016/j.watres.2023.120786

Imran H., Mahmudul, H. A., and Akram Ullah, S.M. (2023). Assessment of Domestic Water Usage and Wastage in Urban Bangladesh: A Study of Rajshahi City Corporation: A Study of Rajshahi City Corporation. *The Journal of Indonesia Sustainable Development Planning*, 4(2), 109–121. https://doi.org/10.46456/jisdep.v4i2.462

Khairulbahri, M. (2021). Analyzing the impacts of climate change on rice supply in West Nusa Tenggara, Indonesia. *Heliyon* 7 e08515. https://doi.org/10.1016/j.heliyon.2021.e08515

Krztoń, W., Walusiak, E., and Wilk-Woźniak, E. (2022). Possible consequences of climate change on global water resources stored in dam reservoirs. *Science of The Total Environment*, 830, 154634. https://doi.org/10.1016/j.scitotenv.2022.154646

Meeks, R. C. (2017). Water Works: The Economic Impact of Water Infrastructure. *Journal of Human Resources*, 52(4), 1119–1153. https://doi.org/10.3368/jhr.52.4.0915–7408R1

Muqsit F. A., Setiadi, R., and LoA. (2024). Waste Management in Heritage Tourism Area: Perspectives from Visitors and Waste Management Operators. *The Journal of Indonesia Sustainable Development Planning*, 5 (1), 15–26. https://doi.org/10.46456/jisdep.v5i1.524

Nurkhaida, R., Hamdani, A., Suriadi, A., and Heryani, N. (2021). Increasing rice productivity and profitability through irrigation water management and biofertilizer in West Nusa Tenggara. *E3S Web of Conferences*, 306, 04012. https://doi.org/10.1051/e3sconf/202130604012

Qiao, N., Fang, L., and Mu, L. (2020). Evaluating the Impacts of Water Resources Technology Prog*ress on Development and Economic Growth Over the Northwest*, China. *PLoS ONE*, 15(3), e0229571. https://doi.org/10.1371/journal.pone.0229571

Rachman, B., Ariningsih, E., Sudaryanto, T., Ariani, M., Septanti, K. S., Adawiyah, C. R., *et al.* (2022). Sustainability status, sensitive and key factors for increasing rice production: A case study in West Java, Indonesia. *PLoS ONE*, 17(12), e0274689. https://doi.org/10.1371/journal.pone.0274689

Schmitt, R. J. P., and Rosa, L. (2024). Dams for hydropower and irrigation: Trends, challenges, and alternatives. *Renewable and Sustainable Energy Reviews* Volume 199. https://doi.org/10.1016/j.rser.2024.114439

Shi, H., Chen, J., Liu, S., and Sivakumar, B. (2019). The role of large dams in promoting economic development under the pressure of population growth. *Sustainability*, 11(10), Article 10. https://doi.org/10.3390/su11102965

Sujianto, Gunawan, E., Saptana, Syahyuti, Darwis, V., Ashari, Syukur, M., Ariningsih, E., Saliem, H., Mardianto, S. and Marhendro. (2022). Farmers' perception, awareness, and constraints of organic rice farming in Indonesia. *Open Agriculture*, 7(1), 284–299. https://doi.org/10.1515/opag-2022-0090

Wichelns, D. (2015). Water productivity and food security: considering more carefully the farm-level perspective. *Food Security*, 7(2), 247–260. doi:10.1007/s12571-015-0426-0

Towards Resilient Societies: The Synergy of Religion, Education, Health, Science, and Technology – Rahiem (Ed.)

Assessing the effectiveness of BisKita Transpakuan's operational policy in achieving sustainable urban transportation in Bogor city

Agus Nugraha
Universitas Islam Negeri Syarif Hidayatullah Jakarta, Indonesia
ORCID ID: 0000-0002-5618-2486

Savira Maharani
Universitas Islam Negeri Syarif Hidayatullah Jakarta, Indonesia

ABSTRACT: This study aims to evaluate the effectiveness of BisKita Transpakuan's operational policy in achieving sustainable urban transportation in Bogor City. A qualitative method with a descriptive approach was employed, with data collected through document analysis, observations, and in-depth interviews with key informants. The findings reveal that BisKita Transpakuan's operational policy has been effective primarily in bus operations, ensuring reliable transportation services. However, the infrastructure aspects, particularly the preparation of safe and comfortable bus stops and terminals, remain ineffective and hinder the overall success of the policy. In conclusion, while there have been operational improvements, BisKita Transpakuan's policy has not fully achieved its objectives in addressing traffic congestion and establishing sustainable urban transportation in Bogor City. The study highlights the need for greater focus on improving supporting infrastructure to maximise policy effectiveness.

Keywords: effectiveness, policy, operation of BisKita

1 INTRODUCTION

Sustainable Development Goal (SDG) 11 is intended to "Make Cities and Human Settlements Inclusive, Safe, Resilient and Sustainable" (United Nations 2015). Sustainable transportation plays a crucial role in creating a sustainable city (Brebbia 2000). Traffic congestion in Bogor City is a serious problem that challenges achieving a sustainable city (Syaiful 2017). The congestion in Bogor City has impacted people's tendency to choose private transportation, making traffic congestion in Bogor City worse (H. P. Sari *et al.* 2019). To overcome this problem, the Bogor City Government officially released the integrated mass public transportation, BisKita Transpakuan, on November 2, 2021. The operational policy of BisKita Transpakuan is required to provide public services following the Minimum Service Standards (MSS) as stipulated in the Minister of Transportation Regulation Number 10 of 2012, which also aligns with the goal of SDG 11, especially on transportation. The MSS includes security, safety, comfort, equality, affordability, and regularity.

Previously, a transportation model similar to BisKita had been realised in several cities throughout Indonesia. The Semarang City Government promotes public transportation through Bus Rapid Transit (BRT) to minimise traffic jams. However, the people of Semarang have not fully utilised public transportation since the distance, i.e., the comfort level, to reach the BRT bus stops is too far (Adibah 2019). The same things are observed in

DOI: 10.1201/9781003645542-7

Bekasi City, where the goals, integration, and adaptation measures were still ineffective. Moreover, socialisation is considered less comprehensive, leaving people in Bekasi unaware of the existence of the BRT (Sahara *et al.* 2022).

Several studies have been done on BusKita Transpakuan operations in Bogor City. Luthfiyyah *et al.* examined the quality of BisKita Transpakuan services using the SERVQUAL theory proposed by Parasuraman *et al.* This study found that the Biskita Transpakuan service has not been running optimally because it has not fulfilled all service quality required in the SERVQUAL theory (Hasna *et al.* 2024). Osly *et al.* (2022) analysed BisKita Transpakuan service performance based on community satisfaction perception on the Bubulak Terminal – Ciawi route. Research results show that user expectations were high, yet the level of satisfaction was low (Osly *et al.* 2022). Murtejo *et al.* evaluated the performance of BisKita Trasnpakuan on the Terminal Bubulak via the Cidangiang – Ciawi route, and the results remained below 70% (Murtejo *et al.* 2023). Those studies were focused on service quality and operational performance. However, no research has yet assessed the effectiveness of BisKita Transpakuan's operational policy in achieving sustainable urban transport in Bogor City.

The main theories employed in the current study were effectiveness and public policy, which are correlated with sustainable urban transportation. Public policy, according to Howlett *et al.*, refers to a government's actions, goals, and statements regarding certain problems that have been decided to be addressed (Howlett and Cashore 2014). The government's actions, in general, are a political decision to solve social problems in the public interest (M. M. K. Sari and Satmoko 2022). Public policy constitutes actions by several elements, individuals, groups, or the government, in realising certain goals (Fajobi and Adesina 2024). Accordingly, a broad outline of government actions is adopted by individuals and groups, which subsequently leads to the behaviour of public policy implementation. Filgueiras *et al.* also explain several criteria in determining policies, namely: effectiveness, efficiency, sufficient, fairness, and accountability (Filgueiras and de Fatima Nascimento de Queiroz 2021).

According to Lestanata and Pribadi, effectiveness is the ability to choose goals by utilising facilities and infrastructure to achieve them appropriately (Ilhami 2023). A policy can be considered effective given that the activity process has achieved the intended goal (Merrill and White 2018). Mardiasmo asserts that policy is related to achieving the policy goals or targets (Virani 2019). Thus, effectiveness refers to the relationship between policy and the goals that must be achieved (Schmidt 2022). A policy can be deemed effective given that the policy goals have been fully achieved. In other words, policy is rendered effective if the policy goals are realised (Mukherjee and Bali 2019).

Sustainable Urban Transport is an SDGs program promoted by the United Nations that emphasises that transportation should be recognised as an important means of achieving various goals, especially in making cities and human settlements inclusive, safe, resilient, and sustainable. By 2030, it is expected to provide access to safe, affordable, accessible, and sustainable transport systems for all, thereby improving road safety, notably by expanding public transport, with special attention to the needs of those in vulnerable situations, women, children, persons with disabilities and older persons (United Nations 2015). This indicator closely correlates with BisKita Transpakuan's operational policy, which requires the provision of public services to comply with the Minimum Service Standards (MSS) outlined in the Minister of Transportation Regulation Number 10 of 2012. This policy also supports the objectives of SDG 11.

2 RESEARCH METHODS

The qualitative method was employed in this study to conduct an in-depth analysis of the effectiveness of BisKita's operational policies in Bogor City (Rogers 1998). The reason for using this approach is to make it easier to understand the examined phenomenon and reach a conclusion (Rundmo and Nordfjærn 2019). Primary data were obtained from observations, documentation, and in-depth interviews with informants selected using purposive sampling. Interviewees consisted of the president director of BisKita Transpakuan, the operation manager of BisKita Transpakuan, staff of Bubulak Station, and community of users. Meanwhile, secondary data were obtained through literature study by analysing several existing literary sources, including books, journals, articles, magazines, newspapers, and the internet. The data analysis techniques used in the current study included descriptive analysis, wherein the researcher conducted analyses by describing, explaining, or validating the phenomenon or problem being studied in detail (Koch *et al.* 2014). Using descriptive research analysis techniques, researchers could describe and explain results relating to the effectiveness of public policy in the operational aspect of BisKita Transpakuan in Bogor City.

3 FINDINGS AND DISCUSSION

The effectiveness of BisKita Transpakuan's operational policy is closely linked to achieving SDG 11. A city that meets sustainable criteria must be supported by a transportation system that ensures smooth mobility for its residents to carry out their activities and is free from traffic congestion issues. Therefore, an assessment was conducted on the effectiveness of BisKita's operational policy, measured by indicators such as security, safety, convenience, equality, regularity, and affordability.

3.1 *Security of BisKita Transpakuan in Bogor City*

Security is a key factor in people's choice of private vehicle or public transportation for their mobility. The lack of security in public transportation certainly impacts people's rational choice to use private vehicles. Thus, it further exacerbates traffic congestion. From field observations and interviews with local users and officers, an overview of the security aspects of BisKita Transpakuan was obtained as follows: (1) BisKita Transpakuan vehicles have clear identity with a number and route affixed to the front and back of the vehicle; (2) the drivers have full identification on by wearing a uniform and displaying the driver's name and company; (3) the lighting on the vehicle functions as a light source to provide security for service users; (4) window film is installed as a coating on vehicle windows to reduce direct sunlight; (5) vehicles have warning lights to provide information about serious issues with the vehicle. This information shows BisKita Transpakuan's security. However, a sense of insecurity is experienced at bus stops where passengers are picked up and dropped off. The condition of open bus stop coverage creates a feeling of insecurity for users. Moreover, no security officers are on guard at each bus stop to help and regulate the inflow and outflow of passengers when BisKita Trasnpakuan arrives. These conditions do not follow the principle of sustainable urban transportation.

3.2 *Safety of BisKita Transpakuan in Bogor City*

In addition to security, safety assurance is a very important factor in public transportation services (Srivastava 2019). The results of research by Sugiyanto *et al.* show that safety factors (28.3%) come first in the results of factor analysis in choosing public transportation modes using the analytic hierarchy process method (Sugiyanto *et al.* 2021). The field research results show that Bogor City has implemented safety standard operating procedures (SOPs) in operating BisKita Transpakuan. The drivers are competent and knowledgeable about the routes they serve, the procedures for transporting passengers, and the traffic procedures; they are also respectful and have friendly attitudes and behaviour toward passengers. The drivers must also be in excellent physical condition when driving the vehicle. They should be physically and mentally healthy and free from any use of drugs and alcohol. However, the safety guarantee provided to citizens using BisKita Transpakuan has not been accompanied by a guarantee of safety at bus stops. The condition of the bus stop, which is open-covered and left unguarded by security officers, causes users to worry about their safety. In fact, BisKita Transpakuan customers are more likely to wait outside the bus stop and enter the bus stop when the bus arrives. This contradicts the target of SDG 11 set by the UN, which emphasises road safety.

3.3 *Convenience and equality of BisKita Transpakuan in Bogor City*

Comfort is key to choosing public transportation (Amador-Jimenez and Christopher 2016). Field research shows that BisKita Transpakukan of Bogor City provides sufficient comfort to their passengers. This is apparent from the condition of the buses, which are still new and well-maintained; the lighting is installed correctly, the room temperature control (AC) facilities are still cold, and the cleanliness of the vehicle is still maintained. There are also instructions to prioritise people with disabilities, children, and pregnant women. However, the situation is very different with the bus stops and terminals where people experience discomfort. The bus stops are dirty, many seats are damaged, and the roof and lighting are poorly maintained. Additionally, the condition of open bus stops is most uncomfortable when it is hot and rainy. Likewise, priority access and facilities are unavailable for passengers with disabilities, children, and pregnant women. This evidence is not relevant to the target of SDG 11, which emphasises inclusivity in sustainable urban transportation.

3.4 *Regularity of BisKita Transpakuan in Bogor City*

The regularity of BisKita Transpakuan during departure and arrival is a very important factor in realising excellent mass transportation services. The observations, documentation studies, and interviews show that the BisKita Transpakuan service is sometimes on time and sometimes late. This is due to the lack of designated bus lanes, resulting in buses being stuck in traffic along with other private vehicles during peak hours. This condition also follows the results of research by Murtejo *et al.* (2023), which shows that the waiting time for the arrival of BisKita Transpakuan is uncertain. Sometimes 5 minutes, 10 minutes, 15 minutes, or even up to 20 minutes, depending on road traffic conditions, with an average value of 15.08 minutes (Murtejo *et al.* 2023). In fact, according to research by Susanto *et al.* (2021) regarding the evaluation of the operational performance of BisKita Transpakuan, waiting times reached an average of 17.36-17.75 minutes (Susanto *et al.* 2021). Such circumstances do not support efficient, sustainable urban transportation.

3.5 *Affordability of BisKita Transpakuan in Bogor City*

Affordability is a minimum standard that must be met to facilitate users accessing BisKita Transpakuan. The true condition can be observed based on the fact that terminals and bus stops, as vital supporting elements, still do not meet the necessary criteria for easy movement of passengers between corridors and the availability of feeder route network integration. This situation certainly influences the interest of Bogor residents in using BisKita as mass public transportation. The lack of minimum standards for the affordability aspect of BisKita Transpakuan's operations also aligns with a study by Novando *et al.* (2024), which shows that bus stops are still difficult for service users to reach. For example, the Ciparigi bus stop, located in a housing complex, is difficult to reach by public transportation. Also, bus stops often become new points of congestion because the bus stops intended as BisKita Transpakuan stops were sometimes occupied by other forms of public transportation

(Novando *et al.* 2024). This situation is not consistent with the target of SDG 11, i.e., ensuring affordable transport systems for all.

4 CONCLUSION

Based on the discussion above, it can be concluded that the effectiveness of public policy in the operation of BisKita Transpakuan Bogor City has not been fully achieved. This is apparent from the fact that the policy objective has not been realised, namely the availability of mass public transportation stops that are secure, comfortable, safe, equal, orderly and affordable. Hence, the current operation of BisKita Transpakuan still has not encouraged the public to switch from using private vehicles to public transportation to overcome the congestion problem. The security, safety, comfort, equality, and affordability indicators in BisKita Transpakuan have been empirically realised only in the bus vehicle units. However, the condition of bus stops and terminals as an integral part of BisKita Transpakuan's operations does not indicate all indicators have been fully satisfied. These conditions result in Bogor City's inability to realise a sustainable urban transportation system. The recommendations for improving transportation systems and infrastructure are: policies should be issued by the central and regional government to provide special lanes for BisKita Transpakuan operation; and terminals and bus stops need to be improved following standards of safety, security, convenience and equality, regularity, and affordability.

REFERENCES

Adibah, A. N. (2019). The Performance of BRT Trans Semarang Services to Achieve Sustainable Transportation. *IOP Conference Series: Earth and Environmental Science*, 328. https://api.semanticscholar.org/CorpusID:211770648

Amador-Jimenez, L., and Christopher, A. D. (2016). A comfort index for public transportation: Case study of Montreal. *2016 IEEE International Conference on Intelligent Transportation Engineering* (ICITE), 3–7.

Brebbia, C. A. (2000). *The Sustainable City: Urban Regeneration and Sustainability*. WIT Press.

Fajobi, M. A., and Adesina, S. P. (2024). Group Dynamics and Public Policy Formation: Insights from Nigerian Society. *International Journal of Emerging Multidisciplinaries: Social Science*. https://api.semanticscholar.org/CorpusID:268826269

Filgueiras, F., and de Fatima Nascimento de Queiroz, L. (2021). The Governance of Public Policy Evaluation Systems: Policy Effectiveness and Accountability. *Organizações & Sociedade*. https://api.semanticscholar.org/CorpusID:236630013

Hasna, P., Luthfiyyah, D., Yuningsih, N. Y., and Afrilia, U. A. (2024). Kualitas Pelayanan Transportasi Publik Bus Rapid Transit (BRT) Melalui Biskita Trans Pakuan Kota Bogor Tahun 2022. *Janitra*, 4(1), 21–30.

Howlett, M., and Cashore, B. (2014). *Conceptualizing Public Policy*. https://api.semanticscholar.org/CorpusID:156513004

Ilhami, R. (2023). Strengthening policy networks to create effective public policies. *Neo Journal of Economy and Social Humanities*, 2(3), 216–222. https://doi.org/10.56403/nejesh.v2i3.143

Koch, L. C., Niesz, T., and McCarthy, H. M. (2014). Understanding and Reporting Qualitative Research. *Rehabilitation Counseling Bulletin*, 57, 131–143. https://api.semanticscholar.org/CorpusID:145382023

Merrill, R. K., and White, L. V. (2018). Bridging Ends and Means: The Centrality of Targets in Comparative Policy Analysis with Illustrations in the United States, New Zealand, and Germany. *Journal of Comparative Policy Analysis: Research and Practice*, 20, 469–485. https://api.semanticscholar.org/CorpusID:157623379

Mukherjee, I., and Bali, A. S. (2019). Policy effectiveness and capacity: two sides of the design coin. *Policy Design and Practice*, 2, 103–114. https://api.semanticscholar.org/CorpusID:199845995

Murtejo, T., Muhajir, A., Alimuddin, A., and Chayati, N. (2023). Evaluasi Kinerja Angkutan Umum Trans Pakuan Trayek Terminal Bubulak via Cidangiang – Ciawi di Kota Bogor. *Jurnal Komposit*, 7(1), 61–68. https://doi.org/10.32832/komposit.v7i1.7367

Novando, Hernawan, D., and Ramdani, F. T. (2024). Pengaruh Kualitas Pelayanan Terhadap Kepuasan Masyarakat pada Biskita Trans Pakuan. *Karimah Tauhid*, 3(2 SE-Articles), 1524–1537. https://doi.org/10.30997/karimahtauhid.v3i2.11823

Osly, P. J., Meutia, W., and Kurniawan, E. (2022). Kinerja Pelayanan "Bis Kita" Trans Pakuan Bogor Berdasarkan Persepsi Kepuasan Masyarakat. *Jurnal ARTESIS*, 2(2), 192–199. https://doi.org/10.35814/artesis.v2i2.4302

Rogers, B. (1998). Descriptive analysis of research data. *Workplace Health & Safety*, 46, 266–267. https://api.semanticscholar.org/CorpusID:41842447

Rundmo, T., and Nordfjærn, T. (2019). Judgement of transport security, risk sensitivity and travel mode use in urban areas. *Sustainability*. https://api.semanticscholar.org/CorpusID:159428227

Sahara, I. A. P., Hidayat, A., and Abdillah, W. S. (2022). Efektivitas Penggunaan Bus Trans Patriot Sebagai Angkutan Umum Massal Dalam Mengatasi Kemacetan Di Kota Bekasi. *Jurnal Inovasi Penelitian*, 3(2 SE-Articles). https://doi.org/10.47492/jip.v3i2.1752

Sari, H. P., Anggraeni, L., and Purnamadewi, Y. L. (2019). Mode Choice Analysis for Commuters in Metropolitan Region A-Case Study of Bogor City, Indonesia. *International Journal of Scientific Research in Science, Engineering and Technology*. https://api.semanticscholar.org/CorpusID:134306615

Sari, M. M. K., and Satmoko, A. (2022). Government Disaster Policy: Dealing Between Political Will and Political Demands. *Proceedings of the International Joint Conference on Arts and Humanities* 2021 (IJCAH 2021), 618. https://doi.org/10.2991/assehr.k.211223.103

Schmidt, R. (2022). A model for choice infrastructure: looking beyond choice architecture in Behavioral Public Policy. *Behavioural Public Policy*. https://api.semanticscholar.org/CorpusID:246511601

Srivastava, G. N. (2019). Understanding perception of commuters for safety and risk mitigation mechanism: the case of public transportation. *International Journal of Applied Management Science*.

Sugiyanto, S., Arnaya, I. W., Ryanto, S. S., and Surya, A. A. B. O. K. (2021). Analisa Faktor Pemilihan Moda Transportasi Menggunakan Metode Analytic Hierarchy Process. *Jurnal Teknologi Transportasi Dan Logistik*, 2(1), 11–18. https://doi.org/10.52920/jttl.v2i1.18

Susanto, B. A., Firdausiyah, N., Rini, I., and Ari, D. (2021). Evaluasi Kinerja Operasional dan Pelayanan Trans Pakuan Koridor 3 Cidangiang – Bellanova Kota Bogor Pada Masa Pandemi Covid-19. *Planning for Urban Region and Environment*, 10(0341), 147–158.

Syaiful. (2017). Engineering model of traffic and transportation safety with pattern of cooperation between sustainable region in Bogor.

United Nations. (2015, September 24). Sustainable Development Goal 11: Make Cities Inclusive, Safe, Resilient and Sustainable. https://sdgs.un.org/goals/goal11

Virani, A. (2019). Notions of policy effectiveness and implications for policy design: insights from public-private partnerships in India. *Policy Design and Practice*, 2, 198–214. https://api.semanticscholar.org/CorpusID:181741084

Towards Resilient Societies: The Synergy of Religion, Education, Health, Science, and Technology – Rahiem (Ed.)

Aligning SDGs with local migrant predeparture policies: Insights from Serang regency (2020–2022)

Ana Sabhana Azmy
Universitas Pembangunan Nasional "Veteran" Jakarta, Indonesia

Mutiara Pertiwi & Holisah Holisah
Universitas Islam Negeri Syarif Hidayatullah Jakarta, Indonesia

ABSTRACT: This research examines the alignment between the Serang Regency Government's pre-departure migrant policies and the principles of the Sustainable Development Goals (SDGs). The study's key contribution lies in analysing local policy dynamics and identifying opportunities to improve the well-being of migrant workers, particularly in relation to SDG 8.5 (decent work and economic growth) and 8.8 (protection of labour rights and safe working environments). Serang Regency, as the largest contributor of migrant workers from Banten Province and one of Indonesia's main migrant sources, faces significant challenges. It ranks 15th in the Indonesian Migrant Complaint Statistics, with deportation emerging as the most common issue in 2019, indicating frequent legal violations committed by migrants abroad. This raises concerns about the effectiveness of pre-departure policies. Since migrant governance aligns with the United Nations SDGs, there is a pressing need for Serang Regency to implement robust and informed predeparture policies. Through interviews with stakeholders, this study evaluates the coherence of Serang Regency's migrant predeparture policies (2020–2022) with SDG principles. The findings reveal that while notable policy reforms have been introduced, further improvements are needed in pre-departure training quality and reducing misinformation among prospective migrants.

Keywords: migrant workers, predeparture policies, sustainable development goals, Serang, Indonesia

1 INTRODUCTION

Migrant workers' governance is a crucial aspect of the United Nations (UN) Sustainable Development Goal (SDG) No. 8. This global commitment recognises the importance of a coherent multilevel dimension of migrant workers' governance to promote sustainable economic growth, full and productive employment, and decent work for all (Degila and Valle 2022).

Research on the compatibility of the SDGs with policies regarding migrant workers is limited, especially in the context of local or sub-national governance. While Sinkovics *et al.* (2024) and Gammage and Stevanovic (2019) have explored the opportunities and challenges related to achieving SDG No. 8, they primarily focus on global, regional, and national policies. Similarly, Hari and Hennebry (2019) examine the situation in Nepal, Low (2024) analyses Malaysia, and Schmitz and Cosman (2021) look at the national context in China.

Indonesia contributed nine million migrant workers in 2016, with one-third of that number working in caregiving and domestic sectors (Bal and Palmer 2020). Various research has revealed the vulnerabilities of Indonesian migrant workers to abuse, exploitation, and trafficking (Azmy 2024; Andrevsky and Lyneham 2014). The driving factors to this precarity include weak predeparture governance, and poor selections of working destinations and professions. Among the few studies addressing local government policy dynamics, Anggia and Teuku (2018) analyse Sumba Barat Daya, a regency in Nusa Tenggara Timur Province. They argue that the regency should prioritise predeparture capacity building and improve social protection provisions to align with SDG No. 8. It is important to investigate whether similar issues exist in other regions of Indonesia that supply migrant workers.

Our study parallels the work of Anggia and Teuku by examining a different case. It aims to investigate how well the SDG framework aligns with the predeparture migrant policies at the local level, with a focus on the policies and practices of Serang Regency from 2020 to 2022. Serang Regency is the primary source of Indonesian migrant workers (pekerja migran Indonesia - PMI) from Banten Province, one of the major suppliers of Indonesian migrant workers (BP2MI 2022). Between 2015 and 2021, Serang Regency sent almost 5000 migrant workers to various countries. Regrettably, PMI from this regency have encountered numerous issues abroad, evidenced by their ranking of 15 in the Indonesian Migrant Complaint Statistics 2019.

It is argued that despite various efforts, the Serang Regency Government has made minimal progress in enhancing the protection and well-being of PMI. The government still needs to focus on providing high-quality predeparture training and addressing misinformation about procedures, processes, and qualifications. The remaining challenges indicate that achieving SDG No. 8, particularly Goals 8.5 and 8.8, in Serang Regency still requires significant efforts.

2 METHODS

This study utilises qualitative analysis through a case study approach. We base our analysis on primary data gathered from field observations, semi-structured interviews, as well as secondary data from policy papers and literature. The field observations and interviews took place in Serang between July and December 2023. The participants include staff from the Serang Regency Regional Office of the Manpower and Transmigration Department (Dinas Tenaga Kerja dan Transmigrasi Kabupaten Serang), the Indonesian Migrant Workers Protection Services (Balai Pelayanan Pelindungan Pekerja Migran Indonesia (BP3MI)) in Banten, and five PMI and prospective Indonesian migrants workers (calon pekerja migran Indonesia (CPMI)) originating from Serang Regency who were assigned abroad between 2015 and 2023. The identity of the five workers is protected with pseudonyms. The interviews inform our analysis of the extent to which the policy implementation on PMI protection in Serang Regency conforms to Sustainable Development Goals No. 8, particularly Goals 8.5 and 8.8.

3 FINDINGS AND DISCUSSION

Before we explore the Serang Regency's pre-departure policies, it is important to note the clarity of the SDG aspirations regarding migrant workers governance. While the SDG framework consists of 17 goals, the most explicit connection to migrant workers issues is articulated by Goal No. 8 on ethical economic growth and decent work, which includes 12 specific targets (sdgs.un.org). The most straightforward targets concerning migrant workers protection are:

> *Target 8.5:* By 2030, achieve full and productive employment and decent work for all women and men, including for young people and persons with disabilities, and equal pay for work of equal value; and *Target 8.8:* Protect labour rights and

promote safe and secure working environments for all workers, including migrant workers, in particular women migrants, and those in precarious employment.

The above indicators will help this research evaluate the compatibility of Serang Regency's migrant predeparture policies with the SDGs framework.

We define public policy as any action or inaction conducted by governmental agencies and officials in regulating specific issues (Birkland's 2020). Our particular interest lies in the role of the Serang Regency Government as the official protector of provisions of CPMI at the pre-departure stage. This actor identification underscores the crucial responsibilities of those regulated by the constitution, in this case, Law No. 18/2017.

Law No. 18/2017 on the Protection of Migrant Workers is a crucial legal framework for addressing issues related to Indonesian migrant workers (PMI) (Azmy 2023; Wisnuwardhani *et al.* 2018). Unlike previous regulations, this law recognises the responsibilities of various stakeholders in safeguarding the rights of migrant workers and in preventing work-related hazards. It offers enhanced protection for PMI and their families from the predeparture to post-placement stages (Kharisma 2023). The Law stipulates that the Regency Government is among the central protection operators at the local level, highlighting the academic rigour of evaluating its policies.

Serang Regency, a part of Banten Province, currently lacks regulations to align its local PMI policies with Law No. 18/2017. The existing Banten Provincial Government Regulation No. 4/2016 on Workforce Services is only partially coherent with the national law. For instance, it does not recognise BP2MI as the key PMI national protection agency and BP3MI as the regional one. The current coordination between the Serang Regency Government and BP3MI is based on a non-binding Memorandum of Understanding (MoU), which has no legal consequences. The need for amending the Banten Regulation is clear for the national law to effectively impact this region. The legal gap hampers efforts to synchronise the national framework with the policies and practices at municipalities and regencies in Banten.

As revealed in the Table 1 of Migrant Complaint Statistics (2019) below, the highest recorded issue faced by Serang Regency origin PMI is related to deportation, which means the PMI have committed legal violations. This situation raises questions about their qualification and adaptation skills, which should have been standardised through pre-departure training at home. Improving the safety and well-being of PMI must begin with the exercise of prudent migrant predeparture policies by the Serang Regency Government.

Table 1. Migrant complaint statistics (2019): Serang regency origin.

Unpaid salary	Deportation	Undocumented	Over stay	Papers issues	Health issues
37	45	–	22	5	10
Human trafficking & smuggling	Personal request to return	Issues with employers	Lay off	Violence Victims	Others
3	7	12	7	8	43

Source: BP2MI Banten 2019.

The findings section below is divided into four parts. The findings suggest that although some coordination has taken place between the local government and BP3MI Banten, there is still a need for tangible protection policy reforms to align with Law No. 8/2017 and the Sustainable Development Goals (SDGs).

3.1 *Administrative protection*

Law No. 18/2017 (Article 8) codified administrative protection by the government to ensure every CPMI possesses a legal identity and adequate documentation, including contracts and agreements with employers in the placement countries. There are two types of agreements here: placement and employment agreements. Both must be signed at the pre-departure stage by the CPMI, the employer, and the Indonesian Embassy where the placement is. During this research interview, the Manpower and Transmigration Office of the Serang Regency explained that the local government is involved in the CPMI predeparture processing through the following stages in Figure 1:

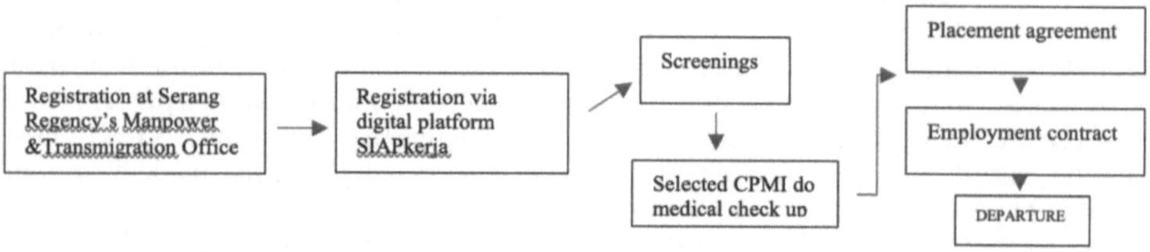

Figure 1. CPMI predeparture processing in Serang regency. *Note.* Manpower and Transmigration Office of the Serang Regency (Personal communication 2024).

The above illustration demonstrates the government's effort to participate in the national framework for migrant protection by facilitating the roles of the Ministry of Manpower and Transmigration and the BP2MI. The BP3MI Banten and the Manpower and Transmigration Office of the Serang Regency must witness the signing process of these agreements. supervising agreements between CPMI and external parties by attending the signing process. Additionally, the local government is involved in supervising the departure of CPMI to their placement countries.

3.2 *Technical protection*

The second dimension of protection at the pre-departure stage is technical protection. Law No. 18/2017 (Article 8) codified this form of protection to include the provisions of (1) reliable and accessible information; (2) capacity building programs; (3) social security; (4) facilities to fulfil migrant rights; (5) enhanced capacities of government officers specialised in migrant assistantship and placement services (pegawai fungsional pengantar kerja) of the Ministry of Manpower and Transmigration; (5) migrant one-stop service; and (6) coaching and monitoring.

The Serang Regency Government has actively contributed to various technical protection aspects in collaboration with relevant institutions such as BP3MI Banten and local P3MI. BP3MI serves as the local implementing agency of BP2MI programs. From 2020 to 2022, the Regency has been a supportive partner to BP3MI, particularly in its efforts to provide the CPMI capacity-building program. Additionally, the Regency established the Migrant Workers One-Stop Services (Lembaga Terpadu Satu Atap or LTSA) in 2020, operated by BP3MI Banten, to streamline administrative processes more effectively. Furthermore, the Regency actively participates in BP2MI Outreach Programs to ensure that accurate information and employment opportunities are accessible to CPMI in remote areas. The majority of these programs from 2020 to 2022 were concentrated in the Ciruas District. Lastly, in line with the pre-departure procedures for migrant workers mentioned previously, the Regency also engages in the coaching and supervision process to safeguard the rights of migrant workers during registration, selection, and departure stages.

3.3 *Obstacles*

This section highlights the key protection barrier in the context of the Serang Regency migrant predeparture policies. Firstly, there is the spread of disinformation about migration and working procedures. A trafficking network is operating in Serang Regency, targeting those who are in economic hardship and desperately need jobs. This criminal network spreads false information about various migrant workers' procedures, manipulating the CPMI to commit fraud or resort to illegal methods that make them vulnerable to harm and future exploitation.

Secondly, there are trust issues. CPMI tends to avoid government officers when encountering problems. They believe the government would be unable to help them and would only make the migration process more difficult. Therefore, the CPMI seeks assistance from local leaders, such as the neighbourhood coordinators (Rukun Tetangga (RT)), rather than directly reporting their issues to the local government.

Thirdly, the issue of weak monitoring is significant. The Regency needs more resources and a stronger institutional capacity to monitor stakeholders' compliance with rules effectively. For instance, during our research interview, we found two female CPMIs in the formal sector who had not yet received training during their nearly one-year departure waiting period. This is a serious concern as job training is a fundamental part of migrant rights in Law No.18/2017.

3.4 *Local-national-global policy coherence*

The analysis indicates that while the policies of the Serang Regency Government partially align with the principles of Law 18/2017 and Sustainable Development Goals No. 8, the quality of its services in protecting migrants is inadequate. This limited support impacts the local government's contributions to SDGs Targets 8.5 and 8.6, which aim to achieve full and productive employment and reduce youth unemployment, as they currently lack a guarantee of decent work and equal reward. At present, the Serang Regency Government does not have specific policies to measure the achievement of Target 8.7 on eradicating forced labour and child exploitation, indicating an area in need of further investment. Additionally, the government's efforts to achieve Target 8.8, focusing on the protection provisions of labour rights, are difficult to measure due to the lack of comprehensive data. There also appears to be a lack of government awareness regarding SDG Target 8.9 concerning the global strategy for youth employment and the Global Jobs Pact of the International Labour Organization.

On a positive note, the Serang Regency Government is in the process of drafting the Regency Government Regulation on Migrant Workers Protection, which is anticipated to have a significant impact once enacted. Furthermore, the Regency is exploring ways to strengthen its partnership with BP3MI, particularly in providing adequate and quality migrant training centres. These initiatives are expected to enhance the Regency's commitment, capacity, and resource allocation for migrant protection.

Restricting legal pathways for migrant workers often drives individuals toward 'illegal' options. The Regency needs to craft a comprehensive policy, including measures to boost the local economy and reduce poverty. It is important to also mark the socio-cultural push factors of women migrant domestic workers in Serang Regency. As also found in other areas of Indonesia, many women become migrant workers to escape the pressure of forced marriages at a young age (Halimatusa'diyah and Arif 2020). Instead of closing off their opportunities to pursue a better life, safer alternatives should be provided to help them thrive.

4 CONCLUSION

The elaboration above reveals a range of institutional efforts made by the Serang Regency Government in exercising protection and supporting well-being of migrant workers at pre-departure stage. Despite the high level of migrant complaints abroad, the local government has not introduced any new policies to improve migrant protection and well-being. The government is also behind in introducing regional regulation to operationalise Law No. 18/2017 on Migrant Workers Protection, let alone regulations to conform to SDG No. 8. This legal vacuum has made the Serang Regency Government operate in conventional ways. Considering these findings, we argue that the Serang Regency Government's migrant pre-departure policy structure and implementation require substantive improvement to be able to contribute to the attainment of SDG no. 8. Without this real reform, the migrant population with Serang Regency origin will experience similar cases of vulnerabilities abroad to their precursors. This reform will be most effective with better collaboration among stakeholders, including civil society organisations and businesses, to improve pre-departure training and information sharing.

REFERENCES

Andrevsky, H. and Lyneham, S. (2014). Experiences of Exploitation and Human Trafficking among a Sample of Indonesian Migrant Domestic Workers. *Trends and Issues in Crime and Criminal Justice*, 471–481.

Anggia, U. D. and Teuku, R. (2018). Indonesian Migrant Domestic Workers and Sustainable Development Goals: Feasible Policy and Practices. *E3S Web Conferences*, 73, Article 09017.

Azmy, A. S. (2023). Pelindungan Negara atas Perempuan Pekerja Migran Indonesia; Tinjauan terhadap Implementasi Kebijakan. Yayasan Pustaka Obor Indonesia.

Azmy, A. S. (2024). Representasi Perempuan Pekerja Migran Indonesia (PPMI) di Parlemen: Peluang dan Tantangan. *Jurnal Perempuan*, 29(2), 127–140.

Bal, C. S., and Palmer, W. (2020). Indonesia and Circular Labor Migration: Governance, Remittances and Multi-directional Flows. *Asian and Pacific Migration Journal*, 29(1), 3–11. https://doi.org/10.1177/0117196820925729

Birkland, T. A. (2020). *An Introduction to the Policy Process: Theories, Concepts, and Models of Public Policy Making*. Routledge.

BP2MI (2022) *Data Pekerja Migran Indonesia tahun 2021*. Pusat Data dan Informasi BP2MI.

Degila, D.E. and Valle, V.M. Multilevel Governance of Migration for Sustainable Development: Perspectives from the Global Souths. (2022). *International Development Policy*, 14, Article 1. https://doi.org/10.4000/poldev.4769

Gammage, S. and Stevanovic, N. (2019). Gender, Migration and Care Deficits: What Role for the Sustainable Development Goals? *Journal of Ethnic and Migration*, 45 (14), 2600–2620.

Halimatusa'diyah, I. and Arif, S. (2020). Gender Equality, Women's Political Representation, and Maternal Mortality: Lessons from Indonesia. *Journal of Development Perspectives*, 4 (1–2), 39–58.

Hari, K. C. and Hennerbry, J. L. (2019). Gender, Labour Migration Governance, and the SDGs: Lessons from the Case of Nepal. *Achieving the Sustainable Development Goals*, 1, 71–84.

Kharisma, D. B. (2023). Optimalisasi Pemerintah Desa dalam Pelindungan Pekerja Migran Indonesia. *Rechts Vinding: Media Pembinaan Hukum Nasional*, 12(1), 77–85.

Low, C. C. (2024). Migration and SDGs in Malaysia: Implementing Policy Reform Toward Migrant Rights Protection. *Asia Pacific Social Science Review*, 24 (3), Article 5.

Schmitz, M. A. and Cosman, H.G. (2021). 'Sustainable Development and Chinese "Domestic": Migrant Workers: A Policy Perspective'. In F. Weider, L. Krainer and M. Karmasin (Eds.), *The Sustainability Communication Reader: A Reflective Compendium* (543–566). Springer.

Sinkovics, N., Sinkovics, R. R, Boussebaa, M., and Fletcher, M. (2024). *International Business and SDG 8: Exploring the Relationship Between IB and Society*. Palgrave Macmillan.

Wisnuwardhani, S., Alwy, B., Awigra, D., Wiratama, O., Dwi, R., Sulisyowati, Y., and Devi, W. (2018). *Buku Saku Memahami Undang Undang Perlindungan Pekerja Migran Indonesia: Kelebihan dan Kelemahan UU PMI*. Jaringan Buruh Migran.

Towards Resilient Societies: The Synergy of Religion, Education, Health, Science, and Technology – Rahiem (Ed.)

Community-based mangrove ecosystem management for sustainable tourism in Sukakerta village, Karawang regency

Angga Kurniawansyah
Ministry of National Development Planning/Bappenas, Indonesia
ORCID ID: 0009-0005-3093-6242

Raden Bonnyswasara
Ministry of National Development Planning/Bappenas, Indonesia

Dewi Susiloningtyas
Department of Geography, University of Indonesia, Indonesia
ORCID ID: 0000-0002-9027-6128

Masita Dwi Mandini Manessa
Department of Geography, University of Indonesia, Indonesia
ORCID ID: 0000-0002-4531-3057

Rasta Aditya
Indonesian College of Business Economics/STIEBI, Indonesia

ABSTRACT: This study explores mangrove ecosystem management through community participation at the Tangkolak Maritime Center (TMC) tourism site in Sukakerta Village, Cilamaya Wetan District, Karawang Regency. A qualitative method was employed, incorporating discussions and interviews with key and primary informants, supported by quantitative assessments using a Likert scale. The findings indicate that community participation in mangrove ecosystem management is considered adequate, with an average score of 3.00 out of 5.00. Participation includes activities such as mangrove planting, maintenance, and sustainable use of timber and non-timber resources, as well as tourism development. Best practices observed include active involvement in conservation initiatives, utilisation of mangroves for eco-tourism, firewood, and non-timber products such as honey, tannin, animal feed, and aquaculture pond activities, particularly when conducted on privately owned land. These activities provide ecosystem services directly benefiting the community. This study offers valuable insights for researchers and suggests pathways for future research to explore deeper aspects of mangrove ecosystem management. Additionally, it highlights the importance of policymaker support in promoting awareness, mobilising community participation, and implementing sustainable mangrove management practices as a practical contribution to climate change mitigation.

Keywords: mangrove, ecosystem, management, community, participation, tourism

1 INTRODUCTION

Mangrove ecosystem management emphasises restoration and reduction of degradation (Adenan 2018). As is known, mangroves are plants that thrive in coastal areas with high salinity and waterlogged regions (Suryana *et al.* 2024) with various functions, such as tourism object (Ewaldo *et al.* 2023), and effectively as a carbon sink (Sjögersten *et al.* 2021). One

DOI: 10.1201/9781003645542-9

of the uses of the existing mangrove ecosystem is located at the Tangkolak Bahari Center (TMC), Sukakerta Village as tourism in the last 6 years (Pin *et al.* 2021). Based on the report by (Kurniawansyah *et al.* 2023), there was ±2 Ha of mangrove degradation in Sukakerta Village in 2019 as a result of an increase in the number of tourists. Based on the Karawang Regency Strategic Plan (Renstra) for 2021–2026, visits to natural tourism object (mountains and beaches) have increased significantly. The increased number of tourists due to the high concentration of tourism activities will harm the ecosystem (Kurniawan *et al.* 2016). If this condition is left unchecked, the mangrove ecosystem will be damaged due to a decrease in the area and quality of the ecosystem.

Therefore, managing the mangrove ecosystem is important to prevent or improve the condition of the mangrove ecosystem from damage due to tourism activities. This management can be carried out in coordination and consolidation with the community around the location of the tourism object regarding efforts as a form of their participation so that ecosystem management is carried out well and mangrove functions are utilised sustainably (Surjanti *et al.* 2020). Macamo *et al.* (2024) suggests that community participation provides awareness in the management and conservation of mangrove ecosystems. This study aims at finding out how to manage the mangrove ecosystems in TMC based in the involvement of the surrounding community, which is a form of implementation of the action plan in preventing climate change. In other words, this considers environmental and sustainability aspects which are also in line with the achievement of SDGs 13 concerning the prevention of climate change. Research on mangrove management found in the last 5 years in coastal areas, bays, deltas, and islands at tourist attraction locations has not connected it with community participation using a Likert scale. The following structure of this paper includes research methodology in the form of data collection and calculation, including sample determination. The results section presents the findings and the discussion explains the findings in the context of existing research. Finally, this study's conclusions provide recommendations for future policymakers and researchers regarding mangrove ecosystem management.

2 METHODS

This research uses qualitative methods through discussions with tourism managers and interviews with local communities, as well as quantitative methods in the form of Likert scale calculations. Discussion is a form of expressing opinions, collecting information, and making group conclusions (Rohimawati *et al.* 2024). Meanwhile, interviews are a way of exchanging ideas and some people try to obtain information, expressions of opinion, or beliefs from other people (Young *et al.* 2018). The questions asked to resource persons/informants will relate to managing mangrove ecosystems, referring to Susilo (2007) regarding aspects of mangrove protection, utilisation, and rehabilitation. Mangrove ecosystem management data from interviews will be processed based on comprehensive calculations and classified according to their respective classes (low, insufficient, sufficient, high, and very high) according to Hunecke *et al.* (2010) using a Likert scale. The classification of community participation using scores 1, 2, 3, 4, and 5. Likert scale has often been used in social science measurements, such as attitudes, norms, understanding, a person's behaviour, and other social characteristics (Magdolen *et al.* 2024). Therefore, the analysis of this research will use a Likert scale and answers from the community (never, rarely, sometimes, often, and always) referring to Cahyadinata *et al.* (2019) which will be calculated based on the class interval (0.8) based on Equation 1 below.

$$Class\ Interval = \frac{Highest\ Score - Lowest\ Score}{Number\ of\ Classes} \tag{1}$$

$$Class\ Interval = \frac{5-1}{5} = 0,8$$

Determining the sample of respondents for interviews is based on non-random sampling, a type of purposive sampling, namely the selection of a group of subjects based on the suitability of the sample and arguments with certain reasons for their selection and providing an adequate basis for experimental surveys and is subjective (Klar and Leeper 2019). Suzanne (2020) suggests that the type of purposive sampling is very important to clarify the justification of the criteria in sampling determined by the researchers. There are also criteria or considerations set by researchers in taking samples, namely having interests or activities/ receiving benefits, and/participating in activities in managing mangrove ecosystems in and around tourism object, and being aged $\geq$ 17 years because according to Law of the Republic of Indonesia No. 22 (2009) is considered capable of taking responsibility for oneself. The interviews conducted took into account the population size based on Memon *et al.* (2020) with a sample size of $\geq$ 50 people for observation and survey recommendations because of the ease of selecting respondents randomly and this study found at least 50 people in the field. Sample selection has nothing to do with the contribution of the research (including bias), but rather about the appropriateness of its use in the research.

3 FINDINGS AND DISCUSSION

3.1 *Main findings: best practices from successful case studies*

The best practices identified from successful case studies of community participation related to the management of the mangrove ecosystem at the TMC tourism object, namely that the community has carried out activities within the scope of management, such as conservation activities carried out only on the coastline and slightly indented towards the mainland, utilisation of mangroves from timber and non-timber, and activities tourism, where it is said that participation is included in the sufficient classification. In the context of utilising mangrove timber, it is usually used by the community for household purposes by cutting it randomly around the TMC tourism object location. If allowed to continue, this will result in the degradation of the mangrove ecosystem at the TMC tourism object.

3.2 *Analysis of findings: Research unit characteristic identify and classification of participation*

In this research, the unit of analysis is the location of a mangrove ecosystem tourism object, namely TMC. So, the key informant is the manager of the TMC tourism object, namely Mr. Yanto. He is the chairman of the Kreasi Alam Bahari (KAB) Tourism Awareness Group (Pokdarwis), which has 11 members in its operational activities and is also the manager of the TMC tourism object. Meanwhile, the main informants in this research were the local community at the TMC tourism object. In total, 50 people were interviewed around the TMC tourism object, with ages ranging from 19 to 72 years. In society, women are dominant compared to men. They have educational levels ranging from elementary school (SD) to university (S1) and varied jobs, such as food and beverage traders, housewives (IRT), fishermen, consultants, and government employees. It is known that the average total community participation has a score of 3.00, which is classified as sufficient. The average community participation in implementing activities in the protection aspect was 2.98 (sufficient classification), the utilisation aspect was 2.50 (poor classification), and the rehabilitation aspect was 3.52 (high classification) (Figure 1).

Figure 1. Community participation in TMC.

3.3 *Interpretation of findings: Participation in mangrove ecosystem management*

The protection aspect is based on the presence of the community in the implementation stages of mangrove planting, mangrove nursery, and mangrove maintenance activities. Based on interviews conducted, they said that the protection aspect in the form of conservation activities was carried out at the coast border and slightly indented towards the mainland but remained at the location of the TMC tourism object based on programs originating from the regional government or academics (with several types of mangroves, including *Avicennia, Rhizopora stylosa,* and *Sonneratia*).

In the utilisation aspect, the community admits that they do not participate in aquaculture activities, only a few use non-mangrove timber products (honey, tannin, animal feed, etc.), and are dominant in tourism activities and mangrove timber products (for firewood). The community's non-participation in aquaculture activities is because the majority of them do not have more land to use and need to learn how to carry out these activities. The local government can provide land grant assistance for this problem. Meanwhile, regarding the utilisation of non-mangrove timberproducts, it is said that only a few people understand further processing, and these people are dominated by mothers (as housewives) who have received some training from tourism managers or the local government voluntarily. This is a challenge that occurs there. They claim to get benefits from this, such as skills in processing for consumption and profits from buying and selling that occurs (from Rp50.000 to Rp200.000 in a day). Non-timber products usually include mangrove juice, crackers, lunkhead, and others. The uneven training received by the community is due to the lack of internal will of the community itself, not because there is no training program there. This is what makes the value of community participation included in the poor classification.

For the rehabilitation aspect, the community has participated by planting mangrove seedlings and caring for man-grove trees around the TMC tourism object. Some of them admitted that they already had several mangrove trees that they bought their seeds (5-10 tree seeds) and cared for until the tree seeds grew big (around 1-2 years). Based on community recognition, these are people whose jobs are directly related to the existence of the mangrove ecosystem, such as Food and Beverage Traders and Fishermen (12,00% and 30,00% of the total people met). According to them, the economic improvement can be felt from the existence of the TMC tourism object. However, purchasing mangrove seeds for planting was only carried out by people who had an income of $\geq$ Rp250.000/day (16,00% of the total people met). They did both of these things because they believed that the presence of mangrove trees would positively impact or benefit on their lives during their stay, such as as a tourism destination, wave barrier, and mangrove timber products.

3.4 *Policy recommendation: Presence of participation and development technologies*

Policies that must be considered to achieve the success of coastal ecosystem management programs are collaboration, a sense of commitment, shared responsibility, and ownership of resources. Based on the research findings of Gultom *et al.* (2021), the community will be more active in planting and caring for mangroves because they are very aware of the benefits of the mangrove ecosystem on the coast and because of their experience in preventing disasters, which collaborate with organisations, student groups, youth organisations, and non-governmental organisations. Seeing the conditions at the TMC tourism object, which has lost several hectares of area, community participation is considered important in managing the mangrove ecosystem in this study and can be continuously improved. Technology such as remote sensing can also be utilised in this study. One such method is the *Participatory Geography Information System* (PGIS), which desires to deliver a clearer understanding of the problems spatially from community's active role.

3.5 *Implication of findings: Implementation community participation in mangrove ecosystem management*

Increasing participation in managing mangrove ecosystems from the utilisation aspect can include firewood, non-timber cultivation (honey, tannin, animal feed, etc.), and tourism activities. The utilisation aspect of pond activities (intercropping/wanamina/silvofishery and open type) can also be carried out if the community has land and further knowledge of implementation mechanisms, which is necessary regarding the role and involvement of the community in the mangrove ecosystem. Increasing community participation is one way that can be done for better management of mangrove ecosystems, such as in the female gender locus in the research of Begum *et al.* (2024) by resolving some social issues that arise. Community-based tourism object management can be a strategy to achieve development goals because decision-making regarding resources and the environment lies with organisations within the community (Sumarmi *et al.* 2021). Sattayapanich *et al.* (2022) also said that community participation in managing mangrove ecosystems is essential to encourage sustainable use of mangrove resources. According to Elwell *et al.* (2018), understanding a community or society is the basis for their participation in a natural resource by identifying what is important to them and the public. Therefore, the existence of a community in an ecosystem is important in terms of management.

4 CONCLUSION

This study highlights the management of mangrove ecosystems based on community participation in TMC tourist attractions for environmental sustainability to achieve climate change mitigation (SDGs 13) and underwater life (SDGs 14) goals presented as output activities, such as planting, seeding, and maintenance of mangroves, utilisation of mangrove ecosystems (such as the use of firewood, non-timber cultivation), pond activities, and for tourist attractions. Of course, the utilisation of these activities must be wise in their implementation by the surrounding community. Policymakers need to conduct socialisation related to the benefits of mangrove ecosystems to the community intensively and periodically. In addition, it is necessary to develop new information regarding the management of mangrove ecosystems in the rehabilitation aspect (related to location details) and in the protection aspect (used as a wave breaker by the community around the tourist attraction).

REFERENCES

Adenan, R. (2018). Restoration Of Mangrove Co-Management To Sequester Coastal Blue Carbon. *Advances in Social Sciences Research Journal*, *5*(2), 27–40. https://doi.org/10.14738/assrj.52.4134

Begum, F., Lobry de Bruyn, L., Kristiansen, P., and Islam, M. A. (2024). What factors influence women's participation in co-management? A case study of Sundarban mangrove forest management in Bangladesh. *Journal of Environmental Planning and Management*, *0*(0), 1–26. https://doi.org/10.1080/09640568.2024.2346600

Cahyadinata, I., Fahrudin, A., Sulistiono, and Kurnia, R. (2019). Perception And Participation of Fishermen In The Sustainable Management of Mud Crabs On The Outermost Small Island (Case Study: Enggano Island, Bengkulu Province, Indonesia). *International Journal on Advanced Science, Engineering and Information Technology*, *9*(4), 1330–1336. https://doi.org/10.18517/ijaseit.9.4.7287

Elwell, T. L., Gelcich, S., Gaines, S. D., and López-Carr, D. (2018). Using People's Perceptions of Ecosystem Services To Guide Modeling And Management Efforts. *Science of the Total Environment*, *637–638*, 1014–1025. https://doi.org/10.1016/j.scitotenv.2018.04.052

Ewaldo, K., Karuniasa, M., and Takarina, N. D. (2023). Suitability of Mangrove Forests for Sustainable Tourism. *Advances in Tropical Biodiversity and Environmental Sciences*, *7*(3). https://doi.org/10.24843/ATBES.v07.i03.p02

Gultom, J. E. I., Hasibuan, H. S., and Patria, M. P. (2021). Local Communities Participation in Mangrove Management for Tsunami Disaster Mitigation at Palu City Coastal. *IOP Conference Series: Earth and Environmental Science*, *940*(1). https://doi.org/10.1088/1755-1315/940/1/012084

Hunecke, M., Haustein, S., Böhler, S., and Grischkat, S. (2010). Attitude-Based Target Groups To Reduce The Ecological Impact of Daily Mobility Behavior. *Environment and Behavior*, *42*(1), 3–43. https://doi.org/10.1177/0013916508319587

Klar, S., and Leeper, T. J. (2019). Identities And Intersectionality: A Case For Purposive Sampling In Survey-Experimental Research. *Experimental Methods in Survey Research: Techniques That Combine Random Sampling with Random Assignment*, 419–433. https://doi.org/10.1002/9781119083771.ch21

Kurniawan, F., Adrianto, L., Bengen, D. G., and Prasetyo, L. B. (2016). Vulnerability assessment of small islands to tourism: The case of the Marine Tourism Park of the Gili Matra Islands, Indonesia. *Global Ecology and Conservation*, *6*, 308–326. https://doi.org/10.1016/j.gecco.2016.04.001

Kurniawansyah, A., Susiloningtyas, D., Dwi, M., and Manessa, M. (2023). Zone-Based Tourism Planning Using Satellite Imagery. *The Journal of Indonesia Sustainable Development Planning (JISDeP)*, *4*(3), 263–283. https://doi.org/10.46456/jisdep.v4i3.441

Macamo, C. da C. F., Inácio da Costa, F., Bandeira, S., Adams, J. B., and Balidy, H. J. (2024). Mangrove Community-Based Management In Eastern Africa: Experiences From Rural Mozambique. *Frontiers in Marine Science*, *11*(February), 1–13. https://doi.org/10.3389/fmars.2024.1337678

Magdolen, M., Von Behren, S., Vallée, J., Chlond, B., and Vortisch, P. (2024). Response Bias In Likert-Style Psychological Items - An Example From A Large-Scale Travel Survey In China. *Transportation Research Procedia*, *76*(2022), 349–360. https://doi.org/10.1016/j.trpro.2023.12.060

Memon, M. A., Ting, H., Cheah, J. H., Thurasamy, R., Chuah, F., and Cham, T. H. (2020). Sample Size For Survey Research: Review And Recommendations. *Journal of Applied Structural Equation Modeling*, *4*(2), i–xx. https://doi.org/10.47263/jasem.4(2)01

Pin, T. G., Supriatna, J., Takarina, N. D., and Tambunan, R. P. (2021). Mangrove Diversity And Suitability Assessments For Ecotourism In Cimalaya Wetan Coast, Karawang District, Indonesia. *Biodiversitas*, *22* (2), 803–810. https://doi.org/10.13057/biodiv/d220234

Rohimawati, F. A., Shohib, M. W., and Apriantoro, M. S. (2024). Exploring Critical Discussion Research: A Bibliometric Analysis of Scopus-Indexed Publications. *ISETH (International Summit on Science, Technology, and Humanity)*, 968–978.

Sattayapanich, T., Janmaimool, P., and Chontanawat, J. (2022). Factors Affecting Community Participation In Environmental Corporate Social Responsibility Projects: Evidence From Mangrove Forest Management Project. *Journal of Open Innovation: Technology, Market, and Complexity*, *8*(4), 1–28. https://doi.org/10.3390/joitmc8040209

Sjögersten, S., de la Barreda-Bautista, B., Brown, C., Boyd, D., Lopez-Rosas, H., Hernández, E., Monroy, R., Rincón, M., Vane, C., Moss-Hayes, V., Gallardo-Cruz, J. A., Infante-Mata, D., Hoyos-Santillan, J., Vidal Solórzano, J., Peralta-Carreta, C., and Moreno-Casasola, P. (2021). Coastal wetland ecosystems deliver large carbon stocks in tropical Mexico. *Geoderma*, *403*(December 2019). https://doi.org/10.1016/j.geoderma.2021.115173

Sumarmi, S., Arinta, D., Suprianto, A., and Aliman, M. (2021). The Development of Ecotourism With Community-Based Tourism (CBT) In Clungup Mangrove Conservation (CMC) of Tiga Warna Beach For Sustainable Conservation. *Folia Geographica*, *63*(1), 123–142.

Surjanti, J., Soejoto, A., Seno, D. N., and Waspodo. (2020). Mangrove Forest Ecotourism: Participatory ecological learning and sustainability of students' behavior through self-efficacy and self-concept. *Social Sciences and Humanities Open*, *2*(1), 100009. https://doi.org/10.1016/j.ssaho.2019.100009

Suryana, T. N., Purnamasari, S., and Ewaldo, K. (2024). Monitoring mangrove forest degradation in mangrove nature tourism Park Angke Kapuk, North Jakarta, Indonesia Using NDVI. *European Journal of Forest Engineering, February*, 29–42. https://doi.org/10.33904/ejfe.1395676

Susilo, F. (2007). *Pengelolaan Ekosistem Mangrove Kabupaten Deli Serdang Sumatera Utara* [IPB University]. https://repository.ipb.ac.id/handle/123456789/10777

Suzanne, D. (2020). Commentary: Purposive Sampling: Complex or simple? research case examples. *Journal of Research in Nursing*, *25*(8), 662–663. https://doi.org/10.1177/1744987120928156

Young, J. C., Rose, D. C., Mumby, H. S., Benitez-Capistros, F., Derrick, C. J., Finch, T., Garcia, C., Home, C., Marwaha, E., Morgans, C., Parkinson, S., Shah, J., Wilson, K. A., and Mukherjee, N. (2018). A Methodological Guide To Using And Reporting On Interviews In Conservation Science Research. *Methods in Ecology and Evolution*, *9*(1), 10–19. https://doi.org/10.1111/2041-210X.12828

Towards Resilient Societies: The Synergy of Religion, Education, Health, Science, and Technology – Rahiem (Ed.)

Sustainable development strategies for Indonesia as a maritime axis

Eska Dwipayana Pulungan
Universitas Islam Negeri Syarif Hidayatullah Jakarta, Indonesia

ABSTRACT: This study examines sustainable development efforts undertaken by the Indonesian government to position Indonesia as a global maritime axis, employing a literature review as its primary research method. The study focuses on identifying key government initiatives and assessing their relevance in achieving this goal, specifically in the areas of maritime-oriented human resource preparation, maritime infrastructure development, and political will. The findings indicate that while these three components are crucial in supporting Indonesia's maritime ambitions, they have not yet succeeded in establishing Indonesia as a global maritime axis. This shortfall is largely due to obstacles such as the ongoing politicisation of the sea toll discourse, which hinders regional connectivity. To address these challenges, the study emphasises the importance of aligning sustainable development strategies with SDG 9, which focuses on industry, innovation, and infrastructure. This alignment is essential for achieving consistent and healthy development progress. The study concludes by recommending the establishment of robust inter-island connectivity as a strategic measure to enhance economic engagement and maximise the potential of Indonesia's vast marine resources.

Keywords: Indonesia, maritime infrastructure, sea toll, sustainable development, world maritime axis

1 INTRODUCTION

Indonesia is built upon the unity of numerous islands and interconnected waters by the concept of Archipelagic State, which was championed by Mochtar Kusumatmadja in 1982. The maritime waters between Indonesia's islands are a unifier, not a separator. Such conditions should make Indonesia a world maritime power and a regional leader (Wicaksana 2017). However, the relationship between the national government and the regions and outer islands has led Indonesian policymakers to see geography as a source of difficulty in developing Indonesia. It is, therefore, necessary to design a strategy to establish Indonesia as a world maritime axis by initially making domestic improvements, and by following the 9th SDG, which focuses on developing industry, innovation, and infrastructure (Butcher and Elson 2017).

The Indonesian government need to achieve several movements in order to be consistent in its vision of realising Indonesia as a world maritime axis based on Sustainable Development Goal (SDG) 9, namely: one, development outside of Java; building maritime infrastructure; international cooperation for integrated regional development; and technological development (Connelly 2015). These are important aspects because in order to achieve a revival, Indonesia needs to prepare human resource expertise in various fields in this context in the marine sector, ranging from technical and technological experts to strategists and specialists in international maritime law (Juliadi *et al.* 2023). Meanwhile, to reach a more strategic level, it is necessary to initiate the strengthening of maritime environmental

DOI: 10.1201/9781003645542-10

awareness, namely Maritime Domain Awareness (MDA) (Chintoan-Uta and Silva 2017). Steps must be taken immediately and instilled in the younger generation by mainstreaming maritime insights into the education process. Second, MDA needs to be supported and carried out to strengthen maritime infrastructure. The third point relates to funds and technology (Agastia and Perwita 2017). This may be difficult for Indonesia to achieve without foreign assistance, hence the need to establish multilateral maritime partnerships. The point is not only to gain inspirations on how to develop an archipelagic nation equitably and sustainably, but also how to obtain the availability of qualified technology in the maritime sector (Lauder and Lauder 2016).

This study is aimed at finding out the obtained benefits given that the three aspects above are carried out properly, moreover the three components are closely related to SDG 9, which are not properly carried out by the government. Geoffrey Till in his book entitled Seapower A Guide for the Twenty-First Century states that the most important measure a country can do to achieve progress and glory is empowering its maritime power, which will automatically improve the welfare and security of the country. If ignored, it will likely result in great losses for the country and even potentially weaken the country (Till 2018). This is a benchmark in making changes to start shifting economic activities from land to sea and also to boost infrastructure development outside Java by empowering the many benefits of Indonesian seas. In meeting societal needs, it should be noted that excessive focus should not be put upon infrastructure development, but it should be balanced and adjusted to sovereignty of food, minerals, and raw materials (Salim and Negara 2018). All of which can be found in Indonesian seas as new natural resources to help build equitable and sustainable infrastructure.

This particular type of deep research using literature review is employed to analyse the actions implemented by the government to address these three components. In particular, this study will evaluate the government's actions and identify better actions by comparing them with Japan's maritime policy as a final recommendation for the government. The research questions the article is intended to address are as follows: (1) what are the government's actions in making Indonesia a world maritime axis; (2) what actions or policies carried out by the government have succeeded in making Indonesia a maritime axis.

The significance of this research is to assist the government in realising its vision of becoming a maritime axis or at least making Indonesia a master of maritime industry in the region. Based on the deep research conducted, the study also addresses the need for synergy between government institutions, instead of just a single institution, as well as the need for policy alignment between the national and subnational level governments to avoid conflicting interests and prevent regulations that slow down action in realising Indonesia's vision of becoming a maritime axis.

2 METHODS

The qualitative research method was specifically used in this study. Qualitative research constitutes research that relies on deep research using data first and subsequently analysing them into new and credible findings or facts that have benefits in scientific development. The triangulation model was used as a lens to see more clearly the policies of the government of the Republic of Indonesia in making Indonesia a world maritime centre. Secondary data sources in the form of media coverage or the President's social media accounts were also used in collecting data because media coverage and social media are currently the most frequent communication tools the government uses for expressing its statements or conveying information relating to new government policies (Creswell and Creswell 2017).

The data collection technique was carried out by conducting a literature review. Books, among others, were used to deepen the research in order to analyse the state of advancement of Indonesia's natural resources, as well as to obtain data on the various kinds of Indonesian maritime wealth. Journal articles were used to examine the research further and determine why President Jokowi was most interested in realising Indonesia as a maritime axis. Additionally, government websites and international online news were also used to analyse the development of sea toll policies, the sustainability of the development of marine food security, and the progress of inter-sea interconnectivity development.

Deep research was then conducted by focusing on the use of books and journal articles published in the past ten years so that data validity is maintained in monitoring the series of implemented policies, particularly during President Joko Widodo's time in office, by using the keywords: maritime axis, archipelagic state, maritime policy, and sustainable development. The sea toll policy is assumed to be successfully implemented; this step can help realise Indonesia as a maritime axis even though it is of little significance; it, at the very least, makes Indonesia into a provider of transportation for logistics between islands instead of being controlled by outsiders (Savin-Baden and Major 2023).

3 FINDINGS AND DISCUSSION

This research aims to discover the government's actions or policies to realise Indonesia as a global maritime axis. The research question is focused on identifying government actions and policies to carry out sustainable development so that the public gains the benefits of a balanced economic development that is not solely concentrated on the island of Java but on other islands, to the outermost regions of Indonesia as well.

3.1 *Understanding the shifting origins of global economic resources and infrastructure development*

The future of economic resources will no longer be on land but in the ocean. The battle for maritime wealth will most likely begin in the next 10 to 15 years. The geopolitical shift from west to east also maintains a strong influence, with a glance at the rise of China and the battle for economic resources that emerged in the Asian region, which no longer relates to the ideological competition between the United States and the Soviet Union in the 20th century (Heriawan 2022). In fact, Indonesia's maritime economic potential reaches US$ 1.33 trillion per annum, yet utilisation remains below 10%, given the composition of Indonesia's territory comprising 79% sea areas, 17,508 islands with 5,707 inhabited islands, 99,000 km of coastline and 275 million people. Thus, launching a maritime axis is an absolute strategy to achieve balanced development among the regions because each region in Indonesia has aquatic areas and is capable of independently developing them. What actually happened to the process of equal development in Indonesia? Upon deeper observation we inquire how far has Indonesia been able to realise the dreams previous leaders have envisioned and fought for, which resulted in their diplomatic victory over the archipelagic state, which led to Indonesia gaining complete sovereignty over its own territorial waters and also as a bridge for implementing equal development throughout Indonesia.

Accordingly, facilities or infrastructure such as ports are crucial for archipelagic countries like Indonesia. On the contrary, the availability of infrastructure such as ports is still very minimal, this has been the cause of development inequality remaining a problem among the regions, causing the spread of logistics in Indonesia to be more expensive compared to other countries. Similarly, the flow of goods in and out of the country, or between the islands is hampered and logistics costs rise. If this is not urgently addressed, then Indonesia must immediately prepare to lose investors, as they generally seek countries that support their security and comfort in investing. In this case, neighbouring countries such as Singapore, Malaysia, Thailand, and even Brunei Darussalam have overtaken Indonesia in port infrastructure development. What actually happened to the process of levelling out development in Indonesia. Upon deeper examination, the extent to which Indonesia is able to realise the dreams previous leaders have fought for, which resulted in their diplomatic victory over the archipelagic state, and led to Indonesia obtaining complete economic sovereignty over its own territorial waters and for the sea to serve as a bridge to level out development throughout Indonesia instead of becoming a barrier (Anna 2020).

Research shows that 70% of a country's foreign relations are economic relations with foreign countries, Indonesia remains sovereign in the international maritime arena because 40% of international trade passes through Indonesian waters (Juliadi *et al.* 2023). If used properly, Indonesia can become a new power in the global political and economic arena, because in the global scope Indonesia lies at the centre of world services, which is an important key in building new economic resources toward becoming a maritime axis (Hastuti *et al.* 2023). Undoubtedly, this must be supported by improving sea transportation. The sea transportation development model is adjusted to Indonesia's geographical location and archipelagic conditions with the main objective of building inter-island connectivity by: (1) optimising growth based on regional unity; and (2) integrating modes with the inter-modal supply chain system method that connects remote areas with economic centres (Ditjen Perhubungan Laut 2020).

In addition to building inter-island connectivity and implementing equitable development, if the maritime axis were carried out to the maximum, it would become a reality instead of remaining simply a discourse. A point to consider is that foreign relations in the world are almost entirely dominated by economic relations, and Indonesia is located in a favourable region in the international economy (Suropati *et al.* 2018). Indonesia needs to learn from Japan, since it is known as one of the strongest maritime nations, Japan implemented numerous measures to reach that position, among them are: conducting high-quality maritime research, building a high-quality and large ocean fleet, utilising abundant marine resources wisely and developing advanced fisheries technology, and utilising good use of its bays by turning it into ports with international standard (Patalano 2014).

3.2 *Political will in the development of sea tolls*

Sea toll means shipping without obstacles (logistics cost suppression), connecting ports throughout Indonesia to support various national interests, especially in eastern Indonesia (Bradford and Kharisma 2024). It is also necessary to develop short sea shipping in coastal areas as an alternative to freight transportation, as well as the construction of 100 non-commercial port tolls in support of the sea tolls. The sea toll also means opening up regional access by building two international-scale ports to serve large commercial vessels above 3,000 TEU. At present, the government has built 65 ferry ports nationwide, with 34 routes. The main sea toll ports are located in Belawan, Makassar, Surabaya, and Sorong. The sea toll development commitment with 65 new ports aims to increase the intensity of national and international shipping. Meanwhile, from a prosperity perspective, increased intensity of shipping will be in line with increased intensity of trade that would encourage economic growth and accelerate the connection of all regions throughout Indonesia, including its outermost and foremost regions, which would also balance the growth of development between the islands to avoid concentrating development only in Java (Talitha *et al.* 2020).

On the other hand, the sea toll development failed during President Joko Widodo's era because it was more of a political strategy to gain people's sympathy for his government, especially when the notion was delivered during his campaign. In fact, it was poorly managed and corruption was observed in the project, there were several issues such as lack of on-field supervision and warehousing facilities at the port, limited facilities and infrastructure for ships loading goods, imbalance of outbound and return cargo, sea toll vessels not fully meeting seaworthiness requirements, and deviations in subsidy calculations (Pikoli 2021). However, it should be recognised that the sea toll discourse is an extremely brilliant idea to raise the community's economic level and reduce social inequality, especially for communities outside Java Island. The most important measure today is the need for the government to ensure strategic actions in developing Indonesia, establish easy access to all remote areas to ensure inter-island connectivity, and it should also be accompanied by political will. The development that must be carried out is the integration between sea transportation modes in achieving cross-region connectivity (Adam 2015). The sea toll development policy was developed to realise Indonesia's vision as a maritime axis, and that significantly aligns with SDG 9 (Sambhi 2015).

Realising Indonesia's ideal as a world maritime axis, which is what Mochtar Kusumatmadja did for Indonesia (i.e., having economic sovereignty over its waters), requires synergy and totality from all parties (both local and national government agencies even political parties), especially their political will and commitment to succeed. In this context, political will is expected to continue from one administration to the next to ensure a successful transfer of national development because political will includes patterns of interaction between political actors, which will influence political culture and subsequently lead to actions or behaviours that form general and specific goals to be achieved and procedural measures considered appropriately applied to achieve a well-executed goal, namely Indonesia's vision as a maritime axis (Sa'adah *et al.* 2019).

4 CONCLUSION

In reality, to achieve equitable and sustainable development, Indonesian leaders and key stakeholders must be able to collaborate with international actors and even global elites in order to maximise profits to protect our national interest, especially defending citizens in vulnerable social and economic classes. To realise this, development must have a clear and far-seeing vision. To become a visionary nation, Indonesians must understand that they were initially born as a maritime nation. In a sense, we must realise that the support for the sustainability of this nation's life is not only on land but in the waters, even the potential of Indonesia's marine resources is greater than that on land because the Indonesian seas offer multiple elements to complete primary and secondary needs for supporting human sustainability in the modern era. Indonesia must make the most out of its fortune in terms of its maritime position, which is a hotbed of global trade transportation flows, instead of being controlled by foreigners due to the government's helplessness to take advantage of the opportunities that God has bestowed. Ultimately, the government should turn to the Japanese model, as a maritime country, in building inter-island connectivity based on SDG 9.

REFERENCES

Adam, L. (2015). Kebijakan konektivitas maritim di Indonesia. *Jurnal Politica*, 6(1), 20–41. https://jurnal.dpr.go.id/index.php/politica/article/view/302

Agastia, I. G. B. D., and Perwita, A. A. B. (2017). Building Maritime Domain Awareness as an Essential Element of the Global Maritime Fulcrum: Challenges and Prospects for Indonesia's Maritime Security. *Jurnal Hubungan Internasional*, 6(1), 113–123. https://doi.org/https://doi.org/10.18196/hi.61109

Anna, Z. (2020). 75 tahun merdeka, Indonesia masih punya banyak potensi kembangkan sektor kelautan dan perikanan. *The Conversation*. https://theconversation.com/75-tahun-merdeka-indonesia-masih-punya-banyak-potensi-kembangkan-sektor-kelautan-dan-perikanan-143188

Bradford, J., and Kharisma, W. (2024). The Maritime Implications of Indonesia's Presidential Election. *The Diplomat*. https://thediplomat.com/2024/01/the-maritime-implications-of-indonesias-presidential-election/

Butcher, J. G., and Elson, R. E. (2017). *Sovereignty and the Sea*. NUS Press. https://doi.org/10.2307/j.ctv1xz0cq

Chintoan-Uta, M., and Silva, J. R. (2017). Global maritime domain awareness: a sustainable development perspective. *WMU Journal of Maritime Affairs*, 16(1), 37–52. https://doi.org/10.1007/s13437-016-0109-5

Connelly, A. L. (2015). Sovereignty and the sea: President joko widodo's foreign policy challenges. *Contemporary Southeast Asia*, 37(1), 1–28. https://doi.org/10.1355/cs37-1a

Creswell, J. W., and Creswell, J. D. (2017). *Research Design: Qualitative, Quantitative, and Mixed Methods Approaches.* SAGE Publications. https://books.google.co.id/books?id=335ZDwAAQBAJ
Ditjen Perhubungan Laut. (2020). Transportasi Laut Pegang Peranan Strategis untuk Merajut Keberagaman Indonesia dan Mendorong Pertumbuhan Ekonomi. *Kementerian Perhubungan Republik Indoensia.* https://hubla.dephub.go.id/home/post/read/7825/transportasi-laut-pegang-peranan-strategis-untuk-merajut-keberagaman-indonesia-dan-mendorong-pertumbuhan-ekonomi
Hastuti, H., Muhidu, A., Rastin, R., and Agus Mokodompit, E. (2023). Indonesia's Marine Economic Potential As A Maritime Country. *International Journal of Science, Technology and Management*, 4(4), 813–825. https://doi.org/10.46729/ijstm.v4i4.897
Heriawan. (2022). The Ever-growing China's Maritime Rise and South China Sea Dispute: A Literature Review. *Jurnal Hubungan Internasional*, 15(1), 74–92. https://doi.org/10.20473/jhi.v15i1.33644
Juliadi, A. A., Oktirellia, A. S., Fadhilah, G. C. N., Shaffiyah, I., and Dwiyanti, V. (2023). Indonesia Toll Roads: Its Effect on The Trade Routes in ASEAN. *ASEAN Journal of Science and Engineering*, 3(1), 57–62. https://doi.org/10.17509/ajse.v3i1.33765
Lauder, M. R. M. T., and Lauder, A. F. (2016). Maritime Indonesia and the Archipelagic Outlook: Some reflections from a multidisciplinary perspective on old port cities in Java. *Wacana*, 17(1), 97–120. https://doi.org/10.17510/wacana.v17i1.428
Patalano, A. (2014). Japan as a Seapower: Strategy, Doctrine, and Capabilities under Three Defence Reviews, 1995–2010. *Journal of Strategic Studies*, 37(3), 403–441. https://doi.org/10.1080/01402390.2014.904788
Pikoli, A. N. P. (2021). Critical Analysis of Indonesia's Global Maritime Fulcrum under Joko Widodo: Problems and Challenges. *Publik (Jurnal Ilmu Administrasi)*, 10(1), 152. https://doi.org/10.31314/pjia.10.1.152-108.2021
Sa'adah, K., Yakti, P. D., and Susanto, S. R. (2019). Evaluating 4 years of Jokowi Sea Toll Policy: The concept of Indonesia-Centric connectivity for economic equality. *Masyarakat, Kebudayaan Dan Politik*, 32 (3), 290. https://doi.org/10.20473/mkp.v32i32019.290-302
Salim, W., and Negara, S. D. (2018). Infrastructure development under the jokowi administration progress, challenges and policies. *Journal of Southeast Asian Economies*, 35(3), 386–401. https://doi.org/10.1355/ae35-3e
Sambhi, N. (2015). Jokowi's 'Global Maritime Axis' Smooth Sailing or Rocky Seas Ahead? *Security Challenges*, 11(2), 39–55. https://www.jstor.org/stable/26465437
Savin-Baden, M., and Major, C. H. (2023). *Qualitative Research: The Essential Guide to Theory and Practice.* Taylor \& Francis. https://books.google.co.id/books?id=uQa6EAAAQBAJ
Suropati, U., Montratama, I., and Sulaiman, Y. (2018). Indonesia Inc: peta jalan menuju poros maritim dunia. *PT Elex Media Komputindo.* https://books.google.co.id/books?id=tXHPwQEACAAJ
Talitha, T., Firman, T., and Hudalah, D. (2020). Welcoming two decades of decentralization in Indonesia: a regional development perspective. *Territory, Politics, Governance*, 8(5), 690–708. https://doi.org/10.1080/21622671.2019.1601595
Till, G. (2018). *Seapower: A Guide for the Twenty-First Century.* Taylor \& Francis. https://books.google.co.id/books?id=jSZhDwAAQBAJ
Wicaksana, I. G. W. (2017). Indonesia's maritime connectivity development: Domestic and international challenges. *Asian Journal of Political Science*, 25(2), 212–233. https://doi.org/10.1080/02185377.2017.1339618

Towards Resilient Societies: The Synergy of Religion, Education, Health, Science, and Technology – Rahiem (Ed.)

Policy effectiveness on abrasion prevention and coastal community development

Haniah Hanafie & M. Zenal Abidin
Universitas Islam Negeri Syarif Hidayatullah Jakarta, Indonesia

ABSTRACT: This study evaluates the effectiveness of policies implemented to prevent abrasion and promote coastal community development in Ketapang Village. The village has experienced severe land loss due to abrasion, with 23.35 hectares lost in 2015 and 44.75 hectares in 2019. Despite being close to abundant biological resources, the coastal community faces economic hardship, evident from the insufficient infrastructure supporting economic development. The study employed field observations, informant selection, and analysis based on the effectiveness theory. The findings indicate that the policies have not been fully effective. Countermeasures were implemented far from the actual abrasion points, replacement houses for affected communities are significantly smaller, and economic empowerment programs have failed to increase community income meaningfully. However, some elements show effectiveness: the right implementer (government and private sector with community involvement), the right target (aligned with planning), the right environment (adequately supported internally), and the right process (executed according to established procedures). Overall, the policies have not delivered the intended outcomes. Greater attention from stakeholders and the Ketapang Village community is needed to improve environmental initiatives' planning, implementation, and supervision in coastal areas.

Keywords: abrasion, effectivity, livability, income

1 INTRODUCTION

Ketapang Village, Mauk Subdistrict, Tangerang District is one of the coastal areas with a focus on housing and settlement development by the Tangerang District Government, as stipulated by Decree No. 050/Kep.380-HUK/2017 on the Establishment of Housing and Slum Settlement Sites, and the Plan for Preventing and Improving the Quality of Urban Slum Settlement in Tangerang District.

These coastal areas often have complex and interrelated environmental problems. One of the flagship programs of Tangerang District is the Ketapang Urban Aquaculture development policy. It is aimed at solving the problems of abrasion disasters, providing basic infrastructure (livable houses, sanitation, and clean water), improving the economy, and empowering the community.

The coastal areas often face damages and disasters (Sherly *et al.* 2015) caused by human hands, settlement patterns, and coastal ecosystem vegetation, resulting in abrasion, flooding, irregular settlements, and deforestation, impacting coastal ecosystems (Pinto 2015). Coastal erosion can also be caused by man-made structures such as groins, harbours, docks, and breakwaters (Abualtayef *et al.* 2013).

Due to the impact of abrasion, Ketapang Village, Mauk Subdistrict, Tangerang District, lost 23.35 ha of land in 2015 and 44.75 ha in 2019, thus impacting the community's aquaculture ponds and the fisher's income (Suwandana 2019). At Kuwaru Beach, Yogyakarta, the cutting down of cypress trees resulted in abrasion, and several businesses went bankrupt (Pinto 2015).

DOI: 10.1201/9781003645542-11

This study aims to examine and analyse policy effectiveness in preventing abrasion, arranging livable houses, and increasing the income of coastal communities.

The significance of the current study is to inform the public that the complexities of problems in the coastal areas, such as disorganised or slum settlements and indiscriminate deforestation, lead to flooding and poverty. This study is therefore required to capture the effectiveness of policies for preventing abrasion, arranging livable housing, and increasing the income of coastal communities. Additionally, the current study can serve as an evaluation material, especially for the benefit of the local government for further development.

Many references discuss coastal areas. Beck *et al.* (2018) says that coral reefs can prevent coastal abrasion, since abrasion can reduce areas for tourism, coastal conservation is, thus, an absolute necessity (Talakua *et al.* 2024).

The community's economy is threatened due to sluggish beach tourism and the felling of trees (Pinto 2015). Meanwhile, beach tourism has economic development potential for the surrounding community (Az Zahra *et al.* 2024).

The vulnerability of fisher households caused by climate change (Winda *et al.* 2023) and sea level changes (Veron *et al.* 2019), therefore, needs to be circumvented by integrating social and environmental systems (Chakraborty and Joshi 2016).

In the discussion of the current study, the theory of policy effectiveness consists of five policy accuracies: right policy, right implementation, right target, right environment, and right process (Nugroho 2017), along with several relevant literary sources.

2 METHODS

A qualitative approach was used in this study to obtain descriptions of the phenomena of abrasion, livable settlements, and income of coastal communities that are in-depth, thorough, accurate and as they are. Primary data were obtained from interviews with the Head of Bappeda of Tangerang District, the Head of the Fisheries and Maritime Service of Tangerang District, the Head of Mauk Subdistrict, community leaders, and representatives of the Ketapang Village community who were selected by using quota and purposive sampling because the respondents were considered representative of the coastal community and had knowledge of the issues under study. Secondary data were obtained through observation at the location of the study before and after the policy was issued and documentation by analysing related books, journals, reports, magazines, newspapers, and information from the internet. Data were analysed descriptively using data reduction, presentation, and conclusion drawing. To obtain data validity, extended research time, extended observation, triangulation, and peer discussion were carried out to provide more accurate data.

3 FINDINGS AND DISCUSSION

Once data were collected, discussions and analyses were carried out by referring to the theory of policy effectiveness. The following passages discuss abrasion prevention, the arrangement of livable houses, and increased income.

3.1 *Abrasion prevention*

Abrasion on the border of Ketapang and Margamulya Villages, Mauk Subdistrict, has become a concern for the Tangerang District Government because it has disrupted the coastal road and can result in being cut off. The Tangerang District Government, thus, made a policy to develop Ketapang Urban Aquaculture, transforming the slum coastal area into a mangrove conservation park (ppid.tangerangkab.go.id, n.d.).

The research results show that the abrasion prevention policy has not met expectations, as the location of the countermeasures is far from the abrasion point. The following is an excerpt from an interview with a Ketapang administration staff:

*"When viewed from the abrasion point with the location of Ketapang Urban Aquaculture Development, it is inappropriate. If the development is done according to the abrasion point, it will clash with other villages. The Ketapang Village was chosen because there are local government assets in the form of approximately 15 hectares of land belonging to the Department of Fisheries and Maritime Affairs, so these assets are utilised" (*Ibnu Su'ud, personal communication, June 1, 2023).

Based on the interview above, which is supported by observations made through Google Earth, it is true that the development of Ketapang Urban Aquaculture is far from the point of abrasion. This is presented in the Figure 1 below:

Figure 1. Position of Ketapang urban aquaculture (Source: google Earth 2023).

3.2 *Livable house arrangement*

Some of the biggest problems in Ketapang Village are found in housing and sanitation management. Concerning this, the Tangerang District Government issued the Coastal Community Development Movement Program to improve the quality of life and basic infrastructure Hayat and Napilah 2022).

The research results show that the more livable houses have not replaced the old ones due to their smaller size. Access to clean water is difficult, and flooding occurs due to inadequate drainage (Darsem and Novi, personal communication, April 4 2023). Livable houses as replacements were initially promised, as stated by Darsem and Novi below:

"During a socialisation meeting at Empang Mang Omat, at the village administration office and the Ketapang Urban Aquaculture construction site, it was promised that when construction is carried out, the houses around the site would be relocated, moved, and they said it would be built better" (Darsem and Novi, personal communication, April 4 2023).

Besides being smaller in size, land ownership status is also unclear. People are expected to buy the land, but the payment method has not been determined. Fiqih, a resident of Ketapang Village stated as follows:

"This house and the land will be bought, but how we pay for it has not yet been decided. In the past, we lived on the land there without ownership status, so we were asked to move here; we were forced and disappointed; what else could we do? Don't let it happen twice. We don't have ownership status, so we don't get stuck" (Faqih, personal communication, April 4 2023).

The Ketapang village community hopes to pay off the land and gain ownership status. Hence, they have the certainty of occupying a livable house without having to feel worried at any time about being evicted again.

3.3 *Increased income*

The Coastal Community Development Movement is a flagship program of the Tangerang District Government that aims to develop and accelerate the economic growth of coastal areas. Program development is centred on coastal areas, considering fishers still struggle with poverty and poor economic conditions despite the sea's abundant potential (Octavia and Akhsaniyah 2023). Therefore, this program is expected to directly impact the community's economy and empower the community to manage the potential and resources of Tangerang District's coastal areas. One of the forms of development is Urban Aquaculture.

Urban Aquaculture as ecotourism development is a manifestation of the Coastal Community Development Movement Program that has succeeded in opening up employment opportunities by way of selling and renting parking lots, as shown in Figure 2 below:

Figure 2. Trader stalls in the Ketapang urban aquaculture area.

The head of the Mauk Subdistrict said that a new economic sector has emerged, and this is a development goal. The following is an excerpt from his interview:

> *"The way I look at it, the economic impact is significant. The community initially worked as fishers. Today, new economic sectors, such as drink and snack sellers, have new income (generator) other than fishing. This is our real goal, so there are changes and improvements in the community's economy"* (Hakim and Cahyono, personal communication, March 30, 2023).

A Coastal Festival is also held in the Urban Aquaculture area, providing fantastic proceeds of billions of rupiahs to the local government and community. Gradually, the increase in community income was not achieved due to expensive ticket prices, lack of visitors, and lack of buyers, which consequently impacted buying and selling activities, and some people had to close their businesses (Faqih, personal communication, April 4, 2023).

Community income was reduced because the Urban Aquaculture area had no visitors and because of the failure to prevent abrasion. Abrasion resulted in residents losing residential, agricultural, and aquacultural land, causing a loss of livelihoods, which ultimately meant reduced income (Damayanti 2013).

A policy can be considered the **right Policy** if it is made to solve problems and policy formulation is aligned with the characteristics of the problem, implemented by authorised, professional institutions (Nugroho 2017) and trust. The results show that abrasion prevention had failed because the location of the countermeasure was not in line with the abrasion point, rendering the abrasion prevention policy ineffective.

Failure to prevent abrasion can damage tourism areas and attractions, as observed at Sigandu Beach, Batang District (Muhammad and Sugiri 2015). The results of a study done in Kulonprogo suggest that abrasion disasters can be prevented by planting mangroves at estuaries, shrimp cypress trees (Sri Wahyuningsih 2016) and coral reefs (Beck *et al.* 2018) in and around the coast. Otherwise, abrasion disasters will eliminate homes, damage agricultural land and ponds, and reduce people's livelihoods (Damayanti 2013).

The policies for developing livable housing with clear land status as a policy target are not ready for intervention because some residential lands do not have legal ownership status, which is also the case for replacement houses. The land ownership status remains indeterminate by the local government, leaving residents worried. People's concerns are quite plausible as they can be subject to eviction again at any time. Additionally, the size of the replacement house is smaller than that of their previous house, and clean water remains unavailable. Accordingly, the arrangement of livable houses cannot be said to be effective regarding the right policy.

Right implementation can be considered effective, given that the implementation is not only carried out by the government but is a concerted effort between the government, the community, and the private sector (Nugroho 2017; Fell and Johnson 2021) or integration of central government, strategy, perception and land supply significantly, thus impacting the selection of affordable housing programs (Cia and Wang 2019). In addition, its sustainability depends on its technical, social, and governance aspects (Silva *et al.* 2024).

The development of the Ketapang Urban Aquaculture area involves not only the government as an implementer but also the private sector and should be supported by community participation and inclusion (Fell and Johnson 2021), especially among the people of Ketapang village. Ketapang villagers' approval of the relocation to new homes shows the community's support for the policies of the Tangerang District Government despite some shortcomings observed in the implementation of the development program.

Regarding the **right target**, it can generally be effective because policies on abrasion prevention, livable housing arrangements, and income generation follow the plan, do not overlap with other interventions, or do not clash with other policy interventions. The target is in harmony, no conflict is observed, and the current policy renews the previous one (Nugroho 2017).

Nonetheless, improvement in environmental quality has not been achieved, as floods and uninhabitable houses remain. Livable houses for the poor are difficult to realise, given the lack of interest from developers. Accordingly, the government should take responsibility (Winarno 2018).

The **right environment** refers to the interactions between formulating institutions, policy implementers and related institutions. In terms of the external environment, the government, as the implementer, conducted regular communication meetings and exchanges of ideas regarding the program with sub-districts and villages. Because communication was well established, implementing the program was easier for public policy actors. Mauk Subdistrict always met with stakeholders to align their views and facilitate institutional coordination (Hakim and Cahyono, personal communication, March 30 2023). Without good administration, finance and communication (Shakeri *et al.* 2021), development cannot be realised, because communication is a human necessity (Octavia and Akhsaniyah 2023). Communication is an important element of the organisation; without communication, an organisation does not operate properly, misunderstanding occurs, and significant impacts are posed upon individuals or groups.

The **right process** is considered effective in cases where a policy is implemented and meets policy acceptance, policy adoption, and strategic readiness (Nugroho 2017), meaning that it has gone through the stages of the planning and socialisation process, making people aware of, understand about, accept, and experience the policy. The policies of abrasion prevention, decent housing arrangement, and increased income can be considered effective in terms of the right process because it has gone through the planning, socialisation, and implementation process properly with well-balanced conditions, without any significant conflict (rejection), despite imperfect results.

4 CONCLUSION

The current study on the effectiveness of policies on abrasion prevention, livable houses development, and income generation of coastal communities can be considered effective in terms of the right implementer, target, environment and process. Meanwhile, the right policy can be considered ineffective since it is not aligned with the expected policy and has not been able to solve the problems coastal communities face.

REFERENCES

Abualtayef, M., *et al.* (2013). Mitigation measures for Gaza coastal erosion. *Coastal Development*, 16(2), 145.

Az Zahra, N. N., *et al.* (2024). Analisis karakteristik sampah laut dan tingkat kebersihan di Pantai Sendang Biru dan Pelabuhan Perikanan Pondokdadap, Kabupaten Malang, Jawa Timur. *Ilmu Lingkungan*, 22(4), 852–860. https://doi.org/10.1829-8907

Beck, M. W., Losada, I. J., Menéndez, P., *et al.* (2018). The global flood protection savings provided by coral reefs. *Nature Communications*, 9. https://doi.org/10.1038/s41467-018-04568-z

Chakraborty, A., and Joshi, P. K. (2016). Mapping disaster vulnerability in India using analytical hierarchy process. *Geomatics, Natural Hazards and Risk*, 7(1), 308–325. https://doi.org/10.1080/19475705.2014.897656

Cia, X., and Wang, W.-N. (2019). Affordable housing policy development: Public official perspectives. *International Journal of Housing Markets and Analysis*, 12(5), 934–951. https://doi.org/10.1108/IJHMA-08-2018-0063

Damayanti, K. (2013). Dampak abrasi pantai terhadap lingkungan sosial (Studi kasus di Desa Bedono, Sayung Demak). *Jurnal Ilmu Lingkungan*, 366.

Fell, T. F., and Johnson, M. (2021). The role of public-private partnerships in housing as a potential contributor to sustainable cities and communities: A systematic review. *Sustainability*, 13(14). https://doi.org/10.3390/su13147783

Hayat, N., and Napilah, S. N. (2022). Dampak pembangunan Gerbang Mapan pada perilaku sosial masyarakat Desa Ketapang. *Jurnal Pendidikan Tambusai*, 6(2), 8805.

Muhammad, M. S., and Sugiri, A. (2015). Bentuk-bentuk adaptasi lingkungan terhadap abrasi di kawasan Pantai Sigandu Batang. *Teknik PWK*, 4(4), 702.

Nugroho, R. (2017). *Public policy: Dinamika kebijakan publik, analisis kebijakan publik, manajemen politik kebijakan publik, etika kebijakan publik, kimia kebijakan publik.* PT Elex Media Komputindo.

Octavia, S. P. N. K., and Akhsaniyah, N. (2023). Manajemen komunikasi pelaku reklamasi ilegal Nggacardi Pesisir Surabaya. *Ilmu Komunikasi*, 20(1), 59–72. https://doi.org/10.24002/jik.v20i1

Pinto, Z. (2015). Kajian perilaku masyarakat pesisir yang mengakibatkan kerusakan lingkungan (Studi kasus di Pantai Kuwaru, Desa Poncosari, Kecamatan Srandakan, Kabupaten Bantul, Provinsi DIY). *Wilayah dan Lingkungan*, 3(3), 170.

PPID Tangerang.(n.d.). Retrieved from https://ppid.tangerangkab.go.id

Shakeri, E., Vizvari, B., and Nazerian, R. (2021). Comparative analysis of disaster management between India and Nigeria. *International Journal of Disaster Risk Reduction*, 63. https://doi.org/10.1016/j.ijdrr.2021

Sherly, M. A., *et al.* (2015). Disaster vulnerability mapping for a densely populated coastal urban area: An application to Mumbai, India. *American Geographer*, 105(6). https://doi.org/10.1080/00045608.2015.1072792

Silva, L. P. P., *et al.* (2024). Sustainable affordable housing: State-of-the-art and future perspectives. *Sustainability*, 16(10). https://doi.org/10.3390/su16104187

Sri Wahyuningsih, D. (2016). Efektivitas upaya mitigasi abrasi berbasis ekosistem di Kabupaten Kulonprogo. *Jurnal Mitigasi Abrasi*, 255.

Suwandana, E. (2019). Dinamika morfologi Pantai Kabupaten Tangerang Banten dan Pantai Indah Kapuk Jakarta melalui analisis citra Google Earth. *Perikanan dan Kelautan*, 9(1), 55–68.

Talakua, E. G., *et al.* (2024). Kesedian membayar wisatawan untuk pelestarian pantai di Desa Hukurila Kota Ambon. *Ilmu Lingkungan*, 22(4), 878–886. https://doi.org/10.1829-8907

Veron, S., *et al.* (2019). Vulnerability to climate change of islands worldwide and its impact on the tree of life. *Scientific Reports*, 9(1), 1–10. https://doi.org/10.1038/s41598-019-51107-x

Winarno, B. (2018). Evaluasi program penyediaan rumah layak huni bagi masyarakat miskin di Kabupaten Belitung. *Pengembangan Kota*, 6(1), 66–74. https://doi.org/10.14710/jpk.6.1.66-74.

Winda, G., *et al.* (2023). Kerentan nafkah rumah tangga nelayan dalam tekanan variabilitas iklim: Studi kasus Desa Dendun, Kabupaten Bintan, Kepulauan Riau. *Ilmu Lingkungan*, 21(1).

Towards Resilient Societies: The Synergy of Religion, Education, Health, Science, and Technology – Rahiem (Ed.)

Democracy and Islam in the reform era of Indonesia: A bibliometric analysis from 1998–2024

Ida Farida, Fahma Rianti, Prisinta Wanastri, Okta Reni Azrina RA & Pungki Purnomo
Universitas Islam Negeri Syarif Hidayatullah Jakarta, Indonesia

ABSTRACT: Indonesia has the largest Muslim population in the world. Following the 1998 reform, the country transitioned from three decades of Suharto's authoritarian rule to a democratic system. The study of democracy is essential for achieving the United Nations Sustainable Development Goals (SDGs), as history demonstrates that democracy fosters good governance, peace, and justice within society. This study aims to map academic trends on democracy in Indonesia by analysing articles indexed in Scopus from 1998 to 2024 using bibliometric analysis. The findings show that Robert W. Hefner is the most prolific author in this field, while Boston University ranks as the top institution publishing related research. UIN Syarif Hidayatullah also appears among the top 30 contributing institutions. The top five keywords identified are "civil society," "political Islam," "authoritarianism," "religion," and "Asia." The most prominent trending topics include "Islamic Law," "Civil Religion," "Identity Politics," "Candidate Behaviour," and "Islamic Movement." Studia Islamika is the leading journal in publication volume, while Civil Islam holds the highest citation count. The recurring emphasis on "civil society" and "civil Islam" underscores their pivotal role in advancing democracy in Indonesia, aligning with the objectives of SDG 16 (Peace, justice, and strong institutions).

Keywords: bibliometric analysis, democracy and Islam, reform, SDGs

1 INTRODUCTION

The fall of Suharto's presidency in 1998 marked a significant turning point in restoring democracy in Indonesia. After some 35 years of centralised state rule, the people could again have a critical voice. As the world's largest Muslim-majority country, Indonesia's democratic journey is closely watched. The concept of democracy does not contradict existing values in Islam, including shura (deliberation), al-'is (justice), ijma' (consensus), ijtihâd (freedom of thought), tasamuh (tolerance), al-hurrîyah (freedom), al-musâwah (egalitarianism), ash-shidqu wal amânah (honesty and responsibility) (Ciftci 2019). Muslims value these principles in society and the state, which contributes to democracy in Indonesia (Hakiki 2016; Widodo 2007).

Post-1998, Prominent Muslim figures from the two largest Islamic organisations, Nahdatul Ulama (NU) and Muhammadiyah, Abdurrahman Wahid (Gus Dur) and Amin Rais, were proactively involved in the leadership change process that became part of building a democratic system in Indonesia (Bourchier 2019). Indonesians generally hold a positive view of democracy, with many agreeing that it is the best form of government for the country (Muhtadi and Warburton 2020). This appreciation peaked in 2004 and remains high today (Mujani *et al.* 2018). Despite economic conditions influencing democracy, Indonesia is considered more democratic than countries like Singapore, based on political freedom and civil liberties (Freedom House 2016).

However, Indonesia faces challenges such as corruption, military control, rule of law issues, and inadequate minority protection (Horowitz 2013; Tambunan 2023). Studying democracy and Islam in Indonesia is crucial, aligning with SDG 16, which aims to promote

DOI: 10.1201/9781003645542-12

peaceful, inclusive societies, provide justice for all, and build effective, accountable institutions by 2030. This study aims to map the trend of democracy in Indonesia through a bibliometric analysis of academic articles in Scopus from 1998 to 2024.

Significant research has been conducted on the intersection of politics and Islam in Indonesia, focusing on democracy. Analysing these studies is crucial to see their development so that they provide valuable insight for future researchers in the same field. For this reason, bibliometric studies are an effective method for mapping research trends on this topic. Additionally, this method can be used to highlight under-researched topics and primary sources and identify potential collaboration opportunities with other researchers in the same field, both nationally and internationally.

A variant of systematic literature reviews known as "bibliometric research" involves the application of quantitative and statistical methods to bibliographic data (such as publications and citations), such as performance analysis, cluster analysis, and descriptive statistics analysis (Linnenluecke *et al.* 2020; Mukherjee *et al.* 2022). Methodologically, it empirically builds theory by grouping knowledge and examining indicators from a collection of papers to reveal scientific components needing explanation. This approach also assesses the quantity and quality of scientific output and its societal impact (Salinas-Ríos and García-López 2022).

Ellegaard and Wallin (2015) emphasise the importance of qualifying and quantifying research outputs for decision-makers. Bibliometric analysis serves as an effective tool in this context. The most prevalent technique in bibliometrics for assessing research quality, especially for individual researchers, university and institution rankings, is citation analysis. These assessments inform researchers about emerging trends, competitive organisations, and collaborative opportunities. The necessity to evaluate scientific output and disseminate the findings to scientists, policymakers, and other stakeholders drives the use of bibliometric techniques. Consequently, Bibliometric analysis has become a widely accepted method for assessing scientific productivity.

Applying bibliometric analysis to the study of Islam and democracy in Indonesia can identify the primary research topics that are often researched in this field, both globally and regionally. It can also reveal the latest research trends, emerging issues, and knowledge gaps that required attention. Furthermore, this method can map the collaboration network among researchers, institutions, and countries that are active in this field, whether in research or publication. Such analysis provides relevant information for shaping the direction of public policy related to Islam and democracy in Indonesia and identifies research questions that are still open and require further investigation.

2 METHODS

This research uses a quantitative approach with a descriptive method type and bibliometric analysis. The aspects studied are co-authorship, co-occurrence, and co-citation. Co-authorship analysis is used to find relationships between various studies based on research documents produced by researchers. This analysis can help overcome and provide substantial contributions to scientific development. Finding social relationships (such as those between writers, institutions, and nations) and social dominance or hidden biases (such as triangulation between contributors—such as those between authors, institutions, and governments) are the goals of co-authorship analysis (Mukherjee *et al.* 2022; Sidiq 2019).

On the other hand, co-occurrence analysis can reveal research topics statistically. It is simply a calculation of paired data within a collection unit. Several word analysis methods can be applied based on the word matrix, which consists of factor analysis, cluster analysis, multivariate analysis, and social network analysis to reveal the structure and development of the research field. These methods are helpful for researchers who want to study an overview of a field. So, it plays an essential role in identifying the value of academic disciplines (Sidiq 2019).

A fundamental method for mapping scientific publications is called citation analysis. It works under the premise that citations show the intellectual connections that arise when two articles quote one another (Donthu *et al.* 2021). Citation analysis is one of the fundamental characteristics of scientific work. The relationship of scientific work citations is the primary basis of citation analysis. Citation analysis uses various methods, including mathematics, statistics, comparison, induction, abstraction, generalisation, and logic. This method is used to analyse various scientific journals, papers, citation objects, and cited phenomena to determine the inherent characteristics of the bibliometric analysis method (Gunawan 2021; Herawati *et al.* 2022).

Data was obtained from the Scopus database, using the keywords "democracy - Islam - Indonesia" all year, capture the unique interaction between Indonesia's democratic governance and Islamic values. From the search process, 215 documents from 1998 until 2024 were obtained and processed using Excel and VOS Viewer to obtain data visualisation describing the author network and productivity, topic mapping and trends, and citations of each author, source, and country. The time frame from 1998 to 2024 marks the significant evolution of Indonesia's political landscape, transitioning from authoritarian rule to democracy. Throughout these years, Indonesia has diligently worked to enhance and strengthen its democratic institutions.

3 FINDINGS AND DISCUSSIONS

3.1 *Findings*

The results of the bibliometric analysis research on the topic "Democracy and Islam in Indonesia" were based on the Scopus database. The keywords "Democracy - Islam - Indonesia" were used to obtain data from 215 publications, with the type of document being article publication. This number is the result of publications from 1998 to 2024.

3.1.1 *Co-authorship analysis*

3.1.1.1 *Authors* Based on the processed data, 247 authors published articles on Democracy and Islam in Indonesia. Of the 247 authors with networks between authors, 10 are divided into 3 clusters. Robert W. Hefner leads with 14 publications, making him the most prolific author in this area. He is followed by Azyumardi Azra, Diego Fossati, Vedi R. Hadiz, and Marcus Mietzner, each with 5 documents. The list also includes prominent authors such as Burhanuddin, Edward Aspinall, Kikue Hamayotsu, Wayne Hudson, and Jeremy Menchik, with contributions ranging from 3 to 4 documents.

3.1.1.2 *Organisations* Based on the processed data, it is known that 255 organisations publish articles on the topic of Democracy and Islam in Indonesia. Of the 255 organisations, 13 organisations are related in 1 cluster. The following are details of the 30 institutions with the most documents. The Institute on Culture, Religion, and World Affairs, Boston University is the institution that has published the most articles on this topic, followed by Boston University, Australian National University, National University of Singapore, and the Asia Institute, University of Melbourne, Australia. From Indonesia, the institutions included in the list of top 30 are UIN Syarif Hidayatullah Jakarta (5 documents), UIN Sunan Kalijaga Yogyakarta (2 documents), University of Indonesia (2 documents), Gadjah Mada University (2 documents), Bina Nusantara University (1 document), and Center for Political Studies, Indonesian Institute of Sciences (LIPI) Jakarta (1 document).

3.1.1.3 *Countries* Based on the processed data, it is known that 26 countries publish articles on the topic of democracy and Islam in Indonesia. Indonesia has the highest number of publications, followed by the United States, Australia, England, and Singapore. Of the 26 countries, there are 12 networked countries divided into 4 clusters.

3.1.2 *Co-occurrence analysis*

Co-occurrence analysis using VOSviewer helps visualise the relationships between keywords in academic publications. By analysing the frequency and co-occurrence of specific keywords, VOSviewer can create a network map highlighting the connections and clustering of concepts within a given research field. From the processed data, there are a total of 524 keywords divided into 44 clusters. These clusters show the relationships between keywords in articles published on Democracy and Islam in Indonesia. The relationships and connections between key themes, trends and gaps in the literature, which can be useful for understanding the context and focus areas. Apart from "Indonesia", "democracy", and "Islam", the five most widely used keywords are "Islamism", "civil society", "political Islam", "authoritarianism", and "religion". The ten most frequently used keywords in published articles. The bottom four within range nine up to 10 mentions are "ethnopolitics", "Asia", "religion", and "authoritarianism". These topics have been widely researched in the field of Islam and Democracy in Indonesia. Keywords that are still rarely used in articles on the topic of democracy and Islam in Indonesia, including "accountability", "administration", "agnostic plurality", "ahlus-sunnah wal jama'ah", "anti-terrorism", and so on. Subsequent researchers could still study these keywords associated with democracy and Islam in Indonesia.

Network visualisation of author keywords using VOSviewer shows the interrelated keywords in articles on the topic "Democracy and Islam in Indonesia". These interrelated are marked with distinct colours, consisting of forty-eight colours/clusters. Author keyword network visualisation provides a graphical visualisation that displays the relationship between author keywords in a network. These author keywords are taken from the title, abstract, or keywords the author gave in their publication. From this visualisation, we can see: 1) the relationship between keywords that shows how often the keywords appear together in one publication, 2) keyword groups that represent a specific theme or topic, and 3) central keywords, namely keywords that have many relationships with other keywords, are usually at the centre of the network. These keywords are considered the most important keywords or represent the main theme of the research. Keywords that have relationships, for example, between "Islam-Democracy-Indonesia" and "Civil Society", "Pluralism", "foreign policy", and others.

This co-currencies analysis also showed the novelty of the keywords used. The brighter, the newer/more recent (yellow colour). Keywords in yellow include "Islamic law", "civil religion", "identity politics", "candidate behaviour", "Islamic movement", "al-sofwah foundation", "criminal law", "science advocacy", "gender quotas", etc. The keywords are the latest articles on "Democracy and Islam in Indonesia". At the same time, dark keywords are outdated keywords in articles.

Keywords that are still rarely used by the authors in the articles include "civil society and Indonesia", "civic education, skills, and Islam", "citizenship", "candidate behaviour", and many more. These keywords can be an opportunity for further research.

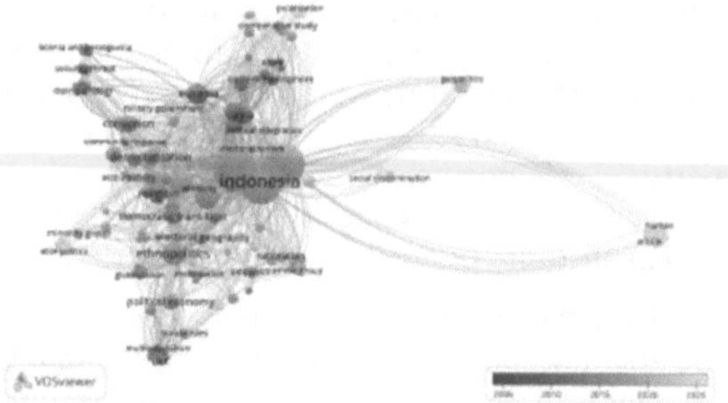

Figure 1. Overlay visualisation of index keywords.

Figure 1 shows the overlay visualisation in VOSviewer, which is a useful feature for examining how the relevance of specific keywords within a research dataset changes over time. It shows recent trends (in the yellow gradient) used in articles on the topic of 'Democracy and Islam in Indonesia' are 'gender issue', 'interpersonal communication', 'human rights', 'elite politics', 'institutional framework', 'military government', etc. There are several topics in the article, including "Democracy and Islam in Indonesia," that have not been studied much. Research topics that have not been studied much are marked with inconspicuous colours and smaller font sizes. These topics that have not been widely studied can be used as input for further research. Keywords that are still rarely used by the authors in the articles include "demographic trend", "drug development", "electoral system", elite politics", "ethnic conflict", etc. These keywords can be an opportunity for further research.

3.1.3 *Citation analysis*

3.1.3.1 *Sources* Based on the processed data, there are 142 journals, and 114 of them have been cited. Of the 114 journals, 41 have networks and are divided into 10 clusters. The ten journals with the most significant number of documents. In first place is Studia Islamika (11 articles), followed by Journal of Indonesian Islam (10 articles), Asian Survey, Indonesian Journal of Islam and Muslim Societies, Islam Beyond Conflict, and Trans (5 articles each).

Civil Islam has the highest number of citations, followed by Asian Survey, The Future of Political Islam, Bulletin of Indonesian Academic Studies, etc. It can be said that the Civil Islam journal is a reference for writers/researchers on the topic "Democracy and Islam in Indonesia".

3.1.3.2 *Authors* Based on the processed data, it is known that there are 247 authors, with citations of 190 authors divided into 140 clusters. Moreover, 58 authors are interrelated and divided into 11 clusters. The ten authors with the most citations are Robert W. Hefner (1037 citations), followed by Graham E. Fuller, Kathryn Robinson, Jeremy Menchik, and others. The 58 authors in the networked authors are organised into 11 clusters. Robert W. Hefner became the author with the most citations because his work provides deep, comprehensive insights into the complex interplay between Islamic traditions and democratic practices in Indonesia, such as "Islam and Citizenship in Indonesia" and "Civil Islam: Muslims and Democratisation in Indonesia".

3.1.3.3 *Countries* Based on the processed data, out of 26 countries, 19 countries have citations divided into 5 clusters. Based on the number of citations, Ranked first is the United States (1389 citations), followed by Australia (860 citations), Indonesia (646 citations), Hong Kong (107 citations), and Canada (93 citations). According to this result, the United States is the nation whose publications on Islam and democracy in Indonesia are most frequently cited by other nations, with Australia and Indonesia coming in second and third, respectively.

3.2 *Discussion*

This study, using the Scopus database, identified 215 articles, 247 authors, 255 organisations, 26 countries, and 524 keywords in Democracy and Islam in Indonesia from 1998 to 2004. Robert W. Hefner and Azyumardi Azra are the top authors. Leading institutions include Boston University, Australian National University, and National University of Singapore. Indonesian institutions in the top 30 are UIN Syarif Hidayatullah Jakarta, UIN Sunan Kalijaga Yogyakarta, University of Indonesia, Gadjah Mada University, Bina Nusantara University, and LIPI Jakarta. Indonesia leads in article count, followed by the USA, Australia, England, and Singapore.

The top most frequently used keywords are "civil society". The top keywords that are still under research is "accountability". The 5 topics that are trending include "Islamic law", "civil religion", "identity politics", "candidate behaviour", "gender issue". Studia Islamika is the journal with the most significant number of documents. Civil Islam has the highest number of citations. United States is the most cited countries. The authors with the most citations include Robert W. Hefner, Graham E. Fuller, Kathryn Robinson, and Jeremy Menchik.

From the co-authorship analysis, we can see the development of research on "Islam-Democracy in Indonesia" based on author, organisation, and country. Robert W. Hefner as an author, leads with 14 publications. The Institute on Culture, Religion, and World Affairs, Boston University, is the institution that has published the most articles on this topic. The research findings also highlight valuable information related to a prominent Muslim scholar, the esteemed late Azyumardi Azra, who once served as a lecturer of faculty Adab and Humaniora from 1992 to 2022 and as a rector from 1998–2006 at UIN Syarif Hidayatullah Jakarta. His contribution to the academic community, especially on democracy and Islam has left a lasting legacy. Furthermore, He also initiated the Studia Islamika journal in 1993, which plays a crucial role in disseminating scholarly works on Islam in Indonesia and South East Asia.

4 CONCLUSIONS

This study showed the significant contribution of bibliometric analysis findings, highlighting the importance of enhancing collaboration among leading institutions, encouraging research on underexplored keywords, and promoting interdisciplinary approaches to emerging topics. This information can be used to recommend potential partners for research collaborations, form research consortia, and ultimately advance the field of Islam and democracy in Indonesia."

Based on the countries' ranking results, Indonesia has the highest number of publications. This information can be used to recommend potential partners to conduct research collaborations between researchers, institutions, and countries. In addition, it can also be used as input to form a research consortium in the field of Islam and Democracy in Indonesia. Various research topics found in this research in Islam and Democracy in Indonesia also provide data to create relationships among topics, topic maps, and topic trends for further exploration.

Identification of topics that link Islam and democracy in Indonesia with "Islamic law", "civil religion", "identity politics", "candidate behaviour", and "gender issues" requires more research to find out the practice of democracy in the world's most populous Muslim country. This comprehensive approach will undoubtedly lead to more understanding and practical solutions for strengthening democracy in Indonesia. By examining these interconnected topics, researchers can uncover the dynamics at play and propose strategies that promote good governance, peace, and justice, aligning with the goals of UN SDG 16.

REFERENCES

Bourchier, D. M. (2019). Two decades of ideological contestation in Indonesia: From democratic cosmopolitanism to religious nationalism. *Journal of Contemporary Asia*, 49(5), 713–733. https://doi.org/10.1080/00472336.2019.1590620

Ciftci, S. (2019). Islam, social justice, and democracy. *Politics and Religion*, 12(4), 549–576. https://doi.org/10.1017/S1755048318000810

Donthu, N., Kumar, S., Mukherjee, D., Pandey, N., and Lim, W. M. (2021). How to conduct a bibliometric analysis: An overview and guidelines. *Journal of Business Research*, 133, 285–296. https://doi.org/10.1016/j.jbusres.2021.04.070

Ellegaard, O., and Wallin, J. A. (2015). The bibliometric analysis of scholarly production: How great is the impact? *Scientometrics*, 105(3), 1809–1831. https://doi.org/10.1007/s11192-015-1645-z

Freedom House. (2016). *Freedom in the World 2016: Indonesia*. https://freedomhouse.org/report/freedom-world/2016/indonesia

Gunawan, I. (2021). Analisis sitasi pada JAMP: Jurnal Administrasi dan Manajemen Pendidikan, Universitas Negeri Malang, 2018–2020. *JAMP: Jurnal Administrasi dan Manajemen Pendidikan*, 4(2), 163–170. http://journal2.um.ac.id/index.php/jamp/

Hakiki, K. M. (2016). Islam dan demokrasi: Pandangan intelektual Muslim dan penerapannya di Indonesia. *Wawasan: Jurnal Ilmiah Agama dan Sosial Budaya*, 1(1), 1–17.

Herawati, P., Utami, S. B., and Karlina, N. (2022). Analisis bibliometrik: Perkembangan penelitian dan publikasi mengenai koordinasi program menggunakan Vosviewer. *Jurnal Pustaka Budaya*, 9(1), 2442–7799. https://journal.unilak.ac.id/index.php/pb/article/view/8599/3633

Horowitz, D. L. (2013). *Constitutional Change and Democracy in Indonesia*. Cambridge University Press.

Linnen-luecke, M. K., Marrone, M., and Singh, A. K. (2020). Conducting systematic literature reviews and bibliometric analyses. *Australian Journal of Management*, 45(2), 175–194. https://doi.org/10.1177/0312896219877678

Muhtadi, B., and Warburton, E. (2020). Inequality and democratic support in Indonesia. *Pacific Affairs*, 93 (1), 31–58. https://doi.org/10.5509/202093131

Mujani, S., Liddle, R. W., and Ambardi, K. (2018). *Voting Behavior in Indonesia Since Democratization*. Cambridge University Press.

Mukherjee, D., Lim, W. M., Kumar, S., and Donthu, N. (2022). Guidelines for advancing theory and practice through bibliometric research. *Journal of Business Research*, 148, 101–115. https://doi.org/10.1016/j.jbusres.2022.04.042

PPIM UIN Syarif Hidayatullah Jakarta. (2024). Alm. Prof. Dr. Azyumardi Azra, CBE pendiri PPIM UIN Jakarta. PPIM UIN Jakarta.

Salinas-Ríos, K., and García-López, A. J. (2022). Bibliometrics, a useful tool within the field of research Bibliometría, una herramienta útil dentro del campo de la investigación. *Journal of Basic and Applied Psychology Research Biannual Publication*, 3(6), 9–16. https://repository.uaeh.edu.mx/revistas/index.php/jbapr/issue/archive

Sidiq, M. (2019). Panduan analisis bibliometrik sederhana. *Universitas Negeri Jakarta Journal*, 10. https://doi.org/10.13140/RG.2.2.15688.37125

Tambunan, D. (2023). Indonesia under threat: The danger of corruption to political legitimacy. *Asian Journal of Comparative Politics*, 8(1), 112–140. https://doi.org/10.1177/20578911221124965

Widodo, S. A. (2007). Islam dan demokrasi pasca orde baru. *UNISIA*, 30(65), 217–230.

Towards Resilient Societies: The Synergy of Religion, Education, Health, Science, and Technology – Rahiem (Ed.)

The effectiveness of SIPOL in ensuring electoral integrity in the 2024 Indonesian general election

Idris Thaha & Muh Revi S.
Universitas Islam Negeri Syarif Hidayatullah Jakarta, Indonesia

ABSTRACT: This study critically examines the effectiveness of the Political Party Information System (Sistem Informasi Partai Politik– SIPOL) used by the Indonesian Solidarity Party (Partai Solidaritas Indonesia – PSI) and the Labor Party (Partai Buruh) during the 2024 general elections in Indonesia, aiming to ensure the integrity of the electoral process. The research investigates how SIPOL was employed in verifying political party candidates. Using a qualitative method with a descriptive approach, data were collected through interviews with political figures and document analysis from both parties. The findings highlight SIPOL's crucial role in facilitating the verification process, encompassing stages such as registration, administrative verification, and the revision and announcement of verification results. Electoral integrity is characterised by four key attributes: accountability, transparency, honesty, and accuracy, all of which were positively acknowledged by PSI and the Labour Party. The study concludes that SIPOL significantly contributes to maintaining electoral integrity. It recommends that policymakers and political stakeholders refine and strengthen SIPOL's implementation to enhance the quality, transparency, and reliability of future electoral processes in Indonesia.

Keywords: electoral integrity, the Labour Party (Partai Buruh), political party verification, the Indonesian Solidarity Party (PSI), the political party information system (SIPOL)

1 INTRODUCTION

The Political Party Information System (SIPOL), used to ensure electoral integrity, is a relatively new application in Indonesia. Although SIPOL was initially designed for use in the 2014 elections, it was first implemented in the 2019 elections. This technology serves various purposes in different countries. For instance, in India, it is known as the Political Parties Registration Tracking Management System (PPRTMS), used to register and manage political parties. Similarly, in Mexico (Sistema de Información sobre Partidos Políticos-SIPP), South Africa (Party Funding Management System-PFMS), and Australia (Election Funding and Disclosure System-EFADS), the technology is utilised to manage, monitor, and track party funds and finances. Despite these differences, the Indonesian SIPOL and the technologies used in other countries share a common objective: ensuring transparency and accountability in managing party funds and finances.

Existing literature on this technology includes studies by Wisanggen (2021), Sari (2023); Alfarabi *et al.* (2023). These studies primarily addressed the implications of political parties' use of this technology in different regions throughout Indonesia. Yet, they often overlooked the critical role of technology in supporting electoral integrity in the country. Effective technology deployment is a significant indicator of electoral integrity in Indonesia.

This research addresses this gap by analysing the effectiveness of SIPOL usage by the Indonesian Solidarity Party (PSI) and the Labor Party. Specifically, the study focused on how SIPOL assisted these parties in navigating the verification process during the 2024 elections. The current study is intended to address the following questions: 1) what do PSI

 DOI: 10.1201/9781003645542-13

and the Labor Party apply the process for utilising SIPOL in the candidate verification process during the 2024 elections? 2) how does using SIPOL by PSI and the Labour Party contribute to ensuring electoral integrity?

The objective of this study is to evaluate how PSI and the Labor Party utilised SIPOL to meet the requirements for the 2024 election. This topic is of particular interest for several reasons. First, comparing SIPOL usage between PSI and the Labor Party provides valuable insights into its application in the 2024 elections. Second, as parties without parliamentary seats, i.e., those that have not met the parliamentary threshold, PSI and the Labor Party must undergo comprehensive verification, including administrative and factual checks. Third, the increasing prevalence of digital technology necessitates digitising all relevant aspects.

The structure of this research is organised into several sections. The first section reviews existing literature relating to the use of SIPOL. The second section outlines the research methodology, including the research design and data collection methods. The third section presents and discusses the research findings. The fourth section affirms that SIPOL can contribute to achieving electoral integrity.

2 METHOD

The present study employed a qualitative research methodology, specifically a qualitative descriptive approach. The aim is to gain comprehensive and in-depth insights into the registration process applied by the Indonesian Solidarity Party (PSI) and the Labor Party (Partai Buruh) in utilizing the Political Party Information System (SIPOL) as a prerequisite for participating in the 2024 elections. The study examined the role of SIPOL in ensuring electoral integrity through four key indicators: accountability, accuracy, honesty, and transparency (Surbakti 2016).

The data were obtained from interviews with six competent politicians from both parties, official documents from political institutions, including the General Elections Commission (KPU), and articles from journals and mass media published in the past two years. The data were then analysed descriptively to provide a thorough and nuanced depiction of the verification process undertaken by PSI and the Labor Party in the use of SIPOL for the 2024 elections.

3 FINDINGS AND DISCUSSION

This study aims to evaluate the effectiveness of the Political Party Information System (SIPOL) as a tool in the registration and verification process for political parties participating in the 2024 elections. The research focused on how SIPOL, particularly by the Indonesian Solidarity Party (PSI) and the Labor Party, can contribute to achieving electoral integrity.

3.1 *SIPOL for document submission*

Through the Political Party Information System (SIPOL) (Figures 1 and 2), all parties intending to participate in the 2024 elections must submit the necessary registration documents manually or electronically. The General Election Commission (KPU) considered SIPOL an essential supporting tool in the 2024 general election. KPU member Idham Holik stated, "We have introduced SIPOL, the Political Party Information System, as an instrumental platform to assist in the registration and verification processes of political parties" (*Detiknews* 2022). The utilisation of SIPOL is underpinned by four legal frameworks: Law Number 7 of 2017 concerning General Elections, General Election Commission Regulation (PKPU) Number 3 of 2022 concerning the Stages and Schedules for Organizing the 2024 General Elections, PKPU Number 4 of 2022 concerning Registration, Verification, and Determination of Political Parties Participating in the General Elections for Members of the House of Representatives and the Regional House of Representatives, and Constitutional Court Decision (MK) Number 55/PUU-XVII/2020.

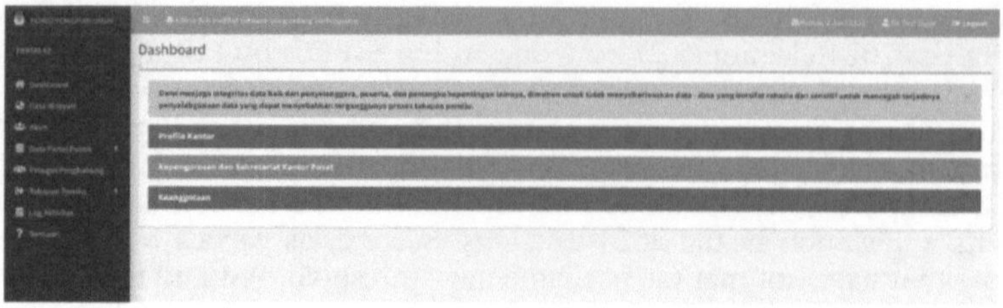

Figure 1. The 2024 general election SIPOL Platform (Source: kpu.go.id 2022).

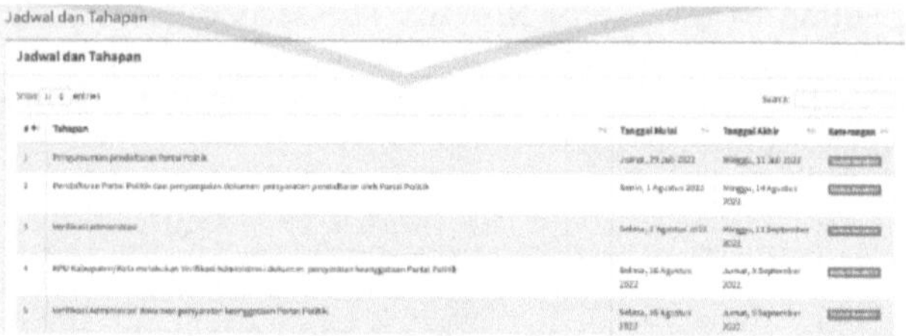

Figure 2. The display of the stage schedule reminder feature in SIPOL, PSI (2022).

3.1.1 *Registration stage*

The Indonesian Solidarity Party (PSI) and the Labour Party (*Partai Buruh*) shared similarities in submitting party requirements for election candidates. Both parties have established internal systems for managing party data: PSI utilises the PSI Administrative Information System (SIAP), while the Labor Party employs SEMATA.

Fauzi (personal communication 2023), Head of the Information and Technology Division at PSI, explained, "SIAP is our internal system designed to facilitate the verification process. It was developed to meet specific verification requirements, streamlining the management of membership and organisational data". SIAP and SEMATA serve as data repositories that support the parties' administrative functions and facilitate the submission of verification documents. Both parties successfully input their management documents at the provincial, district/city, and sub-district levels.

However, as recognised by Diding Sudrajat, Liaison Officer of the Labor Party, the implementation of SIPOL presented significant challenges for his party. He remarked, "The digitalisation process is inherently complex, particularly given the geographical and infrastructural disparities across Eastern and Western Indonesia, where the distances to provincial capitals are considerable. Additionally, maritime and island regions exacerbate these difficulties, especially concerning connectivity and signal reliability" (Sudrajat, personal communication 2023).

3.1.2 *Administrative verification stage*

Administrative verification involved assessing the completeness and validity of submitted documents, including management decision documents, office accounts, office domicile certificates, and documents about women's representation. This stage was divided into the initial verification phase (August 2 - September 14, 2022) and the correction phase (September 15 - October 14, 2022). During the initial verification phase, the General Elections Commission (KPU) designated PSI and the Labor Party with Unqualified (*Belum Memenuhi Syarat*-BMS) status. Following the correction phase, the KPU subsequently declared both parties as Eligible (MS) for the next stage of factual verification.

According to Deviani Nurrismon (personal communication 2023), Regional Coordinator of PSI for SIPOL, the BMS status during the administrative verification stage is quite typical since nearly all political parties seeking to participate in the election were assigned the BMS status. She explained, "Most parties were considered BMS due to the cumulative nature of the requirements. The submission will be flagged as BMS if even one condition is unmet. The reasons for this vary, including typographical errors, illegible documents, or incorrect attachments."

3.1.3 *Factual verification stage*

Factual verification is confirming the accuracy of required documents through field verification. During the initial factual verification phase, the Indonesian Solidarity Party (PSI) and the Labor Party were assigned Unqualified (BMS) status. The General Elections Commission (KPU) conducted site visits to party offices to verify the presence of administrators or members. Both parties identified correcting management and membership data during the factual verification as a significant challenge. PSI and the Labor Party updated their administrative, office, and membership data in the final factual verification phase. Both parties were subsequently declared to have successfully passed the administrative and factual verification stages, as documented in the Official Report Number 303/PL.01.1-BA/05/2022. PSI met the requirements for office and management in 34 provinces and 441 districts/cities, while the Labor Party met the requirements in 414 districts/cities.

3.2 *SIPOL's role in ensuring electoral integrity*

The introduction of SIPOL was expected to play a crucial role in verifying political parties and contributing to realising electoral integrity. For the 2024 general election, the General Election Commission (KPU) refined the technical processes and the system for utilising SIPOL. Political parties have become more receptive to and engaged with SIPOL as a verification tool. Satia Chandra Wiguna (personal communication 2023), Liaison Officer of PSI for SIPOL, remarked, "In the 2024 election, SIPOL was undoubtedly more effective, facilitating the process and providing greater support. In contrast, SIPOL faced challenges in the 2019 election due to inadequate socialisation and an underdeveloped system."

An election with integrity is defined by inseparable and reinforcing principles: accountability, honesty, transparency, and accuracy at every stage of the electoral process, including registering political parties as candidates for the election (Surbakti 2016). PSI and the Labour Party adopt and apply these four criteria.

3.2.1 *Accountability*

The accountability of SIPOL in party verification can be assessed from several aspects, including regulatory compliance, information presentation, and reporting at each stage. Accountability refers to responsibility for the process of using SIPOL. It also encompasses reporting and the presentation of information. The KPU is required to provide clear information to parties and address any technical issues encountered during the use of SIPOL. Accountability also entails the KPU's responsibility for data security and protection against cyber threats.

3.2.2 *Transparency*

Transparency is demonstrated through the enhanced features and functions of SIPOL in the 2019 elections. SIPOL's functions extend beyond electoral process stages and data management, including monitoring progress, archival recordkeeping, continuous data updating, and consultation. Providing access to SIPOL for the Election Supervisory Board (Bawaslu) is a step toward increasing transparency. The public's awareness of SIPOL's use, particularly its impact on party verification, also enhances transparency. Public access to SIPOL allows for the identification of party membership discrepancies and verification of status.

3.2.3 *Honesty*

Honesty is a fundamental principle in elections. It is defined as the accuracy of information without exaggeration or omission. Honesty in using SIPOL is evaluated based on the accuracy of verification results reported by the KPU to the parties. PSI and the Labour Party consider the KPU honest in reporting administrative and factual verification outcomes. The KPU's honesty in using SIPOL is reflected in the provision of written reports and through SIPOL. The KPU reports the results of party verification following PKPU Number 4 of 2022.

3.2.4 *Accuracy*

Accuracy is a critical measure in reviewing party requirement documents. SIPOL features support accurate data verification, particularly in identifying internal and external duplications. However, the KPU's accuracy in document verification through SIPOL requires improvement. For instance, during administrative verification, the KPU assessed PSI as not meeting the requirements despite submitting complete documents, such as copies of ID cards and membership cards. Similarly, discrepancies were noted in the Labor Party's data compared to what was entered into SIPOL. PSI and the Labor Party conducted audits with the KPU to address these issues (News.detik.com 2022).

4 CONCLUSION

PSI and the Labor Party underwent three stages in the party verification process using SIPOL: party registration, administrative verification, and factual verification. SIPOL plays a pivotal role in facilitating an election characterised by integrity. PSI and the Labor Party navigated these stages using SIPOL technology, providing periodic verification results at each electoral stage. Despite the advancements by SIPOL, human involvement remains necessary, as the document verification process still requires manual oversight. While SIPOL has contributed to ensuring electoral integrity, human oversight remains essential. However, even though SIPOL significantly contributes to ensuring electoral integrity, human involvement remains indispensable. Hence, ongoing improvements to the system are essential to enhance the quality and integrity of future elections.

REFERENCES

Alfarabi, A., Warganegara, A., and Maryanah, T. (2023). Implementasi sistem informasi partai politik (SIPOL) pada tahapan pemilihan umum tahun 2024 di Komisi Pemilihan Umum Kabupaten Muara

Constitutional Court(MK) Decision Number 55/PUU-XVII/2020.

General Election CommissionRegulation (PKPU) No. 4 of 2022 on the Registration, Verification, and Accreditation of Political Parties Participating in the General Election for the People's Representative Council and the Regional Representative Council.

Law Number 7of 2017 concerning General Elections.

Partai SolidaritasIndonesia (PSI) mendaftar ke KPU sebagai calon peserta Pemilu tahun 2024. (2022, August 10). Komisi Pemilihan Umum. https://www.kpu.go.id

Puspasari.(n.d.). Prinsip-prinsip good governance transparansi dan akuntabilitas pada KPP Pratama Cikarang Utara. Perspektif, 17(2), 195–201.

Sari, D. P. (2023). Implementasi sistem informasi partai politik (SIPOL) dalam tahapan verifikasi menuju Pemilu 2024 (Studi kasus pada Partai PDA, SIRA, PAS, GABTHAT). *Journal of Political Sphere (JPS*, 4(2), 60–80.

SIPOL socialization materials by the Indonesian KPU. (2022, June 15).

Surbakti, R. (2016). Tata kelola Pemilu sebagai sub kajian Pemilu terapan. Makalah pada inagurasi anggota baru Akademi Ilmu Pengetahuan Indonesia (AIPI).

Tricahyono, I. (2009). *Reformasi Pemilu menuju pemisahan Pemilu nasional dan lokal.* Malang: Intrans Publishing.

Utama, F. (2022, June 30). KPU: 28 partai politik sudah miliki akses SIPOL. Sindonews. https://nasional.sindonews.com.

Wisanggen, A. S. T. (2021). Sistem informasi partai politik (SIPOL) dalam verifikasi partai politik peserta Pemilu tahun 2019. *Electoral Governance*, 2(2).

Towards Resilient Societies: The Synergy of Religion, Education, Health, Science, and Technology – Rahiem (Ed.)

Education and economic inclusivity: A comparative study of Java and non-Java regions in Indonesia

Lucky Mutiara Pindan Rattekarua, Erni Nora M Siahaan & Anindita Wilandari
Ministry of National Development Planning/Bappenas, Indonesia

Deago Fraimon Sirenden
MSCE Student UIUC, USA

ABSTRACT: The lack of an inclusive economy exacerbates inequality, particularly impacting disadvantaged communities in underdeveloped regions. This condition has driven government initiatives to develop tourism sectors as a strategy to stimulate economic growth and reduce regional disparities, especially outside Java. However, empirical studies examining the relationship between population, education, and economic inclusiveness remain limited. This study contributes to the literature on equitable economic development by utilising the Inclusive Development Index (Bappenas). Employing quantitative methods, the study uses regression data analysis from 2016 to 2019 across 514 districts and cities in Indonesia. By incorporating district-level panel data and comparing Java and non-Java regions, the study offers fresh insights into regional disparities. The findings reveal that population size has no significant effect on economic inclusiveness, while education exerts a positive influence. The study recommends further exploration of human resource development to support inclusive economic growth. Special attention should be directed toward non-Java regions, where targeted educational initiatives and strategic investments are crucial for bridging regional gaps and promoting sustainable and equitable development.

Keywords: education, inclusive economy, Indonesia, panel data regression, population

1 INTRODUCTION

The inclusive growth paradigm was first adopted by various international organisations such as the OECD, World Bank, World Economic Forum, African Development Bank and Asian Development Bank with the aim of establishing a growth and development indicators that are broad-based and equitable. The World Bank explains that economic inclusion is growth that focuses not only on growth nationally, but also on reduction poverty that allows people to participate and benefit from economic growth. The inclusive growth paradigm is a deep sustainable development framework and can be seen as a process or policy response (Hasmath 2015) as well as the desired output. However, the theoretical basis of inclusive growth has not been fully developed.

Hasmath (2015) argues that inclusive growth is related to Sen's capability theory who emphasises the role of economics not only as a mechanism for reducing inequality, but also social mechanisms. Inclusive growth involves extending the size of the economy rather than dispersing its resources (Bakker and Messerli 2017) and ensures poverty reduction and reduction inequality (Seth and Alkire 2014). Development can be inclusive only if all segments of society contribute to creating opportunities, share development benefits, and

DOI: 10.1201/9781003645542-14

participate in decision-making (UNDP 2016). This shows that economic inclusiveness is related to alleviating the problem of inequality.

Many countries focus on growth, which has a negative influence on social stability and the economy, contributing to increased inequality. The lack of an inclusive economy leads to increasing inequality in regions, especially among impoverished communities in underdeveloped countries. Previous research by Bulman and Nguyen (2014) indicates that middle-high-income nations like China, Malaysia, and Thailand can no longer accelerate their economic growth due to the inequality that hinders it. This inequality issue is the foundation for nations‘ worldwide commitment to achieving inclusive economic development, which is one of the SDGs' main goals (Bakker and Messerli 2017). The SDGs‘ commitment to Goals 8 and 10 aims to promote inclusive and sustainable economic growth, productive and comprehensive employment opportunities, and decent work for all, as well as to reduce inequalities within and between countries. This serves as a solution to mitigate the trade-off between high economic growth rates and inequality. Indonesia is one nation that is dedicated to accomplishing the SDGs despite having a high degree of economic disparity.

Inequality occurs in Indonesia not only between provinces but also between islands and between rural and urban communities within a region. According to statistics from Indonesia's Central Statistics Agency (BPS 2019), there was a disparity in the Regional Gross Domestic Product (PDRB) between the islands between 2014 and 2018, with the provinces of Java Island (58%) and Sumatra (22%) having the highest PDRB. Only roughly 19–20% of Indonesia's economic growth came from the eastern region; the remaining 80–81% came from the western region. Additionally, there is variation in the inequality between the provinces within the island region; the highest levels are found in Java-Bali and Kalimantan, as well as the highest levels between rural and urban areas in Sulawesi, Java-Bali, and Nusa Tenggara (Bappenas 2020). This inequality demonstrates the existence of marginalised groups with limited access to economic opportunities.

In response to Indonesia's regional disparity concerns, several policies are being introduced on a continuous basis to ensure inclusive development. Tourism is one of the most important areas for developing regional economies in the long-term development plan (RPJMN 2020). The government has begun to develop new tourism growth centres by designating National Priority Destinations and tourist villages. The development of these new tourist destinations is projected to boost the economy in regions, especially those outside of Java Island, which are less developed than Java Island. Nonetheless, the goal of encouraging equitable development through tourism must be examined in terms of its role in fostering a balance of growth and equity. According to Tang and Tan (2018), the impact of tourism on economies might differ among nations based on institutional quality and income levels.

Research is needed to examine the influence of development on the level of inclusiveness in regions that are being significantly driven by new economic growth centres with varying levels of population and education. The research indicates that population size has no effect on the level of inclusiveness, while education positively impacts economic inclusiveness in the region. This research also provides new findings regarding differences between Java and non-Java.

2 DATA

This study draws on data from multiple institutions, including the Central Statistics Agency (BPS), the Ministry of National Development Planning (Bappenas), the Ministry of Public Works and Housing and the Ministry of Finance. This study uses an index that measures inclusive economic growth (Bappenas) which is consistent with study conducted by Fatimah *et al.* (2021) in the case of Malaysia. There are population variables. Perspective of classical economic theory (John Stuart Mill, David Ricardo, Thomas Robert Malthus, and Adam Smith) suggest that population size is one of the driving factors of economic growth, which will subsequently lead to an increase in economic inclusivity in the region. Inclusive growth depends on the average opportunities available to the population and the distribution of opportunities within the population. Another variable, namely education, is also important as the most crucial element in empowering people with skills and knowledge and giving them access to productive work. Pro-poor growth strategies emphasise improving the welfare of the poor to mitigate inequality (Dinda 2014), to anticipate insufficient investment in human capital among the poorer segments of society. Education has an impact to inclusivity in Indonesia (Fitrianasari 2021; Lestari and Rahmawati 2021; Prakoso and Agustina 2022; Sri Hartati 2021).

3 METHOD

This study uses panel data for analysis. The unit of analysis is at the regional level (district/city) from 2016 to 2019. The number of districts in Indonesia is 514 districts, resulting in 2.056 observations over four years. The use of district-level units for Java and non-Java comparisons is a novelty of this study. Panel data can account for variables that do not change over time. In this study, a fixed-effects model was chosen based on the Breusch and Pagan Lagrangian Multiplier test and the Hausman test. The fixed-effects model accommodates unobserved heterogeneity—variables not explicitly captured in the model. These variables, which may influence the level of economic inclusivity, include factors such as area size, religious diversity, cultural and racial composition, and geographical conditions within districts or cities. The decision to exclude variables such as "religious diversity" and "cultural composition" from the analysis was made for several reasons. First, these variables tend to remain relatively stable over time, making them difficult to capture dynamically within the scope of this study. Second, data availability and consistency pose challenges; comprehensive and reliable datasets for such variables at the district level across Indonesia are limited. Lastly, while these factors may influence economic inclusivity indirectly, their effects are likely mediated through other measurable variables, such as education, infrastructure, and governance, which are directly included in the model. By focusing on these more actionable variables, the study ensures a pragmatic approach that aligns with its objective of informing policy recommendations.

The panel data model used to address the research questions is represented in equation (1):

$$\mathrm{IPEI}it = \alpha 0 + \alpha 1\mathrm{EduP}it + \alpha 2\mathrm{pop}it + \alpha 3\mathrm{GRDP}it + \alpha 4\mathrm{Gov}it + \alpha 5\mathrm{Inf}\,it + \varepsilon i \quad (1)$$

To address the significant disparities among variables in the dataset, the model equation was transformed into a natural logarithmic form, except for the government expenditure variable (GovP). This transformation aims to normalise the data distribution, as recommended through normality testing. Additionally, this logarithmic transformation enables the estimated coefficients to be directly interpreted as elasticities. The resulting equation is as follows:

$$\mathrm{lIPEI}it = \beta 0 + \beta 1\mathrm{lEduP}it + \beta 2\mathrm{lpop}it + \beta 3\mathrm{lGRDP}it + \beta 4\mathrm{lGovt} + \alpha 5\mathrm{lInf}\,it + \varepsilon i \quad (2)$$

In this equation, IPEI represents the level of inclusiveness in the region, specified as the dependent variable. The independent variables are education level (Edu), population level (pop), income level (GRDP), infrastructure (Inf), and tourism support (GovP). The inclusion of variables such as "tourism support" reflects its critical role in fostering an inclusive economy. Tourism can serve as a catalyst for economic inclusivity by generating employment opportunities, enhancing local income, and encouraging regional development. Particularly in underdeveloped regions, government support for tourism development can empower local communities and reduce economic disparities.

This regression analysis aims to assess, both partially and simultaneously, the influence of independent variables on the dependent variable. It also determines the proportion of variation in the dependent variable explained by the independent variables. The analysis includes t-tests (Partial Significance), F-tests (Overall Significance), and the Coefficient of Determination (R2).

4 FINDINGS AND DISCUSSION

4.1 *Descriptive statistics*

Table 1 summarises the number of observations, average value, standard deviation, and highest and lowest values of each research variable. It shows that there are six research variables, consisting of the level of inclusive economic development (IPEI), education level (Edu), population rate (Pop), and income level (GRDP), Government Support for the Tourism Sector (GovP), Infrastructure (Inf). Total observations are as many as 2,056, consisting of 514 districts spread across 34 provinces from 2016 to 2019.

Table 1. Summary statistics.

Variable	N	mean	Sd	Min	Max
Inclusive economy index (IPEI) (Indeks 0-10)	2.056	5,31	0,60	2,72	7,53
Education level: average years of schooling (year)	2.056	8,04	1,66	0,7	12,64
Population rate (number)	2.056	514.069	635.441	13.400	5.965.006
Income level (billion rupiah)	2.056	20.080,98	3.792,67	43,792	452.519
Government support for the tourism sector	2.056	0,32	0,46	0	1
Infrastructure (road conditions)	2.056	60.4	21.66	1.8	99.86

Source: author's data processing

4.2 *Fixed effect model analysis*

The relationship between population, education and the level of economic inclusiveness in the region was analysed using a fixed effect model. The estimation results are shown in Table 2. Simultaneously (F test) shows Prob > F = 0.0000 in each research model, which indicates that each equation model can provide a relationship between the independent variables (population, education, GRDP, government support, infrastructure) and dependent variable (level of economic inclusivity/IPEI).

The estimation results, on a national scale, show that education, income & government support for the tourism sector variables have an effect with a significant level of 1 percent and there is no relationship between population & infrastructure variables to IPEI. In Java areas, education shows an effect with a significant level of 10 percent; income shows an effect with a significant level of 1 percent; and government support for the tourism sector, population, and infrastructure variables did not have a significant effect. In non-Java areas, show that education, income & government support for the tourism sector have an effect with a significant level of 1 percent and there is no relationship between population & infrastructure variable to IPEI.

Table 2. Analysis.

	Model		
	Nasional	Java areas	Non-Java areas
lpop	0.017	0.076	0.001
	[0.002]	[0.059]	[0.032]
lEdu	0.144***	0.135*	0.151***
	[0.044]	[0.079]	[0.051]
lGRDP	0.136***	0.206***	0.115***
	[0.015]	[0.030]	[0.018]
lInf	−0.002	0.001	−0.002
	[0.003]	[0.007]	[0.003]
Govp	0.005***	0.001	0.006***
	[0.002]	[0.004]	[0.002]
_cons	0.232	−0.880*	0.558*
	[0.278]	[0.493]	[0.318]
N	2,056	476	1,580
r2	0.86	0.78	0.86

Standard errors in brackets
* $p<0.10$ ** $p<0.05$ *** $p<0.01$
Source: author's data processing

Many studies have demonstrated the influence of education on economic growth. However, significant economic growth does not necessarily guarantee equity. On a global scale, there have been decreases in poverty and inequality between nations, but disparities within individual countries continue to rise (Summers and Balls 2015). A focus on growth alone is one of the causes of these growing disparities, which negatively impact both social stability and the economy. Therefore, it is critical to identify variables that can enhance equality within regions.

This study finds that education significantly influences inclusive economic development. Inclusive growth rates, supported by strong investment activities, create new job opportunities (Fatimah *et al.* 2021), which are heavily dependent on the level of education within the community. The findings also indicate that the impact of education is greater in non-Java regions compared to Java. Conversely, the population variable did not have a significant impact, aligning with the classical economic view that a large population, especially when coupled with a high rate of population growth, is often seen as a hindrance to development.

Furthermore, the study reveals that government support for tourism has had a significant effect in non-Java regions but not in Java. This distinction highlights the unique socio-economic and infrastructural disparities between these regions. Non-Java regions often rely more heavily on government-driven tourism initiatives to stimulate local economies due to limited industrial diversification, underdeveloped infrastructure, and lower levels of private sector investment. Tourism in these areas plays a pivotal role in creating jobs, increasing income, and reducing regional disparities. By contrast, Java benefits from more established infrastructure, a more diversified economic base, and stronger private sector investment, making it less reliant on tourism as a driver of economic inclusivity.

5 CONCLUSION

The study analyses the relationship between population and education and their impact on the level of economic inclusivity in various regions. It finds that population has no significant effect on inclusiveness, while education positively influences it. Developing human resources is critical, as evidenced by the study's findings, which show that the variable "length of school time," an aspect of human resource development through education, has a significant and positive effect on improving the inclusive economy, particularly in non-Java regions, which are more affected than Java regions. The study also finds that government support for tourism has a strong impact in non-Java regions but not in Java. These findings highlight the need for targeted policies. In non-Java regions, improving tourism infrastructure and offering financial support can help boost economic inclusivity. In Java, the focus can shift more to using its industrial and service sectors while making sure education and workforce development promote inclusivity. For policymakers, this emphasises the need to reduce educational gaps and promote inclusive growth. Strategies should focus on non-Java regions, where economic inclusion is low, but tourism potential is high. Support should include training, education, and business capital to help local communities engage in regional development and use tourism as a tool for inclusivity.

REFERENCES

Badan Perencanaan Pembangunan Nasional (Bappenas). (2020). *Indeks Ekonomi Inklusif tahun 2016–2019*. Retrieved February 2022, from http://inklusif.bappenas.go.id//data

Badan Pusat Statistik (BPS). (2019). *Rata-rata lama sekolah tahun 2016–2019*. Retrieved June 2022, from https://www.bps.go.id/indicator/26/415/2/-metode-baru-rata-rata-lama-sekolah.html

Badan Pusat Statistik (BPS). (2019). *Produk domestik regional bruto kabupaten/kota di Indonesia atas dasar harga konstan 2010 tahun 2016–2019*. Retrieved February 2022, from https://www.bps.go.id/publication/produk-domestik-regional-bruto-kabupaten-kota-di-indonesia-2016-2020.html

Badan Pusat Statistik (BPS) Berbagai Provinsi. (2019). *Jumlah penduduk tahun 2016–2019*. Retrieved February 2022, from https://www.bps.go.id/

Bakker, M., and Messerli, H. R. (2017). Inclusive growth versus pro-poor growth: Implications for tourism development. *Tourism and Hospitality Research*, 17(4), 384–391.

Bulman, D., Eden, M., and Nguyen, H. (2014). Transitioning from low-income growth to high-income growth: Is there a middle-income trap? *Policy Research Working Paper No. 7104*. World Bank. https://doi.org/10.1596/1813-9450-7104

Dinda, S. (2014). Inclusive growth through the creation of human and social capital. *International Journal of Social Economics*, 41(10), 878–895.

Fatimah, C. S. N., Fahmee, A. H. M., and Rahim, R. A. (2021). Inclusive growth: Comprehensive dimension of income distribution. *Экономика региона*, 17(1), 301–317.

Fitrianasari, R. N. (2021). Analisis dampak globalisasi, kebijakan fiskal, dan modal manusia terhadap pertumbuhan ekonomi inklusif: Studi kasus dengan data panel pada 9 kabupaten/kota di Provinsi Kalimantan Timur.

Hasmath, R. (Ed.). (2015). *Inclusive growth, development and welfare policy*. Routledge.

Lestari, C. D., and Rahmawati, F. (2021). Analisis pengaruh teknologi, politik, dan sumber daya manusia terhadap pertumbuhan ekonomi inklusif di Indonesia 2017–2019.

Prakoso, A. D., and Agustina, N. (2022). Inclusive growth analysis in Central Sulawesi, the eastern province of Indonesia, 2015–2019. *Asian Journal of Business Environment*, 12(2), 1–12.

RPJMN 2020–2024. (2020). *Rencana Pembangunan Jangka Menengah Nasional* 2020–2024. Retrieved February 2022, from https://www.bappenas.go.id/files/rpjmn/Narasi%20RPJMN%20IV%202020-2024

Seth, S., and Alkire, S. (2014). Did Poverty Reduction Reach the Poorest of the Poor? Assessment methods in the counting approach (No. 77). Queen Elizabeth House, University of Oxford.

Sri Hartati, Y. (2021). Analisis pertumbuhan ekonomi inklusif di Indonesia. *Jurnal Ekonomi dan Bisnis*, 12(1), 79–92.

Summers, L. H., and Balls, E. (2015). *Report of the Commission on Inclusive Prosperity*. Center for American Progress.

Tang, C. F., and Tan, E. C. (2018). Tourism-led growth hypothesis: New global evidence. *Cornell Hospitality Quarterly*, 59(3), 304–311.

UNDP. (2016). *Inclusive Development*. Retrieved from http://www.undp.org/content/undp/en/home/ourwork/povertyreduction/focus_areas/focus_inclusive_development.html.

Towards Resilient Societies: The Synergy of Religion, Education, Health, Science, and Technology – Rahiem (Ed.)

Female circumcision: Cultural practices, Islamic law and implications for SDGs

Marifat Kilwakit, Siti Isnaini & JM Muslimin
Universitas Islam Negeri Syarif Hidayatullah Jakarta, Indonesia

ABSTRACT: The practice of female circumcision, or FGM/C, remains a controversial issue within the Muslim community. While some view it as part of the Prophet's Sunnah and even consider it obligatory, as seen in the Kur Islands, Maluku, others oppose it due to its harmful health impacts. This study examined whether female circumcision in the Kur Islands is purely symbolic or involves harmful cutting practices. This qualitative research utilised primary data from observations and interviews with birth shamans—trusted practitioners of circumcision—religious leaders and community members involved in the practice. Secondary data were sourced from relevant books and documents. The findings reveal that female circumcision in the Kur Islands, conducted without medical professionals and using sharp tools, poses a risk of mutilation. This contradicts maqāṣid al-Sharī'ah, which prioritises protecting life and health, and undermines the achievement of the Sustainable Development Goals (SDGs). Specifically, it endangers women's health (SDG 3), deepens gender inequality (SDG 5), and violates human rights (SDG 16). Addressing this issue requires health education, gender awareness, and supportive policies that promote equality, health, and women's rights in alignment with Islamic values and the SDGs.

Keywords: cultural practices, female circumcision, genealogy, Islamic law, SDGs

1 INTRODUCTION

Female circumcision, or in English it is called *mutilation/mutilation of female genitalia* (FGM/C), or some call it female genital mutilation and injury (P2GP), is a traditional social practice that has been going on for a long time in the realities of today's society. The practice of FGM/C is considered commonplace in all parts of the world, including Indonesia. As shown from the data issued by the *World Health Organization* (WHO), about 100-132 million women worldwide are victims of FGM/C practice. Every year, about 2 million people are at risk of this practice (Irianto 2006). The practice of circumcision today is not only mandatory for men, but women are also required to perform circumcision (Farida *et al.* 2018). Female circumcision is believed to be a religious imperative that has been firmly rooted in traditional beliefs. The logical consequence of this practice is that if a woman refuses to be circumcised, she will be ostracised in society (Mulongo *et al.* 2014). The same thing also happened in the Kur Islands, Maluku Province. On average, the entire Kur community views female circumcision as part of the obligation and responsibility of parents to circumcise their children as early as possible.

Research on female circumcision in Kur society is still limited, especially related to its social, cultural, and health impacts. This study is important to explore whether the practice is symbolic or involves mutilation that injures a woman's sexual organs, which can be categorised as gender-based violence. The study stems from concerns about the physical and psychological effects of female circumcision, such as pain, infection, and trauma, which contradicts the traditional claim of Kur society that the practice aims to maintain the health of women's genitals. The research is also relevant to the SDGs Goal 5 (Gender Equality) and Goal 3 (Good Health), as these practices can violate women's rights and health and support social change for their equality and well-being.

2 METHOD

This research was conducted in the Kur Islands of Maluku by interviewing approximately ten informants: parents (husband and wife), religious leaders, traditional leaders, and birth shamans. Data sources were sampled using purposive and snowball methods. In understanding the phenomenon of female circumcision in Muslim women, the author uses a phenomenological approach to look at female circumcision in the reality of society in the Kur Islands of Maluku (Setyowati *et al.* 2023). Research data was obtained from primary and secondary sources. Primary data is collected through interviews, while secondary data is collected from books, articles, and other relevant documents related to the research topic. The data analysis method used is descriptive-analytical. In the analysis stage, the researcher applied an interactive analysis model that included four main components: data collection, data condensation, data presentation, and conclusion. (Matthew *et al.* 2014). These four components interact with each other in the cycle of the data collection process in the field. During the data collection process, the researcher moved between three components. Once the data collection is complete, the focus shifts between data reduction, data presentation, and conclusions to develop a substantive understanding based on empirical findings. This method is used to understand the practice of female circumcision in the Kur Islands, focusing on its establishment and preservation in socio-cultural and religious contexts. The interactive approach explores the dynamics of relationships between individuals, communities, and traditions and the factors that influence the practice. In this study, the research subject voluntarily gave or expressed his or her desire to participate after explaining the overall study's scope, benefits, and risks.

3 RESULTS AND DISCUSSION

3.1 *Kur society and the practice of female circumcision*

Geographically, Kur is one of the islands in Tual City, Maluku Province, with a population of around 5,003 people. Kur society as a whole embraced Islam and had a social structure that was still traditional, but they also did not close themselves to modernity today. Their socio-religious activities are generally similar to those of other regions. It is just that the Kur people are still thick with local traditions that occur from generation to generation, one of which is female circumcision, which has been preserved to this day. The tradition of female circumcision has been rooted for a long time and is considered an obligation that cannot be let go.

After interviewing BT as a religious leader, he gave information that the Kur community considers female circumcision as a religious obligation to identify a person as a Muslim; Without circumcision, it will be regarded as an infidel (BT, Personal Interview, t.t). Meanwhile, in practice, the UK, as a fertility shaman entrusted with circumcision, gave information that circumcision in Kur society involves cutting off part of the genitals with a razor or needle, accompanied by special rituals and bathing with holy water as a symbol of Islam. Male circumcision is performed in the morning, while female circumcision is done secretly in the morning (UK, Personal Interview, t.t). As for the impact of female circumcision, most informants dismissed the notion that there was a negative impact of female circumcision. This was confirmed directly by seven parents who applied the practice of circumcision to their children. Likewise, it was also strengthened by SY as a traditional figure who argued that since ancient times, the tradition of circumcision in the area did not cause health problems for women, including in the genital area that was cut (SY, Personal Interview, t.t) Therefore, they oppose the prohibition of female circumcision because it is considered an obligation as well as a tradition that must be maintained and preserved.

3.2 *Understanding and genealogy*

In Islam, circumcision is known as the act of cutting the skin that covers the head of the penis in men or a small part of the skin on the top of the female genitals (Farida *et al.* 2018). The WHO classifies female circumcision into several types, ranging from partial or complete clitoral removal to other dangerous procedures on the female genitalia that are not medically intended. This practice includes various actions such as piercing, cutting, scraping, or burning the genital area

(Jannah 2021), as well as inserting herbal substances into the vagina that can cause wounds and bleeding (Alsibiani and Rouzi 2010).

In 1997, the *World Health Organization* (WHO), United Nations Children's (UNICEF), and the United Nations Fund for Population Activities (UNFPA) jointly established a concrete definition of the practice of FGM/C, which is an attempt to remove part or all of a woman's genitals for religious, cultural, or other non-therapeutic reasons. The practice of FGM/C is considered contrary to the basic principles of humanity contained in the Universal Declaration of Human Rights (HAM). Therefore, there needs to be an agreement to eliminate all forms of violence and injustice experienced by women. It is necessary to look further into the various physical and psychological impacts on women, which last in the short and long term (Goldberg *et al.* 2016). Loss of perception of the genitals and overall function and genital space can disrupt the psychophysical balance, self-image, and body perception (O'Dey 2019). Female circumcision generally occurs in girls with an age range of 4-12 years, with only 12% performed by doctors or medical personnel. The rest is performed by non-medical personnel, cutting in sterile conditions and without anaesthesia or painkillers, except for local ointments (Chibber *et al.* 2011).

Research by Dan Reisel and Sarah M. Creighton reveals that FGM/C has adverse health effects, both short- and long-term, including bleeding, trauma, infection, and other serious complications (Ahinkorah *et al.* 2021; Reisel and Creighton 2015). A 2016 study in Africa showed the impact of FGM/C in 57 countries, with causative factors including a lack of antiseptic techniques and the use of anaesthetic agents. Common complications include excessive bleeding, urine retention, and swelling of the genital tissue (Odukogbe *et al.* 2017). In addition to having an impact on the body's health, female circumcision also has the potential to cause death, as happened in Egypt, where a doctor who practised FGM/C on an 11-year-old child later died. (Modrek and Sieverding 2016). The reasons behind the practice of FGM/C are quite diverse. They are based on social, cultural, and religious issues, such as social pressure, rites of transition, religious mandates, acculturation, and views on femininity, decency, and sexuality (Phillips 2024).

Until now, circumcision is still controversial in some circles regarding the recommendation for female circumcision. In Indonesia itself, the Indonesian Ulema Council (MUI), as an institution that has the authority to issue fatwas, allows FGM/C because it is considered fitrah and part of Islamic teachings. The decision to ban female circumcision, which had been fatwa by the MUI in 2008, was again raised in the decision of Commission A Bahtsul Masail of the Java-Madura Islamic Boarding School Consultative Forum (FMPP) XXXVIII on January 21-22, 2023, (Bahtsul Masa'il FMPP XXXVIII 2023). The recommendation for female circumcision is based on the first two Hadiths of the Prophet from Anas bin Malik (ra): *"The Prophet PBUH said to um Atiyah when he asked about circumcision, when you circumcise a woman, cut it a little and do not overdo it, because it brightens the face and pleases the husband."* (HR. Thabrani). In addition, some hadiths come from Um Atiyah: *"At that time there was a woman who was circumcised (a daughter) in Medina, and the Prophet said to her, do not overdo it because it is better for women and more pleasing to husbands."* (HR. Abu Daud).

The above two hadiths, according to Ibn al-Munzir and Syamsul Haq, are the hadiths that *do'if* (weak) because the narrator of the hadith is not strong. Ibn al-Munzir's opinion is also supported by several contemporary scholars, including Muhammad Sayyid Thantawi, former Grand Sheikh al-Azhar, Dr. Ali Gom'ah, Chairman of the Egyptian Fatwa Council Sheikh Yusuf al-Qardhawi and Chairman of the International Islamic Scholars. In addition, Abu Daud also considered this hadith not strong (weak) because Muhammad bin Hasan, one of the narrators, was categorised as *Majhul* (unknown). From some of the opinions above, it can be concluded that these two hadiths cannot be used as *Arguments*. The opinion of fiqh scholars strengthens that the hadith about circumcision is invalid as a postulate for the ability to perform circumcision (Muhammad 2024). Asy Syaukhani also expressed a similar statement. According to Dr. Achmad Asy Syarbashi, if there is evidence that the practice of circumcision on women does not benefit them, then there is no objection to abandoning it, as has been stipulated by the jurists (Azizah 2020). Therefore, the law of female circumcision in Islam requires an in-depth study by paying attention to the principle of gender equality. The validity of the Nash texts on which the law is based is still debated and is even considered "weak" by some experts, as stated by Sayyid Sabiq in "Fiqh al-Sunnah". Thus, female circumcision is included in the issue of ijtihadiyah (Hakim 2017).

3.3 *Challenges and changes in tradition*

Recently, the government has issued a law related to the prohibition of FGM/C, which is contained in article 99 paragraph (2), namely, "reproductive health efforts are carried out through promotive, preventive, curative, rehabilitative or palliative efforts in a comprehensive and integrated manner" (Indoensia 2023). The above opinion is strengthened by the Indonesian Midwives Association (IBI) which took a decisive step by declaring the complete elimination of the practice of female genital mutilation and injury (P2GP) or female circumcision through a Circular Letter of the Chairman of the Central Board of the Indonesian Midwives Association, Dr. Ade Jubaedah on February 22, 2024, which is intended for all Indonesian Midwives Associations (Indonesia 2024). Likewise, the regulation of the Minister of Health of the Republic of Indonesia No. 1636/2010 prohibits female circumcision performed by non-licensed health workers and limits the procedure to the removal of the clitoris skin without damaging it (Hanna *et al.* 2024). This prohibition is also strengthened in the Results of the 2nd Religious Deliberation of the Indonesian Women's Ulema Congress (KUPI) No.08/MK-KUPI-2/XI/2022, concerning the protection of women from mutilation and dangerous female genital injuries (P2GP) without medical reasons ((KUPI) 2023). KUPI seeks to criticise patriarchal thinking in official Islamic interpretations, as well as cultural practices, by reinterpreting religious teachings to realise gender justice, especially for women (Mun'im *et al.* 2024). The prohibition issued by KUPI is based on a fatwa issued by Dar al-Ifta' where female circumcision is not mandatory according to sharia because there is danger as stipulated in the sharia that "there should be no harm and danger." The practice of P2GP has been proven to cause physical and moral harm to women for themselves and their families without any benefit. So, based on the above view, female circumcision is punished as haram (Allam 2021).

3.4 *Sustainable Development Goals (SDGs) and the elimination of female circumcision*

The tradition of female circumcision that has been enshrined in Kur society has not had a bad impact, but procedurally (ordinance) carried out can be classified as violence in the form of female genital mutilation. It has been agreed that all violence (mutilation) committed, including female circumcision (FGM/C), is prohibited in Islam (Alsibiani and Rouzi 2010). In addition, the sustainability of FGM practices is a significant obstacle to achieving several SDGs, including Goal 3, Goal 5, and Goal 10. The adverse health impacts of FGM, such as the risk of severe bleeding, urinary complications, and serious problems during childbirth, directly hamper efforts to ensure healthy lives and support well-being for all, as outlined in SDG 3 (Knipscheer *et al.* 2015). The practice of FGM, specifically in Kur society and at large, has violated women's human rights, exacerbated gender inequality, and negatively impacted their health and empowerment (Mustakimah *et al.* 2023). Various campaigns and programs are designed to increase understanding of the challenges faced by individuals with gender identities while championing the creation of inclusive communities that uphold respect for all gender identities (Ramli *et al.* 2024). Efforts to ensure and promote healthy living for all are necessary (Roa *et al.* 2019). In line with the SDGs, it emphasises the importance of strengthening the health system by increasing service capacity and ensuring sustainability.

From the above presentation, there are two important points regarding health and gender equality issues that focus on all violence against women, including the practice of FGM/C (Sari 2024). The content of the SDGs is a continuation of the Millennium Development Goals, which ended in 2015 and continues with the SDGs, which emphasise gender equality, placing women in an equally important position in development so that all gender inequalities must be eliminated (Zitri *et al.* 2023). The United Nations (UN), through the SDGs, seeks to eliminate the practice of FGM/C as it is considered a serious violation of the rights of girls and women, as well as forms of violence against them (Wogu *et al.* 2019). These harmful practices, preserved through culture and tradition, reinforce the subordination of women in society while legitimising and perpetuating gender-based violence (Adebola 2020). FGM/C also violates human rights and is a form of discrimination based on gender, gender, age, and other grounds (Khosla *et al.* 2017). Therefore, to implement the SDGs goals, it is necessary to design a roadmap for eliminating female circumcision to support the SDGs goals towards world transformation by 2030 (Roa *et al.* 2019). Therefore, legal reforms are needed to address the ongoing issue of FGM/C or P2GP, given the cultural practices that systematically allow FGM/C to discriminate against women. Legal products based on gender sensitivity and victim perspectives are essential as a progressive step toward creating a just and equal society. Such laws are necessary for the prevention, handling, and eradication of violence against women. Regardless of their background, everyone deserves legal treatment prioritising justice (Nurtjahyo and Wicaksono 2022). In addition, according to the author, the prohibition of FGM/C is not only in the form of a written law but needs to be encouraged with maximum efforts to socialise to remote parts of Indonesia so that violent practices in the form of FGM/C and the like can be stopped.

4 CONCLUSION

The practice of female genital mutilation/cutting (FGM/C) in the Kur Islands is a hereditary tradition that has no basis in obligation in Islam but developed as part of a custom that often ignores the impact on women's health. The change in regulations is an important step to protect women from sexual violence, gender discrimination, and reproductive health risks that endanger their well-being. In line with the "Sustainable Development Goals" (SDGs), especially the third and fifth goals, ending FGM/C is a priority to ensure gender equality and reproductive health. Local policymakers and leaders need to raise public awareness through dialogue with religious and traditional leaders, implement strict policies prohibiting FGM/C, and replace harmful traditions with harmless cultural symbols. Ongoing support through health services and counselling is also important to protect women and prevent gender-based violence.

REFERENCES

Adebola, O. (2020). The Nexus between Female Genital Mutilation and Child Marriage in Nigeria: A Cultural Inhibition to Achieving Sustainable Development Goals. *International Journal of Social Sciences (IJSS)*, *10*(2), 1–8.

Ahinkorah, B. O., Ameyaw, E. K., Seidu, A.-A., and Yaya, S. (2021). Predictors of Female Genital Mutilation or Cutting Among Daughters of Women in Guinea, West Africa. *International Journal of Translational Medical Research and Public Health*, *5*(1), 4–13. https://doi.org/10.21106/ijtmrph.319

Allam, S. I. (2021). *Hukmu Khitan al-Ināts fī Al-Syari'ah Al-Islāmiyyah*. Mesir November. https://www.dar-alifta.org/ar/fatawa/16487/حكم-ختان-الاناث-في-الشريعة-الاسلامية

Alsibiani, S. A., and Rouzi, A. A. (2010). Sexual function in women with female genital mutilation. *Fertility and Sterility*, *93*(3), 722–724. https://doi.org/10.1016/j.fertnstert.2008.10.035

Azizah, A. (2020). Status Hukum Khitan Perempuan (Perdebatan Pandangan Ulama dan Permenkes RI No.1636/Menkes/Per/XI/2010. *Musawa Jurnal Studi Gender Dan Islam*, *19*(1636), 173–186.

Bahtsul Masa'il FMPP XXXVIII. (2023). *Hasil Keputusan Komisi A*.

Chibber, R., El-Saleh, E., and El Harmi, J. (2011). Female circumcision: Obstetrical and psychological sequelae continues unabated in the 21st century. *Journal of Maternal-Fetal and Neonatal Medicine*, *24*(6), 833–836. https://doi.org/10.3109/14767058.2010.531318

Farida, J., Elizabeth, M. Z., Fauzi, M., Rusmadi, R., and Filasofa, L. M. K. (2018). Sunat Pada anak Perempuan (Khifadz) Dan Perlindungan Anak Perempuan di Indonesia: Studi Kasus di Kabupaten Demak. *Sawwa: Jurnal Studi Gender*, *12*(3), 371. https://doi.org/10.21580/sa.v12i3.2086

Goldberg, H., Stupp, P., Okoroh, E., Besera, G., Goodman, D., and Danel, I. (2016). Female genital mutilation/cutting in the United States: Updated estimates of women and girls at risk, 2012. *Public Health Reports*, *131*(2), 340–347. https://doi.org/10.1177/003335491613100218

Hakim, L. (2017). Khitan Perempuan Dalam Perspektif Hukum Islam. *Ar-Risalah: Media Keislaman, Pendidikan Dan Hukum Islam*, *15*(1), 138–140.

Hanna, S., Mukri Aji, A., Tholabi, A., and Amin, M. (2024). Woman and Fatwa: an Analytical Study of MUI's Fatwa on Women's Health and Beauty. *Ahkam: Jurnal Ilmu Syariah*, *24*(1), 171–184. https://doi.org/10.15408/ajis.v24i1.37832

Indoensia, P. R. (2023). *Peraturan Pelaksanaan Undang-Undang Nomor 17 Tahun 2023 Tentang Kesehatan*.

Indonesia, P. P. I. B. (2024). Surat Edaran Pengurus Pusat Ikatan Bidan Indonesia Tentang Penghapusan Pelayanan Pemotongan dan Pelukaan Genetalia Perempuan (P2GP). Jakarta, Februari 2024.

Irianto, S. (2006). *Perempuan and Hukum: Menuju Hukum yang Berperspektif Kesetaraan dan Keadilan*. Yayasan Obor Indonesia.

Jannah, R. (2021). Sunat Perempuan Dalam Tinjauan Maqashid Syariah Menurut Al-Ghazali. *International Conference On Syariah and Law 2021 (ICONSYAL 2021), 2021(April)*, 124–132.

Khosla, R., Banerjee, J., Chou, D., Say, L., and Fried, S. T. (2017). Gender equality and human rights approaches to female genital mutilation: a review of international human rights norms and standards. *Reproductive Health*, *14*(1), 59. https://doi.org/10.1186/s12978-017-0322-5

Knipscheer, J., Vloeberghs, E., van der Kwaak, A., and van den Muijsenbergh, M. (2015). Mental health problems associated with female genital mutilation. *BJPsych Bulletin*, *39*(6), 273–277. https://doi.org/10.1192/pb.bp.114.047944

(KUPI), K. U. P. I. (2023). *Hasil Musyawarah Keaagamaan Kongres Ulama Perempuan Indonesia (KUPI) Ke-2*. KUPI (Kongres Ulama Perempuan Indonesia).

Matthew, B., Miles, and Huberman, A. M. (2014). *Qualitative Data Analysis*. Sage Publications.

Modrek, S., and Sieverding, M. (2016). Mother, daughter, doctor: Medical professionals and mothers' decision making about female genital cutting in Egypt. *International Perspectives on Sexual and Reproductive Health*, *42*(2), 81–92. https://doi.org/10.1363/42e1116

Muhammad, H. (2024). *Khitan Perempuan Untuk Apa?* 14 Januari 2024. https://www.kupipedia.id/index.php/Khitan_Perempuan:_Untuk_Apa%3F

Mulongo, P., Hollins Martin, C., and McAndrew, S. (2014). The psychological impact of Female Genital Mutilation/Cutting (FGM/C) on girls/women's mental health: a narrative literature review. *Journal of Reproductive and Infant Psychology*, *32*(5), 469–485. https://doi.org/10.1080/02646838.2014.949641

Mun'im, Z., Nasrudin, M., Suaidi, and Hasanudin. (2024). Revisioning Official Islam in Indonesia : The Role of Women Ulama Congress in Reproducing Female Authority in Islamic Law. *Ahkam: Jurnal Ilmu Syariah*, *24*(1), 135–152.

Mustakimah, L., Huriani, Y., and Zulaeha, E. (2023). Tantangan Mewujudkan Sustainable Development Goals (SDGs) Tentang Kesetaraan Gender pada Masyarakat Desa. *Az-Zahra: Journal of Gender and Family Studies*, *4*(1), 12–29. https://doi.org/10.15575/azzahra.v4i1.25462

Nurtjahyo, L. I., and Wicaksono, M. A. (2022). *Gender-Based Violance In South East*. Springer.

O'Dey, D. mon. (2019). *Vulvar Reconstruction Following Female Genital Mutilation/Cutting (FGM/C and other Acquired Deformities*. Springer.

Odukogbe, A. T. A., Afolabi, B. B., Bello, O. O., and Adeyanju, A. S. (2017). Female genital mutilation/cutting in Africa. *Translational Andrology and Urology*, *6*(2), 138–148. https://doi.org/10.21037/tau.2016.12.01

Phillips, K. M. (2024). *The feminization of freedom: An analysis of love, happiness and freedom from the perspective of childfree, never-married single women of color*. *0*(0), 1–20. https://doi.org/10.1177/02654075241269667

Ramli, M. A., Kasa, A. R., Sharifuddin, N. M., Yusof, Y., and Hassan, P. (2024). The Quest for Third Gender Equality : Challenges and Implications for Islamic Law and Muslim Women' s Sustainability in Malaysia. *Ahkam: Jurnal Ilmu Syariah*, *24*(2), 295–310.

Reisel, D., and Creighton, S. M. (2015). *Maturitas Long term health consequences of Female Genital Mutilation (FGM)*. *Maturitas*, *80*(1), 48–51. https://doi.org/10.1016/j.maturitas.2014.10.009

Roa, L., Jumbam, D. T., Makasa, E., and Meara, J. G. (2019). Global surgery and the sustainable development goals. *British Journal of Surgery*, *106*(2), e44–e52. https://doi.org/10.1002/bjs.11044

Sari, G. K. (2024). Peran PBB Melalui MDGs Dalam Peningkatan Indeks Pembangunan Manusia di Indonesia. *Journal of Multidisciplinary Research and Development*, *1*(2), 102–107. https://doi.org/10.56916/jmrd.v1i2.727

Setyowati, S., Mashuri, Fanggidae, L. W., Nainggolan, F. M. R., Vitrianto, P. N., and Sari, I. K. (2023). *Memahami Fenomologi, Etnografi, Studi Kasus, Dan Metode Kombinasi Dalam Jagad Metode Riset*. DOTPLUSH Publisher.

Wogu, J. O., Amonyeze, C., Oluwasina, R., Folorunsho, B., and Aloh, H. E. (2019). Influence of communication on Sustainable Development Goals (SDGs) campaign against Female Genital Mutilation (FGM) in Ebonyi State, Nigeria. *Idosr.Org*, *11*(14), 80–90.

Zitri, I., Rosiastawa, A., Amil, and Hadi, A. (2023). Mencapai Target Sustainable Development Golas's (SDG's) di Desa Sermong Kecamatan Taliwang Kabupaten Sumbawa Barat Tahun 2021. *Journal of Social and Policy Issues*, *2*, 41–45. https://doi.org/10.58835/jspi.v3i2.101.

Towards Resilient Societies: The Synergy of Religion, Education, Health, Science, and Technology – Rahiem (Ed.)

Assessing just public transport in Indonesia: Disability, mobility justice and Jakarta-Bandung high-speed railway

Mohammad Hasan Ansori
Universitas Islam Negeri Syarif Hidayatullah Jakarta, Indonesia

Luthfy Ramiz, Johari Efendi & Aan Permana
The Habibie Center

ABSTRACT: This study examines the accessibility of persons with disabilities (PWDs) and mobility justice in public services, focusing on the Jakarta-Bandung High-Speed Railway as a case study. Based on Sheller's mobility justice theory, the research develops a holistic, network-wide approach to promoting the social inclusion of PWDs by enhancing their participation in public transport. The findings reveal that although PWDs' involvement in the design and planning stages of public transport remains limited, significant improvements have been made in providing accessible facilities, helping to address mobility injustices for individuals with varying physical abilities along the Jakarta-Bandung route. The study recommends that the government and policymakers actively involve PWDs in the early stages of railway development planning, design, and decision-making processes. Furthermore, these findings contribute to reducing social inequality, aligning with the Sustainable Development Goals (SDGs) objectives.

Keywords: disability, mobility justice, public transport, planning, railway

1 INTRODUCTION

This study is generally aimed at examining the accessibility of persons with disabilities (PWDs) in public transportation, and it specifically highlights the involvement of persons with disabilities in design and planning processes as well as the provision of accessibility facilities. The transportation sector was chosen as the focus of research because it is one of the strategic sectors that has not significantly met the accessibility standards for persons with disabilities. More specifically, this study is intended to put the issue under the broader theoretical framework of mobility justice put forward by Sheller (2018), arguing that forms of mobility for differently positioned individuals should cover accessibility for those differently abled. The mobility Justice framework recognises that mobility injustices emerge beyond the confines of transport infrastructure and are often shaped by unequal power relationships in wider society. It also recognises the importance of including marginalised groups in decision-making processes that shape opportunities for mobility. For this reason, the current study also emphasises public transport's decision-making and design processes.

In Indonesia, the results of the National Socioeconomic Survey (SUSENAS) in March 2019 show that more than 9 percent or about 23.3 million people had some form of disability, with 2.2 percent or 5.7 million people with severe disabilities. Of the total number of persons with disabilities, 55.5 percent or around 12.9 million were women and 44.5 percent or the remaining 10.4 million were men (tnp2k.go.id, accessed on 30 May 2021.) According to the United Nations Convention on the Rights of People with Disabilities (UN CRPD),

DOI: 10.1201/9781003645542-16

persons with disabilities are those who have long-term physical, mental, intellectual, or sensory impairments that, given the various barriers in their interactions, can prevent their full and effective participation in society (UN.org, accessed on 30 May 2021). This definition is reinforced by the Government of Indonesia through Article 1 paragraph (1) of Law No. 8 of 2016 concerning Persons with Disabilities (Law on Persons with Disabilities).

This study mainly builds on the theoretical framework of mobility justice introduced by Sheller (2018). Mobility justice offers a new way to think across the micro, meso, and macro scales of transitioning toward more just mobilities. He argued that "current approaches to transport justice have not spent enough time showing how embodied differences in class, gender, race, ethnicity, nationality, sexual identity, and physical ability influence accessibility and interact with the mobility regimes and control systems that reproduce uneven mobilities". That is, addressing the injustices of our daily unequal mobilities requires that we develop a deeper understanding of how asymmetrical mobility relates not only to how we move but also to gendered and racialised perspectives of the world and global inequalities and liberal and neo-liberal power (Everuss 2019). Given this theoretical backdrop, the present study is intended to assess how the accessibility of public transport development in Indonesia, taking the case of Jakarta-Bandung High-Speed Railway, has not only addressed the injustice of public mobilities but also dealt with racialised perspective in looking at persons with disabilities.

Set against this backdrop, the current study aims to examine the extent to which persons with disabilities (PWDs) are involved in public transport's design and planning process and how it addresses the mobility injustice in public transport in Indonesia. A large number of research and studies have been done on public transport and accessibility, justice and equity in public transport (Everuss 2019; Hananel and Berechman 2016; Holstein and Legacy 2020; Martens 2016; Pereira *et al.* 2016; Sajib 2022). However, only a few examined how space and mobility in public transport interact with how people are differently represented regarding physical abilities, gender, race, etc. Most of the time, the interaction is part of the politics of movement beyond the transport system. The Jakarta-Bandung High-Speed Railway was selected as the prototypical case of this study as it is currently considered one of the most modern and sophisticated public transport in Indonesia.

2 METHOD

The qualitative semi-structured data collection method, instead of the structured one, was adopted in this study. While a structured interview generally maintains a series of formal and limited or previously structured questions, the semi-structured model tends to be more flexible and extensive (Lindlof and Taylor 2002). Given the above goal, some informants were purposively selected based on their characteristics and the study's requirements (Patton 2002). These informants included persons with disabilities who regularly use public transport, policymakers, public transport service providers, civil society organisations, and academics.

In addition to the primary data, the study also used secondary data, including data stored in general data archives, special collections, publishing companies, reports and public documents, academic publications (Heaton 2004), local and national newspapers and magazines, both printed and online. Secondary data can be used for cross-validation to support primary data analysis (Patton 2002) or triangulation. Furthermore, secondary data significantly contributes to what we can learn directly from the data. More importantly, secondary data gives us the stimulus to explore other issues posed by the study's purpose further. The mixed analysis of interviewing and documenting is particularly intended to avoid error vulnerabilities associated with one particular method (Patton 2002; Thorne 1994). Secondary data mainly validates the existing primary data, either confirming or bolstering. The secondary data may also serve as additional elements to the primary ones.

3 FINDINGS AND DISCUSSION

Many studies on mobility justice emphasise the importance of engaging persons with disabilities in the transport planning and design stage. One of the underlying themes of mobility justice is a query on how we can create a more just world in terms of not only freedom of movement but also equality of movement in the public transport domain. Such mobility justice in public transport can only be enabled by engaging PWDs in earlier processes. These studies have been done as case studies in diverse countries, including the USA, Australia, Turkey, etc. (Bascom and Christensen 2017; Cook and Butz 2019; Hananel and Berechman 2016; Kett *et al.* 2020; Romano 2022; Verlinghieri and Schwanen 2020).

It is important to note that people with a range of impairments are included in planning discussions, on accessibility panels, and in audits of existing services and structures. In relation to this, governments need to develop established measures, framework, and appropriate sets of standards, targets, and indicators in collaboration with local disability organisations (e.g., organisations of persons with disabilities) to assess availability, accessibility, affordability, acceptability and quality, as well as safety and security, independence, and autonomy.

However, the data indicate that most informants emphasised that they did not witness any involvement of organisations or groups of persons with disabilities, both in the design, planning, and development stages (Interview with Rachmita Harahap, a person with disabilities, 5 May 2021). Even though the design and planning stage of the Jakarta - Bandung High-Speed Railway did not involve organisations or groups of persons with disabilities, its planning is considered good in terms of inclusiveness-accessibility of public transportation by relevant stakeholders (Interview with Haris Muhammadun, chairperson of the Jakarta City Transportation Council. 7 May 2021).

This finding reinforces that justice considerations have traditionally played a marginal role in transportation decision-making, such as investment in a new infrastructure project. Considerations of justice accentuate the plight of disadvantaged populations and are, thus, aimed at furthering equality among groups and individuals, especially equality of opportunities and mobility (Bierbaum *et al.* 2021; Hananel and Berechman 2016; Juarez 2020).

Regarding facilities, the Jakarta - Bandung High-Speed Railway boasts several facilities that support ease of access for persons with disabilities (Interview with PT. KCIC, 09 June 2021). The train design was adopted from the Chinese and Indonesian railway regulations, namely the Ministry of Transportation Regulation No. 63/2019 and China Code Standard. These facilities include the 11 trains in 1 series, facilities accessible to persons with disabilities are especially provided in passenger carriage 4, i.e., the persons with reduced mobility (PRM) carriage.

There are 4 PRM special priority seats on each train. The PRM special toilet was designed to accommodate wheelchair users and has brailed features for the blind. The access door is a sliding door that PRM can pass through, and the wheel-chair storage area is available on passenger carrier 4 (Figure 1).

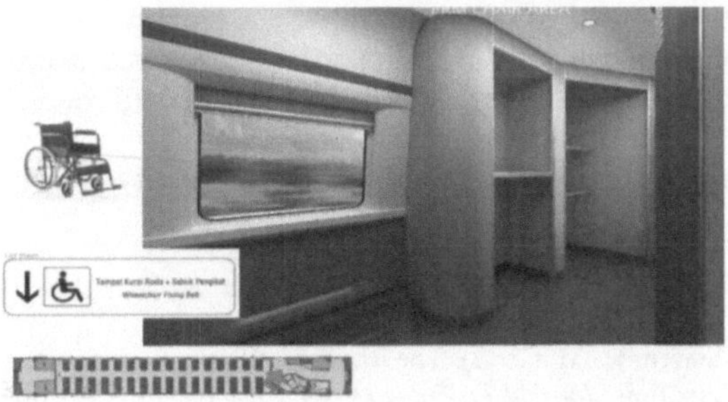

Figure 1. Storage area for wheelchairs on the Jakarta - Bandung high-speed railway.

Meanwhile, the forms of accessibility for persons with disabilities on the Jakarta - Bandung High-Speed Railway infrastructure (i.e., train stations) include special escalators and elevators for persons with disabilities, platform floors parallel to train doors, special officers assisting persons with disabilities, priority ticketing facilities through e-ticketing applications and counters that can accommodate groups of persons with disabilities, manual ticket checking door with larger space, special toilets for persons with disabilities, and special parking space (Interview with PT. KCIC, *Ibid.*)

The data generally indicate that persons with disabilities (PWDs) were not significantly involved in the design and planning process of the Jakarta-Bandung High-Speed Railway. Their lack of engagement in this earlier stage of the Jakarta-Bandung High-Speed Railway development is assumed to be associated with their different physical abilities, as theoretically framed by Sheller (2018). PWDs were prevented from participating during this earlier process since they were seen as having different physical abilities. There was an interaction between this lack of participation and the general policy system of public transport. However, participation at this earlier stage is crucial for granting their just participation and ensuring that the accessibility facilities properly meet their needs. This implies that their participation at this earlier stage would have paved the way to stronger mobility justice. The Jakarta-Bandung High-Speed Railway public transport has strengthened the mobility justice of persons with different physical abilities in Jakarta (Sheller 2018). Accordingly, all these accessibility facilities make it easier for them to move freely between Jakarta and Bandung.

Finally, it is also important to note that the mobility justice framework should be considered within other aspects or dimensions of development, not only public transport and differences in race, ethnicity, age, and gender. This can pertain to education, cycling, energy transition, pandemic, public services, tourism, youth, senior citizens, and women. Studies on these issues in relation to mobility justice have been conducted by a number of researchers in diverse countries as case studies (Bascom and Christensen 2017; Gillot and Rerat 2022; Harada *et al.* 2023; Martini 2024; Verlinghieri and Schwanen 2020; Sheller 2021).

4 CONCLUSION

The legal framework in Indonesia explicitly states that the rights of persons with disabilities are guaranteed in the constitution. However, policies relating to accessibility of public transportation services for persons with disabilities still need strengthening. There are still many obstacles that pose challenges in providing better accessibility to persons with disabilities, access to disability-related law enforcement, and negative stigmas against persons with disabilities. The involvement of persons with disabilities from the early stages of development is a strategic matter for ensuring that the facilities have met the principle of accessibility and followed the needs of persons with disabilities. More importantly, the involvement of persons with disabilities in design and planning cannot be delegated to non-disabled persons.

In addition, while the process of the design and planning stage of the Jakarta-Bandung High-Speed Railway does not guarantee the prerequisite of mobility justice since PWDs were not significantly engaged, the later stage of the development and provision of accessibility facilities was observed to be the opposite. Given the properly developed accessibility facilities of the Jakarta-Bandung High-Speed Railway, which are considered decent and well-developed, PWDs have greater space for moving between Jakarta and Bandung, which is similar to those with normal physical abilities, and this, thereby, grants mobility justice to PWDs.

REFERENCES

Bascom, G. W., and Christensen, K. M. (2017). The impacts of limited transportation access on persons with disabilities' social participation. *Journal of Transport and Health.* http://dx.doi.org/10.1016/j.jth.2017.10.002

Bierbaum, A. H., Karner, A., and Barajas, J. M. (2021). Toward mobility justice: Linking transportation and education equity in the context of school choice. *Journal of the American Planning Association*, 87(2).
Cook, N., and Butz, D. (2019). *Moving toward mobility justice*. In *Mobilities, Mobility Justice, and Social Justice*. New York: Routledge.
Everuss, L. (2019). *Mobility justice: A new means to examine and influence the politics of mobility. Applied Mobilities*, 4(1), 132–137. https://doi.org/10.1080/23800127.2019.1576489
Gillot, M., and Rerat, P. (2022). The plurinational cycling revolution in Santiago de Chile: Demands for mobility justice. *Applied Mobilities*, 9(1), 1–19. https://doi.org/10.1080/23800127.2022.2145078
Hananel, R., and Berechman, J. (2016). Justice and transportation decision-making: The capabilities approach. *Transport Policy Journal*, 42, 78–85.
Harada, T., Birtchnell, T., and Bo Du. (2023). The rush of the rush hour: Mobility justice for seniors on public transport in Sydney, Australia. *Social and Cultural Geography*, 24(2), 212–231. https://doi.org/10.1080/14649365.2021.1922735
Heaton, J. (2004). *Reworking qualitative data*. London: SAGE Publications.
Holstein, E. V., and Legacy, C. (2020). Mobility justice and accessible public transport networks for people with intellectual disability. *Applied Mobilities Journal, November 2020*. Taylor and Francis Group.
Juarez, Z. (2020). The movement towards mobility justice in Los Angeles: *Building a framework grounded in popular education and community knowledge*. Capstone Projects Publication, UCLA.
Kett, M., Cole, E., and Turner, J. (2020). Disability, mobility, and transport in low- and middle-income countries: A thematic review. *Sustainability Journal*, 12, 589. https://doi.org/10.3390/su12020589
Lindlof, T. R., and Taylor, B. C. (2002). *Qualitative communication research methods* (2nd ed.). Thousand Oaks, CA: SAGE Publications.
Martini, A. (2024). Geographies of mobility justice: Post-disaster tourism, recognition justice, and affect in Tohoku, Japan. *Mobilities Journal*, 19(3), 363–378. https://doi.org/10.1080/17450101.2023.2242002
Patton, M. Q. (2002). *Qualitative research and evaluation methods*. Thousand Oaks, CA: SAGE Publications.
Pereira, R., Schwanen, T., and Banister, D. (2016). Distributive justice and equity in transportation. *Transport Reviews Journal*. Retrieved May 12, 2024, from http://dx.doi.org/10.1080/01441647.2016.1257660
Romano, E. F. T. (2022). "*No one should feel like they're unsafe*": Mobility justice photovoice as a youth advocacy tool for equitable community mobility. *Family and Community Health Journal*, 45(4).
Sajib, S. H. (2022). Identifying barriers to the public transport accessibility for disabled people in Dhaka: A qualitative analysis. *Transactions on Transport Sciences Journal*, 1, March.
Sheller, M. (2018). Theorizing justice mobility. *Tempo Social, Revista de Sociologia da USP*, 30(2).
Sheller, M. (2021). Mobility justice and the return of tourism after the pandemic. *Mondes du Tourisme Journal*, 19.
Thorne, S. E. (1994). Secondary analysis in qualitative research: Issues and implications. In Morse, J. M. (Ed.), *Critical issues in qualitative research methods*. London: SAGE Publications.
United Nations. (2016). *Leaving no one behind: The imperative of inclusive development*. Retrieved May 29, 2021, from https://smeru.or.id/sites/default/files/publication/wp_disabilitas_in_0.pdf
Verlinghieri, E., and Schwanen, T. (2020). Transport and mobility justice: Evolving discussions. *Journal of Transport Geography*, 87.

Towards Resilient Societies: The Synergy of Religion, Education, Health, Science, and Technology – Rahiem (Ed.)

Silent treatment in conflict resolution: Communication and uncertainty management for sustainability

Musfiah Saidah
Universitas Islam Negeri Syarif Hidayatullah Jakarta, Indonesia

Ashlaha Baladina Zaimudin
University Of Leeds, United Kingdom

Isna Rahmawati, Fauziah Muslimah, Siti Nurbaya & Mahjoba Hasina Wardak
Universitas Islam Negeri Syarif Hidayatullah Jakarta, Indonesia

ABSTRACT: Communication is a key strategy for mitigating conflict and fostering peaceful interactions, aligning directly with Sustainable Development Goal (SDG) 16. However, in certain contexts, communication may be temporarily avoided through "silent treatment," a behaviour characterised by taking intentional distance to facilitate more effective communication later. This study investigates how silent treatment impacts conflict resolution and uncertainty reduction in millennial relationships, employing qualitative interviews to examine its effectiveness and implications. The findings reveal that silent treatment, when used thoughtfully, can serve as a tactical approach to managing interpersonal conflicts. It allows individuals to regulate emotions, reduce stress, and promote emotional well-being, thus supporting SDG 3 (Good Health and Well-being). Furthermore, it contributes to SDG 16 by fostering respect and promoting peaceful interactions. However, the study underscores the critical importance of follow-up communication to fully address underlying tensions and restore relational harmony. By highlighting both the benefits and limitations of silent treatment, the research offers insights into its role in conflict management and its contribution to creating more peaceful and inclusive environments as envisioned by the SDGs.

Keywords: communication, conflict, consideration, silent treatment, uncertainty

1 INTRODUCTION

Effective communication is essential for good message reception, but disorders like silent treatment can disrupt it, leading to neglect and poor relationship quality. To address this, people seek to reduce uncertainty in communication. Communication, defined by Rogers, is the exchange of messages aimed at behaviour change, requiring cooperation for mutual understanding (Genc 2017). Communication fulfils human needs, aligns interests, and fosters relationships (Dyatmika 2022). Vederber highlights its social function in fulfilling desires for pleasure and relationship-building. Cooley's looking glass self suggests self-concept is shaped by others' perceived thoughts, further supported by studies on self-awareness stimuli (Milyane 2022). Miscommunication, like silent treatment, can lead to conflict (Zolbanin 2021, May 11).

Effective conflict resolution depends on strategies like active listening and clear messaging. The silent treatment, a form of miscommunication, involves deliberate indifference through behaviours like avoiding eye contact, refusing to speak or ignoring others (Agarwal 2022). It is often used to express disapproval, manipulate, or punish, disrupting healthy communication and causing anxiety, stress, and power imbalances, especially in gendered contexts (Wright C. N. 2009). Research links this behaviour to negative outcomes such as heightened relational tension, hindered conflict resolution, and harm to mental well-being, aligning it with challenges addressed by SDGs, including good health (SDG 3), gender equality (SDG 5), reduced inequalities (SDG 10), and peace and justice (SDG 16).

 DOI: 10.1201/9781003645542-17

Traditional gender roles further perpetuate the silent treatment as a negative communication pattern. Addressing this issue requires promoting empathetic and open communication to foster equality and improve relational dynamics. Silent treatment threatens three basic human needs: self-esteem, control, and meaningful existence. This behaviour creates a sense of rejection, loss of control, and feelings of insignificance. Individuals who receive this treatment are more likely to experience confusion, isolation, and mental health issues. In the relationship, silent treatment is often used as a control mechanism by those with low self-esteem and tolerance for conflict, which can ultimately worsen the quality of the relationship and trigger an aggressive response from the victim (Agarwal 2022). However, silent treatment can also have a positive impact by providing individuals with needed space outside of social interaction (He 2022). While the silent treatment is predominantly viewed negatively, it can also provide individuals with necessary space away from social interactions, allowing for personal reflection (Wright 2015).

Neglect in silent treatment creates communication uncertainty by ignoring the behaviour, such as avoiding eye contact. Uncertainty Reduction Theory (URT) explores how people reduce this uncertainty through communication, particularly in relationships (Knobloch 2015). Direct communication is a common strategy, as seen in romantic relationships, to manage uncertainty (Afifi 2004). URT also highlights the importance of verbal and nonverbal cues, as well as computer-mediated communication, in reducing uncertainty (Antheunis 2010). Berger and Calabrese's URT emphasises systematic planning to manage uncertainty (Fatwasuci 2021). Several studies have explored the topic. Dina S.P. highlights URT as a tool for improving communication sensitivity. Mark V.R. defines "environment" in URT as the interaction setting. Yusuf Yuksel suggests building common perceptions through identity in communication. Walid A. Afifi discusses how people manage uncertainty by seeking or avoiding information. Berger and Calabrese note that symbolic interactions often fail to reduce uncertainty effectively. Despite this, few studies have examined the benefits of silent treatment in conflict situations in the media era (Callabrate 2001). Further research is needed to comprehensively understand the role of silent treatment in contemporary communication dynamics. Self-disclosure is sufficiently associated with an individual's intrinsic motivation to see how the individual expresses himself to others (Latifa 2019).

2 METHODOLOGY

This research is a research with a constructivist paradigm. Constructivists state that humans do not obtain or discover knowledge but construct and shape it. This study employs a qualitative research approach to explore the use of silent treatment in relationship conflicts, as it allows for in-depth examination of social and humanitarian issues through direct interviews (Creswell 2003). The research is descriptive, aiming to provide a detailed account of the situation, social settings, and relationships involved. A case study strategy is utilised to investigate contemporary phenomena in depth and in real-world contexts, particularly where the boundaries between phenomena and context are unclear (Neuman 2003). Data collection techniques ensure high-quality data is gathered and stored properly. This qualitative study explores the use of silent treatment in relationship conflicts through an in-depth examination of social dynamics using direct interviews (Creswell 2003). Employing a descriptive and case study approach, it investigates contemporary phenomena within real-world contexts, particularly where boundaries between phenomena and context are unclear (Neuman 2003). Primary data were collected via in-depth interviews, while secondary data came from books, journals, and scientific articles. Participants in this study were millennials aged 20-25 who have experienced silent treatment in their relationships. Inclusion criteria required participants to have firsthand experience with silent treatment as a conflict resolution strategy were analysed thematically, involving transcription, repeated readings, and identifying key themes to understand the issue comprehensively.

3 RESULTS AND DISCUSSION

3.1 *The use of silent treatment in relationship conflict*

Silent treatment involves choosing silence in a relationship, disrupting communication. Although brief silences can interrupt conversations, communication is essential for meeting social needs.

The silent treatment is often used as a colloquial term for social exclusion. Emotions and miscommunication often cause this silence, but individuals may eventually break their silence to preserve the relationship (Koudenburg 2011).

> *"Actually, I feel bad if they do feel guilty because I know how it feels like to be ignored without knowing what we did wrong so sometimes I (give) silent (treatment) because I am simply not in the mood." (AA, personal interview).*

Meanwhile, the second source revealed that in the use of silent treatment, they will see the opponent of communication and be associated with the power they have.

> *"It depends on the power (relation), if it's with someone who has more power (than me) then I avoid to resolve (a problem) by giving the silent treatment. However, if it's with someone closer like with a partner then I'll be silent as an effort to improve the relationship. It is not really about finding out who's guilty." (KF, personal interview).*

Based on the information by informant 2, it is known that there is a degree of consideration toward the level of closeness and power in the use of silent treatment in a relationship. This is part of the preparation of a plan in communicating in a relationship because the more meaningful a relationship is, the more a person will try to maintain the relationship. Meanwhile, Informant 4 chose silent treatment as a temporary effort to solve the problem they had. *"It (silent treatment) is actually not a good way (to solve a problem). However, sometimes being silent can give space to each party. It also gives us time to reduce the possibility of a bigger conflict." (MZ, personal interview).*

Based on the information from Informant 4, it can be said that silent treatment is an effort to resolve conflicts that can be chosen, especially in the short term. According to them, silent treatment can be considered as a 'cool down'. During the silent treatment, both parties can think more clearly. In addition, silent treatment is a step to resolve conflicts in the early stages of a relationship.

Then, Informant 3 said that silent treatment is the safest way to end conflicts. According to them, this step can be started by one party and then followed by the other.

> *"Sometimes this (silent treatment) is the best way. When we are angry and in no position to remain calm, it is better to be silent first. To me, being silent can gradually reduce the conflict. After that, the two parties can continue their relationship as usual (as it was before)." (MS, personal interview).*

Informant 3 explains that the silent treatment serves as a signal to pause a conflict but emphasises the importance of setting limits to allow relationship restoration. Insights from the four sources reveal that while silent treatment can temporarily reduce tension, it is not universally effective. For instance, workplace conflicts with superiors require proactive approaches like coordination and negotiation instead of withdrawal. Silent treatment has both benefits and drawbacks. On the positive side, it provides space for emotional processing, potentially enabling constructive dialogue when communication resumes. However, prolonged silence can lead to resentment, isolation, and misunderstandings, worsening conflicts. Fostering an environment that encourages dialogue helps minimise risks, creating a more balanced approach to conflict resolution.

3.2 *Effects of silent treatment in problem-solving*

Everyone has a reason for using silent treatment to resolve relationship conflicts, but as Berger notes, even well-thought-out plans can fail, as intentions may be misinterpreted (Griffin 2019). Silent treatment may be effective for addressing temporary conflicts, but its long-term use plays a critical role in the sustainability of the relationship. Ultimately, the goal is to maintain and repair relationships.

> *"I am the type of person who is silent when I am angry so that (I can) cool down and (my anger) doesn't drag on, so people won't speculate.". (AA, personal interview).*

During the silent process, they review the events that caused the problem and become introspective about their attitude towards the situation. The emotions from these intersecting forces shape our interpretation. Authors can't foresee every question, so the unspoken elements often reflect the hopes and doubts that enhance the story's meaning. Silence, then, is used as a mediation to help one or both partners, especially those whose emotions are more attached to the problem, to have time to reflect on their own position (Oduro 2011). They will also try to think from the target's point of view to understand whether they are doing the right thing or not. In addition, when staying silent, individuals tend to consider how the relationship should continue, as stated by informant 2.

> *"Say if there was a problem, then a person expressed their emotions. Well, I actually really wanted to be angry as well but it will ruin (the problem) even more. I do introspection and find solutions by asking (my) other friends." (KF, Personal interview).*

In the time of silence, there is the potential for uncomfortable circumstances. However, this needs to be passed as a phase of reactive results, for example how a person behaves towards others. The silent phase is also an individual's reflection on emotional responses when a conflict occurs. This action aims to reduce the threat of uncertainty.

Meanwhile, Informant 4 concluded that silent treatment is the effect of the inability of both parties to resolve the conflict. According to them, silent treatment is a transition that cannot be positioned as a solution. The reason is that the silent treatment does not lead both parties to the cause and effect of the conflict. Therefore, although it can be chosen as an alternative, silent treatment is only a temporary alternative. As the opinion of informant 4 stated below:

> *"In my opinion, silent treatment occurs when both parties do not understand the conflict that is going on. So, that's basically it and (sometimes) just being silent is because there are no other options for the time being." (MZ, Personal interview).*

Therefore, it can be said that silent treatment is an effect that arises from the helplessness of the parties involved in communication. Silent treatment is generally chosen as the first step to provide peace of mind for oneself. However, silent treatment can also be the limit of the end or continuation of a relationship. Meanwhile, self-disclosure and visible signs of anxiety serve as cues that underpin others' evaluations of similarity (Ilma 2019). Of course, being upset with someone does not always translate into conflict, but the literature on conflict is useful in showing that when we actively pay attention to the things that upset us in our relationships, we do have a greater chance of maintaining and improving individual and relational well-being (Rittenour 2019).

3.3 *Uncertainty reduction: communicated displeasure, confession of disappointment, and silence*

Silent treatment is considered to be anger which is expressed with the act of choosing silence as punishment for others. If the target does not get close to the participant or if the participant is not satisfied with the target's effort then the relationship ends. People use silence in various relationships for various reasons and purposes such as time limits, punishment, manipulation, defense, and for rejection (Agarwal 2022). If examined further, the silent treatment is a type of behavioural uncertainty. In this position, there is miscommunication and a feeling of uncertainty that needs to be confirmed immediately. Silent treatment goes against the human tendency to respond and communicate with others and therefore requires more effort to stop an improper situation. At this stage, it is necessary to reduce uncertainty by starting conversations. Based on the informant's statement, the main reason for finally starting to express is when the situation is deemed to be calm. As stated by Informant 1:

> *"The reason to communicate again is because (we) feel that the temperature is back to zero. Cooled down already. Because if (we were) angry (we) became unstable. However, if (we're) calmed down then it is good to talk and smooth things up. Like that was (the problem before) is actually like this and that." (AA, personal interview).*

This is in line with the information given by informant 2. Unpleasant things will be attempted to be communicated when the situation feels better. Informant 2 will still choose to express his frustration.

> *"I believe that when we talk in a comfortable condition like asking them to go to a restaurant, that can make them more willing to talk. So it is good to make the atmosphere as comfortable as possible. After that, (you can) express your grievances, anger, or disappointment but then again not to find out who's wrong or guilty. " (KF, Personal interview).*

As Berger argues, interpersonal communication is the main means of reducing uncertainty. Because the communication process that is finally established is an effort to improve relationships and reduce all uncertainties through conversation. According to informant 3, in the end, silent treatment is an 'introduction' to the next stage. According to them, the period after the silent treatment is carried out, both parties can communicate further, especially in finding steps to solve problems.

> *"Yes, after silent treatment, we do have to communicate again. Slowly but surely starting with (asking) 'why' the previous problems could occur and 'how' are the solutions and what can be done in the future." (MS, Personal interview).*

After silent treatment is deemed to be done, the next step that can be considered is to reopen communication. The opinion of informant 3 goes in line with informant 2 who chooses communication as a reduction of uncertainty due to the silent treatment that occurs. When communication is re-established, the two parties will usually agree on several things that are considered conflicts and then resolve them with an agreement that also comes from both parties. This is in line with the opinion of informant 3 regarding the role of communication in silent treatment. Furthermore, according to informant 4, communication after silent treatment should be done in the form of verbal communication.

The consensus among four informants highlights that communication is vital for reducing uncertainty caused by silent treatment. This process is inherently emotional, encouraging both parties to express feelings and perspectives openly. Emotional reactions to a situation indicate a personal response to disapproval. An interactive communication strategy, where both individuals exchange as much information as possible, effectively reduces relational uncertainty. In modern contexts, including digital interactions, clear and direct communication—via video calls, voice messages, or well-crafted written texts—is essential, particularly when non-verbal cues are limited. Emphasising clarity and emotional expression minimises misunderstandings and fosters better conflict resolution. The findings suggest practical applications in education and workplace settings. Educational programs could incorporate training on effective conflict resolution, such as simulations to practice reopening communication. Similarly, workplaces can implement open communication policies and training on emotional intelligence to enhance dialogue and reduce relational tensions. When viewed through the lens of Uncertainty Reduction Theory, the silent treatment can temporarily diffuse tension but requires clear boundaries and open communication to restore relationships. Integrating these insights with other conflict resolution methods provides a deeper understanding of managing interpersonal conflicts.

4 CONCLUSION

The findings of this research highlight that while silent treatment is often employed to manage relationship dynamics, it can ultimately exacerbate conflicts if underlying issues remain unaddressed. This is particularly relevant in the context of the Sustainable Development Goals (SDGs), especially SDG 3 (Health and Well-being) and SDG 16 (Peace, Justice, and Strong Institutions). Effective communication emerges as a crucial element for conflict resolution and the promotion of mental well-being. Although the silent treatment may offer temporary respite for reflection, the lack of constructive dialogue escalates tension and uncertainty. To counteract the adverse effects of the silent treatment, the study proposes several actionable strategies. First, implementing communication training workshops can enhance interpersonal skills, enabling individuals to articulate their feelings and concerns more effectively, thereby reducing reliance on silence as a conflict resolution tactic. Second, counselling interventions are recommended to create a safe environment for individuals or couples to discuss conflicts and learn healthier communication techniques under professional guidance, fostering better emotional navigation and mutual understanding. Lastly, developing community-based conflict resolution programs can equip individuals with essential skills that encourage open dialogue and diminish the likelihood of resorting to silence during disputes. Fostering open and inclusive communication is essential for developing sustainable solutions and maintaining healthy, peaceful relationships. The insights from this study provide a valuable reference for enhancing interpersonal communication strategies, contributing to broader efforts to cultivate a more inclusive and harmonious society.

BIBLIOGRAPHY

Afifi. (2004). Toward a Theory of Motivated Information Management. *Communication Theory.*

Agarwal. (2022). When Silence Speaks: Exploring Reasons of Silent Treatment from Perspective of Source. *International Journal of Trend in Scientific Research and Development (IJTSRD).*

Antheunis. (2010). Getting acquainted through social network sites: Testing a model of online uncertainty reduction and social attraction. *Computers in Human Behavior.*

Callabrate. (2001). Initial interactions between strangers are characterized by information seeking in order to reduce uncertainty. *Uncertainty is reduced as levels of self-disclosure, nonverbal warmth, and similarity increas.*

Creswell. (2003). Chapter One, "A Framework for Design." *Research Design Qualitative Quantitative and Mixed Methods Approaches.*

Dyatmika. (2022). *Textbook: Introduction to Communication Sciences.*

Fatwasuci. (2021). Development of the Application of Uncertainty Reduction Theory in Face-to-Face Interaction and Online Interaction. *Indonesia Scientific Journal.*

Genc, R. (2017). The Importance of Communication in Sustainability & Sustainable Strategies. *Procedia Manufacturing.*

Griffin. (2019). *The first Look of Communication behaviour.* New York: McGraw-Hill Education.

He. (2022). Exploring the role of communication in missed nursing care: A systematic review. *Journal of Advanced Nursing.*

Ilma. (2019). *Social Anxiety on Instagram Second Account Use.*

Knobloch. (2015). Uncertainty Reduction Theory. *The International Encyclopedia of Interpersonal Communication.*

Koudenburg, N. P. (2011). Disrupting the flow: How brief silences in group conversations affect social needs. *Journal of Experimental Social Psychology.*

Latifa, R. (2019). Self-Disclosure in Indonesia Social Media User: What Leads People to Self-Disclosure in Social Media? *Journal of Engineering and Applied Sciences.*

Milyane. (2022). *Introduction to Communication Science.*

Neuman. (2003). *Social Research Methods: Qualitative and Quantitative Approaches.* Boston: Allyn & Bacon.

O'Grady. (2015). Silence: Because What's Missing Is Too Absent To Ignore. *Journal of Societal.*

Oduro. (2011). Semiotic silence in intimate relationships. *Much silence makes a powerful.*

Rittenour. (2019). Socializing the Silent Treatment: Parent and Adult Child Communicated Displeasure, Identification, and Satisfaction. Socializing the Silent Treatment: Parent and Adult Child Communicated Displeasure, Identification, and Satisfaction. *Journal of Family Communication.*

Wright. (2015). You shouldjust knowwhy i'm upset: expectancy violation theory and the influence of mind reading expectations (mre) on responses to relational problems. *Communication Research Reports.*

Wright, C. N. (2009). Relational Commitment and the Silent Treatment. *Communication Research Report.*

Zolbanin. (2021, May 11). *Miscommunication in the age of communication: A crowdsourcing framework for symptom surveillance at the time of pandemics.*

Towards Resilient Societies: The Synergy of Religion, Education, Health, Science, and Technology – Rahiem (Ed.)

Exploring early marriage and its impact on stunting in West Java, Indonesia

Narila Mutia Nasir
Universitas Islam Negeri Syarif Hidayatullah Jakarta, Indonesia
ORCID ID: 0000-0001-6116-5835

Raihana Nadra Alkaff
Universitas Islam Negeri Syarif Hidayatullah Jakarta, Indonesia
ORCID ID: 0000-0003-2235-3002

Ai Juariah
Badan Penelitian dan Pengembangan Daerah Provinsi Jawa Barat, Indonesia

ABSTRACT: Stunting remains a significant public health issue in Indonesia, with a prevalence of 21.5% in 2023. Early marriage has been identified as a potential risk factor for stunting. This study aims to explore adolescents' knowledge of stunting within the context of early marriage. Using a qualitative phenomenological approach, the research involved adolescents who entered early marriages and have stunted children, as well as their parents. Data were collected through in-depth interviews and focus group discussions and analysed using content analysis. Findings reveal that adolescents in early marriages lack an understanding of its definition, consequences, and the prevention of stunting. This limited knowledge contributes to higher stunting rates by affecting maternal health during pregnancy and the provision of adequate nutrition for infants. Comprehensive interventions are urgently needed to reduce early marriage rates. Furthermore, education on the connection between early marriage and stunting should be consistently delivered to all community groups. Cross-sector collaboration is essential to support these efforts and achieve the Sustainable Development Goals, particularly those related to health and well-being.

Keywords: adolescent, early marriage, Indonesia, stunting, qualitative

1 INTRODUCTION

Early marriage is a complex socio-cultural problem in Indonesia that indirectly has an impact on public health issues. The phenomenon of early marriage- marriage under the age of 19 - becomes a concern in Indonesia since it is found to correlate with the Maternal Mortality Rate (MMR) and Infant Mortality Rate (IMR) in Indonesia (Devi *et al.* 2022; The National Development Planning Agency 2020). In the global context, mortality should be reduced to achieve the target goal of Sustainable Development Goal (SDG) 3 (good health and well-being) (United Nations 2023). Besides mortality, the other important issue related to early marriage is stunting. According to the World Health Organization (WHO), stunting is a condition where the growth and development of children is impaired due to poor nutrition, repeated infections, and inadequate psychosocial stimuli. A child is defined as stunted if their height exceeds two standard deviations below the WHO Child Growth Standards median (WHO 2014). Previous studies show that early marriage is one of the determinants that might cause stunting among children in Indonesia (Ashar *et al.* 2024; Fuada *et al.* 2020).

 DOI: 10.1201/9781003645542-18

Factors associated with stunting are poor maternal health and nutrition, feeding practices, especially complementary foods, and infections. Maternal health before, during, and after pregnancy is crucial since it influences the early growth and development of children from the beginning phase of life. Maternal health factors including low height, too close to childbirth interval, and being too young for childbirth might contribute to the incidence of stunting. Adolescent pregnancy, which usually occurs following early marriage, can interfere with the fetus's growth due to competition with the teenage mother's growth.

Optimal maternal health must be supported by the mother's physical and mental condition and also adequate family support. According to several studies, maternal health conditions cannot be optimal because of early marriage (Wemakor *et al.* 2018). Research in India also states that the incidence of stunting is greater among children whose mothers had an early marriage compared to those who get married >18 years of age (Raj *et al.* 2010). Cultural norms and socioeconomic factors perpetuate the practice of early marriage. Poverty drives early marriage because the parents think it would reduce the financial burden of the family. However, this early marriage often leads to early pregnancy, which contributes to stunting. West Java is one of the provinces in Indonesia that has many early marriages and stunting. The previous studies analysing early marriage and stunting in this province are still limited. Therefore, the objective of this study is to explore early marriage and knowledge on stunting among adolescents who had an early marriage.

2 METHODS

2.1 *Study design*

This study used a qualitative method with a phenomenological approach. This approach is suitable for exploring the phenomena that occur in society more deeply and analysing the causal factors of the phenomena. Using qualitative methods provides the opportunity to understand the practice of early marriage. In this study, the phenomenon to be analysed is the phenomenon of early marriage and the incidence of stunting in the community in the research location in West Java.

2.2 *Data collection*

The data were collected through in-depth interviews and Focus Group Discussions (FGD). As for the main informant, the interviews and FGD were conducted among the adolescents who had an early marriage (marriage before the age of 19) and had a stunting child. The other data were obtained from the parents of adolescents who had early marriages and were involved in FGDs as the supporting informants. Data triangulation was conducted using two methods (in-depth interview and FGDs) and source informants (the adolescent and parent).

2.3 *Data management and analysis*

Data were stored in encrypted digital formats on secure servers, with access restricted to authorised personnel only. Data were anonymised before analysis to protect privacy, and regular backups were performed to prevent data loss. The data analysis was performed using content analysis, where the research results are studied one by one from all the results of interviews and discussions with informants and then analysed to get a complete description.

2.4 *Ethical consideration*

The ethical approval of this study was obtained from the Research Ethics Committee of the Faculty of Health Sciences, Syarif Hidayatullah State Islamic University Jakarta. Informed consent was received from all informants, with assurances of confidentiality, the voluntary nature of participation, and the right to withdraw from the study without any negative consequences.

3 FINDING AND DISCUSSION

Two in-depth interviews and two Focus Group Discussions (FGDs) were conducted in this study. The researchers interviewed 2 adolescents who had early marriages and had stunted children. Two FGDs were performed among the adolescents who had early marriage and had stunting children. Meanwhile, the data from parents of adolescents who had early marriages were obtained from one FGD.

3.1 *Characteristics of informants*

The age of the early married adolescent as the main informants range from 17 to 30 years old and their husband's age is between 20 to 34 years old. The age of informants when they got married ranged from 12 to 18 years old, while their husband's ages at that time were 18 to 42 years old. Meanwhile, according to the Indonesia National Population and Family Planning Board, the ideal marriage age is 21 for women and 25 for men. Most informants' ages when they first gave birth were under 18 years old (13-18 years old; only 1 person gave birth at 19 years old). The national data in Indonesia in 2018 indicated that 1 of 9 girls aged under 18 years old were married. It is estimated that about 1.2 million girls had early marriages in Indonesia, and it places the country as one of the countries with a high number of early marriages in the world (UNICEF *et al.* 2020).

The highest level of education of the informants was vocational school, while the lowest level of education was elementary school. The number of early marriages among adolescents mostly with low education (Marphatia *et al.* 2023). The majority of primary informants who participated in the FGDs were pregnant, and some of them had Chronic Energy Deficiency (CED) conditions. Almost all informants had no work. Only a few of them were helping the family financially with irregular jobs such as selling at the market or receiving food orders (catering). Almost all informants' husbands had work. In another study, it was found that women who married early had fewer prenatal check-ups and were less likely to receive standard checks during their pregnancy. In addition, women who had early marriage had lower labor force participation and were less likely to work in the formal sector for both women and men who had early marriage (Cameron *et al.* 2023).

3.2 *Attitude and experience toward early marriage*

The results of the study found that of the two informants from the in-depth interview stated that they got married without any coercion. However, they actually hoped not to get married early. They said that they wanted to get married at the age of 20 or 21 years old. This result is in line with the research conducted among adolescents in the Pakulon area which showed that 77.8% of adolescents understand that early marriage should be avoided (Yastirin 2019).

Most informants who had early marriages did not get any counselling related to marriage before they got married. Counselling related to marriage is important to do before marriage so the adolescent gets an understanding and knowledge of the impact of early marriage (Aggarwal *et al.* 2023). Parents play an important role in early marriage by doing the matchmaking and approving the marriage proposal. After their children get married, parents still have roles in their children's households, such as taking care of children, participating in the household costs if the marriage children still live together in their house, and a place to return to if their married children quarrel with their husband. In this study, the parents considered that early marriage has a positive side, especially in preventing premarital sex. In another study, it was also found that one of the factors for early marriage was parents who encouraged and supported children to marry at an early age (Kok *et al.* 2023; Wibowo *et al.* 2021). The previous study in Cianjur, West Java also revealed that parents determined the practice of early marriage for economic and religious dogma (Sopyan *et al.* 2023).

Generally, the main informants did not understand the meaning of early marriage and its consequences. They only mentioned the definition literally and did not understand the age that was categorised as early marriage. This opinion arises because early marriage is perpetuated as part of cultural practice. However, their lack of understanding about this issue might imply they do not have the desire to do it. This condition could be considered a forced marriage (Fauzi and Ismayawati 2022). They perceived that getting married at a young age was a positive thing. The positive impacts of early marriage mentioned by the informants were reducing the burden on their parents, their husbands being responsible for their lives, and their belief they are getting healthier. In fact, they have faced various problems after the early marriage, mainly related to economic problems. They also explained that adapting to the conditions of marriage, especially with their spouse, was a problem. The informant felt that they did not have friends in their age who also had children. In addition, the adolescents who had early marriages expressed that after marriage, they are really dependent on their husbands, and they must ask permission from their husbands to do some things. This situation described the phenomenon of gender inequality that conflicts with the target of SDG to achieve gender equality and empower all women and girls.

Based on this research data, the culture of early marriage has become common and usual among the communities in the research area. For parents, marrying off their children at a young age is also to maintain the tradition of marrying at a young age in the community. This tradition has occurred in the West Java region and other areas of Indonesia, such as in Kudus (Indanah *et al.* 2020), West Nusa Tenggara (Rahiem 2021), and South and Central Sulawesi (Ratnaningsih *et al.* 2022). Indanah *et al.* (2020) also found that culture in the family is associated with early marriage. The other factor that might encourage early marriage is that most informant have no intention to continue their education to a higher level. This is because the higher education school is located outside the village, which requires high transportation costs. It is also a burden for a family with low socioeconomic levels to let their children go to the highest level of education, such as senior high school or vocational school. This result aligns with a previous study that showed the relationship between educational background and early marriage (Pourtaheri *et al.* 2023).

Economic factors were a trigger for early marriage in this study. The phenomenon of early marriage in the research area is a recurring cycle closely related to the low economic status of the informant. The parents thought they could give the responsibility to the children's husbands if their children married early. However, some of the marriages unfortunately ended with divorce, and then the children went back to their homes. This situation causes an economic burden for the parents. Early marriage and school drop-out contribute to the poverty in the family (Schlecht *et al.* 2013).

3.3 *Early marriage and its relationship with stunting*

The results of this study described that the main informant (an adolescent who had early marriage) did not know the prevention of stunting. They thought that stunting could only be prevented by feeding their children. It is a concern because the other study explained that good maternal knowledge is related to maternal behaviour to prevent stunting (Halimatunnisa *et al.* 2020). This study also revealed that the main informant did not get comprehensive information about stunting or malnutrition. Some informants stated they ever got information about malnutrition, complementary feeding, vitamins, and eating vegetables and fruits, but the explanation was incomplete. They received health information either from doctors, midwives, the Public Health Center (*Puskesmas*), integrated health post (*Posyandu*), or their mothers. This lack of information has led to the low knowledge of the main informant about stunting. Furthermore, the ability to receive and understand information is generally owned by highly educated people, while adolescent who had early marriages usually do not continue their education to a higher level (Kuswanto *et al.* 2024). The fact that low education level, poverty, early marriage, and stunting are reinforced factors each other. Those with a low level of education will have lower opportunities to increase their economic status and make them remain in poverty. They might think early marriage can help them out from poverty. This early marriage leads to early pregnancy which can increase the risk of stunting. This is the cyclical relationship that should be ended.

Based on FGD, the main informants stated that stunting is stunted, failure to grow and malnutrition. One informant said, "a stunting child will be slow in thinking, it's horrible...".

This shows that the informant who had early married general information about stunting. Although their information about stunting is limited, they were aware that stunting children must visit integrated health post (*Posyandu*). The informant also know that to prevent stunting the child should get nutritious food, vegetables and fruits.

The informant had poor knowledge of the influence of early marriage on the incidence of stunting. They only know that low nutritional intake or poor care of children can cause stunting. They did not think that paying attention to nutrient intake should be done during the pregnancy period to prevent stunting. If the early married mother is giving birth, the child has a higher risk of dying before the age of five years old, being underweight and stunting (Cameron *et al.* 2023).

This study revealed that the informant wanted to delay pregnancy after their early marriage, but in reality they did not use any contraception method. In this case, the informant then became pregnant at a very young age, although they were actually not yet ready physically and mentally to have a child. Those who had early marriage might not have optimal health during pregnancy and put the baby at risk (Fan and Koski 2022). Adolescent pregnancy causes competition in the absorption of nutrients between the mother and the fetus. As a result, the baby is born with low birth weight and has a higher risk of stunting (Hendrawati *et al.* 2023; Wemakor *et al.* 2018).

4 CONCLUSION

Early marriage is a common practice in West Java. Factors that encourage early marriage are cultural factors, the role of parents, and economic status. Parents consider early marriage as shifting responsibility to the children's husband to reduce the economic burden of the family, although it is not always as the expectation. The adolescents who had early marriage mostly have poor knowledge about the risk of stunting and its relationship with early marriage. They also lack information on stunting prevention. The low level of education and economic status have increased the risk of stunting among informants. Thus, it is necessary to educate the community to be more knowledgeable about early marriage and its impact on health, particularly stunting. Reducing early marriage through education and economic empowerment programs is critical for breaking the cycle of poverty and stunting in West Java. A comprehensive intervention should be implemented to decrease the number of early marriages, which might reduce the number of stunting. Interventions should focus on cross-sector collaboration to educate communities about the long-term consequences of early marriage and stunting. It can also support the achievement of the target of Sustainable Development Goals, particularly in ensuring good health and well-being for all, achieving gender equality, and ending poverty.

REFERENCES

Aggarwal, S., Francis, K. L., Dashti, S. G., and Patton, G. (2023). Child marriage and the mental health of adolescent girls: A longitudinal cohort study from Uttar Pradesh and Bihar, India. *The Lancet Regional Health–Southeast Asia*, 8. https://doi.org/10.1016/j.lansea.2022.100102

Ashar, H., Laksono, A. D., Supadmi, S., Kusumawardani, H. D., Yunitawati, D., Purwoko, S., and Khairunnisa, M. (2024). Factors related to stunting in children under 2 years old in the Papua, Indonesia: Does the type of residence matter? *Saudi Medical Journal*, 45(3), 273–278. https://doi.org/10.15537/smj.2024.45.3.20230774

Cameron, L., Contreras Suarez, D., and Wieczkiewicz, S. (2023). Child marriage: Using the Indonesian family life survey to examine the lives of women and men who married at an early age. *Review of Economics of the Household*, 21(3), 725–756. https://doi.org/10.1007/s11150-022-09616-8

Devi, Y. P., Ekoriano, M., Sari, D. P., and Muthmainnah, M. (2022). Factors associated with adolescent birth in Indonesia: A national survey. *Rural and Remote Health*, 22(2). https://doi.org/10.22605/RRH7219

Fan, S., and Koski, A. (2022). The health consequences of child marriage: A systematic review of the evidence. *BMC Public Health*, 22(1), 309. https://doi.org/10.1186/s12889-022-12707-x

Fauzi, I., and Ismayawati, A. (2022). Child Marriage in Indonesia: Sexual Violence or Not?. *De Jure: Jurnal Hukum dan Syar'iah*, 14(2), 288–303. doi:https://doi.org/10.18860/j-fsh.v14i2.17850

Fuada, N., Latifah, L., Yunitawat, D., and Ashar, H. (2020). Assessment of Nutritional Status of Children Under-Five in Families of Adolescent Mothers in Indonesia 2013. *Journal of Nutritional Science and Vitaminology*, 66(Supplement), S425–S431. https://doi.org/10.3177/jnsv.66.S425

Halimatunnisa, M., Indarwati, R., Ubudiyah, M., Trihartuty, Putri, N., Sari, M., and Suhardin, S. (2020). Family Determinants of Stunting in Indonesia: A Systematic Review. *International Journal of Psychosocial Rehabilitation*, 24, 815–822. https://doi.org/10.37200/IJPR/V24I9/PR290099

Hendrawati, L. D., Santoso, P., and Koesbardiati, T. (2023). Early marriage is one of the causes of stunting in Bondowoso Regency. *Jurnal Antropologi: Isu-Isu Sosial Budaya*, 25(2), Article 2. https://doi.org/10.25077/jantro.v25.n2.p194-200.2023

Indanah, I., Faridah, U., Sa'adah, M., Sa'diyah, S. H., Aini, S. M., and Apriliya, R. (2020). FAKTOR YANG BERHUBUNGAN DENGAN PERNIKAHAN DINI. *Jurnal Ilmu Keperawatan Dan Kebidanan*, 11(2), Article 2. https://doi.org/10.26751/jikk.v11i2.796

Kok, M. C., Kakal, T., Kassegne, A. B., Hidayana, I. M., Munthali, A., Menon, J. A., Pires, P., Gitau, T., and van der Kwaak, A. (2023). Drivers of child marriage in specific settings of Ethiopia, Indonesia, Kenya, Malawi, Mozambique and Zambia – findings from the Yes I Do! Baseline study. *BMC Public Health*, 23 (1), 794. https://doi.org/10.1186/s12889-023-15697-6

Kuswanto, H., Oktaviana, P. P., Efendi, F., Nelwati, N., and Malini, H. (2024). Prevalence of and factors associated with female child marriage in Indonesia. *PLOS ONE*, 19(7), e0305821. https://doi.org/10.1371/journal.pone.0305821

Marphatia, A. A., Saville, N. M., Manandhar, D. S., Cortina-Borja, M., Wells, J. C. K., and Reid, A. M. (2023). The role of education in child and adolescent marriage in rural lowland Nepal. *Journal of Biosocial Science*, 55(2), 275–291. https://doi.org/10.1017/S0021932022000074

Pourtaheri, A., Sany, S. B. T., Aghaee, M. A., Ahangari, H., and Peyman, N. (2023). Prevalence and factors associated with child marriage, a systematic review. *BMC Women's Health*, 23(1), 531. https://doi.org/10.1186/s12905-023-02634-3

Rahiem, M. D. H. (2021). COVID-19 and the surge of child marriages: A phenomenon in Nusa Tenggara Barat, Indonesia. *Child Abuse and Neglect*, 118, 105168. https://doi.org/10.1016/j.chiabu.2021.105168

Raj, A., Saggurti, N., Winter, M., Labonte, A., Decker, M. R., Balaiah, D., and Silverman, J. G. (2010). The effect of maternal child marriage on morbidity and mortality of children under 5 in India: Cross sectional study of a nationally representative sample. *BMJ*, 340, b4258. https://doi.org/10.1136/bmj.b4258

Ratnaningsih, M., Wibowo, H. R., Goodwin, N. J., Rezki, A. A. K. S., Ridwan, R., Hadyani, R. N., Minnick, E., Ulum, D. F., Kostaman, T. K., and Faizah, S. N. (2022). Child Marriage Acceptability Index (CMAI) as an essential indicator: An investigation in South and Central Sulawesi, Indonesia. *Global Health Research and Policy*, 7, 32. https://doi.org/10.1186/s41256-022-00252-4

Schlecht, J., Rowley, E., and Babirye, J. (2013). Early relationships and marriage in conflict and post-conflict settings: Vulnerability of youth in Uganda. *Reproductive Health Matters*, 21(41), 234–242. https://doi.org/10.1016/S0968-8080(13)41710-X

Sopyan, Y., Muttaqin, Z. Z., Solihat, C., and Aripin, J. (2023). Child Exploitation by Parents in Early Marriage: Case Study in Cianjur West Java, Indonesia. *Samarah*, 7(3), 1921–1942. https://doi.org/10.22373/sjhk.v7i3.14804

The National Development Planning Agency. (2020). National Strategy on The Prevention of Child Marriage. *In The National Development Planning Agency Republic of Indonesia.*

UNICEF, Statistics Indonesia (Badan Pusat Statistik [BPS]), Bappenas (National Planning Board), and PUSKAPA. (2020). *Prevention of Child Marriage Acceleration that Cannot Wait.* In UNICEF. https://www.unicef.org/indonesia/sites/unicef.org.indonesia/files/2020-06/Prevention-of-Child-Marriage-Report-2020.pdf

United Nations. (2023). *The Sustainable Development Goals Report 2023: Special Edition.* In United Nations. https://unstats.un.org/sdgs/report/2023/

Wemakor, A., Garti, H., Azongo, T., Garti, H., and Atosona, A. (2018). Young maternal age is a risk factor for child undernutrition in Tamale Metropolis, Ghana. *BMC Research Notes*, 11(1), 877. https://doi.org/10.1186/s13104-018-3980-7

WHO. (2014). *Stunting Policy Brief.* In WHO.

Wibowo, H. R., Ratnaningsih, M., Goodwin, N. J., Ulum, D. F., and Minnick, E. (2021). One household, two worlds: Differences of perception towards child marriage among adolescent children and adults in Indonesia. *The Lancet Regional Health - Western Pacific*, 8, 100103. https://doi.org/10.1016/j.lanwpc.2021.100103

Yastirin, P. A. (2019). Persepsi Remaja tentang Pernikahan Usia Anak. *Journal for Quality in Women's Health*, 2(1), 43–49.

Towards Resilient Societies: The Synergy of Religion, Education, Health, Science, and Technology – Rahiem (Ed.)

An evidence-based evaluation framework for Indonesia's priority tourism destinations program

Nawang Anandhini
Directorate of Monitoring and Evaluation Procedures, Ministry of National Development Planning

Dina Maulina
Directorate of Destination Development II, Ministry of Tourism and Creative Economy

ABSTRACT: Since 2019, Indonesia has launched ten new tourism destinations to diversify tourist preferences, which have been heavily concentrated in Bali. However, the effectiveness of the Priority Tourism Destinations Program remains unclear due to the lack of a comprehensive evaluation. This study introduces an evaluation framework to assess the program's implementation and impact. A systematic literature review was conducted to develop this framework, which consists of two key steps: (1) constructing the theory of change for the Priority Tourism Destinations Program and (2) formulating evaluation questions to assess six criteria—relevance, coherence, effectiveness, efficiency, impact, and sustainability—through evidence of difference-making. The proposed framework is designed to align with Indonesia's program structure, ensuring its practical application while providing deeper analysis. This approach facilitates collaborative evaluation among stakeholders and supports informed decision-making and evidence-based policy development.

Keywords: evaluation criteria, evidence-based evaluation, priority tourism destinations program, theory of change, tourism evaluation framework

1 INTRODUCTION

Over the years, tourism has been a main source of income for many developing countries due to its significance in creating job opportunities and accelerating economic prosperity (Dwyer *et al.* 2020). Bali, Indonesia's most popular tourism destination, faces massive urbanisation, land use changes, traffic congestion, waste disposal issues, and water supply deficiency (Cole and Browne 2015). In Kuta, coastal land use has significantly changed and has become a high-value real estate business for tourism (Chin *et al.* 2015). The Indonesian government initiated Priority Tourism Destinations (DPP) Program in 2019 to address these issues. This program aims to disperse tourists concentrated in Bali and accelerate the growth of the tourism sector's share in gross domestic product, tourism foreign exchange, and number of tourist visits.

Through this program, the fulfilment of key tourism components: attraction, accessibility, and amenity in ten targeted areas will be stimulated. However, Westoby *et al.* (2021) discussed how natural degradation and communities in DPP development are excluded. The goal of 10 New Bali would not be achieved, as the key issues are not resolved: accessibility, human capacity, community empowerment, and land use regulation. Yet, there is no comprehensive evaluation towards the Priority Tourism Destinations Program, and as a result, the progress and impacts of the program cannot be identified.

The evaluation framework is formulated to promote Sustainable Development Goals (SDGs). Target 12.b explicitly mentions that the tourism stakeholders must "develop and

 DOI: 10.1201/9781003645542-19

implement tools to monitor sustainable development impacts for sustainable tourism that creates jobs and promotes local culture and products". This action aims to assess DPP program performances and its achievement while ensuring the opportunities in the tourism sector align with principles of sustainable consumption. By implementing target 12.b, the ideal condition of sustainable tourism could be achieved by 2030 as the economic contribution of tourism (GDP) reaches its successful point (target 8.9).

An evaluation should be incorporated into the organisation's decision-making process; thus, evaluation should deliver integrated, credible, useful, and evidence-based information (UNDP 2021). The empirical study by SMERU (2017) points out that the monitoring and evaluation role in the development management cycle is overlooked and less supported. Furthermore, monitoring and evaluation practices are often perceived as mere formalities and regarded as extra duties that add to the employee's workload. SMERU (2017) also stated that monitoring and evaluation reports are focused on budget absorption and physical performance and are less effective in assessing outcomes and impacts of programs/activities.

By theory, contribution analysis has similar foundations yet is complementary to logic analysis and realist evaluation (Brousselle and Buregeya 2018) to measure the achievements of programs. However, the theory of change often fails to depict the anticipated causal connections clearly. Davies (2018) argues that theories of change have six structural problems: 1) unlabelled connections between events in the theory of change diagram; 2) missing connections in chain models; 3) oversimplified symmetric connections; 4) a huge range of possibilities in developing diagrams; 5) feedback loops; and 6) limited articulation of connections to a broader context. Moreover, Sjöö and Callerstig (2023) highlight that a sector-specific evaluation in the government has limited capacity to address program challenges beyond the sector domain, while asking independent evaluators is vulnerable as evaluation risks exposing failure.

Thus, this paper focuses on developing an evaluation framework specified in a tourism program called the 'Priority Tourism Destination Program'. This study developed a theory of change from the program and interlinked it with several evaluation guidelines, resulting in a specific framework applicable for stakeholders to evaluate the program.

2 METHODS

This study adopts a qualitative approach through a systematic literature review to develop an applicable evaluation framework. The literature review is focused on 1) addressing interventions that support the Priority Tourism Destination Program, 2) improving the program theory of change, and 3) developing evaluation questions for the program. A literature review is chosen to effectively synthesise research findings to provide meta-level evidence and identify gaps that require further exploration (Snyder 2019), which is essential for constructing conceptual models. This method emphasises transparency, and criteria must be defined and justified (Shaffril *et al.* 2021). For the first part of the research, The National Medium Term Development Plan (RPJMN) 2020-2024 (Presidential Regulation No. 18/2020) was reviewed to identify the evaluation components such as input, activities, output, outcome, and impact. For the second part, a research paper regarding the theory of change and evaluation guidelines from development institutions was examined to formulate a proposed evaluation framework.

3 FINDINGS AND DISCUSSIONS

UNDP (2021) defines evaluation as a systematic and impartial assessment of an activity, project, program, strategy, policy, topic, theme, sector, operational area, or institutional performance. Evaluation methods have continually emerged (Brousselle and Buregeya 2018; Koudoumakis *et al.* 2022; Sabet and Brown 2018) to meet stakeholders' needs and provide rigorous evidence in development policy making. To choose an evaluation method, it should consider the ability to answer evaluation questions, the available design, and programme attributes (Befani 2020; Stern *et al.* 2012). However, all evaluations should be grounded in clear results frameworks and theories of change (UNDP 2021). The evaluation examines the results chain, process, contextual factors, and causality to analyse the level of achievement (UNDP 2021).

Tourism evaluation highlights the industry's value to economic share. However, tourism evaluation is typically partial and focused on financial and impact indicators (number of jobs in the tourism industry and tourists' beds). Other factors, namely effectiveness, efficiency, and profitability of investment, are not measured (Ballesteros and Hernández 2016).

3.1 *Developing the theory of change*

The theory of change is intended to give a framework for whether intervention is making a difference. The theory of change describes how programs are intended to operate (Vaessen 2024) so that it can be used as a basis for contribution analysis (Befani 2020). Befani (2020) argued that the theory of change is viewed as a dynamic model that is developed and planned but can be revised during program implementation. It depends on stakeholders' perspectives and expectations, the understanding of those involved in the intervention, the experiences of the beneficiaries, and the prior research and evaluation findings.

Theory of Change of Priority Tourism Destinations Program (Figure 1) was built from RPJMN 2020 - 2024 (Presidential Regulation No. 18/2020). Regarding this document, the impact of the DPP program could be identified. Furthermore, Ministerial Regulation Number 12 of 2020 on the Strategic Plan of the Ministry of Tourism and Creative Economy is used to identify other evaluation components, namely input, activities, output, and outcome.

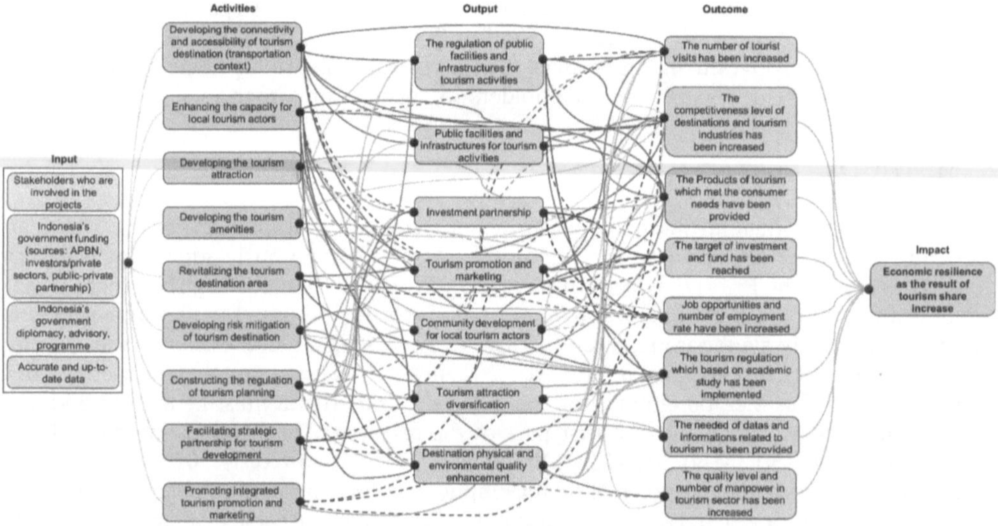

Figure 1. The theory of change of priority tourism destinations program.

3.2 *Formulating evaluation questions*

Formulating evaluation questions is fundamental. It allows evaluators to define the information produced, specify the scope to be examined and decide the evaluation approaches (Befani 2020; UNDP 2021). Evaluation questions are not limited to numbers and types. However, the reliability of the answers should be considered. It is preferred to have smaller scope, precise, well-defined, and more manageable questions (Befani 2020; UNDP 2021).

Formulating evaluation questions should go beyond the intended outcome/impacts, accounting for spillover effects that can underestimate a programme's impact. Angelucci and Maro (2015) recommend developing a theory about what may cause the spillover effects, how it works, and who may be affected. Naldini (2018) suggests explaining the evaluation questions to address the greater meaning of evaluation criteria and build real policy-relevant knowledge. In the tourism context, evaluation should consider externalities that might affect the program (Ballesteros and Hernandez 2016), such as considering various attractions and tourism products, proximity in terms of time and cost to the attraction, presence of complementary services, and the stage reached by tourism destinations. A qualitative approach can offer a comprehensive understanding of: 1) unidentified impacts of tourism interventions on local communities; 2) meaningful changes of the program beneficiaries from their ownership of the evaluation process; 3) engage the policy actors to promote more sustainable tourism; and (4) challenges faced by local authorities and organisations when evaluating tourism interventions using sustainability frameworks (Montano *et al.* 2024; Stoffelen 2018; Vanclay 2015).

The OECD's six evaluation criteria—relevance, coherence, effectiveness, efficiency, impact, and sustainability—are used as a reference to formulate evaluation questions (OECD 2021; UNDP 2021). However, evaluation practice expects that the questions are applied to the overall theory of change (Koudoumakis *et al.* 2021; UNDP 2021; OECD 2021; Vignetti *et al.* 2020). In a public policy context, an assessment of the interlinkages between the theory of change aspects—inputs, activities, outputs, outcomes, and impacts—is needed to illustrate the progress and results of the program. Detailed evaluation answers are used as lessons learned and provide specific feedback for future intervention plans.

This paper formulated evaluation questions using the OECD's six criteria in every theory of change aspects (Table 1). It is not mandatory to use all evaluation questions; they should be adjusted to the evaluators' objectives and resources. These findings support evaluation frameworks developed through the Regulation of the Minister of National Development Planning/Head of Bappenas No. 1/2023 that expect the user—government institutions—to complement quantitative score of program performance with structured qualitative explanations. In addition, focus group discussions between stakeholders and external evaluators are needed to ensure the objectivity of the evaluation process.

Table 1. The evaluation questions mapping.

	Inputs	Activities	Outputs	Outcomes	Impacts
Relevance	Are the planned/existing inputs appropriate & relevant given the political, economic, & social context in the country? Are the planned/existing inputs relevant with the government's/ development partner's policy framework & strategies?	Are the inputs relevant to the activities? Are the outputs/ outcomes/impacts pertinent with the activities? How does the activities address relevant priorities & needs? Do the activities align with beneficiary needs? Are the activities consistent with its design? Are the activities suitable & relevant	Are the planned/ existing inputs & activities adequate & relevant in achieving the desired outputs? Do the outputs respond to beneficiaries' needs & priorities? Are the outputs suitable & relevant within the country's political, economic, & social context? Are the outcomes & impacts aligned with the activities?	Do the inputs, activities, & outputs align with the desired outcomes? Are the planned/existing inputs, activities, & outputs adequate & relevant to achieve the desired outcomes? Do the outcomes address the initial problems?	Are the planned/existing inputs, activities, outputs, & outcomes relevant for achieving the desired impacts? Are the impacts aligned with national/global needs, policies, & priorities?

(*continued*)

Table 1. Continued

	Inputs	Activities	Outputs	Outcomes	Impacts
		within the country's political, economic, & social context, as well as government policies & frameworks? How the activities adapt to any context change?		Are the outcomes appropriate & relevant within the country's political, economic, & social context? Do the outcomes support national/global needs, policies, & priorities? Are the impacts pertinent with the outcomes?	
Coherence	Are the inputs across the same or other programs interconnected? How are the inputs synergised to/support each other?	How are the inputs within the same program synergised to support the activities? Are there any inputs from other programs that support this program's activities?	How the inputs & activities within the same program synergised to achieve the outputs? How might inputs, activities & outputs from other programs affect the program's outputs?	Are the activities within the same program synergised to achieve the outcomes? Do inputs from other programs affect this program's outcomes?	Are the inputs within the same program synergised to achieve the impacts? Do inputs from other programs affect this program's impacts?
Effectiveness	Are the resources available according to plan? Are there any risks that affect inputs/resource availability?	Do the inputs/resources support the activities? How well do the activities address the beneficiary's needs? Are the activities progressing as planned? Are there any problems or issues? Are there any risks that affect the activities?	Do the outputs achieve the targets? How the inputs & activities contribute to achieving the desired outputs? How well the outputs match the beneficiary's needs?	Do the outcomes achieve the intended results? How the inputs, activities, & outputs, contribute to achieve the desired outcomes?	Do the impacts achieve the intended results? How the inputs, activities, outputs, & outcomes contribute to achieve the desired impacts?
Efficiency	Are resources being utilized efficiently to achieve the desired targets/results?	How effectively are resources being utilized to carry out activities from economic, operational, & timeliness aspects?	How effectively are inputs/resources is utilized to achieve the desired outputs considering economic, operational, & timeliness aspects? How well are activities managed to achieve the desired outputs?	How effectively are inputs/resources is utilized to achieve the desired outcomes, considering economic, operational, & timeliness aspects? How well are activities &	How effectively are inputs/resources is utilized to achieve the desired impacts, considering economic, operational, & timeliness aspects? How well are activities &

(*continued*)

Table 1. Continued

	Inputs	Activities	Outputs	Outcomes	Impacts
				outputs managed to achieve the desired outcomes?	outputs managed to achieve the desired impacts?
Impact	Does the program design anticipated to evaluate the planned and differential impacts? Does the program identify the overall impacts both on the target groups or not?	Do the activities generate significant positive/negative impacts both on the target groups or not? Do the activities generate significant intended/unintended impacts both on the target groups or not?	Do the outputs generate significant positive/negative impacts both on the target groups or not? Do the outputs generate significant intended/unintended impacts both on the target groups or not?	Do the activities & outputs generate significant positive/negative & intended/unintended impacts that affect achievement of the outcomes? Do the outcomes generate significant positive/negative & intended/unintended impacts both on the target groups or not?	Do the activities, outputs, & outcomes generate significant positive/negative impacts that affect achievement of the targets? Do the activities, outputs, & outcomes generate significant intended or unintended impacts that affect achievement of the targets? Do the impacts of the program influence other programs' impacts?
Sustainability	–	Do the activities help enhance the enabling environment for development?	To what extent do the benefits of the outputs sustain over time for the beneficiary?	To what extent will the outcomes last after the program has ended in terms of stability & relative permanence? What conditions ensure the continuation of the outcomes, including institutional, economic & financial, environmental, political, social, & cultural sustainability? Do the outcomes contribute to improve the enabling environment for development?	To what extent will the impacts last after the program has ended in terms of stability & relative permanence? What kind of conditions ensure continuation of the impacts, such as institutional, economic & financial, environmental, political, social, & cultural sustainability? Do the impacts contribute to improve the enabling environment for development?

4 CONCLUSION

This study establishes an evaluation framework that is generally applied to government programs, combining OECD's six criteria of evaluation and the Theory of Change framework. The framework allows stakeholders to measure the program's performance based on evidence. Evidence of difference-making is needed to identify the program's activities towards its results. The output of this developing framework is descriptive analysis resulting from the evidence comparison in each criterion and inter.

To specify the scope of The Priority Tourism Destinations Program, the stakeholders should build the theory of change framework based on planning documents, RPJMN 2020-2024. The main goal of this framework is to illustrate how to examine the linkages between the activities of The Priority Tourism Destinations Program and its outputs, results, and outcomes. However, the enhancement of this framework is necessary due to its limitation in concluding the evaluation result. Evaluation questions should address externalities and go beyond the expected outcome/impacts that have been written in planning documents. Stakeholders should expand the topic to cover any possible impacts on local communities, stakeholders' commitment to sustainable tourism, program beneficiaries, and obstacles encountered by local governments and institutions.

REFERENCES

Angelucci, M. and Di Maro, V. (2015). Programme evaluation and spillover effects. *Journal of Development Effectiveness*, *8*(1), 22–43.

Befani, B. (2020). *Choosing Appropriate Evaluation Methods: A Tool for Assessment and Selection* (2nd ed.). Centre for the Evaluation of Complexity Across the Nexus (CECAN). https://www.cecan.ac.uk/wp-content/uploads/2020/11/Final_Choosing-Appropriate-Evaluation-Methods-1.pdf

Ballesteros, J. G. T. and Hernandez, M. (2016). Assessing the impact of EU rural development programs on tourism. *Tourism Planning and Development*, *14*(2), 149–166. 10.1080/21568316.2016.1192059

Brousselle, A., and Buregeya, J.-M. (2018). Theory-based evaluations: Framing the existence of a new theory in evaluation and the rise of the 5th generation. *Evaluation*, *24*(2), 154–168. 10.1177/1356389018765487

Chin, W. L., Fraser, J. H., and Hampton, M. P. (2015, May). Destination competitiveness: evidence from Bali. *Current Issues in Tourism*, 20(12), 1265–1289. https://doi.org/10.1080/13683500.2015.1111315

Cole, S., and Browne, M. (2015, May). Tourism and Water Inequity in Bali: A Social-Ecological Systems Analysis. *Human Ecology*, 43, 439 - 450. https://doi.org/10.1007/s10745-015-9739-z

Davies, R. (2018). Representing theories of change: technical challenges with evaluation consequences. *Journal of Development Effectiveness*, *20*(4), 438–461. 10.1080/19439342.2018.1526202

Dwyer, L., Forsyth, P., and Dwyer, W. (2020). *Tourism Economics and Policy* (2nd ed.). Channel View Publications.

Kementerian PPN/Bappenas. (2023, January). Peraturan Menteri/Kepala PPN Bappenas No. 1 Tahun 2023 tentang Tata Cara Pemantauan, Pengendalian dan Evaluasi Pelaksanaan Rencana Pembangunan. JDIH BPK. Retrieved December 28, 2024, from https://peraturan.bpk.go.id/Details/254887/permen-ppnkepala-bappenas-no-1-tahun-2023

Koudoumakis, P., Botzoris, G., and Protopapas, A. (2022). Cohesion policy evaluation: Guidelines for selection of appropriate methods. *Regional Science Policy and Practice*, *14*(5), 1062–1084. https://doi.org/10.1111/rsp3.12524

Montano, L. J., Font, X., Elsenbroich, C., and Ribeiro, M. A. (2024). Harnessing the power of stories: Evaluating complex tourism interventions through a 'most significant change' approach. *Annals of Tourism Research Empirical Insights*, *5*(2). https://doi.org/10.1016/j.annale.2024.100150

Naldini, A. (2018). Improvements and risks of the proposed evaluation of Cohesion Policy in the 2021–27 period: A personal reflection to open a debate. *Evaluation*, *24*(4), 469–504. 10.1177/1356389018804261

OECD. (2021). *Applying Evaluation Criteria Thoughtfully*. OECD Publishing.

Sabet, S. M., and Brown, A. N. (2018). Is impact evaluation still on the rise? The new trends in 2010–2015. *Journal of Development Effectiveness*, *10*(3), 291–304. 10.1080/19439342.2018.1483414

Shaffril, H. A. M., Samsuddin, S. F., and Samah, A. A. (2021). The ABC of systematic literature review: the basic methodological guidance for beginners. *Quality and Quantity*, *55*. 1319–1346. https://doi.org/10.1007/s11135-020-01059-6

Sjöö, K., and Callerstig, A.-C. (2023). The challenges in integrating horizontal perspectives in sectoral policy evaluation. *Policy Studies*, *44*(2), 174–194. 10.1080/01442872.2021.2015312

SMERU. (2017). Monitoring and Evaluation of Development Programs in Five Ministries: A Study on the System and Implementation. *SMERU Research Report*.

Snyder, H. (2019). Literature review as a research methodology: An overview and methodology: An overview and guidelines. *Journal of Business Research*, *104*. 333–339. https://doi.org/10.1016/j.jbusres.2019.07.039

Stoffelen, A. (2018). Tourism trails as tools for cross-border integration: A best practice case study of the Vennbahn cycling route. *Annals of Tourism Research*, *73*. 91–102. https://doi.org/10.1016/j.annals.2018.09.008

UNDP. (2021). *UNDP Evaluation Guidelines*. UNDP. https://popp.undp.org/document/undp-evaluation-guidelines-revised-edition-june-2021

Vaessen, J., Lemire, S., and Befani, B. (2020). *Evaluation of International Development Interventions: An Overview of Approaches and Methods*. International Bank for Reconstruction and Development.

Vanclay, F. (2015). The Potential Application of Qualitative Evaluation Methods in European Regional Development: Reflections on the Use of Performance Story Reporting in Australian Natural Resource Management. *Regional Studies*, *49*(8). 1326–1339. 10.1080/00343404.2013.837998

Vignetti, S., Giffoni, F., Pancotti, C., and Pagliara, F. (2020). Analytical framework for ex-post evaluation of transport projects: Lessons learnt on retrospective CBA. *Papers in Regional Science*, *99*(3), 683–703. 10.1111/pirs.12501

Westoby, R., Gardiner, S., Carter, R. W., and Scott, N. (2021, April). Sustainable livelihoods from tourism in the "10 New Balis" in Indonesia. *Asia Pacific Journal of Tourism Research*, 26(6), 702–716. https://doi.org/10.1080/10941665.2021.1908386

Towards Resilient Societies: The Synergy of Religion, Education, Health, Science, and Technology – Rahiem (Ed.)

The fragile intersection of human rights and national security: Lessons from Indonesia's democracy

Nawiruddin & Muhammad Arif Maulana
Department of Political Science, Universitas Islam Negeri Syarif Hidayatullah Jakarta, Indonesia
ORCID ID: 0000-0002-5618-2486

ABSTRACT: The dissolution of the Islamic Defenders Front (FPI) and Hizbut Tahrir Indonesia (HTI) through Law No. 16/2017 introduced a new dimension to the discourse on democracy and human rights in Indonesia. This study aims to examine the dissolution mechanisms of these two mass organisations and their implications for upholding democratic values and human rights in the country. Through comprehensive in-depth interviews and an analytical framework grounded in democratic theories, the study reveals a connection between the formulation and implementation of Law No. 16/2017—targeted at disbanding FPI and HTI—and the decline in Indonesia's democracy index. The findings suggest that this regulation restricts freedom of expression and hinders the broader implementation of human rights. Additionally, the study highlights the need for the government to evaluate the enforcement of this law to ensure it aligns with democratic principles and the protection of human rights.

Keywords: democracy, FPI, HTI, human rights, mass organisation

1 INTRODUCTION

In 2017 and 2019, the Indonesian government dissolved two Islamic mass organisations, i.e., the Islamic Defenders Front (FPI) and Hizbut Tahrir Indonesia (HTI). FPI, known for its hardline actions in enforcing Islamic sharia, was disbanded due to its involvement in activities deemed to threaten national order and security. Meanwhile, HTI was dissolved because the organisation sought to replace the state ideology of Pancasila with the ideology of the Caliphate, which stands in direct conflict with the foundation of the Indonesian state. These organisations were disbanded based on Law No. 16 of 2017 concerning Mass Organisations, which replaced Law No. 17 of 2013.

This regulation has triggered extensive debate regarding the limits of freedom of expression and democracy in general (Qodir 2024). Furthermore, the Democracy Index produced by The Economist Intelligence Unit (EIU) asserted that Indonesia's democracy index had declined. In 2019, Indonesia scored 6.48, which dropped to 6.3 the following year. This figure represents Indonesia's lowest democracy index score in the past 14 years and indicates the country is facing significant governance challenges (Fossati 2023; Mietzner 2024; Warburton and Aspinall 2019).

Mass organisations are an essential element that must be considered to improve the quality of democracy (Amal 2020; Liddle 2013; Matesan 2020; Tumin 2018). The existence of mass organisations shows the state's guarantee of the freedom of association its citizens enjoy to organise their ideas (Barton 2021; Hadiz 2017). This study further examined whether the mechanism used for dissolving FPI and HTI based on Government Regulation in Lieu of Law (Perppu) No. 02/2017, which was later passed into Law No. 16/2017, has an impact on weakening democratic values and upholding human rights in Indonesia.

 DOI: 10.1201/9781003645542-20

2 METHODS

This study employed a qualitative research methodology, utilising two types of data sources: primary and secondary. The primary data, which constitutes the main source for this research, were obtained from several informants who provided analyses relevant to the research subject (Table 1). The secondary data, which supports the in-depth understanding of the primary data, were collected from various books, journals, print media, and online sources (Anggito and Johan 2018).

Table 1. The informants.

No.	Name	Description
1.	Habib Abu Bakar Alatas	• Member of Islamic Defenders Front Lawyer Team • Secretary General of the Islamic Defenders Front
2.	Ir. Muhammad Ismail Yusanto, M.M.	• Spokesperson for Hizbut Tahrir Indonesia
3.	Dr. Mardani Ali Sera, M.Eng	• Member of the Indonesian House of Representatives / PKS Deputy Secretary General of PAN
4.	Slamet Nur Achmad Effendy	• Deputy Secretary General of PAN
5.	Prof. Dr. Masykuri Abdillah	• Professor of UIN Jakarta • Special Staff to the Vice President of the Republic of Indonesia
6.	Muhammad Anwar	• Chairman of GP Ansor South Jakarta Branch

The current study is descriptive explanatory, meaning that data collection was linked to relevant theories concerning the issue. This descriptive approach is aimed at providing an overview of the mechanisms involved in the dissolution of the Islamic Defenders Front (FPI) and Hizb ut-Tahrir Indonesia (HTI) organisations under Law No. 16 of 2017, as well as the implications it has on freedom of expression and association to these organisations.

3 FINDINGS AND DISCUSSION

Law No. 16 of 2017, an extension of the Government Regulation in Lieu of Law (Perppu), has undergone a legislative process. The formulation and issuance of a Perppu, which is later enacted into law, typically involve at least two stages. First, as outlined in Article 1, Paragraph 4 of Law No. 12 of 2011 concerning the formation of legislative regulations, creating a Perppu requires urgency or an emergency situation as a primary factor. The Constitutional Court (*Mahkamah Konstitusi* or MK) is responsible for assessing this urgency to determine whether the situation justifies the issuance of a Perppu. Second, according to Article 7 of Law No. 12 of 2011, the proposed Perppu must be tested against the existing legal hierarchy in Indonesia.

Perppu No. 02 of 2017, which forms the basis for Law No. 16 of 2017, should have undergone a series of hierarchical legal checks by the Constitutional Court. This process manifests the procedural mechanisms within a legal state and the judicial function. However, before testing the legal hierarchy of a Perppu, the Constitutional Court must assess its alignment with other laws. A Perppu should not conflict with existing laws and should not contain values inconsistent with or threatening other regulations.

3.1 *Dissolution mechanism for HTI and FPI*

The dissolution of HTI and FPI began with Government Regulation in Lieu of Law (Perppu) No. 02 of 2017, issued on July 10, 2017, and subsequently ratified by the Indonesian House of Representatives (DPR) on November 22, 2017. However, there were slight differences in the dissolution processes for HTI and FPI. HTI was the first organisation to dissolve under this Perppu nine days after its issuance on July 19, 2017. The government used the mechanisms outlined in Perppu No. 02 of 2017 to issue a one-time, seven-day warning to HTI. Subsequently, based on the recommendation given by the Ministry of Law and Human Rights, HTI was deemed suitable for dissolution, and its legal status was revoked. The dissolution mechanism under Perppu No. 02 of 2017 differs from the previous law, i.e., Law No. 17 of 2013. Under the previous law, the government issued periodic warnings to organisations, followed by suspension of grants, temporary suspension, and then dissolution through a judicial mechanism in the District Court.

In response to the dissolution, HTI, through Hizbut Tahrir Statement No. 293 and spokesperson Ismail Yusanto, asserted that the government employed the principle of *contrarius actus*—the same authority that grants an organisation the right to operate also holds the authority to dissolve it. HTI rejected the dissolution order and argued that the applied mechanism goes against democratic principles. Furthermore, HTI contended that the government did not adhere to the due process of law principle, which ensures legal protection and justice. According to HTI, the Perppu did not provide organisations with the right to defend themselves (Interview with Ismail Yusanto, HTI Spokesperson, on April 11, 2022).

Regarding this issue, the parties involved have differing interpretations. Abu Bakar Alatas, the General Secretary of FPI, indicated that the government seemed to have used a shortcut to dissolve organisations based on its perspective without allowing citizens the opportunity to defend themselves (Interview with Abu Bakar Alatas, FPI Lawyer Team, on April 4, 2022). Meanwhile, Special Staff to Vice President Masykuri Abdillah explained that a judicial process is still necessary to maintain democracy. He noted that if an organisation is dissolved, it has the right to challenge the decision in court. According to Abdillah, this process is consistent with the previous system. Despite revoking an organisation's legal status, individuals are still allowed to express their opinions and promote their views (Interview with Masykuri Abdillah, Special Staff to the Vice President of Indonesia, on July 1, 2022).

The dissolution of FPI was unlike that of HTI. While HTI's dissolution involved revoking its legal status as an organisation, FPI was dissolved through the non-issuance of a Certificate of Registered Organization (SKT) (Interview with Abu Bakar Alatas, FPI Lawyer Team, on April 4, 2022). Perppu No. 02 of 2017 introduced new regulations requiring organizations to obtain a recommendation letter from the relevant ministry, in FPI's case, the Ministry of Religious Affairs (MoRA), before applying to the Ministry of Home Affairs (MoHA) for an SKT (Interview with Abu Bakar Alatas, FPI Lawyer Team, on April 4, 2022). Under the previous law, organisations seeking to establish themselves could directly apply to MoHA for registration.

FPI received a recommendation letter from MoRA (Interview with Abu Bakar Alatas, FPI Lawyer Team, on April 4, 2022). However, MoHA did not issue the SKT for FPI. In this context, DPR called FPI to explain discrepancies in their organisational statutes, particularly concerning using terms like "*jihad*" and "*khilafah.*" DPR advised FPI to revise its organisational vision and mission, eliminating terms deemed odd and inconsistent with Pancasila's ideology.

The non-issuance of SKT for FPI, despite receiving a recommendation from MoRA, suggests that the government regarded FPI as problematic or contrary to Pancasila ideology, warranting its dissolution (Interview with Abu Bakar Alatas, FPI Lawyer Team, on April 4, 2022). FPI was seen as affiliated with prohibited ideologies and administrative issues that were not addressed (Interview with Abu Bakar Alatas, FPI Lawyer Team, on April 4, 2022). To address the SKT issue, FPI attempted to amend its organisational statutes through a National Deliberation (Munas) to rectify deficiencies leading to the non-issuance of SKT. However, these efforts were ineffective as the dissolution notice was issued on December 30, 2020.

According to Special Staff to the Vice President Masykuri Abdillah, HTI and FPI did not acknowledge Pancasila and the Constitution as the foundation of their ideological struggle (Interview with Masykuri Abdillah, Special Staff to the Vice President of Indonesia, on July 1, 2022). They disregarded Pancasila ideology as values worth upholding, which led the government to dissolve both organisations. Additionally, the government perceived actors within these organisations as being affiliated with prohibited ideologies, although Abu Bakar Alatas mentioned MoRA's recommendation that FPI was eligible for SKT.

3.2 *Implications of the dissolution of FPI and HTI on human rights and democracy index*

3.2.1 *Implications to human rights*

The enactment of Law No. 16 of 2017 has been interpreted differently by various interest groups and has brought about several implications. According to FPI's legal team, the dissolution mechanism for mass organisations outlined in this law is inconsistent with human rights perspectives, where the state should ensure freedom and security for its citizens. In the context of the dissolution of organisations, FPI's lawyer, Abu Bakar Alatas, asserted that individual rights are inherent to every person. Thus the state should not revoke an individual's right to associate and establish organisations, which must be guaranteed and protected within the constitution (Interview with Masykuri Abdillah, Special Staff to the Vice President of Indonesia, July 1, 2022).

In contrast, Special Staff to the Vice President, Masykuri Abdillah, argued that under Law No. 16 of 2017, the state still accommodates these rights. He asserted that the state provides a judicial process for citizens whose organisations have been dissolved. Those involved in the dissolved organisations are still allowed to engage in activities, including expressing their opinions or establishing new organisations, provided they comply with applicable regulations (Interview with Masykuri Abdillah, Special Staff to the Vice President of Indonesia, July 1, 2022). Abdillah further explained that the implementation and protection of human rights in Indonesia remain intact, citing that many leaders of dissolved organisations continue to engage in their usual activities, such as conducting preaching activities.

The dissolution of FPI and HTI provides an alternative perspective on human rights. In this case, the government did not restrict individual freedoms but primarily acted on the administrative aspects. The government revoked the legal status of the organisations, but no direct dissolution took place; instead, individuals involved were still allowed to resume their activities. This approach is also observed in other countries that have experienced a decline in the quality of democracy, such as Turkey, which is also a majority-Muslim country (Yilmaz *et al.* 2021).

The dissolution of HTI and FPI illustrates how this law can restrict the ability of organisations and individuals to express their views and participate in public discourse. The dissolution of HTI and FPI also raises concerns about religious freedom in Indonesia. Although the law addresses groups deemed radical, it may inadvertently impact religious freedom by limiting the rights of individuals and groups to practice and promote their beliefs. Moreover, this case also has implications for legal protection in Indonesia.

3.2.2 *Implications to democracy*

In democratic principles, constructive communication to determine the direction of policies, the presence of freedom of speech, freedom of the press, and freedom of opinion, as well as the right to assemble and associate, and the absence of destructive conflicts among communities, are key benchmarks (Suteki 2021). The enactment of Law No. 16 of 2017 has caused some mass organisations operating within society to feel uneasy and fearful of engaging in activities and expressing their opinions (Interview with Mardani Ali Sera, Chairman of PKS, April 12, 2022).

The discomfort these organisations experienced reflects the dysfunction of democratic principles (Power 2018). The enactment of Law No. 16 of 2017, which laid the foundation for the dissolution of FPI and HTI, has led to four key impacts contributing to the decline of democracy in Indonesia. First, it restricts freedom of organisation and expression. The dissolution of HTI and FPI demonstrates how this law limits groups from expressing their views, even when they do not align with government ideology, which can undermine civil liberties and democratic practices (Fossati 2023). While the government strives to uphold human rights, as evidenced by the continued activities of members from these organisations, such as preaching, despite their dissolution (Interview with Masykuri Abdillah, Special Staff to the Vice President of Indonesia, on July 1, 2022), this measure still affects freedom of speech and association. Second, it reduces political participation. By dissolving these organisations, which serve as platforms for political engagement, the government risks narrowing down channels for public participation (Nuraniyah 2023).

The third impact is social dissatisfaction. Dissolving organisations like HTI and FPI can intensify public discontent, as affected groups may feel marginalised, deepening distrust in state institutions. Although the State Administrative Court (PTUN) allows citizens to voice their concerns, it may not fully address these issues. Fourth, the dissolution process highlights a decline in political transparency. The government's lack of communication with civil society, coupled with the hasty policy decisions behind the dissolution of these organisations, reflects a failure of democratic principles (Interview with Mardani Ali Sera, Chairman of the PKS, on April 12, 2022). Law No. 16 of 2017 grants substantial governmental authority without adequate oversight, increasing the risk of power abuse and eroding public trust in democratic institutions.

4 CONCLUSION

The dissolution of the societal organisations Front Pembela Islam (FPI) and Hizbut-Tahrir Indonesia (HTI) has had a negative impact on the implementation of the democratic system and the fulfilment of the rights to freedom of expression and association in Indonesia. The dissolution mechanism, in which the principle of *actus contrarius* is applied by the government and the shift in judicial process to post-dissolution, has reduced the effectiveness of democracy and human rights in Indonesia. The enactment of Law No. 16 of 2017, which grants the government the authority to dissolve societal organisations, has raised serious concerns about the decline in democratic quality in Indonesia.

The emerging fear within society to express opinions, individually and collectively through organisations, indicates that the guarantee of the right to free speech and association has been affected by the enactment of Law No. 16 of 2017 and the dissolution of HTI and FPI. Law No. 16 of 2017 is a relatively clear law that provides a legal mechanism to address organisations not aligned with Pancasila and the 1945 Constitution. However, aspects still need to be revised, such as prioritising more humane and democratic values, in line with SDG 10 (Reduced Inequalities) and 16 (Peace, Justice, and Strong Institutions). In this regard, the government should revisit this law to create a more democratic mechanism for the dissolution of organisations while prioritising and protecting the implementation of human rights within society.

REFERENCES

Amal, M. K. (2020). Islamic populism in Southeast Asia: An Indonesian Muslim intellectual's perspective. *Journal of Critical Reviews*, 7, 583–588.

Anggito, and & Johan, S. (2018). *Metodologi Penelitian Kualitatif*. CV Jejak.

Barton, G. (2021). Contesting Indonesia's democratic transition: Laskar Jihad, the Islamic Defenders Front (FPI), and civil society. In A. Vandenberg and & N. Zuryani (Eds.), *Security, democracy, and society in Bali* (Chap. 13, pp. 305–331). Palgrave Macmillan.

Fossati, D. (2023). Illiberal resistance to democratic backsliding: The case of radical political Islam in Indonesia. *Democratization*, 31(3), 616–637. https://doi.org/10.1080/13510347.2023.2210504

Hadiz, V. R. (2017). Indonesia's year of democratic setbacks: Towards a new phase of deepening illiberalism? *Bulletin of Indonesian Economic Studies*, 53(3), 261–278. https://doi.org/10.1080/00074918.2017.1410311

Liddle, R. W. (2013). Improving the quality of democracy in Indonesia: Toward a theory of action. *Indonesia*, 96, 59–80. https://doi.org/10.5728/indonesia.96.0057

Matesan, I. E. (2020). Grievances and fears in Islamist movements: Revisiting the link between exclusion, insecurity, and political violence. *Journal of Global Security Studies*, 5(1), 44–62. https://doi.org/10.1093/jogss/ogz042

Mietzner, M. (2024). The limits of autocratization in Indonesia: Power dispersal and elite competition in a compromised democracy. *Third World Quarterly*, 1–17. https://doi.org/10.1080/01436597.2024.2317970

Nuraniyah, N. (2023). Repressed therefore radicalized? Explaining variation in Islamists' responses to state repression in Indonesia. *Counter Terrorist Trends and Analyses*, 15(3), 1–9. https://www.jstor.org/stable/48732710

Power, T. P. (2018). Jokowi's authoritarian turn and Indonesia's democratic decline. *Bulletin of Indonesian Economic Studies*, 54(3), 307–338. https://doi.org/10.1080/00074918.2018.1549918

Qodir, Z. (2024). Conservative turn and political identity: Challenges to democracy in Indonesia after presidential election 2019. *Identities*, 1–18. https://doi.org/10.1080/1070289X.2024.2415220

Suteki, A. J. (2021). Dissolution of Islamic community organizations (Ormas) in the context of a state of law and a democratic state. *Walisongo Law Review (Walrev*, 3(2). https://doi.org/10.21580/walrev/2021.3.2.10848

Tumin, M. (2018). Civil society and democracy in Southeast Asia and Turkey. *Democracy and Security*, 14(1), 90–93. https://doi.org/10.1080/17419166.2018.1426283

Warburton, E., and & Aspinall, E. (2019). Explaining Indonesia's democratic regression: Structure, agency and popular opinion. *Contemporary Southeast Asia*, 41(2), 255–285. https://www.jstor.org/stable/26798854

Yilmaz, I., Shipoli, E., and & Demir, M. (2021). Authoritarian resilience through securitization: An Islamist populist party's co-optation of a secularist far-right party. *Democratization*, 28(6), 1115–1132. https://doi.org/10.1080/13510347.2021.1891412

Towards Resilient Societies: The Synergy of Religion, Education, Health, Science, and Technology – Rahiem (Ed.)

University museums as catalysts for sustainable cultural communities

Pungki Purnomo & Ida Farida
Universitas Islam Negeri Syarif Hidayatullah Jakarta, Indonesia

ABSTRACT: This study explores the role of university museums in Indonesia as catalysts for academic and off-campus community engagement, with a focus on supporting the Sustainable Development Goals (SDGs), particularly SDG 4 (Quality Education), SDG 5 (Gender Equality), and SDG 16 (Peace, Justice, and Strong Institutions). Employing qualitative research methods, data were gathered through in-depth interviews with university museum managers, academics, and visitors, alongside the analysis of relevant case studies. The findings reveal that various programs initiated by university museums—such as community curator programs, community engagement initiatives, community exhibitions, and seminar series—effectively foster cultured, inclusive, critically thinking communities that value equality and social justice. These outcomes significantly contribute to advancing SDGs 4, 5, and 16. The study highlights the untapped potential of university museums in driving community development and addressing global challenges while offering practical recommendations to enhance their role within a sustainable development framework.

Keywords: gender equality, quality education, strong institution, sustainable development goals, university museum

1 INTRODUCTION

Today, the meaning of museums is not only a repository of artefacts and artworks but also a platform for critical reflection and dialogue on various social issues, including those related to democracy. The community can learn many things through the museum's collection-based learning program, including democratic values. Concrete examples of learning applied democratic values can be found in a study by Hagmayr and Fröschl (2023) regarding learning democracy in museums with the research object at The Museum Arbeitswelt in Steyr, Austria. The democratic learning implemented in the museum shows that everyone at various age levels carries out museum-based democratic learning.

The presence of museums in higher education institutions plays a role as a library partner in promoting culture and supporting sustainable development goals. Therefore, many university museums in leading universities, especially in Indonesia, are not passive and exclusive in their management. The role of university museums is to preserve heritage and foster understanding, increase participation, and promote inclusivity in society (Lipovec 2023). Community participation is vital in forming a strong cultural society for sustainable development goals (Rosetti *et al.* 2022). In line with that, the role of museums and community participation are important factors in maintaining human and natural heritage in the past and for the goals of sustainable community development in the present and future (Song *et al.* 2024).

Museums and community participation are reflected in various community development activities (King *et al.* 2023). Museum visits by people of different ages, education, and social

DOI: 10.1201/9781003645542-21

status backgrounds can potentially democratise their museum visit experience (Coghlan 2018; Stroja 2018). The form of visitor participation is not only providing comments on their visit experience. Participation can trigger conversations and build museum relationships with real and imaginary communities inside and outside the museum (Antón *et al.* 2018).

The central theme of this study is to highlight the intersection between democracy and museum service programs. The democratic values contained in various museum services it is certainly conducive to further empowering a cultural and religious community (Murzina 2020). Public participation, in terms of their interactions, when visiting museums and when they participate or are involved in planning various presentation programs, is a very important discourse (Robinson 2017). This is because various programs such as exhibitions and educational programs will effectively impact the community if they follow their needs. Therefore, museums must ensure that the presentation of programs, both in the form of exhibitions and other educational programs, can enable the visiting public to interact directly or indirectly with the museum (Gata *et al.* 2023).

As spaces for education and public engagement, museums often discuss democratic values such as freedom of expression, diversity, equality, and participation. The museum's service program shapes public discourse and promotes democratic ideals. The presence of museums in all leading universities in Indonesia is due to the realisation that libraries are not the only reliable ones to shape the character of both students and society. Therefore, the initiators and founders of university museums in Indonesia strive to make museums a complementary library partner.

It is very important to introduce the university's identity to its academicians with its various cultural heritages. This effort requires seriousness in improving the quality of their literacy about primary sources of the university's cultural heritage (Öztemiz 2020). The higher their level of literacy, the better their understanding of their identity. University museums, with their collections, can help academicians obtain adequate and comprehensive information according to their needs (Murray 2020). The literacy ability of academicians is a combination of knowledge and skills needed so that they are effectively able to identify their needs, trace and find what they need, then the ability to interpret, evaluate, and use primary sources ethically in the context of a specific discipline, to create new knowledge or to revise existing understanding.

2 METHODS

The research design to explore the significant role of university museums in supporting the Sustainable Development Goals (SDGs) employed a qualitative approach and case study method with observation and in-depth interviews of insights, perceptions and experiences of museum managers and campus community who engaged in advancing sustainable development programs. The approach is suitable for revealing various understandings and practices that can support society's sustainable development efforts that align with the research questions.

An interview guide was developed with in-depth semi-structured interviews to ensure consistency while maintaining flexibility to accommodate unique informants' perspectives. The primary data was collected from museum managers, diverse lecturers, and students from different backgrounds, including experts of cultural organisations, both governmental and non-governmental organisations (NGOs). The sample size of this study consisted of 10 individuals using purposive sampling, and the informants' criteria covered their expertise, experience, involvement, contribution and interest in the sustainable development of the museum's role.

Furthermore, all interview data was recorded and transcribed for analysis. Then, all data findings related to SDGs were analysed using relevant case study methods through data reduction, data presentation, and conclusion to show concrete examples of the university museum's role in supporting the Sustainable Development Goals (SDGs).

Various strategies were used to ensure the reliability and validity of the findings. The triangulation technique by comparing interview findings was used to avoid bias. Therefore, the researcher ensured that informants with different backgrounds of status, expertise, contributions, and interests in their studies, such as museum managers, lecturers, students, experts, and other communities outside the campus, were interviewed. In addition, the researcher also confirmed the interview transcripts with the informants to ensure the truth of the intent of what was previously expressed.

Finally, to gain recognition of the findings from the interviews with these informants, the researcher then asked for their consent by signing a form that ensured they were fully aware of the purpose, procedures, and their right to withdraw at any time. The informants were informed that confidentiality and anonymity were strictly guaranteed.

As we know there are 17 interconnected global goals set by the United Nations in 2015 as part of the 2030 agenda for the Sustainable Development Goals (SDGs), but considering the limitations of researchers both in terms of time and research budget and for this research to be more focused, this research is only limited to 3 SDGs, namely SDG 4 (Quality Education), SDG 5 (Gender Equality) and SDG 16 (Peace, Justice, and Strong Institutions).

3 FINDINGS AND DISCUSSION

3.1 *Findings*

The Sustainable Development Goals (SDGs), consisting of 17 goals (Table 1), replace the Millennium Development Goals that expired in 2015 and have been implemented for about 15 years. About 193 UN member countries in September 2015 agreed and approved the Sustainable Development Goals (SDGs), which are then known as the 2030 agenda.

Table 1. List of 17 SDGs of UN.

No. SDGs	Goals
1	No Poverty
2	Zero Hunger
3	Good Health & Well-Being
4	Quality Education
5	Gender Equality
6	Clean Water & Sanitation
7	Affordable & Clean Energy
8	Decent Work & Economic Growth
9	Industry, Innovation & Infrastructure
10	Reduced Inequalities
11	Sustainable Cities & Communities
12	Responsible Consumption & Production
13	Climate Action
14	Life Below Water
15	Life on Land
16	Peace, Justice, and Strong Institution
17	Partnership for the Goals

Based on Table 1 above, the Sustainable Development Goals (SDGs) are a collection of 17 global goals established by the United Nations (UN) in 2015 as part of the 2030 Agenda for Sustainable Development. These goals address the world's most pressing challenges by fostering sustainable, inclusive, and equitable development (Gigliotti *et al.* 2019).

Table 2. University Museum programs support SGDs.

No.	University museum programs	No of SDGs	Goal of SDGs
1	Community Curator Program	4, 5, & 16	Quality Education, Gender Equality, & Peace, Justice & Strong Institution
2	Community engagement Program	4,5, & 16	Quality Education, Gender Equality, & Peace, Justice & Strong Institution
3	Community Exhibition Program	4, 5, & 16	Quality Education, Gender Equality, & Peace, Justice & Strong Institution
4	Seminar & Conference Program	4, 5, & 16	Quality Education, Gender Equality, & Peace, Justice & Strong Institution

Table 2 above explains that the findings of this study reveal that there are at least four university museum programs that support sustainable community development efforts that are in line with the goals of SDG 4, SDG 5, and SDG 16. The four museum programs are as follows: community curator programs, community engagement programs, community exhibition programs, and seminar and conference programs.

3.1.1 *Community curator programs*

Implementing the community curator program aims to increase community literacy about the importance of heritage. This effort requires seriousness in improving the quality of their literacy about primary sources of the university's cultural heritage (Öztemiz 2020). The higher their level of literacy, the better their understanding of their identity. University museums, with their collections, can help academicians obtain adequate and comprehensive information according to their needs (Murray 2020).

3.1.2 *Community engagement programs*

Community engagement programs, which include workshops, community-guided tours, classes and workshops, collaborative projects, discussions and debates, and special events, aim to build relationships, encourage dialogue, and create a sense of belonging among community members. The programs allow the community to engage in dialogue, collaboration, and mutual learning (Rislow and Nelson 2020).

3.1.3 *Community exhibitions and events*

Museums can host exhibitions and events curated by community members, showcasing their talents, traditions, and stories to a wider audience. The implementation of the community exhibitions and event program provides benefits not only for visitors but also for museum managers.

3.1.4 *Seminar and conference programs*

The museum can organise seminars, scientific conferences and trainings on various issues such as democracy, green universities, and the environment. The seminar and training program invites experts from different disciplines to share their ideas on current societal issues in response to the sustainable development goals (SDGs).

3.2 *Discussion*

The findings of this study demonstrated four museum programs, which include community curator programs, community engagement programs, community exhibitions programs, and seminar and conference programs, have been practised in museums by involving the community in the decision-making process related to the management and development of the university museum. The community engagement programs provide space for various voices and opinions to be heard and valued (Nikonanou and Misirloglou 2023). Encouraging people to participate in a community development program at a museum requires thoughtful planning, engagement strategies, and ongoing communication (Ibrahim 2022). The museum should actively engage with the local community by listening to their needs and interests.

Community curator programs reflect the museum's efforts to embrace an inclusive approach to curation and exhibition programs. Collaboration with local communities allows people to be directly involved in the creative process and decision-making. Community participation can be done by creating programs appropriate to the community's specific interests and needs (Schuch *et al.* 2018) .

The community engagement program highlights the importance of museums as a space for dialogue and collaboration at various levels of society. These activities usually involve local communities to learn and participate in certain creative and learning activities to create a learning society that can anticipate various developments in science and technology (Matuk 2016). Museums can also act as hosts for discussion and debate activities on topics related to culture, history, and social issues that are relevant to the community (Johnson-Cunningham 2018). The goal of museum community engagement programs is to make museums more accessible, relevant, and inclusive to diverse audiences (Springinzeisz 2024). Community engagement also helps museums to build trust, strengthen partnerships, and create positive social impact within their local communities (Emifoniye 2023).

Through the organisation of the exhibition, the museum can convey various messages to visitors about what is needed by the community. Organising a museum exhibition is one of the most important activities and is a demand that museum managers cannot ignore (Wang *et al.* 2021). Organising an exhibition by a museum is a form of openness of the university museum to the wider community which is not limited to its campus community (King *et al.* 2023).

The seminar and conference programs demonstrate the role of museums as intellectual centres and places of multidisciplinary dialogue. By hosting seminars on democracy and the environment, museums can facilitate important discussions involving the public, academics, and professionals from various fields.

4 CONCLUSION

This study has revealed the significant contributions of University Museums as catalysts for sustainable cultural communities. Key findings highlight the success of four university museum programs practised, and the programs are reflected in community curator programs, community engagement programs, community exhibitions and events, and seminar and conference programs to achieve sustainable development goals no. 4 (quality education), no. 5 (gender equality), and no. 16 (peace, justice and strong institution). The implementation of four museum programs of community development directly answers research questions by identifying effective strategies and best practices that leverage the role of university museums as catalysts for sustainable cultural communities.

The practical implications of implementing the four museum programs of community development allow it to be adopted as something that becomes material for policymakers and provides the necessary support. This study shows the importance of future research in increasing the role of museums in the long term, especially in supporting sustainable development goals.

REFERENCES

Antón, C., Camarero, C., and Garrido, M. J. (2018). Exploring the experience value of museum visitors as a co-creation process. *Current Issues in Tourism*, 21(12), 1406–1425.

Coghlan, R. (2018). 'My voice counts because I'm handsome.' Democratising the museum: the power of museum participation. *International Journal of Heritage Studies*, 24(7), 795–809.
Emifoniye, A. (2023). Heritage Management and Community Engagement: The Story of the National Museum, Benin City, Nigeria. *International Journal of Current Research in the Humanities*, 26(1),
Gata, D., Valakos, E., and Georgiou, M. (2023). Engaging and assessing students via a museum educational program. *Eurasia Journal of Mathematics, Science and Technology Education*, 19(10), 1–13.
Gigliotti, M., Schmidt-Traub, G., and Bastianoni, S. (2019). The Sustainable Development Goals. In *Encyclopedia of Ecology*: Volume 1-4, Second Edition (2nd ed., Vol. 4). Elsevier Inc.
Hagmayr, M., and Fröschl, F. (2023). Democracy Needs Democratically-Engaged Citizens. The Politics Workshop (Politikwerkstatt) at the Museum Arbeitswelt as Learning Environment for Civic Education. *Journal of Museum Education*, 48(3), 283–292.
Ibrahim, A. (2022). The role of museums, design accessibility and community concerns: A case study of the State Bank of Pakistan Museum and Art Gallery. *Journal of Community Archaeology and Heritage*, 9(1), 57–67.
Johnson-Cunningham, S. (2018). Beyond gallery walls and performance halls: five essential steps museums and other cultural institutions must take to center people, communities, and cultivate effective societal change. *Museums and Social Issues*, 13(1), 2–7.
King, E., Smith, M. P., Wilson, P. F., Stott, J., and Williams, M. A. (2023). Creating Meaningful Museums: A Model for Museum Exhibition User Experience. *Visitor Studies*, 26(1), 59–81.
Lipovec, N. Č. (2023). 'Memory Talk' in a Museum in a Contested Land. *Ethnologia Fennica*, 50(2), 75–102.
Matuk, C. (2016). The learning affordances of augmented reality for museum exhibits on human health. *Museums and Social Issues*, 11(1), 73–87.
Murray, C. (2020). Functions of art museums: what visitors and museum staff believe. *The Museum Review*, 5 (1), 1–16.
Murzina, I.Y. (2020). Museum As Information and Communicative Educational Institution. *KnE Social Sciences*, 2020, 159–166.
Nikonanou, N., and Misirloglou, T. (2023). 'Together We Curate': Cultural Participation and Collective Curation. *Museum and Society*, 21(1), 31–44.
Öztemiz, S. (2020). Cultural heritage literacy: A survey of academics from humanities and social sciences. *Journal of Librarianship and Information Science*, 52(3), 818–831.
Rislow, M., and Nelson, A. (2020). Artist Files and Art Students: Expanding Audience Through Collaboration Between an Art Museum Library and a Contemporary Art History University Course. *Journal of Museum Education*, 45(2), 166–175.
Robinson, H. (2017). Is cultural democracy possible in a museum? Critical reflections on Indigenous engagement in the development of the exhibition Encounters: Revealing Stories of Aboriginal and Torres Strait Islander Objects from the British Museum. *International Journal of Heritage Studies*, 23(9), 860–874.
Rosetti, I., Cabral, C. B., Roders, A. P., Jacobs, M., and Albuquerque, R. (2022). Heritage and Sustainability: Regulating Participation. *Sustainability* (Switzerland), 14(3), 1–27.
Schuch, J. C., Harden, S. B., Bostick, K., and Smith, H. A. (2018). Museums engaging diverse Millennials in community dialogue. *Museums and Social Issues*, 13(2), 58–77.
Song, Y., Gilardi, F., and Lam, C. (2024). Building culturally sustainable communities. Community museums and transmedia storytelling. *Museum Management and Curatorship*, 39(1), 2–19.
Springinzeisz, K. (2024). Social Inclusion In Museum Education: A Literature Review. 11(2), 4–20.
Stroja, J. (2018). My history, your history, our history: Developing meaningful community engagement within historic sites and museums. *Queensland Review*, 25(2), 300–321.
Wang, Y. C., Chen, C. L., and Deng, Y. Y. (2021). Museum-authorization of digital rights: A sustainable and traceable cultural relics exhibition mechanism. *Sustainability* (Switzerland), 13(4), 1–25.

Towards Resilient Societies: The Synergy of Religion, Education, Health, Science, and Technology – Rahiem (Ed.)

Sustainable urban design for local economic development: Agrarian reform in Subang regency

Syarifah Muslimat
Ministry of National Development, Indonesia
ORCID ID: 0009-0009-8642-3699

Endah Kartika Syahri
Ministry of National Development, Indonesia
ORCID ID: 0009-0007-4158-8823

Nur Azizah Ferawati
Ministry of National Development, Indonesia
ORCID ID: 0009-0000-1437-7165

ABSTRACT: Land conflicts in Indonesia are prevalent in high-poverty regions. This study investigates the impact of Agrarian Reform Village initiatives on economic development in Subang Regency. Employing a mixed-methods approach, the research integrates qualitative analysis with spatial modelling based on Sustainable Urban Design (SUD) and Participatory Planning. Data were collected through direct observation, socio-economic mapping, interviews, and Focus Group Discussions (FGDs) with key stakeholders. A distinctive feature of this study is the application of SUD—typically used in urban contexts—to rural areas experiencing land conflicts. The findings reveal that community leaders and local businesses play a critical role in driving development, while support from government institutions, academia, and the private sector enhances planning and implementation efforts. The successful establishment of a local economic development centre in 2023 highlights the model's effectiveness, generating a 'spread effect' that inspires similar initiatives in neighbouring areas. These outcomes support Sustainable Development Goal (SDG) 1 (No Poverty), demonstrating how land reform can enhance community welfare. Future recommendations include evaluating the long-term impact of SUD in fostering inclusive settlements and promoting local economic growth through sustainable industries.

Keywords: agrarian reform, land conflict, local economic development, participatory planning, sustainable urban design

1 INTRODUCTION

Land issues, dating back to Ancient Greece, have long disadvantaged low-income populations (Kartodihardjo and Cahyono 2021). These challenges are global, prompting the development of Agrarian Reform regulations. Addressing land issues requires in-depth research into evolving social structures and expert perspectives to meet a country's development needs.

Agrarian Reform covers multiple dimensions: rights, security, structure, egalitarianism, gender, compensation, macroeconomic, environmental, and process (Gafuraningtyas *et al.* 2024). Countries have used reform to achieve these goals, aligning with SDGs like No Poverty and Decent Work, also Sustainable Cities and Communities. Effective land management must promote justice and community welfare (De Vries 2024).

DOI: 10.1201/9781003645542-22

Indonesia, an agrarian nation with a long history of land conflicts dating back to colonial times (Darmawan *et al.* 2023), recognised the need to reform land ownership long before the enactment of the Basic Agrarian Law. At that time, the government understood the crucial role of balanced land ownership in shaping economic development. However, many land issues, particularly in rural areas, remain unresolved.

Land issue resolution in Indonesia centres on conflicts over National Strategic Project acquisitions and cultivation rights (*Hak Guna Usaha* or HGU). The HGU scheme, a partnership between companies and the government for up to 25 years, impacts the local economy by creating jobs and business partnerships, particularly in plantations (Parihah *et al.* 2022; Herwati and Sumarlan 2016). However, challenges remain, including unproductive land, contract violations, and companies continuing land use after contracts expire, leading to community conflicts. High dependence on agriculture prompts locals to claim problematic HGU land, as outlined in Presidential Regulation No. 86 of 2018, which mandates 20% of HGU land be reserved for public use.

Agrarian Reform is a key goal in President Jokowi's "Nawacita" development agenda, focusing on balancing asset redistribution and land access. As a result, it significantly shapes regulations, particularly at the local government level. Subang Regency is one of the regions in Indonesia, and a significant portion of its plantations are identified as HGU land. Since 2021, Agrarian Reform efforts have focused on resolving conflicts, particularly regarding HGU land owned by PT. PG Rajawali II is a sugarcane plantation company. These efforts have allocated land for housing and public facilities for the community. However, some land remains abandoned and unproductive, requiring a utilisation scheme to ensure fair and sustainable land access management.

This study aims to develop a land management model to boost the local economy, often called a Pilot Project in Agrarian Reform contexts. The research integrates Sustainable Urban Design with local potential, using land released from HGU in Subang Regency. Sustainable Urban Design focuses on social, economic, and environmental factors, implemented through collaboration with local communities as key drivers. The goal is to empower communities to sustain their livelihoods and effectively manage local assets. This model can be adapted to other regions and tailored to their unique conditions and needs. The spreading effect occurs because almost all areas have the same land problems in Subang Regency. So that other areas can adopt the successful example of the implementation of the SUD concept.

Since the 1950s, Urban Design principles have evolved alongside the dynamic growth of cities and societies (Nag and Ghosh 2022). Urban and Landscape Design, a key aspect of urban planning, focuses on a city's aesthetics, layout, and form (Karandeev *et al.* 2021; Tian *et al.* 2014). It is closely linked to the urban environment, addressing interconnected spatial, ecological, economic, and social factors. Research by Karandeev *et al.* (2021) highlights the importance of parks and green spaces in sustainable urban development, including suburban areas. Similarly, Jamei *et al.* (2020) found a significant relationship between Urban Design elements and air temperature levels, demonstrating the impact of design on climate change.

Urban Design elements are inherently linked to sustainability (Ameen *et al.* 2015; Das and Das 2022; Rodríguez *et al.* 2014). Sustainable Urban Design integrates sustainability components into city planning, focusing on social, economic, and environmental aspects (Lange and Waal 2016; Mersal 2016). To support urban dynamics and holistic development, urban planning must embrace sustainable principles. This approach can be adapted to diverse regions, addressing local needs and conditions, and is essential for ensuring community resilience amidst growing pressures on urban life.

The most common urban design elements, as outlined by Shirvani in Sari and Alhamdani (2020), include land-use patterns, circulation and parking, building mass forms, pedestrian paths, open spaces, activity support elements, signage, and preservation. These elements can evolve and adapt to urban science and planning needs. The use of urban design elements can be tailored to the needs and existing conditions of the study area (Dias *et al.* 2014; Tangestanizadeh and Pari 2017).

2 METHODS

The research was conducted on land released from the Right to Cultivate by PT. PG Rajawali II, located in Pasirbungur Village, Purwadadi District, Subang Regency. This location underwent a by-name, by-address data collection process in 2022 and was designated as redistributed certified land in 2023. According to existing conditions, the research site is situated in a public facility, specifically a Sports Field in Rawasari Hamlet, covering a total area of 6 hectares and surrounded by residential areas. This location was chosen as a Pilot Project for Agrarian Reform because the land conflict has been going on for decades, and the local community has been very participatory in the resolution process.

Based on Figure 1, data collection for this research began in 2022 by conducting direct observation, social-economic mapping, and interviews with community leaders to understand the area's potential and issues thoroughly. To deepen the application of theory and design concepts, a Focus Group Discussion (FGD) process was conducted with various stakeholders, which resulted in the following inputs and support: 1) The government supports the land acquisition process and regulations; 2) Academics provide input on the application model of the theory and concept of community development; 3) Private sector plays a role in the physical implementation model of infrastructure and funding assistance.

In the second stage, a Pre-modelling Theory Approach was conducted by synthesising the theories of Sustainable Urban Design and Participatory Planning and reviewing various regulations that are relevant to the design analysis process. The results of the synthesis of various theories and policies and the results of field data collection are used for site analysis with SUD as the theoretical basis. This step identified development needs, forming the basis for adaptive urban design components aligned with current conditions. The process involves diagrammatic and qualitative methods to outline the design requirements comprehensively.

The study's final output is a SUD Model (in the Modelling Process), serving as a prototype for the Agrarian Reform Access Arrangement Program in Subang Regency. The process uses visual tools like ArcGIS for socio-economic mapping, AutoCAD for 2D design visualisation, and SketchUp for 3D modelling. The outcome will offer design recommendations that align with the area's potential, challenges, and community needs. The final product of this activity is a Masterplan document and a financial and management scheme for the area.

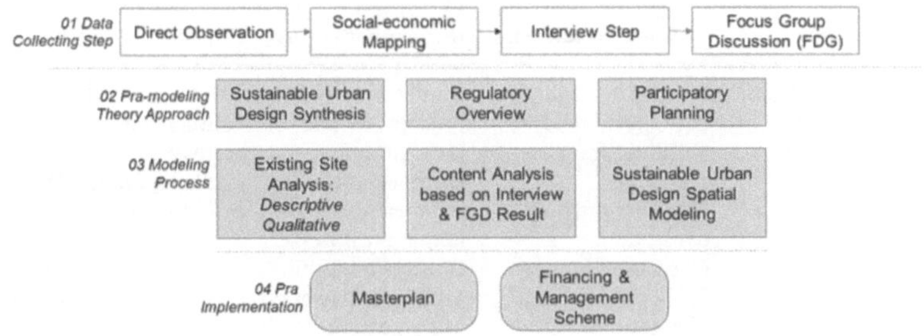

Figure 1. Design analysis process.

3 RESULT AND DISCUSSION

The socio-economic mapping identifies both physical and non-physical conditions in the area. Physical observations are made of infrastructure that directly supports social and economic activities around the area, such as land conditions, roads, drainage, etc. Meanwhile, non-physical factors include individual livestock and fishery activities, such as residents grazing sheep and occasionally allowing them to roam Rawasari Field. Additionally, Rawasari Hamlet has the potential for undeveloped cattle and poultry farming, with local markets as the primary distribution method.

A unique economic opportunity in this area is the emergence of a resident practising home-scale freshwater lobster aquaculture, including broodstock maintenance, seedling, and lobster rearing. This activity has been identified as a potential focus for the Agrarian Reform Access Arrangement Program in Subang Regency. However, significant improvements are needed, particularly in aquaculture infrastructure and human resources to manage the operations effectively.

Focusing on local development in Rawasari Hamlet is essential, as many retired rubber plantation workers from PTP XXX live below the poverty line. Programs are needed to help optimise their skills and ensure their livelihood in later years. This aligns with the Agrarian Reform Access Arrangement's goal of boosting the community's economy and improving environmental quality.

Based on the sustainability components of the Sustainable Urban Design concept, the design recommendations and visualisations integrate social, economic, and environmental aspects. As shown in Figure 2, the design accommodates the community's social activities while retaining the sports arena function, with improved land use to better meet local needs. Additionally, supporting spaces such as tourism areas and SME outlets promote economic sustainability. Information centres, educational facilities, and R&D hubs are designed to provide sustainable community development and activity spaces.

The circulation and drainage systems of the streets will be reorganised to improve environmental sustainability. As new activities and the area's development emerge, the predicted increase in density makes circulation and parking arrangements essential. The concept of sustainability is closely related to urban design components, ensuring that all recommendations are aligned with sustainability principles (Ameen *et al.* 2015; Rodríguez *et al.* 2014). Although the research location is rural, sustainable urban design can be effectively adapted to meet its needs, as rural areas often require more specific planning than urban areas. Table 1 shows the design's implementation process along with its impact and implications on the SDGs.

Table 1. Analysis of urban design component needs at rawasari field.

Component	Description	Existing conditions	Design recommendations	Impact	SDGs support
Land use	Allocation of land use and control of land utilisation	Used only as a regular sports field without clear planning Used for grazing livestock Occasionally used for special events like national holidays	Implement comprehensive land use planning Optimise land use for social and economic purposes Control land utilisation post-asset Management	Resolving land conflicts and increasing local economic activity	SDG's 1: No-poverty SDG's 8: Decent Work and Economic Growth SDG's 11: Sustainable Cities and Communities
Open spaces	Optimisation of activity spaces and availability of green areas	Lack of green open spaces or effective water catchment areas Lack of comfortable and safe social interaction spaces	Establish interaction zones Develop sports zones Provide appropriate vegetation Create inclusive, open spaces	Landscaping, reforestation, and increasing social interaction venues	SDG's 11: Sustainable Cities and Communities

(continued)

Table 1. Continued

Component	Description	Existing conditions	Design recommendations	Impact	SDGs support
		The vegetation planning is not suited to the needs and existing environmental conditions. Inefficient water catchment function leading to frequent flooding during rain			
Circulation and parking	Traffic and parking management, optimisation of road networks and their components	There is no parking arrangement, and the roads are still dirt paths without integrated drainage.	Install environmental drainage systems Designate parking zones Manage parking zones	Reduce flood risk and improve circulation	SDG's 11: Sustainable Cities and Communities
Signage	Directional and location/activity signage	There are no directional signs within the area.	Install road signs Place location markers Establish area landmarks Provide location maps and information centres	Strengthening the area's image and tourist attractions	SDG's 11: Sustainable Cities and Communities
Supporting elements	Economic, tourism, educational, research, and other centres	There is no professional and managed planning for economic centres. There are no dedicated spaces for education and research There is no tourism management in the area	Develop economic zones Establish tourism zones Create research and development zones Set up educational zones	Improving the quality of life of the community and the local economy	SDG's 1: No-poverty SDG's 8: Decent Work and Economic Growth SDG's 11: Sustainable Cities and Communities

Source: Shrivani Data Processing in Sari and Alhamdani (2020)

This area design aims to implement Access Management that aligns with the community's needs and existing conditions. A professional management system is essential for sustainable and effective execution. The scheme follows the Pentahelix Theory of regional development, involving key stakeholders, the government, the private sector, academics, and media to drive economic development and area planning (Susanti *et al.* 2022).

The key stage in this management system is forming a professional team from within the local community, including leaders, youth groups, and women's organisations. Strengthening this team requires targeted training and skill development to build professional management that can drive the local economy by addressing challenges and innovating in response to market demands. The government oversees the regulatory process, academics contribute to theoretical studies and community empowerment, while the private sector supports funding through initiatives like Corporate Social Responsibility (CSR).

With strong internal management and effective coordination with the external sector, an empowered community capable of managing the area independently is expected. This will create internal and external funding opportunities, while local businesses such as fisheries, livestock, and agriculture will find their markets. Over time, the production and distribution chain will strengthen, encouraging collaboration with larger companies and boosting regional tourism. However, potential challenges, such as increased waste, air pollution and socio-cultural changes, must be anticipated. Preventive measures should be planned from now on to prevent these problems.

Figure 2 illustrates the design model based on SUD theory, integrating economic, environmental, and socio-cultural factors for a sustainable community solution. A key example is the "*Rumah Belajar*" Educational House, which began with a site analysis to assess potential and challenges. The collaborative design process involved multiple stakeholders, ensuring the master plan met local needs. Using the Pentahelix model, it engages the government, private sector, community, media, and academics. The project is a replicable pilot, with the media playing a key role in spreading the initiative. Challenges include ensuring ongoing community involvement and securing funding for non-profitable areas.

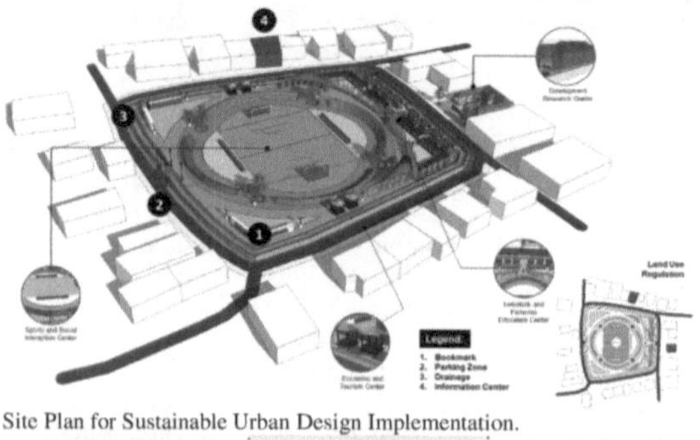

(a) Site Plan for Sustainable Urban Design Implementation.

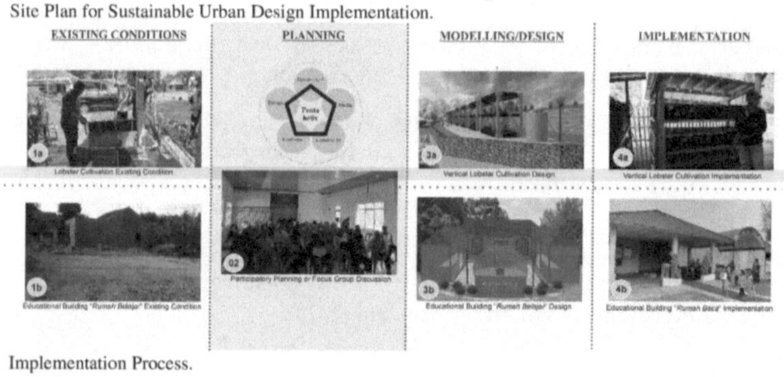

(b) Implementation Process.

Figure 2. Progress of sustainable urban design implementation in rawasari hamlet.

4 CONCLUSIONS

Agrarian Reform seeks to restructure land ownership and resolve disputes, particularly in rural areas with high poverty. While past efforts have focused on Asset Management through land titling, the program's core is Access Management for community economic empowerment. This research outlines a Pilot Project designed for Agrarian Reform, addressing a land conflict over the use rights of PT PG Rajawali II, beginning in 2022. Implementing Sustainable Urban Design offers an effective solution for small-scale area management, transforming regions into social, economic, and environmental hubs. This results in an Area Management Master Plan that guides future development. While land titling provides short-term benefits, it doesn't meet the long-term needs of rural residents. Therefore, transparent Access Management with lasting economic impact is crucial to prevent land sales post-asset Management. The program aims to serve as a model for other regions in Indonesia, emphasising local economic development for sustainable Agrarian

Reform. Further research is needed to assess the impact of Sustainable Urban Tourism on communities. To ensure consistency, government regulations should formalise the Master Plan and prevent rapid land sales, with annual monitoring and evaluation. This project could become a flagship program for the Subang Regency Government.

REFERENCES

Ameen, R. F. M., Mourshed, M., and Li, H. (2015). A critical review of environmental assessment tools for sustainable urban design. *Environmental Impact Assessment Review*, 55, 110–125. https://doi.org/10.1016/j.eiar.2015.07.006

Darmawan, D. A., Soetarto, E., Asnawi, Y. H., and Mahasari, J. (2023). The History of Agrarian Reforms in Indonesia: A Sociological Perspective, *Journal of Law and Sustainable Development*, 11(11): e2196, 1–27. Doi:10.55908/sdgs.v11i11.2196

Das, S., and Das, S. (2022). Sustainable urban design guidelines for different types of urban ponds: a case study of sylhet, Bangladesh. *Current Urban Studies*, 10, 593–610. https://doi.org/10.4236/cus.2022.104035

De Vries, W. T. (2024). How to know if land management is responsible? evaluating the 8r framework of responsible land management. *The Journal of Indonesia Sustainable Development Planning*, 5(1), 76–80. https://doi.org/10.46456/jisdep.v5i1.546

Dias, N., Professor, S. C., and Professor, E. B. (2014). The current approach of urban design, its implications for sustainable urban development. *Procedia Economics and Finance*, 18, 497–504. https://doi.org/10.1016/S2212-5671(14)00968-X

Gafuraningtyas, D., Koestoer, R. H., Setiadi, H., and Lessy, M. (2024). Transformative pathways of agrarian reform: comparative bibliometric insights of southeast asia and indonesia cases. *The Journal of Indonesia Sustainable Development Planning*, 5(1), 38–53. https://doi.org/10.46456/jisdep.v5i1.534

Herwati, S. R. M., and Sumarlan, Y. (2016). Peasants' land rights claims over plantation companies' sites in Central Java, Indonesia (1998–2014). *Indonesia Law Review*, 6(1), 111–133. Doi: http://dx.doi.org/10.15742/ilrev.v6n1.164

Jamei, E., Ossen, D. R., Seyedmahmoudian, M., Sandanayake, M., Stojcevski, A., and Horan, B. (2020). *Renewable and Sustainable Energy Review*, 134, 110362. https://doi.org/10.1016/j.rser.2020.110362

Karandeev, A., Sedykh, O., and Yartseva, E. (2021). Accessing landscape design of small towns in russia. *Proceedings of the Institution of Civil Engineers-Urban Design And Planning*, 174(3), 116–130. https://doi.org/10.1680/jurdp.20.00078

Kartodihardjo, H., and Cahyono, E. (2021). Agrarian reform in Indonesia: analyze concepts and their implementation from a governance perspective. *Jurnal Manajemen Hutan Hujan Tropika*, 27(te), 1–8. Doi: 10.7226/jtfm.27.te.1

Lange, M. D., and Waal, M. D. (2016). *Owning the city: new media and citizen engagement in urban design. Apple Academic Press* (pp. 87–110), London.

Mersal, D. A. (2016). Sustainable urban future: environmental planning for sustainable urban development. *Procedia Environmental Sciences*, 34, 49–61.

Nag, S., and Ghosh, M. (2022). Urban Design-Introduction, History, Characteristics, Parameters. *Encyclopedia of Design* (pp. 364–369), London. Doi:10.5040/9781472596154-BED-U009

Parihah, V. S., Prastiawan, M. A., and Hapsari, R. A. (2022). Penyelesaian sengketa tanah dengan hak guna usaha dalam penertiban tanah terlantar [Settlement of land disputes with land use rights in the regulation of abandoned land]. *Administrative Law and Governance Journal*, 5(3): 205–215.

Rodríguez, M. I., Cuevas, M. M., Martínez, G., and Moreno, B. (2014). Planning criteria for water sensitive urban design. *WIT Transactions on Ecology and the Environment*, 191, 1579–1591. https://doi.org/10.2495/SC141342

Sari, D. P., and Alhamdani, M. R. (2020). The urban design guidelines of sungai pinyuh street corridor. *The 1st International Conference on Urban Design and Planning*, 409: 012028. Doi:10.1088/1755-1315/409/1/012028

Susanti, R., Purwandari, S., and Prilosadoso, B. H. (2022). Penta helix as strategy of tourism village development in karangasem village, bulu district, sukoharjo regency. *International Journal of Social Science (IJSS)*, 2(4), 1979–1984. Doi: https://doi.org/10.53625/ijss.v2i4.4261

Tangestanizadeh, N., and Pari, I. (2017). Sustainable urban design with an approach in sustainable urban development. *4th International Conference of Sustainable Architecture and Urbanism.*

Tian, Y., Jim C. Y., and Wang, H. (2014). Assessing the landscape and ecological quality of urban green spaces in a compact city. *Landscape and Urban Planning*, 121, 97–108.

Towards Resilient Societies: The Synergy of Religion, Education, Health, Science, and Technology – Rahiem (Ed.)

Poverty and marginalisation in Abdo Khal's Tarmi Bi Sharar

Vania Sabila Maharani Jauhari & Rizqi Handayani
Universitas Islam Negeri Syarif Hidayatullah Jakarta, Indonesia

ABSTRACT: This article examines poverty, family disintegration, and the violation of societal norms in Abdo Khal's novel *Tarmi Bi Sharar*, focusing on how marginalised communities in Saudi Arabia respond to government policies affecting natural resources. Using qualitative methods and Alan Swingewood's sociological framework, the study uncovers the resistance of marginalised characters, particularly Hameed and Salem, to government actions that restrict access to natural resources and eliminate vital income opportunities. The article highlights the Saudi Arabian government's tendency to employ the death penalty as retaliation for acts of disobedience. Policies limiting access to the sea in impoverished areas have intensified poverty, family breakdown, prostitution, theft, alcoholism, and environmental degradation. This research underscores the urgent need for comprehensive policies prioritising social welfare and expanding economic opportunities, particularly for marginalised communities in Jeddah. Implementing policies that support the poor and disenfranchised can foster greater security and social harmony in Saudi Arabia while reducing the risk of political unrest resulting from oppressive and inhumane policies.

Keywords: Abdo Khal, Arabic novel, family disorganisation, marginal society, poverty

1 INTRODUCTION

Inadequate living conditions worsen the quality of life for vulnerable communities by contributing to environmental problems like pollution and slums. Abdo Khal addresses how a marginalised culture is portrayed in society in his book Tarmi Bi Sharar. The social issues that the community faces and the reality of Saudi society are both reflected in the book (Ahlam bin Asy-Syaikh 2018). Tarmi Bi Sharar, the novel, was initially released in Arabic by Al-Jamal Publications in 2009 (Abdo Khal, Saudi Arabia, International Prize for Arabic Fiction, n.d.). Abdo Khal's novel delves into the effects of unchecked wealth on Saudi society, specifically focusing on the role of corruption. The story, set in Jeddah, is told in an opulent castle where the residents strive for prosperity and work. However, because of the owner's autocratic behaviour, they find the palace more brutal and depressing than anticipated. Thariq, the main character, has been enslaved in the palace for twenty years. Because of its sensitive examination of politics, sexuality, and religious topics that are not typical of Arab society, the book was banned in the Arab world. The author discusses the intersection of two domains, double standards, and their interactions. Consequently, he can vociferously articulate the malevolent aspects of his society (Dante 2010).

Tarmi Bi Sharar, written by Abdo Khal, won the 2010 International Fiction Prize. It narrates the tale of 50-year-old Tariq Fadel, who uproots his modest upbringing to work at the Master's Palace in Jeddah, Saudi Arabia, during the 1980s (Khal 2011). The novel captures Saudi reality from the seaside city of Jeddah, combining truth and imagination to depict contrasting scenarios of human relationships with work, the sea, sex, religion, and social class (Wadi 2021). Saudi Arabia has undergone tremendous social, political, economic, and cultural transformations since the 1930s (Hidayat and Machmudi 2017). Legitimacy has shifted from tribal to state, with the Al Saud family emerging as the dominant social class. Natural gas wealth has revolutionised the nation's politics and economy, but it also has the potential to breed inequality in the community, corruption, and power struggles. Among the countries with the highest execution rates is Saudi

 DOI: 10.1201/9781003645542-23

Arabia, where 1,243 people were put to death between 2010 and 2021, mostly women, children, and foreigners (Al-Hatloul 2023). This emphasises how important it is to change the law to protect the nation's justice and human rights.

The research conducted by Husam and Mansoer focuses on linguistic characteristics, especially those seen in narrative repetition. The repetition indicates a state of hesitation and intensifies feelings of regret (Al-Mahwas 2021; Muhibis, n.d.). Sulaimani and Saadiyah research focuses more on the rebellion carried out by marginalised communities shown in the novel (Sulaymani and Saadiyah 2022). In addition, Sameh and Ahmed's review, "Throwing Sparks," also discusses translation techniques related to translating culturally unique taboos. This review focuses on maintaining the same metaphor when the intended audience is aware of cultural nuances or changing it when the culture is misinterpreted (Yousef and Albarakati 2023). Finally, Albalawi's writing focuses on death and the process of dying in Abdo Khal's fiction. The writing explores several manifestations of death but fails to address the issues of marginalised communities (Albalawi 2023). Existing studies only focus on aspects of narrative structure, translation techniques, and the rebellion problems of marginalised communities. There has been no research that discusses the social realities that occur in the Marginal community of Jeddah, such as poverty, family disorganisation, and violations of community norms in Tarmi Bi Sharar by Abdo Khal as an impact of the bad policies of the Saudi Arabian government. In the novel Tarmi Bi Sharar, Abdo Khal attempts to criticise the economic and political policies taken by the Saudi Arabian government to reduce social inequality so that poverty, family disorganisation, and social crime can be minimised.

2 METHOD

The methodology of this research focuses on qualitative content analysis research, which examines the response of marginalised communities to government policies in the novel Tarmi Bi Sharar by Abdo Khal. This study uses primary data in the form of narratives in the novel Tarmi Bi Sharar by Abdo Khal. In contrast, secondary data is taken from books and scientific journals, both printed and online. Data collection is carried out using the technique of reading and identifying narratives in the novel in the form of sentences or paragraphs related to the response of marginalised groups to Jeddah government policies, then classified based on the social problems described in the novel. The available data is analysed using a sociology of literature approach to view literature as a social mirror of society underlies it. Finally, the social reality in the novel is reconstructed in relation to the reality of Jeddah society.

3 RESULTS AND DISCUSSIONS

3.1 *The portrayal of marginalised communities in the novel*

A concept that is often associated with marginality is poverty. The view that globalisation will impact reducing poverty and social inequality needs to be reviewed because, in reality, globalisation in several countries, including Saudi Arabia, has impacted the emergence of various social problems (Chand *et al.* 2017). Acute poverty is a common occurrence in marginalised communities, and it leads to a host of social problems like dysfunctional families, social norm violations, and deteriorating environmental problems. In the novel Tarmi bi Sharar, marginalised communities experience the following effects:

3.1.1 *Poverty*

Poverty is a complicated and multifaceted problem. Several factors contribute to poverty, including hunger, psychological problems, illiteracy, a lack of access to basic infrastructure, illness, and the inability to manage resources to deal with vulnerability rather than discussing income (Narayan *et al.* 2000). Regardless of a nation's level of development, poverty is a serious problem. Its severity is increasing, impacting development programs and creating income gaps between affluent and low-income groups. The severity of this problem increases in developing cultures, underscoring the significance of tackling poverty. This is exemplified in the subsequent quotation:

ونفوس بعثرهتا الفاقة فخرجت تبحث هلا عن مكان داخل الجنة. هكذا، وبسرعة متناهية تكونت طحالب الفقر، واستشرت في ذلك الحي البائس، ولمن أراد القفز للجهة الأخرى عليه أن يتخفف من حمولة الأيام، حمولات الضمير، والأخلاق، هذا إن حملها أحدهم أصلا. (عبده خال, ترمي بشرر, ص: 38)

The novel does a good job of depicting poverty by drawing a contrast between the luxurious lifestyle of a palace and the deplorable conditions in slums. For those who live inside the palace, it is a paradise; however, it is a place of suffering for those who live outside. The sea, the main source of income for the underprivileged population, is overfished, which depletes natural resources. Their economic well-being is negatively impacted by the unchecked growth of concrete infrastructure, which limits their capacity to use these resources. The downtrodden depend on the generosity of the wealthy in the palace, and they compete to win over the rulers' compassion and support.

3.1.2 *Family disorganization*

Broken family relationships are a sign of disorganised families (Swamy 2021). Interpersonal conflicts and psychological strain can result from financial limitations, existential uncertainty, and family educational disparities. Social injustice makes these problems worse and breeds dissatisfaction and hopelessness. Rising crime rates and a general lack of security in neighbourhoods worsen these problems. Inadequate social assistance is a consequence of dysfunctional families, particularly for children who are more likely to participate in criminal or undesirable activities. The subsequent quotation exemplifies this:

تلك الأم الحية لم أرها منذ أن انتقلت لبيت زوجها، فمع قبولها بالزواج من جمال المهندس شعرت بأنها خانتني. خانت أمومة كان عليها أن تبقيها داخلي كأم لا تمنح جسدها لرجل آخر، ينهش ثدييها اللذين وهباني الحياة (عبده خال, ترمي بشرر, ص: 157)

The story's protagonist, Thariq, experiences betrayal and abandonment from his mother following her father's departure. Because of her dysfunctional upbringing, his aunt Khairiyyah harbours a strong desire for vengeance. After Thariq's mother is severely injured and becomes mute, he becomes determined to exact revenge for her murder. This act of retaliation is an outward sign of his anger and mental instability, which are caused by dysfunctional families and a diminished sense of trust in his loved ones. Thariq's journey illustrates how familial chaos can significantly impact a person's psychology and behaviours, even at the most basic level of interpersonal relationships.

3.1.3 *Violation of societal norms*

Intense Family strife and financial strain can cause people to look for quick fixes, which loosens social norms. Family dysfunction is becoming more common as a result of people deviating from social norms due to a lack of support. The deviant behaviour of younger generations reinforces these norms. Environments that promote undesirable behaviours and doubt identity give rise to subcultures that reject or disregard established norms. Differing moral standards, inequality, family conflicts, and insufficient guidance all impact deviance from societal norms.

3.2 *The Portrayal of transgressions in the novel*

3.2.1 *Prostitution*

Poverty, unemployment, inadequate reintegration services, stigma, unfavourable social perceptions, familial expectations, and the need to adapt to a new lifestyle are all factors that contribute to prostitution (Bhat 2022). Its rise is partly attributed to low education levels, economic hardship, and broken families. Ignorance of Islamic principles, the desire for financial independence, and religious emancipation can also be driving forces behind prostitution. For women, this practice is a serious problem. The following quote evidence this:

والصغيرات حين يتعلمن أن ابتسامتهن لها مقابل مادي يكن قد وضعن أقدامهن على طريق البغاء، ويعشن بقية أعمارهن عاهرات يف شبابهن، وقوادات مع غروب جمالهن (عبده خال, ترمي بشرر, ص: 98)

The work highlights the connection between poverty and the need for survival while examining the harsh realities of prostitution for young women in crowded urban areas. The young female must make difficult decisions while navigating a barren environment full of hunger and suffering. Although these women would prefer not to, they are forced to become prostitutes to support their families or survive. Nevertheless, the results of their deeds are deep sadness, despair, and a vicious cycle of exploitation and poverty. The reader is forced to acknowledge that prostitution is their only chance of survival as the narrative progresses.

3.2.2 *Juvenile delinquency*

The term "juvenile delinquency" describes a range of legal and social norm violations that frequently arise from development and maturation. The United Nations Guidelines for the Prevention of Juvenile Delinquency emphasise that between two-thirds and three-quarters of all juvenile offences are committed by members of gangs, highlighting the fact that juvenile delinquency is frequently a group phenomenon (Almasi 2003). Aggression and violence are used as a form of conflict resolution by some groups and subcultures rather than labelling antisocial behaviour or young conduct as juvenile delinquency. When someone under the age of eighteen commits crimes against society, including drug abuse, assault, sexual promiscuity, adolescent delinquency, and other offences, it is considered juvenile delinquency. Peer pressure, unstable family environments, erratic social and economic circumstances, and inadequate parental supervision are all common causes of this behaviour. Examples of larceny, burglaries, physical assault, moral or ethical failings, drug misuse, and reckless driving are all examples of juvenile delinquency. These behaviours demonstrate a lack of adherence to current social and legal norms, resulting in negative consequences for individuals and society. It can be seen in the following text:

أول مرة أتسلل إلى مخدع أبي، وأمد يدي لجيبه، وقبل أن أنعم بنجاح أول سرقة في حياتي، تعلقني من أذني (عبده خال, ترمي بشرر, ص: 98)

The main character, Thariq, represents juvenile criminality by acting abnormally from a young age. He starts having bad relationships and doing things that are against the law. His upbringing and maltreatment mould his conduct, causing him to follow a route that defies social and legal norms. Thariq is a living example of the complicated problem of teenage misbehaviour and how it affects both the individual and the environment.

3.2.3 *Alcoholism*

Social and economic challenges in Arab societies frequently provoke alcoholism concerns. Challenging economic circumstances and intense social pressures can lead individuals to experience a sense of being confined in a repetitive pattern of stress and unpredictability. As a means of dealing with stress, certain individuals resort to alcohol as a kind of escape. This is demonstrated by the quote that follows:

احتجت لكمية وفرية من اخلمور، لتعينين على أداء املهمة اليت جلبت كي أهنيها كما ينهي جزار إراقة دماء ذبيحته على أحسن حال من غري أن يتهم بسوء أداء مهته (عبده خال, ترمي بشرر, ص: 121)

Within marginalised populations, alcohol is frequently perceived as a means of enjoyment and alleviation of stress resulting from the burdens of everyday life. Numerous persons experience a sense of being confined in repetitive patterns of uncertainty and financial constraints, desiring to break free from this state of affairs. Alcohol is readily available and is often seen as a source of temporary gratification.

3.2.4 *Homosexuality*

This research defines homosexuality as the experience of pleasurable emotions, motivations, and sexual attractions towards others of the same gender, typically accompanied by a desire to engage in sexual activities or partake in related behaviours. Homosexuality is demonstrated by the quote that follows:

كنت أنقض على فريستي لإثبات رجولة، وليس لإفراغ شهوة ولكي لا أخسر هذه الشهرة بين أقراني كنت أقدم على اقتناص فرائسي لإبقاء سيرتي مهابة بني أترابي، وبهذه الوسيلة أبعد بقية الصياديين عن التهامي (عبده خال, ترمي بشرر, ص: 43)

In the story, the depicted environment does not view sodomy activities as abnormal behaviour. However, it is perceived as a means to establish a reputation for masculinity and is regarded as an indication of an individual's sexual desire rather than a sexual inclination. This method facilitates categorising individuals as either predators or prey within the social setting outlined in the story.

3.2.5 *Environmental issues*

Due to societal expectations and financial limitations, people who live in slums frequently overlook environmental issues and endure poor living conditions as well as limited access to public goods and services (Hung 2024). Living conditions that are unclean, littering, and a lack of interest in sustainability result from this. In India, some causes of environmental damage due to urbanisation lie in the legislation and the country's regulating agencies (Uttara and Aggarwal 2012). Urbanisation does not always harm the environment, as several measures can be taken to effectively mitigate them, namely improving urban planning and design, increasing resource utilisation efficiency, and promoting green technology and innovative financing (Gao 2024). The situation is made more difficult by elements like chaotic families and teenage delinquency. These problems are made worse by ignorance and disdain for environmental laws. This is seen in the subsequent quotation:

عابراً أزقة ملتوية تكدست بها القمائم. وفاحت منها روائح شتى مختمرة تجبره على سد فتحتي أنفه بأداء مسرحي متأفف لمن يرقب تنقلات خطواته (عبده خال, ترمي بشرر, ص: 136)

The marginalised environment is portrayed in the work as a dirty, unclean, and miserable place with broken windows, crumbling debris, and dilapidated buildings. It draws attention to the social injustice and personal hopelessness that residents of this harshly impoverished environment go through. Because literary works are frequently shaped by the experiences and circumstances of their authors, they frequently reflect social reality. For instance, a writer in a marginalised neighbourhood with environmental concerns and poverty may be inspired to tackle these topics in their stories, mirroring the neighbourhood's social difficulties. The difficulties that people who live in this setting face are highlighted in this depressing depiction.

4 CONCLUSION

Based on research findings, this paper shows that the government's decision to close the main marine resources has caused unpleasant reactions from various stakeholders, especially fishermen who feel significant losses. As a result, isolated communities, including prominent figures such as Hamed and Saleem, expressed significant rejection of construction projects that blocked their access to the sea. However, the rejection and resistance the characters in the novel carry out result in death. The deaths of the two characters show that the government does not protect human rights and, with its power, dominates the marginalised. Thus, the government's policy of restricting access to the sea in slum areas has exacerbated social inequality, resulting in poverty, family problems, social crimes (such as prostitution, juvenile delinquency, theft, alcoholism, and deviant sexual behaviour), and environmental damage. Based on this, this paper recommends that the government, especially the Jeddah government, formulate comprehensive policies covering social, economic, and environmental aspects to reduce negative impacts and social inequality. In addition, the government also needs to carry out sustainable urban development and equitable distribution of natural resources.

REFERENCES

Abdo Khal, Saudi Arabia, International Prize for Arabic Fiction. (n.d.). *International Prize for Arabic Fiction.* Retrieved January 31, 2024, from Https://www.arabicfiction.org/En/Abdo-Khal

Ahlam bin Asy-Syaikh. (2018). Al-wāqi'ah wa sinā'atu riwāyah al-muhammisyīn fī al-manẓūrīn al-ijtimā`ī wa an-naqdī. Majallatu Maqaliyyah, 14.

Albalawi, M. H. (2023). Death And Dying In The Fiction Of Abdo Khal. *Theory And Practice In Language Studies*, 13(7), 1730–1737.

Al-Hatloul, L. (2023). Bloodshed And Lies Mohammed Bin Salmans Kingdom Of Executions. *Retrieve European Saudi Organisation Of Human Right*, 3.

Al-Mahwas, M. bin Àbdul A. (2021). Al-Tawātur As-Sardiy fī Riwāyah "Tarmi Bisharar" li Àbduh Khal. *Jāmiàh Al-Majmaàh*, 419–377.

Almasi, J. N. (2003). *World Youth Report Chapter 7 Juvenile DELINQUENCY* (p. 191). https://childhub.org/sites/default/files/library/attachments/332_397_EN_original.pdf

Bhat, R. M. (2022). *A Historical Study of Prostitution.* 2(04), 4–5. https://doi.org/10.55529/jmcc24.1.6

Chand, R., Nel, E., and Pelc, S. (2017). *Societies, Social Inequalities, and Marginalization Raghubir Chand Etienne Nel Stanko Pelc Editors Marginal Regions in the 21st Century.* Springer.

Dante, D. (2010, March 10). 'Terrifiying' Saudi Novel Wins Arabic Booker. 1.

Gao, X. (2024). Research on the Interrelationship Between Urbanisation and Sustainable Development, *Proceedings of the 3rd International Conference on Art, Design and Social Sciences.* 200–203. https://doi.org/10.54254/2753-7064/44/20240028

Hidayat, R., and Machmudi, Y. (2017). Perubahan Sosial Politik Di Arab Saudi 1932-1975. *Jurnal Middle East and Islamic Studies*, 4(1).

Hung, J. (2024). *Sociological and Policy Examination of Poverty_Led Crime in Indonesia: Case Study of (Child) Prostitution.* https://doi.org/10.1080/10875549.2024.2379766

Khal, A. (2011). Tarmi Bi Sharar. Mansūrāt Aljamal.

Muhibis, H. A. (n.d.). Al-Khiṭāb Al-Waṣif fī Al-Riwāyah "Tarmi Bisharar" li `Abduh Khal. Dirāsāt Tarbawiyyah, 13(52), 101.

Narayan, D., Raj Patel, K. Schafft, Rademacher, A., and Koch-Schulte, S. (2000). *Voices Of The Poor: Crying Out For Change.* Oxford University Press for the World Bank.

Sulaymani, R., and Saadiyah, N. (2022). Tamaẓurat At-Tamarrud fi Ar-Riwāyah As-Su`ūdiyyah, Riwāyatu "Tarmi Bisharar" li `Abduh Khal anmūżajuhā. *Jusūr Al-Ma`rifah*, 8(2), 416–413.

Swamy, C, C. (2021). Family Disorganization—A Social Problem in Karnataka. *SHANLAX International Journal of Art, Science and Humanities*, 8(1), 236. https://doi.org/10.34293/sijash.v8iS1-Feb.3958

Uttara, B., N., and Aggarwal, V. (2012). Impacts Of Urbanization On Environment. *Ijreas*, 2(1).

Wadi, F. (2021, February 16). Riwāyah "Tarmi Bisharar". *Tanāquḍāt Hayah Al-Qusūr wa al-Hawāmisy Bi `ain Su`ūdiyyah.* https://www.aljazeera.net/midan/intellect/literature/2021/2/169

Yousef, S. S., and Albarakati, M. A. (2023). Strategies for Translating Culture-Specific Methafor on Taboos in Abdo Khal's "Throwing Sparks." *Theory and Practice in Language Studies*, 13(6), 1403–1412

Towards Resilient Societies: The Synergy of Religion, Education, Health, Science, and Technology – Rahiem (Ed.)

Local wisdom, conflict resolution, and sustainable peace: A systematic literature review

Zulkifli Zulkifli
Universitas Islam Negeri Syarif Hidayatullah Jakarta, Indonesia
ORCID ID: 0000-0002-3231-5352

Nuryaman Nuryaman
Universitas Islam Negeri Syarif Hidayatullah Jakarta, Indonesia

ABSTRACT: This study explores the role of local wisdom in conflict resolution and sustainable peace in Indonesia. Employing a Systematic Literature Review (SLR) method, the research reviewed and analysed sixteen relevant articles from the Scopus database. The study identified twenty-one local wisdom practices addressing conflict resolution and sustainable peace, categorised into inter-ethnic tensions, religious disputes, crimes, territorial conflicts, and community welfare. While most practices serve latent functions in conflict resolution, only a few directly and actively resolve conflicts. These practices align with the Sustainable Development Goals (SDGs) by promoting freedom from violence, non-discrimination, and inclusivity. The study recommends that policymakers and practitioners develop strategies to integrate local wisdom with human rights and legal frameworks while enhancing its value through education and collaboration with community leaders. Given the study's limitation to published literature, future research should adopt ethnographic approaches to examine local wisdom practices in greater depth and contribute to developing sustainable peacebuilding models.

Keywords: conflict resolution, local wisdom, sustainable peace, systematic literature review

1 INTRODUCTION

Indonesia is globally notorious for its astonishing cultural heterogeneity (Nursabrina *et al.* 2023). Each folkloric has local and regional languages with the state's officially recognised religions: Islam, Christianity, Catholicism, Hinduism, Buddhism, and Confucianism (Budiwanti 2021), excluding local religious beliefs. However, these differences typically provoke conflicts between majority and minority groups, which can escalate into tensions or peaceful resolutions through debates and negotiations (Garing *et al.* 2023). Fortunately, in this rich tapestry of sociocultural landscape lies a profound reservoir of local wisdom, an identity of local culture fashioned by communal activities and transmitted orally from generation to generation (Hunaepi and Firdaus 2017; Pattiasina *et al.* 2024). Thus, local wisdom varies from language and literary systems, technology, livelihood, social organisations, knowledge, religion, and art (Syamsuddin and Purnama 2021). As these cultural practices are inherently wedged into Indonesian civilisation, the rising role of local wisdom in overcoming conflicts is unveiled in current literature, showing local wisdom as the fundamental value to sustaining peacefulness in today's societies (Ibo 2023; Meilinar *et al.* 2023; Ratnasari *et al.* 2023; Pasaribu *et al.* 2024). Despite the growing recognition of local wisdom in peacebuilding, studies remain unspecified on the potential of local wisdom practices in Indonesia to achieve the Sustainable Development Goals (SDGs). Therefore, this research aims to review Indonesian local wisdom dealing with conflict resolution and the functions of sustainable reconciliation. This study fills that gap in existing studies by offering insights that enrich the global discourse on unique local wisdom practices in Indonesia and their alignment with SDGs.

 DOI: 10.1201/9781003645542-24

2 METHODS

SLR (Systematic Literature Review) is a protocol-based literature review method (Muka *et al.* 2020), including identification, screening, eligibility, and inclusion as part of the PRISMA Model (Haddaway *et al.* 2022). Boolean logic was formulated (Local AND wisdom AND conflict AND Indonesia and Local AND wisdom AND peace AND Indonesia) to gather relevant documents from the Scopus Database. These keywords retrieved seventy-five papers in various languages, including journal articles, book chapters, and other sources. The results were screened to include only open-access, fully English to obtain more global perspectives and ensure language convenience for researchers, and peer-reviewed journal articles, resulting in 52 documents. Subsequently, through analytically examining titles, abstracts, subtitles, and key findings, 46 irrelevant articles to the research question were excluded. Finally, sixteen journal articles were suitably included in the study of conflict resolutions (Figure 1).

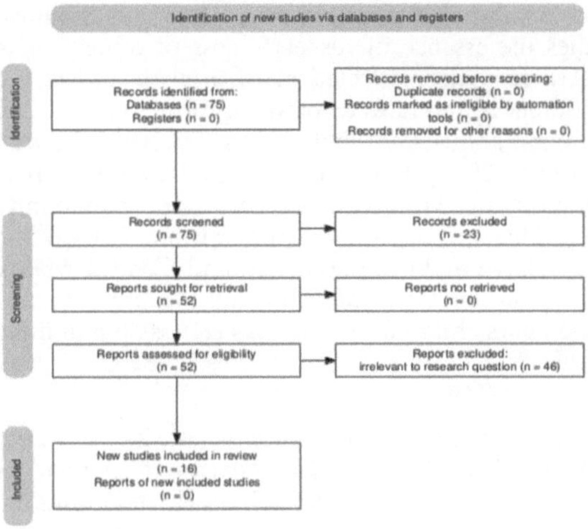

Figure 1. PRISMA Flow (Haddaway *et al.*, 2022).

3 FINDINGS AND DISCUSSION

As Shown in Table 1, three types of local wisdom are found in the reviewed articles, namely language expressions, rituals, and artefacts which have developed in various parts of Indonesia. Each type has connections with certain functions in conflict resolution, as explained in succeeding sections.

Table 1. Types of local wisdom and their contributions to conflict resolution.

Conflicts	Language expressions	Rituals	Artifacts
Inter-ethnicities	*Ale Rasa Beta Rasa, Samama Baraya, Meutingkue Tijiek, Pela Gandong, and Kadeudeuh*	-	-
Religion disputes	*Satu Tungku-Tiga Batu and Ahimsa*	*Nyadran and Kaneka Hagasir*	-
Criminalities	*Belom Bahadat and Silih Hampura*	*Tepung Tawar*	-
Territorialities	*Purpusage and Rewang*	*Rewang and Molo*	*Kalosara*
Community welfare	*Seloko Adat*	*Tingkeban, Dewi Sri and Khanduri*	-

3.1 *Ethnic conflict resolutions*

Sociocultural differences, such as religion, ideology, and language, contribute to tensions of ethnic conflicts (Markhulia 2024), becoming regional or international crises when foreign powers get involved (Reuter 2023). In today's unavoidable multicultural circumstances, traditional language expressions from Indonesian local wisdom are universal values, including mutual care and respectfulness. Fostering these principles is a reasonable antidote to conflict-related civilian mortality, supporting SDGs' peace and inclusive humanitarian goals on a country scale. In parallel with cosmic war narratives, this actual praxis encompasses equality, which extends beyond mere tolerance to sustain respect and empathy, regardless of ethnicity or religious beliefs. Eventually, these localised wisdom practices underscore the universality of community-based reconciliation, offering scalable models to address broader global conflicts.

The rooted expression of *Ale Rasa Beta Rasa* (You Feel, I Feel) in Maluku embodies mutual care within communities (Matakena *et al.* 2022). In the sociocultural fabric of the Maluku people (Moluccans), this linguistic expression philosophically transcends to foster inclusive and supportive communities. Similar to *Samama Baraya* (Equal as Siblings) for Balinese religious diversity, which embodies the essence of respectfulness of other beliefs (Arjawa and Zulkifli 2021; Rasyid *et al.* 2023). It is also reflected in *Meutingkue Tijiek* (Caring Attitude), which essentially minimises differences and hatreds between former GAM (Gerakan Aceh Merdeka - Free Aceh Movement) combatants and civil society, evolving the sustainability of peace in Aceh Province (Nirzalin *et al.* 2023). On smaller scales, like organisations or institutions, *Pela Gandong* was implemented by the Moluccans as a means of interaction and conflict resolution (Pattipeilohy *et al.* 2023). This language-based local wisdom is used in hospitals' working culture and organisational behaviour in Ambon City and Central Maluku Regency. In family relationships, the *Kadeudeuh* or *Kanyaah* (affection) represents an agreed-upon sharing arrangement between the heir's children and siblings (Khosyi'ah and Rusyana 2022).

3.2 *Religious conflict resolutions*

As global religious conflict has gained concern since the late 20th century, religious harmony can be elevated by "One Stove, Three Stones" (*Satu Tungku, Tiga Batu*) philosophies among the Papuan residing on the West Coast (Cahyo and Indriasari 2021). This metaphor exemplifies the importance of dialogue and shared cultural values in maintaining peace and unity amidst religious diversity in the Fakfak region. Reflecting on the notable Poso (2001) and Ambon (2002) incidents in Indonesia, religious sentiments and manipulations can destabilise the community trust (Hanumbhawono *et al.* 2024; Harahap *et al.* 2023). To prevent this bloodshed war in the next civilisation, the universal human values and guiding principles of *Ahimsa* and *Atat Twam Asi* (I am You) embedded in Bali societies underscore the principle of opposing harming oneself and mitigating radicalism (Rasyid *et al.* 2023). *The Nyadran* ceremony, on the other hand, can build harmony among religious groups by honouring the founding fathers and offering prayers for the communal feast (Agung *et al.* 2024). Muslims and Hindus perform this ritual during 'Id al-Adha and Nyepi Day. The folks sense this rite will make their neighbourhood harmless. Similarly, the *Kaneka Hagasir* ceremony (Cahyo and Indriasari 2021) serves as a gratitude feast with the spirit of *Tormormarge* for family welfare and diverse religious affiliations in Papua.

3.3 *Criminal and legal conflict resolutions*

Scrutinising from legal cases in customary Palangkaraya Central Kalimantan, *Belom Bahadat* (living with inherent norms) is a criminal procedural law that intertwines beliefs in God, horizontal human relationships, and consensus-based deliberation (Wulandari *et al.* 2022). Central to this system is the role of *Penatua*, or customary elders, who are trusted arbiters tasked with resolving disputes and upholding individual rights. Although this method handled complex criminal cases of sexual harassment, domestic violence, traffic accidents, and land burning in Dayak communities, the justice framework's implementation could challenge secular regions or states as they might disregard God/religion. These practices suggest that while customary law offers valuable insights into the United Nations to end abuse, exploitation, trafficking, and violence as SDG's agenda, its integration into broader legal systems requires watchful negotiation to balance tradition with modern legal standards. The comparable contest possibly applies to the principle of *Silih Hampura* (forgiving each other) in Baduy society, Banten Province. *Silih Hampura* plays an initial role in conflict resolution, and the community habitually turns the cases to the traditional ritual of *Ngabokoran* (self-purification) when conflicts cannot be resolved. A similar scheme of removing unwanted menace is exhibited in the *Tepung Tawar* (Fresh Flour) ceremony, aiming to reconcile disputing parties, especially those involved in conflicts or murder (Alfitri *et al.* 2014; Alfitri and Hambali 2013). Led by a traditional leader acting as a mediator, the ceremony brings together the conflicting parties, facilitating peace and resolution in Lahat, South Sumatra. The concern manifests in how this local wisdom merely ends grave offences like killings and rapes by prioritising forgiveness to foster peacebuilding and familial unity, which is difficult to tolerate in modern-day criminal law reform and human rights.

3.4 *Territorial conflicts resolutions*

This enduring resilience of local wisdom across generations reflects a culture of mutual support and cooperation, aligning with the SDGs by promoting and enforcing non-discriminatory laws and policies for sustainable development. However, its practical implications may be limited to certain ethnic contexts and require robust legitimacy and endorsement from government authorities regarding people's lives and personal choices.

The disputes among land authorities in Karo Regency, North Sumatra Province, created social clashes between households and police officers after the volcanic eruption disaster. The customary conflict resolution model, *Purpusage*, has been passed to uphold equal life, promoting heart-to-heart communication to resolve social conflicts adjusted to typologies and actor characteristics (Lestari *et al.* 2019). Besides underpinning the values of latent function from language expressions, forms of rituals and ceremonies have the potential to resolve territoriality affairs. Most global issues, such as immigrant interrelations, are reflected in the *Rewang* tradition of Javanese immigrants who moved to Langsa, Aceh Province, due to Dutch colonialism in the 1910s (Rahman *et al.* 2022). This practice embodies a collective effort to assist relatives or neighbours.

Furthermore, *Kalosara*, as the only local wisdom cultural object, also addressed conflicts that often arise due to the invasion of migrants who have yet to acclimate to social conditions, exacerbated by primordialism focusing on group identities (Diab *et al.* 2022). *Sara Wonua*, a ritual material made of rattan, gold, silver, thread, and others, facilitates problem-solving, with all involved parties ultimately accepting decisions wholeheartedly. This stage marks the shape of an eight, meaning the customary leader's decisions harm no one. Utilised ritually, Ambon and Central Maluku communities undertake the *Molo* (diving) ceremony when conflicts over territorial rights face deadlocks (Garing *et al.* 2023). This customary law rules both parties to diving. The final justification for who floats quickly is that he is wrong and right if he is still drowning.

3.5 *Community welfare resolutions*

To preserve prosperous life cycles and community welfare, *Tingkeban,* and *Mitoni* rituals constitute a range of rituals and activities to bestow blessings on newborns and safeguard their well-being and prosperity. *Tingkeban* is a ceremony during the seventh month of pregnancy and centres on the expecting mother's welfare and health in Malang, East Java (Agung *et al.* 2024). At the same time, *Mitoni* strikes in the seventh month following the birth of a newborn, concentrating explicitly on the postnatal time to celebrate the influx of a neonatal. While nurturing the primary needs for living, Malang believes that the rituals associated with *Dewi Sri* as Rice Goddess (Agung *et al.* 2024). Apart from *Dewi Sri*, *Sri Kuncung* (tradition of corn plants) and *Sri Manis* (sugar cane plants) are particularly used in these ceremonies. In Acehnese tradition, *Khanduri* is deeply community-entrenched and is a unifying force across various segments of society (Rahman *et al.* 2022). It can be the *Khanduri Laot* for the fishing community, the *Khanduri Blang* for rice farmers, or *the Khanduri Maulid* commemorating the Prophet Muhammad's birthday. To further strengthen social relations, *Seloko Adat* is a repository of traditional literature and poetry in the Jambi language that guides residents (Halim and HM 2020), ensuring the community's commitment to fair and inclusive decision-making processes.

These principles embody profound teachings of togetherness, friendship, and mutual assistance, particularly maintaining peace and balance with environmental sustainability for world communities with similar social and land structures.

4 CONCLUSION

Amidst Indonesia's historical complexities, the significance of local wisdom in conflict resolution and the pursuit of sustainable peace has become a crucial area of academic research and practical application. The types of local wisdom including language expressions, rituals, and artifacts demonstrate the power of mutual respect, forgiveness, and dialogue in resolving conflicts, providing scalable and adaptable models that echo the principles of Sustainable Development Goals (SDGs), predominantly in fostering inclusive societies and peacebuilding. While most exhibit latent functions for conflict resolution, only a few possess manifest functions directly aiding in conflict resolution, such as *Molo, Kalosara,* and *Purpusage.* Considering the challenges of integrating local wisdom into broader frameworks, policymakers should develop mechanisms to ensure compatibility between human rights and legal standards and increase awareness of the value of local wisdom through educational initiatives and partnerships with religious and community leaders, further advancing its relevance in addressing contemporary conflicts and sustaining peace. By implementing these practices, Indonesia can be a global model for utilising cultural heritage to achieve sustainable peace. In the future, with its immersive approach, ethnographic research will capture the lived experiences, narratives, and interactions underpinning these practices within their cultural and environmental settings in the framework of formulating an applicable model of sustainable peacebuilding.

REFERENCES

Agung, D. A. G., Nasih, A. M., Sumarmi, Idris, and Kurniawan, B. (2024). Local wisdom as a model of interfaith communication in creating religious harmony in Indonesia. *Social Sciences and Humanities Open*, 9, 100827. https://doi.org/10.1016/j.ssaho.2024.100827

Alfitri, A., Awang, M. M., and Ahmad, A. R. (2014). Culturally-Responsive Strategies for Resolving Social Conflict in Rural Community. *Mediterranean Journal of Social Sciences*. https://doi.org/10.5901/mjss.2014.v5n20p2267

Alfitri, A., and Hambali, H. (2013). Integration of National Character Education and Social Conflict Resolution through Traditional Culture: A Case Study in South Sumatra Indonesia. *Asian Social Science*, 9 (12), p125. https://doi.org/10.5539/ass.v9n12p125

Arjawa, I. G. P. B. S., and Zulkifli, Z. (2021). The Social Integration of Hindu and Muslim Communities: The Practice of "Menyama-Braya" in Contemporary Bali. *Studia Islamika*, 28(1), Article 1. https://doi.org/10.36712/sdi.v28i1.10914

Budiwanti, E. (2021). Religion and public sphere: Revisiting balinese religious tradition in the island of thousand mosques. *Religion, Culture, and State Journal*, 1(1), Article 1. https://journal.unram.ac.id/index.php/rcs/article/view/3

Cahyo, P., and Indriasari, D. T. (2021). Preventing Religious Conflict in Papua Land: Adopting Cultural Traditions of Peacebuilding. *Asian Journal of Peacebuilding*, 9(2), 331–356. https://doi.org/10.18588/202108.00a119

Diab, A. L., Pabbajah, M., Nurina Widyanti, R., Muthalib, L. M., and Fajar Widyatmoko, W. (2022). Accommodation of local wisdom in conflict resolution of Indonesia's urban society. *Cogent Social Sciences*, 8(1), 2153413. https://doi.org/10.1080/23311886.2022.2153413

Garing, J., Firdaus, W., Herianah, H., Ridwan, M., Erniati, E., Budiono, S., and Pariela, T. D. (2023). Identifying and Resolving Conflicts Using Local Wisdom: A Qualitative Study. *Journal of Intercultural Communication*, 23(4), 69–81. https://doi.org/10.36923/jicc.v23i3.156

Haddaway, N. R., Page, M. J., Pritchard, C. C., and McGuinness, L. A. (2022). PRISMA2020: An R package and Shiny app for producing PRISMA 2020-compliant flow diagrams, with interactivity for optimised digital transparency and Open Synthesis. *Campbell Systematic Reviews*, 18(2), e1230. https://doi.org/10.1002/cl2.1230

Halim, A., and HM, P. (2020). The study of Seloko Adat as safety-valve to prevent religious conflict in Jambi City, Indonesia. *Jurnal Ilmu Sosial Dan Ilmu Politik*, 23(3), 223. https://doi.org/10.22146/jsp.50532

Hanumbhawono, W., Maarif, S., Swastanto, Y., and Madjid, M. A. (2024). Potential For Religious Exploitation of Conflict in Indonesia. *International Journal of Religion*, 5(10), Article 10. https://doi.org/10.61707/6hfhjv52

Harahap, A. S., Mulyono, H., Nuzul, A. N. A., Milhan, M., and Siregar, T. (2023). Dalihan Na Tolu as a Model for Resolving Religious Conflicts in North Sumatera: An Anthropological and Sociological Perspective. *Samarah: Jurnal Hukum Keluarga Dan Hukum Islam*, 7(3), Article 3. https://doi.org/10.22373/sjhk.v7i3.13091

Hunaepi, and Firdaus, L. (2017). Integrating local wisdom of sasak tribe in ecology learning to develop scientific attitude. *Proceeding 14th ADRI*, 478–482. https://doi.org/10.31227/osf.io/p36bc

Ibo, J. M. (2023). Revitalizing Peace Values in the Dimensions of Local Wisdom in Papua (Case Study: Sentani). *Migration Letters*, 20(S9), 1130–1140. https://doi.org/10.59670/ml.v20iS9.4952

Khosyi'ah, S., and Rusyana, A. Y. (2022). Inheritance settlement of descendants of children and siblings in Islamic law with local wisdom in Indonesia. *Cogent Social Sciences*, 8(1), 2126615. https://doi.org/10.1080/23311886.2022.2126615

Lestari, P., Kertamukti, R., and Ruliana, P. (2019). Use of Local Wisdom (Purpusage) through Heart-to-Heart Communication in Settling of Social Conflicts in Karo, North Sumatra Indonesia. *Jurnal Komunikasi: Malaysian Journal of Communication*, 35(3), 163–181. https://doi.org/10.17576/JKMJC-2019-3503-10

Markhulia, G. (2024). Geopolitical motives of interethnic conflicts in Sri Lanka. *Georgian Academy of Business Sciences "Moambe"*, 50, Article 50. https://doi.org/10.52340/gbsab.2024.50.08

Matakena, F., Watloly, A., Agustang, A., Adam, A., and Alim, A. (2022). The Self-Concept of Ale Rasa Beta Rasa in the Orang Basudara Community in Ambon (Studies on the Community of Passo and Batumerah Country). *International Journal of Criminology and Sociology*, 9, 1306–1311. https://doi.org/10.6000/1929-4409.2020.09.150

Meilinar, F., Syahrin, A., and Khairani, C. (2023). Analysis of Local Wisdom Values in the Seumapa Tradition in Society in Bireuen Regency. *Edumaspul: Jurnal Pendidikan*, 7(2), 3223–3231. https://doi.org/10.33487/edumaspul.v7i2.6892

Muka, T., Glisic, M., Milic, J., Verhoog, S., Bohlius, J., Bramer, W., Chowdhury, R., and Franco, O. H. (2020). A 24-step guide on how to design, conduct, and successfully publish a systematic review and meta-analysis in medical research. *European Journal of Epidemiology*, 35(1), 49–60. https://doi.org/10.1007/s10654-019-00576-5

Nirzalin, N., Bachri, N., Fakhrurrazi, F., Yunanda, R., Ilham, I., and Muchlis, M. (2023). Strengthening Reintegration through Social Capital: Learning from Aceh, Indonesia. *Jurnal Ilmu Sosial Dan Ilmu Politik*, 27(2), 141. https://doi.org/10.22146/jsp.82251

Nursabrina, A., Palevi, A., Robbyanti, D. R., and Meha, R. R. (2023). *The Influence of Cultural Diversity on Human Perspective*. 3(1).

Pasaribu, I., Siahaan, J., and Ginting, R. (2024). Local Wisdom Traditions of Ulaon Matumona Toba Batak Ethnic. *International Journal of Research and Review*, 11(3), 496–504. https://doi.org/10.52403/ijrr.20240358

Pattiasina, P. J., Maspaitela, M., and Gommies, C. C. (2024). Bentuk dan Fungsi Kearifan Lokal Cerita Rakyat Waisalaka di Negeri Waai. *Journal on Education*, 6(3), Article 3. https://doi.org/10.31004/joe.v6i3.5506

Pattipeilohy, W. S. E., Suharti, L., Ihalauw, J. J., and Dwiatmadja, C. (2023). The Role of Pela Gandong Local Wisdom Value on Organizational Behavior: A Case Study in Dr. M. Haulussy Regional Public Hospital, *Indonesia. Quality-Access to Success*, 25(198). https://doi.org/10.47750/QAS/25.198.20

Rahman, A., Wasino, Arsal, T., and Shintasiwi, F. A. (2022). Local wisdom and strengthening social integration in multiethnic society post-aceh conflict. *Kasetsart Journal of Social Sciences*, 43(3). https://doi.org/10.34044/j.kjss.2022.43.3.06

Rasyid, A., Lubis, R. F., Hutagalung, M. W. R., Lubis, M. A., Nor, M. R. M., and Vinandita, A. (2023). Local Wisdom Recognition in Inter-Ethnic Religious Conflict Resolution in Indonesia from Islah Perspective. *JURIS (Jurnal Ilmiah Syariah)*, 22(1), 13. https://doi.org/10.31958/juris.v22i1.8432

Ratnasari, D., Sidiq, F., and Saputra, J. (2023). Local Wisdom, Environmental Management and Sustainable Traditions in Naga Village, *Tasikmalaya Regency*. 3(4).

Reuter, T. K. (2023). Ethnic conflict—Identity, Power, Inequality | Britannica. *In Encylopeda of Britannica*. https://www.britannica.com/topic/ethnic-conflict/Dynamics-of-ethnic-conflict

Syamsuddin, S., and Purnama, R. (2021). Local wisdom in the novel of the snking of the van der wijk by buya hamka. *Prosiding Seminar Nasional Dan Internasional HISKI*, Article 0. https://doi.org/10.37905/psni.v0i0.31

Wulandari, C., Pujirahayu, E. W., Hiariej, E. O. S., Hassan, M. S., and Kambuno, J. A. (2022). Penal Mediation: Criminal Case Settlement Process based on the Local Customary Wisdom of Dayak Ngaju. *Lex Scientia Law Review*, 6(1), 69–92. https://doi.org/10.15294/lesrev.v6i1.54896

Towards Resilient Societies: The Synergy of Religion, Education, Health, Science, and Technology – Rahiem (Ed.)

Post-ISIS dynamics and the evolution of terrorism in Indonesia

Mujaddid Akbar
Republic Indonesia Defense University, Jakarta, Indonesia
ORCID ID: 0009-0009-5371-7420

Fauzia Gustarina Cempaka Timur
Republic Indonesia Defense University, Jakarta, Indonesia
ORCID ID: 0000-0001-9273-2249

Priyanto Priyanto
Republic Indonesia Defense University, Jakarta, Indonesia
ORCID ID: 0009-0008-1164-6911

M. Afifuddin
Republic Indonesia Defense University, Jakarta, Indonesia

ABSTRACT: The Islamic State of Iraq and Syria (ISIS) and its affiliates remain one of the most significant terrorist threats worldwide. Originally declaring the establishment of an Islamic Caliphate in the Middle East, ISIS has expanded its influence to various countries, including Indonesia. The collapse of the ISIS Caliphate weakened terrorist movements both in Indonesia and globally. This study employs a qualitative approach using secondary data from diverse sources, including academic literature, journals, brochures, and newspapers. The findings reveal that: first, several terrorist groups in Indonesia are affiliated with ISIS. Second, the destruction of ISIS strongholds in multiple regions has significantly impacted terrorist activities in Indonesia and beyond. Third, terrorist groups continue to use asymmetric strategies, notably mass propaganda on social media platforms. The future of terrorism poses new threats with the adoption of advanced technologies such as drones, artificial intelligence (AI), and cyberterrorism. Fourth, widespread conflict zones contribute to the uncontrolled circulation of weapons, increasing the risk of them falling into terrorist hands. Fifth, monitoring the potential rise of new global terrorist organisations is crucial. Terrorism often stems from injustice, inequality, and institutional weaknesses, directly contradicting Sustainable Development Goal 16, which promotes peace, justice, and strong institutions worldwide.

Keywords: asymmetric strategy, Indonesia, post-ISIS, roots of terrorism

1 INTRODUCTION

The terrorist movement in Indonesia has a long history, even from the early days of Indonesian independence until now. Sulastiana (2018) categorises terrorism trends in Indonesia into three distinct periods. The first period was during the Darul Islam/Indonesian Islamic Army (DI/TII) era from 1949 to 1954. The second period was the Jemaah Islamiyah (JI) era. JI was popular in Indonesia until the early 2000s and had succeeded in carrying out acts of terror in Bali and Jakarta. The third period is the era of the Islamic State of Iraq and Syria (ISIS), which continues to develop to this day.

DOI: 10.1201/9781003645542-25

As it continues to develop, terrorism in Indonesia continues to spread and diversify among various groups. At the end of 2021, the National Counterterrorism Agency (BNPT) said that there were several terrorist groups still operating in Indonesia. At least six terrorist groups remain active in Indonesia, including Jamaah Islamiyah (JI), Jamaah Ansharut Daulah (JAD), Jamaah Ansharut Khilafah (JAK), Jamaah Ansharut Syariah (JAS), Indonesian Islamic State (NII), and Indonesian Mujahideen East (MIT). The number of terrorist groups that still exist are suspected of having affiliations with global terrorist groups like Al Qaeda and ISIS.

Terrorism in Indonesia continues to evolve, adapting to geopolitical conditions. Various complex factors, such as economic inequality, social marginalisation, ideological radicalisation and armed conflict, often trigger the increasing trend of terrorism. At a time when the influence of the ISIS and Al Qaeda groups is weakening, it is important to examine further what the real direction of terrorist groups' movements in Indonesia will be in the future after the ISIS Caliphate. In addition, technological advancements, such as the Internet of Things (IoT), undoubtedly offer terrorist groups the potential to carry out acts of terror against a nation. This research focuses on the development of terrorist groups in Indonesia affiliated with ISIS or Al Qaeda and what asymmetric methods they may use in the future.

In the face of this threat, global partnerships play a crucial role in counterterrorism efforts. International cooperation allows countries to share intelligence information, technology, and resources needed to identify and stop terrorist networks. In addition, global collaboration helps build the capacity of national institutions, especially in developing countries, to prevent violence and combat terrorism effectively. This aligns with sustainable development goals (SDGs), particularly Goal 16, which relates to peace, justice, and strong institutions. Strengthening national institutions through international cooperation is a strategic step. For example, global initiatives such as the United Nations Global Counter-Terrorism Strategy provide a platform for countries to work together to build a coordinated response to terrorism while promoting approaches that prevent radicalisation through social inclusion and education.

2 METHODS

The author employed a qualitative approach in this research that aims to understand social phenomena or human behaviour from participants' perspectives in a natural context. In this research, the author employs secondary data gathered from various sources. The author quotes and analyses various documents, such as academic texts or literature, journals, brochures, newspapers, etc. This helps researchers validate the validity of the data and its relevance to the research. The data is then analysed by focusing on exploring the context, relationships, and meaning behind the data to produce a holistic understanding of the phenomenon under study. The analysis results are then presented in a narrative so that they can describe the research findings in depth.

3 FINDINGS AND DISCUSSION

3.1 *ISIS and terrorist groups in Indonesia and the struggle of the Islamic Caliphate*

The current discourse on global terrorism cannot be separated from discussions about ISIS, which is affiliated with a network of terrorist groups in a number of other countries in the world. In 2023, the Institute of Economics and Peace (IEP) released the 2023 Global Terrorism Index. Based on these findings, ISIS remains the world's deadliest international terrorist organisation. ISIS continues to expand beyond the territorial boundaries of nation-states by utilising advances in information and communications technology. Since March 2023, the IS media administration centre has reported being responsible for 1,121 attacks. Based on their published data, these attacks left around 4,770 people dead or injured.

After the defeat of Al Qaeda in military attacks carried out by the US in several regions in the Middle East and the death of Al Qaeda leader Osama bin Laden, ISIS emerged as a new force that threatened global security. The emergence of ISIS has awakened sleeper cells of Islamic State or Islamic Caliphate fighters throughout the world, including in Indonesia. Several Islamic terrorist groups in Indonesia have also pledged their allegiance to ISIS.

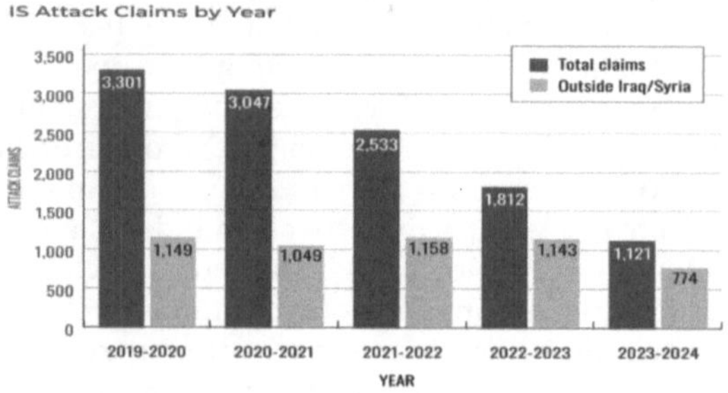

Figure 1. IS attack claims by year (Source: Zelin 2024).

3.2 *The collapse of the ISIS Caliphate*

Attacks by the ISIS terrorist group in various regions of the world continue to decline from year to year (Figure 1). According to data collected by BBC Monitoring (2024), ISIS activity has decreased by more than 53% compared to last year. The Global Terrorism Index 2024 also noted that terrorist attacks against Western countries fell by 55% compared to the previous year. Even Iraq, once a yearly target of terrorist attacks, saw a significant decline. The death rate due to terrorist attacks in Iraq from its peak in 2007 to 2023 has decreased by 99%. Likewise, terrorist attacks from 2007 to 2023 decreased by up to 90%. The epicentre of the global terrorist movement currently appears to be shifting from the Middle East to the Sahel and Sub-Saharan Africa.

Similarly, the terrorist movement in Indonesia is facing significant setbacks. The BNPT and related stakeholders run the deradicalisation program, in which several terrorist convicts have also participated. One of the terrorist convicts who participated in the deradicalisation program was Umar Patek, who was one of the perpetrators of the Bali Bombing I. In Indonesia, the Global Terrorism Index 2024 report shows a significant decrease in terrorist attacks compared to previous years. Based on GTI data, Indonesia is currently ranked 31st globally (Figure 2). Compared to the rankings in previous years, this position signifies positive progress, as the decline in rankings indicates that the incidence of terrorism in the country continues to decrease.

RANK	COUNTRY	SCORE	RANK CHANGE	RANK	COUNTRY	SCORE	RANK CHANGE	RANK	COUNTRY	SCORE	RANK CHANGE
1	Burkina Faso	8.571	↑1	28	Thailand	4.219	↓5	55	Spain	1.669	↑5
2	Israel	8.143	↑24	29	Türkiye	4.168	↓8	56	Lebanon	1.562	↓5
3	Mali	7.998	↑1	30	United States of America	4.141	↓1	57	Australia	1.475	↓2
4	Pakistan	7.916	↑3	31	Indonesia	3.993	↓7	58	Italy	1.447	↓4
5	Syria	7.890	↔	32	Bangladesh	3.317	↑14	59	Central African Republic	1.445	↓2
6	Afghanistan	7.825	↓5	33	Sri Lanka	3.072	↓2	60	Saudi Arabia	1.366	↑5
7	Somalia	7.814	↓4	34	Greece	3.028	↓1	61	Argentina	1.274	↓3
8	Nigeria	7.575	↔	35	Russia	3.016	↑8	62	Ethiopia	1.272	↓3
9	Myanmar	7.536	↔	36	Tunisia	2.914	↑4	63	Kosovo	1.218	↑28
10	Niger	7.274	↔	37	Germany	2.782	↓2	64	Japan	1.189	↔
11	Iraq	7.078	↓5	38	France	2.647	↓4	65	Venezuela	1.174	↓13
12	Cameroon	6.98	↓1	39	Libya	2.469	↓7	66	Slovakia	1.092	↓5
13	Democratic Republic of the Congo	6.514	↑1	40	Burundi	2.434	↓4	67	Mexico	1.04	↑1
14	India	6.324	↓1	41	United Kingdom	2.373	↑3	68	Austria	0.953	↓5
15	Mozambique	6.267	↓3	42	Tanzania	2.267	↓3	69	Tajikistan	0.871	↓16
16	Colombia	6.188	↓1	43	Angola	2.254	↑48	70	Sweden	0.735	↓4
17	Chile	5.679	↓1	44	Algeria	2.197	↓6	71	Switzerland	0.627	↓4
18	Kenya	5.616	↑1	45	Nepal	2.163	↓8	72	Cyprus	0.616	↑3
19	Philippines	5.383	↓1	46	Côte d'Ivoire	2.06	↓5	73	China	0.582	↑21
20	Egypt	5.221	↓3	47	Peru	2.045	↓6	74	Netherlands	0.577	↓4
21	Chad	4.967	↓1	48	Djibouti	2.035	↓3	75	Jordan	0.455	↓4
22	Palestine	4.966	↑6	49	Brazil	1.968	↓2	=76	Armenia	0.423	↑18
23	Yemen	4.951	↓1	50	New Zealand	1.947	↓2	=76	Uzbekistan	0.423	↓4
24	Benin	4.898	↑3	51	Belgium	1.904	↑11	78	Paraguay	0.241	↓5
25	Togo	4.67	↑5	52	Canada	1.753	↑4	=79	United Arab Emirates	0.233	↓2
26	Iran	4.464	↓1	53	Norway	1.747	↓3	=79	Iceland	0.233	↓2
27	Uganda	4.377	↑22	54	Ukraine	1.686	↑20	81	Malaysia	0.192	↓5

Figure 2. Global terrorism index 2024 (Source: Resky 2024).

3.3 *The threat of post-ISIS caliphate*

In such a situation, terrorist groups will use asymmetrical warfare strategies to fight for their existence and interests. The post-ISIS Caliphate presents a terrorist threat that requires consideration and anticipation of at least five factors. *Firstly*, terrorist propaganda in social media. Terrorist groups heavily utilise social media to propagate their propaganda in Indonesia's terrorism context (Aththaariq Rizki 2021). Data from the National Counterterrorism Agency (BNPT) indicates that 600 social media sites or accounts contain elements of radicalism in 2022. The BNPT report indicates that terrorist groups primarily use Facebook among several social media platforms. Similarly, in 2023, the BNPT reported that terrorist groups were using 2,670 digital content to spread radicalism and terrorist propaganda. Based on the latest data from BNPT, in 2024, there were 180,954 contents containing radicalism, intolerance and extremism. The social media platforms used also adjust to the trends used by the Indonesian people; in this case, the 3 highest social media platforms are Instagram (86,203 content), then Facebook (45,449 content) and TikTok (23,595 content). Stricter supervision needs to be carried out from the upstream; in this case, the cooperation of the Indonesian Government with the providers of social media platforms needs to be carried out as a preventive measure.

Secondly, terrorist groups have adopted artificial intelligence (AI) and drones as new threats. Currently, the world is facing a new challenge: the proliferation of autonomous warfare systems (including AI-based drones), which are not yet strictly regulated by international law. The proliferation of small, inexpensive drones is particularly affecting options on the battlefield. Small commercial drones, initially primarily used for reconnaissance, have become a crucial tool for effectiveness due to the widespread use of FPV (First Person View) drones as of 2023. It is plausible that terrorist groups, including those in Indonesia, may develop this technology to carry out their operations. The asymmetric strategy by terrorist groups that utilise autonomous warfare systems needs to be anticipated early on by all stakeholders in Indonesia, including, in this case, the need to adjust legal regulations related to the development and use of autonomous technology so that terrorist groups do not misuse it.

Thirdly, cyberterrorism. Although it seems that terrorist groups have not yet reached the capacity to carry out cyber-attacks on certain government infrastructures, as technology continues to develop and information becomes more open, terrorist groups may have this capability to shake up a country's government and create a sense of security. This could cause worry, fear, and terror among the country's citizens. Cyberterrorism, to put it simply, involves the use of the internet and information technology to carry out acts of terrorism. This presents significant challenges for law enforcement and security agencies tasked with preventing cyber-terrorist activities and bringing perpetrators to justice (Lee 2024).

Fourthly, the world is witnessing a massive circulation of weapons. The global political and security situation is currently experiencing turmoil in various places, starting with the Russian and Ukrainian wars. The Israeli and Palestinian conflict continues to grow and involves several other countries. The ongoing events undoubtedly create vulnerabilities by fostering the growth of extremism and radicalism, which in turn can lead to acts of terrorism. Apart from that, the large number of conflict points also creates vulnerabilities in the massive circulation of weapons. If it is not controlled correctly, these weapons may fall into the hands of terrorists and be used to carry numerous high-tech weapons. Upon closer inspection, the weapons the United States left in Afghanistan in 2021 also present their vulnerabilities, as it is unclear if the Taliban seized all of them or if terrorist groups took control of some of them (Omar 2024). Indonesia also needs to watch out for the global circulation of weapons.

Fifthly, new terrorist organisations. Meanwhile, the decline of ISIS as a global terrorist organisation is expected to impact terrorist movements in other countries, including Indonesia (Risa Rizania 2023). Terrorist groups sometimes feel the need to carry out physical attacks to demonstrate their existence. On a global scale, although terrorist movements currently tend to be decentralised in each region or country, the movement will have more global influence if there is a large organisation that houses it and acts as a patron for terrorist groups spread across various countries. Naturally, such a large organisation needs substantial and independent funding to sustain its activities. The largest and most well-known terrorist organisations in the world are ISIS and Al-Qaeda, but new patrons from the global terrorist movement may replace or reincarnate them.

4 CONCLUSION

The declaration of the founding of the Islamic Caliphate has drawn the attention of numerous radical Islamic groups in various countries, leading them to support and pledge allegiance to ISIS. Consequently, the perception of ISIS has expanded beyond its origins in Iraq and Syria, now encompassing various global regions. In its development, both globally and nationally, the ISIS Caliphate appears to have experienced a significant setback or even collapsed. This situation undoubtedly impacts terrorist groups operating in Indonesia and other countries.

The terrorist groups changed their strategy in response to the collapse of the ISIS Caliphate. Terrorist groups, including ISIS and its networks in other countries such as Indonesia, primarily employ asymmetric warfare techniques and methods in their movement patterns. The asymmetric strategies include: *first,* using popular social media in Indonesia as a propaganda tool, including Instagram, Facebook and TikTok. *Second,* there is the threat of the use of autonomous warfare systems by terrorist groups. *Third,* the threat of cyber warfare can cause chaos in society and the state. Additionally, it is crucial to enhance the surveillance of weapon distribution, particularly in areas of conflict, and closely monitor the rise of new terrorist groups.

REFERENCES

Aththaariq Rizki, F. G. (2021). Security threats of radicalism through social media amid COVID-19 pandemic: Indonesia's perspective. *Connections: The Quarterly Journal*, 20(3–4).

Badan Nasional Penanggulangan Terorisme. (2021). Modul pengetahuan dasar ancaman dan perkembangan terorisme di Indonesia. *Jakarta*: *BNPT.*

Global Terrorism Index 2023: *Key findings in 5 charts.* Retrieved from https://www.visionofhumanity.org/global-terrorism-index-2023-key-findings-in-5-charts/

Lee, D. D. (2024). *Ukraine cyberwar: Digital warfare in the Russian invasion.* Columbia: SkyCuration.

Omar, A. A. (2024). *A framing analysis of the United States government counter-terrorism messaging strategies during the rise and fall of ISIS (Master's thesis).* University of Leicester, United Kingdom.

Risa Rizania, F. G. (2023). The role of teachers in facing radicalism threats at Al-Kautsar Labschool Middle School, Bandung. *International Journal of Humanities Education and Social Sciences (IJHESS)*, 3(3), 1544–1554.

Zelin, A. Y. (2024, March 20). *One year of the Islamic State Worldwide Activity Map.* Retrieved from https://www.washingtoninstitute.org/policy-analysis/one-year-islamic-state-worldwide-activity-map

Towards Resilient Societies: The Synergy of Religion, Education, Health, Science, and Technology – Rahiem (Ed.)
© 2026 The Author(s), ISBN: 978-1-041-08471-6
Open Access: www.taylorfrancis.com, CC BY-NC-ND 4.0 license

Indonesia's public diplomacy for al-Quds and al-Aqsa mosque

Masyrofah Masyrofah
Universitas Islam Negeri Syarif Hidayatullah Jakarta, Indonesia
ORCID ID: 0000-0001-9657-0443

Amany Lubis
Universitas Islam Negeri Syarif Hidayatullah Jakarta, Indonesia

Nahrowi Nahrowi
Universitas Islam Negeri Syarif Hidayatullah Jakarta, Indonesia
ORCID ID: 0000-0001-5771-3667

ABSTRACT: This paper examines the efforts of international communities to support the Palestinian people and protect Al-Quds and Al-Aqsa from Israeli occupation. Indonesia has consistently supported the Palestinian struggle, aligning with United Nations Security Council (UNSC) Resolution No. 338 (1973), which calls for the unconditional return of all Arab territories occupied by Israel and recognises the legitimate rights of the Palestinian people to establish an independent state with Al-Quds in East Jerusalem as its capital. This qualitative study draws data from document analysis to explore Indonesia's role in this cause. The findings reveal that Indonesia has engaged in public diplomacy through major organisations such as Nahdlatul Ulama and Muhammadiyah. The Indonesian government also encourages support from various societal sectors, including community organisations, educational institutions, and non-governmental organisations (NGOs). Indonesia is urged to continue its advocacy for the Palestinian people and the protection of Al-Aqsa Mosque across local, national, and global platforms.

Keywords: al-Quds and al-Aqsa mosque, international communities, public diplomacy

1 INTRODUCTION

The existence of a pluralistic Indonesian society provides nuances for diversity in the state and religious life. The obedience of the Indonesian Muslim population in carrying out worship is reflected in their life behaviour guided by the values of the Islamic religion *Rahmatan Lil' aalamiin,* which is indicated by the increasing percentage of the Indonesian population who perform Umrah every year and the waiting period of years for those who have registered for Hajj. In addition, religious tourism in Muslim countries, including pilgrimages to the Al Aqsa Mosque, the first Qiblat of Muslims in Palestine, is also a destination with many devotees in Indonesia.

Al-Quds and Al-Aqsa Mosque in East Jerusalem are still under Israeli occupation. While peace has not been realised in Palestine, the two holy places have not received essential security. UN Resolution No. 271 of September 15, 1969, states that any act of destroying or polluting sacred places, religious buildings, and sites in Jerusalem endangers international peace and security. The OIC strongly condemns the Israeli attack on Al-Quds and the Al-Aqsa Mosque in 2021, as stated in the Resolution of the Extraordinary Meeting of the OIC Executive Committee at the Foreign Ministerial Level Virtual on May 16, 2021, which discussed Al-Quds in East Jerusalem. The existence of Al Quds is a breath of Muslims on Earth.

DOI: 10.1201/9781003645542-26

2 METHODS

This study was qualitative, with the data source obtained through a document study. The position of Al Quds can be analysed in several aspects, namely philosophical, historical, political, and humanist. These three aspects influence each other and play a role in maintaining the existence of Al Quds as holy places for the three religions. Some principles must be maintained so that the sustainability of Al Quds becomes a hope for peace on earth. Security factors cannot be realised in the Al Quds area. Therefore, it is still limited to Muslims who worship in the Al Quds. According to Constance Duncombe (Duncombe 2019), public diplomacy is increasingly facilitated through social media, with government leaders and diplomats using platforms such as Twitter, Facebook, and Instagram to communicate with foreign audiences.

On the other hand, public diplomacy through some international communities has played an essential role in opening the way for communication and cooperation with people directly and how the technology with social media could share the activities of international communities for fundraising and distributing humanitarian aid to Palestinian people and saving al Quds from Israeli occupation. This article has significantly contributed to strengthening the global partnership between the government and civil society on a local, national, and global scale to unite the collaboration and cooperation through international communities to stand with Palestine.

3 FINDINGS & DISCUSSION

3.1 *Al-Quds is a pluralistic city in international politics and law*

Muslims refer to the Al Quds as al-Haram al-Sharif or the Holy Place. The Jews call this the Temple Mount. The complex is located in Jerusalem's Old City, designated a World Heritage Site by the United Nations (UN) cultural agency UNESCO, and a holy place for the three Abrahamic religions. The site has been the most contested territory in the Holy Land since Israel occupied East Jerusalem, including the Old City, in 1967, along with the West Bank and Gaza Strip. However, the conflict began even further before Israel's state existed. In 1947, the United Nations created a partition plan. The goal was to separate the history of Palestine, which was then under British control, into two states: one for Jews, mainly from Europe, and one for Palestine. The Jewish state was given 55 percent of the land and the remaining 45 percent to the Palestinian state. Jerusalem, which houses the Al-Aqsa compound, belongs to the international community under UN administration. The site was given a special status because of its importance to the three religions (Puspaningrum, kompas.com 2021) The first Arab-Israeli war broke out in 1948 after Israel declared statehood, controlling about 78 percent of the land. The rest of the West Bank, East Jerusalem, and Gaza are controlled by Egypt and Jordan. Israeli encroachment on land intensified in 1967. After the second Arab-Israeli War, Israel occupied East Jerusalem and eventually illegally annexed Jerusalem, including the Old City and Al-Aqsa. Israel's illegal control of East Jerusalem, including the Old City, violates several principles of international law. It is determined that the occupying power does not have sovereignty in its territory. For Muslims, it is the third holiest site in Islam: the Al-Aqsa Mosque and the Dome of the Rock. The seventh-century building is believed to be where Prophet Muhammad ascended to heaven. Jews thought that the complex was the site of the Biblical Jewish temples. However, Israeli Jewish and Rabbinate law forbids Jews from entering the compound and praying there, as it is considered too holy to be stepped on. The complex's Western Wall, known as the Wailing Wall to Jews, is believed to be the last remnant of the Second Temple (Eddeh 2001). At the same time, Muslims refer to it as the al-Buraq Wall and think that it was there that the Prophet Muhammad tied al-Buraq, the animal used to ascend to the sky and speak God (Puspaningrum, kompas.com 2021) (Puspaningrum, kompas.com 2021).

In the 1967 Six-Day War, Israel annexed East Jerusalem, which triggered the Israeli-Palestinian conflict. Israel wanted to make Jerusalem its capital after the birth of the Jewish state. It was corroborated by the statement of Israel's first Prime Minister, David Ben-Gurion, on December 5, 1949. At that time, Jerusalem was divided into two parts: East Jerusalem, controlled by Jordan, and West Jerusalem, controlled by Israel. Israel's official government was in West Jerusalem. However, in the 1967 war, Israel also controlled the eastern part of Jerusalem and established its status as the state's capital, which was confirmed in the 1980 Constitution of Israel. Palestinians also consider East Jerusalem to be the capital of a future Palestinian state after being freed from Israeli occupation. It proves that Israel's intention to control Jerusalem was to occupy Palestinian land. Therefore, a significant consequence of the 1967 War was the take-over of the PLO by Palestinian factions dominated by Fatah in 1969. Perhaps the most important political outcome of the 1967 War was reframing the struggle's final resolution. Before 1967, there had been a regional consensus to reject the legitimacy of the Zionist regime and define the goal of the struggle as the liberation of Palestine and the return of the 1948 displaced Palestinian refugees (Al-Arian 2024).

On November 12, 2004, the body of Palestinian leader Yasser Arafat was buried in Ramallah in the Judean Hills. The Grand Imam of Jerusalem Ikrema Sabri conveyed through the Reuters News Agency on May 11, 2004, that Yasser Arafat wishes to be buried in Jerusalem near the Al Aqsa Mosque. However, Israel has rejected this desire. The Israeli government believes that Arafat's funeral in Jerusalem will strengthen the Palestinian people's struggles. Israel says that Arafat's burial in Jerusalem will strengthen Palestinian claims to the Arab sector of Jerusalem as a traditional Palestinian territory and as the future Palestinian capital. Although Arafat grew up in Jerusalem and Cairo, Egypt, Arafat's clan, Al Kidwas, is native to Gaza. Arafat's extended family owns 25 to 30 grave plots in Khan Younis, southern Gaza (Kuncahyono 2017).

Observers argue that the settlement of the Palestinian-Israeli conflict depends on resolving Jerusalem's status. In ancient times, Europeans made ancient maps, placing Jerusalem as the centre of the spiritual map of the world and as the centre of the earth and the universe. Jerusalem is also still the centre of world attention, feuds, and conflicts that have not ended until now. West Jerusalem is the capital of Israel, and East Jerusalem is the capital of future Palestine. Nevertheless, until now, Israel still claims Jerusalem as its capital. Some Palestinians feel that they have a personal connection to Jerusalem, even though they have not visited the city (Souri 2006) (Souri 2006).

Jerusalem's uniqueness stems from its spiritual significance and not from its long historical heritage. There are three major religions: Judaism, Christianity, and Islam. However, all three differ in their fundamental concepts of God in their faith and worship rituals, united by their love for Jerusalem. In her article titled Jerusalem, Leah Sullivan, The Three Religions of the Temple Mount, states that Jews, Christians, and Muslims view Jerusalem as "the door to heaven." Jerusalem is sanctified by religion, tradition, history, theology, holy places, and houses of worship. They reflect the zeal and piety of three monotheistic religions. For Muslims, Jerusalem is where Prophet Muhammad SAW made his night journey (Isra) from the cities of Mecca and Jerusalem in Palestine, followed by Sidratul Muntaha. The trip strengthened the relationship between Mecca and Jerusalem. The religious dimension has given more weight to Al Quds Jerusalem but also emphasises that it is a heritage of humanity and is a holy city and a city where people live for Palestinians, Israelis, Jews, Christians, and Muslims. In this city, two nations and three religions live in mutual respect. It makes Jerusalem different from other cities globally, but the Palestinian population has not felt peaceful. The statement "Al Quds is sacred to the three monotheistic religions and has not made the city's status the subject of negotiations. The legal aspects of the Jerusalem issue are numerous and complicated by political, religious, and historical factors.

3.2 *Indonesia's public diplomacy for Al-Quds*

Diplomacy, in its conventional form, is characterised by state-level interactions that include the establishment of official ties to achieve foreign policy aims. Simultaneously, 'public' diplomacy involves diverse interactions, such as cultural exchanges, education programs, and athletic events that seek to elevate a nation's standing and reshape the host country's attitudes (Ayoub 2023) (Ayoub 2023). According to Wang and Chang, public diplomacy is a form of international public relations (Wang 2004) (Wang 2004). Mohan J. Dutta-Bergman maintained that both public diplomacy and public relations strive to engage target audiences to preserve and oversee perceptions, showcasing numerous shared strategies and tactics (Dutta-Bergman 2006).

Under uncertain conditions in the Middle East, Indonesia has become a great hope for Palestine. Compared to other countries with the same Muslim majority population, support from the Indonesian people for Palestine is no less extensive (not to say, the largest). Indonesia is also relatively stable both economically and politically. Moreover, Indonesia has good relations with several influential countries at the global level, which can be used to resolve the Palestinian-Israeli conflict entirely and permanently. Of course, there are more things that Indonesia must do to increase the capacity of its foreign policy stance than just political statements (Sastrawi 2020) (Sastrawi 2020).

Al Aqsa Mosque does not only belong to Palestinians but also to Muslims in the world, who are obliged to maintain and glorify the place. Nevertheless, as long as there is no peace between Israel and Palestine, the existence of Al Aqsa is still threatened by Israeli occupation, so the conflict is still ongoing. Indonesia's position fully supports the Palestinian state's future struggle, independence, and sovereignty. The Indonesian government supports all elements of society, including community organisations, educational institutions, and non-governmental organisations (NGOs). Indonesia is called upon to support the Palestinian people and protect Al Aqsa on local, national, and global scales.

Indonesia has a unique influence on the Palestinian issue, with Indonesia being the largest Muslim country located far from the Middle East. In terms of the population of Indonesia, which is predominantly Muslim outside the Middle East, Indonesian Muslims struggle to achieve independence. Religious issues have significant weight as a form of religious identity in Indonesia. Therefore, the Palestinian issue has always had a place in the hearts of the Indonesian people. (Wahab 1997) (Wahab 1997) The government and people of Indonesia cooperate in voicing the independence of Palestine and providing assistance to humanity to the Palestinian people. The issue of Palestine has always been a hot topic of discussion in domestic and international forums. When the Presidential and Vice-Presidential Candidate Debates were held in 2014, on the theme of Indonesian Foreign Policy, the material asked to the candidate pairs was related to Indonesia's attitude towards Palestine and Indonesia's foreign policy towards Palestine. It indicates Indonesia's great concern in fighting for the rights of the Palestinian people to gain independence. The Palestinian issue is an integral part of Indonesia's public diplomacy. Diplomacy can be carried out in various forms, including humanitarian diplomacy involving the Indonesian people, who work hand-in-hand with the Palestinian people to realise their hope that they can live everyday life as citizens in the country they love. Humanitarian assistance from money, cash, food, clothing, and medical treatment to construct infrastructure in hospitals and schools continues to flow from Indonesian people to Palestinian people.

Indonesia's role in supporting Palestine has been proven by hosting conferences, namely the Asia-Africa Summit (AA Summit) and the Conference of the Organisation of the Islamic Cooperation (OIC). At the extraordinary OIC Summit, President Joko Widodo conveyed a firm stance and an invitation to other OIC countries to increase support for Palestine by seeking the following steps: strengthening political support to revive the peace process, increased pressure on Israel, including the boycott of products; Israel generated in the occupied territories; increased pressure on the UN Security Council to provide protection international law to Palestine; and the deadline for ending the Israeli occupation of the Palestinian territories. Humanitarian factors are highly prioritised in Indonesia's foreign policy. It can be seen in Indonesia's support for Palestine, diplomatic and humanitarian assistance, and capacity building. The Indonesian government initiated 1,800 capacity-building programs in Palestine covering the housing, telecommunications, and media sectors and seeking humanitarian aid during Israeli aggression in the West Bank and Gaza Strip. Humanitarian assistance is sought through the government (*government to government*) and the active participation of the Indonesian people, who are enthusiastic about defending Palestine's independence through non-governmental organisations, community organisations, and individuals who provide humanitarian assistance for Palestine (*people to people*).

As a country that has experienced the bitterness of colonialism, Indonesia deeply feels the suffering experienced by the Palestinian people with all the limitations in meeting the necessities of life in the West Bank, Gaza, and refugee camps. Israeli military attacks often brutally attack the Palestinian people, so millions of Palestinians, both adults, parents, and children, have become victims of the conflict. Thus, to meet the basic needs of the Palestinian people, Indonesians provide continuous humanitarian assistance. During the current financial crisis that hit the United Nations Agency for Palestine Refugees (UNRWA), Indonesian people's concern was again expressed through the distribution of humanitarian assistance to Palestinian refugees. Among other things, assistance in the form of school supplies and snacks was provided to students at Amman New Camp (Kamp Wihdat) Elementary School in South Amman. The distribution of humanitarian aid was welcomed by UNRWA, especially when the UNRWA experienced a financial crisis due to cuts in aid funds by the United States. According to Davis, this assistance from the Indonesian people is concrete proof of Indonesia's support and commitment to the independence and welfare of the Palestinian people. UN member countries must follow Indonesia for humanitarian assistance (Palestina 2024) (Palestina 2024).

The NU is consistent in supporting the struggle of the Palestinian people, both through humanitarian assistance and in expressing its condemnation of Israeli aggression. However, NU refused to voice its aspirations for solidarity action against Palestine through street demonstrations. The NU stated that it would defend Palestine through diplomacy with other religious leaders as an act of solidarity that criticised the US policy of recognising Jerusalem as the capital of Israel at Mona's on December 17, 2017. NU prefers to defend Palestine by strengthening its diplomacy and garnering support. Therefore, on December 15, 2017, the NU met with the faith leaders. They produced humanitarian points at the meeting as a form of solidarity between the leaders toward Palestine and the realisation of a peaceful world (Dewi, tempo.co, 2017) (Dewi, tempo.co, 2017).

Table 1. 2023 Confirmed Pledges towards UNRWA's programmes as of December 31, 2023.

	Program Budget			Non Programme Budget							
Donor	Program Budget	In-kind	Total Program Budget	EA OPT	Syria Appeal	Projects	Syria Flash Appeal-Earthquake	Early Recovery Appeal for Jenin	Ein El Hilweh Flash Appeal	Gaza Flash Appeal	Total
Andorra	0	0	0	0	0	0	0	0	0	49.308	49.308
Australia	14.542.856	0	14.542.856	0	0	0	0	0	0	0	14.542.856
Austria	3.744.237	0	3.744.237	0	1.108.773	0	0	0	0	0	4.853.010
Azerbaijan	0	0	0	0	0	0	0	0	0	2.000.000	2.000.000
Bahrain	50.000	0	50.000	0	0	3.000.000	0	0	0	0	3.050.000
Bangladesh	50.000	0	50.000	0	0	0	0	0	0	0	50.000
Belgium	8.402.028	0	8.402.028	0	3.043.854	3.380.435	0	0	0	0	14.826.317
Government of Flanders	177.013	0	177.013	0	0	0	0	0	0	0	177.013
Bulgaria	0	0	0	0	0	0	0	0	0	107.875	107.875
Brazil	75.000	0	75.000	0	0	0	0	0	0	0	75.000
Brunel Dar-ussalam	114.712	0	114.712	0	0	0	0	0	0	100.000	214.712
Canada	20.070.078	0	20.070.078	2.243.270	1.409.496	1.112.760	0	0	0	14.599.660	39.435.264
Chile	25.000	0	25.000	0	0	0	0	0	0	200.000	225.000
China	5.000.000	0	5.000.000	0	0	0	0	0	0	1.000.000	6.000.000
Colombia	0	0	0	0	0	0	0	0	0	160.000	160.000
Croatia	0	0	0	0	0	0	0	0	0	269.687	269.687
Cyprus	107.000	0	107.000	0	0	0	0	0	0	550.000	657.000
Denmark	19.324.615	0	19.324.615	0	0	193.240	0	0	0	6.014.166	25.532.021
Djibouti	0	0	0	0	0	0	0	0	0	2.094.747	2.094.747
EU	98.492.986	0	98.492.986	0	2.157.497	0	0	0	0	0	100.650.483
ECHO	0	0	0	2.072.944	2.217.787	0	0	0	0	15.224.530	19.515.261
Estonia	85.288	0	85.288	0	0	0	0	0	0	36.549	121.837
Finland	5.458.515	0	5.458.515	0	0	184.808	0	0	0	0	5.643.323
France	53.356.349	64.759	53.421.108	1.097.695	6.777.168	299.570	0	0	736.773	88.790	62.421.104
Germany	40.381.401	0	40.381.401	39.869.258	30.072.873	51.815.529	0	0	0	48.593.673	210.732.734
Greece	42.508	0	42.508	0	0	0	0	0	0	43.150	85.658
Holy See	20.000	0	20.000	0	0	0	0	0	0	0	20.000
Iceland	1.982.384	0	1.982.384	0	0	255.082	0	0	0	0	2.237.466
India	5.000.000	0	5.000.000	0	0	0	0	0	0	0	5.000.000
Indonesia	600.000	0	600.000	0	0	0	0	0	0	2.000.000	2.600.000

Muhammadiyah's attitude and view of conflict in Palestine is a religious issue. It was conveyed by Abdul Mu'ti, general secretary of PP Muhammadiyah, who assessed that the Palestinian people were getting more and more limited due to the expansion of territory by Israel. Indiscriminate attacks on Palestinian civilians exacerbated Israeli army attacks and violence. PP Muhammadiyah reiterates that this problem is not merely a political issue but also a religious issue. However, it is not a religious conflict because it is the religion that teaches us always to do something if there is a crime, munkar, or human actions that arbitrarily do earth damage. Even if religious people, especially Muslims, are not sensitive to colonialism and oppression on Earth, then that person is included in the awful faith category. In other words, their faith is fragile (Dahlan 2018) (Dahlan 2018) Expressing concerns and condolences for victims of the Al Aqsa Mosque conflict. Muhammadiyah strongly condemns the actions of The Israeli government and the army's violence against civilians.

4 CONCLUSION

As one of the largest Muslim countries in the world, Indonesia should build and develop several aspects of civil society to strengthen Indonesian public diplomacy for Palestine. That's why international support is also significant; the more support from the international community, the more substantial and decisive the status of Palestine in the United Nations as a non-observer state. All elements of society synergise with their respective capacities to realise the independence of Palestine and the Al Aqsa Mosque. The research has found that historically and religiously, Jerusalem is a holy city for three monotheistic religions. The involvement and attitudes of the public, such as NU and Muhammadiyah, have a significant influence on raising funds and distributing them so that they are beneficial for the Palestinian people. NGOs collected and distributed humanitarian funds for Palestine, where the critical role of community organisations and NGOs synergised with the government in public diplomacy for Palestine. It proves that Indonesia has an active role in protecting the human rights of the Palestinian people and safeguarding the Al-Quds and Al-Aqsa mosques for world peace.

REFERENCES

Al-Arian, S. (2024). Palestine and The Arab Regional Order. Insight Turkey, SPRING Vol. 26 No.2, Published by: SET VAKFI Iktisadi Isletmesi, SETA VAKFI.

Ayoub, H. H. (2023). Assesing Palestinian Public Diplomacy: Realities, Challenges, and Opportunities. *Global Media and Communication, Volume 19, Issue 2, Sage Journals*, 207–223.

Dahlan, I. A. (2018, May 26). Retrieved from itb-ad.ac.id: https://www.itb-ad.ac.id/2021/05/26/muhammadiyah-konflik-palestina-juga-permasalahan-agama/

Dewi, C. M. (2017, December 16). Retrieved from tempo.co: https://fokus.tempo.co./read/1042742/2017/12/16/alasan-penolakan-nahdlatul-ulama-turun-ke-jalan-bela-Palestine

Duncombe, C. (2019). Digital Diplomacy: Emotion and Identity in The Public Realm. *The Hague Journal of Diplomacy*, 14(1–2), 102–116. https://doi.org/10.1163/1871191x-14101016.

Dutta-Bergman, M. J. (2006). U.S. Public Diplomacy in the Middle East A Critical Cultural Approach. Journal of Communication Inquiry Volume 30 Number 2 April 2006 102–124 © 2006 Sage Publications 10.1177/0196859905285286 http://jci.sagepub.com .

Eddeh, M. (2001). The Zionist Claims to al-Haram al-Sharif and al-Aqsa Mosque: A Biblical Perspective . Islamic Studies, Autumn-Winter 2001, Vol.40, No. 3/4, Special Issue; Jerusalem, Research Institute, International Islamic University, Islamabad, https://www.jstor.org/stable/20837147, 649–654.

Kuncahyono, T. (2017). *Kesucian, Konflik, dan Pengadilan Akhir*. Jakarta: Gramedia.

Palestina, K. B. (2024, October 2024). *KBRI Amman Sambut Kedatangan Bantuan Kemanusiaan untuk Gaza, Palestina*. Retrieved from kemlu.go.id: https://www.kemlu.go.id/amman/berita/kbri-amman-sambut-kedatangan-bantuan-kemanusiaan-untuk-gaza-palestina?type=publication

Puspaningrum, B. A. (2021, August 5). Retrieved from kompas.com: https://internasional.kompas.com/read/2021/05/08/091212370/masjid-al-aqsa-dot-pertikaian-panjang-palestina -Israel?page=all

Sastrawi, H. (2020, September 15). *Peran Indonesia untuk Palestina*. Retrieved from media indonesia.com: https://mediaindonesia.com/opini/344936/peran-indonesia-untuk-palestina

Souri, H. T. (2006). Palestine, Israel and The Politics of Popular Culture. *Journal of Palestine Studies*, Vol. 36, No. 1 (Autumn), University of California Press on behalf of The Institute for Palestine Studies., 92–93.

Wang, J. (2004). Strategic public diplomacy and local press: How a high profile "head-of- state". *Public Relations Review*.

Wahab, S. A. (1997). *Analisis Kebijakan: Dari Formula ke Penyusunan Model-Model Implementasi Kebijakan Publik*. Jakarta: Bumi Aksara.

Towards Resilient Societies: The Synergy of Religion, Education, Health, Science, and Technology – Rahiem (Ed.)
© 2026 The Author(s), ISBN: 978-1-041-08471-6
Open Access: www.taylorfrancis.com, CC BY-NC-ND 4.0 license

Achieving social sustainability through corporate social responsibility: Modelling impediments of CSR in India

Suha Bilquis & Ayesha Farooq
Aligarh Muslim University, India

ABSTRACT: This study aims to identify and rank barriers to corporate social responsibility (CSR) implementation in Indian manufacturing companies and examine the interactions among these barriers. Key obstacles include inadequate recycling infrastructure, a lack of sustainable supply chain management, and resistance to adopting eco-friendly technologies. By aligning CSR initiatives with sustainable development goals (SDGs), Indian companies can address these barriers, fostering economic inclusion and environmental stewardship to achieve long-term social and environmental impact. The study employs the interpretive structural modelling (ISM) approach to analyse how these barriers interact, offering a systematic framework to map and prioritise their interrelationships. ISM effectively identifies CSR challenges, uncovers underlying causes, and visually represents problem structures, simplifying complex CSR issues for better strategic planning. The study identifies eleven barriers to CSR implementation based on an extensive literature review and expert opinions. Key challenges include resource constraints, knowledge gaps, limited top management commitment, and insufficient CSR skills, exacerbated by required time investment and consumer apathy. This research offers valuable insights for academics and practitioners by presenting an ism-based model tailored to the Indian context. It enables managers to address critical barriers and enhance CSR implementation in their organisations.

Keywords: barriers, CSR, ISM, manufacturing

1 INTRODUCTION

In recent decades, corporate social responsibility (CSR) and related practices like sustainability, corporate citizenship, and corporate social performance have become strategic priorities across various industries and countries (Ghosh 2020). CSR yields numerous benefits, including enhanced organisational performance, increased customer satisfaction, reduced legal risks, and improved employee engagement and skill development(Mukherjee and Nuñez 2019).CSR's popularity stems from its comprehensive approach to sustainable development, unlike other methods with limited scope (K. Kumar 2022). Due to differing socio-economic contexts, CSR in developing economies faces unique challenges compared to developed economies. In developing economies, issues such as poverty, illiteracy, and inadequate healthcare require CSR efforts to focus on basic needs, whereas developed economies often emphasise environmental sustainability and ethical practices. Limited regulatory frameworks and weak enforcement in developing countries create challenges in ensuring accountability and transparency. Additionally, resource constraints, lack of awareness, and resistance to change among businesses and communities hinder CSR initiatives. Conversely, developed economies benefit from robust legal systems, mature markets, and higher stakeholder expectations, enabling more strategic and structured CSR practices.

1.1 *Literature review*

This study aims to identify and analyse the significant CSR barriers in Indian manufacturing industry. The objective is achieved by initially collecting relevant studies on CSR barriers in different contexts and making them available in different journals related to sustainability, organisational and business strategy, and sustainable practices. The most authenticated and widely used databases, including Scopus, Web-of-Sciences, Google Scholar, Taylor and Francis, Wiley Online, Science Direct, Emerald Insight, and Springer, were searched by using the following keywords: "Corporate social responsibility adoption, "Corporate social responsibility barriers", "CSR barriers", "CSR obstacles', "CSR implementation", "CSR in manufacturing industry", "CSR in India", and "CSR adoption in developing economies."

1.1.1 *Corporate social responsibility*

Despite numerous definitions of CSR in business literature, the European Commission provides the most explicit and widely referenced definition: "The responsibility of companies for affecting local communities, and therefore, companies should show more affiliation with society by including social, environmental, ethical, moral, and human rights aspects into their business agenda and also following the law". Barnett (2007) defined CSR as "The free allocation of company resources to advance social well-being as a way of strengthening associations with dominant stakeholders." CSR is also associated with ethics, environmental safety, excellent corporate citizenship, social stability, transparency, and stakeholder satisfaction. Recently, companies worldwide have adopted CSR to integrate economic, social, and environmental responsibilities and compete globally (Mukherjee and Nuñez 2019).

1.1.2 *Barriers of corporate social responsibility implementation*

The factors that influence and hinder the implementation of a system are known as barriers. Researchers have identified several barriers to CSR implementation across various sectors (Table 1). Major obstacles to CSR adoption in the textile industry include a lack of knowledge, insufficient stakeholder awareness, inadequate training, and a poor CSR reputation.

Table 1. List of barriers of corporate social responsibility.

S.No.	Barriers	Source
1	Complexity of CSR Implementation	(S. P. Kumar and Garg 2019) (Bux *et al.* 2020)
2	Time-Intensive Processes	(K. Kumar 2022) Verma and Singh 2016)
3	Lack of Policy Support	(Fahad and Rahman 2020)
4	Stakeholder Awareness Deficit	(Schlegelmilch and Szőcs 2020)(Nag and Bhattacharyya 2016)
5	Doubt About CSR Effectiveness	(Gatti *et al.* 2019) (Bux *et al.* 2020)
6	Insufficient Commitment from Top Management	(Ghosh 2020)
7	Inadequate CSR Training	(Verma and Singh 2016)(Fahad and Nidheesh 2021)
8	Knowledge Deficiency	(K. Kumar and Prakash 2019)
9	Resource Shortage	(Boodoo 2020) Verma and Singh 2016)
10	Absence of a CSR Strategy	(Sharma *et al.* 2020) (Ghosh 2020)
11	Low Concern for Reputation	(Bux *et al.* 2020)(Verma and Singh 2016)

1.1.3 *Research gaps*

The general impression is that many countries face the same challenges while implementing CSR. However, the degree of importance of each barrier might not be the same owing to the country-specific scenarios. The extensive literature review showed a significant interest among academia and scientific researchers; more essentially, they propose strategies to overcome these challenges. Therefore, based on the previous sections, this study identified the following research gaps:

- Most of the earlier studies on CSR implementation focussed on developed countries. However, very few studies used established methodologies to address CSR implementation barriers in developing countries.

- Several researchers have investigated the barriers. However, very few studies exist in the manufacturing industry with serious CSR issues; most studies did not consider the Indian context.
- Developing countries, such as India, continue to lack policies and strategies while facing problems in implementing the CSR approach.

2 RESEARCH METHODOLOGY

Interpretive structural modelling (ISM) finds and summarises relationships among specific variables. The ISM methodology works on the experts' opinions and group discussions to develop the interrelationships among the variables under study (Warfield 1974). The number of experts in most of the studies varied from five to fifteen. Structural Modelling (ISM) is better than Structural Equation Modelling (SEM) for modelling CSR challenges as it provides a structured, hierarchical framework that identifies and analyses interrelationships among challenges. ISM is particularly effective in dealing with complex, qualitative factors and organising them into levels based on their driving and dependence power. Unlike SEM, which relies on statistical data, ISM emphasises expert judgment, making it suitable for exploring subjective, context-specific issues like CSR challenges, where quantitative data may be limited or unavailable.

3 ANALYSIS

For this study, data were collected from companies in six types of manufacturing industries: Textiles, automobiles, oil and gas, steel, pharmaceuticals, and beverages; all of them are from India.

3.1 *Structural self-interaction matrix*

The relationships among the variables (e.g., i and j) were established by conducting a pair-wise comparison of variables. The associated direction of the relation, for example, whether i leads to j or vice versa, was questioned. Four symbols were deployed to find the existing contextual link between the two concerned sub-variables (i and j).

They are as follows:

V = Variable i will help achieve variable j
A = Variable j will help achieve variable i
X = Variable i and j will help achieve each other
O = Variables i and j are unrelated.

The Structural Self Interaction Matrix (SSIM) was developed that was based on the contextual relationships for the ten barriers of CSR. The SSIM is depicted in Table 2.

Table 2. Structural self-interaction matrix (SSIM).

S.No.	Barriers	16	14	12	9	8	7	5	4	3	2	1
B1	Insufficient Commitment from Top Management	V	A	A	O	A	V	V	A	V	X	
B2	Knowledge Deficiency	X	A	O	O	V	V	V	A	A		
B3	Inadequate CSR Training	O	A	O	V	X	V	A	A			
B4	Resource Shortage	V	O	O	O	O	O	V				
B5	Absence of a CSR Strategy	O	A	V	V	V	V					
B6	Low Concern for Reputation	A	A	O	O	A						
B7	Doubt About CSR Effectiveness	O	A	A	A							
B8	Complexity of CSR Implementation	O	A	V								
B9	Time-Intensive Processes	V	O									
B10	Lack of Policy Support	V										
B11	Stakeholder Awareness Deficit											

3.2 *Level partition*

After compiling the final reachability matrix, the next step, partitioning, was done (Table 3). The objective of partitioning was to find the hierarchy of each variable.

Table 3. Level partition.

S.No.	Barriers	Level
B1	Insufficient Commitment from Top Management	III
B2	Knowledge Deficiency	II
B3	Inadequate CSR Training	III
B4	Resource Shortage	V
B5	Absence of a CSR Strategy	III
B6	Low Concern for Reputation	I
B7	Doubt About CSR Effectiveness	III
B8	Complexity of CSR Implementation	III
B9	Time-Intensive Processes	III
B10	Lack of Policy Support	IV
B11	Stakeholder Awareness Deficit	III

3.3 *Development of digraph*

Based on the conical form of the reachability matrix, the initial digraph, including transitive links, was developed (Figure 1).

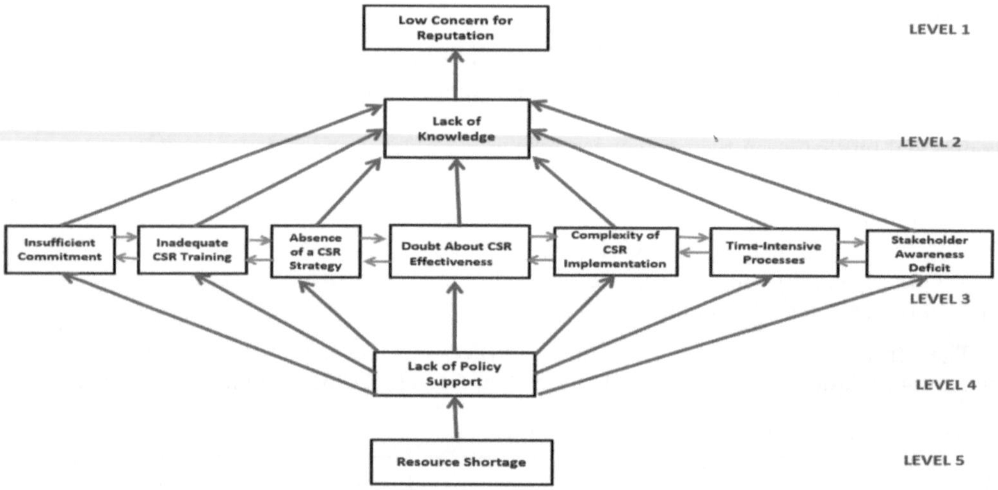

Figure 1. Diagram depicting levels.

4 RESULT AND DISCUSSION

The ISM methodology results identified five barriers to CSR adoption, as shown in Table 4. The primary barrier is resource shortage (B4). To address this, companies should initiate efforts to increase financial resources, seeking government support through subsidies, tax relief, funding, and duty reductions on importing and exporting environmentally friendly machinery.

At Level II, the critical barrier is the lack of policy support (B10). The government should enhance the enforcement of regulations and standards and create policies that promote CSR, fostering an environment conducive to responsible business practices. Once regulations are strictly enforced, companies must adopt more socially and environmentally responsible behaviour.

Level III includes eight barriers: B1, B3, B12, B5, B7, B8, B9, and B11. Insufficient commitment from top management (B1) is particularly crucial. Research in India indicates that top management plays a significant role in CSR implementation. However, CSR often receives less attention from top management than other departments, suggesting that CSR is not yet integrated into the core corporate strategy and that its role in achieving long-term objectives is poorly understood.

At Level IV, the identified barrier is knowledge Deficiency (B2). To overcome this, CSR activities should be organised by the government, industry, and third parties, such as universities, educational institutions, or NGOs, to enhance CSR knowledge.

The sole barrier at Level V is Low Concern for Reputation (B6). This can be mitigated by increasing knowledge and customer awareness about CSR, subsequently improving the company's reputation.

This study aids industrial managers in understanding the obstacles to CSR implementation. By identifying these critical barriers, managers can work towards eliminating them, facilitating CSR adoption in their organisations.

4.1 *Practical implications*

Companies need to implement practical strategies to effectively address CSR challenges in India while collaborating with government bodies. Below are key recommendations and suggestions for overcoming the identified challenges. Companies can effectively address CSR challenges in India by focusing on these practical recommendations and collaborating with government bodies: 1. Enhance Transparency and Accountability in CSR Activities, 2. Align CSR Initiatives with National Development Priorities, 3. Engage Local Communities in CSR Planning and Execution, 4. Address Skill Development and Employment, 5. Ensure Proper Utilization of CSR Funds, 6. Promote Innovation and Technology in CSR Initiatives, 7. Foster Collaboration Between Corporations and NGOs, 8. Address Environmental Sustainability, 9. Improve Education and Awareness on CSR, 10. Establish Stronger Corporate Governance.

5 CONCLUSION

Although corporate social responsibility (CSR) has been successfully adopted in developed countries, its implementation in developing nations like India is challenging due to various obstacles. This study identifies and analyses the key barriers to CSR implementation within the Indian manufacturing sector. It identifies these barriers from previous research and examines their interrelationships using the Interpretive Structural Modelling (ISM) methodology.

The findings highlight the current state of CSR implementation in developing countries. Notably, the "lack of resources (B4)" barrier is foundational in the ISM hierarchy model, indicating that financial resources are essential for CSR adoption in a developing country. This resource gap can be bridged in India with appropriate financial policies and government subsidies, accelerating CSR implementation in the manufacturing industry. However, the research has limitations. The ISM-based model developed in this study outlines a hierarchical structure of barriers but does not assess their overall impact on CSR adoption.

Recently, the total interpretive structural model (TISM) has gained traction among researchers interested in understanding the impact of various factors (Jena *et al.* 2017). Future studies could employ the fuzzy-TISM technique to include more barriers, clarifying the impact of each barrier on CSR implementation in the manufacturing sector. While ISM was used to create an initial model, other methods like structural equation modelling could also be utilised. Additionally, this study primarily focuses on identifying barriers rather than solutions. Future research could explore enablers of CSR implementation. The results are based on a case study of the Indian manufacturing industry, and future research could test their applicability in different contexts and countries.

REFERENCES

Barnett, M. L. (2007). Stakeholder influence capacity and the variability of financial returns to corporate social responsibility. *Academy of Management Review*, *32*(3), 794–816. https://doi.org/10.5465/AMR.2007.25275520

Boodoo, M. U. (2020). Does Mandatory CSR Reporting Regulation Lead to Improved Corporate Social Performance? *Academy of Management Proceedings*, *2020*(1), 17332. https://doi.org/10.5465/ambpp.2020.17332abstract

Bux, H., Zhang, Z., and Ahmad, N. (2020). Promoting sustainability through corporate social responsibility implementation in the manufacturing industry: An empirical analysis of barriers using the ISM-MICMAC approach. *Corporate Social Responsibility and Environmental Management*, *27*(4), 1729–1748. https://doi.org/10.1002/csr.1920

Fahad, P., and Nidheesh, K. B. (2021). Determinants of CSR disclosure: an evidence from India. *Journal of Indian Business Research*, *13*(1), 110–133. https://doi.org/10.1108/JIBR-06-2018-0171

Fahad, P., and Rahman, P. M. (2020). Impact of corporate governance on CSR disclosure. *International Journal of Disclosure and Governance*, *17*(2–3), 155–167. https://doi.org/10.1057/s41310-020-00082-1

Gatti, L., Vishwanath, B., Seele, P., and Cottier, B. (2019). Are We Moving Beyond Voluntary CSR? Exploring Theoretical and Managerial Implications of Mandatory CSR Resulting from the New Indian Companies Act. *Journal of Business Ethics*, *160*(4), 961–972. https://doi.org/10.1007/s10551-018-3783-8

Ghosh, S. (2020). Mandated corporate social responsibility: evidence from India. In *CSR, Sustainability, Ethics and Governance*.

Jena, J., Sidharth, S., Thakur, L. S., Kumar Pathak, D., and Pandey, V. C. (2017). Total Interpretive Structural Modeling (TISM): approach and application. *Journal of Advances in Management Research*, *14*(2), 162–181. https://doi.org/10.1108/JAMR-10-2016-0087

Kumar, K. (2022). Emerging phenomenon of corporate sustainability reporting: Evidence from top 100 NSE listed companies in India. *Journal of Public Affairs*, *22*(1). https://doi.org/10.1002/pa.2368

Kumar, K., and Prakash, A. (2019). Examination of sustainability reporting practices in Indian banking sector. *Asian Journal of Sustainability and Social Responsibility*, *4*(1). https://doi.org/10.1186/s41180-018-0022-2

Kumar, S. P., and Garg, M. C. (2019). Relationship between corporate social reporting (CSR) practices and company characteristics in indian companies. *Finance India*, *33*(4), 1001–1014.

Mukherjee, A., and Nuñez, R. (2019). Doing well by doing good: can voluntary CSR reporting enhance financial performance? *Journal of Indian Business Research*, *11*(2), 100–119. https://doi.org/10.1108/JIBR-07-2018-0199

Nag, T., and Bhattacharyya, A. K. (2016). Corporate Social Responsibility Reporting in India: Exploring Linkages with Firm Performance. *Global Business Review*, *17*(6), 1427–1440. https://doi.org/10.1177/0972150916653032

Schlegelmilch, B. B., and Szőcs, I. (2020). *Rethinking Business Responsibility in a Global Context.*

Sharma, P., Panday, P., and Dangwal, R. C. (2020). Determinants of environmental, social and corporate governance (ESG) disclosure: a study of Indian companies. *International Journal of Disclosure and Governance*, *17*(4), 208–217. https://doi.org/10.1057/s41310-020-00085-y

Verma, P., and Singh, A. (2016). Fostering Stakeholders Trust through CSR Reporting: An Analytical Focus. *IIM Kozhikode Society and Management Review*, *5*(2), 186–199. https://doi.org/10.1177/2277975215618473

Towards Resilient Societies: The Synergy of Religion, Education, Health, Science, and Technology – Rahiem (Ed.)

Challenges in addressing urban sprawl and housing affordability in Indonesian cities through land development policy

Nuraini Safitri & Rizal Galinato Junior
Ministry of National Development Planning, Indonesia

Fadhila Almapuspita
University of New South Wales, Australia

Anri Ariq Rafif
Ministry of National Development Planning, Indonesia

ABSTRACT: Indonesia has experienced significant urbanisation and population growth in recent decades, leading to increased land consumption and housing demand in cities, contributing to urban sprawl. This rapid expansion has strained land availability, increased property prices and limiting strategic development areas. As a result, developers are increasingly turning to suburban areas with scattered development patterns to provide more affordable housing options. This article aims to propose policy recommendations to address the housing affordability challenges caused by urban sprawl, aligning with Sustainable Development Goal (SDG) 11, which seeks to ensure access to adequate, safe, and affordable housing. The study employs a qualitative case study approach, synthesising data from academic literature, government reports, and case studies to analyse urban sprawl's characteristics and economic impacts. Findings reveal that urban sprawl significantly affects housing affordability, exacerbated by fragmented spatial planning, inadequate urban infrastructure, and lack of land banking mechanisms. These challenges limit the government's ability to provide affordable housing, particularly for lower-income communities. The proposed recommendations aim to support policymakers in fostering sustainable, inclusive, and liveable communities in alignment with SDG 11.

Keywords: housing affordability, land use, land management, urban sprawl

1 INTRODUCTION

As populations grow and economic opportunities concentrate in urban centres, the demand for housing has intensified, leading to significant challenges in housing affordability. Particularly for housing affordability, the impact of urbanisation creates pressure on land utilisation to accommodate activities in urban areas, where there is an increasing demand for land but decreasing in land supply. This phenomenon induces competition between settlement and agricultural areas, resulting in land conversion and shifting land use patterns.

The urban landscape of Indonesian cities is undergoing rapid transformation, driven by population growth, economic development, and social dynamics. The urbanisation rate in Indonesia is expected to increase by approximately 57% in 2030 (UN – Dept of Economic and Social Affairs and World Bank 2021). According to the World Bank (2023), in the 1960s, the Indonesian population lived in urban areas only around 15%, and this steadily increased each year until it peaked at 59% in 2023.

DOI: 10.1201/9781003645542-28

The particular issue tackled in this research is unprecedented increases in the urban population, which affected the housing deficit. Notably, this population growth also affects urban land use patterns (Olanrejawu and Adegun 2021) (Olanrejawu and Adegun 2021). Regarding urban sprawl, land use decisions are largely influenced by financial considerations, which affect the degree of sprawl in urban areas (Chapple 2018). Parallel with housing affordability and urban sprawl issues, land control and ownership inequality play a critical role. It can be seen from the gini ratio of land tenure in Indonesia which only 0,58, it means that 1% of the population controls 58% of the resources of agrarian power, land, and space (Ministry of Agrarian and Spatial Planning 2022). The concentration of land in a handful of parties will limit small actors or marginal groups from accessing credit, which impacts limited access to their land market.

While significant research has been conducted on urban sprawl and housing affordability in various contexts, a notable gap exists in the specific relationship between these challenges in Indonesian cities and the policies that address them. In particular, existing studies have not sufficiently examined the intersections of urban sprawl, housing affordability, and the spatial distribution of land ownership, especially for the lower-income community. This journal explores the relationship between urban sprawl, housing affordability, and land policy in Indonesian cities. By examining current trends, main drivers, and the impact of existing policies, it seeks to identify strategies that promote sustainable urban growth while ensuring everyone has the same chance to get housing. Ultimately, this article aims to recommend a land management policy to mitigate the housing affordability challenges posed by urban sprawl.

The paper is structured as follows: the methods section details the research design and studied cities. The findings and discussions from secondary datasets, such as previous research and government data interpret and offer insights about challenges in addressing urban sprawl and housing affordability. Finally, the conclusion summarises the key points and suggests directions for future research.

2 METHODS

2.1 *Research design*

This study employs a qualitative case study method, leveraging a secondary approach to explore the intricate issues of urban sprawl and housing affordability in Indonesia. This methodology is ideal for in-depth analysis of complex phenomena using existing data and prior studies (Yin 2017). Data collection involved a comprehensive literature review, drawing from academic articles, books, government reports, and previously documented case studies, particularly focusing on urban sprawl and housing affordability in Indonesian cities. Key sources included peer-reviewed journal articles, which offered detailed analyses and findings across various contexts, enriching the study's insights (Takahashi and Araujo 2019). By synthesising these diverse perspectives, the research delineates urban sprawl's characteristics and economic implications, providing a strong foundation for policy recommendations to address the identified housing affordability challenges in Indonesian cities.

2.2 *The studied cities*

In recent years, Indonesian cities have experienced significant growth and development rates caused by rapid urbanisation. This rapid expansion of urban economic activities due to urbanisation has burdened these urban centres substantially (Silver 2024). Some major cities in Indonesia, such as Jakarta, Surabaya, Bandung, and Makassar, exemplify this trend. The more people move to urban areas, the more housing is needed to accommodate them.

However, the increasing population and limited land availability have led to housing issues (Alzamil 2020), presenting a significant challenge that must be addressed.

Table 1 depicts that DKI Jakarta has Indonesia's highest population and density, corresponding with a GDP of 3,442.98 trillion rupiahs in 2023, significantly exceeding other major cities. The substantial population and limited land area in Jakarta have resulted in significant housing availability issues (Alzamil 2017). According to the report for DKI Jakarta (2016), there was a housing backlog of at least 302,319 units in 2015. This extensive backlog has driven developers and the government to focus on affordable housing, particularly for low-income residents. The high demand for housing has further led to the development of residential areas on the outskirts of Jakarta, utilising available land and contributing to urban sprawl (Silver 2024). This study employs DKI Jakarta as a case study because it illustrates many urban challenges other Indonesian cities face, including rapid population growth, urbanisation, and poverty. Additionally, Jakarta's expansion into surrounding cities like Bogor, Depok, Tangerang, and Bekasi (Bodetabek) provides valuable lessons on how urbanisation impacts metropolitan governance and regional connectivity. These characteristics make Jakarta an ideal focus for understanding the interplay between urban sprawl and housing affordability in Indonesia.

Table 1. Indonesian major cities characteristics.

	Population	Population Growth 2020–2023	Area	Density	GDP	Prices of Flat by (IDR)		
Cities	P	%	Km^2	P/Km^2	Trillion Rupiah	District	(m^2)	(unit)
DKI Jakarta	10.672.100	0,38	660,98	16.165	3.442,98	West Jakarta	8.900.000	320.400.000
						South Jakarta	9.200.000	331.200.000
						East Jakarta	8.800.000	316.800.000
						North Jakarta	9.600.000	345.600.000
						Centre Jakarta	9.300.000	334.800.000
Bandung	2.506.603	0,92	167,31	14.982	351,28		7.300.000	262.800.000
Surabaya	3.009.286	0,42	326,81	8.958	715,29		7.900.000	284.400.000
Makassar	1.434.393	-0,29	175,77	8.388	226,90		7.300.000	262.800.000

Source: BPS-Statistics Indonesia, 2024; Decree of Minister of Public Works and Housing No. 552/KPTS/M/2016.

3 FINDINGS AND DISCUSSIONS

3.1 *Main findings: the fundamental challenge of housing affordability significantly provokes urban sprawl*

Urban sprawl, marked by the uncontrolled expansion into rural areas, increases infrastructure costs and social segregation. Housing affordability, shaped by economic, demographic, and policy factors, impacts access for different income groups (Daniel and Rosario

2019). Before the housing policy, the government produced about 200,000 units (1%), the private sector had 2 million units (14%), and individual owners, through self-help efforts, had 9.8 million units (85%). The government mandates that private developers follow a 1:3:6 ratio for high, middle, and low-income housing (UN-HABITAT 2018).

However, it often fails to provide affordable housing options for lower-income earners within urban centres, leading to a mismatch between where affordable housing is available and where it is needed. Providing proper infrastructure for poor households, such as drinking water facilities and latrines for the poor, is highly expected. The government's housing and public facilities allocation can be maximised to provide decent housing for the poor (Novitasari and Iskandar 2022). On the other hand, numerous private entities have acquired land by leveraging financial resources, driving up prices. For instance, land prices in Setiabudi have reached Rp 19-29 million per square meter, with central Jakarta locations being even more expensive. In Bumi Serpong Damai (BSD) and Cibubur, prices are Rp 10-12 million per square meter, respectively. These high prices make it difficult to build affordable housing for low-income communities, even with apartment developments (rusun). Ideally, land prices for such projects should be below Rp 8 million per square meter, but suitable land in Jakarta at this price is scarce.

3.2 *Analysis of findings: Indonesia's housing and land use policies are characterised by inefficient implementation compared to Singapore and Australia*

In Indonesia, the housing market predominantly operates on a free trade concept, facilitating marketisation that empowers certain groups while marginalising others. Over time, as dominant property forms are embraced as family assets, they gradually integrate into daily life, subtly co-opting individuals. This process often subjugates many people, making adequate housing inaccessible for middle to low-income individuals, who then choose more affordable options in peripheral areas. Indonesia's government has committed to addressing housing issues through programs like *Tapera*, established by Law No. 4 of 2016 and Government Regulation No. 25/2020, aiming to collect sustainable funds for housing finance. However, *Tapera's* funding, sourced from the public, has been inefficiently implemented because of the uncertain regulation. The government also created District/City Spatial Plans or *Rencana Tata Ruang Wilayah* (RTRW) and Detailed Spatial Plans or *Rencana Detail Tata Ruang* (RDTR) with zoning regulations to optimise land use and mitigate sprawl. Despite these efforts, land utilisation often does not comply with spatial regulations, driven by limited land resources and allocation for housing.

A major obstacle to these policies in Indonesia is the issue of land ownership. In contrast, Singapore's government owns 76.2 percent of its land, managed by agencies like JTC, which facilitates affordable housing through the HDB program by avoiding speculative land pricing and providing substantial subsidies, resulting in a homeownership rate of 87.9 percent. Likewise, First Home Owner (New Homes) Grant and Australia's Affordable Rental Housing SEPP (State Environmental Planning Policy) support an increase in housing supply, contributing to a homeownership rate of 71.4 percent as part of a comprehensive approach to curb speculative activities in the housing market. Both countries emphasise the importance of accessibility by developing housing with well-connected transport systems in mixed-use areas.

3.3 *Policy recommendation: enhancement of land banking and land consolidation to support the integration of vertical development strategies*

Innovative strategies to address urban sprawl and align with SDG 11 include land banking, vertical development, and land consolidation. These approaches promote sustainable urbanisation, optimise land use, and reduce costs. Land banking allows authorities to control land for planned growth and affordable housing. In Indonesia, land banking is established by Government Regulation No 64/2021, and the land banking organisations have the authority to conduct land acquisition and prepare the master plan. Vertical development maximises land use with high-rise buildings, addressing housing shortages. Land consolidation reorganises fragmented plots, improving efficiency and access for low-income populations. The Government of DKI Jakarta's Vertical Land Consolidation program in Palmerah expanded liveable home areas from $10m^2$ to $18m^2$, supporting SDG 11 for sustainable urban environments (Ministry of Agrarian Affairs 2024). However, implementing land consolidation and vertical development faces several challenges, such as community resistance to vertical housing due to their preference for landed houses and lack of supporting infrastructure because vacant land with affordable prices is located in suburban areas that are not accessible.

3.4 *Implications and next steps: implementation of land banking, land consolidation, and vertical development in rapidly growing areas*

The parameter of successful implementation to tackle urban sprawl is converting existing landed houses into vertical housing, reducing the need for new residential areas on the city's outskirts. Additionally, to connect multiple nodes, transit-oriented development for affordable housing (TODAH) projects, such as Tanjung Barat, Pondok Cina, and Rawabuntu Stations, leverage public transportation infrastructure to provide affordable housing and curb urban sprawl. Local governments assist in administrative and traffic management (National Water and Sanitation Information Services 2020). Furthermore, land banking—a strategy involving the acquisition and management of vacant land—transforms these areas into productive assets, stabilising land prices and supporting sustainable urban planning (Sasu *et al.* 2022; Alexander 2015) presents as another viable strategy. This proactive approach is crucial for managing urban expansion and promoting sustainable development (Syed Abu Bakar and Jaafar 2018).

4 CONCLUSION

Appropriate land management is vital for improving urban planning. In cities where urban sprawl is becoming a complex issue, local authorities should reassess building regulations and zoning laws and encourage the development of highly populated cities. This paper aims to endorse future research in land management to address the issues of urban sprawl and housing affordability. Despite having a relatively functional housing market, reform of land-use regulation in Indonesia is important. Some clear policy reforms were raised from the case studies of Jakarta (Indonesia) and other countries. Based on the study, it can be concluded that the allocation of land for housing cannot be fully entrusted to the free-market mechanism because it does not ensure equal access for the lower-middle-income groups. That, resolving legal issues about land banking is important by preparing a land acquisition masterplan, mechanisms, and incentives and disincentives. The government can collaborate with the private sector in land banking mechanisms to anticipate future low-income housing needs. Afterwards, in the medium term, the government should implement a series of pilot vertical housing projects, starting with the most populous urban centres, such as Jakarta, in partnership with local governments and private developers. Furthermore, integrating Transit-Oriented Development (TOD) projects into new residential areas should be prioritised to ensure equitable access to transportation and housing. Over the next 5 years, a robust monitoring framework should be established, tracking the success of these policies with clear KPIs. This approach will not only mitigate urban sprawl but also align with SDG 11 by providing affordable, safe, and adequate housing, especially to low-income groups in Indonesia.

REFERENCES

Alexander, F. S. (2015). *Land Banks and Land Banking. Michigan, USA: Center for Community Progress.* ISBN-10: 0692405127. ISBN-13: 978-0-692-40512-3

Alzamil, W. S. (2017). The urban features of informal settlements in Jakarta, Indonesia. *Data in Brief*, 993–999. https://doi.org/10.1016/j.dib.2017.10.049

Alzamil, W. S. (2020). Housing Challenges for Urban Poor: Kampungs in Jakarta, Indonesia. *Current Politics and Economics of South, Southeastern, and Central Asia, 29*(1), 81–112. ISBN: 978-1-53618-715-1

Chapple, K. (2018). The fiscal trade-off: Sprawl, the conversion of land, and wage decline in California's Metropolitan Regions. *Landscape and Urban Planning*. https://doi.org/10.1016/j.landurbplan.2018.01.002

Daniel, C.-H. 1., and Rosario, A.-C. (2019). Urban Sprawl and Sustainable Urban Policies A Review of the Cases of Lima, Mexico City and Santiago de Chile. *Sustainability MDPI*, 3–4. https://doi.org/10.3390/su11205835

DPRKP Provinsi DKI Jakarta. (2016). RP3KP. *Jakarta: Government of Jakarta Province.*

Ministry of Agrarian Affairs. (2024, 07 04). *Bersama Kementerian ATR/BPN*, Pemprov DKI Jakarta Resmikan Program Perbaikan Rumah dan Konsolidasi Tanah Vertikal. Retrieved 07 23, 2024, from Kementerian Agraria dan Tata Ruang/Badan Pertanahan Nasional: https://wwwv2.atrbpn.go.id/siaran-pers/detail/14349/bersama-kementerian-atrbpn-pemprov-dki-jakarta-resmikan-program-perbaikan-rumah-dan-konsolidasi-tanah-vertikal

National Water and Sanitation Information Services. (2020, August 25). *Alternatif Mekanisme Penyediaan Tanah bagi Perumahan Masyarakat Berpenghasilan Rendah (MBR) Di Perkotaan.* Retrieved from PerkimPedia: https://www.nawasis.org/perkimpedia/wiki/alternatif-mekanisme-penyediaan-tanah-bagi-perumahan-masyarakat-berpenghasilan-rendah-mbr-di-perkotaan/10

Novitasari, M., and Iskandar, D. A. (2022). Spatial Spillover Impact of Sectoral Government Expenditure on Poverty Alleviation in South Kalimantan Province. *Journal of Indonesia Sustainable Development Planning*, 3(3), 207–221. https://doi.org/10.46456/jisdep.v3i3.361

Olanrejawu, S. D., and Adegun, O. B. (2021). Urban Sprawl and Housing: A Case of Densification in Nigerian Cities. In *Housing and SDGs in Urban Africa* (pp. 287–299). Springer. https://doi.org/10.1007/978-981-33-4424-2

Sasu, A., Squires, G., and Javed, A. (2022, November 2). Land banking and land markets: A literature review. *Habit International*, 130, 1–13. https://doi.org/10.1016/j.habitatint.2022.102698

Silver, C. (2024, April 6). Rapid Urbanization: The Challenges and Opportunities for Planning in Indonesian Cities. *The Indonesian Economy and the Surrounding Regions in the 21st Century*, 76; 35–48. https://doi.org/10.1007/978-981-97-0122-3_3

Syed Abu Bakar, S. P., and Jaafar, M. (2018). Achieving business success through land banking and market analysis: Perspectives of Malaysian private housing developers. *Property Management*, 36(5), 562–574. https://doi.org/10.1108/PM-10-2017-0059

Takahashi, A. R., and Araujo, L. (2019). Case study research: opening up research opportunities. *RAUSP Management Journal*, 55, 100–111.

UN-HABITAT. (2018). *The Role of Government in the Housing Market: The Experiences from Asia. Nairobi: United Nations Human Settlement Programme.* ISBN: 978-92-1-132025-1

Yin, R. K. (2017). *Case Study Research and Applications: Design and Methods.* California: Sage.

Towards Resilient Societies: The Synergy of Religion, Education, Health, Science, and Technology – Rahiem (Ed.)
© 2026 The Author(s), ISBN: 978-1-041-08471-6
Open Access: www.taylorfrancis.com, CC BY-NC-ND 4.0 license

Leveraging WhatsApp support to improve well-being of families with autistic children: An Indonesian virtual ethnographic study

Dhia Damarisa Qariru
Ministry of National Development Planning, Indonesia

Herna Herna
Universitas Gunadarma, Indonesia

Hidayaturahmi Hidayaturahmi
Politeknik STIA LAN Jakarta, Indonesia

Hiswanti Hiswanti
Institut Bisnis dan Informatika Kosgoro 1957, Indonesia

ABSTRACT: The global prevalence of autism spectrum disorder (ASD) is increasing, including in Indonesia, where approximately 2.4 million individuals are affected. Families of children with ASD often experience social stigma, isolation, and discrimination, particularly those with limited resources. As Indonesia's most widely used social media platform, WhatsApp presents a promising avenue for social support and health communication. This study aims to explore the satisfaction levels of parents of autistic children within virtual communities. Employing a qualitative approach with a virtual ethnographic method, data were collected through eight months of participant observation (116,238 chat messages) and in-depth interviews with members of the "Anak Istimewa" virtual community. The findings reveal that virtual communities provide significant support across several dimensions: emotional (enthusiasm, empathy), informational (problem-solving, early expert intervention), social (reduced isolation, increased solidarity), and spiritual (emotional comfort through religious encouragement). This study highlights the potential of virtual communities to improve health and well-being, aligning with Sustainable Development Goal (SDG) 3 (Good Health and Well-Being) and SDG 10 (Reduced Inequalities). Virtual communities can help reduce stigma, increase access to information and health education, and engage health experts to offer accurate and practical guidance, thereby enhancing support for families of children with ASD.

Keywords: autism, communication, social support, virtual community, WhatsApp

1 INTRODUCTION

The prevalence of autism spectrum disorder (ASD) continues to increase worldwide. Autism is a neurodevelopmental disorder affecting communication, social interaction, and behaviour. According to (Hodges *et al.* 2020), one in 160 people is identified with autism, occurring across all racial, ethnic, and socioeconomic groups. In Indonesia, although precise data is lacking, it is estimated that around 2.4 million people have autism, with an increase of 500 individuals per year. Autism significantly impacts family life, especially for elderly parents who often experience stress due to their child's behaviour. Other issues include high medical costs, lengthy treatment processes, the absence of effective medication, and the

DOI: 10.1201/9781003645542-29

uncertainty of the child's future (Ilias *et al.* 2018). Research indicates that parents of autistic children experience higher stress levels compared to parents of children without disorders (Pepperell *et al.* 2018). Therefore, social support is crucial for parents of autistic children. Another study found that parents of autistic children require more support compared to parents of children without disorders or children with Down syndrome (McIntyre and Brown 2018).

In Indonesia, autism is often perceived as karma or a result of the parents' past sins (Riany *et al.* 2016). This leads to pervasive social stigma, isolating families and hindering access to crucial support and services, particularly for those with limited resources. The resulting discrimination negatively impacts the health and well-being of both mothers and children, exacerbating the challenges they face. While the Indonesian government has made strides in protecting the rights of children with disabilities through legislation like Law No. 8 of 2016 and Ministerial Regulation PPPA No. 4 of 2017, significant barriers to accessing adequate health and education services persist. This gap in access underscores the urgent need to explore innovative and accessible solutions, particularly in the context of achieving the Sustainable Development Goals (SDGs) 3 with an emphasis on good health and well-being by promoting mental health and psychosocial well-being for individuals with ASD.

Furthermore, it aligns with SDG 10 (Reduce Inequalities) by exploring accessible and affordable technology to bridge communication gaps and empower individuals with ASD to participate more fully in society. However, existing research often overlooks the link between autism and the achievement of these SDGs in developing countries. This study aims to bridge this gap by investigating the potential of WhatsApp, a widely accessible and user-friendly platform, to support individuals with autism and their families in Indonesia, ultimately contributing to a more inclusive and equitable society.

Social support is crucial for people with autism and their families. With technological advancements, social support can be obtained both offline and online. Social media platforms like WhatsApp are potential options for health communication due to their high participation rate in Indonesia (84%). The internet allows people to access information and social support that they may not obtain offline, allowing them to join virtual communities to discuss autism-related issues. Virtual communities have become very popular in the last decade. Platforms like Facebook, WhatsApp, Instagram, and Twitter play significant roles in people's daily lives (Gazit *et al.* 2019). Discussion groups on these platforms often evolve into communities where participation is voluntary, and members feel they benefit from the experience (Bruckman and Jensen 2002). Online communities let people freely express their opinions, share feelings, and engage by posting, commenting, and liking (Gazit and Amichai-Hamburger 2020).

Previous research states that virtual communities for health communication provide informational and emotional support (Cole *et al.* 2017; Doty and Dworkin 2014; Mohd Roffeei *et al.* 2015; Purnamasari 2020; Schiavo 2014). Parents with autistic children tend to use virtual communities to obtain information related to childcare. This study explores the motivations of the WhatsApp group member "Anak Istimewa." This research can contribute to understanding the factors that motivate parents to join virtual communities and the satisfaction achieved. The Uses and Gratifications Theory (UGT) approach is a theoretical foundation for understanding members' motivations. Observations of the group and in-depth interviews are qualitatively analysed to gain a deeper understanding. By addressing these questions, this research illuminates how virtual communities enhance the well-being of families with autistic children, ultimately contributing to Indonesia's progress toward achieving the SDGs, particularly those related to the health and well-being of individuals with autism in Indonesia through informational support within the groups addresses their need for practical knowledge about autism care, emotional support provides much-needed empathy and a sense of solidarity with others facing similar challenges, and entertainment support offers stress relief through humour and shared experiences.

2 METHODS

This study employs a qualitative methodology with a virtual ethnographic approach, adapting traditional ethnographic techniques to study artifacts, culture, and communities in online contexts. (Kozinets 2015). This method involves the researcher's active presence and participation in the observed community, including participant observation and elicitation interviews to understand the differences between what participants say and do (Boellstorff *et al.* 2012). The "Anak Istimewa" WhatsApp Group was selected as the unit of analysis because it is relevant, active, interactive, diverse, and data-rich. The research aims to address questions about the community's support and interactions. Before data collection, ethical consent was obtained from group administrators and participants, ensuring confidentiality and voluntary participation.

Data from the "Anak Istimewa" WhatsApp Group includes texts, visuals, and a combination. The researcher recorded 116,238 chats, which were condensed to 48,221 data points and analysed using NVivo 12 Plus. This included statements, responses, and replies within the group. Offline interviews were conducted in the Jabodetabek area, and online interviews were conducted using Zoom or video calls for those outside Jabodetabek.

Purposive sampling was used to select informants based on the severity of the disorder, income level, and gender, ensuring a representative sample of the community's demographics. The sample included parents with autistic children aged 2 to 18. The research process involved several steps, including selecting the community, obtaining permission from the admin, conducting initial observations, and interacting with the community from January to December 2020. Data condensation resulted in 387 meaningful chats, illustrating patterns of reciprocal communication. Data analysis focuses on continuously reflecting on the data to answer the research questions, including posing analytical questions and taking brief notes throughout the study (Creswell and Creswell 2018).

3 FINDINGS AND DISCUSSION

The interaction dynamics within support groups, particularly among mothers of children with severe autism from low-income backgrounds, provide valuable insights into the multifaceted nature of support systems in these communities (Figure 1). These mothers frequently encounter societal stigmatisation, behavioural challenges, and a lack of offline support, yet

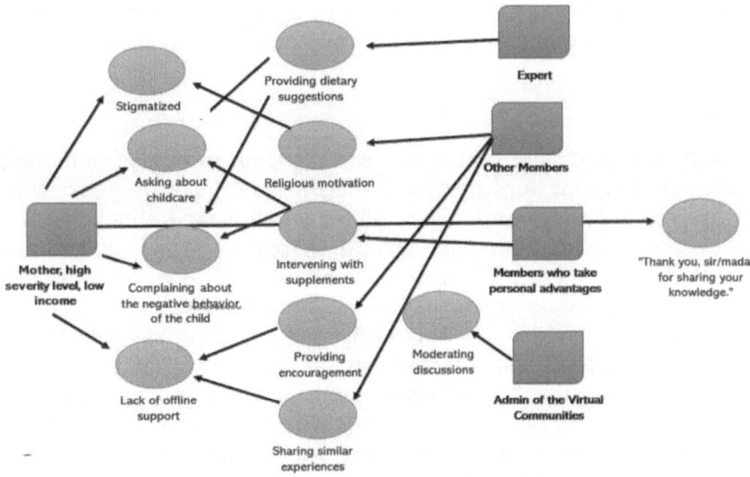

Figure 1. Interaction dynamics in a support group for mothers of children with severe autism.

find empowerment through shared experiences and mutual assistance within the group. The intersection of health, technology, and community engagement is key to understanding the effectiveness of these support systems. Online platforms, such as WhatsApp, facilitate the exchange of information, emotional support, and expert advice, thereby bridging the gap between healthcare needs and accessible social resources. This technology integration enables continuous engagement, ensuring that mothers receive real-time guidance and solidarity, regardless of geographic constraints.

The following subsections examine key findings from the study, including information Support, which highlights the group's role in providing autism-related knowledge; Social Support, focusing on peer connections that reduce isolation; Spiritual Support, which fosters resilience through shared beliefs; and emotional support, which emphasising empathy and encouragement. The study also addresses the need for entertainment through stress-relieving interactions, the presence of experts offering tailored guidance, and accessibility and limitations, which discuss the challenges of internet connectivity and platform features. Together, these dimensions reveal how online communities enhance the well-being of mothers and their children.

3.1 *Information support*

Parents of autistic children join the "Anak Istimewa" WhatsApp community driven by the need for information that is hard to access in their local environment due to the fear of stigma. The group, primarily composed of mothers of autistic children, shares problems, solutions, and early interventions from credible experts. These mothers use the community to seek information related to autism, including causes, symptom reduction, education, therapy tips, and diet. One participant shared, *"The group provides so much information, like which foods can trigger tantrums and should be avoided. I have learned a lot about how to support my child better."* Another example is members actively engaging in discussions, posing questions about their children's behaviour, and receiving advice from peers and experts within the group. This collaborative approach has created a trusted platform for learning and mutual support, empowering parents to better care for their children.

Social support plays a vital role in improving the life satisfaction of families with autistic children. Therefore, interventions that focus on building positive social support for them in the future are needed (Lu *et al.* 2015). Without sufficient information or understanding of autism, parents may experience delays in addressing their child's needs. Therefore, information exchange and discussions in the group aid the knowledge transfer process (Mohd Roffeei *et al.* 2015). The support provided by the group includes receiving expert advice, learning from the experiences of other members, and obtaining information shared by members.

3.2 *Social support*

Many mothers with autistic children experience social stigma and isolation in society. Online communities like WhatsApp groups provide a space where they can speak without fear of judgment. Parents join online communities to share emotions and obtain social support. In the Anak Istimewa virtual community, mothers can share experiences and gain support from fellow group members. For example, one mother hesitated to discuss the challenges of raising her child with neighbours or friends but felt comfortable expressing her concerns in the WhatsApp group. However, in the WhatsApp group, she could freely express her concerns, receive advice from other mothers with similar experiences, and feel less alone. This helps reduce feelings of isolation and provides significant moral support, as illustrated by participant quotes like, *"When my child was diagnosed with autism, I felt like my world was falling apart. But in this group, I do not feel alone. The mothers here support me, and now I have someone to talk to."* or *"Just reading a message in the group saying, 'We are chosen and strong mothers,' makes me feel empowered."*

Words of encouragement reflecting emotional support often appear in the communication within the virtual community. For instance, a member venting about stigma receives support from other members with words of encouragement. Communication in the virtual community provides emotional support such as sympathy, solidarity, motivation, advice, prayers, or sharing similar experiences.

Social support in this virtual community helps members feel they are not alone and gain gratification from their involvement in the community. This encouragement shows that members providing support often go through the same stage, allowing them to offer more meaningful support.

3.3 *Spiritual support*

The "Anak Istimewa" community also provides spiritual support, with members sharing prayers, motivational verses, and inspirational stories. This support dimension strengthens faith and provides a sense of hope and resilience. One participant shared, *"When I feel hopeless, the mothers in the group remind me to surrender to God. It gives me strength. The phrase 'our children are heaven's children' motivates me to accompany them with more enthusiasm."* This highlights the multifaceted role these communities play in supporting the overall well-being of families.

3.4 *Need for entertainment*

Parents of autistic children use WhatsApp groups to socialise and seek entertainment, which helps them reduce stress and feel happier. They share funny stories, videos, and happy moments that can reduce stress and provide moments of joy. This supports the Uses and Gratifications Theory (U&G), which suggests that people use media to entertain and reduce stress (Gazit and Amichai-Hamburger 2020). The study found that the desire for social connection drives people to engage in social media groups. The study indicates that this need for community and support frequently motivates parents of autistic children.

3.5 *Presence of experts*

One significant finding of this study is the presence of experts in the virtual community who provide direct information and guidance. This differs from information obtained from the internet, which is general and not specific. The presence of experts becomes a credible source of information, offering more relevant guidance to individual problems, as highlighted by an informant, HL. HL explained that while information from the internet and social media can be abundant, it often lacks the specificity needed for individual cases. In contrast, the ability to interact with experts within the group provides parents with credible and reliable information that addresses their unique concerns more effectively than the often vague and broad suggestions found online.

Virtual communities on WhatsApp can serve as a forum for parents to obtain information related to autism care and emotional support, which are solutions to their social problems. These communities help solve social problems by providing a platform for personalised consultations and tailored advice specific to each child and family. Health professionals' proactive involvement in such groups is crucial, as they can educate mothers and the community, as suggested by (Indhy Prastyo 2021). Personalised consultations allow for tailored advice and intervention strategies specific to each child and family, or facilitating support groups and offering counselling services can help families cope with emotional stress and anxiety.

3.6 *Accessibility and limitations*

While WhatsApp is generally accessible in rural areas for text-based communication, limitations in network infrastructure can affect features requiring a stronger internet connection, such as video calls or media sharing. One participant noted, *"WhatsApp is easier to use, but video calls often lag or do not open properly. Text messages, though, usually work smoothly."* This highlights the need for continued improvements in internet access to ensure equitable access to all features of these virtual communities.

4 CONCLUSION

Virtual communities on WhatsApp have proven to be highly effective in health communication by offering crucial support to their members. These communities provide valuable information, emotional support, social interaction, and expert access. This sets them apart from general internet use, focusing on interpersonal connections, seeking information, convenience, and entertainment. For example, the "Anak Istimewa" WhatsApp group provides significant help to parents of autistic children in Indonesia. By sharing advice on early interventions, suitable education, and managing autism, these communities help ensure that autistic children get the inclusive and proper education they need, improving the quality of education for children with disabilities.

Additionally, these virtual communities allow members to find information that is hard to get elsewhere, share their experiences, and receive emotional support. This helps parents deal with social stigma and feelings of isolation. The support provided not only reduces stress and boosts the psychological well-being of parents but also strengthens their knowledge and sense of community. As a result, these communities enhance the well-being of autistic individuals and their families and support achieving sustainable development goals in Indonesia.

Given their significant impact, integrating these virtual communities into Indonesia's health and social support systems is crucial for enhancing care, fostering inclusivity, and addressing existing service gaps. This requires a multi-faceted approach, including updating legislation to formally recognize these communities, ensuring legal protections for participants, and integrating them into programs like BPJS Kesehatan to expand access. Furthermore, a comprehensive national strategy and engagement from stakeholders such as Kementerian Kesehatan and Kementerian Sosial, supported by adequate funding and Nationwide training programs for facilitators, should be implemented to provide high-quality, culturally sensitive support in virtual communities. Additionally, virtual communities should be incorporated into hybrid care models, particularly for families facing geographic or financial barriers to in-person therapies. By embracing and investing in virtual communities, Indonesia can create a more inclusive and accessible autism care system that genuinely supports the well-being of autistic individuals and their families.

REFERENCES

Boellstorff, T., Nardi, B., Pearce, C., and Taylor, T. L. (2012). *Ethnography and virtual worlds: A handbook of method.* In *Ethnography and Virtual Worlds: A Handbook of Method.* Princeton University Press. https://doi.org/10.5860/choice.50-5679

Bruckman, A., and Jensen, C. (2002). The mystery of the death of MediaMOO: seven years of evolution of an online community", in Renninger, K.A. and Shumar, W. (Eds), *Building Virtual Communities.* Cambridge University Press, Cambridge, pp. 21–33. https://doi.org/doi:10.1017/cbo9780511606373.006

Cole, L., Kharwa, Y., Khumalo, N., Reinke, J. S., and Karrim, S. B. S. (2017). Caregivers of School- aged Children with Autism: Social Media as a Source of Support. *Journal of Child and Family Studies.* https://doi.org/10.1007/s10826-017-0855-9

Creswell, J. W., and Creswell, J. D. (2018). *Research Design: Qualitative, Quantitative, and Mixed Methods Approaches.* Fifth edition. Thousand Oaks, California, SAGE Publications, Inc.

Doty, J. L., and Dworkin, J. (2014). Online Social Support for Parents: A Critical Review. *In Marriage and Family Review.* https://doi.org/10.1080/01494929.2013.834027

Gazit, T., Aharony, N., and Amichai-Hamburger, Y. (2019). Tell me who you are and I will tell you which SNS you use: SNSs participation. *Online Information Review*, 44(1), 139–161. https://doi.org/10.1108/OIR-03-2019-0076

Gazit, T., and Amichai-Hamburger, Y. (2020). Factors Underlying Engagement in Facebook Support Groups of Female Infertility Patients. *Psychological Reports*, 124(3), 1150–1173. https://doi.org/10.1177/0033294120934703

Hodges, H., Fealko, C., and Soares, N. (2020). Autism spectrum disorder: Definition, epidemiology, causes, and clinical evaluation. In *Translational Pediatrics* (Vol. 9, pp. S55–S65). AME Publishing Company. https://doi.org/10.21037/tp.2019.09.09

Ilias, K., Cornish, K., Kummar, A. S., Park, M. S. A., and Golden, K. J. (2018). Parenting stress and resilience in parents of children with autism spectrum disorder (ASD) in Southeast Asia: A systematic review. *In Frontiers in Psychology*. https://doi.org/10.3389/fpsyg.2018.00280

Indhy Prastyo, J. (2021). The Influence of Father and Mother's Joint Decision Making on Children's Basic Immunization:Empirical Research on Indonesia's Demographic and Health Survey 2017. *The Journal of Indonesia Sustainable Development Planning*, 2(3), 234–256. https://doi.org/10.46456/jisdep.v2i3.175

Kozinets, R. V. (2015). *Netnography: Redefined* (2nd Edition). SAGE Publications.

Lu, M., Yang, G., Skora, E., Wang, G., Cai, Y., Sun, Q., and Li, W. (2015). Self-esteem, social support, and life satisfaction in Chinese parents of children with autism spectrum disorder. *Research in Autism Spectrum Disorders*, 17, 70–77. https://doi.org/10.1016/j.rasd.2015.05.003

McIntyre, L. L., and Brown, M. (2018). Examining the utilisation and usefulness of social support for mothers with young children with autism spectrum disorder. *Journal of Intellectual and Developmental Disability*. https://doi.org/10.3109/13668250.2016.1262534

Mohd Roffeei, S. H., Abdullah, N., and Basar, S. K. R. (2015). Seeking social support on Facebook for children with Autism Spectrum Disorders (ASDs). *International Journal of Medical Informatics*. https://doi.org/10.1016/j.ijmedinf.2015.01.015

Pepperell, T. A., Paynter, J., and Gilmore, L. (2018). Social support and coping strategies of parents raising a child with autism spectrum disorder. *Early Child Development and Care*. https://doi.org/10.1080/03004430.2016.1261338

Permen PPPA Nomor 4 Tahun 2017 Tentang Perlindungan Khusus Bagi Anak Penyandang Disabilitas, 12 Peraturan Menteri Pemberdayaan Perempuan dan Perlindungan Anak Nomor 4 Tahun 2017 tentang Perlindungan Khusus bagi Anak Penyandang Disabilitas 145 (2017).

Purnamasari, A. (2020). Perilaku Komunikasi pada Komunitas Virtual Terkait Diet MAkanan Sebagai Dukungan Sosial Bagi Orang tua yang Memiliki Anak Autism. IPB University.

Riany, Y. E., Cuskelly, M., and Meredith, P. (2016). Cultural Beliefs about Autism in Indonesia. *International Journal of Disability, Development and Education*. https://doi.org/10.1080/1034912X.2016.1142069

Schiavo, R. (2014). Health Communication from Theory to Practice. In *Journal of Chemical Information and Modeling*. San Fransisco (US): Jossey-Bass. https://doi.org/10.1017/CBO9781107415324.004

Towards Resilient Societies: The Synergy of Religion, Education, Health, Science, and Technology – Rahiem (Ed.)

Fostering green innovation through adaptive regulations for sustainable development

Fathudin
Universitas Islam Negeri Syarif Hidayatullah Jakarta, Indonesia
ORCID ID: 0000-0002-3692-0154

Ahmad Tholabi Kharlie, Hilmy Firdausy & Bilqis Khoiriyyah
Universitas Islam Negeri Syarif Hidayatullah Jakarta, Indonesia

Nur Ihsan Hinna
Narathiwat School, Thailand

ABSTRACT: This article examines how adaptive regulatory frameworks can foster green innovation for sustainable development. It analyses key Indonesian policies—Law No. 17/2023 on Health, Presidential Regulation No. 98/2021 on Carbon Economic Value, and Presidential Regulation No. 55/2019 on Battery Electric Vehicles—and their alignment with SDGs 9 (Industry, Innovation, and Infrastructure) and 13 (Climate Action). Using a normative legal research approach, findings suggest that despite promoting environmental sustainability, gaps persist in integrating harm reduction and incentivising green innovation. For instance, revisions to Law No. 32/2009 reduce penalties for environmental violations, potentially weakening enforcement. Comparative insights from Germany and Norway show that robust policy ecosystems, including tax incentives and international collaboration, accelerate green technological advancements. This research underscores the urgency for Indonesia to refine its regulatory frameworks, ensuring they adapt and drive innovations that balance economic growth with environmental preservation.

Keywords: climate action, green innovation, Indonesia, regulatory frameworks, sustainable development

1 INTRODUCTION

The rapid development of technology has significantly impacted the global economy, creating efficiencies across socio-cultural and economic fields. Countries focus on boosting their economic competitiveness through innovation, which is a key factor in improving economic positions. Innovation is a driving force for economic progress, and the relationship between innovation and technological development is a key driver of growth. The concept of innovation has since evolved, leading to extensive research on its applications in various sectors.

Countries with high innovation levels tend to have stronger economic competitiveness, enabling them to produce more sophisticated products, enhance productivity, and create job opportunities (Sorin MS Krammer 2017). South Korea, for example, has improved the competitiveness of small and medium enterprises (SMEs) through technological advancements. Similarly, countries like the United States, China, Japan, and Germany leverage innovation to strengthen their global economic positions. These nations prioritise innovation as a central strategy for boosting market competitiveness.

Innovation must also integrate sustainability to avoid crises like environmental degradation. The United Nations' Sustainable Development Goals (SDGs), particularly Goal 9 (Industry,

DOI: 10.1201/9781003645542-30

Innovation, and Infrastructure) and Goal 13 (Climate Action), stress the need to balance economic growth with environmental conservation. Global examples, such as Germany's renewable energy initiatives and Ecuador's constitutional recognition of the rights of nature, show how green innovation can drive sustainability (Fraccascia *et al.* 2018). Developing countries like Malaysia also demonstrate how technological innovation can promote economic growth and environmental sustainability, highlighting the importance of supportive policies aligning innovation with environmental goals.

In Indonesia, innovation plays an important role in overcoming various economic development challenges, especially as the country strives to become a developed country by 2045. In addition to encouraging economic growth, innovation has the potential to significantly improve people's welfare by improving the quality of life and public health. The Indonesia Vision 2045 document, published by the Ministry of National Development Planning/National Development Planning Agency, outlines four main development goals that need careful attention: (1) fostering a superior and cultured society skilled in science and technology, (2) building an advanced and sustainable economy, (3) promoting equitable and inclusive development, and (4) building a democratic country and transparent government (Bappenas 2021). In the same vision, he to position Indonesia among the top 10 global economies by 2030, with a focus on technological advancements across key industrial sectors (Ministry of Industry 2018).

However, encouraging green innovation within this framework presents challenges, especially in aligning industrial growth with environmental goals. Current policies, such as Presidential Regulation Number 98 of 2021 concerning the Economic Value of Carbon and Presidential Regulation Number 55 of 2019 concerning Electric Vehicles, signal progress. However, there are still gaps in encouraging innovation and comprehensively addressing hazard reduction (Bappenas 2021). To support green innovation, governments need to prioritise adaptive regulatory frameworks that can facilitate sustainable development. The need for these adaptive regulations is aligned with the growing global emphasis on sustainable living and the creation of a low-emission environment. As countries realise the importance of sustainability, they are increasingly encouraged to develop and implement innovative strategies, especially in the industrial sector. This sector is a key player in shifting towards a green economy, emphasising environmental preservation and sustainable living.

This article explores how an adaptive regulatory framework can drive green innovation in Indonesia, ensuring alignment with national and global sustainability goals. By examining the strengths and weaknesses of current policies, the study offers insights into strategies to increase Indonesia's capacity to innovate sustainably while addressing environmental and public health challenges.

2 METHODOLOGY

This study employs a normative legal research approach to evaluate the effectiveness of Indonesia's regulatory frameworks in fostering green innovation. Normative legal research involves analysing laws, regulations, and policies within their legal and sociopolitical contexts (Creswell 2017). The statutory approach forms the core of this analysis, focusing on key legislations such as: Law No. 17/2023 on Health, particularly Article 149 Paragraph 4 which integrates harm reduction strategies, Presidential Regulation No. 98/2021 that emphasises carbon economic value and its alignment with Indonesia's commitment to the Paris Agreement, and Presidential Regulation No. 55/2019 which accelerates the adoption of battery electric vehicles (BEVs). The policies analysed in this study were selected based on their relevance to the following criteria: first is alignment with SDGs or policies directly supporting Goals 9 (Industry, Innovation, and Infrastructure) and 13 (Climate Action), second is the impact on Green Innovation or regulations that incentivise the development and adoption of sustainable technologies, and the last is potential for public-private collaboration or policies fostering partnerships between the government, private sector, and academia.

Primary data sources were primary legal documents, including statutes and presidential regulations. Secondary data were obtained from journal articles, policy reports, and case studies. The analytical framework involved a comparative analysis with international best practices, such as Germany's Energiewende and Norway's EV policies, to identify potential gaps and areas for improvement in Indonesia's regulatory framework (Fraccascia *et al.* 2018; IEA 2022). This study also examines challenges in implementing green innovation policies, such as inconsistencies in enforcement and a lack of measurable indicators for policy outcomes. Addressing these issues requires a deeper understanding of the socio-economic and institutional barriers that hinder the effectiveness of Indonesia's regulatory frameworks (Bappenas 2021).

3 FINDINGS AND DISCUSSION

3.1 *Indonesia's roadmap for industrial revolution 4.0*

The advent of Industry 4.0 is marked by a digital revolution, which is characterised by the integration of digital technologies into industrial processes. This revolution encompasses various technological advancements, including 3D printing and robotics, which are anticipated to enhance productivity (Hasan Tinmaz 2020). Industry 4.0 represents a technological shift driven by the concept of smart production, aiming to foster a more modern, informative, personalised, and environmentally friendly manufacturing sector as a foundation for achieving sustainable living. The definition and concept of Industry 4.0 are grounded in the convergence of knowledge, communication systems, and industrial technologies, with a strong reliance on developing cyber-physical systems (CPS) to establish digital factories and advance the concept of smart production. (Sharma *et al.* 2021).

The industrial sector has experienced a significant transformation in the current era, driven by advancements in information and communication technology. These changes have profoundly affected business models within the production process and the entire industrial value chain. Technological progress is an unavoidable reality that all countries must confront. Consequently, nations adopting Industry 4.0 recognise the necessity of comprehensive government policy support as a critical element for effectively implementing digital economy strategies. For Indonesia, the onset of the fourth industrial revolution (Industry 4.0) is an inescapable reality, necessitating strategic measures to ensure the country's successful adaptation to this new era.

The Indonesian government has initiated efforts to accelerate the fourth industrial revolution through the "Making Indonesia 4.0" roadmap. This initiative encourages the manufacturing sector to adopt digital technologies throughout its industrial value chain. Given its strong demand factors, supportive institutional framework, and favourable conditions for global trade and investment, Indonesia is poised to leverage this opportunity. The "Making Indonesia 4.0" roadmap outlines integrated strategies to position Indonesia among the world's top 10 economies by 2030. Additionally, the relatively strong resilience of the Indonesian economy is seen as a potential asset that can be further enhanced through the adoption of Industry 4.0 technologies.

The "Making Indonesia 4.0" Roadmap was developed to provide a strategic framework for Indonesia's future industrial development, focusing on five key sectors and ten national priorities to strengthen the country's industrial base. The Fourth Industrial Revolution (4IR) is expected to encompass a range of advanced technologies, including artificial intelligence (AI), the Internet of Things (IoT), wearables, advanced robotics, and 3D printing. The initial implementation will target five primary sectors: (i) food and beverage, (ii) textiles and apparel, (iii) automotive, (iv) chemicals, and (v) electronics. These sectors were chosen based on economic impact, feasibility of implementation, GDP contribution, trade significance, potential effects on other industries, investment levels, and market penetration rates. Indonesia plans to review the strategy for each sector every three to four years to assess progress and address any challenges in implementation (Ministry of Industry).

Among the ten national priorities outlined in the "Making Indonesia 4.0" roadmap to advance the manufacturing sector in Indonesia is the development of an innovation ecosystem. This ecosystem is crucial for the success of the initiative. Efforts to build this ecosystem include establishing a national innovation centre blueprint, setting up pilot innovation centres, and optimising relevant regulations. These regulations encompass intellectual property protection and fiscal incentives designed to foster cross-sector collaboration between private enterprises, state-owned enterprises (SOEs), and universities (Ministry of Industry). Incentives are recognised for their potential to stimulate innovation and technology adoption; thus, additional incentives are provided for technology investments as part of the government's support for accelerating Industry 4.0. Consequently, the Indonesian government has revised its technology adoption incentive plans to include subsidies, corporate tax rebates, and import duty exemptions for companies committed to integrating 4IR technologies. Furthermore, the government will introduce a state investment fund to provide supplementary financial support for investment and innovation in advanced technologies (Ministry of Industry).

The roadmap "Making Indonesia 4.0" provides a strategic framework for Indonesia's industrial transformation, emphasising the integration of advanced technologies such as artificial intelligence, the Internet of Things (IoT), and renewable energy systems. While the roadmap highlights the need for digital transformation to enhance productivity, its alignment with sustainability outcomes remains underdeveloped. For instance, the focus on manufacturing sectors like electronics and automotive presents an opportunity to embed green innovation, particularly through the adoption of electric vehicles (EVs) and energy-efficient production methods (Ministry of Industry 2018). However, the absence of explicit sustainability metrics within the roadmap limits its potential to fully contribute to Sustainable Development Goals (SDGs), particularly Goals 12 (Responsible Consumption and Production) and 13 (Climate Action).

3.2 *Adaptive regulation and recognition of environmentally friendly products*

The Fourth Industrial Revolution (Industry 4.0) offers an opportunity to integrate product innovation to achieve the "Making Indonesia 4.0" objectives. This is essential for addressing global challenges: expanding economic opportunities while mitigating environmental pressures. The key solution lies in fostering environmentally friendly, sustainable economic growth, often referred to as "green growth," which emphasises economic development while ensuring the preservation of environmental resources (Fraccascia *et al.* 2018).

International examples, like Germany's Energiewende policy, show how adaptive regulatory frameworks promote green innovation. Germany's renewable energy commitments and emission targets have spurred green technology development (Fraccascia *et al.* 2018). Similarly, Norway's tax incentives for electric vehicles (EVs) have made it a global leader in sustainable transportation, with EVs constituting over 50% of new vehicle sales in 2022 (IEA 2022). These cases highlight the importance of fiscal incentives and public-private partnerships in regulatory frameworks for green innovation.

Environmental issues like global warming, energy crises, and pollution have driven companies to adopt environmentally friendly product innovations. These companies are changing production methods and creating products designed to minimise environmental impacts. Eco-friendly products help address environmental challenges, establish competitive advantages, and foster future growth. Eco-friendly practices are critical to sustaining the environment and improving public health by reducing emissions, conserving resources, and minimising waste.

Green innovations can be promoted through strategies like decarbonisation, encouraging cleaner production methods. In the automotive sector, electric vehicles are seen as a solution to air pollution caused by carbon emissions. Indonesia's Presidential Regulation Number 55 of 2019 aims to accelerate the Battery Electric Vehicle Program, offering incentives for domestic manufacturers to develop the electric vehicle industry. The regulation includes both fiscal and non-fiscal incentives, such as subsidies for electric vehicle purchases.

Indonesia targets 23% of its energy mix from New and Renewable Energy (NRE) by 2025, aiming for a 29% emissions reduction by 2030. Presidential Regulation Number 98 of 2021 supports this, focusing on reducing greenhouse gas emissions in line with the Nationally Determined Contribution (NDC). Indonesia also aims for net-zero emissions by 2060, with potential emissions reductions of up to 50% compared to business-as-usual scenarios, particularly with international support.

In alignment with these efforts, Article 149, Paragraph 4 of Law No. 17 of 2023 on Health addresses the production and use of low-risk tobacco products. The law encourages innovation in the tobacco industry (IHT) to create alternative tobacco products (HPTL) with a harm reduction approach, aiming to reduce health and environmental risks associated with traditional cigarettes. These products aim to mitigate the adverse effects of tobacco use amidst the continuing high prevalence of smoking in Indonesia.

The concept of harm reduction in public health aims to provide better alternatives to behaviours or substances that pose health and social risks. In tobacco, harm reduction strategies focus on reducing risks through products that heat rather than combust, such as vapour products. These alternatives help reduce cigarette waste, which contributes to environmental issues, with reports noting that 766,571 metric tons of cigarette butts are produced globally each year. Alternative tobacco products are a better option for smokers and contribute to environmental preservation by reducing litter.

The rise of sustainable entrepreneurship, where companies directly address environmental issues and the Sustainable Development Goals (SDGs), is key to fostering green innovation. Companies are expected to prioritise ecological, economic, and social sustainability, with products designed to reduce environmental impact throughout their lifecycle. Research shows that green sector development in countries like China has substantially reduced industrial pollution while boosting economic growth (Luca Fraccascia *et al.* 2018). Sustainable product innovations help mitigate the negative impacts of production and consumption, promoting a greener and healthier world.

4 CONCLUSION

Indonesia has implemented policies such as Law No. 17 of 2023 on Health, Presidential Regulation No. 98 of 2021 on Carbon Economic Value, and Presidential Regulation No. 55 of 2019 on Battery Electric Vehicles to address environmental challenges. However, additional regulations incorporating harm reduction as a strategic approach are needed. The revision of Law No. 32/2009 on Environmental Protection and Management through the Job Creation Law in March 2023, which shifted penalties from criminal to administrative sanctions, weakens the enforcement of green innovation. While existing policies demonstrate a commitment to green innovation, they lack mechanisms for long-term sustainability. To address these gaps, there needs to be enhanced integration of sustainability metrics into industrial policies, stronger fiscal incentives for green technology adoption, and improved monitoring mechanisms to ensure compliance with environmental standards.

REFERENCES

Bappenas. (2021). *Vision Indonesia 2045: Sustainable Development Plan*. Bappenas.

Creswell, J. W. (2017). *Research design: Qualitative, quantitative, and mixed methods approaches* (4th ed.). Sage.

Fraccascia, L., Giannoccaro, I., and Albino, V. (2018). Green product development: What does the country product space imply? *Journal of Cleaner Production*, *170*, 1076–1088. https://www.sciencedirect.com/science/article/pii/S0959652617321911

IEA. (2022). *Global EV outlook 2022: Securing supplies for an electric future*. International Energy Agency.

Katsikeas, C. S., Leonidou, C. N., and Zeriti, A. (2016). Eco-friendly product development strategy: Antecedents, outcomes, and contingent effects. *Journal of the Academy of Marketing Science*, *44*, 660–684.

Krammer, S. M. S. (2017). Science, technology, and innovation for economic competitiveness: The role of smart specialization in less-developed countries. *Technological Forecasting and Social Change*, *123*, 95–107.

Ministry of Industry. (2018). *Making Indonesia 4.0*. Ministry of Industry.

Sharma, A. K., Bhandari, R., Pinca-Bretotean, C., Sharma, C., Dhakad, S. K., and Mathur, A. (2021). A study of trends and industrial prospects of Industry 4.0. *Materials Today: Proceedings*, *47*, 2364–2369.

Stiglitz, J. E. (2016). How to restore equitable and sustainable economic growth in the United States. *American Economic Review*, *106*(5), 43–47.

Tinmaz, H. (2020). History of industrial revolutions: From Homo sapiens hunters to Bitcoin hunters. *In* Blockchain Technology for Industry 4.0: Secure, Decentralized, Distributed and Trusted Industry Environment (pp. 1–26).

Towards Resilient Societies: The Synergy of Religion, Education, Health, Science, and Technology – Rahiem (Ed.)

Analysis of social capital in the family hope program in Baktijaya village, Depok City

Aliska Putri Maulida & Muhtadi Muhtadi
Universitas Islam Negeri Syarif Hidayatullah Jakarta, Indonesia

ABSTRACT: Social capital is a resource derived from interactions within a community or group, encompassing emotional bonds such as trust, reciprocal relationships, social networks, values, and norms that facilitate coordination and cooperation to achieve shared goals. This study aims to analyse social capital and the implementation of the Family Hope Program (Program Keluarga Harapan/PKH) in Baktijaya Village. A descriptive quantitative method was employed, with data collected through interviews and questionnaires. The study targeted the Beneficiary Group (*Kelompok Penerima Manfaat*/KPM) as recipients of the Family Hope Program. Using the Slovin formula, a sample of 202 respondents was drawn from a population of 410. The findings reveal that: (1) the social capital within the Family Hope Program in Baktijaya Village achieved an average score of 3.34, categorised as high, indicating that social capital among the Beneficiary Group is well-established; and (2) the implementation of the Family Hope Program received an average score of 3.49, also categorised as high, demonstrating that the program is being effectively carried out in Baktijaya Village.

Keywords: community empowerment, family hope program, social capital

1 INTRODUCTION

As one of the efforts to overcome poverty, the government has realised many community empowerment programs. One of the empowerment programs realised by the government to overcome poverty and improve community welfare is the Family Hope Program. The Depok City Government also plays a role in implementing poverty alleviation in its area under the auspices of the Ministry of Social Affairs to overcome poverty. One of the Ministry of Social Affairs programs run by the Depok City Social Service is the Family Hope Program, a social protection program in the form of social assistance. This assistance is given to poor and vulnerable families. The Family Hope Program can reduce the expenditure burden on poor families and increase understanding of the importance of education, health and family financial management in improving the quality of life of their families in the future. Through the Family Hope Program, beneficiary groups are encouraged to have access to and utilise basic social services for health, education, food and nutrition, care, and assistance, including access to various other social protection programs, which are complementary programs on an ongoing basis.

Community empowerment does not always have to be given capital through revolving funds or other direct community assistance funds. However, the most important thing is that the community can always participate in all activities, trust each other, uphold existing norms and values, and actively carry out empowerment programs. Without involving the existing community, the program will not run optimally. One of the capitals that the community must own is social capital. Social capital is a resource as a result of interactions in a community or group, either between individuals and individuals, individuals and groups, or groups and groups that make emotional bonds in the form of trust, reciprocal relationships, social networks, values and norms that form a community structure that is useful for coordination and cooperation in achieving common goals. Social capital is also important in a community development process, as the existence of social capital in a community becomes the glue between one community group and another. The social capital referred to here is the norms and rules that bind the citizens in it and regulate the behaviour patterns of its citizens, as well as elements of trust and networking between citizens or community groups. Existing norms and rules also regulate individual behaviour in terms of internal behaviour (internal group) and external behaviour (external relations with other community groups).

However, not all community empowerment programs have elements of social capital that grow well in a community. In reality, various efforts made by various parties in the context of empowerment are less able to grow and develop the potential of existing social capital, even not infrequently, because of errors in the approach; what happens is the opposite, reducing the potential of social capital. The problem that usually occurs is the lack of trust that is owned because there are no values and norms that bind a group and community, so there is no relationship between each other. This study aims to provide an overview of the results of the analysis of social capital and the implementation of the Family Hope Program in Baktijaya Village, Sukmajaya District, Depok City.

2 METHOD

The study was conducted using a descriptive quantitative method with a population of 410 people who were the beneficiary coverage of the Family Hope Program in Baktijaya Village and determined the sample using the Slovin formula and set an error of 5%. Based on the Slovin formula, the following sample size was obtained: n = 202,469 (rounded to 202 people). The sampling technique was the probability sampling technique with the simple random sampling type. Data was obtained from primary data sources through questionnaires distributed directly to the beneficiary family group of the Family Hope Program in Baktijaya Village. While secondary data is in the form of literature related to social capital and the Family Hope Program. The data collection technique used in this study was a questionnaire.

The data validity test was conducted using Pearson bivariate correlation, and the reliability test using Cronbach's alpha. The data analysis technique used in this study is descriptive statistics, which is used to analyse data by describing or depicting the data that has been collected as it is without intending to make conclusions that apply to the public or generalisation. Descriptive statistics can be used if the researcher only wants to describe the sample data, and does not want to make conclusions that apply to the population from which the sample was taken. The data processing technique uses Microsoft Excel 2010 and the IBM SPSS Statistics Version 25 application. Then, determine the average of each respondent's score per statement indicator. After that, the results are categorised according to the predetermined category level. This study's assessment category level consists of three categories: high, medium and low.

3 RESULT

The Family Hope Program is one of the social protection programs in Indonesia in the form of social assistance. This assistance is given to poor and vulnerable families with certain requirements. The Family Hope Program was first implemented in Depok City in 2014 based on the Decree of the Director of Social Security No. 252/SK/JS/082014. Depok City consists of 11 sub-districts and 63 villages. At the beginning, the coverage of Family Hope Program beneficiary families in Depok City was 5,219 beneficiary families spread across 10 sub-districts and 59 villages in Depok City. Then, in 2015, the Family Hope Program in

Depok City received an additional coverage of 377 beneficiary families, spread across 1 sub-district and 4 villages, namely Cinere District. So, the number of beneficiary families in 2015 was 5,375 beneficiary families. Every year the number of Family Hope Program beneficiary families in Depok City has increased so that in 2022 the coverage of beneficiary families of the Family Hope Program in Depok City will total 30,444 spread across 11 sub-districts and 63 villages throughout Depok City.

One sub-district included in the beneficiary families' Family Hope Program is Baktijaya Sub-district, located in Sukmajaya District. Baktijaya Sub-district has an area of around 295 hectares with 29 RW and 229 RT with a population of 1,851,878 people. Based on the results of interviews with Family Hope Program facilitators in Baktijaya Sub-district, at the beginning of distribution (phase IV) Baktijaya Sub-district had 101 Family Hope Program with one facilitator. In 2022, the Family Hope Program in Baktijaya Sub-district had a beneficiary coverage of 441 beneficiary families Family Hope Program (phase IV). Then, in 2023, there was a decrease in the coverage of beneficiary families in the Family Hope Program by 31 beneficiary families. In 2023 (phase I), the number of beneficiary families in the Family Hope Program in Baktijaya Sub-district was 410. In phase II, beneficiary families Family Hope Program in Baktijaya Village received an additional coverage of 46 beneficiary families Family Hope Program so that in phase II, the number of beneficiary families Family Hope Program in Baktijaya Village was 456 beneficiary families spread throughout Baktijaya Village.

3.1 *Analysis of social capital*

In social capital, the dimensions used in this study come from Adi's theory (2013), which explains that social capital consists of norms and rules that have rule indicators, trust that has an indicator of trust attitudes and networks that have indicators of social interaction and social relationships. The following in Table 1 are the results of the analysis of social capital in the Family Hope Program:

Table 1. Analysis of social capital in family hope program.

Dimension	Average	Category
Norm	3,2	High
Trust	3,38	High
Networking	3,46	High
Average	3,34	High

Based on the results of the social capital analysis of the Family Hope Program in Table 1, it is known that the norm dimension obtained an average of 3.2, so it is included in the high category. However, if we look at the results of the social capital analysis above, the norm dimension obtained a lower percentage than the trust and networking dimension. Then, the trust dimension obtained an average of 3.38 to be included in the high category. The trust dimension has a slightly lower average than the networking dimension.

Furthermore, an average of 3.46 was obtained by the networking dimension included in the high category. The networking dimension has the highest average among norms and trust. Overall, the average social capital dimension is 3.34. So, the analysis of social capital consisting of norms, trust and networking in the Family Hope Program is in the high category.

3.2 *Analysis of the implementation of the family hope program*

The Family Hope Program variables used in this study come from the 2021 Family Hope Program Implementation Guidelines of the Indonesian Ministry of Social Affairs, which include planning, implementation and evaluation of the Family Hope Program. The following are the results of the analysis of the Family Hope Program indicators:

Table 2. Results of the PKH implementation analysis.

Dimension	Result	Category
Planning	3,35	High
Implementation	3,45	High
Evaluation	3,69	High
Average	3,49	High

Based on the analysis of the implementation of the Family Hope Program in Baktijaya Village in Table 2, it is known that the planning dimension obtained an average of 3.35, which is included in the high category. However, if seen in the table of the analysis results of implementing the Family Hope Program in Baktijaya Village, the planning dimension obtained an average lower than implementation and evaluation. Then, the implementation dimension obtained an average of 3.45, which is in the high category. The implementation dimension has an average value slightly lower than the evaluation dimension. Furthermore, an average of 3.69 was obtained by the evaluation dimension, which is in the high category. The evaluation dimension has the highest average between planning and implementation. Overall, the average result of the Family Hope Program dimension is 3.49. So, the implementation of the Family Hope Program, which consists of planning, implementation and evaluation, is in the high category.

4 DISCUSSION

Based on the results of the study, it shows that the social capital variables consisting of norms, trust and networking obtained an average of 3.34, which is included in the high category, meaning that the application of social capital in each beneficiary group of the Family Hope Program is implemented very well. According to Adi (2013), the existence of social capital in a society becomes the glue between one community group and another. The social capital referred to here is the norms and rules that bind the citizens in it and regulate the behaviour patterns of its citizens, as well as elements of trust and networking between citizens or community groups.

In the norm dimension, some rights and obligations must be fulfilled and implemented as beneficiary groups of the Family Hope Program. In addition, separate norms are compiled and agreed upon by the group, both written and unwritten, to be obeyed together. The norm itself is a set of rules expected to be obeyed and followed by members of society in a particular social entity (Simbolon 2018). In the trust or attitude that each beneficiary group of the Family Hope Program has towards their companions and fellow group members and between other beneficiary groups. Trust in a relationship grows because of a common view, which can be manifested in the presence of a common goal to achieve something, which then gives rise to a desire to do it together, shown in the form of involvement, both involvement of time, funds, energy, thoughts and voice (decisions) (Anjani *et al.* 2018). Networking includes interactions and social relationships with companions, fellow group members, and other Family Hope Program beneficiary groups. According to Anjani *et al.* (2018), a network is said to be effective if there is an exchange of knowledge and mutually beneficial relationships (reciprocity).

The results of the analysis of social capital in the Family Hope Program are also in line with previous research, namely research stating that social capital in the community empowerment program can be categorised as high. High social capital is seen in the components of social capital theory, including networks, trust, and norms. Then, in the overall research, the role of social capital of the Joint Business Group (KUBE) in the Family Hope Program is categorised as high because social capital plays a very good role in achieving the group's goals. Based on the study results for the Family Hope Program variables consisting of planning, implementation, and evaluation obtained an average of 3.49, which is included in the high category, meaning that the implementation of the Family Hope Program in Baktijaya Village is implemented very well.

In another context, the Family Hope Program indirectly relates to women's empowerment. Women's empowerment is important for migrant women workers (Tandos *et al.* 2022), Aisyiyah women's economic empowerment (Hidayah 2020), and women in rural areas (Bain *et al.* 2018). The sustainability of the Keluaga Harapan Program requires empowered women. Even rural women entrepreneurs who are digitally literate are also important (Fauzia and Hidayati 2023), and Islam has given women the right to be self-employed (Hidayah 2023). Because those who are digitally literate and aware of their economic rights can accelerate and help the success of community empowerment programs.

5 CONCLUSION

The results of the analysis of social capital in the Family Hope Program in Baktijaya Village, Sukmajaya District, Depok City, with an average of the overall dimensions of social capital, namely norms, trust and networking, of 3.34 and can be categorised as high. This proves that the application of social capital in the Family Hope Program, especially in the beneficiary group of the Family Hope Program, has been implemented very well. The results of the analysis of the implementation of the Family Hope Program in Baktijaya Village, Sukmajaya District, Depok City, with an average of the overall dimensions of the Family Hope Program, namely planning, implementation and evaluation, of 3.49 and can be categorised as high. It can be seen that the Family Hope Program, especially in Baktijaya Village, is implemented very well.

REFERENCES

Adi, I. R. (2013). *Intervensi Komunitas & Pengembangan Masyarakat Sebagai Upaya Pemberdayaan Masyarakat* (Edisi Revisi 2012). Jakarta: PT Raja Grafindo Persada.

Anjani, F. D., Ariwibowo, and Subarkah, A. (2018). Modal Sosial Masyarakat Dalam Pengelolaan Pariwisata Pantai Muarareja Indah di Kelurahan Muarareja Kecamatan Tegal Kota Tegal. *Ilmiah Perlindungan dan Pemberdayaan Sosial*, 2(2).

Bain, C., Ransom, E., and Halimatusa'diyah, I. (2018). 'Weak winners' of Women's empowerment: The gendered effects of dairy livestock assets on time poverty in Uganda. *Journal of Rural Studies.*

Fauzia, A., and Hidayati, S. (2023). Improving Digital Literacy of Rural Women Entrepreneurs in Indonesia. 2023 11th International Conference on Cyber and IT Service Management (CITSM), 1–5.

Hidayah, N. (2020). Women's Movement in the Framework of Modernist Muslims in Indonesia: Aisyiyah And Economic Empowerment. *Hamdard Islamicus*, *43*(5). https://doi.org/10.57144/hi.v43iSpecialIssue.218.

Hidayah, N. (2023). Gender, Economy, and the Law: Women Entrepreneurs in Indonesian and Islamic Legal Perspectives. *Samarah: Jurnal Hukum Keluarga dan Hukum Islam*, 7 (2). DOI: 10.22373/sjhk.v7i2.17944

Simbolon, A. K. (2018). *Analisis Modal Sosial untuk Kesejahteraan Masyarakat Lokal* (Studi pada Wisata Petik Jeruk di Dusun Borogragal, Desa Donowarih, Kecamatan Karangploso, Kabupaten Malang). Cakrawala, 12(1). https://doi.org/10.32781/cakrawala.v12i1.266.

Tandos, R., Zhu, R., and Krever, R. (2022). The protection and empowerment of Indonesian female migrant domestic workers: Proposals from a multi-stage analysis. *Asian Journal of Women's Studies*, 28, 205–227.

Session 2: Education and psychology in building resilient and inclusive societies

Towards Resilient Societies: The Synergy of Religion, Education, Health, Science, and Technology – Rahiem (Ed.)

Risk & resilience in early childhood education: Strategies for social and emotional development

Maila Dinia Husni Rahiem
Universitas Islam Negeri Syarif Hidayatullah Jakarta, Indonesia
ORCID ID: 0000-0002-5618-2486

Husni Rahim
Universitas Islam Negeri Syarif Hidayatullah Jakarta, Indonesia
ORCID ID: 0000-0003-1797-5018

Marhamah Lidinia
Unisadhuguna International College (UIC College), Indonesia

Supriyanto Suharman
Dharmaila Pradita Foundation, Indonesia

Krey Amiros
Institute An-Nikmah Al-Islamiyah Phnom Penh, Cambodia

ABSTRACT: This study explores Early Childhood Education (ECE) teachers' perspectives on fostering social and emotional development in young children, focusing on strategies to build resilience, empathy, and collaboration. The research addresses the importance of these skills in equipping children to navigate challenges and regulate emotions, particularly within the diverse socio-cultural context of Greater Jakarta. Using a qualitative exploratory design, data were collected from 108 ECE teachers with over five years of teaching experience through open-ended questionnaires. Data analysis followed Saldaña's coding framework, identifying four key themes: peer support mechanisms, resilience strategies, emotional development activities, and environmental factors. Findings reveal that structured peer interactions, stress management exercises, positive reinforcement, and supportive classroom environments are essential for nurturing social-emotional growth. The study recommends professional development programs for teachers and collaborative efforts among educators, parents, and policymakers. Future research should explore broader regions and conduct longitudinal studies to examine long-term impacts.

Keywords: children resilience, early childhood education, emotional development, peer support, social-emotional learning

1 INTRODUCTION

Social and emotional development during early childhood forms the foundation for lifelong skills in empathy, resilience, and collaboration. Early Childhood Education (ECE) settings provide critical opportunities to foster these attributes through intentional classroom strategies and supportive environments. Research highlights the significance of fostering social-emotional learning (SEL) in young children to promote skills such as empathy, problem-solving, and emotional regulation, all of which contribute to their success in academic and social settings (Mondi *et al.* 2021; Pollak *et al.* 2024).

DOI: 10.1201/9781003645542-32

The focus on resilience in early childhood is especially important given the rising concerns about mental health in Indonesia. Reports indicate that suicide rates are alarmingly high, with increasing cases among young people. Additionally, societal discourse often highlights that the current generation of children is perceived as less resilient and more prone to giving up when faced with challenges. This underscores the urgency of equipping children with the tools to manage adversity effectively from an early age. Resilience, as a critical component of character development, helps children build the perseverance and emotional regulation necessary to navigate life's difficulties.

The Greater Jakarta (Jabodetabek) area was selected as the focus of this study due to its diversity in ECE institutions, which range from public to private, religious to secular, and serve families from various socio-economic backgrounds. This variety offers a broad perspective on the practices and strategies used to foster social and emotional development in early learners. Given the emphasis in early childhood education on character building, understanding how resilience is nurtured within these diverse settings provides valuable insights for educators and policymakers.

Despite the recognised importance of resilience and social-emotional learning in early childhood, limited research has explored how ECE teachers, particularly in diverse urban contexts like Greater Jakarta, perceive and implement strategies to foster these skills. This study seeks to address the following research questions: (1) How do ECE teachers perceive their role in fostering resilience, empathy, and collaboration in young children? (2) What specific strategies do ECE teachers use to develop these skills in their classrooms? Addressing these questions will contribute to a deeper understanding of how resilience and social-emotional competencies can be effectively supported during early childhood.

2 METHODS

This study utilised a qualitative exploratory research design to explore early childhood education (ECE) teachers‘ perspectives on fostering social and emotional development in young children. This approach was chosen for its ability to delve deeply into participants' experiences, capturing rich and nuanced insights into complex and under-researched phenomena (Creswell and Poth 2016). The flexibility of this method allowed for contextual interpretations of the participants‘ narratives, making it particularly suitable for this study.

The sample comprised 108 ECE teachers from the Greater Jakarta area (Jabodetabek), selected through purposive sampling. Participants were required to have a minimum of five years of teaching experience, ensuring they could provide meaningful, reflective, and detailed accounts of their classroom practices. Notably, 98% of the participants were female, which aligns with the broader demographic composition of ECE teachers, a field predominantly occupied by women. Despite this gender imbalance, the participants' extensive professional experience ensured their responses were rich and valuable.

The data were collected using open-ended questionnaires, which allowed participants to articulate their thoughts freely and provide detailed descriptions of their strategies and practices. This approach ensured the collection of broad yet in-depth responses. The completed questionnaires were systematically organised and entered into Excel, where they were prepared for analysis. Each response was assigned a unique identifier for easy tracking and categorisation. This initial organisation facilitated the subsequent coding and theme development process.

The analysis followed Saldaña's (Saldaña 2016, 2021) approach, which was iterative and systematic. The first step involved open coding, where key concepts, phrases, and ideas were identified and labelled to capture the core meaning of each response. These codes were then examined for patterns and relationships through axial coding, leading to the grouping of related codes into broader categories. Finally, these categories were synthesised into overarching themes that encapsulate the central findings of the study. The key themes identified—peer support mechanisms, resilience strategies, emotional development activities, and environmental factors—reflected the multifaceted approaches ECE teachers used to foster social and emotional growth.

To ensure the reliability and validity of the analysis, researchers independently conducted the coding process. The research team then convened in a focused group discussion (FGD) to review, refine, and confirm the final themes. This collaborative process ensured the findings were robust and accurately represented the participants' perspectives. By combining systematic management and analysis of data, the study provides meaningful insights into the strategies employed by experienced ECE teachers to support young children's social and emotional development.

3 RESULTS AND DISCUSSION

3.1 *Results*

Table 1. Strategies for ECE social and emotional development.

Codes	Buddy system	Group tasks	Praising teamwork	New students paired with buddies
	Pairing children to each other	Team games	Acknowledging acts of kindness	Helping shy children connect
	Older children guiding younger ones	Collaborative play	Highlighting peer help in class	Promoting inclusiveness
Categories	***Buddy systems***	***Team activities***	***Encouragement and Recognition***	***Collaborative play***
Theme 1	**Peer support mechanisms**			
Codes	Relaxation exercise	Working together to solve puzzles	Teaching children to identify feelings	Rewarding resilience
	Calming breathing techniques	Finding solution during group activities	Encouraging reflection in emotions	Praising efforts to overcome challenges
	Mindfulness breaks	Teacher-guided conflict resolution	Discussing how to handle frustation	Highlighting perseverance
Categories	***Stress management activities***	***Problem-solving skills***	***Emotional regulation support***	***Positive reinforcement***
Theme 2	**Resilience strategies**			
Codes	Sharing stories about feelings	Building structure together	How do you feel today charts	Discussing how others might feel
	Discussing characters' emotions in books	Sharing toys during group tasks	Talking about facial expressions	Imagining others' points of view
	Role-playing emotional scenarios	Rotating roles in group games	Labelling emotions during discussions	Encouraging acts of kindness
Categories	***Empathy building exercises***	***Collaboration in play***	***Recognition of emotions***	***Social perspective-taking***
Theme 3	**Emotional development activities**			
Codes	Creating cosy spaces for group works	Teacher modelling empathy	Establishing rules for mutual respect	Encouraging participation from all children
	Accessible and safe materials for all	Teacher reinforcing positive behaviours	Scheduling collaborative activities	Addressing cultural differences
	Visual cues for emotional regulation	Building warm and supportive relationships	Routine discussions on feelings	Adjusting activities for diverse needs
Categories	Classroom setup	Teacher role	Classroom rules and routines	Inclusive practices
Theme 4	**Environmental factors**			

Based on the analysis of qualitative data collected through open-ended questionnaires, four key themes emerged, each highlighting critical strategies and considerations for fostering social and emotional development in early childhood education: 1) peer support mechanisms, 2) resilience strategies, 3) Emotional development activities, 4) Environmental factors (Table 1).

3.1.1 *Peer support mechanisms*

The importance of structured peer interactions in fostering social integration and emotional comfort among children was consistently emphasised by participants. Many highlighted the role of buddy systems and group activities in encouraging teamwork and building a sense of mutual support. One participant noted, *"Pairing children to support each other helps shy students connect and feel included in group activities."* Others shared similar insights, such as using group tasks to foster collaboration: *"I divide children into teams for activities like solving puzzles or cleaning up, so they learn to help and rely on each other."* Another participant stressed the significance of recognising children's contributions, stating, *"Praising teamwork and acknowledging acts of kindness encourages children to repeat these behaviours."* These approaches not only create an inclusive classroom environment but also teach children the foundational values of empathy, cooperation, and shared responsibility, which are essential for their social development.

3.1.2 *Resilience strategies*

Participants consistently emphasised the importance of helping children develop resilience through structured classroom activities. Several noted the value of stress management exercises and emotional regulation techniques. One participant shared, *"I teach children calming breathing exercises to help them handle frustration and reduce anxiety."* Similarly, problem-solving tasks were frequently mentioned as tools to build perseverance: *"I use group activities like assembling puzzles or constructing block towers to encourage children to work through challenges together."* Positive reinforcement was also highlighted as a key strategy, with one participant explaining, *"I always praise children who show persistence, saying things like 'Great job for not giving up!' to motivate them."* These approaches were identified as crucial in equipping children to adapt to challenges, manage emotions, and grow into confident, resilient individuals capable of facing difficulties both inside and outside the classroom.

3.1.3 *Emotional development activities*

Participants highlighted the importance of activities that nurture emotional intelligence, focusing on helping children understand and express their feelings. One participant shared, *"I read stories about kindness and discuss how the characters feel, asking questions like, 'How would you help a friend in need?'"* Such storytelling was seen as an effective way to teach empathy. Group discussions were also noted as essential, with one respondent explaining, *"I encourage children to talk about their emotions by asking, 'How do you feel today?' and let them share with their friends."*Additionally, collaborative play was described as a way to foster empathy and teamwork: *"I organise group games like building structures together, where children learn to support and listen to each other."* These activities not only help children recognise their own emotions but also build their capacity to empathise and form meaningful relationships with peers.

3.1.4 *Environmental factors*

Participants highlighted the crucial role of a positive and inclusive classroom environment in supporting children's social and emotional growth. Many emphasised that creating a safe and welcoming space is essential. One participant shared, *"I ensure the classroom feels safe and comfortable so children feel accepted and valued."* Teachers' actions were also seen as critical, with one noting, *"I model empathy by helping children and encouraging them to do the same."* Clear classroom rules and routines were frequently mentioned as foundational, as another respondent explained, *"We create simple rules together, like helping friends or taking turns, to build a supportive atmosphere."* Group-oriented activities, such as collaborative play, further enhanced this dynamic: *"I organise team games and projects to encourage cooperation and mutual respect."* Together, these environmental factors create a foundation for fostering care, collaboration, and emotional well-being among young learners.

3.2 *Discussion*

The results of the study highlight four key themes that are critical in fostering social and emotional development in early childhood education: peer support mechanisms, resilience strategies, emotional development activities, and environmental factors. These themes underscore the multifaceted approaches educators use to nurture children's emotional well-being, empathy, and social skills.

Participants highlighted the importance of structured peer interactions, such as buddy systems and collaborative group activities, in promoting social integration and emotional comfort among children. Encouraging teamwork and acknowledging peer contributions were noted as effective ways to instil a sense of belonging and mutual support. This approach is corroborated by recent research indicating that social-emotional learning (SEL) programs enhance peer connectedness and classroom climate, thereby fostering a sense of belonging and mutual support among students (Pollak *et al.* 2024; Rahiem 2024). Similarly, a study by Wentzel and Watkins (Wentzel and Watkins 2002) emphasised that cooperative learning strategies, where students work together towards shared goals, significantly improve their ability to collaborate and support one another. Furthermore, Vygotsky's sociocultural theory supports the idea that social interactions are critical to learning and emotional development, suggesting that peer collaborations provide opportunities for children to develop empathy, communication skills, and a shared sense of responsibility (Eun 2010; Morcom 2014).

Developing children's resilience was seen as pivotal, with strategies including stress management exercises, problem-solving tasks, and positive reinforcement. These approaches were recognised as essential for helping children adapt to challenges, regulate emotions, and build perseverance. The American Psychological Association advocates for such approaches, emphasising the importance of building resilience in children through supportive relationships and adaptive skill-building (American Psychological Association 2012). Structured interventions, combined with nurturing and supportive environments, empower children to overcome adversities and develop lasting resilience (Doll and Song 2023; Rahiem 2021; Rahiem *et al.* 2018; Tabibnia and Radecki 2018). An integral part of this process involves cultivating positive emotions and a sense of control, which are foundational to resilience (Sciaraffa *et al.* 2018). These elements not only enhance children's ability to face and adapt to difficulties but also align seamlessly with the strategies designed to promote emotional regulation and perseverance (Sciaraffa *et al.* 2018; Smith and Ascough 2016).

Programs designed to foster emotional intelligence, such as empathy-building exercises, storytelling, and group discussions about emotions, were found to be crucial in helping children recognise, understand, and express their feelings. These activities not only enhanced children's emotional awareness but also strengthened their capacity to empathise with peers and build meaningful social connections. Research supports this, indicating that early childhood education programs focusing on social-emotional development contribute significantly to improved social skills and school readiness (Mondi *et al.* 2021). The development of empathy appears closely intertwined with children's ability to adopt others' perspectives (Lithoxoidou *et al.* 2017). As children engage in social interactions, they begin to exhibit empathic behaviours, such as recognising and responding to the emotions, perspectives, and needs of others (Frydenberg 2017; Silver and Stafford 2017). This growing awareness encourages children to demonstrate altruistic behaviours, such as showing interest, offering care, and providing assistance to their peers (Garnett 2017; Lithoxoidou *et al.* 2017).

The classroom environment plays a pivotal role in shaping children's social and emotional development. A positive and inclusive atmosphere, reinforced by teacher modelling, clear rules, and group-oriented routines, establishes a strong foundation for nurturing care, collaboration, and emotional well-being. Research supports the idea that such environments are instrumental in fostering resilience and promoting social-emotional health in young children (Darling-Hammond and Cook-Harvey 2018; Osher *et al.* 2021; Schonert-Reichl 2017). Teachers contribute significantly by using effective teaching methods, leveraging resources, and employing thoughtful classroom management strategies to turn challenges into opportunities for creative problem-solving (Buettner *et al.* 2016; Javed 2023; Jennings *et al.* 2017; Schonert-Reichl 2017). By encouraging children to confront, rather than avoid, less-than-ideal situations, educators equip them with the skills to build resilience that extends beyond the classroom (Collet 2017). These experiences enable children to develop perseverance and effectively utilise available resources to rise above disappointments and setbacks, strengthening their capacity to navigate challenges in diverse contexts (Frydenberg 2017; Silver and Stafford 2017; Skinner *et al.* 2020).

4 CONCLUSION

This study highlights the critical role of Early Childhood Education (ECE) teachers in fostering social and emotional development through strategies such as peer support mechanisms, resilience-building activities, emotional development programs, and creating supportive classroom environments. These approaches are essential for equipping young children with the skills to navigate challenges, regulate emotions, and build meaningful relationships. Teachers play a pivotal role in shaping resilient and empathetic individuals, underscoring the importance of integrating social-emotional learning into early education curricula. The study recommends professional development programs for ECE teachers, focusing on effective strategies for resilience and emotional intelligence development. Additionally, collaboration between educators, parents, and policymakers is essential to create supportive systems for young learners. The findings are limited to the Greater Jakarta context and reflect the perspectives of predominantly female teachers. Future research should include diverse regions, a balanced gender sample, and longitudinal studies to explore long-term impacts of these strategies.

REFERENCES

American Psychological Association. (2012). *Resilience guide for parents and teachers.* American Psychological Association. https://www.apa.org/topics/resilience/guide-parents-teachers?utm_source=chatgpt.com

Buettner, C. K., Jeon, L., Hur, E., and Garcia, R. E. (2016). Teachers' social–emotional capacity: Factors associated with teachers' responsiveness and professional commitment. *Early Education and Development, 27*(7), 1018–1039.

Collet, V. S. (2017). "I Can Do That!" Creating Classrooms That Foster Resilience. *YC Young Children, 72* (1), 23–32.

Creswell, J. W., and Poth, C. N. (2016). *Qualitative inquiry and research design: Choosing among five approaches.* Sage publications.

Darling-Hammond, L., and Cook-Harvey, C. M. (2018). Educating the Whole Child: Improving School Climate to Support Student Success. *Learning Policy Institute.*

Doll, B., and Song, S. Y. (2023). Enhancing resilience in classrooms. In *Handbook of resilience in children* (pp. 525–537). Springer.

Eun, B. (2010). From learning to development: A sociocultural approach to instruction. *Cambridge Journal of Education, 40*(4), 401–418.
Frydenberg, E. (2017). *Coping and the challenge of resilience.* Springer.
Garnett, H. (2017). *Developing empathy in the early years: A guide for practitioners.* Jessica Kingsley Publishers.
Javed, M. (2023). The Effectiveness of Different Teaching Methods in Education: A Comprehensive Review. *Journal of Social Signs Review, 1*(1), 17–24.
Jennings, P. A., Brown, J. L., Frank, J. L., Doyle, S., Oh, Y., Davis, R., Rasheed, D., DeWeese, A., DeMauro, A. A., and Cham, H. (2017). Impacts of the CARE for Teachers program on teachers' social and emotional competence and classroom interactions. *Journal of Educational Psychology, 109*(7), 1010.
Lithoxoidou, L. S., Georgopoulos, A. D., Dimitriou, A. T., and Xenitidou, S. C. (2017). "Trees Have a Soul Too!" Developing Empathy and Environmental Values in Early Childhood. *International Journal of Early Childhood Environmental Education, 5*(1), 68–88.
Mondi, C. F., Giovanelli, A., and Reynolds, A. J. (2021). Fostering socio-emotional learning through early childhood intervention. *International Journal of Child Care and Education Policy, 15*(1), 1–43.
Morcom, V. (2014). Scaffolding social and emotional learning in an elementary classroom community: A sociocultural perspective. *International Journal of Educational Research, 67*, 18–29.
Osher, D., Cantor, P., Berg, J., Steyer, L., and Rose, T. (2021). Drivers of human development: How relationships and context shape learning and development 1. In *The science of learning and development* (pp. 55–104). Routledge.
Pollak, I., Stiehl, K. A. M., Birchwood, J., Schrank, B., Zechner, K. A., Wiesner, C., and Woodcock, K. A. (2024). Promoting Peer Connectedness Through Social-Emotional Learning: Evaluating the Intervention Effect Mechanisms and Implementation Factors of a Social-Emotional Learning Programme for 9 to 12-Year-Olds. *Journal of Youth and Adolescence, 53*(1), 89–116.
Rahiem, M. D. H. (2021). Remaining Motivated despite the Limitations: University Students' Learning Propensity during the COVID-19 Pandemic. *Children and Youth Services Review*, 105802. https://doi.org/10.1016/j.childyouth.2020.105802
Rahiem, M. D. H. (2024). Early childhood education contingencies for sustaining learning during school closures: Lessons from preschool remote education home visits in Indonesia during the COVID-19 pandemic. *Children and Youth Services Review, 166*, 107955.
Rahiem, M. D. H., Krauss, S. E., and Rahim, H. (2018). The Child Victims of the Aceh Tsunami: Stories of Resilience, Coping and Moving on with Life. *Procedia Engineering, 212*, 1303–1310. https://doi.org/10.1016/j.proeng.2018.01.168
Saldaña, J. (2016). *The Coding Manual for Qualitative Researchers (No. 14). Sage.*
Saldaña, J. (2021). The coding manual for qualitative researchers.
Schonert-Reichl, K. A. (2017). Social and emotional learning and teachers. *The Future of Children*, 137–155.
Sciaraffa, M. A., Zeanah, P. D., and Zeanah, C. H. (2018). Understanding and promoting resilience in the context of adverse childhood experiences. *Early Childhood Education Journal, 46*, 343–353.
Silver, D., and Stafford, D. (2017). *Teaching kids to thrive: Essential skills for success.* Corwin Press.
Skinner, E. A., Graham, J. P., Brule, H., Rickert, N., and Kindermann, T. A. (2020). "I get knocked down but I get up again": Integrative frameworks for studying the development of motivational resilience in school. *International Journal of Behavioral Development, 44*(4), 290–300.
Smith, R. E., and Ascough, J. C. (2016). *Promoting emotional resilience: Cognitive-affective stress management training.* Guilford Publications.
Tabibnia, G., and Radecki, D. (2018). Resilience training that can change the brain. *Consulting Psychology Journal: Practice and Research, 70*(1), 59.
Wentzel, K. R., and Watkins, D. E. (2002). Peer relationships and collaborative learning as contexts for academic enablers. *School Psychology Review, 31*(3), 366–377.

Towards Resilient Societies: The Synergy of Religion, Education, Health, Science, and Technology – Rahiem (Ed.)
© 2026 The Author(s), ISBN: 978-1-041-08471-6
Open Access: www.taylorfrancis.com, CC BY-NC-ND 4.0 license

A comprehensive analysis of administrative issues in Islamic Senior High Schools in Lampung: Striving for improved school performance management

Juanda Naim
Raden Intan State Islamic University, Lampung, Indonesia

Siti Patimah
UIN Sultan Maulana Hasanuddin Banten, Indonesia

Agus Pahrudin & Meriyati
Raden Intan State Islamic University, Lampung, Indonesia

Imam Subchi, Dzikri Rahmat Romadhon & Maila Dinia Husni Rahiem
Universitas Islam Negeri Syarif Hidayatullah Jakarta, Indonesia

Muhammad Sirajuddin Bin Ahmad Khairani
Organisasi Pendidikan Keagamaan "Jabatan Agama Islam Selangor, Malaysia

ABSTRACT: This study explores challenges in performance management faced by administrative staff in Islamic Senior High Schools in Lampung, Indonesia, and proposes improvement strategies. Using a qualitative exploratory approach, in-depth interviews with seven participants (two principals, two heads of administration, and three administrative staff) were analysed through content analysis to identify key obstacles and practical solutions. The findings highlight five major challenges: suboptimal planning, insufficient training, inconsistent evaluations, heavy workloads, and ineffective communication. Addressing these requires a comprehensive management strategy, including systematic planning, continuous training, transparent evaluations, balanced workload distribution, and improved communication. This study provides valuable insights for administrators and policymakers in faith-based institutions, contributing to educational management discourse by integrating Islamic values with effective organisational practices.

Keywords: administrative staff, Islamic education, performance evaluation, performance management, professional development, school management, senior high schools

1 INTRODUCTION

Performance management is crucial for educational excellence, impacting staff productivity and institutional success. This study examines performance management among administrative staff in two Islamic Senior High Schools in Lampung, Indonesia, an underexplored area in education. Faith-based institutions face unique challenges balancing academic rigor and spiritual commitments, requiring innovative management strategies.

This research is internationally relevant, highlighting cultural practices and resource limitations in Islamic schools. It provides insights for administrators, policymakers, and researchers, offering a comparative framework for optimising management strategies.

The study contributes to educational management discourse, emphasising Islamic values in school policies. It aligns with comparative religious education, exploring the link between faith, education, and management. School-Based Management (SBM) promotes institutional

 DOI: 10.1201/9781003645542-33

autonomy and community involvement, enhancing collaboration and student achievement (Bandur 2012; Fatayan *et al.* 2019). SBM is particularly relevant in Islamic schools, where school committees play a key role in maintaining quality.

Leadership also fosters excellence, with kyai in pesantren influencing both culture and management, creating inclusive, effective educational environments (Mutohar 2021). However, administrative staff in religious schools continue to face performance challenges, requiring further study (Zine 2001).

This research identifies challenges and recommends strategies to enhance performance management. It offers valuable insights for educational leaders, policymakers, and practitioners, contributing to educational equity and excellence. By integrating practical management strategies with religious values, it presents a balanced approach to spiritual and academic commitments.

2 METHODS

This study used an exploratory qualitative approach to examine administrative performance management in two Islamic Senior High Schools in Lampung. By analysing local challenges and strategies, it provides an in-depth understanding relevant to the educational context. A qualitative method was chosen for its ability to capture complex issues through participant insights (Teherani *et al.* 2015).

Purposive sampling selected seven participants (two principals, two administrative heads, and three staff members) to ensure diverse perspectives. Data collection involved 60-minute face-to-face interviews with structured, open-ended questions. The interview guide, developed from literature, was pilot tested and reviewed by education and Islamic management experts to ensure reliability and validity.

Data were analysed using qualitative content analysis (Mayring 2004), with systematic transcription, coding, and theoretical linkage to generate evidence-based recommendations. The study received ethics committee approval, ensuring participant confidentiality and informed consent.

3 RESULTS

The data revealed key challenges in administrative performance management. Suboptimal performance planning led to poorly structured appraisals with no effective monitoring team. Limited training meant staff primarily engaged in scientific activities rather than receiving direct managerial guidance. Inconsistent performance evaluations were mostly oral and undocumented. Heavy workloads and resource shortages forced staff to take on excessive tasks. Lastly, ineffective communication and coordination led to operational misunderstandings and decision-making errors.

3.1 *Suboptimal performance planning*

Interviews revealed that performance planning for administrative staff lacks structure, often responding to immediate needs without a clear framework. L stated, "The performance appraisal activities for administrative staff are conducted, but the planning is merely in the form of issuing invitations for performance monitoring without forming an effective monitoring team," highlighting weak oversight. R noted, "The agenda for conducting performance monitoring has never been prepared," and J added, "The scheduling and detailed aspects of what will be assessed have never been properly planned," indicating poor preparation. Administrative staff also reported execution issues. F stated, "The performance appraisal, including the schedule, instruments, venue, funding, infrastructure, and assessors, has never been implemented." These deficiencies, especially for non-civil servant staff, underscore the need for better organisation and systematic planning in performance evaluations.

3.2 *Lack of development and training*

Islamic Senior High Schools in Lampung lack targeted professional development for administrative staff. Instead of management-focused training, schools offer unrelated workshops and seminars, limiting operational improvement. S stated, "They are involved in scientific activities such as workshops, seminars, and others." Staff also lack technical and administrative skill development. AH noted, "Performance guidance for administrative staff is given individually by the head of the school." H added, "Efforts by the head of the school to improve the discipline of administrative staff performance are rarely made." J further highlighted, "The participation of administrative staff in professional organisations is still very rare," reflecting limited professional growth opportunities. These findings stress the urgent need for structured, systematic training to improve administrative performance.

3.3 *Inconsistent performance evaluation*

Performance evaluations in Islamic Senior High Schools in Lampung are mostly informal and oral, with little formal documentation or record-keeping. This lack of structure hinders systematic performance tracking, objective assessment, and long-term professional development. R stated, "Performance evaluations are conducted orally. They are rarely done in writing using evaluation forms." Evaluations prioritise attendance discipline over work outcomes or collaboration. J noted, "Evaluations are also conducted based on the progress of tasks completed by administrative staff." However, key aspects like cooperation and coordination are often overlooked, as S mentioned, "Cooperation and coordination skills are not given much attention in performance evaluations." AH added, "Technical skills are also evaluated, assessing the ability of administrative staff to use the equipment and facilities available at our school." This inconsistent evaluation process fosters unfairness and demotivation, limiting efforts to improve staff performance.

3.4 *Administrative issues and workload*

Administrative staff face excessive workloads due to staff shortages and limited resources, negatively affecting efficiency, service quality, and stress levels. H noted, "There are administrative tasks assigned to teachers who should be focused on the learning process but are instead handling the school's administration." FH added, "The lack of administrative staff hinders the administrative service process." Staff often handle tasks beyond their roles, as J stated, "The tasks often assigned are outside their primary roles." S described the workload burden, "Tasks assigned to administrative staff often pile up before one task is completed. Another one arrives and must be immediately addressed." Without adequate support or training, routine tasks like document handling and data entry further strain staff, reducing efficiency and service quality.

3.5 *Ineffective communication and coordination*

Ineffective communication and coordination among staff and principals cause misunderstandings, errors, and inefficiencies. R emphasised, "All teams must coordinate with each other to avoid misunderstandings and errors in implementing activities and making policy decisions," stressing the need for clear communication channels. External professional engagement is lacking. J noted, "Communication between the school principals and administrative staff has never involved participation in a professional organisation." H added, "The lack of communication between the principal and administrative staff" exacerbates operational challenges by delaying information exchange. S linked poor service quality to communication issues, stating, "The lack of excellent service provided by administrative staff to the school stakeholders." These findings highlight how poor communication affects task execution and decision-making. Schools face delays, conflicts, and inefficiencies without effective frameworks and collaboration. Strengthening internal communication and professional engagement is essential for better coordination and performance.

4 DISCUSSION

4.1 *Suboptimal performance planning*

Administrative staff at Islamic Senior High Schools in Lampung struggle with inadequate performance planning, causing inefficiencies in academic and operational areas, reducing service quality, and leaving schools unprepared for challenges. Structured planning is crucial for staff commitment, productivity, and institutional success (Bandur 2012). However, in Lampung, even well-designed plans fail due to insufficient support, resources, and stakeholder coordination (Zhang 2021). School-Based Management (SBM) supports decentralisation and local autonomy, enabling schools to adapt management to their needs. However, SBM in Lampung lags behind West Java, where structured planning has improved performance management, underscoring the need for context-specific adaptations (Bandur 2012).

4.2 *Lack of development and training*

The lack of continuous professional development for administrative staff hinders skills, innovation, and effectiveness in schools. Without training, abilities stagnate, reducing academic service efficiency. Ongoing development is essential for maintaining staff performance (Forman *et al.* 2009; Griffith 2004). While religious values shape administration, technical training in management and technology is equally vital (Forman *et al.* 2009; Griffith 2004).

Comparative studies show that structured training in Central Java leads to higher staff performance and job satisfaction, emphasising the need for continuous development (Beames *et al.* 2021; Forman *et al.* 2009). Training also improves staff behaviour and response to challenges, enhancing the school environment (Allen *et al.* 2008).

In Lampung, financial constraints and limited support hinder professional development. Collaboration with local educational authorities is needed to improve staff competencies.

4.3 *Inconsistent performance evaluation*

Inconsistent and subjective performance evaluations add to the challenges faced by administrative staff. Structured assessments are essential for constructive feedback and continuous improvement (Helal 2022). However, schools in Lampung rely on informal evaluations, causing unfairness and demotivation among staff. Effective SBM frameworks advocate for objective evaluations and school committee involvement to enhance transparency and accountability (Dieleman *et al.* 2003). Yet, minimal stakeholder participation in Lampung reduces evaluation effectiveness, highlighting a key area for improvement.

4.4 *Administrative issues and workload*

Heavy workloads and administrative inefficiencies are common educational challenges (MacDonald 2003). Excessive workloads contribute to stress and reduced productivity, highlighting the need for effective leadership to manage administrative tasks (Funke *et al.* 2016). In Lampung, staff are often overburdened, affecting efficiency and well-being.

Strong leadership is essential for streamlining processes and balancing workloads. While leadership is crucial, enhancing administrative systems and fair task distribution is equally important (Funke *et al.* 2016). A structured, supportive environment can significantly reduce stress and improve productivity.

Research links high workloads to job stress and dissatisfaction, negatively impacting performance and staff retention (Simanjorang and Wahyanti 2021). Overworked staff struggle to maintain efficiency, leading to declining administrative quality and mental and physical health issues.

Schools must implement workload management strategies, including hiring additional staff, training programs, and automation to address this. Providing regular feedback and support can enhance job satisfaction and performance. These steps create a sustainable, productive environment, improving institutional success and staff well-being.

4.5 *Communication and coordination*

Effective communication and coordination are essential for task execution, decision-making, team cohesion, and operational efficiency. Poor communication between principals and administrative staff leads to misinformation, conflicts, and inefficiencies, hindering educational objectives (Akram and Muhammad Khan 2020; Fatayan *et al.* 2019).

Enhancing internal communication improves stakeholder engagement and performance management (Akram and Muhammad Khan 2020; Fatayan *et al.* 2019). Research highlights clear communication as key to preventing conflicts, increasing productivity, and improving performance (Akram and Muhammad Khan 2020; Fatayan *et al.* 2019).

Studies also show that better communication fosters a harmonious work environment, enhances community participation, and strengthens school-based management (Fatayan *et al.* 2019; Katuuk *et al.* 2018). It prevents conflicts, supports decision-making, and ensures active stakeholder involvement in educational success (Fatayan *et al.* 2019; Katuuk *et al.* 2018).

5 CONCLUSION

This study used qualitative content analysis to examine performance management practices in Islamic Senior High Schools in Lampung, Indonesia. In-depth interviews with principals, administrative heads, and staff identified five key challenges: inadequate performance planning, insufficient training, inconsistent evaluations, heavy workloads, and poor communication. The findings highlight the need for context-specific management strategies that align with studies in Java and Sumatra. This research provides insights into faith-based school administration and offers evidence-based recommendations.

Schools should implement clear planning, regular professional development, and structured evaluations to improve performance management. Technology and workload distribution can enhance efficiency, while open communication through regular meetings fosters coordination.

The study offers valuable insights into Islamic education administration but has limitations, including small sample size and potential researcher bias. Future research should include larger-scale quantitative studies to validate findings across different regions.

Further exploration should cover longitudinal studies on training impact, comparative research on evaluation methods, and technology-driven communication strategies. Involving students, parents, and stakeholders would provide a holistic view of performance management, ensuring all factors are considered for improving school administration.

ACKNOWLEDGMENTS

We sincerely thank the administrative staff, principals, and heads of administration at the Islamic Senior High Schools in Lampung for their participation. We also extend our gratitude to colleagues, staff, lecturers, and university leadership at Universitas Islam Negeri Raden Intan Lampung and Universitas Islam Negeri Syarif Hidayatullah Jakarta for their support, encouragement, and guidance. Their cooperation and expertise were invaluable to this study.

REFERENCES

Akram, M., and Muhammad Khan, A. (2020). Exploring E-Leadership of Principals: Increasing School Effectiveness by Learning Management System. *Journal of Education and Social Sciences*, 8(1), 15–30. https://doi.org/10.20547/jess0812008102

Allen, K., Cornell, D., Lorek, E., and Sheras, P. (2008). Response of school personnel to student threat assessment training. *School Effectiveness and School Improvement*, 19(3), 319–332. https://doi.org/10.1080/09243450802332184

Bandur, A. (2012). School-based management developments: Challenges and impacts. *Journal of Educational Administration*, 50(6), 845–873. https://doi.org/10.1108/09578231211264711

Beames, J. R., Johnston, L., O'Dea, B., Torok, M., Christensen, H., Boydell, K. M., and Werner-Seidler, A. (2021). Factors That Help and Hinder the Implementation of Digital Depression Prevention Programs: School-Based Cross-sectional Study. *Journal of Medical Internet Research*, 23(8), e26223. https://doi.org/10.2196/26223

Dieleman, M., Cuong, P. V., Anh, L. V., and Martineau, T. (2003). Identifying factors for job motivation of rural health workers in North Viet Nam. *Human Resources for Health*, 1(1), 10. https://doi.org/10.1186/1478-4491-1-10

Fatayan, A., Hanafi, İ., Sari, E., and Ghani, A. R. A. (2019). The Implementation of School Based Management: School Committee Involvement in Islamic Schools. *International E-Journal of Educational Studies*, 4(7), 106–110. https://doi.org/10.31458/iejes.608131

Forman, S. G., Olin, S. S., Hoagwood, K. E., Crowe, M., and Saka, N. (2009). Evidence-Based Interventions in Schools: Developers' Views of Implementation Barriers and Facilitators. *School Mental Health*, 1(1), 26–36. https://doi.org/10.1007/s12310-008-9002-5

Funke, G. J., Warm, J. S., Baldwin, C. L., Garcia, A., Funke, M. E., Dillard, M. B., Finomore, V. S., Matthews, G., and Greenlee, E. T. (2016). The Independence and Interdependence of Coacting Observers in Regard to Performance Efficiency, Workload, and Stress in a Vigilance Task. *Human Factors: The Journal of the Human Factors and Ergonomics Society*, 58(6), 915–926. https://doi.org/10.1177/0018720816646657

Griffith, J. (2004). Relation of principal transformational leadership to school staff job satisfaction, staff turnover, and school performance. *Journal of Educational Administration*, 42(3), 333–356. https://doi.org/10.1108/09578230410534667

Helal, I. (2022). The Impact of Performance Appraisals on Employee Productivity: The Case of the Lebanese Retail Sector. *European Journal of Business and Management Research*, 7(5), 109–117. https://doi.org/10.24018/ejbmr.2022.7.5.1616

Katuuk, D. A., Pasandaran, S., and Pangalila, T. (2018). Effectiveness of School-Based Management Practices in Increasing Community Participation and Implementing School Programs. *Proceedings of the Annual Civic Education Conference (ACEC 2018). Proceedings of the Annual Civic Education Conference (ACEC 2018)*, Bandung, Indonesia. https://doi.org/10.2991/acec-18.2018.32

MacDonald, W. (2003). The impact of job demands and workload on stress and fatigue. *Australian Psychologist*, 38(2), 102–117. https://doi.org/10.1080/00050060310001707107

Mayring, P. (2004). Qualitative content analysis. *A Companion to Qualitative Research*, 1(2), 159–176.

Mutohar, A. (2021). Leading for Diversity: How Islamic School Leaders View Diversity. Proceedings of the First International Conference on Islamic History and Civilization, ICON-ISHIC 2020, 14 October, Semarang, Indonesia. *Proceedings of the First International Conference on Islamic History and Civilization*, ICON-ISHIC 2020, 14 October, Semarang, Indonesia, Semarang, Indonesia. https://doi.org/10.4108/eai.14-10-2020.2303848

Simanjorang, W. P., and Wahyanti, C. T. (2021). The Effect of Leadership Style and Workload on Work Stress at Pt Bank Perkreditan Rakyat Buana Agribisnis Saribudolok. *International Journal of Social Science and Business*, 5(4), 451. https://doi.org/10.23887/ijssb.v5i4.41433

Teherani, A., Martimianakis, T., Stenfors-Hayes, T., Wadhwa, A., and Varpio, L. (2015). Choosing a Qualitative Research Approach. *Journal of Graduate Medical Education*, 7(4), 669–670. https://doi.org/10.4300/JGME-D-15-00414.1

Zhang, X. (2021). Planning the Structure of University Teaching Staff Based on Multiobjective Optimization Method. *Scientific Programming*, 2021, 1–10. https://doi.org/10.1155/2021/1773561

Zine, J. (2001). Muslim Youth in Canadian Schools: Education and the Politics of Religious Identity. *Anthropology and Education Quarterly*, 32(4), 399–423. https://doi.org/10.1525/aeq.2001.32.4.399.

Towards Resilient Societies: The Synergy of Religion, Education, Health, Science, and Technology – Rahiem (Ed.)

Community resilience and healing narratives in the novel al-Wabā'

Teguh Luhuringbudi & Rizqi Handayani
Universitas Islam Negeri Syarif Hidayatullah Jakarta, Indonesia

ABSTRACT: This study, employing socio-ecological resilience theory and a therapeutic narrative approach, aims to explore how the narrative in literature can strengthen community resilience during an epidemic. Data were collected through textual and historical context analysis of the novel *al-Wabā'* by Ḥānī al-Rāhib and a review of relevant literature on the applied theories. This research significantly contributes to the field by highlighting the role of Arabic literature in supporting sustainability goals and enhancing the understanding of public health challenges through storytelling. However, the study's focus on a single Arabic literary work limits its scope. Further research is recommended to validate these findings across other Arabic literary texts and broader contexts. The study concludes that literature can function as a therapeutic medium for addressing collective trauma and reinforcing community resilience in times of crisis.

Keywords: Arabic literary novel, community resilience, healing narratives, socio-ecological resilience, therapeutic narrative

1 INTRODUCTION

Infectious diseases such as typhus and tuberculosis continue to represent a significant public health concern in numerous developing countries, including those in the Middle East. In the Levant, the prevalence of these diseases has been exacerbated by the confluence of poor socio-economic conditions (Blanco 2020; Bruce-Chwatt 1973; Crosby 1991; de Graaf 2020; Helman 2004) and limited access to health facilities (Arnold 1991; Blecher 2004; Bölükbaşı 2023; Jacobson 2010; Méouchy and Sluglett 2004; Norrie 2016; Post 1976; Zylberman 2007), particularly in areas affected by conflict during World War I and II. Epidemiological studies have demonstrated a high prevalence of typhus and tuberculosis in the region (formerly Lebanon, Syria, Jordan, Israel, and Palestine), reflecting the impact of poor environmental conditions and a lack of public health awareness. A considerable number of families were compelled to reside in unsanitary conditions, which served to accelerate the transmission of these diseases (Biselli *et al.* 2022; Evered and Evered, n.d.; Jensen 1990; Kazemi 2018; Kunitz 1983; Loddenkemper and Konietzko 2018; Mercer 2021; Nachtigal 2023; Šimkutė and Gudienė 2022).

Even though past investigations-research have inspected the economic, political, and social impacts of typhus and tuberculosis in influenced zones, there is a crevice in understanding how literary narratives can contribute to community resilience. The novel al-Waba by Hani al-Rahib (1981) is a literary work that demonstrates the resilience of Arab society, which managed to survive in the midst of the most traumatic events between the World Wars and the deadly epidemic. A few consider having illustrated the value-esteem of narrative approaches in healing trauma (Martínez 2017; Simoni 2000) and progressing mental health (Bürgschwentner *et al.* 2014; Duraković 2014; Lee 2003; Mackenbach 2020; McKnight Nichols *et al.* 2022; Trubeta *et al.* 2021; Weindling 2021) in communities influenced by infectious diseases. In any case, ponders setting up an interface between Arabic literary narratives and community resilience are rare. Ḥānī al-Rāhib's novel *al-Wabā'* (1981), set against the background of a destroying plague epidemic, allowed looking at the potential of literary narratives to encourage the management of public health issues.

 DOI: 10.1201/9781003645542-34

This paper aims to look at the concept of community resilience and the role of healing narratives-stories within the novel *al-Wabā'* (1981). In doing so, this research paper provided modern writing and literary experiences that support the sustainable development goals (SDGs), especially within the health setting.

2 METHODS

This study employed a qualitative-analytical design, analysing community resilience and healing narratives. This design was selected because it permitted researchers to comprehensively examine the experiences of individuals afflicted with typhus and typhoid, as well as tuberculosis, within the context of the epidemic and its impact on the community. This qualitative analysis provided in-depth experiences into the capacity of literary narratives to reflect and impact social and health realities.

The essential source of this paper is the novel *al-Wabā'* by Hānī al-Rāhib (1981). This Arabic novel was chosen for its complex portrayal of an epidemic and a community's resilience, situated within the context of public health. Additionally, the Arabic literary novel may be noteworthy commitment to Living Arabic Literature, linking narrative art with modern issues of sustainable development (Dhika *et al.* 2023; Fajriati *et al.* 2020; Neumeyer *et al.* 2020; Rahman 1996). The data collection techniques utilised in this research paper about were based on the strategies portrayed by Esubalew Aman Mezmir, identifying thematic frameworks, data reduction, and memoing (Mezmir 2020).

The data analysis procedures in this research paper were carried out in two stages. The starting step of the analysis included categorising the discoveries from data collection into three fundamental categories. The primary was the Social-Ecological Resilience Theory, as proposed by Carl Folke and colleagues, which emphasised the adaptive and transformative capacity of communities in building resilience (Ebbesson and Folke, n.d.; Folke 2016; Folke *et al.* 2011, 2016; Galaz *et al.* 2008; Miller *et al.* 2010; Schlüter *et al.* 2019). The second step was the Therapeutic Narrative Approach, created by Nelya Koteyko and Dimitrinka Atanasova, which centred on the management of trauma and the construction of new meanings (Atanasova *et al.* 2017; Koteyko and Hunt 2018; Koteyko and Atanasova 2018). The thematic moments in the novel *al-Wabā'* have been analysed, including the suffering of the disease that befalls the characters, the narratives about it, the suffering of the epidemic for the community, the efforts and processes of healing, and the memories and reflections of the epidemic after the healing.

3 FINDINGS & DISCUSSION

3.1 *Findings*

3.1.1 *Community resilience and healing narratives through socio-ecological resilience*

Applying the socio-ecological resilience theory encouraged understanding how social and ecological frameworks adapted and transformed in reaction to altering and disturbance. The literary-printed data on page 9 of the novel *al-Wabā'* outlines that the Levantine society was seriously influenced by war and scourge-epidemics. The devastation caused harm to the physical infrastructure and brought about noteworthy human suffering. Those who remained were constrained to take off their homes, look for food, and shelter somewhere else.

The textual data on page 73 of the novel *al-Wabā'* outlines how the Levantine society delineated within the novel reacted to tuberculosis (TB) scourge-epidemic. In this setting, the character Ḥasan ⊠ghā, who has endured noteworthy misfortunes, is constrained to recognise ways to supply care for his wiped-out wife. He procured a metal container and water from the market to give his wife something to drink and clean up the mess she had made. Hasan's decision to sue his affluent brother-in-law appeared to be due to his desperate situation of seeking equity. This adaptation reflected how people and communities looked to distinguish solutions to the health challenges they were confronting, the significance of solidarity, and its role in fortifying social resilience.

The literary data on page 10 of the novel *al-Wabā'* showed how Sheik Abdel Jawad's family adjusted to the challenges posed by the typhus plague epidemic. The family was forced to look for a shield beneath a partially collapsed stone bridge, looking for protection from the overwhelming rain. In this circumstance, they were constrained to make troublesome choices concerning their debilitated children. The discussion between Sheik Abdel Jawad and his wife-spouse about whether or not to desert their wiped-out children reflected the ethical and emotional situations in an emergency. Their adaptation to unforgiving environmental conditions and a dangerous infection showed that community resilience included physical survival strategies and moral decisions.

3.1.2 *Community resilience and healing narratives through a therapeutic narrative approach*

The therapeutic narrative approach permitted an analysis of how the narratives within the Arabic literary novel, *al-Wabā'*, capacitated as a vehicle for healing and restoration for the characters and communities portrayed. By describing encounters of hardship and adversity, the Arabic literary novel allows readers to get it and reflect on traumatic events (Dhika *et al.* 2023; Fajriati *et al.* 2020; Neumeyer *et al.* 2020; Rahman 1996). For this case, Ḥasan ⊠ghā employed the narrative as an implication of seeking equity or justice for his expired wife. In contrast, Shaikh 'Abd al-Jawwād's family looked to secure their relatives from a fatal illness-sickness.

In expansion, the novel *al-Wabā'* outlined how therapeutic or helpful narratives could fortify community resilience by representing resilience and adaptation. The characters within the the novel *al-Wabā* illustrated fortitude, relentlessness, and inventiveness in confronting impressive adversity. The narratives reflected the significance of solidarity, fellowship, and back or support in fortifying community resilience. Consequently, the Arabic literary novel served as a conduit for person healing and a source of learning for the more extensive community.

This analysis appeared that Ḥānī al-Rāhib's novel *al-Wabā'* (1981) could be a compelling outline of the capacity of literary narrative to encourage the examination and reinforcing of community resilience within the setting of a scourge or epidemic. Utilising a therapeutic narrative approach permits an understanding how the Levantine society portrayed within the Arabic literary novel adjusted and changed in the face of crisis. The narrative within the Arabic literary novel served as an implication to encourage healing and recovery, allowing characters and readers to reflect on traumatic experiences and discover hope. The Arabic novel, *al-Wabā'* too illustrated the significance of solidarity, fellowship, and social support in reinforcing community resilience.

3.2 *Discussion*

The findings of this paper show that Ḥānī al-Rāhib's novel *al-Wabā'* effectively portrays the challenges confronted by communities in keeping up their resilience amid infection episodes. This can align with Martínez's and Simoni's (Martínez 2017; Simoni 2000) investigation, which emphasises the vital role of narrative approaches in recuperating trauma. The narrative examination in this consideration uncovers how communities adjust and recuperate-recovery from the collective trauma caused by the flare-up, reinforcing the contentions of Bürgschwentner and Duraković (Bürgschwentner *et al.* 2014; Duraković 2014; Lee 2003; Mackenbach 2020; McKnight Nichols *et al.* 2022; Trubeta *et al.* 2021; Weindling 2021) regarding the work of narratives in moving forward the mental health of communities affected by infectious disease.

Within the socio-ecological resilience theory proposed by Folke *et al.* (Ebbesson and Folke, n.d.; Folke 2016; Folke *et al.* 2011, 2016; Galaz *et al.* 2008; Miller *et al.* 2010; Schlüter *et al.* 2019), the findings of this consider show that the novel *al-Wabā'* portrays the adaptative and transformative capacity of communities in building resilience. The novel's narrative demonstrates how communities can learn from past encounters and create techniques to manage future emergencies, in line with Galaz and Miller's papers (Ebbesson and Folke, n.d.; Folke 2016; Folke *et al.* 2011 2016; Galaz *et al.* 2008; Miller *et al.* 2010; Schlüter *et al.* 2019), who emphasise the significance of learning and adaptation in building community resilience. The

Socio-Ecological Resilience	Interaction	Therapeutic Narratives
Adaptation to change, the literary text: residents abandoning homes due to war (on page 9)		Mechanism for Healing, the literary text: Ḥassān Āghā seeking justice through storytelling (on page 9 to 10)
وفي الشوارع كان صوتهم المسموع الوحيد: خبز! خبز! ومع متاعهم القليل البخس سرعان ما حملوا أمراضًا قاتلة، كان التيفوس أكثرها تمسكًا بهم. لقد وجد القمل في رؤوسهم وآباطهم مزارع مسمدة، فتكاثر بأسرع مما كانوا يموتون. ولذلك ماتوا بأسرع مما عرفوا.		في اليوم السادس أوصل التيفوس أم كحلة إلى الموت، وكانوا يظنونها خالدة. لكنهم ظلوا مطمئنين، الا من صمت خيم عليهم. وقبيل صلاة الفجر سمع بعضهم صيحة عظيمة فأفاق، وبعض بعضهم رأى انفجارًا نورانيًا شق بطن السماء وفاض على الكون: لم تكن لديهم ساعات ليعرفوا كم طال، لكنه طال بما يكفي لأن يشاهد بعض بعض بعضهم شيخ السنديان صاعدًا نحو الأبراج وقد استحال انفجار النور حوله إلى تسبيحة.
Responses to Health Crises, the literary text: Ḥassān Āghā caring for his sick wife during an epidemic (on page 73)	Reinforcement of Community Resilience (healing narratives contribute to community strength, emphasizing togetherness), Example: Shared stories of survival foster social bonds	Community Inspiration, the literary text: collective experiences bolster social cohesion (on page 105)
وقد استطاع تدبير وجبة أو وجبة ونصف له ولزوجته. وسرق ابريقا معدنيا من حارة القلعة، صار يملؤه ماء من حنفية السوق، فيسقي مريم ويغسل لها أوساخها......ومريم تموت بالسل		هذه الأسطورة المنفجرة _ قد لا تموت أبدًا. ربما لأنها استوطنت، ربما لأن لها معنى لا يموت. - ألا تخافين أن يصيبك مرض السل، يا بنت الشيخ عبد الجواد؟ - لا. لا أخاف. الله يعطيك العافية. نهضت مريم بلأي، واتكأت على مرفقها. حدقت إلى خولة قبل أن ترد التحية: ((الله يعافيك.)) وظلت تحدق. لم يبد أنها سترحب بزائرتها الى الجلوس، ولا أن الترحيب نفسه قد خطر لها.
Moral and Ethical Decisions, the literary text: Shaikh 'Abdel Jawād's family deciding wheather to abandon sick children (on page 10)		Promotion of Solidarity, the literary text: depictions of support systems in navigating health crises (on page 106)
أحمد مريض. أخاف منه..لكن ...يا رب! ما هذه التجربة؟ تعبنا عليه خمس سنوات..نتركه؟ هل نترك داوود؟ يمكن أن يكون التيفوس. نأخذه إلى المدينة. ويموت. ونتركهنا واحدًا ويموت. يا رب! يا رب! ذنب كبير ارتبكته. لا أعرف ما هو، ولكنك الآن تعاقبني عليه. نترك داوود؟	Reflecting and Learning (communities reflect on experiences through narratives, enhancing adaptive capacities). Example: Learning from past crises to improve future responses	خبئي منديلك يا بنت الشيخ عبد الجواد. وابعدي يا مجنونة. لئلا يصيبك السل. ماذا جاء بك إلى هنا؟ وإن يكن الانسان قادرًا على الإنتقام؟ الانتقام ضعف. أنا أقوى منه. لا أريد خبزه ولا أريد ماله. ماذا جاء بك إلى هنا؟ مجنون. هذا ما كرهته فيه.

Figure 1. The interaction between socio-ecological resilience and therapeutic narratives.

therapeutic narrative approach created by Koteyko and Atanasova (Atanasova *et al.* 2017; Koteyko and Hunt 2018; Koteyko and Atanasova 2018) is additionally important, as the novel *al-Wabā'* appears how narratives can be utilised to oversee trauma and develop modern implications, supporting the discoveries of Atanasova (Atanasova *et al.* 2017; Koteyko and Hunt 2018; Koteyko and Atanasova 2018) on the role of narratives in making a difference people and communities adapt with traumatic encounters.

This paper highlights the significance of therapeutic narratives in community resilience methodologies. The novel *al-Wabā'* outlines the potential of narratives as an implies to record collective encounters and foster basic awareness. Within the setting of sustainable development and the SDGs, narratives can assist communities in understanding the impact of health crisis and creating sustainable resilience. This is often in line with Trubeta and Weindling's studies (Bürgschwentner *et al.* 2014; Duraković 2014; Lee 2003; Mackenbach 2020; McKnight Nichols *et al.* 2022; Trubeta *et al.* 2021; Weindling 2021), who emphasise the significance of authentic and narrative perspectives in understanding and tending public health crisis. It is evident that the present novel deviates from the narrative of the Spanish flu pandemic on the novel *Amīrikā* by Rabī' Jābir (2009) with the setting of World War II; Ebola outbreak on *Ībūlā* by Amīr Tāj al-Sirr (2012) with the setting of South Sudan and Congo; Covid outbreak in *Kūfīd-7* by Sharf al-Dīn 'Ibābanah (2020) with the fictional setting of a human woman's tragic friendship in a dwarf environment; Covid-19 pandemic in *Hurriyyah fī* Zamān al-Kūrūnā by Sandis al-Shāwyy (2020) with the setting of the revolution from the tyranny of the Iraqi dictatorship and Covid-19 pandemic. Besides, this ponders the have to be therapeutic narrative approaches to open public health to improve communities' resilience, supporting the contentions of Lee and Mackenbach (Bürgschwentner *et al.* 2014; Duraković 2014; Lee 2003; Mackenbach 2020; McKnight Nichols *et al.* 2022; Trubeta *et al.* 2021; Weindling 2021) on the significance of considering social variables in public health approaches.

4 CONCLUSION

This research paper pondered the foremost critical findings concerning how the novel *al-Wabā'* reflected the social and public health emergencies of the early 20th century, with a specific centre on the annihilating plagues-epidemics of typhus and tuberculosis. The findings indicated that Ḥānī al-Rāhib employed narratives as a vehicle for commemorating and critiquing public health conditions and the policies of political authorities, reflecting local aspects of Islam and sustainability within the SDG framework. This paper, which is in line with SDG3 (health) and the discussion of narratives in the novel al-Wabā, not only addresses cases of health and community resilience in the midst of a health crisis, but therapeutic narratives are also educational tools that empower individuals and communities to understand and reflect on their experiences as stated in SDG4 (education). Thus, through storytelling, learning about health, resilience, and solidarity can be preserved at the local to national levels, thereby contributing to a culture of learning in society. In this way, the Arabic literary novel demonstrated the continuing relevance of Arabic literature in the advancement of sustainability goals and the addressing of global health issues.

A notable contribution of this study is the integration of socio-ecological resilience theory and the therapeutic narrative approach in literary analysis. This study makes a novel contribution to the field by emphasising the potential of narratives as tools for critical reflection and adaptation strategies relevant to the SDGs. Furthermore, it demonstrates how literary texts serve as dynamic sources of knowledge, enhancing comprehension of the interconnections between health, policy, and culture in local and global contexts.

The limitation of this study was its narrow focus on a single literary work, namely the novel *al-Wabā'*. Further research could expand the scope by comparing *al-Wabā'* with other Arabic literary works that discuss similar issues, as well as conducting comparative analysis with literature from other cultures to understand global and local influences in health and sustainability narratives. We will add specific recommendations for higher education institutions, specifically Arabic Language and Literature, Tarjamah, *Dirāsāt Islāmiyyah*, and Arabic Language Education programs, to integrate learning about the Arabic Literature genre of Infectious Diseases into their syllabi and curricula. This could include group analysis of themes of resilience and healing in Arabic literature novels and poetry that facilitate discussions that encourage students to connect their personal or community experiences to the narratives found in the literature.

REFERENCES

Arnold, D. (1991). The Indian Ocean as a disease zone, 1500–1950.†. *South Asia: Journal of South Asian Studies, 14* (2), 1–21. https://doi.org/10.1080/00856409108723152

Atanasova, D., Koteyko, N., Brown, B., and Crawford, P. (2017). Representations of mental health and arts participation in the national and local British press, 2007–2015. *Health, 23*(1), 3–20. https://doi.org/10.1177/1363459317708823

Biselli, R., Nisini, R., Lista, F., Autore, A., Lastilla, M., De Lorenzo, G., Peragallo, M. S., Stroffolini, T., and D'Amelio, R. (2022). A Historical Review of Military Medical Strategies for Fighting Infectious Diseases: From Battlefields to Global Health. In *Biomedicines* (Vol. 10, Issue 8). https://doi.org/10.3390/biomedicines10082050

Blanco, M. Á. del A. (2020). Famine in Spain During Franco's Dictatorship, 1939–52. *Journal of Contemporary History, 56*(1), 3–27. https://doi.org/10.1177/0022009419876004

Blecher, R. (2004). *Desert medicine, ethnography, and the colonial encounter in mandatory Syria* (pp. 249–268). Brill. https://doi.org/https://doi.org/10.1163/9789047402695_018

Bölükbaşı, O. (2023). *Historical Lessons from the Pandemics BT - Headache Disorders in Pandemic Conditions* (A. Özge, D. Uludüz, H. Bolay, and Ö. Karadaş, Eds.; pp. 21–33). Springer International Publishing. https://doi.org/10.1007/978-3-031-26309-5_3

Bruce-Chwatt, L. J. (1973). *Global Problems of Imported Disease* (B. B. T.-A. in P. Dawes, Ed.; Vol. 11, pp. 75–114). Academic Press. https://doi.org/https://doi.org/10.1016/S0065-308X(08)60185-2

Bürgschwentner, J., Egger, M., and Barth-Scalmani, G. (2014). *Other Fronts, Other Wars?: First World War Studies on the Eve of the Centennial.* Brill. https://doi.org/https://doi.org/10.1163/9789004279513

Crosby, A. W. (1991). Infectious Disease and the Demography of the Atlantic Peoples. *Journal of World History, 2* (2), 119–133.

de Graaf, B. (2020). Taming Pandemics in International Relations: Why Do States Cooperate in Transboundary Crises? An Applied History Perspective. *Journal of Applied History, 2*(1–2), 36–61. https://doi.org/https://doi.org/10.1163/25895893-bja10011

Dhika, M. A., Khairani, D., Masruroh, S. U., Fiade, A., Arifin, V., and Tsaqofi, W. A. (2023). Comparing GraphQL and ReST Architecture in Arabic Learning Games: A Quality of Service (QoS) Approach. *2023 11th International Conference on Cyber and IT Service Management (CITSM)*, 1–5. https://doi.org/10.1109/CITSM60085.2023.10455108

Duraković, I. (2014). *11. Serbia as a Health Threat to Europe: The Wartime Typhus Epidemic, 1914–1915* (pp. 259–279). Brill. https://doi.org/https://doi.org/10.1163/9789004279513_013

Ebbesson, J., and Folke, C. (n.d.). *Social-Ecological Resilience and Law* (A. S. Garmestani and C. R. Allen, Eds.; pp. 265–292). Columbia University Press. https://doi.org/doi:10.7312/garm16058-011

Evered, K. T., and Evered, E. Ö. (n.d.). Of hovels and homes: consumption, class, and domestic space in early republican Turkey. *Middle Eastern Studies*, 1–19. https://doi.org/10.1080/00263206.2023.2294775

Fajriati, R. P., Khairani, D., Rozy, N. F., Husin, N., Wiyartanti, L., and Rosyadi, T. (2020). Towards the Implementation of Arabic Language Mobile Apps Learning: Designed by User Insight. *2020 8th International Conference on Cyber and IT Service Management (CITSM)*, 1–5. https://doi.org/10.1109/CITSM50537.2020.9268901

Folke, C. (2016). Resilience (Republished). *Ecology and Society, 21*(4).

Folke, C., Jansson, Å., Rockström, J., Olsson, P., Carpenter, S. R., Chapin, F. S., Crépin, A.-S., Daily, G., Danell, K., Ebbesson, J., Elmqvist, T., Galaz, V., Moberg, F., Nilsson, M., Österblom, H., Ostrom, E., Persson, Å., Peterson, G., Polasky, S., . . . Westley, F. (2011). Reconnecting to the Biosphere. *AMBIO, 40*(7), 719–738. https://doi.org/10.1007/s13280-011-0184-y

Galaz, V., Olsson, P., Hahn, T., Folke, C., and Svedin, U. (2008). The Problem of Fit among Biophysical Systems, Environmental and Resource Regimes, and Broader Governance Systems: Insights and Emerging Challenges. In O. R. Young, L. A. King, and H. Schroeder (Eds.), *Institutions and Environmental Change: Principal Findings, Applications, and Research Frontiers* (p. 0). The MIT Press. https://doi.org/10.7551/mitpress/7920.003.0011

Helman, A. (2004). Cleanliness and squalor in inter-war Tel-Aviv. *Urban History, 31*(1), 72–99. https://doi.org/10.1017/S0963926804001798

Jacobson, R. L. (2010). Leishmaniasis in an Era of Conflict in the Middle East. *Vector-Borne and Zoonotic Diseases*, *11*(3), 247–258. https://doi.org/10.1089/vbz.2010.0068

Jensen, J. H. (1990). The Impact of War on Southeastern Europe, 1912—1923. *War and Society*, *8*(1), 29–46. https://doi.org/10.1179/106980490790305702

Kazemi, R. (2018). Epidemics and Revolutions: The Case of Iran in the Late Nineteenth-Century Middle East. *The Historian*, *80*(4), 749–790. https://doi.org/10.1111/hisn.13001

Koteyko, and Hunt. (2018). Special issue: Discourse analysis perspectives on online health communication. *Discourse, Context and Media*, 25. pp. 1–4. ISSN 2211-6958

Koteyko, N., and Atanasova, D. (2018). Mental health advocacy on Twitter: Positioning in Depression Awareness Week tweets. *Discourse, Context and Media*, *25*, 52–59. https://doi.org/https://doi.org/10.1016/j.dcm.2018.04.007

Kunitz, S. J. (1983). Speculations on the European Mortality Decline. *The Economic History Review*, *36*(3), 349–364. https://doi.org/10.2307/2594969

Lee, K. (2003). *Globalization and Health: A Historical Perspective BT - Globalization and Health: An Introduction* (K. Lee, Ed.; pp. 30–60). Palgrave Macmillan UK. https://doi.org/10.1057/9781403943828_2

Loddenkemper, R., and Konietzko, N. (2018). Tuberculosis in Germany before, during and after World War II. In J. F. Murray and R. Loddenkemper (Eds.), *Tuberculosis and War: Lessons Learned from World War II* (Vol. 43, p. 0). S.Karger AG. https://doi.org/10.1159/000481475

Mackenbach, J. P. (2020). Understanding Trends in Population Health. In *A History of Population Health* (pp. 57–117). https://doi.org/10.1163/9789004429130_004

Martínez, F. J. (2017). Mad at Tangier: Hygienist discourses, mental disorders, and European competition for hegemony in early twentieth-century Morocco. *The Journal of the Middle East and Africa*, *8*(1), 57–73. https://doi.org/10.1080/21520844.2017.1290501

McKnight Nichols, C., Ewing, E. T., Gaston, K. H., Marinari, M., Lessoff, A., and Huyssen, D. (2022). What Came Next?: Reflections on the Aftermath(s) of the 1918–19 Flu Pandemic in the Age of COVID. *The Journal of the Gilded Age and Progressive Era*, *21*(2), 111–149. https://doi.org/10.1017/S1537781421000682

Méouchy, N., and Sluglett, P. (2004). *The British and French Mandates in Comparative Perspectives/Les mandats français et anglais dans une perspective comparative*. Brill. https://doi.org/https://doi.org/10.1163/9789047402695

Mercer, A. (2021). Protection against severe infectious disease in the past. *Pathogens and Global Health*, *115*(3), 151–167. https://doi.org/10.1080/20477724.2021.1878443

Mezmir, E. A. (2020). Qualitative Data Analysis: An Overview of Data Reduction, Data Display and Interpretation. *Research on Humanities and Social Sciences*, *10*(21), 15–27. https://doi.org/10.7176/rhss/10-21-02

Miller, F., Osbahr, H., Boyd, E., Thomalla, F., Bharwani, S., Ziervogel, G., Walker, B., Birkmann, J., van der Leeuw, S., Rockström, J., Hinkel, J., Downing, T., Folke, C., and Nelson, D. (2010). Resilience and Vulnerability. *Ecology and Society*, *15*(3).

Nachtigal, R. (2023). Russian Public Health during the First World War. *The Journal of Slavic Military Studies*, *36*(1), 73–95. https://doi.org/10.1080/13518046.2023.2210938

Neumeyer, X., Santos, S. C., and Morris, M. H. (2020). Overcoming barriers to technology adoption when fostering entrepreneurship among the poor: The role of technology and digital literacy. *IEEE Transactions on Engineering Management*, *68*(6), 1605–1618.

Norrie, P. (2016). *How Disease Affected the End of the Bronze Age BT - A History of Disease in Ancient Times: More Lethal than War* (P. Norrie, Ed.; pp. 61–101). Springer International Publishing. https://doi.org/10.1007/978-3-319-28937-3_5

Post, J. D. (1976). Famine, Mortality, and Epidemic Disease in the Process of Modernization. *The Economic History Review*, *29*(1), 14–37. https://doi.org/10.2307/2594505

Rahman, Y. (1996). The Miraculous Nature Of Muslim Scripture: A Study Of 'Abd Al-Jabbār's 'I'jāz Al-Qur'ān'. *Islamic Studies*, *35*(4), 409–424.

Schlüter, M., Haider, L. J., Lade, S. J., Lindkvist, E., Martin, R., Orach, K., Wijermans, N., and Folke, C. (2019). Capturing emergent phenomena in social-ecological systems. *Ecology and Society*, *24*(3).

Šimkutė, V., and Gudienė, V. (2022). Epidemic typhus in the Lithuanian army from 1919 to 1923. In *Vojnosanitetski Pregled* (Vol. 79, Issue 6, pp. 628–635). https://doi.org/10.2298/VSP201114139S

Simoni, M. (2000). At the roots of division: a new perspective on Arabs and Jews, 1930–39. *Middle Eastern Studies*, *36*(3), 52–92. https://doi.org/10.1080/00263200008701318

Trubeta, S., Promitzer, C., and Weindling, P. (2021). *Medicalising borders: Selection, containment and quarantine since 1800*. Manchester University Press. https://doi.org/10.7765/9781526154675

Weindling, P. (2021). *Sealing borders and containing prisoners: From free movement of migrants to containment in concentration camps*. Manchester University Press. https://doi.org/10.7765/9781526154675.00015

Zylberman, P. (2007). *Civilizing the State: Borders, Weak States and International Health in Modern Europe BT - Medicine at the Border: Disease, Globalization and Security, 1850 to the Present* (A. Bashford, Ed.; pp. 21–40). Palgrave Macmillan UK. https://doi.org/10.1057/9780230288904_2.

Towards Resilient Societies: The Synergy of Religion, Education, Health, Science, and Technology – Rahiem (Ed.)
© 2026 The Author(s), ISBN: 978-1-041-08471-6
Open Access: www.taylorfrancis.com, CC BY-NC-ND 4.0 license

Exploring the hidden curriculum in religious-based elementary schools in Indonesia

Fauzan Fauzan
Department of Elementary Education, Universitas Islam Negeri Syarif Hidayatullah Jakarta, Indonesia
ORCID ID: 0000-0002-1441-2757

Mardhatillah Mardhatillah
Primary Education, Universitas Negeri Malang, Malang, Indonesia
ORCID ID: 0000-0002-4610-1531

Faiz Biamrillah
SD Islam Kamila Insancita, Jakarta, Indonesia

Fatkhul Arifin
Department of Elementary Education, Universitas Islam Negeri Syarif Hidayatullah Jakarta, Indonesia
ORCID ID: 0000-0001-9439-5856

Sobareeyah Yaya
Thamavitya Mulniti School, Thailand

ABSTRACT: A hidden curriculum fosters moral education and values implicitly embedded in students' daily habits and activities. However, its constructive impact remains inconsistent and often lacks measurable effectiveness. In Indonesia, the prevalence of criminal behaviour among school-age children highlights the limited success of moral cultivation emphasised through the hidden curriculum. Therefore, research on its implementation is essential. This study, employing a qualitative method through interviews, observations, and document analysis, investigates the application of the hidden curriculum in achieving the vision of religious-based schools. The findings are discussed across three key aspects: awareness among school members, the role of school management in designing a curriculum that supports the hidden curriculum, and the interdependence of these elements in achieving the school's vision. Particular attention is given to two essential areas: improving the quality of religious facilities and enhancing library resources as a foundation for literacy. These facilities serve as central spaces for interaction and habit formation, significantly influencing the success of the hidden curriculum. This study underscores the importance of a well-structured hidden curriculum in fostering moral values and achieving educational goals in religious school environments.

Keywords: hidden curriculum, religious-based school, school management, school vision

1 INTRODUCTION

Educators need to create a positive and impactful hidden curriculum since the students will naturally learn much from the hidden curriculum because the aims and values are addressed implicitly. Moreover, this curriculum can enhance the hidden components of the success of education by involving leadership, awareness of being truthful. For some countries, the official curriculum could tentatively be changed, including Indonesia. The recent changes in

DOI: 10.1201/9781003645542-35

the Indonesian curriculum, namely the 2013 curriculum, have inserted morals and character explicitly through the learning process. It was stated that the Indonesian students' graduation was no longer determined by their national examination scores but by the accumulated scores of affective and cognitive aspects (Akmalia 2021). It means that students should succeed in both academic matters and non-academic matters (e.g. life skills and morals).

Some studies stated that the hidden curriculum is essential and even more influential than the official curriculum in strengthening students' norms and morals (Mulder *et al.* 2019). Regarding the powerful impact of the hidden curriculum, some previous studies have been conducted in Indonesia, such as the value of anti-corruption, leadership mindset, and skills, character building on self-motivation, and integration of national insight.

The application of hidden curriculum occurs through educational processes that are not explicitly taught but are embedded in the practices, rules, and daily interactions within schools. For instance, in countries like the United Kingdom, the hidden curriculum is used to reproduce power relations through school structures that reinforce certain social norms, such as gender roles and class hierarchies (Skelton 1997; Wals 2020). Students unconsciously learn about compliance, order, and the dominant societal norms through school rules, classroom arrangements, and teacher-student interactions. Meanwhile, in the United States, the hidden curriculum helps instill capitalist values by reflecting hierarchical work structures in school activities, such as grading systems and task assignments (Laiduc and Covarrubias 2022). Although invisible, the hidden curriculum plays a crucial role in shaping students' mindsets to align with existing social.

The previous studies were primarily in the Islamic boarding school, junior high school, senior high school, and vocational high school. However, the implementation, effectiveness, and success of a hidden curriculum still need to be investigated in the earlier stage of education due to the urgency of building morals and norms in social life. The students must be introduced and given a role early to upgrade their attitudes through the hidden curriculum implementation. Moreover, for internalising the norms and goodness, religious-based schools take more concern in addressing those aspects of students' activities as identity and cultural aspects because religion is considered the functional dimension of culture (Hannigan *et al.* 2022).

Besides the larger capacity of the religious-based schools to internalise the norms, this research also considered that Indonesia, in which this research was conducted, has been categorised as the largest Muslim country in the world (Chia 2022), with 87.2% of the whole population are Muslim (Hefner 2020). Recently, among the Indonesian Muslim community, Islamic education appeared in the formal education context, leading to the increasing number of formal education institutions that attached the 'Islamic' name to emphasise the basis of Islamic values that will be taught. With such an enormous hegemony, attention to formal Islamic-based schools should be more centralised. Therefore, the current study investigated the implementation of the hidden curriculum in a religious-based elementary school.

This study was conducted to determine the answers to the three research questions: a) To what extent did participatory awareness level among stakeholders in implementing its hidden curriculum? b) What was the role of school management in designing a relevant curriculum that supports the hidden curriculum? c) how did those two factors affect the implementation of the hidden curriculum itself?

2 METHOD

This study employs a qualitative approach to understanding social phenomena in a faith-based elementary school in Banten, Indonesia, as it allows flexibility in exploring the social and cultural complexities of the research site (Creswell 2012). The data sources include individuals as informants (school management, teachers, and students), the school environment (facilities and learning resources), and supporting documents such as regulations, school programs, activity reports, and observations of events during learning processes and social systems among school members. The sampling method is purposive, selecting informants deemed to have in-depth knowledge and high relevance to the studied phenomena, such as school principals, teachers with more than five years of experience, and students actively involved in school programs. Data collection techniques involve three primary methods: in-depth interviews to explore informants' perspectives and experiences, participatory observation to directly observe learning processes and interactions within the academic community, and document analysis to examine policies and practices supporting character development. Additionally, closed questionnaires are used to reach more respondents during the pandemic, which limits field observations. The researcher acts as the main instrument, designing interview and observation guidelines, ensuring data validity through triangulation of various sources, and systematically analysing data using a thematic analysis approach. This methodology is expected to provide a comprehensive understanding of the hidden curriculum and character development processes in faith-based elementary schools.

3 RESULTS

From the on-site research activities (observation, interview, and school document analysis), we would like to present and discuss some key findings.

3.1 *The awareness of school stakeholders in implementing hidden curriculum*

The researchers found that the stakeholders performed well in implementing the hidden curriculum based on the structured and unstructured activities (Figure 1). Social relationships were also well established by involving all school staff. At the same time, teachers tried to reduce the gap with students by using familiar greetings to develop harmonious relationships.

They were highly aware of the school's vision, leading to an excellent school institution that considers the students' 21st-century skills to be mastered. We also found that most school members perform routine activities, namely mandatory prayer, maintaining cleanliness, and practising 5S values (Indonesian term for Senyum, Sapa, Salam, Sopan, Santun, or Smiling, Greeting, and Being Polite). From the above findings, the level of awareness among the school stakeholders could be categorised as high. They share a common understanding of being agents of the successful implementation of the curriculum, which undoubtedly

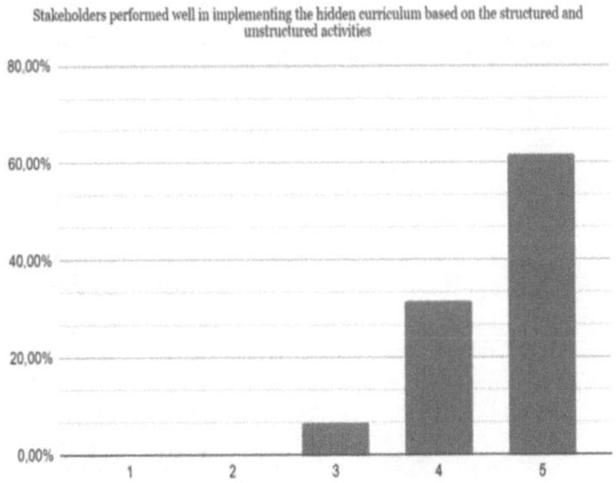

Figure 1. Stakeholders performed.

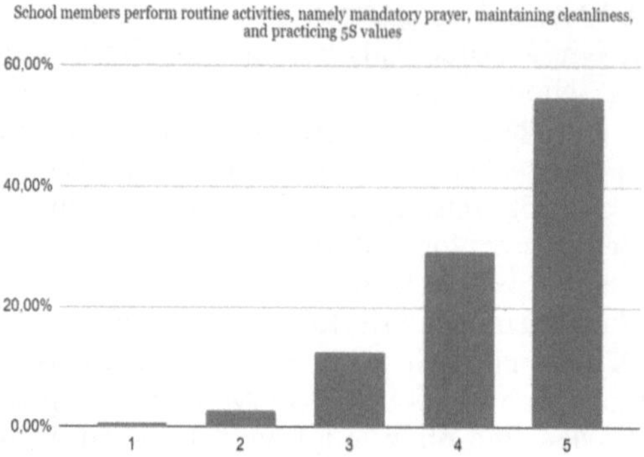

Figure 2. School member performed.

supports the implementation of the hidden curriculum. High awareness is needed in the entire academic community of the school (Villanueva *et al.* 2018).

In the Figure 2 above shows awareness of the school's vision, the awareness of the academic community of the observed school could also be reflected in obeying school regulations and implementing the agreed school culture. These findings aligned with the previous research conducted by Mossop *et al.* (2013), who found that the awareness of all the staff to follow the agreed culture strictly could determine the success of hidden curriculum implementation. Successful changes in social structure in the hidden curriculum are essential to increase teachers' awareness of social structure factors in the hidden curriculum. Lastly, we could also see that the awareness level of the school vision was well articulated to parents as they were actively involved in their children's activities, such as doing parent-student joint projects and other events promoting the school vision (Figures 3 and 4).

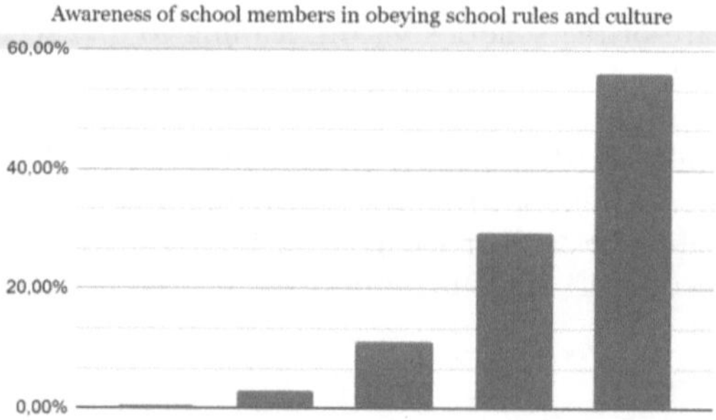

Figure 3. Awareness of school member in obeying school rules and culture.

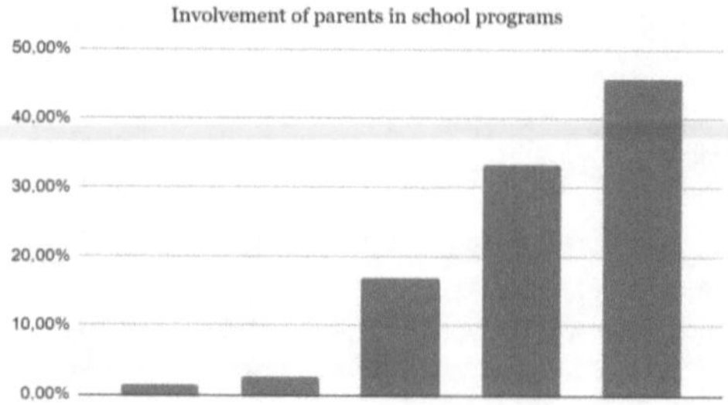

Figure 4. Involvement of parents in school programs.

3.2 *The role of school management in designing the activities which support the implementation of hidden curriculum*

Three crucial roles of school management in implementing the hidden curriculum are observed: mapping activities into the curriculum, scheduling learning time, and integrating it with the national curriculum. The first role focuses on curriculum aspects, including student activities that incorporate elements such as active learning models, Islamic values, bilingual programs, life skills, ICT classes, arts, sports, and field trips. The hidden curriculum reflects students' responses to experiences during learning, with active learning providing a full and positive learning experience.

In religious activities, the school encourages daily worship routines such as reading the Qur'an, performing sunnah prayers, congregational prayers, and memorising Qur'anic verses. Specific religious lessons include Islamic Education, private Iqra lessons, Quranic recitation (murottal), and Quran memorisation. These activities aim for students to not only absorb Islamic values but also practice them in their daily lives as devout Muslims.

For bilingual practice, the school collaborates with Cambridge to enhance students' language skills. With a vision focused on mastering foreign languages and 21st-century skills, the school offers activities like robotics classes to develop ICT skills. The robotics program also encourages teamwork and problem-solving, addressing the needs of students to gradually improve these skills (Boz and Allexsaht-Snider 2022).

The second role involves scheduling learning time. Research on educational effectiveness highlights the importance of time allocation to optimise learning outcomes (Scheerens 2013). Considering this, the school management has carefully structured additional time before and after classes to provide implicit learning opportunities through the hidden curriculum.

The third role is integrating both local and national content to maximise learning outcomes. The school also considers subjects that can effectively support the hidden curriculum. The long-standing research on the hidden curriculum emphasises its focus on values. The school suggests several elements of the hidden curriculum: value comprehension, life skills development, fostering a democratic society, social control mechanisms, and enhancing motivation and academic achievement.

3.3 *Achieving the school's vision by implementing a hidden curriculum*

Three key variables closely related to social interaction in the school environment are organisation, social systems, and culture. In the organisation variable, three interrelated aspects were observed: active, student-centred learning models, inclusive class composition across various learning models and intelligence levels, and the arrangement of facilities and learning resources. Active learning provides better experiences for students, as previous studies show it positively impacts student achievement in STEM subjects (Lugosi and Uribe 2022).

In terms of student organisation, aligned with the hidden curriculum's inclusiveness and non-discriminatory atmosphere (Warren *et al.* 2019), students are grouped heterogeneously, promoting tolerance, cooperation, and mutual support. Additionally, student performance is evaluated not only by theoretical knowledge but also by their attitude development. The third aspect of the organisation variable is the availability of facilities. These play a significant role in implementing the curriculum (Chaudhary 2015). The observed school provided facilities like English development tools, sports grounds, a pool, and 21st-century skills training spaces, all of which were well-maintained.

However, The current prayer space is insufficient, leading to congregational prayers being held in multipurpose rooms, often requiring shifts due to shared space with the middle school. Lower-grade students pray in classrooms. Despite the symbolic importance of religious spaces in Islamic character development (Alamdari *et al.* 2017), this school lacks adequate capacity. While Islamic values are instilled through scheduled activities, students need a larger space for prayer. Plans for mosque development exist, but progress has been delayed, as parents prefer improving teaching quality over infrastructure. Parents feel the current facilities are sufficient, encouraging students to patiently wait their turn for prayer.

The library's location is another concern. To foster literacy, a well-placed library is crucial (Rojas-Estrada *et al.* 2023). However, the main library is far from the elementary school, though there are mini-libraries in classrooms. The main library should be more accessible to elementary students to encourage interaction and discussion (Anderson 2019).

The social system variable is vital to the success of the hidden curriculum, as it implicitly builds awareness of proper attitudes. This includes teacher communication with colleagues, parents, staff, and the community (Maryani and Sulisworo 2015). These relationships are categorised into four key points: teacher-colleague, teacher-parent, teacher-student, and student-student interactions. Weekly meetings and non-formal events foster teacher-colleague relationships. Communication with parents is maintained through mobile technology, updating them on school news, Islamic values, activities, student achievements, and behaviour.

Regarding cultural variables, which involve belief systems, values, and cognitive structures (Cheung and Xu 2016), the school promotes agreed-upon cultural values and internalises them through routine activities. Cultural values are symbolised by the school logo, featuring a mosque for Islamic identity and a globe for global awareness. Daily student routines include English communication, diary writing in English, prayers before and after activities, Quran recitation, congregational prayers, Islamic holiday celebrations, proper waste disposal, and public speaking practice. Such routines promote cultural internalisation through meaningful interactions (Suriadi and Mursidin 2019). Finally, consistent reward and punishment systems reinforce these cultural norms, serving as a social control mechanism in the hidden curriculum (Wu *et al.* 2022).

4 CONCLUSION

The implementation of the hidden curriculum in faith-based elementary schools encompasses three main aspects: the awareness of the school community, the role of school management in designing a curriculum that supports hidden curriculum practices, and its application in achieving the school's vision. The effectiveness of the hidden curriculum in realising this vision is influenced by three variables: organisational structure, social systems, and culture. To optimise the impact of the hidden curriculum, it is recommended that schools consistently evaluate their active learning strategies and inclusivity practices to ensure alignment with the institution's vision. Additionally, enhancing teacher training programs in communication and social skills, as well as explicitly integrating cultural values into daily routines, can further strengthen the role of the hidden curriculum in character development and academic excellence.

REFERENCES

Akmalia, R. (2021). Effect of elimination of national examination on students' graduation standard assessment. *Al-Irsyad: Jurnal Pendidikan Dan Konseling*, *11*(1), 46–57. https://doi.org/10.30829/al-irsyad.v11i1.9332

Alamdari, A. H., Mousavi, S. E., Sheikholeslami, H. K., and Saadatmand, M. (2017). The typology of the mosque-schools 1 of Iran based on the method of access. *Architecture & Urbanism*, *14*(53), 57–68.

Anderson, R., and Helmane, I. (2019). Towards media literacy in the education curriculum: Experience of Latvia. *Rural Environment Education Personality*, 12, 15–21. https://doi.org/10.22616/REEP.2019.001

Boz, T., and Allexsaht-Snider, M. (2022). How do elementary school teachers learn coding and robotics? A case study of mediations and conflicts. *Education and Information Technologies*, *27*(3), 3935–3963. https://doi.org/10.1007/s10639-021-10736-4

Chaudhary, G. K. (2015). Factors affecting curriculum implementation for students. *International Journal of Applied Research, 1*(12), 984–986. www.allresearchjournal.com

Cheung, C. K., and Xu, W. (2016). Integrating media literacy education into the school curriculum in China: A case study of a primary school. In X. Liu and M. L. Tan (Eds.), *Media Literacy Education in China* (pp. 133–146). Springer. https://doi.org/10.1007/978-981-10-0045-4_10

Chia, P. S. (2022). Pancasila and covenantal pluralism in Indonesia: A historical approach. *Transformation, 39* (2), 91–98. https://doi.org/10.1177/02653788211069971

Cresswel, J. W. (2012). *Educatioanal research: Planning, conducting and evaluating quantittaive and qualitative research.* Pearson.

Hannigan, A., Faas, D., and Darmody, M. (2022). Ethno-cultural diversity in initial teacher education courses: the case of Ireland. *Irish Educational Studies, April,* 1–16. https://doi.org/10.1080/03323315.2022.2061559

Hefner, R. W. (2020). Islam and covenantal pluralism in Indonesia: A critical juncture analysis. *Review of Faith and International Affairs, 18*(2), 1–17. https://doi.org/10.1080/15570274.2020.1753946

Laiduc, G., and Covarrubias, R. (2022). Making meaning of the hidden curriculum: Translating wise interventions to usher university change. *Translational Issues in Psychological Science*, 8(2), 221–233. https://doi.org/10.1037/tps0000309

Lugosi, E., and Uribe, G. (2022). Active learning strategies with positive effects on students' achievements in undergraduate mathematics education. *International Journal of Mathematical Education in Science and Technology, 53*(2), 403–424. https://doi.org/10.1080/0020739X.2020.1773555

Mossop, L., Dennick, R., Hammond, R., and Robbé, I. (2013). Analysing the hidden curriculum: Use of a cultural web. *Medical Education, 47*(2), 134–143. https://doi.org/10.1111/medu.12072

Maryani and Sulisworo (2015).?

Mulder, H., ter Braak, E., Chen, H. C., and ten Cate, O. (2019). Addressing the hidden curriculum in the clinical workplace: A practical tool for trainees and faculty. *Medical Teacher, 41*(1),

Rojas-Estrada, EG., Aguaded, I. and García-Ruiz, R. Media and Information Literacy in the Prescribed Curriculum: A Systematic Review on its Integration. *Educ Inf Technol* 29, 9445–9472 (2024). https://doi.org/10.1007/s10639-023-12154-0

Scheerens, J. (2013). The use of theory in school effectiveness research revisited. *School Effectiveness and School Improvement, 24*(1), 1–38. https://doi.org/10.1080/09243453.2012.691100

Skelton, A. (1997). Studying hidden curricula: Developing a perspective in the light of postmodern insights. *Curriculum Studies*, 5(2), 177–193. https://doi.org/10.1080/14681369700200007

Suriadi, S., and Mursidin, M. (2019). Internalisation of cultural values in schools early steps in building nation characters. Proceedings of the Proceedings of 1st Workshop on Environmental Science, Society, and Technology, WESTECH 2018, December 8th, 2018, Medan, Indonesia, 20. https://doi.org/10.4108/eai.8-12-2018.2283966

Villanueva, I., Di Stefano, M., Gelles, L., and Youmans, K. (2018). Hidden Curriculum Awareness: A Comparison of Engineering Faculty, Graduate Students, and Undergraduates. *World Education Engineering Forum Conference*, 1–6.

Wals, A. E. (2020). Transgressing the hidden curriculum of unsustainability: Towards a relational pedagogy of hope. *Educational Philosophy and Theory*, 52(8), 825–826. https://doi.org/10.1080/00131857.2019.1676490

Warren, K., Mitten, D., D'Amore, C., and Lotz, E. (2019). The gendered hidden curriculum of adventure education. *Journal of Experiential Education, 42*(2), 140–154. https://doi.org/10.1177/1053825918813398

Wu, J., Luan, S., and Raihani, N. (2022). Reward, punishment, and prosocial behavior: Recent developments and implications. *Current Opinion in Psychology, 44*, 117–123. https://doi.org/10.1016/j.copsyc.2021.09.003

Towards Resilient Societies: The Synergy of Religion, Education, Health, Science, and Technology – Rahiem (Ed.)

Role of the merdeka curriculum in enhancing critical thinking for addressing global challenges: A study on environmental change education

Abdulloh Yahya
Emer Islamic Boarding School, Lampung, Indonesia
ORCID ID: 0009-0002-2126-0330

Nengsih Juanengsih
Universitas Islam Negeri Syarif Hidayatullah Jakarta, Tangerang Selatan, Indonesia
ORCID ID: 0000-0001-8035-7111

Fatoumata Sonko
Nemasu Senior Secondary, The Gambia

ABSTRACT: The Merdeka Curriculum aims to develop students' critical thinking skills, aligning with the Pancasila Student Profile, which emphasises critical reasoning. These skills are essential for achieving SDG 13 (Climate Action) and SDG 4 (Quality Education), empowering individuals to understand complex issues like climate change, propose practical solutions, and act reflectively based on evidence. Quality education that fosters critical thinking equips students to address global challenges effectively and contribute to a sustainable future. This study evaluates the critical thinking skills of Grade 10 students studying environmental change topics under the Merdeka Curriculum in South Tangerang. Using a quantitative descriptive method, the research involved 113 students from three public senior high schools. The instrument, based on Facione's six indicators—interpretation, analysis, evaluation, inference, explanation, and self-regulation—consisted of nine valid questions. Data were analysed as average indicator achievement percentages. The results show that all schools fall into the low critical thinking category, with no significant differences across schools. The study recommends discussion techniques, self-reflection, problem-solving, and constructive feedback to enhance students' critical thinking skills effectively.

Keywords: critical thinking skills, environmental change, 21st-century skills, Daily practice questions or tests should use questions that hone students' critical thinking skills

1 INTRODUCTION

The national curriculum in Indonesia has undergone eleven times post-independence until 2019 (Alhamuddin 2019). Changes in curriculum aspects include design and models. The government changed the curriculum to improve the level of education in Indonesia. Besides that, globalisation and rapid technological advances force adjustments that need to be made to adapt.

At the end of the COVID-19 pandemic, Indonesia's Ministry of Education and Culture announced the existence of a new national curriculum known as the Merdeka Curriculum. This action is an effort to overcome learning loss during a pandemic and increase student competency. As an innovation, the Merdeka Curriculum brings a new colour to the education perspective in Indonesia. One new thing that appears is the Project for Strengthening the Profile of Pancasila Students (P5), which focuses on problem-solving skills.

DOI: 10.1201/9781003645542-36

The Merdeka curriculum determines objectives in each subject. Biology at the high school level has nine objectives. One is understanding biological issues within the individual, family, surrounding, and global environment and applying Biological knowledge to overcome these problems. To reach these goals, students need to have critical thinking skills. This ability will help them to describe the problem they face and then analyse it to make decisions (Lin 2018). The PISA study emphasises the importance of improving education quality in Indonesia, particularly emphasising 21st-century skills, including critical and creative thinking (Astawan *et al.* 2023).

Critical thinking is a 21st-century student skill (Rosidin *et al.* 2019). It is essential as a provision for life to help solve their problems. Critical thinking involves higher-order thinking skills, creative thinking, problem-solving, and decision-making (Narmaditya *et al.* 2017). Students' development of critical thinking skills must focus on activities to teach content (knowledge) and methods or ways to find the process and evaluate information (Nhat *et al.* 2018). Students can think critically when learning, but this skill sometimes needs to be developed (Isnaeni *et al.* 2021). Apart from that, rapid changes due to globalisation encourage every human being to compete globally (Lau 2011). Students need good thinking skills to help them make realistic and logical decisions to succeed in their chosen field. Critical thinking is only sometimes associated with the ability to solve math problems perfectly or be good at debating. However, taking the right attitude towards something also includes critical thinking. Many problems in human life, such as global warming, floods, and plastic waste, must be solved with innovative ideas. The learning activities determine critical thinking skills (Rapi *et al.* 2022). Argumentation or reasoning is an indicator of critical thinking skills. Students will try to convey information and knowledge to others through scientific argumentation (Wibowo *et al.* 2024).

Cultivating critical thinking in students contributes significantly to building a resilient society that can face social, economic, and environmental challenges in an adaptive, innovative, and sustainable manner. Students who are taught to think critically can identify problems more clearly and independently. They can find solutions based on evidence and reason and make the right decisions without relying entirely on the opinions of others.

A resilient society needs individuals who can adapt to change, whether in terms of technology, economy, or environment. Critical thinking teaches students to understand change and see it as an opportunity, not a threat. They can assess the possibilities and their impacts and adapt to new situations. Critical thinking and creativity are becoming increasingly important in the labour market and contribute to a better personal and civic life (Vincent-Lancrin *et al.* 2019).

Quality education is a compilation of the 17 sustainable development objectives and policies to ensure inclusive and equal education for all and foster opportunities for life-long learning for everyone (Rulandari 2021). Critical thinking skills are closely linked to SDG (Sustainable Development Goal) 4: Quality Education. SDG 4 aims to ensure inclusive, equitable, quality education and promote lifelong learning opportunities. Critical thinking is important in achieving this goal for several reasons: improving problem-solving skills, encouraging more independent learning, increasing social and ethical awareness, fostering reflective thinking skills, preparing for global challenges, and encouraging innovation and creativity.

Various stakeholders' awareness and engagement play a crucial role in the successful implementation of climate policy and Sustainable Development Goals (SDGs). SDG 13, which refers to climate action, has three targets for combating climate change and its impact (Hwang *et al.* 2021). Critical thinking is critical to SDG 13: Action on Climate Change, which aims to take urgent steps to combat climate change and its impacts. Critical thinking helps us understand the issue of climate change and leads us to find more effective and sustainable solutions to address this challenge. Critical thinking plays a significant role in supporting the achievement of SDG 13: Action on Climate Change. With the ability to analyse data, evaluate policies, find innovative solutions, and educate and spread awareness about climate change, critical thinking drives more effective and impactful actions to address the challenges of climate change.

Critical thinking is doing more identifying and evaluating. These skills can be formed by making a personal perception, putting forward good reasons to support the conclusion, or making a logical decision about what to do in some conditions. So, cognitive and intellectual skills are needed. The necessary intellectual standards are clarity, correctness, accuracy, relevance, consistency, logic, completeness, and fairness (Bassham *et al.* 2011).

At the end of the 19th century, Peter Facione assembled a group of 46 educators who were experts in critical thinking. Then, he used the Delphi method to form a view about critical thinking and its constituent dispositions. The results show that critical thinking skills comprise six abilities: interpretation, analysis, evaluation, inference, explanation, and self-regulation. He concludes that these six abilities are the core of critical thinking. Aside from that, he highlighted that a person's knowledge or cognitive abilities are also required to produce rational decisions in all problems (Bassham *et al.* 2011).

The data from the PISA (Program for International Student Assessment) in 2018 told us that the average score for Indonesian students in science has decreased compared to that of the PISA in 2015. The average PISA score in 2015 for Indonesia in Science was 406; in 2018, it decreased to 396. The ability in the field of science measured in PISA includes using logic to solve problems. As reported by Anita and Firmansyah, only one in four students can fulfill the four indicators of critical thinking (Anita and Firmansyah 2022). It happened, as explained by Prasetyo and Firmansyah, because students were not used to high-order thinking skills (HOTS) questions (Prasetyo and Firmansyah 2022).

Based on the urgency of critical thinking skills, it is necessary to research the results of implementing the Merdeka Curriculum in schools that have implemented it in developing these abilities. This research was conducted by collecting data on the critical thinking abilities of grade 10 from public schools that have implemented the Merdeka Curriculum in South Tangerang City.

2 METHODS

Quantitative descriptive research was conducted to identify the profile of students' critical thinking on environmental change. Grade 10 students were sampled from three public schools in South Tangerang, Indonesia, and there were 113 students for the 2023/2024 academic year. They were selected purposively. All schools are accredited A, which means they are good schools. So, the researcher used school rankings from the school's average score in UTBK 2022 for categorisation. School A is ranked 412th nationally, school B is ranked 765th nationally, and school C is not among the top 1000 schools. Sampling that includes low, medium, and high groups is essential to ensure diversity, representativeness, and fairness in research or evaluation results.

To get the data, we used a test instrument with questions representing six aspects of critical thinking skills from Peter Facione and conducted interviews with three students from each school. Before data collection, the test instrument was tested at the eleventh-state high school in South Tangerang. There are two cases: environmental pollution and global warming. Each case contains six questions. Then, from twelve questions, only nine were valid using the validity and reliability test, and it had a 0.713 score on the reliability test using Cronbach's Alpha. A high Cronbach's alpha value indicates that the measurement instrument is reliable and can be relied on to produce consistent and accurate data. The validity test uses the Pearson Correlation test. There are nine valid questions. The data is presented as indicator achievement percentages by finding the average value of each indicator. According to Facione (Fitriani *et al.* 2019), the determination of achievement indicators.

After collecting all the data, the researcher interviewed three students from each school. With the provisions of one person from the upper class, one from the middle class, and one from the lower class on the critical thinking skills test results.

3 FINDINGS AND DISCUSSION

The results of calculating the critical thinking ability data for each school are shown in Table 1. The results of obtaining data from the field regarding critical thinking skills scores show that the highest average score is school C, amount 62.61. Then, the data from three schools in each indicator of critical thinking is presented in Table 2.

Table 1. Data calculation results for each school.

	School		
Data	A	B	C
Total students	34	43	36
Lowest score	39	36	39
Highest score	83	86	86
Mean	60.24	61.65	62.61
Median	62.50	61	64
Mode	61	64	50
Variance	138.31	143.57	145.33
Standard deviation	11.76	11.98	12.06

Table 2. Percentage of critical thinking each indicator in all school.

	School					
	A		B		C	
Critical thinking indicator	Percentage (%)	Category	Percentage (%)	Category	Percentage (%)	Category
Interpretation	72,79	Moderate	68,60	Weak	71,87	Moderate
Analysis	59,92	Not achieved	70,35	Moderate	72,22	Moderate
Evaluation	57,35	Not achieved	60,47	Not achieved	51,39	Not achieved
Inference	55,15	Not achieved	49,42	Not thinking critically	54,17	Not achieved
Explanation	46,32	Not thinking critically	40,70	Not thinking critically	45,14	Not thinking critically
Self-regulation	59,19	Not achieved	63,08	Weak	62,15	Weak
Mean	60,24	Not achieved	61,65	Not achieved	62,61	Weak

The interpretation indicators for each school show a score above 65. However, only schools A and C are categorised as moderate. The analysis indicator score for school A is far below that of schools B and C, categorised as moderate. The explanation indicator for each school cannot reach a score of 50, which means it needs to be revised. Only school A scored below 60 for the self-regulation indicator. All their scores were low, with school categories A and B not achieved and school C weak.

After the data for each student has gone through the prerequisite tests and declared that the data is regular and homogeneous, it continues with the one-way ANOVA test. The results of the ANOVA test that was carried out are presented in Table 3.

Table 3. ANOVA test result.

	Sum of Squares	df	Mean Square	F	Sig.
Between Groups	99.701	2	49.850	.350	.706
Within Groups	15680.441	110	142.549		
Total	15780.142	112			

The calculated Fvalue is 0.350 < Ftable 3.08, and the significance level is 0.706 > 0.05, so H0 is accepted, meaning there is no difference in critical thinking skills between schools A, B, and C. It means their scores are the same. All their scores were low, with school categories A and B not achieved and school C weak.

The descriptive data from interviews in each school is summarised. School A and C students are more often given multiple-choice Biology questions rather than essays, while school B students, apart from multiple choice, have short answer questions. Gustianingrum emphasised that daily test questions must be developed to support critical thinking skills according to the aim of the Pancasila student profile (Gustianingrum *et al.* 2023).

The critical reasoning dimension in the Pancasila Student Profile requires students to find solutions to problems. Therefore, school B obtained the highest self-regulation indicator score compared to other schools because it has much activity in P5 on this topic. Merdeka curriculum opens up space for educators and students to increase learning effectiveness, one of which is through P5 activities. Because in P5, students will learn from various sources, not just books, and relate to students' daily lives (Nadhiroh and Anshori 2023). Prasetyo and Firmansyah mentioned that students' habits in working on questions or problems that make them think at a higher level will develop their critical thinking skills (Prasetyo and Firmansyah 2022).

Interpretation, analysis, and self-regulation indicators in this study consist of two questions, one in each case. In the interpretation question, a graph of river water pollution in the first case and a picture of changes in the shape of mountain glacier ice in the second case are data presented for students to interpret. To clarify the meaning of the information, students must find and describe the outline (Facione 2013). A few of them could not find the red lines of information and just repeated the stimulus of the question.

In the analysis questions, students are given an article that supports the data interpretation question. Then, students are asked to analyse the relationship between the article and the data. They must examine the idea and relate issues or problems to each other's concepts (Facione 2013). Most students in school need to understand the instructions. Most say it has a relation, but the argument must be more relevant. Students are given three articles in the evaluation question. They must choose one article related to the discussed problem and strengthen the previous information. In assessing claims, they must give logical and clear reasons (Facione 2013). The way students give a reason needs to be improved. Many of them need to be put in the topic.

Inference questions force students to conclude the problems. To conclude, their arguments must be supported by the data that has been presented (Facione 2013). Only around fifty percent of the students in each school answered with a reasonable conclusion. Half answered by adding irrelevant information.

For the explanation question, students must give a complete description. To state the results means constructing accurate descriptions correctly (Facione 2013). They clarify the explanation with the data presented in the problem raised. Not all of the schools reach a score of 47% in this indicator. It means they need clarification about the questions and the issues surrounding this matter.

In self-regulation questions, students are asked to provide at least three solutions to the problems faced in each case. To self-monitor, they must reflect on their motivations or attitudes regarding an issue (Facione 2013) to self-monitor. Only school A can not score 60%, but not far behind the others.

Many factors can influence low critical thinking in the explanation and self-regulation aspects, including lack of training and education focusing on critical thinking skills, limitations in cognitive abilities, social influences, and psychological and emotional factors. Teachers should address this issue, as it is important to provide education that encourages critical reflection, self-management, and more open and analytical thinking habits.

Teachers should use an approach that involves active practice, reflection, feedback, and good thinking models to improve explanation and self-regulation in critical thinking. These strategies will help students understand how to provide clear and evidence-based explanations and learn how to manage their thinking processes effectively and independently. In this way, both aspects can develop optimally, facilitating students to become more critical and independent thinkers. In order to train and improve critical thinking skills, teachers are advised to use a guided inquiry learning model. Based on what was stated in previous research, the learning process that integrates all stages of the guided-inquiry learning model with SETS is suitable for all critical thinking skills indicators (Diana *et al.* 2023). Critical thinking could be integrated into any class and was able to find new techniques for involving students in discussion and deepening their critical thinking (Defianty and Wilson 2024). Teachers must facilitate the development of the student's critical thinking skills by providing relevant activities embedded within any subject, with some creativity from the teachers (Husna 2019).

4 CONCLUSION

Based on the results of the critical thinking skills of grade 10 from the three schools studied, there were no significant differences regarding students' average scores. More often, students solve higher-order thinking skills questions, which will help improve their critical thinking. Improving critical thinking skills requires a diverse and consistent approach to learning. Students can develop more profound and structured thinking skills through discussion techniques, self-reflection, problem-solving, and constructive feedback.

REFERENCES

Alhamuddin. (2019). Politik Kebijakan Pengembangan Kurikulum di Indonesia: Indonesia: *Sejak Zaman Kemerdekaan Hingga Reformasi* (1947–2013). Prenadamedia.

Anita, and Firmansyah, D. (2022). Analisis Kemampuan Berpikir Kritis Matematis Siswa SMA pada Materi Barisan Aritmetika. *Jurnal Matematika Ilmiah*, 8(1), 30–44.

Astawan, I. G., Suarjana, I. M., Werang, B. R., Asaloei, S. I., Sianturi, M., and Elele, E. C. (2023). STEM-Based Scientific Learning and Its Impact on Students' Critical and Creative Thinking Skills: An Empirical Study. *Jurnal Pendidikan IPA Indonesia*, 12(3), 482–492. https://doi.org/10.15294/jpii.v12i3.46882

Bassham, G., Irwin, W., Nardone, H., and Wallace, J. M. (2011). *Critical Thinking: A Student's Introduction.* McGraw Hill.

Defianty, M., and Wilson, K. (2024). Using collaborative action research to promote critical thinking pedagogies in ELT in Indonesia. *Asia-Pacific Journal of Teacher Education*, 52(5), 607–621. https://doi.org/10.1080/1359866X.2024.2410837

Diana, N., Sahania, A., Supriadi, N., Irwandani, Rahmayanti, H., Ichsan, I. Z., Alatas, F., and Sison, M. H. (2023). Guided inquiry based on science, environment, technology, and society is needed to improve critical thinking skills in physics learning. *YOUNG SCHOLAR SYMPOSIUM ON SCIENCE AND MATHEMATICS EDUCATION AND ENVIRONMENT.*

Facione, N. (2013). *California Critical Thinking Skills Test: Test Manual (Issue 650)*. California Academic Press.

Fitriani, H., Asy'ari, M., Zubaidah, S., and Mahanal, S. (2019). Exploring the prospective teachers' critical thinking and critical analysis skills. *Jurnal Pendidikan IPA Indonesia*, 8(3), 379–390. https://doi.org/10.15294/jpii.v8i3.19434

Gustianingrum, R. A., Murni, A., and Maimunah. (2023). Analisis Kemampuan Berpikir Kritis Peserta Didik dalam Menunjang Penguatan Profil Pelajar Pancasila. PRISMA, *Prosiding Seminar Nasional Matematika*, 6, 465–471.

Husna, N. (2019). Developing students' critical thinking through an integrated extensive reading program. *Teflin Journal*, 30(2), 212–230. https://doi.org/10.15639/teflinjournal.v30i2/212-230

Hwang, H., An, S., Lee, E., Han, S., and Lee, C. H. (2021). Cross-societal analysis of climate change awareness and its relation to SDG 13: A knowledge synthesis from text mining. *Sustainability*, 13(5596), 1–21. https://doi.org/10.3390/su13105596

Isnaeni, W., Sujatmiko, Y. A., and Pujiasih, P. (2021). Analysis of The Role of Android-Based Learning Media in Learning Critical Thinking Skills and Scientific Attitude. *Jurnal Pendidikan IPA Indonesia*, 10(4), 607–617. https://doi.org/10.15294/jpii.v10i4.27597

Lau, J. Y. F. (2011). An Introduction to Critical Thinking and Creativity: Think More, Think Better. In *An Introduction to Critical Thinking and Creativity: Think More, Think Better*. John Wiley and Sons, Inc. https://doi.org/10.1002/9781118033449

Lin, Y. (2018). Developing Critical Thinking in EFL Classes: An Infusion Approach. In *Developing Critical Thinking in EFL Classes: An Infusion Approach*. Springer Nature Singapore Pte Ltd. https://doi.org/10.1007/978-981-10-7784-5

Nadhiroh, S., and Anshori, I. (2023). Implementasi Kurikulum Merdeka Belajar dalam Pengembangan Kemampuan Berpikir Kritis pada Pembelajaran Pendidikan Agama Islam. *Fitrah: Journal of Islamic Education*, 4(1), 56–68. https://doi.org/10.53802/fitrah.v4i1.292

Narmaditya, B. S., Wulandari, D., and Sakarji, S. R. B. (2017). Does Problem-Based Learning Improve Critical Thinking Skills? *Cakrawala Pendidikan*, 37(3), 378–388.

Nhat, H. T., Lien, N. T., Tinh, N. T., Vu, N., Hang, T., and Trang, N. T. (2018). The Development of Critical Thinking for Students in Vietnamese Schools: From Policies to Practices. *American Journal of Educational Research*, 6(5), 431–435. https://doi.org/10.12691/education-6-5-10

Prasetyo, N. H., and Firmansyah, D. (2022). Analisis Kemampuan Berpikir Kritis Matematis Siswa Kelas VIII dalam Soal High Order Thinking Skill. *Jurnal Educatio*, 8(1), 271–279. https://doi.org/10.31949/educatio.v8i1.1958

Rapi, N. K., Suastra, I. W., Widiarini, P., and Widiana, I. W. (2022). The Influence of Flipped Classroom-Based Project Assessment on Concept Understanding. *Jurnal Pendidikan IPA Indonesia*, 11(3), 351–362. https://doi.org/10.15294/jpii.v11i3.38275

Rosidin, U., Kadaritna, N., and Hasnunidah, N. (2019). Can Argument-Driven Inquiry Models Impact Critical Thinking Skills for Students with Different Personality Types? *Cakrawala Pendidikan*, 38(3), 511–526. https://doi.org/10.21831/cp.v38i3.24725

Rulandari, N. (2021). Study of Sustainable Development Goals (SDGs) Quality Education in Indonesia in the First Three Years. *Budapest International Research and Critics Institute (BIRCI-Journal): Humanities and Social Sciences*, 4(2), 2702–2708. https://doi.org/10.33258/birci.v4i2.1978

Vincent-Lancrin, S., González-Sancho, C., Bouckaert, M., de Luca, F., Fernández-Barrerra, M., Jacotin, G., Urgel, J., and Vidal, Q. (2019). *Fostering Students' Creativity and Critical Thinking.*

Wibowo, A. M., Utaya, S., Wahjoedi, W., Zubaidah, S., Amin, S., and Prasad, R. R. (2024). Critical Thinking and Collaboration Skills on Environmental Awareness in Project-Based Science Learning. *Jurnal Pendidikan IPA Indonesia*, 13(1), 103–115. https://doi.org/10.15294/jpii.v13i1.48561

Towards Resilient Societies: The Synergy of Religion, Education, Health, Science, and Technology – Rahiem (Ed.)

Teacher well-being and professionalism within pesantren

Alfina Pasca Khaira
Universitas Islam Negeri Syarif Hidayatullah Jakarta, Indonesia
ORCID ID: 0009-0004-5450-7875

Muhammad Zuhdi
Universitas Islam Negeri Syarif Hidayatullah Jakarta, Indonesia
ORCID ID: 0000-0002-3015-8322

Sururin Sururin
Universitas Islam Negeri Syarif Hidayatullah Jakarta, Indonesia
ORCID ID: 0000-0003-2030-950X

ABSTRACT: Teacher well-being remains a critical area of focus due to its significant influence on professionalism in the workplace. This study employs a qualitative method with field research, drawing on primary data from five teachers and secondary sources, including books, journals, and school websites. Data were collected through in-depth interviews with the participating teachers. The findings reveal that teacher well-being plays a vital role in enabling teachers to work professionally. From a Western perspective, well-being is typically measured through three aspects: hedonic, eudaimonic, and chaironic. At Dayah Darul Ihsan, teachers' well-being is not solely derived from material or external factors but is strongly influenced by the spiritual (chaironic) dimension. As all participants are Muslim and reside in an Islamic boarding school environment, the hedonic and eudaimonic elements are integrated and overshadowed by spiritual well-being. It is this spiritual strength, derived from chaironic well-being, that empowers teachers at Dayah Darul Ihsan to maintain professionalism, serve as role models, and contribute to producing high-achieving graduates.

Keywords: pesantren, professionalism, teacher well-being

1 INTRODUCTION

The success of educational services can be seen, among other indicators, from the professionalism of teachers in carrying out their duties. One of the government's efforts to enhance teacher professionalism is through certification. Teacher certification policies are supposed to play an important role in improving the quality of education by ensuring professional conduct. In reality, it appears that in Indonesia, teachers with professional certificates have not been able to effect significant changes in their teaching processes and, in many cases, show no difference from their pre-certification conditions (Tambak 2020).

Therefore, evaluating teachers' performance should not be based solely on certification. The subjective well-being of teachers is also crucial and must be considered. According to Zuhdi's research, well-being refers to a state in which an individual possesses balanced physical and mental health, enabling them to perform their life duties effectively. Well-being also allows individuals to realize their potential, practice virtue, and enjoy the various roles they play in life (Zuhdi 2023).

DOI: 10.1201/9781003645542-37

UNESCO states that the world needs to recruit 69 million new teachers to achieve educational goals by 2030. Unfortunately, 80% of current teachers are considering leaving the profession (Mann 2021). Another issue is the limited number of madrasah teachers in pesantren, with uneven and generally low teaching qualifications and competencies. Additionally, there are mismatches between teaching subjects and teachers' educational qualifications (Hayadin 2019).

Bick har-LAM's research shows that in countries like the United States, Australia, Spain, and Asia-Pacific nations, teacher suicide rates are alarming. Increasing numbers of teachers contemplate suicide and view it as a solution to the confusion they face in teaching (har-LAM 2019). In South Korea, a 23-year-old teacher committed suicide due to bullying by parents over their child's misconduct (Mackenzie 2023). In Indonesia, a 55-year-old teacher at a junior high school in Bantul was found dead by hanging at his home, allegedly due to stress from numerous school tasks (Hasanudin 2022). 23% of teachers in America consider leaving their jobs due to difficulties managing student behaviour, lost teaching time to administrative work, low salaries, and long hours, while 77% want to stay because they feel they can still positively impact students (Sy Doan 2023). If teachers feel unhappy and force themselves to keep teaching, it negatively affects students and the teachers' professional conduct (Bosso 2017). This study aims to use not only Western perspectives but also Islamic perspectives based on the Qur'an as indicators of teacher well-being to enhance teacher professionalism. Thus, this research is expected to explore these aspects to fill the knowledge on teacher well-being and professionalism.

This research was conducted in Aceh Besar. In Aceh Besar, parents favour pesantren for their children's education at the junior high and high school levels. Consequently, many public schools face a shortage of students due to the high demand for pesantren education. If the teachers are not qualified and professional, a large number of children in Aceh will not receive a proper education. Therefore, the author has undertaken this study to deeply explore and find solutions to the implications of teacher well-being on teacher professionalism.

2 METHOD

This study is qualitative research conducted as field research at Dayah Darul Ihsan Krueng Kalee in Aceh Besar. The author also uses interview techniques. Primary data are the main sources that provide direct information to the author, involving five teacher informants. The selected teachers have over five years of teaching experience, hold functional positions within the dayah, and are registered in Simpatika (the Management Information System for Educators and Education Personnel of the Ministry of Religious Affairs). Secondary data are obtained from various written sources such as related books and articles. The author chose these data sources to obtain direct information and facts, ensuring strong research analysis results.

In this study, the researcher employs a four-step data analysis process using the Miles and Huberman model. First, the researcher interviews informants by directly listening to their perspectives on the concept of well-being they experience and linking it to their professionalism in performing their professional duties. Second, the data processing stage (data reduction). Third, data presentation allows the researcher to obtain the types and forms of data to be presented in narrative or table form. Fourth, drawing conclusions or verification is carried out to present the research findings.

3 FINDINGS AND DISCUSSION

From the interviews with the teachers at Dayah Darul Ihsan, it was found that they assume several roles as outlined in the Quran. First, the role of *al-murobbi*, which involves developing the potential of students (Qur'an, Al-Isra' verse 24). At Dayah Darul Ihsan, every Thursday and Saturday afternoon, teachers have scheduled sessions to develop students' talents and interests. Additionally, the teachers understand the learning methods and potential of the students and assist them in developing that potential, including opening vocational schools to meet the students' skills needs. Second, the role of *al-mu'allim*, which involves imparting knowledge and skills (Qur'an, Al-Baqarah verse 151). The students not only learn religious and general subjects from their teachers but also develop their skills, such as writing Quranic art, reciting the Quran, reading classical texts, sports, writing, participating in science olympiads, and other activities (Maisarah 2024).

Third, the role of *al-muzakki*, which involves nurturing mental and character development (Qur'an, Al-Baqarah verse 129). Teachers understand the changing times and student character evolution. They do not force students to follow the curriculum of the Islamic senior high school (MA) but also provide platforms to fulfill their technological skills in vocational schools. They create a disciplined environment, foster religious activities, independence, and teach a simple lifestyle. Fourth, the role of *al-ulama*, which involves possessing deep knowledge and piety (Qur'an, Fathir verse 8). Teachers have a routine weekly study schedule to enhance their religious understanding, maintain consistency, and unify the vision and mission in educating students at the dayah. They habituate punctual prayer, memorising the Qur'an, and fasting on Mondays and Thursdays. This not only meets the spiritual needs of a teacher (*ruhul mudarris*) but also serves as a role model for the students (Murtadha 2024).

Fifth, the role of *ahl al-dhikr*, which involves being a source of guidance (Qur'an, An-Nahl verse 43). Teachers motivate and assist students when facing problems, serving as a place for questions and even confiding. Teachers also engage in discussions on various matters. Sixth, the role of *al-muaddib*, which involves nurturing future leaders. Students at Dayah Darul Ihsan are trained and educated to become scholars and leaders for the nation and religion. Thus, teachers must possess intellectual, spiritual, and emotional intelligence. Seventh, the role of *al-mursyid*, which involves instilling upright conduct and honest and commendable character (Qur'an, Al-Baqarah verse 186). Teachers continuously improve their personal competence by strengthening their faith in God, managing their hearts and minds to create a good *ruhul mudaris*, visible in their attitude, speech, and actions, which can then be emulated by the students (Rahmawati 2024).

These seven roles of teachers can be realised if they have good well-being. Spirituality can serve as a source of inspiration that continually helps us live and work with integrity, compassion, connection, enthusiasm, meaning, and vision (Holmes 2005: Nurochim 2022). We will examine teacher well-being at Dayah Darul Ihsan in both objective and subjective aspects to ensure these roles are carried out professionally.

3.1 *Objective well-being*

Tgk. Muakhir Zakaria said, "Teacher well-being at Dayah Darul Ihsan is determined by various factors. Salaries differ slightly among those with S-1, S-2, and S-3 qualifications, paid based on their teaching hours, and those holding functional positions receive additional compensation. Teachers with over five or ten years of teaching experience generally see salary increases (Zakaria 2024). This aligns with Veenhoven's view that highly educated individuals have better opportunities to earn good incomes and tend to secure better jobs (Veenhoven 1984). Surah Al-An'am, verses 76-79, recounts the story of Prophet Ibrahim searching for his God and contains implicit meanings about intelligent and sincere work. There are three characteristics of sincere work: working without being instructed by anyone (eliminating a slave mentality), striving for perfection (regardless of salary, always doing the best possible job), and always innovating.

Ustazah Rahmawati also explained that "Sense of sufficiency makes us feel prosperous" (Rahmawati 2024). As Freedman stated, income only affects the level of extreme poverty, but once basic needs are met, income no longer influences a sense of prosperity or happiness (Freedman 1978), and everyone needs to understand how they should feel content without comparing their situation to others (Diener 1984). Well-being is also influenced by gratitude (Akhmad Baidun 2019).

Tgk. Ataillah explained that "If a teacher joins the management, some are given housing and free meals" (Ataillah 2024). Anne Konu stated that well-being also includes satisfying basic needs (Rimpela 2002), so most teachers want to continue working because they still receive attention from the school (Sy Doan 2023). At Dayah Darul Ihsan, teachers are allowed to continue their studies to the Master's and Doctoral levels while still teaching. Education will expose teachers to a wiser sense of well-being, not just seen from the material benefits of higher education but also from intellectual and emotional intelligence.

3.2 *Subjective well-being*

In the literature, there are three conceptualisations of subjective well-being: hedonic, eudaimonic, and chaironic perspectives. The main difference between eudaimonic and hedonic conceptualisations of well-being is that hedonic is driven by the desire to gain pleasure and avoid pain. The eudaimonistic concept, people can live a good life only when they actualise their potential, not by pursuing pleasures generated by good feelings or bodily satisfaction (Joshanloo 2014). The chaironic perspective, a relatively new addition to the conceptual discussion on well-being, places special emphasis on spiritual aspects (Zuhdi 2023). This perspective shows a sense of comfort in the soul, not feeling lacking, and feeling relatively sufficient.

3.2.1 *Hedonic*

From an Islamic perspective, hedonic well-being is found in Surah Quraysh verse 4, which mentions alleviating hunger. This means that people should work and earn a sufficient livelihood, not excessively, and certainly not by taking others' rights (Athiyyah 1992). Islam teaches that wealth is not the sole indicator of well-being, as it is merely a tool for worshiping Allah SWT (Yusuf 2017).

In Surah An-Nisa' verse 9, this verse encourages people to avoid poverty by working hard as a form of effort and reliance on Allah (Qardhawi 1995). In Ibn Kathir's tafsir, happiness is described as "*falahun*" [QS. An-Nahl: 97] (Katsir 1998). In Al-Qurthubi's tafsir, happiness referred to in QS. Hud verse 105, "Sa'id," is interpreted as a good life (Al-Qurthubi 2007). A good life here is one filled with the blessings of health, well-being, blessings, comfortable housing, and peace of heart and soul (Puspita 2023), so that there is a sense of optimism in life as stated by Tgk Murtadha in Darul Ihsan. Optimism can enhance cognitive, emotional, and motivational aspects. People with high optimism tend to have better moods, more perseverance and success, and good physical and mental health (Csikszentmihalyi 2000; Stuhr 2013; Santos 2022). Latifa also said well-being is often associated with happiness (Latifa 2020) (Latifa 2020).

3.2.2 *Eudaimonic*

The term "Eudaimonic" is based on the meaning and development of virtue (Annas 2009) and refers to optimal psychological functioning and experiences (Deci 2001). The mental well-being of teachers is crucial for the future of education, as attention to this aspect will enable the production of competent teachers (Hercz 2022) and build close relationships with students (Virtanen 2019).

Interviews with teachers at Dayah Darul Ihsan reveal their perception and experience of these six aspects (Hughe 2003) (Hughe 2003) as follows: First, self-acceptance involves surrendering to what Allah has decreed (Supriatna 2021). The Qur'an, in Surah Ghafir, verse 44, surrendering to Allah can broaden one's mind, relieve emotional tension, and soothe the soul when facing any problem (Hamka 2003) (Hamka 2003). Second, positive relationships with others. Third, autonomy. Fourth, environmental mastery. Fifth, purpose in life. Sixth, personal growth. These six aspects significantly impact (Ryff 2003) teacher professionalism. A characteristic of a professional teacher is possessing good intellectual abilities and effectively transferring knowledge (cognitive), as well as understanding the psychology of development and students' learning methods (Anwar 2018).

3.2.3 *Chaironic*

The chaironic aspect posits that God is the supreme force above material matters. Surah Quraish verse 3, represents mental development and shows that if all material well-being aspects are fulfilled, it does not guarantee happiness (Athiyyah 1992). Teachers at Dayah Darul Ihsan believe that spiritual well-being can sustain their *ruhul mudaris*, enabling them to teach with sincerity, patience, and dedication, thereby enhancing their professionalism and contributing to education in Indonesia, particularly within pesantren. As an Arabic proverb says:

الْمَادَّةُ مُهِمَّة وَلَكِن الطَّرِيْقَة أَهَمُّ مِنَ الْمَدَّة الطَّرِيْقَة مُهِمَّة وَلَكِن الْمُدَرِّس أَهَمُّ مِنَ الطَّرِيْقَة وَ رُوحُ الْمُدَرِّس أَهَمُّ مِنَ الْمُدَرِّس

Learning material is important, but the method of learning is far more crucial than the material itself. The learning method is significant, yet the teacher surpasses the method, and the spirit (Ruh) of a teacher surpasses the teacher itself (Saepudin 2022).

4 CONCLUSION

Viewing teacher well-being at Dayah Darul Ihsan using three indicators, namely hedonic, eudaimonic, and chronic. It can be concluded that the well-being of teachers at Dayah Darul Ihsan is fulfilled not solely by material means but by a stronger aspect, the Chaironic (spiritual) aspect. The hedonic and eudaimonic aspects are supported by spiritual well-being (Chaironic). The implications of teacher well-being on their professionalism are reflected in the self-reports of teachers, illustrating how their well-being impacts their professional duties. Teacher well-being serves as a tool, not an end in itself; rather, the ultimate goal is to produce graduates who excel intellectually, emotionally, and spiritually from this pesantren. This chronic strength enables teachers at Dayah Darul Ihsan to perform professionally and fulfill the seven roles of a teacher as outlined in the Qur'an.

REFERENCES

Akhmad Baidun, d. (2019). Effect of Psychological Capital and Gratitude on Subject Well-Being Young Mother of Hijrah Communities in Jakarta. *ICRMH: Proceedings of the 1st International Conference on Religion and Mental Health*, 9. doi:10.4108/eai.18-9-2019.2293469

Al-Qurthubi. (2007). *Tafsir Al-Qurthubi*. Jakarta: Pustaka Azzam.

Annas, C. L. (2009). Feeling good and functioning well: Distinctive concepts in ancient philosophy and contemporary science. *Journal Positive Psychology*, 4, 197–201. doi: 10.1080/17439760902844228

Anwar, M. (2018). *Menjadi Guru Profesional*. Jakarta: Penadamedia.

Ataillah, T. (2024, Februari 7). *Hasil wawancara*. di Dayah Darul Ihsan Aceh Besar.

Athiyyah, M. (1992). *al-Kasyaf al-Iqtishadi li Ayati al-Qur'an. Riyadh: Dar al-Ilmi li Kutub Islamiyah.*

Bosso, D. (2017). Teacher Morale, Motivation and Professional Identity; The National Network of State Teachers of the Year. *Teacher-Researcher Policy Paper Series*, 7.

Chen, J. (2023). The multidimensional teacher well-being: : a mixed-methods approach. *Teachers and Teaching*, 1–21. doi:10.1080/13540602.2023.2282483

Csikszentmihalyi, M. E. (2000). *Positive Psychology: An Introduction. The American Psychological Assosiation Inc*, 55, No. 1, 5–14. doi:DOI: 10.1037/0003-066X.55.1.5

Deci, R. M. (2001). On Happiness and Human Potentials: A Review of Reseach On Hedonic and Eudaimonic Well-Being. *Annual Rev. Psychology(Online)*, 142. Retrieved from http://psych415.class.uic.edu

Diener, E. (1984). *Subjective Well-Being. Bycbological Bulletin: the American Psychological Asociation. Inc.*, 95, 17.

Freedman, J. L. (1978). *Happy people: What happiness is, who has and why*. New York: Harcourt Brace Jovanovich.

Hamka. (2003). *Tafsir Al-Azhar (jilid 8 ed.)*. Singapura: Pustaka Nasional Pte Ltd,.

har-LAM, B. (2019). *Social Support, Well-Being, and Teacher Development*. Singapore: Springer Nature Singapore Pte Ltd.

Hasanudin, U. (2022, Februari 20). *Harian Jogja*. Retrieved from jogjapolitan.harianjogja.com: https://jogjapolitan.harianjogja.com/read/2022/02/20/511/1095657/diduga-stres-karena-banyak-tugas-sekolah-guru-smp-di-bantul-tewas-gantung-diri

Hayadin. (2019). *Balitbangdiklat Kemenag*. Retrieved from https://balitbangdiklat.kemenag.go.id/berita/menyambut-kehadiran-direktorat-guru-madrasah

Hercz, G. T. (2022). The role of positive psychological capital in the prediction of teachers' well-being mediated through motivation. *Athens Journal of Health & Medical Sciences*, 9(4), 245–264.

Holmes, E. (2005). *Teacher Well-Being*. London & New york: Routledgefalmer Taylor dan Francis Group.

Hughe, C. D. (2003). Carol D. RyfStatus inequalities, perceived discrimination, and eudaimonic well-being: Do the challenges of minority life hone purpose and growth. *Journal of Health and Social Behavior*, 44(3), 275–291.

Joshanloo, M. (2014). Eastern conceptualizations of happiness: Fundamental differences with western views. *Journal of Happiness Studies*, 15, 475–493. doi:10.1007/s10902-013-9431-1

Katsir, I. (1998). *Al-Quran al-Azhim*. Al-Dimasyqi: Dar. Al-Kottob al-Ilmiyah.

Latifa, R. (2020). Religious Resilent School: A Mental Health School Study. ICRMH: Proceedings of the 1st International Conference on Religion and Mental Health, 254.

Mackenzie, J. (2023, September 4). bbc.com. Retrieved from bbc.com Indonesia: https://www.bbc.com/indonesia/articles/cd1gy1w4ny1o

Maisarah, U. (2024, Februari 2). Hasil wawancara. di Dayah Darul Ihsan Aceh Besar.

Mann, D. F. (2021). *Practical Applications for Building Teacher Well-Being in Education*. Teacher Wellbeing Pty Ltd, Sydney, NSW Australia, 175–191. doi:10.1007/978-981-15-5963-1_11

Murtadha. (2024, Januari 30). *Hasil Wawancara*. di Dayah Darul Ihsan Aceh Besar.

Nurochim. (2022). Investigating the Relationship Between Spiritual Intelligence, Emotional Regulation and Stress Coping Strategies in the Russian Education Industry. *International Journal of Work Organisation and Emotion*, 13(3), 212–229. Diambil kembali dari https://doi.org/10.1504/IJWOE.2022.126952

Puspita, I. (2023). Tafsir Sa'adah (Kebahagiaan) dalam Surah Hud ayat 105, Jurnal Anwarul; Pendidikan Agama Islam. *Jurnal Anwarul; Pendidikan Agama Islam*, 3, No. 6, 1092–1100. Retrieved from https://doi.org/10.58578/anwarul.v3i6.1599

Qardhawi, Y. (1995). *Al Iman Wa al-Hayah*. Beirut: Muassasah Risalah.

Rahmawati, U. (2024, Januari 29). *Hasil Wawancara*. di Dayah Darul Ihsan Aceh Besar.

Rimpela, A. K. (2002). *Well-Being In Schools*. Health Promotion International, Oxford University Press, 17, 82.

Ryff, Carol D. and Corey L. M. Keyes, Deci L. Hughe. Status Inequalities, Perceived Discrimination, and Eudaimonic Well-being: Do the Challenges of Minority Life Hone Purpose and Growth. *Journal of Health and Social Behavior*. 44(3). 2003, 275–291

Saepudin. (2022). *Pembelajaran Bahasa Arab di Era Postmethode*. Parepare: IAIN Parepare Nusantara Press.

Santos, S. O. (2022). Happiness in Islam The Role of Religion and Spirituality in Muslims Well-Being. *Jurnal from researchgate, Selected Proceedings from the 1st International Conference on Contemporary Islamic Studies (ICIS2021)*, 207–215. doi:10.1007/978-981-19-2390-6_19

Stuhr, P. T. (2013). The Importance of Emotions in Teaching: Developing Positive Emotions in Physical Education. *International Journal of Human Movement Science*, 7, No. 1, 5–29.

Supriatna, E. (2021). Aspects of the Psychological Well-being of Muslim Students in Bandung from the Perspective of Islamic Psychology during the Covid-19 Pandemic. Ilmu Dakwah; *Academic Journal for Homiletic Studies*, 15, No. 1, 149–168.

Sy Doan, E. D. (2023). Sy Doan, Elizabeth D. S. Teacher Well-Being in Intentions to Leave: Findings From The 2023 State of the American Teacher Survey. Jurnal Rand: American Educator Panel, 9–19.

Tambak, S. (2020). *Model Pengembangan Profesionalisme Guru Madrasah.* . Yogyakarta: Graha Ilmu.

Veenhoven, R. (1984). *Indicators of Happiness*. The Netherlands: Erasmus University Rotterdam. doi: 10.1007/978-94-009-6432-7_4

Virtanen, T. E. (2019). Associations between observed patterns of classroom interactions and teacher well-being in lower secondary school. *Teaching and Teacher Education*, 240–252.

Yusuf, M. (2017). *Kesejahteraan Perspektif Islam*. Mataram: Kanhaya Karya.

Zakaria, T. M. (2024, Januari 29). *Hasil Wawancara*. di Dayah Darul Ihsan Aceh Besar.

Zuhdi, M., and Kustiwan Syarief. (2023). *Constructing the Concept of Student Well-Being within Indonesian Islamic Higher Education. Religions* 14, no. 9: 1140. https://doi.org/10.3390/rel14091140

Towards Resilient Societies: The Synergy of Religion, Education, Health, Science, and Technology – Rahiem (Ed.)

Parental expectations, self-concept, and student adjustment in university

Chadijah Saraswati Sumanang, Achmad Syahid & Gazi Saloom
Universitas Islam Negeri Syarif Hidayatullah Jakarta, Indonesia

Rofiatul Hosna
Universitas Hasyim Asy'ari Tebuireng Jombang, Indonesia

Kristel P. Diansay
University of Mindanao, Matina Campus, Matina, Davao City, Davao del Sur, The Philippines

ABSTRACT: This study investigates the impact of parental expectations and self-concept on first-year university students' adjustment in Indonesia and the Philippines, with achievement goal orientation as a mediating factor. Using path analysis on data from 353 students, the findings reveal that self-concept and achievement goal orientation significantly and directly influence adjustment, while parental expectations exert an indirect effect through achievement goal orientation. The study highlights that a strong self-concept and goal-oriented behaviours are pivotal for effective adaptation to university life. Although parental expectations do not directly impact adjustment, they positively influence achievement goal orientation, subsequently enhancing adjustment. The results underscore the importance of fostering self-concept and achievement-oriented mindsets to support students' academic and social transitions. Practical recommendations include integrating resilience and self-concept development into orientation programmes, providing parental education on realistic and supportive expectations, and strengthening university support systems. These insights align with Sustainable Development Goal 4 by promoting equitable education and supporting students' academic success in diverse cultural contexts.

Keywords: university student adjustment, parental expectation, self-concept, achievement goal orientation

1 INTRODUCTION

Studies on first-year university student adjustment (Rahayu and Arianti 2020; Rodríguez *et al.* 2017; van Rooij *et al.* 2017) highlight the challenges of adapting to university life, particularly in ASEAN countries (Rahiem 2020; Tuga *et al.* 2022). Poor adjustment can result in academic dishonesty, social issues, and reduced success (Clinciu *et al.* 2021). Adjustment involves aligning student needs with the educational environment (Schonfeld *et al.* 2020) and supports SDG 4 for equitable education.

Parental expectations and self-concept play key roles—positive parental support enhances academic success (Prasetyo 2022) as long as it doesn't get stuck on typical poisonous parenting style (Syahid *et al.* 2023), and a strong self-concept fosters goal orientation (Goñi *et al.* 2011). Achievement goal orientation further supports adaptation (Elliot *et al.* 2015). This study examines how achievement goal orientation mediates the effects of parental expectations and self-concept on student adjustment in Indonesia and the Philippines.

DOI: 10.1201/9781003645542-38

2 METHODS

This study employed a quantitative approach with 353 first-year university students: 195 from UIN Syarif Hidayatullah Jakarta, Indonesia, and 158 from Campus X in Davao City, Philippines. Convenience sampling was used to streamline data collection during the pandemic-to-normal transition, with participants from diverse faculties to minimise bias. Criteria included first-year students in semester 1 or 2 at the specified universities.

Data collection occurred offline and online (via Google Forms) from May to June 2023. Instruments included: College Adjustment Questionnaire (O'Donnell *et al.* 2018), based on Baker and Siryk's SAQC (1989), with 12 items across three dimensions.; Perception of Parental Expectation Inventory (PPEI) (Sasikala and Karunanidhi 2011), 30 items.; Personal Self-concept Scale (Goñi *et al.* 2011), 22 items across four dimensions; Achievement Goal Orientation-Revised (AGO-R) (Elliot *et al.* 2011), 12 items on performance and mastery goals. Path analysis using MPLUS 8.0 was conducted to assess direct and indirect effects, enabling a comprehensive evaluation of factors influencing student adjustment.

3 FINDING AND DISCUSSION

The respondents who participated in this study were 353 people. The following is a general description of the characteristics of the subjects explained in Table 1 below:

Table 1. Research subject overview.

	Amount	Percentage	Total
Sex			
Male	109	30.9%	353 (100%)
Female	244	69.2%	
University			
UIN Syarif Hidayatullah Jakarta	195	55.2%	353 (100%)
U Mindanao	158	44.8%	
Domicile status			
Migrating	114	32.3%	353 (100%)
Not migrating	239	67.7%	

At this stage, we conducted a descriptive analysis of all true score data from a sample of 353 first-year students. This descriptive analysis provides an overview of the data of a variable used by the researcher to facilitate understanding. Based on data characteristics such as mean, standard deviation, maximum and minimum values, as seen in Table 2 below:

Table 2. Descriptive analysis.

	N	Minimum	Maximum	Mean	Std. Deviation
Adjusment	353	18.96	78.29	50.00	9.03
Parental expectatiton	353	17.13	71.16	50.00	9.49
Self-concept	353	8.70	71.22	49.83	9.12
Achievement Goal orientation	353	24	74.00	50.00	8.58
Valid N	353	353	353	353	353

Table 2 provides statistical descriptions of the variables based on 353 respondents. For student adjustment, the scores range from 18.96 to 78.29, with a mean of 50.0003 and a standard deviation of 9.03463. Parental expectations have scores ranging from 17.13 to 71.16, with a mean of 50.0002 and a standard deviation of 9.49141. The self-concept variable shows scores between 8.70 and 71.22, a mean of 49.83, and a standard deviation of 9.115. As the mediator variable, achievement goal orientation ranges from 24 to 74, with a mean of 50.00 and a standard deviation of 8.576. In this case, the norms are established in Table 3 below.

Table 3. Categorisation of variable scores.

	Frequency %		
Variable	Low	Average	High
Self-adjustment	46 (13%)	270 (76.5%)	37 (10.5%)
Parental expectation	49 (13.9%)	247 (70.0%)	57 (16.1%)
Self-Concept	28 (7.9%)	283 (80.2%)	42 (11.9%)
Achievement goal orientation	37 (10.5%)	270 (76.5%)	46 (13.0%)

Note:
RMSEA: Root Mean Square Error of Approximation
CI: Confident interval
TLI: Tucker Lewis Index
CFI: comparative fit index

Table 3 shows most students exhibit medium levels of adjustment (76.5%), parental expectations (70%), self-concept (80%), and goal orientation (76.5%). Adjustment trends medium to low, while self-concept and goal orientation lean medium to high.

Statistical analysis indicates parental expectations influence adjustment indirectly through goal orientation, while self-concept directly impacts adjustment, with reduced effects via goal orientation. Goal orientation is a crucial mediator between parental expectations and adjustment (Figure 1).

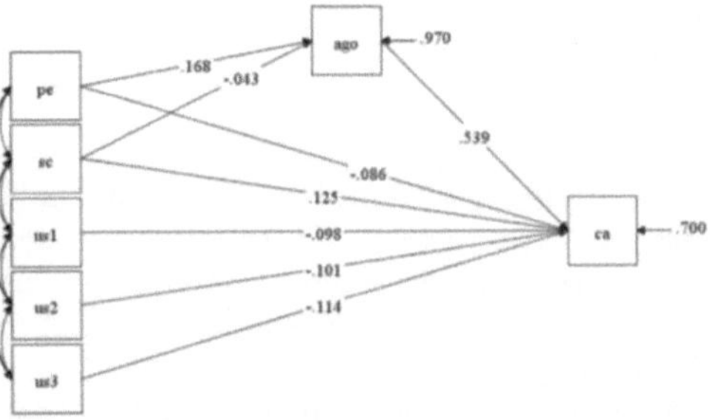

Figure 1. Path analisis diagram.

Based on the test result image and from Figure 1 above, it can be seen that the chi square value shows a value of 3.228, a value that is greater than before being modified, namely 0.000, the df value which was originally 0 became 3. So it can be interpreted that the model is fit, supported by the CFI value = 0.998, TLI = 0.993, and RMSEA = 0.015. The values of the model that has been fitted can be seen in Table 4 below.

Table 4. Model fit hypothesis test scheme.

Index	Adjustment Index	Model Index	Model Evaluation
RMSEA	<0.05	0.015	Fit to criteria
90 Percent C.I	<0.05	0.000 – 0.092	Fit to criteria
CFI	>0.9	0.998	Fit to criteria

(continued)

Table 4. Continued

Index	Adjustment Index	Model Index	Model Evaluation
TLI	>0.9	0.993	Fit to criteria
SRMR	<0.08	0.011	Fit to criteria

Note:
RMSEA: Root Mean Square Error of Approximation
CI: Confident interval
TLI: Tucker Lewis Index
CFI: comparative fit index

Table 5. Direct effect of independent variables on dependent variables.

Impact	Koefisien	S.E	T-Value	P-Value
		Final Model		
PE → CA	−0.086	0.045	−1.901	0.056
SC → CA	0.125	0.044	2.813	0.005
AGO→CA	0.539	0.039	13.956	0.000
US1→CA	−0.098	0.065	−1.505	0.132
US2→CA	−0.101	0.059	−1.723	0.085
US3→CA	−0.114	0.060	−1.901	0.057
PE→AGO	0.168	0.052	3.251	0.001
SC→AGO	−0.043	0.052	−0.815	0.415

Note:
CA: college/university adjustment
PE: parental expectation
SC: self-concept
Ago: achievement goal orientation
Us: Age

Table 5 shows that self-concept directly influences student adjustment (t-value > 1.96), and goal orientation also directly impacts adjustment, indicating significant effects. However, parental expectations and age do not significantly affect student adjustment (t-value < 1.96). Age further shows a negative t-value, meaning the average adjustment is higher for the group coded as 0.

The findings from the analysis offer a comprehensive understanding of both the direct and indirect factors influencing student adjustment at university. Directly, self-concept and achievement goal orientation (AGO) play significant roles, while parental expectations and age do not directly impact adjustment. However, the picture becomes more nuanced when considering indirect effects, as mediated by AGO.

3.1 *Direct effects*

Self-concept demonstrates a significant positive influence on student adjustment at university, supported by a coefficient of 0.125, a T-value of 2.813 (greater than 1.96), and a p-value of 0.0005 (less than 0.05). This indicates that students with a strong self and confidence adapt more effectively to university life. Similarly, achievement goal orientation (AGO) shows a robust positive effect, with a coefficient of 0.539, a T-value of 13.956, and a p-value of 0.000. This underscores the importance of goal-oriented behaviours in facilitating student adjustment, where a higher focus on achieving academic goals correlates with better adaptation.

On the other hand, parental expectations (PE) and age were not found to have direct significant effects on adjustment. The T-value for PE was -1.901 (below 1.96), and the p-value was 0.057 (above 0.05), indicating no direct influence. Similarly, age showed no significant effect, with negative T-values ranging from -1.505 to -1.901 and p-values between 0.057 and 0.132.

3.2 *Indirect effects*

The analysis revealed significant indirect effects for parental expectations on student adjustment, mediated by AGO. A positive relationship with a coefficient of 0.091, a T-value of 3.114 (greater than 1.96), and a p-value of 0.002 (less than 0.05) shows that parental expectations significantly enhance AGO, which in turn improves adjustment Table 6). This suggests that while parental expectations might not directly affect how students adapt, they can positively influence adjustment by fostering goal-oriented behaviours.

Table 6. Indirect effect of independent variable on dependent variable.

Impact	Coeficient	S.E	T-value	P-value	Sig.
Effect from PE to CA					
Sum of indirect	0.091	0.029	3.114	0.002	
Specific indifrect					
PE→AGO→CA	0.091	0.029	3.114	0.002	√
Effect from SC to CA					
Sum of indirect	−0.023	0.028	−0.811	0.418	
Specific indirect	−0.023	0.028	−0.811	0.418	X
SC→AGO→CA					

Note:
CA: college/university adjustment
PE: Parental Expectation
SC: self-concept
Ago: achievement goal orientation
US: age

In contrast, self-concept does not show a significant indirect effect on adjustment through AGO. The relationship was negative, with a coefficient of -0.023, a T-value of -0.811 (below 1.96), and a p-value of 0.418 (above 0.05). This indicates that AGO does not mediate the influence of self-concept on adjustment, suggesting that the effect of self-concept is direct rather than mediated. These findings contribute to a broader understanding of student adjustment in cross-cultural contexts. They emphasise the importance of equipping educators and university systems with strategies to foster achievement goal orientation and self-concept while balancing parental influence to support the diverse needs of students in Indonesia and the Philippines. These efforts align with SDG 4 by supporting students' overall development and ensuring that all individuals have the opportunity to succeed academically. Future research could further explore these relationships in non-pandemic contexts and among students at different academic levels to deepen our understanding of their implications.

4 CONCLUSSION

The study reveals that parental expectations and self-concept significantly influence university adjustment in first-year students, mediated by achievement goal orientation. Achievement goal orientation and self-concept directly impact adjustment, while parental expectations influence adjustment indirectly via achievement goal orientation. These findings highlight the importance of fostering achievement-oriented mindsets and self-concepts to support student adjustment effectively.

Recommendations

(1) For Academic Advisors: Promote self-concept development through mentoring and personal growth programs. It also offers workshops on achievement goal orientation, emphasising realistic and mastery-oriented goal setting.
(2) For Universities: Strengthen student support systems, including counselling services that build resilience and positive self-concept. Integrate goal-setting and self-regulation skills into orientation programs for first-year students.
(3) For Policymakers: Advocate for parental education initiatives to encourage realistic and supportive expectations. Fund socio-emotional learning programs in university curricula to reinforce adaptive behaviours and goal orientations.

By implementing these strategies, stakeholders can create a nurturing environment that supports students' successful transition and adjustment to university life.

REFERENCES

Baker, R. W., and Siryk, B. (1989). *SACQ student adaptation to college questionnaire manual.* Western Psychological Services

Clinciu, A. I., Ives, B., and Cazan, A. M. (2021). Academic Dishonesty and Academic Adjustment Among the Students at University Level: An Exploratory Study, *SAGE Open.* 11(2), DOI: 10.1177/21582440211021839

Elliot, A., Murayama, K., Kobeisy, A., and Lichtenfeld, S. (2015). Potential-based achievement goals. *British Journal of Educational Psychology*, 85 (2), 192–206. doi: 10.1111/bjep.12051

Elliot, A. J., Murayama, K., and Pekrun, R. (2011). A 3 x 2 achievement goal model. *Journal of Educational Psychology, 103*(3), 632–648. doi: 10.1037/a0023952.

Goñi, E., Madariaga, J. M., Axpe, I., and Goñi, A. (2011). Structure of the Personal Self-Concept (PSC) Questionnaire. *International Journal of Clinical and Health Psychology, 11*(3), 509–522. https://doi.org/10.1016/S1697-2600(11)70064-6

O'Donnell, M. B., Shirley, L. A., Park, S. S., Nolen, J. P., Gibbons, A. M., and Rosén, L. A. (2018). The college adjustment questionnaire: A measure of students' educational, relational, and psychological adjustment to the college environment. *Journal of College Student Development. 59*(1), 116–121. doi: 10.1353/csd.2018.0009.

Prasetyo, A. (2022). The role of family in child development and education. *Journal of Child Development and Education, 34*(1), 45–60. https://doi.org/10.1016/j.jcde.2022.100123

Rahayu, M. N. M., and Arianti, R. (2020). Penyesuaian Mahasiswa tahun Pertama di Perguruan Tinggi: Studi pada Mahasiswa Fakultas Psikologi UKSW. *Jurnal Psikologi Sains dan Profesi.* Vol. 4, No. 2. 73–84. DOI:10.24198/jpsp.v4i2.26681

Rahiem, M. D. H. (2020). The Emergency Remote Learning Experience of University Students in Indonesia amidst the COVID-19 Crisis. *International Journal of Learning, Teaching and Educational Research.* Vol. 19, No 6, 1–26. https://doi.org/10.26803/ijlter.19.6.1

Rodríguez, M. S., Tinajero, C., and Páramo, M. F. (2017). Pre-entry characteristics perceived social support, adjustment and academic achievement in first-year Spanish university students: A path model. *The Journal of Psychology, 151*(8), 722–738. DOI: 10.1080/00223980.2017.1372351.

Sasikala, S., and Karunanidhi, S. (2011). Development and validation of parentalexpectations inventory. *Journal of the Indian Academy of Applied Psychology, 37*(1). 114–124. https://jiaap.in/wp-content/uploads/2024/07/12-Sasikala.pdf

Schonfeld, D. J., Demaria, T., and Kumar, S. A. (2020). Supporting Young Children after Crisis Events. *YC Young Children, 75*(3), 6–15. https://www.jstor.org/stable/26979160

Syahid, A., Irvan, M., Randraputri, R. R. R. Y., and Rahmawati, S. (2023). The Indonesian Version of The Poisonous Parenting Style Scale: Development of Measuring Tools and Adaptation. *Jurnal Ilmu Sosial Mamangan.* Vol 12 No. 02. 776–793. DOI: https://doi.org/10.22202/mamangan.v12i02.6997.g3335

Tuga, B., Jocson, J., and Mabunga, R. A. (2022). The impact of COVID-19 on a Philippine university: Challenges and responses towards a new normal in education. *AsTEN Journal of Teacher Education.* Retrieved from https://po.pnuresearchportal.org/ejournal/index.php/asten/article/view/1777

van Rooij, E. C. M., Jansen, E. P. W. A., and van de Grift, W. J. C. M. (2017). First-year university students' academic success: The importance of academic adjustment. *European Journal of Psychology of Education. 33*(4), 749–767. https://doi.org/10.1007/s10212-017-0347-8

Towards Resilient Societies: The Synergy of Religion, Education, Health, Science, and Technology – Rahiem (Ed.)

Effectiveness of rice subsidy programs on child nutrition and education in Indonesia

Felicia Putri Anggrasari & Muhammad Reza Pahlevi
Kementerian PPN/Bappenas

Fikriyah
Dinas Perumahan Rakyat, Kawasan Permukiman dan Cipta Karya Provinsi Jawa Timur

ABSTRACT: Approximately 8–9% of school-age children in Indonesia suffer from undernutrition. To address this issue, the government introduced the Rice for the Poor (Raskin) program to assist low-income households in meeting their daily nutritional needs. However, research evaluating the effectiveness of social assistance programs like Raskin remains limited and often inconclusive. This study aims to assess the impact of Raskin on the nutrition and education of school-aged children, focusing on disparities between urban and rural areas. Using the propensity score matching method, the study reveals that households benefiting from the Raskin program in 2007 did not show significant improvements in their children's nutritional status. However, the program positively influenced children's years of schooling, particularly at the primary school level. These findings highlight the importance of integrating food assistance initiatives with a stronger emphasis on meeting the recommended daily intake (RDA) of essential nutrients. The government should prioritise.

Keywords: nutritional status, propensity score matching, Raskin (Rice for the Poor), school-aged children, school lunch program

1 INTRODUCTION

Global hunger levels are still much higher than before the pandemic. In 2022, it is estimated that between 690 and 783 million people worldwide will face hunger, an increase of 122 million people relative to the pre-COVID-19 period. However, the rise in hunger in the past two years seems to have stopped, with 3.8 million less estimated people having hunger in 2022 compared to 2021 (FAO 2023). Due to hunger, many people around the world are malnourished. Childhood nutrition is crucial for growth and development, affects physical and mental health, therefore future human resources and labour wages through cognition, knowledge and skills (Bütikofer *et al.* 2018). In addition, school-age children's nutrition and children's years of schooling is critical to breaking the intergenerational cycle of poverty. Therefore, a relevant question is how much benefit children receive from welfare programs aimed at them.

United States has a national school breakfast and lunch program to cut family food spending to meet children's nutrition (including vitamins and minerals) and health needs (Ettinger de Cuba *et al.* 2023; Zimmerman *et al.* 2024). While China has a free school meal program for rural primary and secondary schools, it includes nutritious breakfasts and lunches to boost children's nutritional status. Implementation of these programs has an impact on increasing the educational achievement and cognitive and noncognitive abilities of the children (Zhou *et al.* 2024). Through providing school children with lunches equivalent

DOI: 10.1201/9781003645542-39

to 450 kcal per serving, the Indian Government is trying to boost the enthusiasm of children to be in school and decrease the dropout rate (Kaur 2021).

Indonesia prioritises tackling hunger and malnutrition to create a skilled labour force that will contribute to the Indonesia Emas 2045 initiative. Limited access to proper nutrition, diverse foods, and public awareness leads to various nutritional issues in Indonesia. So, nutritious food is essential to promoting balanced nutrition and achieving the SDG goal 2 of eradicating hunger. Social assistance programs to fight hunger and malnutrition are direct cash transfers (BLT) and food assistance programs. Food aid, particularly targeted at impoverished individuals, enables them to consume staple foods at a frequency comparable to those who are not impoverished (Cunningham *et al.* 2021).

Improving the nutritional status of children in other countries requires providing essential foods such as vitamins and balanced nutrition. Raskin, a government program providing subsidised rice to economically disadvantaged families in Indonesia, aimed to assist poor and vulnerable households in meeting their daily food needs and reducing their spending on staple foods. The program operated from 2007 to 2014, and its long-term effects are worth evaluating. This study aims to answer the following research questions: (1) does the Raskin food aid program impact children's nutritional status and education? and (2) is there any difference between this condition in rural and urban areas?

By identifying existing problems and synthesising similar cases, the researcher attempts to analyse the impact of Raskin on nutritional status to provide insights into what efforts can be implemented in Indonesia to improve the nutrition of children, particularly school-aged children. The study is significant as it will guide policymakers to develop programs that aim to end hunger and boost children's nutrition. The government and academics will be able to measure the success of targeted policy practices based on the results of this research.

2 METHODS

This study utilised IFLS data from 2007 to 2014. It included as many elementary school kids (seven to twelve years) and parents as 1,317. Bali, Bangka Belitung, Banten, Central Java, Yogyakarta, East Java, West Java, Jakarta, Lampung, South Kalimantan, Central Kalimantan, North Sumatera, South Sumatera, West Sumatera, Riau, South Sulawesi, and West Nusa Tenggara became study locations.

Propensity Score Matching was used as the analysis technique in this study. This analysis determines the likelihood of an individual being assigned to a specific treatment group based on observed baseline characteristics (Lalani *et al.* 2020). Furthermore, using propensity scores is the most straightforward and rigorous method for reducing bias. Notably, in observational studies, the propensity score approach produces more statistically valid results than traditional multivariable regression.

It should be remembered that the propensity score matching method is very good if the basic data is distributed evenly across all groups in the study (Elze *et al.* 2017). For each unit that is the object of observation, the trend can be seen through the resulting score. This metric facilitates substitution, thereby reducing the number of non-participating individuals who are utilised as counterfactuals (Vairetti *et al.* 2024). The model used in this study can be represented by the equation below.

$$\mathbf{Y}_{\mathbf{it}} = \alpha + \boldsymbol{\beta}\ \mathbf{X}_{\mathbf{ht-1}} + \gamma\ \mathbf{Z'}_{\mathbf{it}} + \mathbf{e}_{\mathbf{it}} \tag{1}$$

Explanation:

- i: Child recipient
- h: Household recipient
- X: Raskin participation of parents (dummy variable) (1 if household receives Raskin and 0 if the household does not receive social assistance)

- Y: Child's malnutrition status
- Z: Control variables

The control variables in this model are chosen based on socio-economic factors associated with a child's nutritional condition. These variables encompass the socio-economic attributes of parents, such as their education level, occupations, and ages (Farooq *et al.* 2019), as well as regional factors like access to household drinking water, sanitation facilities (Dodos *et al.* 2017). These variables serve as benchmarks for assessing the likelihood of receiving specific treatments or interventions. Their selection is informed by prior research that examines their connections with both dependent and independent variables.

Table 1. Data validation and reliability.

Variable	Obs	Mean	Min	Max
Parents' education	1.317	7.6583140	0	19
Parent's occupation	1.317	0.8929385	0	1
Age of parents	1.317	44.1017500	23	98
Household drinking water	1.317	3.1936220	1	10
Household sanitation	1.317	2.1928630	1	11

Table 1 shows the results of the validation and reliability of the data used in this study, the data used in this study can be considered valid and reliable.

3 FINDINGS & DISCUSSION

3.1 *Impact of Raskin on nutritional status and educational children*

Raskin is a social protection that aimed at tackling poverty and improving food security. The program aims to provide low-income and vulnerable people with affordable food. Equal access to food is crucial, particularly considering the persistently poor nutritional and educational conditions among children in Indonesia.

Table 2. Impact of Raskin on child nutrition.

	Propensity score matching	
	Child malnutrition	(log) years of schooling child
Raskin for parents (1 = beneficiary families; 0 = non-beneficiary families)	0.060^{*} (0.024)	0.040^{***} (0.018)
Observation	1.317	1.317

Table 2 shows that beneficiary children who receive Raskin still have a probability of experiencing poor nutritional status of 6.0 percentage points. Providing poor and vulnerable children with food aid in Raskin fails to fully address child malnutrition. Various factors contribute to malnutrition, including low income, household size, and household environment characteristics, these factors are considered risk factors for food insecurity (Brucker *et al.* 2015). Nevertheless, Raskin had a positive impact in increasing children's years of education by 4.0 percent. Raskin can lighten the burden of household expenses so that the budget allocation can be allocated to children's educational needs.

Assessing Raskin's impact on a subset of rural and urban areas is critical for understanding the program's effectiveness in various contexts. Evaluating impact in rural and urban areas makes identifying community's various needs and challenges possible. By implementing policies that are both focused and efficient, it is possible to ensure that the assistance provided is truly beneficial to the individuals who are experiencing poverty in Indonesia.

Table 3. Impact of Raskin on child nutrition: rural and urban differences.

	Urban	Rural
Raskin for parents (1 = beneficiary families; 0 = non-beneficiary families)	-0.040^{**}	0.030
	(0.022)	(0.020)
N	1,493	1,594

Table 3 shows that Raskin has significant impact in urban areas, with an estimated result of 4.0 percentage points. Although with a low probability, it is indicated that beneficiary children in urban areas may experience malnutrition. In contrast to urban areas, in rural areas the nutritional status of beneficiary children is not influenced by Raskin.

3.2 *Food subsidy does not totally affect nutritional status of beneficiary children*

Raskin, a rice assistance program aimed primarily at people experiencing poverty, continues to face several obstacles. Poor rice quality is one of the major issues impeding Raskin's efforts to improve children's nutritional status. According to the SMERU Research Institute (2008), of 220 villages that were sampled, 62.3 percent of areas received low-quality Raskin rice and only 37.7 percent received good quality. In some areas, the rice received was unfit for consumption due to odour, colour, and fleas.

The sole aim of Raskin is to provide rice to the poor and the underprivileged so that they can meet their daily needs. Still, Raskin does not provide other vital nutrition to children, such as protein, vitamins, and minerals, which are needed by their bodies. Malnutrition can impact a child's health and growth if he or she is dependent on rice alone without proper nutrition. It is also reported in research that carbohydrate dependency can lead to malnutrition and obesity in school-aged children. Raskin's focus is also solely on the quantity of assistance provided, with impoverished and vulnerable communities receiving approximately 10-20 kg of rice per month. According to research by Anderson *et al.* (2018), improving child nutrition influences food quality rather than quantity.

The imbalance in the quality and nutrition of Raskin has encouraged the government to increase food assistance programs, Sembako. Sembako program provides non-cash food assistance to the poor and vulnerable while focusing on food diversification. Food diversification is anticipated to offer recipients a more well-rounded nutritional consumption. Various foods from different sources (animal and plant) have a variety of micronutrients that are difficult fullfiled from Raskin that only subsidy for rice (Khamis *et al.* 2019). For instance, adding protein-rich foods like eggs can significantly enhance protein intake, which is essential for cognitive and physical development in children. Nevertheless, the Sembako program does not specifically focus on children of school age. Ensuring that school children have a well-balanced diet and meet their nutritional needs is crucial for enhancing academic performance (Lundborg *et al.* 2022).

Even though Raskin has not produced a substantial improvement in nutritional health of children, this program has potential to extend years of schooling children. By decreasing spending on food, households can reallocate their income to school fees and other essential educational expenses. Therefore, beneficiary households can allocate their income to fulfill other children's needs, such as improving education so that children will have a better future. Nevertheless, the estimated impact of Raskin on enhancing education remains relatively low.

Hence, it is imperative to implement supplementary measures and foster collaboration with other initiatives to promote children's educational achievement.

Suppose it is related to impacts based on regional aspects. In that case, urban beneficiaries tend to rely solely on the food provided by the aid program, as they do not have food reserves such as those from gardens or agricultural produce (Zhong *et al.* 2023). The low estimation result shows that Raskin has not fully improved children's nutritional status. As food barns, the rural areas are self-sufficient in providing food, particularly staple foods, by growing their rice so that the price of local rice is very accessible (Rozaki 2021). As a result, Raskin is less relevant in rural areas because people already have adequate access to rice. In contrast, urban residents depend on the market for their food supplies, including rice, which is primarily available through urban markets. Therefore, Raskin can help ease the financial burden of purchasing rice for households in urban areas.

3.3 *Policy recommendation: rules of balanced nutrition*

In addition to providing food assistance to vulnerable and impoverished individuals, the government should also prioritise the quality of food distribution. Ensuring the provision of high-quality, nourishing food that contains a well-balanced combination of essential nutrients will enhance overall food consumption. Beneficiary children will receive sufficient calories and nutrition. This can be implemented using Isi Piringku concepts that consist of 50% fruit and vegetables, and the remaining 50% consisting of carbohydrates and protein. This concept is in line with the oldest guidelines in Minister of Health Regulation Number 41 of 2014 which states that food assistance must include carbohydrates, protein, vegetables, and fruit.

The program should also enhance the scheme for supplying nutritious and wholesome meals to technical personnel who are responsible for carrying out the activities. School nutritionists and parents should be fully educated on the importance of a balanced daily diet for children. Providing children with nutritious food can improve their physical health and facilitate their academic performance. Higher levels of education are predicted to promote better child well-being, thus preventing the ongoing cycle of poverty across generations.

4 CONCLUSION

According to the findings, Raskin has been ineffective in improving the nutritional status of students. The main reason is a lack of food diversification, as Raskin only provides rice, whereas children require a wide range of nutrients that rice cannot provide. Calorie intake and important nutrients cannot be met by low quality rice so government should have social assistance that include food diversification to improve child nutrition. One of the best ways that the government can take is providing aid packages, including foodstuffs with complete and balanced nutrition for children (nuts, eggs, milk, vegetables and fruit). Besides that, it is important to remember that food assistance for school-age children in collaboration with all levels of education is also needed.

Given the limitations of existing research, it is necessary to carry out a detailed interpretation regarding the public's perception of the Raskin program. For future research, a more complex analysis could be the answer in capturing the long-term impact of the food subsidy program, especially on school-aged children. Additionally, examining the link between childhood nutrition and adult employment, income levels, or productivity could provide valuable insights. These findings could contribute to designing more effective food programs that not only alleviate immediate food insecurity but also enhance long-term human capital and economic development. More detailed, accurate, and comprehensive data is also needed to predict the effectiveness of the free nutritious meal program (Makan Bergizi Gratis) as a flagship program for the President and Vice President in improving the nutrition of school-age children.

REFERENCES

Anderson, M. L., Gallagher, J., and Ramirez Ritchie, E. (2018). School meal quality and academic performance. *Journal of Public Economics*, *168*, 81–93. https://doi.org/10.1016/j.jpubeco.2018.09.013

Brucker, D. L., Mitra, S., Chaitoo, N., and Mauro, J. (2015). More Likely to Be Poor Whatever the Measure: Working-Age Persons with Disabilities in the United States. *Social Science Quarterly*, *96*(1), 273–296. https://doi.org/https://doi.org/10.1111/ssqu.12098

Bütikofer, A., Mølland, E., and Salvanes, K. G. (2018). Childhood nutrition and labor market outcomes: Evidence from a school breakfast program. *Journal of Public Economics*, *168*, 62–80. https://doi.org/https://doi.org/10.1016/j.jpubeco.2018.08.008

Cunningham, S. A., Shaikh, N. I., Datar, A., Chernishkin, A. E., and Patil, S. S. (2021). Food subsidies, nutrition transition, and dietary patterns in a remote Indian district. *Global Food Security*, *29*. https://doi.org/10.1016/j.gfs.2021.100506

Dodos, J., Mattern, B., Lapegue, J., Altmann, M., and Aissa, M. A. (2017). Relationship between water, sanitation, hygiene, and nutrition: What do Link NCA nutrition causal analyses say? *Waterlines*, *36*(4), 284–304. https://doi.org/10.3362/1756-3488.17-00005

Elze, M. C., Gregson, J., Baber, U., Williamson, E., Sartori, S., Mehran, R., Nichols, M., Stone, G. W., and Pocock, S. J. (2017). Comparison of Propensity Score Methods and Covariate Adjustment: Evaluation in 4 Cardiovascular Studies. *Journal of the American College of Cardiology*, *69*(3), 345–357. https://doi.org/https://doi.org/10.1016/j.jacc.2016.10.060

Ettinger de Cuba, S., Bovell-Ammon, A., Ahmad, N., Bruce, C., Poblacion, A., Rateau, L. J., Coleman, S. M., Black, M. M., Frank, D. A., Lê-Scherban, F., Henchy, G., Ochoa, E., Sandel, M., and Cutts, D. B. (2023). Child Care Feeding Programs Associated With Food Security and Health for Young Children From Families With Low Incomes. *Journal of the Academy of Nutrition and Dietetics*, *123*(10), 1429–1439. https://doi.org/10.1016/j.jand.2023.06.003

FAO. (2023). The State of Food Security and Nutrition in the World 2023. In *The State of Food Security and Nutrition in the World 2023*. FAO; IFAD; UNICEF; WFP; WHO; https://doi.org/10.4060/cc3017en

Farooq, M. U., Majeed Nadeem, A., and Ali, F. (2019). The Impact of Parents' Education on Child Health from the Perspective of Demographic and Health Survey. *European Online Journal of Natural and Social Sciences*, *8*(1), 158–174. http://www.european-science.com

Kaur, R. (2021). Estimating the impact of school feeding programs: Evidence from mid day meal scheme of India. *Economics of Education Review*, *84*. https://doi.org/10.1016/j.econedurev.2021.102171

Khamis, A. G., Mwanri, A. W., Ntwenya, J. E., and Kreppel, K. (2019). The influence of dietary diversity on the nutritional status of children between 6 and 23 months of age in Tanzania. *BMC Pediatrics*, *19*(1). https://doi.org/10.1186/s12887-019-1897-5

Lalani, N., Jimenez, R. B., and Yeap, B. (2020). Understanding Propensity Score Analyses. *International Journal of Radiation Oncology, Biology, Physics*, *107*(3), 404–407. https://doi.org/10.1016/j.ijrobp.2020.02.638

Lundborg, P., Rooth, D. O., and Alex-Petersen, J. (2022). Long-Term Effects of Childhood Nutrition: Evidence from a School Lunch Reform. *Review of Economic Studies*, *89*(2), 876–908. https://doi.org/10.1093/restud/rdab028

Rozaki, Z. (2021). Chapter Five - Food security challenges and opportunities in indonesia post COVID-19. In M. J. Cohen (Ed.), *Advances in Food Security and Sustainability* (Vol. 6, pp. 119–168). Elsevier. https://doi.org/https://doi.org/10.1016/bs.af2s.2021.07.002

Vairetti, C., Gennaro, F., and Maldonado, S. (2024). Propensity score oversampling and matching for uplift modeling. *European Journal of Operational Research*, *316*(3), 1058–1069. https://doi.org/10.1016/j.ejor.2024.03.024

Zhong, T., Crush, J., Song, Y., Si, Z., Scott, S., and Peng, Y. (2023). Urban food insecurity and the impact of China's affordable food shop (AFS) program: A case study of Nanjing City. *Applied Geography*, *154*. https://doi.org/10.1016/j.apgeog.2023.102924

Zhou, Y., Ren, J., and Zheng, X. (2024). Feeding for a brighter future: The long-term labor market consequences of school meals in rural China. *Economics and Human Biology*, *52*. https://doi.org/10.1016/j.ehb.2023.101335

Zimmerman, T., Rothstein, M., Dixit-Joshi, S., Vericker, T., Gola, A. A. H., and Lovellette, G. (2024). USDA Summer Meals Programs: Meeting the Nutritional Needs of Children. *Journal of the Academy of Nutrition and Dietetics*, *124*(3), 379-386.e1. https://doi.org/10.1016/j.jand.2023.09.010

Towards Resilient Societies: The Synergy of Religion, Education, Health, Science, and Technology – Rahiem (Ed.)

Integrating Islamic values and culture in mathematics education: A literature study

Gelar Dwirahayu
Universitas Islam Negeri Syarif Hidayatullah Jakarta, Indonesia
ORCID ID: 0000-0002-4098-7562

Gusni Satriawati
Universitas Islam Negeri Syarif Hidayatullah Jakarta, Indonesia
ORCID ID: 0009-0003-3983-2445

Finola Marta Putri
Universitas Islam Negeri Syarif Hidayatullah Jakarta, Indonesia
ORCID ID: 0000-0002-2374-8316

Dindin Sobiruddin
Universitas Islam Negeri Syarif Hidayatullah Jakarta, Indonesia
ORCID ID: 0009-0002-7155-3713

Firdausi
Universitas Islam Negeri Syarif Hidayatullah Jakarta, Indonesia
ORCID ID: 0000-0002-6145-313X

ABSTRACT: The integration of Islamic values, culture, and sciences is a growing topic of discussion in Indonesia, as these values serve as essential guidelines for Indonesian society, particularly Muslims. This study explores the integration of Islamic values, mathematics, and culture to enhance the quality of education, support sustainable development, and promote global citizenship, especially among Muslims, while preserving cultural and natural heritage. Using a qualitative literature study method, this research examines the process of incorporating Islamic and cultural values into mathematical concepts. The findings reveal two key points: first, the conceptualisation of Islamic values is achieved through Islamic education subjects in the school curriculum, including Aqidah Akhlak, the Qur'an and Hadith, the History of Islam, and Fiqh; second, the conceptualisation of cultural values is carried out through the study of local wisdom. As a universal religion promoting peace, Islam encourages all teachers to integrate Islamic values into their lessons, fostering both faith in Allah and a deeper understanding of knowledge. Simultaneously, embedding cultural values in education nurtures students‘ appreciation and love for their cultural heritage, reinforcing their identity as responsible citizens.

Keywords: culture, education, integration, islamic value, mathematical knowledge

1 INTRODUCTION

Learning is transferring knowledge from teachers to students to change the students' knowledge/skills from do not know to know, from cannot become able, and from do not understand become understand. Teachers should have innovative teaching strategies that

DOI: 10.1201/9781003645542-40

can help students learn. Improving teachers' pedagogical knowledge is one of the purposes of SDG: Substantially increase the supply of qualified teachers. Regarding the sustainability of education in Indonesia, the curriculum at school, in this case, is Islamic-based schools we call madrasa, refers to the National Education Curriculum and the Ministry of Religion Curriculum. All madrasas should follow and implement these regulations. There is no difference in students' abilities in cities and villages regarding developing professionalism, knowledge, and noble character as the SDG aim is to ensure inclusive and equitable quality education and increase lifelong learning opportunities for all.

The national education curriculum contains general subjects that teach by utilising the potential of their local wisdom (Ministry of Education 2022), while the Ministry of Religion Curriculum (2022) adds subjects about Islamic concepts, such as Qur'an & Hadist, Aqidah Akhlak, Fiqh, and Islamic Culture. In this case, the learning process at Madrasa needs to connect the concepts taught with local culture and Islamic values. These two approaches need to be used by teachers to strengthen students‘ belief in Islam and introduce culture to encourage them to love their own local culture. It follows the purpose of SDG, which is education for sustainable development and global citizenship and efforts to protect and safeguard the world's cultural and natural heritage. On the other hand, The vision of UIN Syarif Hidayatullah Jakarta is "integrating Islam values, Indonesian culture, and science". Therefore, the mathematics education study program at UIN Syarif Hidayatullah must align with this vision to produce prospective mathematics teachers who can integrate mathematics, Islamic values, and culture.

Many previous studies have been done about the integration of Islam, culture, and mathematics, including research on finding the relationship between the Qur'an or the context of daily life and mathematics (Nafi'an *et al.* 2023; Nuryami and Fadilah 2023), qualitative research on various models of integration of Islam and culture in Mathematics Learning (Mutijah 2018), research on improving the competence of mathematics pre-service teachers through the integration of Islam or culture in the teaching-learning process (Inganah *et al.* 2023), developing a learning module (Satriawati *et al.* 2023), and developing assessment of learning (Rensiana *et al.* 2024). The other research is about the integration of culture and mathematics. For example, Fouze (2017) used traditional games for children in Israel to teach math, Sunzuma (2019) researching the improvement of the ability of mathematics teachers in South Africa to teach geometry using ethnomathematics, and Zhang (2021) researched the use of the Temple of Heaven in Beijing, China, to teach geometry.

This research is different from previous research. Using literature study, we develop the conceptualisation of integrating Islamic values (Suwendi *et al.* 2024) and culture into mathematical content to produce a theoretical framework that can be applied in teaching and learning mathematics. This theoretical framework will provide a reference for teachers to develop learning or assessment indicators (Dwirahayu *et al.* 2022) where the concept of integration is built from Islamic and school mathematics materials in accordance with the curriculum in the madrasah, while cultural integration is built with Indonesian cultural information that needs to be preserved from various sources and school mathematics. Therefore, this study discusses 1) how to design the integration of Islamic values with mathematics concepts, 2) how to design the integration of cultural with mathematics concepts, 3) how to implement the integration of Islamic value and cultural to mathematical concepts.

1.1 *Conceptualisation of integration of Islamic value, culture, and mathematics*

Islam does not separate science and religion (Abdussakir 2018; Inganah *et al.* 2023). It is necessary to restore science to its centre, namely tauhid (only belief in Allah SWT), so that human faith in Allah increases. The Indonesian government is trying to implement the education system in madrasas by integrating Islamic values and other subjects. Mathematical content can also be integrated with Islamic values in mathematics.

As explained, the integration of Islamic values and mathematics can take many forms, such as integration with tajweed (the rules of reading the Quran), history of the Qur'an, Hadith, *aqidah* (belief in God), *akhlak* (spiritual attitude), use of Islamic terms, visualisation of illustrations with Islamic nuances, and Islamic knowledge in daily life (Rahmah *et al.* 2024).

Abdussakir (2018) and Nuryami and Nurfadilah (2023) mention the four models of integration of mathematics with the Quran: 1) developing mathematics from the Quran, 2) using mathematics to implement the values in the Qur'an, 3) using mathematics to understand the miracles of the Quran, and 4) teaching mathematics with the values of the Qur'an (Abdussakir 2018; Nuryami and Fadilah 2023).

Mathematics is a branch of science that is related to the reasoning process. The concepts in mathematics can be integrated with Islamic values while maintaining the uniqueness of each field of knowledge (Nafi'an *et al.* 2023) to balance mastery of mathematical content with the spiritual (Triana *et al.* 2023), understand and know the history of the glory of Islam (Mutijah 2018). Integration between Islamic values can support the achievement of mathematics learning objectives and positive character (Prasetyo and Halimatusa'diyah 2024; Rahmah *et al.* 2024).

Mathematics is a symbolic language that describes patterns found in nature, in the human mind, or created by the human imagination (Nuryami and Fadilah 2023). The patterns in mathematics can be synergised with the verses of the Qur'an. The integration of Islamic values and science in Indonesia aims to improve mathematical abilities and build awareness that all knowledge begins from one source, Allah SWT.

Research on culture and mathematics was introduced by Ubiratan D'Ambrosio around 1960. The combination of culture and mathematics is known as ethnomathematics. Ethnomathematics examines aspects of mathematics related to culture; it is concerned with the comparative study of different human cultures, particularly concerning how mathematics has been shaped and is, in turn, shaped by the values and beliefs of groups of people (Nuryadi *et al.* 2023; Sari *et al.* 2023) Integrating traditional culture into education is essential today because many foreign cultures have come to Indonesia, making children more familiar with foreign cultures than their own.

This study interprets the integration of culture and mathematics as a mathematics learning process taught using Indonesian cultural food, clothes, historical buildings, food, traditional dances, and musical instruments. Similar to Islamic integration, the conceptualisation of culture and mathematical integration is carried out by conducting a literature review.

2 RESEARCH METHOD

The research method is a qualitative literature study, which describes the integration of Islamic values and math, culture, and math through studying the mathematics education curriculum, Islamic context, and culture context. Data analysis techniques include four stages (Badawi *et al.* 2023; Creswell and Creswell 2018): organise, synthesise, identify, and elaborate. The first step, *organise*, is to organise the literature that will be used. The literature was reviewed to make it relevant to the problem. At this stage, the author searches for ideas, goals, and conclusions from several literatures. Second, *synthesise*, which is uniting the results of the literature organisation into an orderly summary and looking for connections between the literature. Third, Islamic values and culture relevant to mathematics concepts must be identified. Fourth, *elaborate* the literature about Islamic value and mathematical knowledge and elaborate information about the culture and mathematical knowledge become the framework of integration of Islamic value and mathematics framework of integration culture and mathematics; further, it can construct indicators in the mathematics learning process. The sources of information used in this study include school mathematics curricula, Islamic religious education curricula in madrasahs, books on culture, and various literature published in nationally indexed Sinta journals as well as reputable international journals.

The elaboration process consists of the integration of mathematical concepts with Indonesian culture and the integration of mathematical concepts with Islamic values. In cultural integration, we collect data from one of the selected cultures that will be integrated with mathematical concepts. For example, we choose the Sundanese tribe. In the Sundanese tribe, we can learn about many cultures, such as traditional food, traditional art tools, traditional dances, historical buildings, clothes, traditional cooking utensils that Sundanese people commonly use, etc. Because of the many cultures, we can focus on one culture, for example, traditional dance. In traditional dance art, we learn many things, and then we find unique forms in traditional Sundanese dances that can be connected with one of the mathematical concepts.

Meanwhile, in Islamic values integration, first, we analyse the school curriculum for Islamic education in madrasa, second, integrated the material in religion education into mathematics concept, therefore mathematics teachers can learn Islamic subjects and develop mathematics learning that is integrated with Islamic values. From the indicators made by religion teachers, we can adopt them in mathematics lessons. Figure 1 is the elaboration stage carried out by the researcher to combine Islamic and cultural values in mathematics lessons.

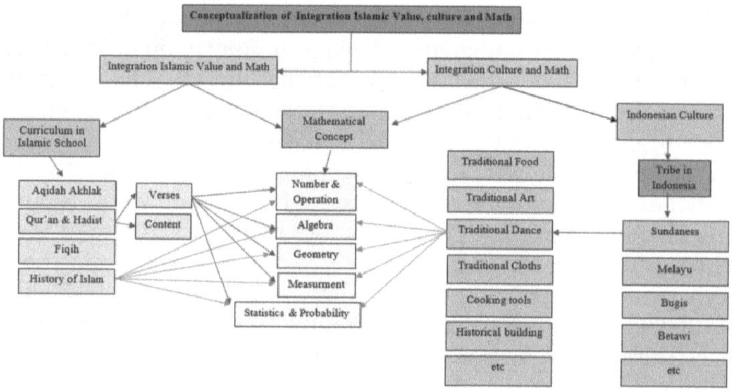

Figure 1. Elaboration process on integration Islamic value, culture, and mathematics.

3 RESULT AND DISCUSSION

3.1 *Integration of Islamic value in mathematics*

Integrating Islamic values in mathematics learning is accessible to convey but challenging to implement. First, it is necessary to make a conceptual framework to facilitate the process of learning mathematics integrated with Islamic values. When Abdussakir (2018) mentions the integration model in mathematics teaching and learning with the Qur'an; in this study, we integrate the concept of mathematics with religious subjects in madrasa. The religious curriculum in Madrasah is divided into four sub-lessons; they are '*akidah akhlak*' (moral theology and belief in God), Qur'an and Hadith, '*Fiqh*' (the science of worship in Islam), and Islamic History (history of Islam, and the story of the prophets and his companions) (Nuryadi *et al.* 2023), while the subject of mathematics consists of numbers, algebra, geometry, measurement, and statistics and probability (Ministry of Education 2022).

Integrating Islam and mathematics in this study is different from the research from Abdussakir (2018), where the integration of mathematics and Islam uses the four models of mathematics from-for-to-with the Qur'an. The conceptual framework of integration in this study offers five model integrations based on religious education. Five model integrations are:

(1) Mathematical integration based on Qur'an verses. It means utilising the content of Qur'an verses to build mathematical knowledge or synergise with some patterns in mathematics
(2) Mathematical integration-based Qur'an content, using Qur'an literacy to build mathematical knowledge

(3) Mathematical integration into the history of Islam (SKI), studying the history of the prophet and her friends to build mathematical knowledge
(4) Mathematical integration into Fiqih, studying social laws based on the Qur'an and Hadist to build mathematical knowledge
(5) Mathematical integration into Aqidah Ahlak, deepening the science of moral theology and belief in God or affective in religion to build mathematical knowledge

3.2 *Integration culture in mathematics*

Indonesia is a country rich in culture. Indonesia has 34 provinces with their peculiarities. Indonesia's culture needs to be preserved. Currently, the flow of technology dramatically influences the pattern of human life. Modern games are available on gadgets, information from abroad is easily accessible, and food from abroad has come to Indonesia (Setiana 2020). Culture in this study is defined as the collection of spiritual, material, intellectual, and emotional characteristics typical of society. It includes, in addition to art and literature, lifestyle, ways of living together (Murtadlo *et al.* 2024), systems of values, traditions, and beliefs (Zuliana *et al.* 2023). Indonesia has more than 1000 cultural heritages that must be preserved (Dihni 2021) through learning mathematics.

For example, teaching mathematics using traditional games Engklek. Learning begins with introducing students to how to play *engklek*, and then, students are invited to play. As long as the games, the teacher gives a case. If five students are playing, and if the students who are allowed to play are even numbers, then how many combinations of players are there? So, the game of *engklek* is a real case for students to understand the differences in probability, permutation, and combination. An example of the assessment of the integration of culture and mathematics can be seen in Rensiana (2024).

4 CONCLUSIONS

The conclusion of this study arranges the theoretical framework for the integration of Islamic values and mathematics and the theoretical framework for the integration of culture and mathematics. Integration can practically be done by determining learning indicators in religious subjects and certain mathematical content. This way, teaching and learning mathematics is more straightforward because religious teachers have taught the religious context. Meanwhile, to integrate culture and mathematics, we must find a culture close to students so mathematics becomes more accurate because it is felt directly by students. Mathematics learning integrated with Islamic values and culture is expected to foster students‘ critical thinking and cultural appreciation.

The implication is that teachers are given training, especially to develop learning that is integrated with Islamic values and culture, following the framework built in this study so that they understand the concept of integrating Islamic values and culture in each subject and improve their pedagogical knowledge and skills.

REFERENCES

Abdussakir. (2018). Integrating mathematics and religious teachings and values in elementary and secondary school. *Contextualization of Mathematics Through Islamic Values Integration*, 4–6.

Badawi, A., Herman, T., and Juandi, D. (2023). Pedagogical content knowledge analysis of mathematics teachers in developing lesson plans on ratio topic. *Hipotenusa: Journal of Mathematical Society*, *5*(1), 29–43. https://doi.org/10.18326/hipotenusa.v5i1.8934

Creswell, J. W., and Creswell, J. D. (2018). *Research design: qualitative, quantitative, and mixed methods approaches* (5th ed.). Sage Publications.

Dihni, V. A. (2021). *Indonesia Miliki 1.239 Warisan Budaya Tak benda [Indonesia has 1.239 intangible cultural heritage].*

Dwirahayu, G., Handayani, I. D., Suhyanto, O., Musyrifah, E., and Sobiruddin, D. (2022). Development of mathematics teaching-learning material with metaphors approach. *Journal of Physics: Conference Series, 2157*(1), 012045. https://doi.org/10.1088/1742-6596/2157/1/012045

Fouze, A. Q., and Amit, M. (2017). Development of mathematical thinking through integration of ethnomathematic folklore game in math instruction. *Eurasia Journal of Mathematics, Science and Technology Education, 14*(2), 617–630. https://doi.org/10.12973/ejmste/80626

Inganah, S., Rizki, N., Choirudin, C., Farooq, S. M. Y., and Susanti, N. (2023). Integration of Islamic values, mathematics, and career readiness competencies of prospective teachers in Islamic Universities. *Delta-Phi: Jurnal Pendidikan Matematika, 1*(1), 11–14. https://doi.org/10.61650/dpjpm.v1i1.31

Ministry of Education. (2022). *Content Standards in Early Childhood Education, Primary Education Levels, and Secondary Education Levels.*

Ministry of Religion. (2022). Implementation of "Merdeka" Curriculum in Madrasa.

Murtadlo, M., Albana, H., Helmy, M. I., Libriyanti, Y., Izazy, N. Q., and Saloom, G. (2024). Preserving the gotong royong character for Indonesian Gen-Z in the digital era. *International Journal of Evaluation and Research in Education (IJERE), 13*(3), 1631. https://doi.org/10.11591/ijere.v13i3.27175

Mutijah. (2018). Model integrasi matematika dengan nilai-nilai Islam dan kearifan lokal budaya dalam pembelajaran matematika [Model of Integration between mathematics, Islamic values, and local cultural wisdom in mathematics learning]. *Jurnal Pendidikan Matematika (Kudus), 1*(2), 51–75. https://doi.org/10.21043/jpm.v1i2.4878

Nafi'an, M. I., Ahmadi, B., Muttaqin, M. N., and Afifah, D. S. N. (2023). Internalization of Islamic moderation value in mathematics problem solving. *Athena: Journal of Social, Culture and Society, 1*(3), 138–143. https://doi.org/10.58905/athena.v1i3.139

Nuryadi, Fitiradhy, A., Marhaeni, N. H., Purwoko, R. Y., and Rumasoreng, M. I. (2023). The effects of puppet ethnomathematics applications as mathematics teaching materials for character education-based. *Pegem Journal of Education and Instruction, 13*(2), 153–160. https://doi.org/10.47750/pegegog.13.02.19

Nuryami, and Fadilah, Y. (2023). Integration of culture and Islam in learning mathematics in The independent curriculum. *Journal of Scientific Research, Education, and Technology (JSRET), 2*(3), 1407–1416. https://doi.org/10.58526/jsret.v2i3.241

Prasetyo, H., and Halimatusa'diyah, I. (2024). Examining Muslim Tolerance Toward Ordinary Non-Muslims: Social, Religious, and Political Tolerance in Indonesia. *International Journal of Sociology, 54*(2), 112–131. https://doi.org/10.1080/00207659.2024.2301881

Rahmah, H., Turmudi, T., and Ghifari, M. T. (2024). Research and development of mathematics teaching material integrated with Islamic values: A systematic literature review. *Prisma Sains: Jurnal Pengkajian Ilmu Dan Pembelajaran Matematika Dan IPA IKIP Mataram, 12*(2), 306. https://doi.org/10.33394/j-ps.v12i2.9519

Rensiana, M., Dwirahayu, G., and Putri, F. M. (2024). Pengembangan instrumen tes pemahaman konsep transformasi geometri berbasis etnomatematika [Development of a test conceptual understanding of geometric transformations based ethnomathematics]. *Math Educa Journal, 8*(1). https://doi.org/http://dx.doi.org/10.15548/mej.v8i1.8226

Sari, N., Saragih, S., Napitupulu, E. E., Rakiyah, S., Sari, D. N., Sirait, S., and Anim, A. (2023). Applying ethnomathematics in learning mathematics for middle school students. *Acta Scientiae, 25*(5), 250–274. https://doi.org/10.17648/acta.scientiae.7690

Satriawati, G., Kholis, N., Dwirahayu, G., and Sobiruddin, D. (2023). Pengembangan bahan ajar transformasi geometri berbantuan website: Pendekatan project-based-learning mozaik geometri [Development of website-assisted geometric tansformation teaching materials: Geometric mosaic project-based-learning approach]. *JINoP (Jurnal Inovasi Pembelajaran), 9*(1), 1–15. https://doi.org/10.22219/jinop.v1i1.2441

Setiana, D. S. (2020). The role of ethnomathematics in welcoming the society 5.0. *Proceedings of the SEMANTIK Conference of Mathematics Education (SEMANTIK 2019) The, 467*, 130–134. https://doi.org/10.2991/assehr.k.200827.130

Sunzuma, G., and Maharaj, A. (2019). Teacher-related challenges affecting the integration of ethnomathematics approaches into the teaching of geometry. *Eurasia Journal of Mathematics, Science and Technology Education, 15*(9). https://doi.org/10.29333/ejmste/108457

Suwendi, S., Mesraini, M., Azka, F. L., and Gama, C. B. (2024). Implementation of Knowledge Integration in Islamic Higher Education. *Jurnal Pendidikan Islam, 10*(1), 41–52. https://doi.org/10.15575/jpi.v10i1.35385

Triana, A. Y., Supono, A., and Aini, A. N. (2023). *Integrating Islamic values on math learning in welcoming the society 5.0: How it works? 1*, 203–211. https://doi.org/10.2991/978-2-38476-044-2_19

Zhang, C., Wijaya, T. T., Zhou, Y., Chen, J., and Ning, Y. (2021). Ethnomathematics values in Temple of Heaven: An Imperial Sacrificial Altar in Beijing, China. *Journal of Physics: Conference Series, 2084*(1), 012015. https://doi.org/10.1088/1742-6596/2084/1/012015

Zuliana, E., Dwiningrum, S. I. A., Wijaya, A., and Purnomo, Y. W. (2023). The Geometrical Patterns and Philosophical Value of Javanese Traditional Mosque Architecture for Mathematics Learning in Primary School: An Ethnomathematic Study. *Journal of Education Culture and Society, 14*(2), 512–532. https://doi.org/10.15503/jecs2023.2.512.532.

Towards Resilient Societies: The Synergy of Religion, Education, Health, Science, and Technology – Rahiem (Ed.)
© 2026 The Author(s), ISBN: 978-1-041-08471-6
Open Access: www.taylorfrancis.com, CC BY-NC-ND 4.0 license

The impact of self-regulated learning and environment on mathematics achievement

Gusni Satriawati
Universitas Islamd Negeri Syarif Hidayatullah Jakarta, Tangerang Selatan, Indonesia
ORCID ID: 0009-0003-3983-2445

Wiwi Nova & Afidah
Islam Negeri Syarif Hidayatullah Jakarta, Tangerang Selatan, Indonesia

ABSTRACT: Mathematics learning outcomes are influenced by intelligence and are also potentially affected by self-regulated learning and the learning environment. This study aims to analyse the relationship between self-regulated learning, the learning environment, and mathematics learning outcomes. A quantitative descriptive research method was employed, conducted at a junior high school in South Tangerang during the 2020/2021 academic year. The research utilised a self-regulated learning questionnaire, a learning environment questionnaire, and a mathematics learning outcomes test as instruments. The results indicate a positive influence of self-regulated learning on mathematics learning outcomes, contributing 45.8%. Additionally, the learning environment positively affects mathematics learning outcomes by 21.6%. A positive relationship of 47.4% was observed between self-regulated learning and the learning environment, indicating mutual influence, represented by the regression equation $Y = -3.201 + 0.989X1 + 0.182X2$. Furthermore, self-regulated learning and the learning environment together contribute 21% to mathematics learning outcomes, with the remaining 79% influenced by other factors. These findings conclude that self-regulated learning and the learning environment significantly affect student performance, aligning with the Sustainable Development Goal of providing a safe and effective learning environment for all students.

Keywords: learning environment, mathematics learning outcome, self-regulated lèarning

1 INTRODUCTION

Mathematics is a systematic discipline that can be used by humans to help solve everyday problems that exist around them. Mathematics has an important role in life, but in reality there are still many students who find it difficult to understand mathematics (Sharma 2021). Students still have difficulty solving math problems related to mathematical concepts and principles (Angelo A. Legarde 2022). Students' higher order thinking skills in mathematics are still in the low category (Satriawati *et al.* 2021). Of course this will affect student learning outcomes in learning mathematics.

Learning outcomes are statements describing what students should know, understand, and be able to do after completing a module or qualification (Edwards 2015). They provide clarity for both tutors and students, linking learning achievements to assessment criteria and methods (Melvyn Dodridge 1999). Learning outcomes can encompass subject-based knowledge, personal transferable skills, and generic academic outcomes (Allan 1996). While associated with rational curriculum planning, they are not synonymous with behavioural objectives and offer a more flexible approach to curriculum design (Allan 1996). Learning outcomes are crucial for subject review in higher education (Melvyn Dodridge 1999) and are intended to empower

DOI: 10.1201/9781003645542-41

students by clearly outlining expected achievements (Edwards 2015). To enhance their utility in intelligent e-learning applications, a more detailed definition and structure may be necessary (Kalou *et al.* 2012). Overall, learning outcomes serve as important indicators for students in the learning process and guide the design of educational experiences in higher education.

One of the internal factors that can affect student learning outcomes is the attitude of student self-regulated learning. Self-regulated learning is self-awareness, self-driven, the ability to learn to achieve its goals (Stephen Brookfield 2000). Schunk and Zimmerman (2013) argue that self-regulated learning is a learning process that occurs due to the influence of thoughts, feelings, strategies, and behavior of someone who has an orientation towards achieving goals. Students who have a conscious effort with the intention and motivation to fulfill their learning goals with full responsibility without depending on others. Indicators of self-regulated learning in this study include: a) learning initiatives; b) diagnosing learning needs; c) setting a target and learning goals; d) view adversity as a challenge; e) using and looking for relevant learning resources: g) choosing and determining learning strategies, and h) evaluating the learning process and learning outcomes (Sumarmo 2004).

The improvement of student quality can be seen from high student learning outcomes, student interest in learning, and self-regulated learning itself. Some studies show that most students' efforts in finding solutions depend on students' self-regulated learning. Hargis (2002) concluded that the more independent a student is in organizing his learning, the higher his chances of success. Fauzi's and widjajanti's research results (2018) concluded that there is a positive effect of self-regulated learning on students' mathematics learning achievement. This statement is also in line with the purpose of national education in Indonesia, which is to build a foundation for the development of students' potential to become independent human beings (Pemerintah Indonesia 2003).

In addition, self-regulated learning, the learning environment is also one of the factors that can affect student learning outcomes (Abbas 2019). The learning environment, including physical, social, and technological aspects, greatly influences educational practices and child development in early education (Florescu 2020). The learning environment encompasses physical, intellectual, and emotional factors that impact student development (Thompson and Wheeler 2008). The learning environment is everything that is around the individual that can affect individual behaviour directly or indirectly. Social and non-social learning environments reveals significant impacts on cognitive abilities across species. Environmental interventions, particularly asocial ones, show larger effects on learning than social interventions alone, with longer interventions yielding stronger outcomes (Lambert and Guillette 2021). The social environment, comprising societal beliefs, customs, and behaviors, influences student behavior and learning (Srivastava 2020)

The indicators of the learning environment in this study are social learning environment family, friends hanging out, school, learning situation, learning place, and learning tools (Srivastava 2020). Based on these conditions, the purpose of this study is 1) to determine the effect of self-regulated learning on students' mathematics learning outcomes, 2) to determine the influence of the learning environment on students' mathematics learning outcomes, 3) how significant is the relationship between self-regulated learning and student learning environments?, and 4) to determine how significant the influence of self-regulated learning is learning and learning environment on students' mathematics learning outcomes.

2 METHODS

This research was a quantitative descriptive study. This study was conducted in one of the junior high schools in South Tangerang in the academic year 2020/2021. The research instrument used was self-regulated learning, learning environment, and the learning of mathematics outcomes test. Data collection techniques used in this study were tests and questionnaires. The test to assess student achievement. The test questions tested consisted of 5 mathematical descriptions on immaterial patterns of numbers and cartesian coordinates. The score was then analysed using the interpretation of the Table 1 score based on the minimum completeness criteria (KKM) of 75.

Table 1. Categories of interpretation of learning outcomes.

Score Interval	Category
X > 75	High
X = 75	Medium
X < 75	Low

To determine the ability of self-regulated learning and student learning environment, were used Questionnaires. The questionnaire consisted of 34 statements regarding self-regulated learning and 30 information regarding the student learning environment. Alternative answers are presented in the form of a checklist using a Likert scale, namely Very Agree (VA), Agree (A), Not Agree (NA) and Very Not Agree (VNA). Scores for positive statements and negative statements can see in Table 2.

Table 2. Scoring of the questionnaire.

	Score	
Alternative statement	Positive	Negative
VA	4	1
A	3	2
NA	2	3
VNA	1	4

The data on the self-regulated learning assessment questionnaire and the student learning environment were analysed using the score interpretation in Table 3.

Table 3. Criteria for interpretation of questionnaire scores.

Score interval	Category
X > Mi +Sdi	High
Mi – Sdi < Mi +Sdi	Medium
X < Mi – Sdi	Low

The situation in this study is to examine whether there is an effect of self-regulated learning and student learning environment on students' mathematics learning outcomes on the material pattern of cartesian numbers and cartesian coordinates. The researcher was collected all the data from the tests and questionnaires. A prerequisite test for data processing techniques was carried out by calculating data scores-applying analysis techniques through normality test, linearity test, multi-collinearity test, hypothesis testing. The data is processed to find answers to the problems that have formula data. Next, the hypothesis was tested using Spearman's Rho correlation, comprehensive, partial, and coefficient of determination (Ghozali 2006).

3 RESULT AND DISCUSSION

3.1 *Students' self-regulated learning, students learning environment, and students' learning outcomes*

Figure 1 shows the results data in the results of the students' learning environment on mathematics based on the indicators. The results data in the results of students' self-regulated learning levels on mathematics learning based on indicators. The average level of students' self-regulated learning towards mathematics is 69,75 with medium criteria. The result is in line with Erni Puji Astuti's research (2016) which concludes that student learning independence is still excellent. Several factors influence the learning rate of self-regulated learning: the learning initiative, diagnosing learning needs, setting targets and learning goals, perceiving learning difficulties as challenges, using learning and searching for learning resources, learning relevant, and evaluating learning resources as learning challenges. The result shows that students still lack the initiative to look for other learning resources related to the studied material. There is no courage from students to express their opinions when discussing the subject matter.

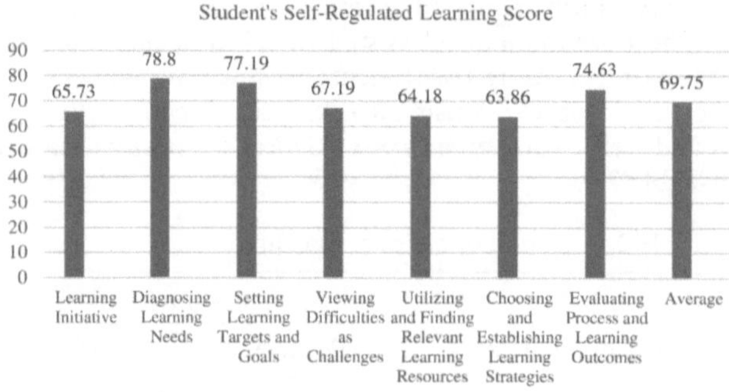

Figure 1. Students' self-regulated learning.

Figure 2 shows that the average level of students' learning environment in mathematics learning is 68,26 with medium criteria. The story of the student learning environment is influenced by several factors, including the family environment, learning place, learning situation, learning tools, friends, and school environment. The result is that the student learning environment is good enough to support students' self-regulated learning and to obtain maximum learning outcomes.

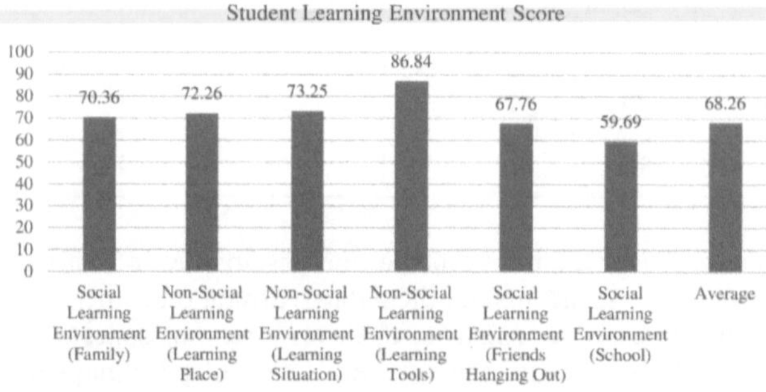

Figure 2. Students' learning environment score.

The results students' mathematics learning outcomes has an average value of 57,77, which means that the matter is smaller than the school's KKM score of 75. So that student learning outcomes in mathematics are relatively low. To determine the effect of self-regulated learning and learning environment on students' mathematics learning outcomes, use The normality test.

Based on statistical test with the SPSS 17.0 program, the asymptotic significance value is 0,200 > 0,05; concluded that the data is a normal distribution. The linearity tests show that the significant value of self-regulated learning variables to learning outcomes is 0,303. The value of significance > 0,05, then the relationship of self-regulated learning with learning outcomes is linear. The result means that the pattern of association between self-regulated learning and learning outcomes will form a straight line. Likewise, with the significance of the student's learning environment to the learning outcome of 0,648, which means the value of relevance >0,05, then the relationship of learning outcomes with the learning environment is linear. The result shows that the pattern of association between learning outcomes and the student's learning environment sets up a straight line. The multicollinearity tests' results show no symptoms of multicollinearity in regression models.

Table 4. Summary of multiple regression test results.

		Unstandardized coefficients	
Model		B	Std. error
1	(Constant)	-3,201	17,604
	Self-regulated learning (X_1)	0,989	0,296
	Learning environment (X_2)	0,182	0,063

Based on Table 4, the multiple linear regression model can be expressed in the following equations:

$$Y = -3,201 + 0,989X_1 + 0,182X_2 \quad (1)$$

The constant value of -3,201. Constant values are negative values meaning the value of self-regulated learning variables (X_1) and learning environments (X_2), if it is considered constant or equal to zero, the student's math learning outcome will be less or lower. The value of coefficient X_1 is 0,989. The value of the X_1 coefficient is positive, meaning that the influence of self-regulated learning on learning outcomes is positive and quite strong. If self-regulated learning is high, then the student's math learning results will be higher. X_2 coefficiency value of 0,182. The value of the X_2 coefficient is positive, meaning that the influence of the learning environment on the student's mathematical learning outcomes is positive and quite strong. If the student's self-regulated learning increases, then the student's math learning outcomes will increase.

Based on statistical test with the SPSS 17.0 program. The results showed that there was a positive influence between self-regulated learning on math learning outcomes of 45.8%. There is a positive influence between the learning environment on math learning outcomes of 21.6%. There is a positive relationship between self-regulated learning and student learning environment, which is 47.4%. The magnitude of the coefficient of determination (R^2) is 0,210 or equal to 21%. The figures mean that the variables of self-regulated learning and the student learning environment simultaneously affect 21% of students' math learning outcomes. At the same time, the remaining 79% are other variables that are not studied. The analysis results showed a partial influence between the student's self-regulated learning variables and the student's math learning outcome variables. The analysis results are in line with the results of the research of Fauzi and Widjajanti (2018), which showed that self-regulated learning has a positive and significant influence on the learning outcomes of mathematics well as partial learning environments.

The self-regulated learning and the student's learning environment simultaneously affect the student's mathematical learning outcomes. This result is in line with Abbas' opinion which states that internal factors that can affect student learning outcomes are self-regulated learning of the students themselves and external factors in the form of the learning environment faced (Abbas 2019). The learning environment also plays an important role in supporting regulated learning. Lahdenperä *et al.*, (2022) found that a student-centred learning environment can encourage regulated learning through scaffolding and a positive social environment. The results of this study are in line with the target of the sustainable development goals in the field of education (SDGs 4), namely providing a safe and effective learning environment to achieve quality education.

4 CONCLUSION

Based on the research results conducted in Junior High School in South Tangerang and referring to the data that researchers have obtained, they are: 1) there was a positive influence between self-regulated learning against outcomes of learning math, 2) there was a positive influence between the learning environment against the results of learning math, 3) there was a positive relationship between self-regulated learning and students' learning environment. This study recommends increasing self-related learning by getting used to solving problems independently, being automated in the learning process, not depending on friends, and instilling self-regulation in students. Provide facilities that support student learning activities to learn independently and maximise their learning outcomes.

ACKNOWLEDGMENT

The researchers would like to thank the Centre for Research and Publication (Puslitpen) LP2M, UIN Syarif Hidayatullah Jakarta for funding this publication.

REFERENCES

Abbas, A. B. (2019). *The Effect of Learning Environment on Learning Discipline and Self-Regulation on Students' Mathematical Learning Outcomes.*

Allan, J. (1996). Learning Outcomes in Higher Education. *Studies in Higher Education*, 21(1), 93–108. https://doi.org/10.1080/03075079612331381487

Angelo A. Legarde, M. (2022). Working With Mathematical Problems: An Analysis of Students Misconseptions And Its Impact On Mathematics LEARNING. *International Journal of Advanced Research*, 10(03), 25–33. https://doi.org/10.21474/IJAR01/14358

Edwards, C. (2015). Learning Outcomes at the Open University. In Institute of Educational Technology (Issue 1). Institute of Educational Technology, *The Open University*. https://doi.org/https://oro.open.ac.uk/46378/

Fauzi, A., and Widjajanti, D. B. (2018). Self-regulated learning: The effect on student's mathematics achievement. *Journal of Physics: Conference Series*, 1097(1). https://doi.org/10.1088/1742-6596/1097/1/012139

Florescu, D. M. (2020). Implications Of The Learning Environment On Educational Practices In Early Education. *In Journal Plus Education*: Vol. XXVII (Issue 2). E-ISSN.

Ghozali, I. (2006). *Aplikasi Analisis Multivariate dengan Program SPSS* (4th ed.). Universitas Diponegoro.

Hargis, J. (2002). *The Self-Regulated Learner Advantage: Learning Science on the Internet.*

Kalou, A., Solomou, G., Pierrakeas, C., and Kameas, A. (2012). An ontology model for building, classifying and using learning outcomes. *Proceedings of the 12th IEEE International Conference on Advanced Learning Technologies*, ICALT 2012, 61–65. https://doi.org/10.1109/ICALT.2012.45

Lahdenperä, J., Rämö, J., and Postareff, L. (2022). Student-centred learning environments supporting undergraduate mathematics students to apply regulated learning: A mixed-methods approach. *The Journal of Mathematical Behavior*, 66, 100949. https://doi.org/10.1016/J.JMATHB.2022.100949

Lambert, C. T., and Guillette, L. M. (2021). The impact of environmental and social factors on learning abilities: a meta-analysis. *Biological Review of The Cambridge Philosophical Society*, 96(6), 2871–2889.

Melvyn Dodridge, by. (1999). *Learning Outcomes their Assessment in Learning Outcomes their Assessment in Education and Higher.*

Satriawati, G., Dwirahayu, G., Afidah, A., and Hafiz, M. (2021). Students' profile on higher order mathematical thinking skill at Islamic Junior High School (MTsN) in Indonesia. In *Emerging Trends in Technology for Education in an Uncertain World* (1st Edition). Taylor and Francis Group. https://www.taylorfrancis.com/chapters/edit/10.1201/9781003219248-12/students-profile-higher-order-mathematical-thinking-skill-islamic-junior-high-school-mtsn-indonesia-satriawati-dwirahayu-afidah-hafiz?context=ubx&refId=c3241cc9-6332-49f0-9df7-b23c5ae5e854

Schunk, D. H., and Zimmerman, B. J. (2013). *Self-Regulation and Learning.*

Sharma, P. (2021). Importance and application of mathematics in everyday life. *International Journal for Research in Applied Science and Engineering Technology*, 9(11), 868–879. https://doi.org/10.22214/ijraset.2021.38869

Srivastava, G. P. (2020). Review of the impact of various dimensions of social environment and social learning theory (SLT) on student's behaviour. *International Journal of Scientific Development and Research (IJSDR)*, 5(10), 100–104. www.ijsdr.org

Stephen Brookfield. (2000). *Understanding and Facilitating Adult Learning.* Josey-bass Publiser.

Sumarmo, U. (2004). *Pembelajaran Matematika untuk Mendukung Pelaksanaan Kurikulum Berbasis Kompetensi.* UPI.

Thompson, N. E., and Wheeler, J. P. (2008). Learning environment: Creating and implementing a safe, supportive learning environment. *In Journal of Family Consumer Sciences Education* (Issue 2).

Undang-Undang No. 20 Tahun 2003 Tentang Sistem Pendidikan Nasional. Jakarta, *Pusat Data dan Informasi Pendidikan*, Balitbang - Depdiknas (2003). https://jdih.kemdikbud.go.id/sjdih/siperpu/dokumen/salinan/UU_tahun2003_nomor020.pdf

Towards Resilient Societies: The Synergy of Religion, Education, Health, Science, and Technology – Rahiem (Ed.)

Integrating SDGs into Arabic education: Curriculum design for higher education

Kisno Umbar
Universitas Islam Negeri Syarif Hidayatullah Jakarta, Indonesia
ORCID ID: 0000-0002-8896-8117

Maswani
Universitas Islam Negeri Syarif Hidayatullah Jakarta, Indonesia
ORCID ID: 0009-0000-3025-6718

Muhbib Abdul Wahab
Universitas Islam Negeri Syarif Hidayatullah Jakarta, Indonesia

ABSTRACT: This study explores lecturers' and teaching staff's understanding of integrating sustainability principles into the curriculum and identifies the challenges and opportunities encountered during implementation. A qualitative research design with a case study approach was employed, with data collected through in-depth interviews and focus group discussions. The findings reveal that stakeholders generally understand the Sustainable Development Goals (SDGs) and recognise the importance of integrating them into Arabic language education to enhance linguistic skills, global awareness, and social responsibility. However, the integration of SDGs into Arabic language education in Indonesia remains in its early stages. Key challenges include limited teaching resources, insufficient time allocated in the syllabus, and resistance from some lecturers towards adopting new approaches. Despite these challenges, the integration offers significant opportunities, such as improving graduates' language proficiency, fostering their social responsibility, and nurturing a global mindset. These outcomes align with the broader objectives of sustainable development, emphasising the need for continued support, capacity-building, and collaboration among stakeholders to embed SDG principles into Arabic language education successfully.

Keywords: Arabic education curriculum, sustainable development, higher education, Indonesia, outcome-based education

1 INTRODUCTION

In the global context, Arabic is not only important as a means of understanding Islam but also a strategic asset in various fields such as education, business, diplomacy, economy, and culture (Rababah 2023; Maswani 2022). With a large Muslim population in the world, Indonesia has a special interest in ensuring effective and applicative mastery of the Arabic language at various levels of formal and informal education, ranging from primary education to higher education. Especially in pesantren that are the basis for the introduction of Islamic religious teachings (Jamhari 2009; Suwendi *et al.* 2024). The Arabic curriculum in higher education in Indonesia should address global challenges, especially sustainable development. Arabic education should focus on improving language skills and helping students understand their role in promoting sustainability.

DOI: 10.1201/9781003645542-42

The discourse of sustainable development in the study of Arabic education can be found in limited literature, for example, Almahdawi (2024), which explores the effectiveness of Arabic textbooks for grade 12 in the UAE in promoting sustainable development. Almaliki *et al.* (2023) examine the important role of Arabic education in achieving the Sustainable Development Goals (SDGs), but the research is limited to literature studies. In Central Asia, Sergey *et al.* (2024) point out students' low knowledge of the SDGs and emphasise the need for better integrating sustainable development principles into the higher education curriculum. Furthermore, Thohri (2024) noted the need for an inclusive and adaptive pedagogical approach and the importance of support from educators, curriculum developers, and policymakers to overcome challenges in implementing an inclusive curriculum. The above narrative reflects various perspectives and challenges in integrating the principles of sustainable development in education, both locally and internationally. Moreover, Cardiff *et al.* (2024) point out that the integration of SDGs into higher education, particularly in foreign language curricula, has been lacks practical implementation frameworks.

This gap becomes evident in the context of Arabic education in Indonesia, where research on curriculum design that incorporates sustainability principles remains scarce. While existing studies emphasise general curriculum development, they do not specifically address how Arabic education can actively contribute to sustainable development. To address this gap, this study aims to investigate how stakeholders—such as program heads, lecturers, and educators—understand and integrate sustainability principles into the Arabic education curriculum. Furthermore, it seeks to identify the challenges and opportunities encountered in the process, offering a practical framework for integrating the SDGs into Arabic education in Indonesian universities. By doing so, this research provides a novel perspective on aligning foreign language education with sustainability goals, bridging the gap between theory and practice.

2 METHOD

This study uses a qualitative research design with a case study approach. Case studies allow researchers to deepen their understanding of the concept of sustainable development from stakeholders in Indonesia (Yin 2018). This research is limited to the Jakarta area and its surroundings because Jakarta is often the centre of national education policy implementation (Fitramadhana 2023). This research was conducted in 2024 (May-July), which coincides with the transition period of higher education in Indonesia to implement OBE by the Regulation of the Minister of Education, Culture, Research, and Technology Number 53 of 2023 concerning Higher Education Quality Assurance. The respondents in this research were selected by *purposive sampling* by considering institutions that have Arabic Education or Arabic programs with "superior" or A accreditation in the Jakarta area and its surroundings, such as the Syarif Hidayatullah State Islamic University Jakarta (participant A), the University of Indonesia (participant B), and the State University of Jakarta (participant C).

Data was collected through in-depth interviews and focus group discussions (FGD) with study program heads and lecturers. It was supported by analysis of curriculum documents and participatory observation while implementing teaching and learning activities.

Data management is carried out by recording and transcribing interviews and group discussions, and then the data is analysed using thematic analysis methods by Terry *et al.* (2017). These procedures consist of familiarisation, coding, theme generation, theme review, theme definition and naming the themes. To ensure the reliability and validity of the data, data sources are triangulated by comparing the results of interviews, FGDs, observations, and document analysis. In addition, *member checking* is carried out by confirming the initial findings to the resource persons to ensure accuracy.

Ethical considerations in this study include obtaining consent from all participants before participating, maintaining the confidentiality of the information provided, and ensuring that participation is voluntary.

3 FINDINGS AND DISCUSSION

3.1 *Integration of SDGs principles in the arabic education curriculum*

Integrating Sustainable Development Goals (SDGs) principles is one of the main focuses in developing an Arabic education curriculum more responsive to global challenges. Based on interviews with the speakers and notes from the focus group discussion (FGD), it was revealed that the understanding of the SDGs among lecturers and study program managers has developed quite well. See the following interview responses regarding the question, "What are SDGs, and how can their values be integrated into Arabic language teaching?"

"The SDGs are an important framework for language learning that can connect students with global issues and positively contribute to society. I believe sustainability values can be integrated by analysing texts discussing themes such as environmental conservation, social justice, and ethics. However, we also need to ensure that students understand the relevance of these themes to their daily lives, not just as abstract concepts." - Participant A.

The understanding above is also relevant to Mambu's (2023) research that examines the teaching of the United Kingdom. In addition, language teaching has been shown to promote sustainability-focused education in some countries, such as Colombia, emphasising community engagement and global citizenship (Ruiz *et al.* 2024).

Furthermore, integrating the SDGs into the Arabic education curriculum is still early. This step cannot be separated from the intervention of the Indonesia government through the regulation of the Minister of Education, Culture, Research, and Technology Number 53 of 2023 concerning Higher Education Quality Assurance, which was issued to direct higher education to apply the OBE curriculum, which emphasises learning outcomes that are relevant to sustainability goals. Implementing OBE in higher education is a paradigm that actively handles the SDGs agenda (Stoltenberg and Michelsen 2023). In addition, OBE also reflects a paradigm shift from conventional education approaches to sustainable education. Furthermore, the comprehensive understanding of the integration of sustainable development principles and the Arabic education curriculum in Indonesia can be seen in Figure 1 below.

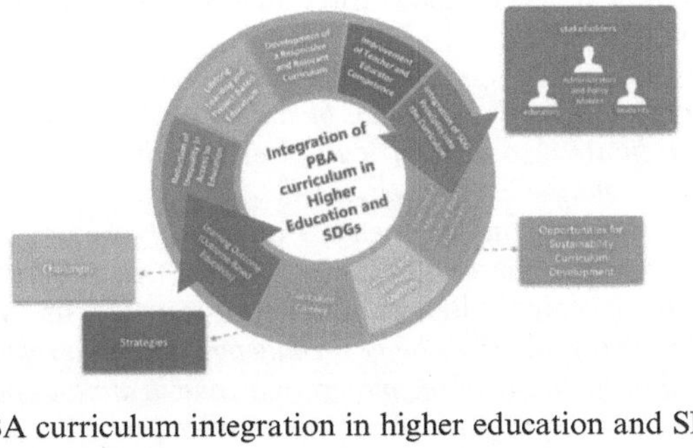

Figure 1. Pattern of PBA curriculum integration in higher education and SDGs.

Figure 1 above is classified into the first two groups related to the principles of the SDGs related to quality education and also components of the curriculum. The two parts are combined in the form of a circle that is interconnected with each other. First, it is necessary to understand the relevant SDGs principles referring to (UNESCO 2017) including 1) integration of SDG principles into the curriculum. Developing a curriculum incorporating SDG principles ensures relevant, high-quality education focused on environmental sustainability, gender equality, and social justice; 2) improvement of teacher and educator competence. Continuous training enhances lecturers' ability to teach sustainability topics, enabling effective integration of SDGs into their teaching (Herlanti *et al.* 2024); 3) development of a responsive and relevant curriculum. Encourages student involvement in practical sustainability projects; and 4) reduces inequality in access to education through institutional support and interdisciplinary collaboration, ensuring that the curriculum developed is inclusive and provides equitable access to all students.

The second is a component of the curriculum. In this context, researchers follow Tyler's (2013) idea of defining a curriculum as a framework that organises goals, content, methods, and assessments in the learning process. Tyler (2013) emphasised that all teaching elements must be arranged systematically and logically to achieve the desired educational goals. The integration process can be understood through each of the following components: 1) OBE-oriented learning outcomes must include linguistic skills and global awareness by the principles of the SDGs. The results of the interviews and FGDs emphasise the importance of directing learning outcomes to not only master the Arabic language but also understand global sustainability issues. Research by Barth *et al.* (2007) shows that outcome-based learning related to the SDGs can improve sustainability competencies among students; 2) Curriculum content. The curriculum must be adapted to integrate materials related to the SDGs, such as texts that discuss social and environmental issues in the Arab world. In this context, lecturers highlight the need for relevant teaching materials to teach sustainability topics; 3) Teaching and learning methods, namely relevant teaching methods such as project-based learning, can be used to connect theory with sustainability practices, and these have been applied to increase students' social awareness through relevant projects. Wiek *et al.* (2011) support this approach by showing that project-based learning is highly effective in promoting a deep understanding of sustainability; and 4) Assessment methods, namely outcome-based assessments, must include an evaluation of students' ability to apply sustainability concepts in the context of the Arabic language. This emphasises the importance of an assessment system that can measure learning outcomes related to the SDGs. According to Lozano *et al.* (2019) the OBA, which measures sustainability achievements, it can strengthen learning and the implementation of the SDGs in higher education.

3.2 *Challenges, opportunities, and strategies in the sustainability curriculum development*

In-depth interviews and discussions with lecturers reveal that integrating sustainability into the curriculum faces significant challenges in higher education. The main issues are limited educational resources, such as relevant teaching materials and competent lecturers. There are no basic guidelines for developing sustainable teaching materials nor sufficient training programs to enhance educators' competence in addressing the SDGs program. The following is an excerpt of the participant's answer.

"The main challenge is the lack of appropriate resources. Many of the materials available still focus on traditional language teaching and do not include a sustainability context. In addition, the time in the syllabus is also limited, so it is difficult to add additional topics without reducing the time for learning the core language." – Participant C.

This problem is not only found in Indonesia, but Wan Norazmi *et al.* (2024) found this problem in Malaysia. It even Madiha *et al.* (2023) revealed this problem in Punjab, where the commitment of students to participate in sustainable programs is also very minimal.

Moreover, Participant B also said, "*The biggest challenge is to find relevant material and by the level of understanding of students. Many of the materials available are too specific or general, making it difficult to adapt them to the existing curriculum.*" In addition, some lecturers resist incorporating sustainability into the curriculum, possibly due to a lack of understanding of its relevance in teaching Arabic. This problem is also confirmed by Weiss *et al.* (2021). His statement is that implementing the sustainability curriculum in higher education faces challenges such as resistance to change and various interests of policy stakeholders, so a transformative approach is needed.

Sustainable curriculum development is possible, even with challenges. When stakeholders understand the principles of the Sustainable Development Goals (SDGs), they can work together to include these in the Arabic Education curriculum through teachers. Support from institutions, such as helpful policies and plans, also makes this integration easier. Furthermore, when students show interest in sustainability issues, it motivates the creation of a more inclusive and globally relevant curriculum. This interest lays a strong foundation for engaging students in interdisciplinary sustainability projects. Stefani *et al.* (2024) added that integrating SDGs content into the curriculum can increase student involvement in the SDGs. Furthermore, interdisciplinary collaboration is also cited as a promising strategy, allowing the integration of sustainability through a more holistic and cross-sectoral approach (Liu *et al.* 2022). In addition, Sok *et al.* (2023) show how developmental psychology in higher education institutions in Cambodia can play a role in achieving SDG targets related to quality education, considering factors such as gender.

To include SDG principles in education, we can create learning modules that combine different subjects and offer ongoing training for teachers. Using project-based learning helps students participate in sustainability activities. This approach improves their language skills, raises social awareness, and links theory with real-world practice on a global scale.

In addition, Cardiff *et al.* (2024) added a content and language-integrated learning course (CLIL) merger strategy that embeds global issues into the curriculum. According to him, this method allows students to engage with various social and environmental topics, expanding their understanding of sustainability beyond traditional environmental issues. The same thing is also applied by Liu *et al.* (2022) only he specifically classifies it in two parts, namely strategies for combining sustainability themes in various disciplines and focusing on developing special programs or courses focusing on sustainability. Although the research settings of Cardiff and Liu are different, the empirical facts presented in their research should be considered for application in Indonesia.

4 CONCLUSION

This study concludes that stakeholders—including program heads, lecturers, and educators—possess a well-developed understanding of sustainability principles (SDGs) and recognise the importance of incorporating these principles into the Arabic education curriculum. They view the SDGs as a crucial framework that enhances students' linguistic competencies while fostering global awareness and social responsibility. Various efforts have been made, such as utilising relevant texts, employing project-based learning methods, and integrating materials that address sustainability issues, including social justice and environmental conservation.

However, implementing these initiatives faces several challenges, including a lack of appropriate teaching materials, limited time allocated in the syllabus, and some lecturers' resistance to adopting changes. Furthermore, inadequate training and resources to support educators in integrating sustainability are significant obstacles.

Despite these challenges, the study identifies substantial opportunities, such as institutional support, government policies advocating for OBE, and increasing student awareness of sustainability issues. Strategies like developing interdisciplinary learning modules, providing ongoing training, and implementing project-based learning demonstrate considerable potential for successful integration.

By adopting a collaborative and cross-sectoral approach, integrating SDGs into Arabic education can be a pivotal step toward creating a more inclusive and globally relevant educational framework, ultimately equipping graduates to tackle sustainability challenges effectively.

REFERENCES

Almahdawi, A. J. (2024). The Extent to Which Grade12 Arabic Textbooks Meet the United Arab Emirates Goals of Sustainable Development in the United Arab Emirates: Grade12 Arabic Textbooks. *American Journal of Multidisciplinary Research and Innovation,* 3(4), Article 4. https://doi.org/10.54536/ajmri.v3i4.3080

Almaliki, M. F., Fahraini, S., and Muslimah, M. (2023). Integrity of Arabic Language Education in The Sustainable Development Goals (SDGs): Opportunities and Challenges in The Indonesian Education Context. *Annual International Conference on Islamic and Science Integration (AICCII)*, 1(1), Article 1.

Barth, M., Godemann, J., Rieckmann, M., and Stoltenberg, U. (2007). Developing key competencies for sustainable development in higher education. *International Journal of Sustainability in Higher Education*, 8(4), 416–430. https://doi.org/10.1108/14676370710823582

Cardiff, P., Polczynska, M., and Brown, T. (2024). Higher education curriculum design for sustainable development: Towards a transformative approach. *International Journal of Sustainability in Higher Education,* 25(5), 1009–1023. Scopus. https://doi.org/10.1108/IJSHE-06-2023-0255

Fitramadhana, R. (2023). Education in the Midst of Indonesia's Development Agenda. *Jurnal Sosiologi Pendidikan Humanis*, 8(1), 55. https://doi.org/10.17977/um021v8i1p55-81

Herlanti, Y., Nobira, S., Kuboki, Y., and Qumilaila, Q. (2024). Online lesson study design: Integrating environmental issues with science learning to enhance students' environmental literacy. *International Journal for Lesson and Learning Studies*. https://doi.org/10.1108/IJLLS-08-2024-0169

Jamhari, J. (2009). New Trend of Islamic Education in Indonesia. *Studia Islamika,* 16(2), Article 2. https://doi.org/10.15408/sdi.v16i2.482

Liu, J., Watabe, Y., and Goto, T. (2022). Integrating sustainability themes for enhancing interdisciplinarity: A case study of a comprehensive research university in Japan. *Asia Pacific Education Review*, 23(4), 695–710. https://doi.org/10.1007/s12564-022-09788-z

Lozano, R., Barreiro-Gen, M., Lozano, F. J., and Sammalisto, K. (2019). Teaching Sustainability in European Higher Education Institutions: Assessing the Connections between Competences and Pedagogical Approaches. *Sustainability*, 11(6), 1602. https://doi.org/10.3390/su11061602

Madiha, K., Naima, Q., and Khan, A. M. (2023). Challenges of Implementing Education for Sustainable Development: University Teachers' Perspectives. *PJE*, 39(1). https://doi.org/10.30971/pje.v39i1.876

Mambu, J. E. (2023). Embracing Sustainable Development Goals Critically to Explore Life Purposes in English Language Teacher Education. *TEFLIN Journal - A Publication on the Teaching and Learning of English*, 34(2), 264–282. https://doi.org/10.15639/teflinjournal.v34i2/264-282

Maswani, M. (2022). Bahasa arab qurani (model pembelajaran bahasa arab berbasis religiusitas pada orang dewasa) [Disertasi, Sekolah Pascasarjana UIN Syarif Hidayatullah Jakarta]. https://repository.uinjkt.ac.id/dspace/handle/123456789/71150

Rababah, H. A. (2023). The Advent of Islam has launched the Greatest Arabic Language Planning Process Ever-made in History for a Language. *European Journal of English Language and Literature Studies*, 11(4), 1–20.

Ruiz, M. R., Bula, L. M., Meza, L. G., and Luna, A. G. (2024). Perspectives for The Implementation of The Sdgs as An Innovative Strategy for Teaching English in Professional Practice Scenarios. *Journal of Law and Sustainable Development*, 12(2), e3402. https://doi.org/10.55908/sdgs.v12i2.3402

Sergey, B., Alfiya, A., Gulmira, A., Ainura, A., and Alexandr, P. (2024). Education for sustainable development: Comparative analysis and prospects at universities in Kazakhstan, Tajikistan, Kyrgyzstan and Uzbekistan. *Discover Sustainability*, 5(1), 140. https://doi.org/10.1007/s43621-024-00343-x

Sok, S., Chhinh, N., Cheb, H., Bo, C., and Nguonphan, P. (2023). Developmental psychology within Cambodian higher education institutions: How gender influences the achievement of the SDG targets related to quality education. *International Journal of Educational Management*, 37(6/7), 1498–1522. https://doi.org/10.1108/IJEM-09-2022-0350

Stefani, S. R., Souza, M. D. de, Silveira, A. R., Liz, L. P. de, and Zampier, M. A. (2024). Assessing the awareness and engagement of management students with the sustainable development goals: A case study at Unicentro. *Review of Sdgs in Emerging Countries,* 6(sdgs), Article sdgs. https://doi.org/10.37497/2965-7393.SDGs-Countries.v6.nsdgs.64

Stoltenberg, U., and Michelsen, G. (2023). Theoretical Reflections on Education for Sustainable Development and Digital Technologies. *In Digitalization, New Media, and Education for Sustainable Development* (pp. 1–13). IGI Global. https://doi.org/10.4018/978-1-7998-5033-5.ch001

Suwendi, S., Gama, C. B., Farabi, M. F. F., Fuady, F., and Arman, A. (2024). Roles and Challenges of Pesantren Intellectual Networks. *Jurnal Ilmiah Islam Futura*, 24(2), 453. https://doi.org/10.22373/jiif.v24i2.23134

Terry, G., Hayfield, N., Clarke, V., and Braun, V. (2017). Thematic analysis. *The SAGE Handbook of Qualitative Research in Psychology*, 2(17–37), 25.

Thohri, M. (2024). Development of Arabic Language Curriculum Focused on Diversity and Inclusion. *An Nabighoh*, 26(1), 31–50. https://doi.org/10.32332/annabighoh.v26i1.31-50

Tyler, R. W. (2013). *Basic Principles of Curriculum and Instruction*. Univ. of Chicago Press.

UNESCO. (2017). *Education for Sustainable Development Goals: Learning objectives*. UNESCO. https://doi.org/10.54675/CGBA9153

Wan Norazmi, W. N. A., Syed Azman, S. M., and Abd Wahid, N. H. (2024). Exploring Issues and Challenges in Integration of Sustainable Development Goals (SDGs) in The Agricultural Curriculum in National Secondary Schools (SMK) Malaysia. *International Journal of Modern Education*, 6(20), 161–175. https://doi.org/10.35631/IJMOE.620013

Weiss, M., Barth, M., Wiek, A., and Von Wehrden, H. (2021). Drivers and barriers of implementing sustainability curricula in higher education—assumptions and evidence. *Higher Education Studies*, 11(2), 42. https://doi.org/10.5539/hes.v11n2p42

Wiek, A., Withycombe, L., and Redman, C. L. (2011). Key competencies in sustainability: A reference framework for academic program development. *Sustainability Science*, 6(2), 203–218. https://doi.org/10.1007/s11625-011-0132-6

Yin, R. K. (2018). *Case Study Research and Applications: Design and Methods* (Sixth edition). SAGE.

Towards Resilient Societies: The Synergy of Religion, Education, Health, Science, and Technology – Rahiem (Ed.)

Spiritual intelligence and the Islamic education worldview: Insights from Tafsir *Fi Zhilalil Qur'an*

Lalu Muhammad Nurul Wathoni
Mataram State Islamic University, Mataram, Indonesia
ORCID ID: 0000-0003-4168-6224

Abdul Halim
Sekolah Tinggi Agama Islam Ibnu Sina Batam, Indonesia
ORCID ID: 0000-0002-5618-2486

Zainal Arifin Haji Munir
Mataram State Islamic University, Mataram, Indonesia
ORCID ID: 0000-0002-7915-0065

ABSTRACT: This research aims to provide a fundamental, holistic, and comprehensive understanding of Spiritual Quotient (SQ) as a basis for developing Islamic education rooted in the worldview paradigm, as interpreted through Sayyid Qutb's *Fi Zhilalil Qur'an*. Conducted as a qualitative descriptive library study, the research examines Qutb's thoughts on SQ, focusing on primary sources, including the six-volume *Fi Zhilalil Qur'an* and *Ma'alim fi Ath-Thariq*. A historical and philosophical approach guides the study, with data collected using documentary techniques and analysed through content analysis, emphasising descriptions, consequences, and implications. The findings highlight that SQ, as conceptualised in Qutb's tafsir, encompasses the spirit of *rahmatan lil 'alamin* (mercy to all creation), universality, theological purity (*qauliyah*), and a balance between theological and ecological values (*qouliyah* and *kauniyah*). SQ represents an internal intuitive capability that activates theological values, fostering environmental and humanistic principles through harmonious and dialogical relationships within society and the universe. This perspective positions SQ as a critical element in actualising Islamic education's humanistic and ecological objectives.

Keywords: spiritual quotient, basic, Islamic education, worldview

1 INTRODUCTION

Spiritual Quotient (SQ) is a concept developed by scholars and has been growing rapidly universally in recent times. As Kuhn (Ushuluddin *et al.* 2021) stated about the paradigm shift (IQ), (EQ), and (SQ) as a scientific revolution.

Spiritual Quotient (SQ) is one of the important dimensions that underlie various human crises today. As Nasr sees the fundamental emptiness in all aspects of life due to the loss of spiritual dimension (Ushuluddin *et al.* 2021). The fading of this spiritual dimension has had a terrible impact on various principles of life. Emptiness will understand the essence of oneself, usefulness to others, and awareness of the evidence of divine truth makes man negligent, cramped, stiff and not good at gratitude or what Hossein Nasr used to call positivistic-anthropocentric (Nasr 1968, 2020).

DOI: 10.1201/9781003645542-43

An interesting fact shows that the incessant discussions such as seminars, health, social, economic, psychology, religion and even education and learning actively dissect the SQ paradigm. As Jane I. Lu (2021) seeks to develop students' self-awareness based on spirituality is an important indicator. Wiggles Worth (Mafuzah Mohamad 2020) views SQ as able to maximise student learning achievement. The same thing is also intended by Hensey (2019) and Cseh *et al.*, (2024). Rajabi *et al.* (2023) stated the importance of the SQ dimension in today's educational performance.

From this fact, SQ is important to be considered as a worldview Islamic education. As with the 1977–1982 Islamic education world conference, Muslim students must "think precisely and logically through the spirituality of the truth (Tan and Ibrahim 2017)

Unfortunately, SQ still leaves problems in the review of Islamic education that have not touched the theological-qouliyah values (Ushuluddin *et al.* 2021). This sparked Fitriyah *et al.* (2021) to introduce the Islamic model of spirituality and its foundations and moral aspects.

Various views of Muslim thinkers and scholars also do not want to be left behind in dissecting the SQ paradigm (Abdalla *et al.* 2022; Dodi and Abitolkha 2022; Feizi *et al.* 2020; Khalil 2020; Oparin 2020; Sukmawati *et al.* 2022; Wahab 2022). For example, Najati (Muhammad Utsman Najati 1992; Utsman Najati 2008) refers to it as al-nafs al-mutmainnah. Al-Ghazali said "tazkiyatun nafs" (Hanefar *et al.* 2016; Sukmawati *et al.* 2022; Wahab 2022; Xavier and Dickson 2021). SQ with a broader meaning, and able to reach fundamental values (Bensaid *et al.* 2014; Dodi and Abitolkha 2022; Ghazal 2021; Khalil 2020; Mawardi 2022; Oparin 2020; Rababa and Al sabbah 2023; Sukmawati *et al.* 2022; Wahab 2022; Xavier and Dickson 2021).

Such an SQ paradigm is urgently needed to develop Islamic education with a worldview that emphasises human values and the universe in complete unity. One of the phenomenal Muslim intellectuals, Sayyid Quthb, stated that Islam is not just a string of words and rhetoric but is present in life (Sayyid Quthb 1979; William E. Shepard 1996).

This research focuses on exploring SQ in the interpretation of Fi Zhilalil Qur'an which is urgently needed by the education of the Muslim generation Worldview. As Jamil and Matondang (2017) insist that Muslim education must become a "total world".

Various studies discuss Quthb's thoughts, for example, Umbar and Istiqomah (2019) "Ali Syari'ati's Perspective of Humanism Value in Diwan Sayyid Quthb". Jamil and Matondang (2017) "The Eduction Thoughts of Sayyid Qutb in the Tafsir of Fi Zhilal Al-Qur'an, "The Concept of Family Education According to Sayyid Quthb (Study of Surat Thaha Verse 132 in Tafsir Fi Zhilalil Qur'an). Suheri Sahputra Rangkuti (2018) "The Values of Islamic Education in the Tafsir of the Ajat Jihad."

2 METHODS

This research is library research with qualitative description. This approach is used to get messages that contain meaning and value in the book Fî Zhilâlil Qur'ân to develop theories and describe reality and the complexity of the phenomenon under study. The object of the research explores Quthb's thoughts in Fî Zhilâlil Qur'ân about SQ. The subject of the study is Quthb as a figure in da'wah and the Egypt movement who was born on October 9, 1906 AD (1324 H). Meanwhile, the tafsir of Fî Zhilâlil Qur'ân used is published by Darusy Suruq, Beirut, in 1142 H/2004 AD 4th/5th edition.

Primary data sources are cited directly through Sayyid Quthb's books which include; First, Fî Zhilâlil Qur'ân, there are 6 volumes, published by Darusy Suruq, Beirut, in 1142 H/ 2004 AD 4th/5th edition. Second, Ma'âlim Fî Ath-Thariq, published by Darusy Syuruq, Beirut 1399 H / 1979 AD. This book has been translated by Mahmud Harun Muchtarom with the title, Ma'âlim Fî Ath-Thariq; Road Directions that Shake Faith, published in Yogyakarta, by Darul Uswah, in 2009. Secondary data sources through sinta.kemdikbud.

go.id searches, Scopus.com., Google Socolar, DOAJ (Directory of Open Access Journal), Garuda, Mendeley.Com., and other media.

This research uses a historical and philosophical approach. Data collection techniques using documentary techniques. Data analysis techniques using Content Analysis with descriptions, consequences and implications.

3 RESULT

3.1 *Biography Sayyid Quthb*

Sayyid Quthb Ibrahim Husaini Shadhilli's full name. He was born on October 9, 1906 AD (1324 H) in the village of Musyah, Ashut City, Egypt. In 1933 Quthb graduated with a bachelor's degree in education and worked as a school superintendent in the education department.

In 1949, Quthb continued his studies at Wilson's Teacher College and Stanford University, and in 1953, Quthb joined the Muslim Brotherhood. In 1954 Quthb was temporarily detained, in 1955 Quthb was accused of subversive, agitation and others until he was sentenced to 15 years in prison. (Musallam, n.d.) In 1964 Quthb was released with the help of the Iraqi president, in 1965 Quthb was again detained and on August 29, 1966 Quthb was sentenced to be hanged (Sayyid Quthb, n.d.).

Quthb's writing style in his works marked the break with classical scientific conventions. KHis alimat is short, the use of exclamation marks that often makes his work dramatic and urgent. The tendency of the tafsir of Fî Zhilâlil Qur'ân uses the pattern of da'wah wal harokah which means stepping and moving in continuity through systematic tactics and strategies in calling people to Islam with wisdom and advice of goodness and virtue, so that they stay away from thaghût and believe only in Allah SWT.

3.2 *SQ: a theological fact*

In this section, it is necessary to bring up the theological facts of SQ. As De Klerk-Weyer and Le Roux and Goleman view, SQ is part of emotional intelligence and not self-contained intelligence that can be measured. Dallas Willard (Dallas Willard 1988) said that the content of faith contributes significantly to SQ. Frank MacHovec called a new theory of personality that identifies spiritual transcendence as the ultimate goal of humanity. MacHovec views SQ as universal and finds evidence of it in sacred religious texts as well as in the humanities in general (Frank MacHovec 2002).

Several studies have emerged on the relationship between religion and spiritual intelligence, for example Wahab, (Wahab 2022) Sukmawati *et al.*, (Sukmawati *et al.* 2022) and Cseh *et al.*, (Cseh *et al.* 2024) reveal life-changing spiritual experiences trigger religious attitudes and behaviors.

3.3 *SQ: a harmonisation of theological, sociological and ecological facts*

SQ is a person's inner ability (intuition) that entrusts the creation of a harmonious relationship for IQ and EQ to access divine values (faith), which are manifested in social, universal (worship) and manifested in wise and wise behaviour (morals). As Noegroho and Wulansari (Noegroho and Wulansari 2020) view SQ as the ability to interpret life correctly. SQ is the ability to adapt to emotional and social awareness of the surrounding environment. (Khamida *et al.* 2019; Mitsea *et al.* 2023; Skrzypińska 2021) As Tony Buzan (Buzan 2001) stated SQ as the natural potential of human beings through social awareness and appreciation for the universe. Buzan refers to SQ as a "spirit" that is connected in humans, society and the universe.

3.4 *SQ as the base of islamic education worldview*

Departing from the harmonious relationship of theological, humanist and ecological facts above, it shows that SQ is a need to be maximised in the development of Islamic education that is Rahmatal Lil 'Alamin. As Abdul Halim, (Abdul Halim 2022) views SQ as the ability to activate theological-qouliyah values towards humanist-insaniyah and ecological-kauniyah through harmonious dialogical relations within the scope of society, society, and the surrounding universe.

This SQ is intended as the base of Islamic education with the face of Worldview, which is able to penetrate space and time and has universal awareness. As Vargas Herrera and Moya Marchant (Vargas Herrera and Moya Marchant 2020) stated that the goal of education is the spiritual development of all residents who are schooled. Jane Hensey (Hensey 2019) makes SQ an important dimension in equipping hard skills well.

3.5 *Spiritual quotient as the basic of islamic education worldview: study of tafsir Fi Zhilalil Qur'an by Sayyid Qutb*

Departing from the understanding of SQ above, it is important to dig deeper into SQ through Quthb's thoughts in the tafsir of Fî Zhilâlil Qur'ân. Quthb, as a contemporary Muslim intellectual, understands Islam very well, not just a string of words, and rhetoric, but present in life (Sayyid Quthb 1979).

Quthb displays spiritual maturity with solid and applicative principles of faith,

"It is a big mistake in the parameters of Islam. When faith only crystallises in a "theory" format, it is limited to intellectual studies or cultural epistemes. Even this kind of view is very dangerous to Islam." (Sayyid Quthb 1979).

3.5.1 *SQ as the spirit of rahmatal Lil 'Alamin*

This is SQ which hints at the religious command by entrusting its relationship to all its creations.

"Allah has sent His Messenger as a mercy for all mankind to lead him to His guidance. This last treatise appeared when the human intellect was mature. He came with a book that was open to all minds in every generation, covering all the unchanging points of human life. He is always ready to meet all the continuous, new human needs, known to God the Creator of man" (Sayyid Quthb 1425). From this it is affirmed that Quthb views the maturity of human reason as associated with a mind that is open to all aspects of life, including religion, humanities, humanities and scientist-ecology that is not limited by space and time. Islam provides breadth to humans in developing their potential in accordance with the needs of the developing times.

3.5.2 *SQ as a universal spirit*

Islam is very open, gentle, voluntary and firm in introducing the values of the faith by presenting information and explanations first, after which it is completely surrendered to make a choice, not forcing and pressing (Sayyid Quthb 1425).

"There is no coercion in religion (Islam)" (Sayyid Quthb 1425). "Freedom of religion is a human right. So a person who strips man of his freedom of freedom of faith means that he has stripped him of his glory" (Sayyid Quthb 1425).

This is what SQ means in the tafsir Fi Zhilalil Qur'an Sayyid Quthb, where Islam is not a harsh teaching, dogma, backwards as the term is embedded today, radical, intolerant, terrorist and others. For Quthb, Islamic principles always test human rationality through attitudes and behaviours of openness, politeness, gentleness, and voluntarily providing breadth.

3.5.3 *SQ as the spirit of theological purity-qauliyah*

In this rhythm, Quthb opens the dish by displaying the end for those who are mukhlis (Sayyid Quthb 1425). This is where the important point of attitude and behaviour of "mukhlis" as SQ which guides every human behaviour by eliminating resentment in the heart and will give birth to a sense of brotherhood that displays attitudes and behaviours of mutual love, mutual embrace and mutual glory through relationships of equality and togetherness in the social order.

3.5.4 *SQ as a theological-qouliyah spirit towards humanist-insaniyah*

In this sub-discussion, a harmonious dialogical relationship between religion and humanity will emerge as the spirit of SQ, as Quthb opens the dish about the principle of structuring the human side that is always connected to the spirit of religion, making humans to one origin, one family and making the humanist-humanist integrity which incidentally is a "nafs" (soul) that directs piety to their god and the maintenance of family. It is intended to uphold the great principles in the social, social, cultural and religious order that prioritise solidarity attitudes and behaviours, mutual love in a family, and establish human relations that are full of truth and virtue (Sayyid Quthb 1425). Quthb emphasised that Islam highly upholds the human side, which has been untouched in general arrangement of people's lives.

From this, it is clear that an intelligent soul is the ability to view humanist social reality as the integrity of religious commands that value human values by displaying polite attitudes and behaviours, being gentle, open, prioritising the needs of others, not hurting the feelings of the weak and glorifying the poor, women and orphans. This is spiritual intelligence (SQ) which entrusts a harmonious relationship from the spirit of religion to the spirit of humanity by seeking attitudes and behaviors that complement and strengthen each other. The unity between the two is a new spirit for the worldview of Islamic education paradigm, which is considered very much needed in this contemporary era.

3.5.5 *SQ as a theological-qouliyah spirit towards ecological-kauniyah*

In the discussion of the Attributes of Ulul Albab, Quthb gave a briefing on the harmonious relationship between the creation of the universe as the embodiment of the spirit of religion that is integrated in a whole, complementary and mutually reinforcing. It is intended that the universe is an open expanse with the guidance of faith in it, there is the power of the hand that arranges behind it wisely and there is the afterlife, hisab and retribution.

"Isn't human awareness of the majesty and beauty of the expanse of the universe in the form of the alternation of night and day as a unique phenomenon that shows the rotation of the earth in front of the sun, the regularity of the sky and the earth has attracted attention to the existence of attraction (gravity) or not. These are some of the natural wonders of kauniyah along with the law of its careful and careful arrangement in regulating and maintaining it" (Sayyid Quthb 1425).

This paints a picture so beautiful that it shines like a pearl that is felt for healthy and intelligent minds. As Quthb called it, the mind (feeling) "Ulul Al-Bab" is the ability to perceive the impressions of the universe through the design of the creation of heaven and earth and alternate night and day (Sayyid Quthb 1425).

4 DISCUSSION

Departing from the understanding of SQ above, the author views how Islam actually pays great attention to the values of humanity, humanities and scientist-ecology as an important part that should not be ignored in religious understanding and religious behavior anywhere and anytime. Likewise, religious values as a guide and enlightenment,, as well as filtering,, should be included in the structuring, search and development of humanity, humanities and scientific-ecological values through dialogical harmonisation, openness, criticality, mutual penetration (semi verbal), complementing and strengthening each other.

This SQ dimension is urgently needed as the ability to activate and connect religious values in the arrangement and search for the values of humanity, humanities and universal values. The integrity of SQ, which considers the importance of humanist-humanist and ecological-kauniyah spirit, is a fundamental need as the basis of the Islamic education Worldview, which is currently under sharp scrutiny by contemporary Muslim thinkers. The impasse of Islamic education that presents a jumud, antagonist, and ritualistic face needs to be ended by opening a more contextual learning space that involves humanist and ecological spirits.

As Karim, (Karim 2022) found facts in the community in Tambakromo, Pati Regency, Central Java, that the community's religious understanding has not touched environmental issues.

Likewise, Dodi and Abitolkha, (Dodi and Abitolkha 2022) stated that spiritual teachings that focus on the practice of riyadhah (inner movement) will be able to reduce and resolve religious conflicts through several principles.

This is where the importance of spiritual intelligence (SQ) lies as the basis of Islamic education Worldview by strengthening in two directions of search, namely the function of spiritual intelligence (SQ) as a guide, motivation, inspiration and the function of spiritual intelligence (SQ) as a basis, limitation as well as strength. The integration of the direction of the search for spiritual intelligence (SQ) will produce the performance of Islamic education both substantively philosophically and methodologically, which entrusts the harmonious relationship of religious behaviour, humanity, and the universe.

The openness of the spiritual intelligence (SQ) concept is a solution, prospective, applicative and productive effort in relation to Islamic education in this contemporary era. S that Islamic education will display a new face that is more polite, cool, harmonious, critical, and 'wise' in terms of its relationship with human behavior and the surrounding environment. As Amroeni Drajat, *et al.*, in their research results stated that the concept of children's education in the Qur'an emphasizes the element of "affection" like the upbringing of parents to children in the concept of tarbiyah (Ridwan *et al.* 2022).

Spiritual intelligence (SQ) by uniting theological-qouliyah, humanist-insaniyah and ecological-kauniyah values is a solution for improving and renewing Islamic education that is natural, humanist and religious. The necessity of Islamic education that makes spiritual intelligence (SQ) a dialogical partner must be formulated in such a way at the learning level that habituation exercises are no longer limited to religious commandments but are able to touch the human side and the wisdom of the environment as a holistic and comprehensive order. Likewise, the search and development of human behavior and the products of scientists and technology also do not leave the spirit of religion as the basis of its strength.

5 CONCLUSION

Spiritual Quotient (SQ) is a product of scientists and scientists that has been growing rapidly universally in recent times. Spiritual intelligence (SQ) is a person's internal ability (intuition) that entrusts the creation of a harmonious relationship for IQ and EQ to access divine values (faith) that are manifested in social, universal (worship) and manifested in wise and wise behaviour (morals). Spiritual intelligence (SQ) as the basis of Islamic education Worldview in the tafsir of Fi Zhilalil Qur'an by Sayyid Quthb includes; the spirit of rahmatal Lil 'Alamin, the spirit of universa, the spirit of theological purity-qauliyah, the spirit of theology-qouliyah towards humanists-humanists, the spirit of theology-qouliyah towards ecological-kauniyah. This is SQ an internal ability (intuition) that activates theologian-qouliyah values as an effort to actualise them to humanist-humanist and ecological-kauniyah values through harmonious dialogical relations within the scope of society, society and the surrounding universe.

REFERENCES

Abdalla, M., Aisyah, N., Aziz, A., Syairah, N., Rani, A., Waqiyuddin, M., Hakimi, M. Rashid, M., Syazwanee, M., Iman, N., Amiri, M., Ghani, A., Reza, F., and Muzaimi, M. (2022). A review of the holy

Quran listening and its neural correlation for its potential as a psycho spiritual therapy Mo-ham med. *Heliyon*, 8(July), e12308. https://doi.org/10.1016/j.heliyon.2022.e12308
Abdul Halim. (2022). Konsep Spiritual Quotient dalam Tafsir Fî Zhilâlil Qur'ân Karya Sayyid Quthb dan Implikasinya Terhadap Pendidikan Islam. *Disertasi*, 227.
Bensaid, B., Machouche, S. B. T., and Grine, F. (2014). A Qur'anic framework for spiritual intelligence. *Religions*, 5(1), 179–198. https://doi.org/10.3390/rel5010179
Buzan, T. (2001). The Power of Spiritual Intelligence: 10 Ways To Tap Into Your Spiritual Genius. Thorsons.
Cseh, O., Karsai, I., and Szabo, A. (2024). The relationship of life changing spiritual experiences to current religious/spiritual attitudes and practices: A pilot study. *Pastoral Psychology*, 73(2), 227–238. https://doi.org/10.1007/s11089-023-01120-9
Dodi, L., and Abitolkha, A. M. (2022). From Sufism to resolution: Examining the spiritual teachings of Tarekat Shiddiqiyyah as the theology of peace in Indonesia. *Qudus International Journal of Islamic Studies*, 10(1), 141–174. https://doi.org/10.21043/qijis.v10i1.11260
Feizi, S., Nasiri, M., Bahadori, H., and Hosseini, M. (2020). The relationship between spiritual well being and happiness among healthcare students: Application of the spiritual health questionnaire for the Iranian population. *Heliyon*, 6(11), 1–7. https://doi.org/10.1016/j.heliyon.2020.e05448
Fitriyah, F. K., Hidayah, N., Muslihati, M., Hambali, I., and Ibad, M. (2021). The role of demographic characteristics and spiritual dimensions in predicting empathy: A study in muslim pre service teachers. *Islamic Guidance and Counseling Journal*, 4(2). https://doi.org/10.25217/igcj.v4i2.1553
Ghazal, A. (2021). Ibadi Muslims of North Africa: Manuscripts, mobilization, and the making of a written tradition by Paul M. Love Jr. *Journal of Islamic Studies*, 32(2), 305–308. https://doi.org/10.1093/jis/etab001
Hanefar, S. B., Sa'ari, C. Z., and Siraj, S. (2016). A synthesis of spiritual intelligence themes from Islamic and Western philosophical perspectives. *Journal of Religion and Health*, 55(6), 2069–2085. https://doi.org/10.1007/s10943-016-0226-7
Hensey, J. (2019). EQ + SQ = ST2: A spiritually intelligent approach to actor training. Part two: Principles in practice. *Stanislavski Studies*, 7(2), 219–228. https://doi.org/10.1080/20567790.2019.1644749
Jamil, S., and Matondang, A. Y. (2017). The eduction thoughts of Sayyid Qutb in the tafsir of Fi Zilal Al Qur'an. *International Journal on Language, Research and Education Studies*, 1(1), 53–66. https://doi.org/10.30575/2017081203
Karim, A. (2022). Integration of religious awareness in environmental education. *Qudus International Journal of Islamic Studies*, 10(2), 415–442. https://doi.org/10.21043/qijis.v10i2.14404
Khalil, A. (2020). Ethics and spirituality in Islam: Sufi adab, edited by Francesco Chiabotti, Eve Feuil lebois Pierunek, Catherine Mayeur Jaouen, and Luca Patrizi. *Journal of Islamic Ethics*, 4(1–2), 268–273. https://doi.org/10.1163/24685542-12340039
Khamida, K., Mubarak, R., and Budury, S (2019). Relationship between spiritual quotient and self adjustment of students at Jabal Nor Islamic Boarding School, Sidoarjo, Indonesia. *Journal of Public Health in Africa*. https://doi.org/10.4081/jphia.2019.1184
Lu, J. I. (2021). Educational models of spiritual formation in theological education: Introspection based spiritual formation. *Teaching Theology and Religion*, 24(1), 28–41. https://doi.org/10.1111/teth.12560
MacHovec, F. (2002). *Spiritual intelligence, the behavioral sciences, and the humanities* (p. 291). Lewiston, New York.
Mafuzah Mohamad. (2020). Lurking on the essential attributes required in Industrial Revolution 4.0. *Global Business and Management Research*, 12(4), 570.
Mawardi, K. (2022). Nationalism and spiritualism of Javanese tarekat: Study of Tarekat Rinjani in Banyumas Central Java. *Qudus International Journal of Islamic Studies*, 10(1), 75–108. https://doi.org/10.21043/qijis.v10i1.12509
Mitsea, E., Drigas, A., and Skianis, C. (2023). VR gaming for meta skills training in special education: The role of metacognition, motivations, and emotional intelligence. *Education Sciences*, 13(7), 639. https://doi.org/10.3390/educsci13070639
Muhammad Utsman Najati. (1992). al Hadith al Nabawi wa 'Ilm al Nafs (5th ed.). *Dar al Shuruq*.
Musallam. (n.d.). No Title.
Najati, U. (2008). The ultimate psychology; Psikologi sempurna ala Nabi SAW. Pustaka Hidayah.
Nasr, S. H. (1968). *The encounter of man and nature: The spiritual crisis of modern man. George Allen and Unwin.*
Nasr, S. H. (2020). The environmental crisis in the Islamic world pertinence of the teachings of traditional Islam. *In Handbook of Ethics of Islamic Economics and Finance* (pp. 156–165). https://doi.org/10.1515/9783110593419-008
Noegroho, A. D., and Wulansari, N. A. (2020). Religiosity as a moderation of the effect of intellectual and spiritual quotient. *Management Analysis Journal*, 9(2), 113–121.

Oparin, D. (2020). Spiritual authority and religious introspection among Muslim migrants in Western Siberia. *Problems of Post Communism*, 67(4–5), 362–374. https://doi.org/10.1080/10758216.2019.1616566
Rababa, M., and Al sabbah, S. (2023). The use of Islamic spiritual care practices among critically ill adult patients: A systematic review. *Heliyon*, 9(3), 1–10. https://doi.org/10.1016/j.heliyon.2023.e13862
Rajabi, R., Eslami Aliabadi, H., Javad Mahdizadeh, M., and Azzizadeh Forouzi, M. (2023). A comparative study of religious beliefs, spiritual intelligence and spiritual well being in two therapies based on education (anonymous drug user) and methadone in drug user in Iran. *BMC Research Notes*, 16(1), 4–9. https://doi.org/10.1186/s13104-023-06377-0
Rangkuti, S. S. (2018). Suheri Sahputra Rangkuti.
Ridwan, A., Pasaribu, S., Drajat, A., and Salminawati, S. (2022). The concept of children's education from the Qur'an perspective. *Jurnal Basicedu*, 6(2), 2231–2241.
Sayyid Quthb. (n.d.). Sayyid Quthb.
Sayyid Quthb. (1425). Fi Zhilalil Qur'an, Jilid 1, Juz 4 (4/5). *Darusy Suruq.*
Sayyid Quthb. (1979). Ma'alim Fi Ath Thariq. *Darusy Syuruq.*
Shepard, W. E. (1996). *Sayyid Quthb and Islamic activism. Bril.*
Skrzypińska, K. (2021). Does spiritual intelligence (SI) exist? A theoretical investigation of a tool useful for finding the meaning of life. *Journal of Religion and Health*, 60(1), 500–516. https://doi.org/10.1007/s10943-020-01005-8
Sukmawati, F., Sumin, S., Istiyono, E., Widhiastuti, W., and Mardiyati, I. (2022). Muslim spiritual happiness scale: The instrument development and validation. *Islamic Guidance and Counseling Journal*, 5(2), 135–151. https://doi.org/10.25217/igcj.v5i2.2754
Tan, C., and Ibrahim, A. (2017). Humanism, Islamic education, and Confucian education. *Religious Education*, 112(4), 394–406. https://doi.org/10.1080/00344087.2016.1225247
Umbar, K., and Istiqomah, H. (2019). "Ali Syari'ati's perspective of humanism value in Diwan Sayyid Quthb." *Al Arabi: Journal of Teaching Arabic as a Foreign Language*, 3, 155–156.
Ushuluddin, A., Madjid, A., Masruri, S., and Affan, M. (2021). Shifting paradigm: From intellectual quotient, emotional quotient, and spiritual quotient toward ruhani quotient in ruhiology perspectives. *Indonesian Journal of Islam and Muslim Societies*, 11(1), 139–162. https://doi.org/10.18326/IJIMS.V11I1.139-162
Vargas Herrera, F., and Moya Marchant, L. (2020). Analyzing the concept of spiritual development based on institutional educational projects from schools located in Valparaíso, Chile. *Religious Education*, 115(2), 201–214. https://doi.org/10.1080/00344087.2019.1677986
Wahab, M. A. (2022). Islamic spiritual and emotional intelligence and its relationship to eternal happiness: A conceptual paper. *Journal of Religion and Health*, 61(6), 4783–4806. https://doi.org/10.1007/s10943-021-01485-2
Willard, D. (1988). *The spirit of the disciplines: Understanding how God changes lives. Harper Collins Publishers.*
Xavier, M. S., and Dickson, W. R. (2021). Between Islam and the new age: The Jerrahi Order and categorical ambiguity in the study of Sufism in North America. *Religion Compass*, 15(1), 1–10. https://doi.org/10.1111/rec3.12383

Towards Resilient Societies: The Synergy of Religion, Education, Health, Science, and Technology – Rahiem (Ed.)

A meta-analysis of the effectiveness of problem-based learning on science literacy

Lalu Usman Ali
Universitas Sebelas Maret, Indonesia
Universitas Islam Negeri Mataram, Indonesia
ORCID ID: 0000-0001-7635-2588

Suranto & Meti Indrowati
Universitas Sebelas Maret, Indonesia

Suhirman
Universitas Islam Negeri Mataram, Indonesia

ABSTRACT: This study presents a meta-analysis assessing the impact of Problem-Based Learning (PBL) on scientific literacy in science education. The research followed the PRISMA procedure to synthesise findings from 10 empirical studies between 2015 and 2024. Using a random effects model with Hedges' g statistic, the study found an overall effect size of 0.93 (95% CI: 0.15–1.71), indicating a moderate to large positive effect of PBL on scientific literacy compared to traditional teaching methods. Heterogeneity analysis revealed an I^2 value of 94%, suggesting substantial variability across the studies. Variations in effect size were influenced by educational attainment and geographic location, with tertiary students showing more significant improvements in scientific literacy than secondary school students. Additionally, the Trim and Fill analysis confirmed no publication bias, reinforcing the validity of the findings. These results underscore the effectiveness of PBL in enhancing scientific literacy and improving educational quality. The study recommends broader adoption of PBL methodologies across educational levels to support the development of critical thinking and scientific understanding in students.

Keywords: digital literacy, meta-analysis, problem-based learning, science learning, science literacy

1 INTRODUCTION

Education is very important to achieving the fourth Sustainable Development Goals (SDGs) goal because it facilitates learning and shapes individual attitudes and behaviours to benefit the environment (Krayneva *et al.* 2021; McCaw 2022). Quality education enables individuals to actively address challenges such as climate change, social injustice, and global health crises (Donkoh *et al.* 2023; Grabau *et al.* 2024). 21st-century education requires alterations in methodology and pedagogical approaches to prepare youth for emerging difficulties and opportunities, highlighting the importance of scientific literacy (Muchtar *et al.* 2018; Perdana and Rudibyani 2020). The revolution in information and communication technology has markedly enhanced the capture, processing, and application of information in the modern period (Anttila 2018). Thus, an individual's ability to understand scientific concepts and grasp technology is critically important.

These qualities affect how individuals engage with the world and dictate their ability to tackle the complex difficulties confronting society today (Tasquier *et al.* 2022). Therefore, it

DOI: 10.1201/9781003645542-44

is imperative to integrate high-quality science education into the curriculum to equip future generations with the knowledge and skills essential for innovation and adaptation in a constantly changing environment. Effective science instruction must integrate interactive and problem-based approaches, allowing students to actively engage with applying scientific principles in everyday life (Turiman *et al.* 2012).

Besides, science learning should also be a core subject to prepare students with critical and creative thinking and problem-solving skills (Saputro *et al.* 2021). However, science learning is often stuck in traditional methods that tend to be teacher-centred and do not involve students actively in the learning process.

Problem-Based Learning (PBL) is an educational approach centred on students, emphasising solving real-world problems to facilitate knowledge acquisition (Ali *et al.* 2023; Putranta 2019). PBL has been shown to enhance student engagement and foster the development of higher-order thinking skills, including analysis, synthesis, and evaluation. Research findings indicate that PBL positively impacts learning outcomes. Amaringga *et al.*, (2021) demonstrated that problem-based modules significantly enhanced students' science literacy. Furthermore, the study by Scholkmann *et al.*, (2023) corroborated these findings by demonstrating a strong correlation between PBL and science literacy. Multiple studies have corroborated the efficacy of PBL in enhancing comprehension of complex concepts, critical thinking skills, learning motivation, and problem-solving abilities (Ernawati *et al.* 2022, 2023; Hmelo-Silver 2004).

Studies examining the efficacy of PBL in educational settings yield varied outcomes. Certain studies indicate notable advancements in scientific literacy and computer skills, whereas others present less favourable outcomes. A meta-analysis is necessary to synthesise the diverse findings for a more comprehensive understanding. This study reviewed 225 empirical articles in science education to evaluate the impact of PBL on improving science literacy. The meta-analysis highlighted the relationship between PBL and science literacy and identified factors influencing its effectiveness, such as education level, country, learning media and evaluation methods. The findings are expected to enrich educational theory and practice and provide strategic guidance for educators in designing more optimised learning models.

2 METHOD

This research employs the PRISMA protocol to facilitate systematic and structured data collection (Hoque *et al.* 2022; Malapane *et al.* 2022; White and Delaney 2021; Yu and Xu 2022). Figure 1 showed identification of literature from the Scopus database for 2015– August 8, 2024.

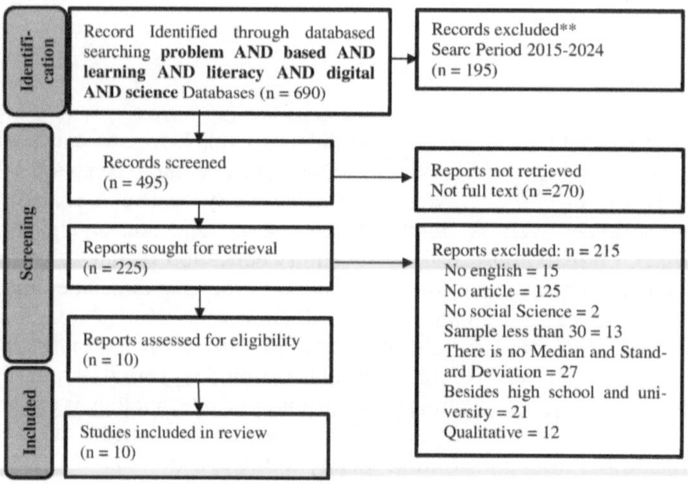

Figure 1. PRISMA diagram.

This research encompasses publicly accessible documents in English released during that timeframe. The search focuses on science literacy, PBL, education, and science, with these keywords appearing in the title, keywords, or abstract of articles from the selected journals. Irrelevant data is filtered using exclusion criteria. R Studio software is used to conduct the meta-analysis. The data comprises effect sizes, standard deviations, sample sizes, and moderator variables like education level, assessment methodologies, media, and study sites.

The analysis findings are shown as a forest plot, illustrating the range of effect sizes and confidence intervals. The I^2 statistic is employed to assess heterogeneity, while moderator tests are utilised to analyse the impact of certain variables. The limited maximum likelihood (REML) approach is employed to derive precision estimates, while effect size is computed using Hedges'g.

3 RESULT AND DISCUSSION

3.1 *Effects and variation across studies*

Forest plots are an important visual tool in meta-analysis that displays the results of each study while providing a picture of the aggregate impact of the entire study (Todd *et al.* 2017). It is useful for identifying between-study variation and measuring the overall effect using a random-effects model approach. This model was chosen because the I^2 value of 94% indicates high heterogeneity. The forest plot results showed that applying the problem-based learning (PBL) model provided 93% higher learning outcomes than the conventional method. With a 94% confidence interval and a summary effect range between 0.15 and 1.71, which does not include a zero value, it is concluded that there is a significant difference between the two learning methods.

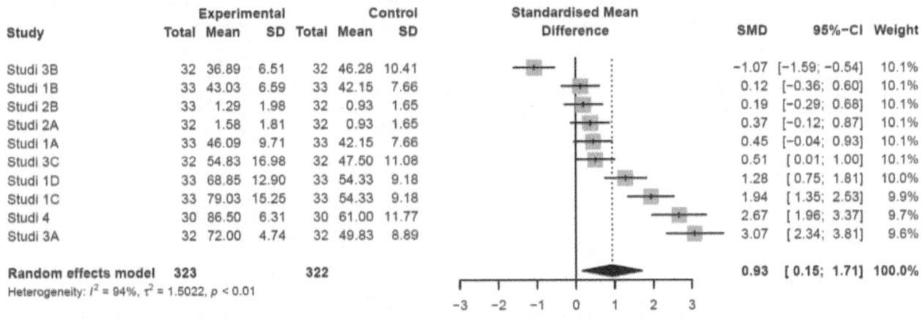

Figure 2. Forest plot of the effect of problem-based learning on students' science literacy.

The study comprised 10 trials, each with a sample size of 30 to 33 participants. Consequently, each group had a total of 322 to 323 participants. The experimental and control groups' data were presented as means and standard deviations (SD) in the experiments designated Study 1A to Study 4. The comparison analysis demonstrated that the SMD varied between research. Some research found superior outcomes in the control group, while others revealed benefits in the experimental group. The standardised mean difference for Study 3B was 0.93, with a confidence interval of [0.15; 1.71]. This signifies that the effect was moderate; still, some doubt remained. In contrast, study 3A demonstrated the highest SMD of 3.07, with a confidence interval of [2.34; 3.81].

Figure 2 demonstrates that the experimental group regularly surpassed the control group but with fluctuating confidence levels in the outcomes. The results show that problem-based learning (PBL) helps students learn more about science and helps them reach SDG 4.1 and SDG 4.7. Yuliati *et al.*, (2020) demonstrated that PBL positively affects various learning dimensions, whereas Hung *et al.*, (2008) asserted that PBL enhances students' problem-solving abilities. Consequently, PBL is recognised as an effective approach to enhancing educational quality concerning global issues, aligning with the objectives of SDG 4.

3.2 *Moderator analysis*

Moderator analysis is an important tool in meta-analysis to identify factors that influence intervention effectiveness. This study shows that various factors influence intervention outcomes, such as education level, skills, research location, sample size, discipline, media, and assessment method.

Education levels ranged from elementary school to college, while assessment methods included multiple choice, essay, questionnaire, peer assessment, and authentic assessment. The studies analysed cover a variety of locations, including Indonesia, with a focus on science fields such as physics and biology, and sample sizes varying between small and large (Fahrezi *et al.* 2020; Todd *et al.* 2017).

An I value of 64% and a value of 0.2617, representing moderate variability among the studies, indicate the model's heterogeneity. We employed the Standardized Mean Difference (SMD) to determine the effect sizes. The SMD and 95% confidence intervals are shown in the graphic's boxed area and horizontal line. Vertical lines at 0 denote statistically insignificant outcomes. Figure 3 presents a standardised mean difference (SMD) of 0.93 with a 95% confidence interval (CI) of [0.15; 1.71], indicating that problem-based instruction positively and significantly enhances science literacy skills despite variances across specific subgroups (Mukagihana *et al.* 2022).

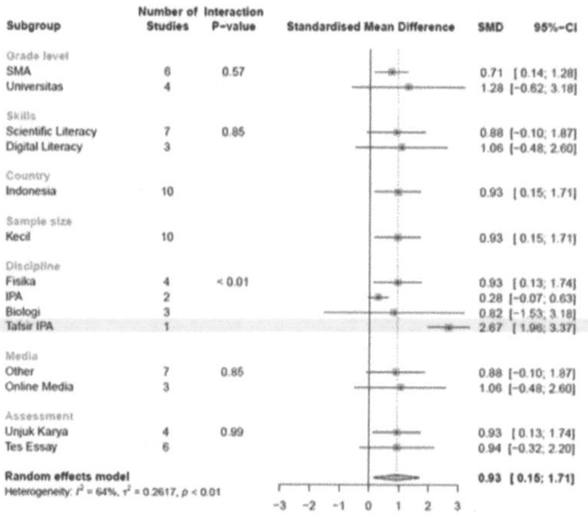

Figure 3. Analysis of the moderators of problem-based learning.

3.3 *Publication bias*

A funnel plot is a graphical tool used in meta-analysis to identify potential publication bias or heterogeneity among studies (Lo and Feng 2020; Yu and Xu 2022). The correlation between the precision of the results (SE or sample size) and the effect size (SMD) is illustrated in this plot. The horizontal axis denotes the SMD value, whilst the vertical axis signifies the SE. Studies utilising extensive samples generally produce more accurate results and correspond more closely with the average effect, while studies with limited samples frequently lead to less reliable conclusions. The funnel plot images exhibited a symmetrical distribution, suggesting no evidence of publication bias or significant heterogeneity.

The analysis results show the effectiveness of PBL over conventional approaches without significant bias. Although there was a slight imbalance due to the smaller number of studies addressing negative impacts, the overall distribution of the data remained symmetrical. This funnel plot helps to evaluate the validity of the data and identify possible biases, although further analysis is still needed to understand the context of each study in depth.

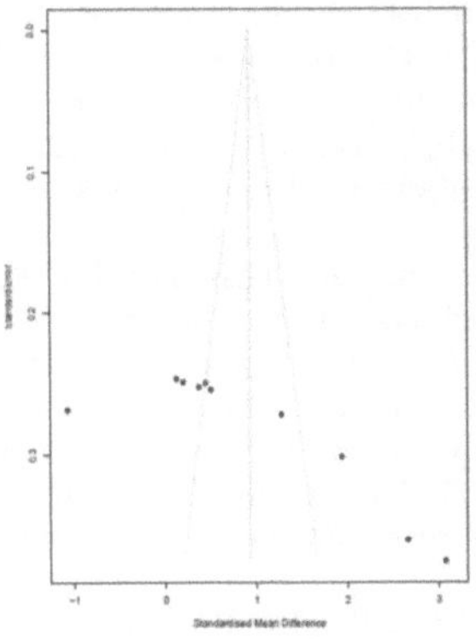

Figure 4. Funnel plot publication bias.

Figure 4 presents an examination of the efficacy of diverse educational programs concerning Sustainable Development Goal 4 (SDG 4), which seeks to guarantee inclusive, equitable, and high-quality education while fostering lifelong learning opportunities for all (Adipat and Chotikapanich 2022; Muñoz-Alcón *et al.* 2022).

4 CONCLUSION

Applying the Problem-Based Learning (PBL) model is significantly more effective than conventional methods in improving science literacy. The PBL model has a positive impact without any indication of publication bias, as analysed using the Trim and Fill method. These findings emphasise the need to integrate PBL within the educational framework. To make this model work better, teachers need to be trained, the curriculum needs to be changed, a competency-based assessment framework needs to be set up, support facilities need to be built, and businesses and the government need to work together.

ACKNOWLEDGMENTS

We would like to thank the Science Education Doctoral Program at the Faculty of Teacher Training and Education, Sebelas Maret University, Surakarta, Indonesia. We also extend our gratitude to BIB LPDP RI for their support in providing educational scholarships.

REFERENCES

Adipat, S., and Chotikapanich, R. (2022). Sustainable Development Goal 4: An Education Goal to Achieve Equitable Quality Education. *Academic Journal of Interdisciplinary Studies, 11*(6). https://doi.org/10.36941/ajis-2022-0159

Ali, L. U., Azmar, Wahyuni, Jumawal, and Fitriana, I. M. (2023). Improving Science Learning Outcomes by Applying Problem-Based Learning Model. *Jurnal Pendidikan Fisika, 11*(2), 173–182. https://doi.org/10.26618/jpf.v11i2.9913

Amaringga, N. G., Amin, M., and Irawati, M. H. (2021). The effect of problem-based learning module containing research result to improve students' scientific literacy. *AIP Conference Proceedings, 2330*. https://doi.org/10.1063/5.0043529

Anttila, J. (2018). Enabling "Quality Society" through quality education for sustainable development. *The Journal of Quality in Education, 8*(12). https://doi.org/10.37870/joqie.v8i11.154

Donkoh, R., Lee, W. O., Ahoto, A. T., Donkor, J., Twerefoo, P. O., Akotey, M. K., and Ntim, S. Y. (2023). Effects of educational management on quality education in rural and urban primary schools in Ghana. *Heliyon, 9*(11). https://doi.org/10.1016/j.heliyon.2023.e21325

Ernawati, M. D. W., Damris, M., Nevriansyah, E., Fitriani, R., and Putri, W. A. (2022). How Scaffolding Integrated With Problem Based Learning Can Improve Creative Thinking in Chemistry? *European Journal of Educational Research*, *11*(3), 1349–1361. https://doi.org/10.12973/eu-jer.11.3.1349

Ernawati, M. D. W., Rini, E. F. S., Aldila, F. T., Haryati, T., and Perdana, R. (2023). Do creative thinking skills in problem-based learning benefit from scaffolding? *Journal of Turkish Science Education*, *20*(3), 399–417. https://doi.org/10.36681/tused.2023.023

Grabau, L., Galand, B., Lafontaine, D., Lavonen, J., Ólafsson, R. F., Trudel, L., and Yoon, S. (2024). What is the association between schoolwork-related anxiety and science literacy proficiency? A comparison between Southeast Asia and Northwest Europe. *Frontiers in Education*, *9*. https://doi.org/10.3389/feduc.2024.1414423

Hmelo-Silver, C.E. (2004). Problem-based learning: What and how do students learn? *Educational Psychology Review*, *16*(3), 235–266. https://doi.org/10.1023/B:EDPR.0000034022.16470.f3

Hoque, F., Yasin, R. M., and Sopian, K. (2022). Revisiting Education for Sustainable Development: Methods to Inspire Secondary School Students toward Renewable Energy. *Sustainability (Switzerland)*, *14*(14). https://doi.org/10.3390/su14148296

Hung, W., Jonassen, D. H., and Liu, R. (2008). Problem-Based Learning, Handbook of Research on Educational Communications and Technology. *Based Learning*.

Krayneva, R., Rudenko, A., and Motylev, R. (2021). Role of education in implementing the sustainable development strategy. *E3S Web of Conferences*, 250. https://doi.org/10.1051/e3sconf/202125007008

Lo, C. O., and Feng, L. C. (2020). Teaching higher order thinking skills to gifted students: A meta-analysis. *Gifted Education International*, *36*(2). https://doi.org/10.1177/0261429420917854

Malapane, O. L., Musakwa, W., Chanza, N., and Radinger-Peer, V. (2022). Bibliometric Analysis and Systematic Review of Indigenous Knowledge from a Comparative African Perspective: 1990–2020. *Land*, *11*(8). https://doi.org/10.3390/land11081167

McCaw, C. (2022). Education for Sustainability: Learning from and with living systems. *Futures of Education, Culture and Nature - Learning to Become*, *1*, 240–246. https://doi.org/https://doi.org/10.7146/fecun.v1i.130249

Muchtar, H., Suparman, M. A., Tamuri, A. H., and Susanto, E. (2018). The influence of learning models and learning reliance on students' scientific literacy. *Jurnal Pendidikan IPA Indonesia*, *7*(4), 458–466. https://doi.org/10.15294/jpii.v7i4.12489

Mukagihana, J., Nsanganwimana, F., and Aurah, C. M. (2022). Effect of instructional methods on pre-service science teachers learning outcomes: a meta-analysis. *Education and Information Technologies*, *27*(2). https://doi.org/10.1007/s10639-021-10696-9

Muñoz-Alcón, A. I., Tejedor-Hernández, V., and Lafuente-Nafría, M. B. (2022). Preparing Vulnerable Populations for Science Literacy and Young Adults for Global Citizenship through Service Learning. *Sustainability (Switzerland)*, *14*(11). https://doi.org/10.3390/su14116775

Perdana, R., and Rudibyani, R. B. (2020). The effectiveness of inquiry social complexity to improving critical and creative thinking skills of senior high school students. *International Journal of Instruction*, *13*(4), 477–490. https://doi.org/10.29333/iji.2020.13430a

Putranta, H. (2019). Development of physics-tier tests (PysTT) to measure students' conceptual understanding and creative thinking skills: A qualitative synthesis. *Journal for the Education of Gifted Young Scientists*, *7*(3), 747–775. https://doi.org/10.17478/jegys.587203

Saputro, B., Tortop, H. S., Zuhri, M., and Saerozi, M. (2021). The effectiveness of the learning management system of saqural learning application on the scientific interpretation learning outcomes. *Jurnal Pendidikan IPA Indonesia*, *10*(1), 111–120. https://doi.org/10.15294/jpii.v10i1.27677

Scholkmann, A., Stegeager, N., and Miller, R. K. (2023). Integrating the Integration: The Role and Challenges of Problem-Based Learning in Bringing Together Social Science and Humanities (SSH) and Science, Technology, Engineering and Mathematics (STEM). *Journal of Problem Based Learning in Higher Education*, *11*(1). https://doi.org/10.54337/ojs.jpblhe.v11i1.7371

Tasquier, G., Knain, E., and Jornet, A. (2022). Scientific Literacies for Change Making: Equipping the Young to Tackle Current Societal Challenges. *Frontiers in Education*, 7. https://doi.org/10.3389/feduc.2022.689329

Todd, E. M., Watts, L. L., Mulhearn, T. J., Torrence, B. S., Turner, M. R., Connelly, S., and Mumford, M. D. (2017). A Meta-analytic Comparison of Face-to-Face and Online Delivery in Ethics Instruction: The Case for a Hybrid Approach. *Science and Engineering Ethics*, *23*(6). https://doi.org/10.1007/s11948-017-9869-3

Turiman, P., Omar, J., Daud, A. M., and Osman, K. (2012). Fostering the 21st Century Skills through Scientific Literacy and Science Process Skills. *Procedia - Social and Behavioral Sciences*, 59. https://doi.org/10.1016/j.sbspro.2012.09.253

White, D., and Delaney, S. (2021). Full STEAM ahead, but who has the map for integration? - A PRISMA systematic review on the incorporation of interdisciplinary learning into schools. *LUMAT*, *9*(2). https://doi.org/10.31129/LUMAT.9.2.1387

Yu, Z., and Xu, W. (2022). A meta-analysis and systematic review of the effect of virtual reality technology on users' learning outcomes. *Computer Applications in Engineering Education*, *30*(5). https://doi.org/10.1002/cae.22532

Yuliati, L., Hermanto, F. M., and Ali, M. (2020). A case study on comparison of high school students' scientific literacy competencies domain in physics with different methods: PBL-stem education, PBL, and conventional learning. *Jurnal Pendidikan IPA Indonesia*, *9*(2), 159–168. https://doi.org/10.15294/jpii.v9i2.23894

Towards Resilient Societies: The Synergy of Religion, Education, Health, Science, and Technology – Rahiem (Ed.)
© 2026 The Author(s), ISBN: 978-1-041-08471-6

Religiosity and academic resilience in Pesantren students

Maya Defianty
Universitas Islam Negeri Syarif Hidayatullah Jakarta, Indonesia
ORCID ID: 0000-0003-0330-1661

Yatni Fatwa Mulyati & Siti Zulfa
Universitas Islam Negeri Syarif Hidayatullah Jakarta, Indonesia

ABSTRACT: This study examines the relationship between religiosity and academic resilience in Indonesia. Employing the PRISMA (Preferred Reporting Items for Systematic Reviews and Meta-Analyses) methodology, the research systematically reviewed articles from Sinta 1–4 indexed journals using the keywords "resilience" and "academic resilience." The review identified 25 relevant papers, categorised into four themes: 18 focused on students, three on teachers, three on Islamic settings, and one on study instruments. However, only four papers directly explored the connection between religiosity and academic resilience and their impact on academic life. The findings reveal that religious teachings and supportive environments in Islamic settings, such as Pesantren, significantly contribute to students' academic resilience and spiritual development. Integrating religious values into the curriculum can play a crucial role in fostering resilience, as teachings often emphasise ethical principles like compassion, forgiveness, and perseverance. These values help students navigate challenges, encouraging responsible and ethical decision-making. The study recommends that educators and policymakers incorporate religious teachings and practices that promote resilience into educational curricula and school activities to strengthen students' personal and academic growth.

Keywords: academic resilience, resilience, religious

1 INTRODUCTION

Recent developments in education have heightened the need for students to be academically resilient in dealing with learning challenges. Almulla (2024) defined academic resilience as the ability to seek aid thoughtfully and adaptively and effectively manage negative emotions. Resilience is crucial in academia as it helps students find inner strength amidst the pressures of the academic setting, especially in the complex world we live in today (Morales 2008; Santaoja 2024). Resilience is increasingly considered a critical life skill (Yuliyanti and Shodiq 2024). Importantly, religiosity can significantly contribute to the development of resilience in students. Many religious beliefs and practices offer frameworks for coping with adversity, finding meaning in suffering, and cultivating hope for the future (Tindaon *et al.* 2024). For example, when students face difficulties, such as academic pressures or social challenges, resilience helps them persevere. Religiosity contributes to this by providing a sense of purpose and meaning, essential for maintaining motivation and hope. For instance, religious beliefs can help students interpret challenges as opportunities for growth or as part of a larger divine plan, fostering a constructive perspective. Religious practices, such as prayer, meditation, and participation in rituals, contribute to emotional regulation, an essential aspect of resilience (Gan *et al.* 2023). These activities help students manage stress, cultivate mindfulness, and find inner peace, making it easier to cope with the demands of their daily lives. The explanation above, emphasis that academic resilient is one of the contributing factors to academic success as revealed by a meta-analysis study conducted by (Nimisha and Deepak 2018). Several current studies have postulated that academic resilient enhances students' learning adjustment, self-efficacy, and motivation (Abdolrezapour *et al.* 2023; Lee and Jung 2024; Lee 2023; Rayyan *et al.* 2023; Yuberti *et al.* 2024).

As academic resilience is not innate, several attempts have been made to identify some contributing factors to students' academic resilience. Nakhostin-Khayyat *et al.* (2024), in their cross-sectional study in the Iranian context, found self-regulation as a mediated factor to academic resilience besides cognitive flexibility. The study corroborates (Ragusa *et al.* 2023) study survey research conducted in Spain involving 991 high school students. The study found that self-regulation can improve academic resilience and thus positively affect academic performance. Previously, Cassidy's (2015) correlational study, which involved students, revealed that academic self-efficacy is a significant predicted factor of academic resilience. A qualitative case study by Rojas (2015) in Bogota, Columbia, has shown that other factors contribute to academic resilience. Drawing out data from document analysis and interviews with teachers and parents, the study revealed that supportive family and individual characteristics contribute to academic resilience.

In addition, several studies have also documented religiosity as one of the contributing factors to students' academic resilience. For example, Javanmard (2013) found that data from the questionnaire distributed to 160 Iranian higher-education students showed a significant correlation between religiosity and academic resilience. Similarly, Bakhtiarvand *et al.*'s (2023) study in the context of medical students also revealed that spirituality was a positive predictor of academic resilience. Caton and Waters (2023), who conducted their study in the USA context, revealed that spirituality affects resilience among race and ethnicity groups; further, the researchers also found that higher spirituality equals higher resilience. The interconnectedness between religiosity and academic resilience has also been documented in several studies in the Indonesian context. This may be because Indonesians are required to inform their religious affiliation on their identity cards. Presumably, religiosity has been an inseparable part of Indonesian. Nevertheless, studies only focused on religiosity and resilience instead of academic resilience (Arifiana *et al.* 2022; Ubaidilah *et al.* 2022). Departing from the aforementioned discussion, this paper aims to review literature that focuses on the interplay between religiosity and academic resilience in the Islamic education institution in Indonesia. The result of the study will shed light on how future researcher in the same landscape should focus their study.

2 METHODOLOGY

This study employed PRISMA (Preferred Reporting Items for Systematic Reviews) as the research design to review studies that discussed the interplay between religiosity and academic resilience in the Indonesian Islamic education context. The papers reviewed in this study were collected from Sinta. Sinta is an indexed system for journals published in Indonesia. For this investigation, the methodology proposed by Schwalm *et al.* (2022) is employed to evaluate the article as research data associated with those phrases through the application of PRISMA. As mentioned previously, the purpose of the study is to review studies that discussed the interplay between religiosity and academic resilience in the context of Indonesian Islamic education. This study utilised the PRISMA framework (Preferred Reporting Items for Systematic Reviews) to systematically identify, evaluate, and select relevant literature. The process consisted of several sequential stages. The stages in implementing PRISMA for this study are illustrated in Figure 1 below:

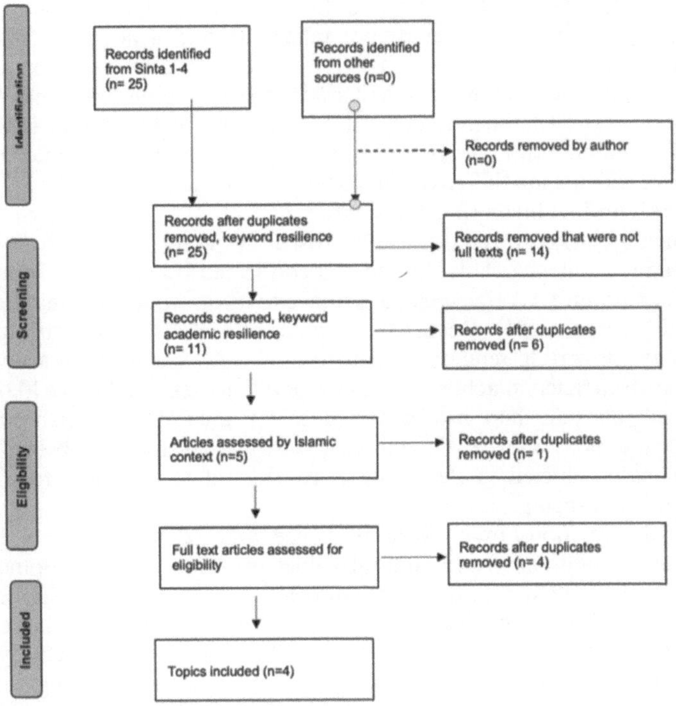

Figure 1. PRISMA diagram.

3 FINDINGS AND DISCUSSION

As illustrated in the previous section, this study only included four studies to be thoroughly reviewed. Two studies conducted in the context of pesantren (Islamic boarding school) found that resilience has been embodied in pesantren education. First is a case study conducted by Mawardi and Ruhiyah (2022). The researchers included teachers and students from three pesantrens to participate in their study, and data was gathered from interviews with teachers and meticulous observation in the research context. Based on the collected data, the researchers claimed that resilience had been part of a hidden curriculum in pesantren education systems. Another study in pesantren was conducted by Jaenullah *et al.* (2022). The study highlighted that resilience in these schools is profoundly rooted in students‘ communal and religious life, known as santri. The pesantren milieu, including religious teachings, community support, and social dynamics, shapes santri's resistance to scholastic and personal problems.

Another study that revealed the correlation between religiosity and academic resilience was conducted in the context of higher education. Although this quantitative has a slightly different focus from the other two studies mentioned previously, the interconnectedness between religiosity and academic resilience can be directly identified. The researchers focused on the impacts of social support and resilience against procrastination. Drawing on data from a questionnaire, the researchers concluded that procrastination is affected by social support and resilience. Further, the researchers stated that "social support from family, friends, and the school can reduce academic procrastination. Resilience reduces and exerts a significant and indirect effect on academic procrastination after mediation by the religious character" (Madjid *et al.* 2021).

Among the collected papers, it can be said that Ali *et al.*'s (2023) is the only article that specifically aimed to identify whether or not religiosity and academic resilience are connected. This mixed-method study involved 341 students in higher education as the research participants. The study revealed that resilient students are religious students. As the study is situated in the COVID-19 Pandemic era, the researcher concluded that students' faith in God helps them overcome difficult times in learning during the Pandemic. Findings from this study have revealed that the environment can foster learning resilience. Studies conducted by Mawardi and Ruhiyah (2022) and Madjid *et al.* (2021) have shown that a learning environment, such as in Pesantren, can be a prospective context to cultivate learning resilience and academic resilience. This finding has confirmed other studies in a similar landscape that argue communities offer social assistance to assist students in managing academic stress Smith and Denton (2005). Research on religious schools across various cultures has demonstrated that integrating religious beliefs into education can create a nurturing atmosphere that enhances academic achievement and psychological welfare (Alfian *et al.* 2021). These findings indicate that there has been insufficient research conducted on the topics of religiosity and academic resilience. Therefore, research about religious beliefs' interconnection with resilience conducted in Pesantren is considered to enhance students‘ academic success and overall well-being.

Furthermore, the relationship between resilience and religious identity highlights the broader influence of community and cultural values on individual development (Aprilianti 2024). Religious settings often provide a sense of belonging, purpose, and support, crucial for developing resilience. Prayer, meditation, and community engagement can serve as coping mechanisms that enable students to handle stress and setbacks more effectively. In addition, Egbaria (2024) depicted the importance of integrating religious teachings and character-building activities into educational curricula. Such efforts can include structured lessons on ethical decision-making, discussions on the relevance of faith in addressing life's challenges, and opportunities for students to engage in service-learning projects that align with their religious principles. These activities not only reinforce students' sense of identity but also provide them with tools to manage adversities in a constructive and faith-driven manner.

Moreover, the research discussion by Amir and Rumiani (2023) invites further exploration of how different dimensions of religious character—such as faith, hope, and compassion—specifically contribute to resilience. For instance, hope derived from religious teachings might empower students to maintain a positive outlook during times of hardship. Similarly, compassion fostered through religious principles could encourage supportive relationships, which are critical for building resilience (Hidayah and Munawaroh 2024). Therefore, the intertwined nature of resilience and religious identity offers valuable insights for educators and policymakers in religious educational settings. By prioritising the development of religious character, institutions can equip students with the skills to succeed academically and the inner strength to thrive amidst life's challenges.

In conclusion, in-depth involvement of explicit integration of religious teachings into discussions about coping strategies, perseverance, and managing preoccupations. Furthermore, this research highlights the need for a more comprehensive exploration of how religious character can be systematically developed alongside resilience to create well-rounded individuals who are both academically successful and spiritually grounded. In addition, although the majority of the analysed research papers are quantitative, additional research that is more in-depth is required to acquire a better and more profound understanding of the topic.

4 CONCLUSION

This literature review-based article reveals that the connection between academic resilience and religious cities is underexplored, with only four articles addressing the topic. The concept of a "hidden curriculum" in religious educational institutions, such as pesantren (Islamic boarding schools), is crucial in fostering academic resilience. This hidden curriculum helps students develop resilience, moral integrity, and a strong moral compass, enabling them to navigate modern challenges. In addition, this research implies that resilience and religiosity are deeply interconnected. Religiosity is a foundation that empowers students to develop and sustain resilience, helping them thrive academically, socially, and personally despite life's inevitable challenges. Therefore, specific actionable recommendations for educators and policymakers in religious education are suggested. Educators and policymakers in religious education play a crucial role in fostering resilience and religiosity in students, equipping them to navigate academic, social, and personal challenges. To achieve this, religious curricula should integrate activities that encourage students to reflect on adversity and develop coping strategies rooted in faith, such as exploring stories of resilience from religious texts and facilitating discussions on the role of spirituality in overcoming hardships. Creating inclusive and supportive religious communities is equally vital. Safe spaces where students can share struggles, peer mentorship programs, and training for educators to recognise stress signals are key to building a nurturing environment. To enhance effectiveness, educators should receive professional training on integrating resilience-building approaches, understanding the psychological aspects of resilience, and modelling these behaviours.

REFERENCES

Abdolrezapour, P., Jahanbakhsh Ganjeh, S., and Ghanbari, N. (2023). Self-Efficacy And Resilience As Predictors Of Students' Academic Motivation In Online Education. *PLOS ONE*, 18(5), E0285984. Https://Doi.Org/10.1371/Journal.Pone.0285984

Alfian, A., Yusuf, M., and Nafiah, U. (2021). Integrating Islamic Values In Teaching English: Lessons Learned From An Integrated Islamic School. *Elsya: Journal Of English Language Studies,* 4(1). Https://Doi.Org/10.31849/Elsya.V4i1.7322

Ali, N., Umar, M., and Mopoliu, A. Z. (2023). Academic Resilience And Religiosity Among College Students Facing Post-Pandemic Learning. *IJORER: International Journal Of Recent Educational Research*, 4(5), 585–596. . Https://Doi.Org/10.46245/Ijorer.V4i5.410

Almulla, M. O. (2024). Academic Resilience And Its Relationships With Academic Achievement Among Students Of King Faisal University In Saudi Arabia. *Revista De Gestão Social E Ambiental*, 18(9), E07391. Https://Doi.Org/10.24857/Rgsa.V18n9-134

Amir, Z. A., and Rumiani, R. (2023). The Relationship Between Religiosity And Resilience In Medical Students. *Psychosophia: Journal Of Psychology, Religion, And Humanity*, 5(2), 53–72.. Https://Doi.Org/10.32923/Psc.V5i2.3375

Aprilianti, E. (2024). Integrating Islamic Psychological Principles In Enhancing Students' Academic Resilience. *Nusantara Journal Of Behavioral And Social Sciences*, 3(2), 63–72. . Https://Doi.Org/10.47679/202246

Egbaria, H. (2024). The Contribution Of Emotional Intelligence And Religiosity In Predicting Psychological Resilience Among Israeli Palestinian High School Students. *European Journal Of Social Sciences Studies*, 9 (6). Https://Doi.Org/10.46827/Ejsss.V9i6.1683

Gan, S. K.-E., Wong, S. W.-Y., and Jiao, P.-D. (2023). Religiosity, Theism, Perceived Social Support, Resilience, And Well-Being Of University Undergraduate Students In Singapore During The COVID-19 Pandemic. *International Journal Of Environmental Research And Public Health,* 20(4), 3620. Https://Doi.Org/10.3390/Ijerph20043620

Hidayah, N., and Munawaroh, M. (2024). The Influence Of Religiosity On Academic Achievement With Resilience As An Intervening Variable. *EDUKASIA: Jurnal Pendidikan Dan Pembelajaran*, 5(2), 1–10. Https://Doi.Org/10.62775/Edukasia.V5i2.770

Jaenullah, J., Utama, F., and Setiawan, D. (2022). Resilience Model Of The Traditional Islamic Boarding School Education System In Shaping The Morals Of Student In The Midst Of Modernizing Education. *Jurnal Kependidikan: Jurnal Hasil Penelitian Dan Kajian Kepustakaan Di Bidang Pendidikan, Pengajaran Dan Pembelajaran*, 8(4), 931. Https://Doi.Org/10.33394/Jk.V8i4.6013

Lee, G.H., and Jung, M.K. (2024). The Moderated Mediating Effect Of Resilience And Social Support In The Relationship Between Academic Self-Efficacy And College Adjustment. *Korean Association For Learner-Centered Curriculum And Instruction*, 24(10), 67–81. Https://Doi.Org/10.22251/Jlcci.2024.24.10.67

Lee, J.Y. (2023). The Effect Of Resilience On Self-Efficacy In University Students. *The K Association Of Education Research*, 8(2), 171–186. Https://Doi.Org/10.48033/Jss.8.2.10

Madjid, A., Sutoyo, D. A., and Shodiq, S. F. (2021). Academic Procrastination Among Students: The Influence Of Social Support And Resilience Mediated By Religious Character. *Cakrawala Pendidikan*, 40 (1), 56–69. Https://Doi.Org/10.21831/Cp.V40i1.34641

Mawardi, M. M., and Ruhiyah, F. (2022). The Relevance Of Positive Education Concept For Pesantren In Indonesia. *Jurnal Pendidikan Islam*, 8(2), 163–176. . Https://Doi.Org/10.15575/Jpi.V8i2.19855

Morales, E. E. (2008). The Resilient Mind: The Psychology Of Academic Resilience. *The Educational Forum*, 72(2), 152–167. Https://Doi.Org/10.1080/00131720701805017

Nakhostin-Khayyat, M., Borjali, M., Zeinali, M., Fardi, D., and Montazeri, A. (2024). The Relationship Between Self-Regulation, Cognitive Flexibility, And Resilience Among Students: A Structural Equation Modeling. *BMC Psychology*, 12(1), 337. Https://Doi.Org/10.1186/S40359-024-01843-1

Nimisha, B., and Deepak, K. (2018). Predictors Of Academic Resilience Among Students: A Meta Analysis. *I-Manager's Journal On Educational Psychology*, 11(4), 37. Https://Doi.Org/10.26634/Jpsy.11.4.14220

Ragusa, A., González-Bernal, J., Trigueros, R., Caggiano, V., Navarro, N., Minguez-Minguez, L. A., Obregón, A. I., and Fernandez-Ortega, C. (2023). Effects Of Academic Self-Regulation On Procrastination, Academic Stress And Anxiety, Resilience And Academic Performance In A Sample Of Spanish Secondary School Students. *Frontiers In Psychology*, 14. Https://Doi.Org/10.3389/Fpsyg.2023.1073529

Rayyan, M., Zidouni, S., Abusalim, N., and Alghazo, S. (2023). Resilience And Self-Efficacy In A Study Abroad Context: A Case Study. *Cogent Education*, 10(1). Https://Doi.Org/10.1080/2331186X.2023.2199631

Santaoja, M. (2024). Fifty Shades Of Academic Resilience. *Journal Of Praxis In Higher Education*, 6(2), 13–22. Https://Doi.Org/10.47989/Kpdc482

Schwalm, F. D., Zandavalli, R. B., De Castro Filho, E. D., and Lucchetti, G. (2022). Is There A Relationship Between Spirituality/Religiosity And Resilience? A Systematic Review And Meta-Analysis Of Observational Studies. In *Journal Of Health Psychology* (Vol. 27, Issue 5, Pp. 1218–1232). SAGE Publications Ltd. Https://Doi.Org/10.1177/1359105320984537

Smith, C., and Denton, M. L. (2005). *Soul Searching.* Oxford University Press new York. Https://Doi.Org/10.1093/019518095X.001.0001

Tindaon, R., Fathiyah, K. N., and Paji, Y. E. (2024). Religiosity And Resilience In Students Of The Faculty Of Catholic Religious Education. *Journal Of Educational, Health And Community Psychology*, 13(2), 480. Https://Doi.Org/10.12928/Jehcp.V13i2.27930

Yuberti, Y., Susilowati, N. E., Velina, Y., and Latifah, S. (2024). Academic Self-Efficacy Inventory: Using Rasch Analysis To Develop And Evaluate Self-Efficacy Rating Scales. *Jurnal Pendidikan IPA Indonesia*, 13(1), 162–171. Https://Doi.Org/10.15294/Jpii.V13i1.47091

Yuliyanti, Y., and Shodiq, S. F. (2024). The Impact of Religious Practices on the Resilience and Post-Traumatic Growth of Adolescents. *Buletin Konseling Inovatif*, 4(3), 182–189. https://doi.org/10.17977/um059v4i32024p189-189.

Towards Resilient Societies: The Synergy of Religion, Education, Health, Science, and Technology – Rahiem (Ed.)
© 2026 The Author(s), ISBN: 978-1-041-08471-6
Open Access: www.taylorfrancis.com, CC BY-NC-ND 4.0 license

Inclusive educational leadership: Promoting tolerance in multireligious schools

Mohamad Iwan Fitriani, Ahmad Sulhan, Ridwan Ridwan, M. Sobry Sutikno, Saparudin Saparudin & Usman Usman
Universitas Islam Negeri Mataram, Indonesia

ABSTRACT: This study aims to identify and analyse the characteristics of students' comprehensive religious tolerance and examine leadership strategies that support its development. Employing a qualitative approach with a case study method, data were collected through in-depth interviews, participatory observation, and document analysis. The findings reveal that fostering comprehensive religious tolerance among students requires a holistic approach, encompassing active and passive tolerance, internal and external tolerance, and both empirical and non-empirical tolerance. Inclusive educational leadership is vital in translating these values into tangible behaviours through supportive policies and programmes. Collaboration with the community, parents, and religious leaders is essential in creating a tolerant school environment. This study highlights the critical role of educational leaders in cultivating a culture of tolerance through well-structured and sustainable strategies. A culture of tolerance is fundamental to fostering peace, justice, and strong institutions while advancing the Sustainable Development Goals (SDGs).

Keywords: comprehensive tolerance, culture of tolerance, educational leadership, inclusiveness, leadership strategies, student behaviour

1 INTRODUCTION

Tolerance is one of the key components that promotes harmony and peace in a multicultural and multireligious community (Suprapto 2017). In addition, tolerance contributes to building community resilience from all actions that disrupt societal harmony, such as radicalism (Muhtar and Jihad 2019). According to UNESCO (1995) , tolerance is "respect, acceptance and appreciation of the rich diversity". Students need to understand tolerance comprehensively, not partially, to foster tolerant behaviour. Comprehensive tolerance is needed to avoid the "trap" of the term tolerance. The trap occurs when (1) there are people who consider themselves tolerant people, but their tolerance is still in their hearts and is not realised in real behaviours; (2) there are people who claim to be tolerant, but their tolerance is still limited to people of the same school (*madzhab*), organisation or religion; (3) there are people who express their tolerance in the real world (of-line community), but intolerance in cyberspace (cyber community) (Fitriani 2023a).

The illustration above shows that; (1) tolerance needs to be transformed from passive tolerance to active tolerance, (2) tolerance is needed in a multi-religious society both in the form of intra and interreligious tolerance; and (3) tolerance needs to be transformed from negative tolerance to positive tolerance. That is why tolerance must be understood broadly to construct true tolerance or what Menchik (2016) calls full tolerance. Menchik mentions five typologies of tolerance : full intolerance, semi-intolerance, neutral, semi-tolerance and full tolerance (Menchik 2016). True tolerance is a concept that needs to be understood and practised in everyday life. Thus, full tolerance can also be understood as active tolerance (Kemenag 2022), while semi-tolerance is passive tolerance (Fitriani and Naamy 2019).

Then value-based leadership (VBL) is one type of leadership that is concerned with values (Sulhan 2020). VBL refers to various types of leadership because values are found in various

DOI: 10.1201/9781003645542-46

leadership typologies (Sulhan 2020). One theory that focuses on values is Kraemer's values-based leadership theory. Kraemer (2011) states that VBL consists of self-reflection, balance, true self-confidence and genuine humility. Furthermore, relevant values transferred by leaders in supporting comprehensive tolerance behaviours are multicultural values. According to Baidhawy (2007), among the multicultural values are living together, love, egalitarianism, *co* and pro-existence, mutual understanding, *mutual* respect, fair competition, mutual trust, positive thinking, tolerance, forgiveness, reconciliation, conflict resolution; peace, anti-violence culture and justice. Misrawi (2013) adds that the multicultural values above are intended to build multicultural behaviours and avoid monocultural behaviours.

Furthermore, this study contributes to providing a typology of tolerance among students and helps educational leaders to build peace and harmony in multi-religious schools in particular and multi-religious communities in general.

Based on the problems presented above, this study is guided by two research questions: what are the characteristics of students' comprehensive religious tolerance, and (2) what are leadership strategies to support students' comprehensive religious tolerance?

2 METHOD

This research uses a qualitative approach with a case study design to explore in depth the phenomenon of tolerance in several madrasas and schools in West Lombok, West Nusa Tenggara. Data were collected through in-depth interviews, document analysis and observation to gain a comprehensive perspective on students' tolerance behaviours and the role of educational leadership. Data analysis was conducted using the Miles and Huberman model, which involves three stages: data collection, condensation, presentation, and conclusion drawing/verification. Data validity was ensured through the triangulation method, which compares information from different sources and methods to ensure consistency and accuracy of findings.

3 RESULTS AND DISCUSSION

This point will describe and analyse two main things, namely, (1) characteristics of comprehensive religious tolerant students and (2) leadership efforts or strategies to support comprehensive religious tolerance.

3.1 *Ideal characteristics of full tolerance students*

Empirical data shows that inclusiveness in the educational environment can help change intolerant attitudes into more open ones. This aligns with UNESCO's idea that tolerance is the basis of a culture of peace (UNESCO 1995) and UNESCO's recommendation on peace education and human rights (UNESCO 2023).

First, regarding nature, tolerance is passive and active; passive tolerance means tolerance (acknowledging and respecting) for all existing differences but is not followed by real behaviours. It can also be called hidden tolerance. This type of tolerance needs to be actualised in real action. Despite its shortcomings, passive tolerance is better than passive intolerance, harbouring intolerant feelings towards other people or different groups. Active tolerance is tolerance that is manifested in real behaviours.

Second, in terms of scope, tolerance is internal and external. In the context of Islam-for example this internal tolerance can take the form of: (a) tolerance towards other Muslims who have different religious organisations such as Nahdlatul 'Ulama, Nahdlatul Wathan, Muhammadiyah and so on; (b) tolerance towards groups that have different school (*madhabs*), for example there are followers of Imam Syafi'i, Maliki, Hambali, Hanafi or there are those who follow all of them selectively and thoughtfully, and so on; (c) tolerance towards Muslim groups who still dialogue local culture with Islamic teachings, and so on; (d) while external tolerance is tolerance towards different religions or groups. For example, Muslim tolerance towards Hindus and vice versa.

Remember, tolerance is reciprocal (*musyarakah*). If you want to be appreciated, respected, loved, then appreciate, respect and love others. The differences that exist will be united by universal values to create peace and harmony; (e) *third*, in terms of empirical and non-empirical, tolerance is divided into two: tolerance in the real world and tolerance in cyberspace. Cyberspace is a world filled with cyber or online communities. In the declaration of the principles of tolerance, UNESCO (1995) states that the essence of tolerance is respect, acceptance and appreciation of differences. Based on the core of tolerance, cyber tolerance culture is a culture that creates respect, acceptance and appreciation of differences to realise a culture of peace in the virtual world. Conversely, "cyber intolerance culture" is a culture of virtual society that does not have respect, acceptance and appreciation of differences that lead to a "culture of war".

To achieve comprehensive tolerance, students are taught to avoid intolerant behaviour. *First*, in terms of nature, intolerance is passive and active. Passive intolerance can be seen in a person's feelings or thoughts that do not accept differences, but the non-acceptance is not expressed in words or real actions. Among the examples are: (1) students who disagree with the differences in other people's organisations, but they do not express it in words or do not spread their dislike to the wider community; (2) students who disagree with other people's sects. This is usually caused by the mono-religious education that is dominantly implemented in educational institutions and reinforced by mono-religious schools of thought. However, this condition can be changed through tolerance education, including tolerance education based on religious education. Besides, an inclusive school environment is needed to eliminate this intolerance. Active intolerance is the intolerance shown in words or actions. Persecution, destruction of places of worship, terror, and suicide bombings are examples of this active intolerance. Other intolerances that need to be avoided are internal-external intolerance and intolerance in the empirical world and cyberspace. Based on this view, the majority and minorities will help each other to realise peace and harmony (Suprapto 2017).

The illustration above shows that tolerant education is an important element for realising peace and harmony in a global society characterised by a multireligious or multicultural society. In its recommendations, UNESCO (2023) emphasises education as an effort to create peace and harmony by preventing all negative actions such as racism, xenophobia, hatred and all forms of intolerance. Then, one of the tolerant education practices implemented by some students in Lombok, West Nusa Tenggara is celebrating diversity and similarity. In this case, an educational leader with inclusive insight is needed. This insight is based on inclusive values such as justice, equality, tolerance, peace and harmony. These values are translated into relevant vision, mission, goals and strategies. Among the characteristics of relevant strategies are inclusive policies and programs.

3.2 *Educational leadership practices to support comprehensive tolerance*

Empirical data shows that this practice successfully builds an inclusive culture where students of different religions can celebrate diversity and strengthen brotherhood through joint activities following the reciprocal principle of fostering tolerance.

Educational leadership practices that support comprehensive tolerance are: (a) identifying common values and principles among different religions to promote unity and understanding; (b) disseminating values to all members to be shared values; (c) transforming shared values into behaviours by identifying behaviours that support interreligious behaviours; (d) modelling behaviours that reflect interreligious tolerance; (e) providing students with religious education according to their religion; (f) providing resource materials that support inter-religious harmony such as books and so on; (g) providing space for differences where leaders create an environment to ensure comfort for all students in expressing their religious beliefs without fear of discrimination or harassment and so on.

The values needed to build comprehensive tolerance are: (1) *ta'aruf,* it relates to values that knowing each other between different students (such as Muslims and Non-Muslims is a potential capital to realize tolerance; (2) *tafahum,* it relates to values that mutual understanding between students about differences is a God design that cannot be avoided, *tafahum* is a value to avoid negative thinking; (3) *tala'um*, it is a value that students need to be selective in receiving any information received to avoid *hoaxes*, *hate speech* and provocations. These values will be maximised by *tabayyun (confirmation)*; (4) *tashaduq* that reflects mutual trust. The basis of *tashaduq* is *al-shidqu (honesty)*: (5) *taakhy*, it is a value related to brotherhood between Muslims and non-Muslims; this *taakhy* can give birth to various types of *ukhuwwah* (broterhood); (6) *ta'athuf*, it is a value related to the process of building sympathy between different religious groups; (7) *tarabuth*, it relates to how to strengthen each other for a common goal. Finally, everything is aimed at *tasamuh (true tolerance)*, between different students, such as tolerance of Muslim students towards non-Muslims and vice versa.

The next task of the leader is to disseminate these values to all members. Without dissemination, the values are only known and owned by certain people.

Leaders also transform the shared values into behaviour to build comprehensive tolerance. The transformation of interreligious values into interreligious behaviour, which builds a multireligious culture and comprehensive tolerance, takes the following form: *Muta'arif* shows inclusive behaviour through the desire to get to know each other between different students. *Mutafahim* shows inclusive behaviours indicated by the desire to learn about others as a prerequisite for mutual understanding between students; *Mutabayyin* shows inclusive behaviour indicated by students' selective behaviour in accepting any information received to avoid *hoaxes, hate speech,* and provocations; this behaviour can also be called appropriation ability." *Mutashadiq* shows mutual trust behaviours; *Mutaakhy* shows inclusive behaviour indicated by brotherhood between all different groups, such as Muslims and non-Muslims; *Muta'athif* is sympathetic behaviour between different religious groups, tribes, and others. *Mutarabith* is a behaviour that relates to binding and strengthening togetherness; *Muta'awin* is the behaviour of helping each other, relieving the burden of others; *Mutasamih* reflects tolerance behaviour between different students, like the tolerance of Muslim students towards non-Muslims and vice versa.

The next role of educational leaders is to build inclusive policies, such as providing students with religious Education following their religion by integrating mono, multi and interreligious education, providing learning materials that support interfaith dialogue, and preparing space for diversity. The important role of the principal is to build inclusive schools that respond to various types of diversity found in schools (DeMatthews *et al.* 2021).

Furthermore, inclusive programs aim to transform inclusive policies into inclusive programs, namely, *celebrating diversity, an event involving* different religions to celebrate differences. For example, Muslim and non-Muslim students agreed to: (a) hold an interfaith event. They gather at school and each religion performs their respective religious traditions; (b) celebrating diversity can be in the form of inviting or attending the invitation of friends of different religions. For example, a Muslim student invited a non-Muslim friend to have a meal together after the Muslim student celebrated the Prophet's birthday. Conversely, a Muslim student also attends the invitation of his/her non-Muslim friends when they celebrate their religious events (such as Christmas, Nyepi or Imlek); (3) saying Happy Holidays to other friends of different religions; (4) inviting multireligious religious leaders; (5) doing community service (*bakti sosial*) involving various religions; (7) interreligious visits. Second, celebrating similarity, it means an event that is carried out to strengthen brotherhood by focusing on the similarities between all students such as; (1) flag ceremony every Monday, (2) the 17-August parade to celebrate the independence day of the Republic of Indonesia; (3) using Sasak traditional clothes for all students (Sulhan and Hakim 2023).

Furthermore, all of the above practices are supported by various parties, including: community involvement, parental involvement, School committee involvement, Religious leader involvement and Government involvement. For the effectiveness of the above values and behaviours, the characteristics of the values and behaviours applied are based on four things: *reciprocal, conditional-proportional, penetrating boundaries and* integrating *self, co-and pro-existence.* Reciprocal means that two parties apply tolerance behaviour (Fitriani 2023b). For example, Muslims are tolerant to Catholics and vice versa.

4 CONCLUSION

This research highlights the importance of building students' characters with comprehensive tolerance through a holistic approach that includes passive and active, internal and external, and empirical and non-empirical tolerance. Comprehensive tolerance creates harmony in diversity and a culture of peace in the real and virtual worlds. Strong and inclusive educational leadership is a key success in transforming the values of tolerance into real behaviour in the school environment through policies and programs that encourage dialogue, respect for diversity, and cooperation across religions and cultures.

Effective leaders play an important role in identifying and transforming values into behaviours. Furthermore, implementing inclusive educational leadership through inclusive policies and programs, such as celebrating diversity and celebrating similarity, shows how schools can become a space for all religions and beliefs to dialogue, respect and work together. The participation of various parties, including the community, parents, school committees, religious leaders and the government, is needed to support the values and behaviours that are built.

For policymakers, values-based leadership training programs must be implemented systematically at all levels of education. Furthermore, for educators, it is important for them to integrate multicultural values in the curriculum as an effort to instil tolerance as part of school culture.

REFERENCES

Baidhawy, Z. (2007). Building harmony and peace through multiculturalist theology-based religious education: An alternative for contemporary Indonesia. *British Journal of Religious Education*, *29*(1), 15–30. https://doi.org/10.1080/01416200601037478

DeMatthews, D. E., Serafini, A., and Watson, T. N. (2021). Leading Inclusive Schools: Principal Perceptions, Practices, and Challenges to Meaningful Change. *Educational Administration Quarterly*, *57*(1), 3–48. https://doi.org/10.1177/0013161X20913897

Fitriani, M. I. (2023a). Islamic Religious Education and Interreligious Tolerance in a Multi-Religious Country: Challenges, Typological Implications, and the Proposed Strategy. *Ulumuna*, *27*(1), Article 1. https://doi.org/10.20414/ujis.v27i1.765

Fitriani, M. I. (2023b). *Pendidikan Agama Islam Dan Toleransi Agama Di Indonesia: Karakteristik Ideal, Model Dan Peran Kepemimpinan Pendidikan (Orasi Ilmiah Pengukuhan Guru Besar Dalam Bidang Ilmu Manajemen Pendidikan Islam dan Disampaikan pada Sidang Senat Terbuka Universitas Islam Negeri (UIN) Mataram Kamis, 19 Sya'ban 1445 H/ 29 Februari) 2024 M.* Sanabil.

Fitriani, M. I., and Naamy, N. (2019). *Islam Wasathiyah; Solusi Alternatif Kontestasi Agama di Pulau Lombok.* Lombok Press.

Kemenag, B. (2022). *Halalbilhalal dan Penguatan Persatuan.* http://balitbangdiklat.kemenag.go.id/berita/halalbilhalal-dan-penguatan-persatuan

Kraemer, H. M. (2011). *From Values to Action: The Four Principles of Values-Based Leadership* (1st edition). Jossey-Bass.

Menchik, J. (2016). *Islam and Democracy in Indonesia: Tolerance without Liberalism.* Cambridge University Press.

Misrawi, Z. (2013). Kesadaran Multikultural dan Deradikalisasi Pendidikan Islam: Pengalaman Bhinneka Tunggal Ika dan Qabul Al-Akhar. *Jurnal Pendidikan Islam*, *2*(1), Article 1. https://doi.org/10.14421/jpi.2013.21.197-215

Muhtar, F. R., and Jihad, S. (2019). Ketahanan Umat Beragama Dalam Mencegah Radikalisme di Kota Mataram Nusa Tenggara Barat. *Jurnal Ketahanan Nasional*, *25*(1), Article 1. https://doi.org/10.22146/jkn.41443

Sulhan, A. (2020). *Teori Kepemimpinan dalam Organisasi Pendidikan Islam.* Sanabil. https://scholar.google.com/scholar?cluster=989129090011947575&hl=en&oi=scholarr

Sulhan, A., and Hakim, L. (2023). Emancipating Islamic Education Management through Good-Quality Santri Character Cultures: Insights from Indonesia. *Eurasian Journal of Educational Research*, *103*(103), 197–214.

Suprapto, S. (2017). Membina Relasi Damai Antara Mayoritas Dan Minoritas (Telaah Kritis atas Peran Negara dan Umat Islam dalam Mengembangkan Demokrasi di Indonesia). *Analisis: Jurnal Studi Keislaman*, *12*(1), 19–48.

UNESCO. (1995). *Declaration of principles on tolerance: Adopted by the General Conference of UNESCO at its twenty-eighth session, Paris, 16 November 1995—UNESCO Digital Library.* https://unesdoc.unesco.org/ark:/48223/pf0000151830

UNESCO. (2023). *Recommendation on Education for Peace and Human Rights, International Understanding, Cooperation, Fundamental Freedoms, Global Citizenship and Sustainable Development—Legal Affairs.* https://www.unesco.org/en/legal-affairs/recommendation-education-peace-and-human-rights-international-understanding-cooperation-fundamental.

Towards Resilient Societies: The Synergy of Religion, Education, Health, Science, and Technology – Rahiem (Ed.)

Ethnographic study of multiculturalism and religious moderation in Jakarta's homeschooling communities

Mohammad Siddiq
Universitas Islam Negeri Syarif Hidayatullah Jakarta, Tangerang Selatan, Indonesia
ORCID ID: 0000-0002-6491-4245

Hartini Salama
Universitas Ibnu Chaldun, Jakarta, Indonesia

Ahmad Juma Khatib Nur Ali
Universitas Gunadarma, Indonesia

Julia Fasha & Fatimah Azzahro
Universitas Ibnu Chaldun, Jakarta, Indonesia

Neneng Nurjanah & Didah Nurhamidah
Universitas Islam Negeri Syarif Hidayatullah Jakarta, Tangerang Selatan, Indonesia

ABSTRACT: This ethnographic study utilises participant observation and interviews to explore community-based homeschooling in Jakarta, focusing on integrating multicultural education and religious moderation. Engaging 19 families and 45 students from diverse religious, ethnic, and social backgrounds, the research highlights the role of families as primary educators in cultivating inclusive and cooperative learning environments. Findings reveal that homeschooling communities promote religious tolerance and multiculturalism through family-led education and project-based learning. Initiatives such as the Indonesian Boy Scouts further enhance children's cultural identity and global readiness. These practices align with Sustainable Development Goals (SDG 4: Quality Education and SDG 16: Peace, Justice, and Strong Institutions) by advancing inclusive education and strengthening community resilience. The study emphasises the need for robust legal frameworks and educational standards to support the positive development of homeschooling, alongside mechanisms to monitor and prevent radicalisation. This model offers significant potential for broader implementation within informal education to foster social cohesion and cultural diversity. Recommendations include enhancing parental involvement, improving monitoring systems, and reinforcing legal safeguards to optimise educational outcomes and community resilience.

Keywords: community-based education, ethnographic study, homeschool family

1 INTRODUCTION

In the era of globalisation, multiculturalism has become an unavoidable reality at international, national, and local levels. Intolerance toward multiculturalism, particularly in diverse nations like Indonesia, can result in societal issues such as fanaticism and hatred (Ham *et al.* 2023). Khisbiyah (2002) identifies two primary drivers of this intolerance: an education system prioritising uniformity over cultural diversity and a lack of mutual respect and dialogue regarding differences. Addressing these challenges requires deconstructing outdated educational paradigms and reconstructing models that emphasise tolerance and the appreciation of multiculturalism (Kassem and Murphy 2023).

DOI: 10.1201/9781003645542-47

Reforming education to nurture human resources that value diversity is essential for strengthening Indonesia's culture and civilisation (Huda *et al.* 2024; Fatimah *et al.* 2024). This objective aligns with Sustainable Development Goals (SDG) 4 and 16, which promote inclusive, equitable education and peaceful, inclusive societies. Integrating multicultural education into Indonesia's formal, nonformal, and informal educational pathways (Dib 1988; Cheng and Hamlin 2023) is critical to achieving this goal.

The failure to integrate multicultural education across these pathways risks perpetuating intolerance, potentially escalating into radicalism or terrorism (Evans and Nelson 2021; Lindsey *et al.* 2023). For instance, the 2018 Sidoarjo bombing involved children indoctrinated under the guise of homeschooling (Detikcom; Liputan6), illustrating the challenges faced by the Ministry of Education and Culture (Kemendikbud) in monitoring radicalism within informal education, particularly in single-family homeschooling (BBC News Indonesia). However, it is crucial to avoid generalising homeschooling as inherently linked to radicalism based on isolated cases. Homeschooling has a long history in Indonesia, dating back to the pre-independence era, and gained prominence in 2005 as a response to disparities in formal education access and the need to accommodate gifted children's unique needs (Holt and Farenga 2003; Illich 2000). Modern homeschooling practices in Indonesia include institutionalised programs and family-based communities offering flexible and adaptive learning opportunities (English 2021; Siddiq and Salama 2021).

Despite its growing popularity, homeschooling in Indonesia faces significant challenges related to inclusivity and regulatory oversight. The absence of robust monitoring frameworks raises concerns about whether homeschooling effectively promotes diversity and multiculturalism (Syahrin *et al.* 2023; Suwendi *et al.* 2024; Turhamun 2024). This study investigates a prominent Jakarta-based homeschooling community known for its diverse membership, frequent activities, and substantial media coverage. The community's heterogeneity—including various beliefs, professions, education levels, ethnicities, and social statuses—provides an ideal context for examining the implementation of multicultural education and religious moderation. This research analyses the community's culture, interaction patterns, and educational practices through an ethnographic approach. The findings aim to construct a family-based education model for informal learning environments, bridging gaps in Indonesia's informal education literature. The study highlights how diverse homeschooling communities can advance SDG 4 by delivering quality education and SDG 16 by fostering peace, tolerance, and inclusivity, providing a replicable blueprint for strengthening social cohesion and cultural appreciation.

2 METHODS

2.1 *Research design*

This study utilised a qualitative research design with an ethnographic approach, following Spradley's Developmental Research Sequence (2006). It explored the implementation of multicultural education and religious moderation in Jakarta homeschooling communities. Data collection involved participant observation, semi-structured interviews modelled as "friendship conversations" to elicit authentic responses, and secondary sources, including media reports. The exploratory design aimed to generate hypotheses through iterative questioning and natural interactions, emphasising the community's lived experiences and practices.

2.2 *Sampling*

Participants were selected using purposive sampling to ensure maximum diversity in terms of religion, ethnicity, education, occupation, and marital status. The sample included 19 families and 45 students, representing a wide range of sociocultural backgrounds. This deliberate selection strategy allowed for a comprehensive exploration of the community's educational perspectives, particularly regarding multiculturalism and religious moderation. To enhance clarity, participant demographics have been summarised in a table (see Table 1).

Table 1. Participant demographics.

Category	Subcategories	Frequency
Religion	Islam, Christianity, Catholicism, Buddhism, Other	5 groups
Ethnicity	Javanese, Chinese, Mixed Ethnic, Other	4 groups
Education	High school, Bachelor's, Master's	3 levels
Occupation	Teachers, Entrepreneurs, Homemakers, Employee	4 categories
Marital Status	Married, Single	2 categories

2.3 *Data collection methods*

Data collection relied on three primary sources: participant statements, participant actions, and artifacts used by participants. Data were gathered through detailed field notes from participant observation, verbatim transcripts of interviews, and documentation of community activities. The ethnographic process was iterative, with data collection and analysis occurring concurrently. This dynamic process enabled the continuous refinement of research questions and hypotheses.

2.4 *Data management and analysis*

The collected data were systematically organised and analysed following Spradley's four levels of ethnographic analysis: domain analysis, taxonomy analysis, componential analysis, and cultural theme analysis (Spradley 2006; Spradley and McCurdy 2012). This multi-tiered approach revealed the cultural meaning systems underpinning the community's practices in multicultural education and religious moderation. The analysis highlighted how families navigate and integrate these values into their daily educational practices, fostering tolerance and cooperation among diverse social groups.

2.5 *Reliability and validity*

To enhance reliability and validity, triangulation was conducted by cross-verifying data from observations, interviews, and documentation. Member checking was performed by sharing interview transcripts and preliminary findings with participants for feedback, ensuring accuracy and credibility. This iterative process reinforced the trustworthiness of the study.

2.6 *Ethical considerations*

Ethical standards were rigorously adhered to throughout the research. Approval was obtained from the relevant institutional review board. Participants provided informed consent and were assured of their rights, including the ability to withdraw at any point. Confidentiality and anonymity were maintained, with all data securely stored and used exclusively for research purposes.

3 FINDINGS AND DISCUSSION

3.1 *Community-based education empowers families as the core of the educational process*

The analysis of observation and interview data reveals that community-based homeschooling empowers families as the cornerstone of the educational process. In this model, parents assume dual roles as primary educators and moral guides, integrating formal education with the holistic development of their children. This personalised approach tailors learning to each child's unique needs and interests, combining academic goals with character-building activities. By emphasising moral and intellectual growth, the model provides a comprehensive framework that aligns education with family values, fostering well-rounded development within a supportive home environment.

The communal aspect further strengthens this model by promoting collaboration and mutual support among families. Activities such as joint educational projects, field trips, and community service initiatives encourage resource-sharing, idea exchange, and collective problem-solving. Programs like "Eksplorasi" exemplify this collaborative spirit, offering educational trips to cultural hubs and coastal regions, where children, guided by parent mentors, explore local traditions, livelihoods, and artistic practices. These experiences blend experiential learning with cultural appreciation, fostering teamwork and mentorship. The community's decentralised "Starfish" organisational structure enhances inclusivity, allowing all members to contribute ideas. As one participant explained, "Our club operates using the Starfish organisational structure, where there is no designated leader but activity coordinators." This flexible, inclusive approach reinforces the community's commitment to shared responsibility and cooperative learning.

3.2 *How family-based multicultural education and religious moderation awareness is implemented in the community?*

Family-based multicultural education within the community is implemented through informal practices that emphasise respect for and understanding of cultural diversity. Central to this approach is the cultivation of a collective identity, where families engage in activities that promote mutual respect and appreciation for diverse cultural backgrounds. This shared identity fosters a sense of belonging within a diverse and inclusive community, encouraging individuals to celebrate rather than merely acknowledge cultural differences. Key initiatives, such as storytelling sessions featuring folk tales from various Indonesian regions and collaborative projects like creating multicultural cookbooks with recipes from families' ethnic backgrounds, provide meaningful opportunities for children to explore and appreciate cultural traditions beyond their own.

Another critical aspect of family-based multicultural education is the emphasis on building strong inter-family relationships, which play a vital role in fostering tolerance and mutual understanding. Regular interactions with families of different cultural, ethnic, and religious backgrounds deepen appreciation for the richness of multiculturalism while strengthening social bonds. Religious moderation is also prioritised through interfaith dialogues, where families discuss shared ethical values like kindness, honesty, and respect. Visits to places of worship, including mosques, churches, and temples, allow families to observe religious practices, engage in respectful dialogues with religious leaders, and broaden their understanding of various belief systems. These initiatives promote unity and mutual respect while addressing the complexities of multicultural and interfaith dynamics through an adaptive and flexible educational approach that meets the community's unique needs.

3.3 *Multiculturalism and religious moderation values in community-based education homeschooling families*

The study of community-based homeschooling families identified core values of multiculturalism and religious moderation as essential for enhancing educational outcomes. A shared identity amidst diversity is fostered through activities celebrating various cultural traditions and emphasising commonalities, promoting belonging and appreciation for diversity. Friendship and mutual respect are cultivated via collaborative learning and social events, strengthening trust among families from different cultural and religious backgrounds. Religious moderation is emphasised through a balanced approach, focusing on shared ethical teachings and avoiding extremes, thereby fostering tolerance and understanding. Scouting activities serve as a platform for these initiatives, offering structured opportunities to practice teamwork, cultural appreciation, and ethical values in a dynamic and engaging setting. These practices collectively build a cohesive, inclusive, and harmonious learning environment, reinforcing community resilience (see Table 2).

Table 2. The intersection of multicultural education, religious moderation, and community resilience.

Key Domains	Core Values	Outcomes	Examples of Practices
Multicultural Education	Inclusivity, Appreciation of Diversity, Cultural Competence	Increased understanding and respect for cultural differences	Cultural Exchange Day: Families share traditional attire, food, and art from their cultural backgrounds. Storytelling Sessions: Parents narrate folk tales from various regions to instil cultural appreciation. Pramuka Activities: Children engage in collaborative outdoor activities that teach teamwork and respect for others.
Religious Moderation	Tolerance, Interfaith Dialogue, Balanced Religious Perspectives	Reduced prejudice and extremism; strengthened interfaith relationships	Interfaith Dialogue: Families discuss shared ethical values like honesty and kindness. Inclusive Celebrations: Religious holidays are celebrated with explanations and active participation from all families. Pramuka Oath and Values: Activities emphasizing moral and ethical principles shared across religious traditions.
Community Resilience	Cooperation, Mutual Support, Adaptability	Creation of cohesive communities that navigate diversity effectively	Collaborative Learning Projects: Families work together on group projects that require teamwork and diverse input. Community Support Networks: Families provide resources, encouragement, and mentorship to each other. Pramuka Leadership Training: Teaching children leadership, problem-solving, and cooperation through structured tasks.
Shared Values (Overlap)	Inclusivity, Tolerance, Cooperative Learning	Promotes a unified educational approach that prepares children for multicultural global societies	Role-Playing Exercises: Simulating real-world scenarios to practice tolerance and teamwork. Multicultural Cooking Projects: Children and parents create a recipe book featuring dishes from diverse backgrounds.

Finally, the study underscores the importance of adaptability in educational practices. Homeschooling families often employ flexible teaching methods that cater to their children's unique needs and interests. This adaptability extends to incorporating multicultural and religious moderation values into the educational experience. By using an eclectic mix of teaching materials and approaches, these families ensure that children receive a well-rounded education that reflects their diverse world. This adaptability enhances educational outcomes and equips children with the skills and mindset necessary to thrive in a multicultural and interconnected global society.

4 CONCLUSION

This ethnographic study highlights the multifaceted dynamics of community-based homeschooling in Jakarta, focusing on integrating multiculturalism and religious moderation into educational practices. Families from diverse backgrounds foster cohesive communities through shared educational objectives, promoting collective identity, cultural appreciation, and mutual respect. Religious moderation is emphasised to prevent extremism through inclusive teachings and interfaith dialogue, cultivating tolerance and understanding. The findings extend beyond homeschooling, offering insights for integrating multicultural and religious values into formal and nonformal education. Policymakers are urged to strengthen homeschooling frameworks with legal safeguards and structured curricula that promote critical thinking, cultural awareness, and interfaith understanding, ensuring inclusivity and global preparedness.

REFERENCES

BBC News Indonesia. (2019, December 3). *Penindakan radikalisme di sekolah: Pemerintah sulit awasi homeschooling*. BBC Indonesia. https://www.bbc.com/indonesia/indonesia-50641555

Cheng, A., and Hamlin, D. (2023). Contemporary homeschooling arrangements: An analysis of three waves of nationally representative data. *Educational Policy, 37*(5), 1444–1466. https://doi.org/10.1177/08959048221103795

Detikcom. (2018, May 14). *Bomber Surabaya tak sekolahkan anak, ngakunya homeschooling*. Detik News. https://news.detik.com/berita/d-4019604/bomber-surabaya-tak-sekolahkan-anak-ngakunya-homeschooling

Dib, Claudio Zaki. (1988). *Formal, Non-Formal and Informal Education: Concepts/Applicability*, Brazil: Institute of Physics University of São Paulo.

English, R. (Ed.). (2021). *Global perspectives on home education in the 21st century. IGI Global*. https://doi.org/10.4018/978-1-7998-6681-7

Evans, A. L., and Nelson, J. K. (2021). The value of adapting counseling to client's spirituality and religion: Evidence-based relationship factors. *Religions, 12*(11), 951. https://doi.org/10.3390/rel12110951

Fatimah, R., Pahrudin, A., Syafi'i, I., Romlah, and Akmansyah, M. (2024). Developing a religious moderation-based multicultural education model for religious nationalist student formation. *Educational Administration: Theory and Practice, 30*(5), 1506–1517. https://doi.org/10.53555/kuey.v30i5.3110

Ham, S.-H., Cha, Y.-K., and Lee, S. (2023). The global institutionalization of multicultural education as an academic discourse. *Societies, 13*(8), 191. https://doi.org/10.3390/soc13080191

Holt, J., and Farenga, P. (2003). *Teach Your Own: The John Holt book of Homeschooling*. Perseus Publishing.

Huda, M., Saeful, A., Rahim, H., Rosyada, D., and Zuhdi, M. (2024). The implementation of multicultural education in Indonesia: Systematic literature review. In M. D. H. Rahiem (Ed.), *Religion, education, science and technology towards a more inclusive and sustainable future* (pp. 83–89). Taylor and Francis. https://doi.org/10.1201/9781003322054-13

Illich, Ivan. (2000). *Bebaskan Masyarakat dari Belenggu Sekolah, judul asli Deschooling Society*, terjemahan A. Sonny Keraf, Jakarta: Yayasan Obor Indonesia.

Kassem, D., and Murphy, L. (2023). Addressing multicultural societies: Lessons from religious education curriculum policy in Indonesia and England. *Journal of Religious Education, 71*(3), 224–236. https://doi.org/10.1007/s40839-023-00164-x

Khisbiyah, Y. (2002). Core issues fueling intolerance: Education and multiculturalism. *Indigenous Journal of Multicultural Studies, 3*(2), 45–60. https://journals.ums.ac.id/index.php/indigenous/article/view/4623

Lindsey, T., Makruf, J., and Pausacker, H. (2023). *Islam, Education and Radicalism in Indonesia: Instructing piety*. Routledge.

Liputan6. (2018, May 15). *Mengaku homeschooling, anak bomber Surabaya hanya terima doktrin orangtua. Liputan6 News*. https://www.liputan6.com/news/read/3519159/mengaku-home-schooling-anak-bomber-surabaya-hanya-terima-doktrin-orangtua

Siddiq, M., and Salama, H. (2021). Homeschooling as a form of informal education: Legality and its various learning approaches. *Jurnal Penelitian Kebijakan Pendidikan, 14*(2). https://doi.org/10.24832/jpkp.v14i2.508

Spradley, J. P. (2006). Metode etnografi. Tiara Wacana.

Spradley, J. P., and McCurdy, D. W. (2012). *Conformity and Conflict: Reading in Cultural Anthropology* (14th ed.). Pearson Education.

Suwendi, S., Mesraini, M., and Fuady, F. (2024). Transforming religious moderation in the education world. *In Proceedings of the 7th International Colloquium on Interdisciplinary Islamic Studies (ICIIS)*. European Union Digital Library. https://doi.org/10.4108/eai.17-6-2024.2349071

Syahrin, A. A., Yasa, I. W. P., Wirawan, I. G. M. A. S., and Nur, I. (2023). Multicultural education and religious moderation education: Strategies for harmony education in schools. *In Proceedings of the 5th International Conference on Law, Social Sciences, and Education (ICLSSE)*. European Union Digital Library. https://doi.org/10.4108/eai.1-6-2023.2341378

Turhamun. (2024). Implementation of religious moderation in the non-formal education in forming a multicultural society. *International Journal of Social Science and Religion, 5*(1). https://doi.org/10.53639/ijssr.v5i1.196

Towards Resilient Societies: The Synergy of Religion, Education, Health, Science, and Technology – Rahiem (Ed.)

The relationship between school feeding programs and school well-being: A systematic review

Muhammad Syukri Pulungan, Gazi Saloom & Maswani
Universitas Islam Negeri Syarif Hidayatullah Jakarta, Indonesia

ABSTRACT: In July 2024, SDG advocates led by UN Secretary-General Jeffrey Sachs emphasised that free lunch programmes are a vital investment in achieving the Sustainable Development Goals (SDGs), particularly in combating hunger and malnutrition, which significantly impact physical and mental development. Addressing physical, psychological, and social aspects is essential for fostering school well-being. To explore this, a comprehensive literature review was conducted on school feeding programmes and their relation to school well-being in the context of the SDGs, using databases such as Google Scholar and Scopus. Through a multi-step exclusion process, 11 relevant documents were selected from 213 Scopus articles and 315 Google Scholar articles. The findings offer valuable insights into promoting health and well-being in educational settings through school feeding initiatives. The literature reveals a strong connection between school feeding programmes and school well-being, demonstrating that school meals, directly and indirectly, influence the dimensions of having, loving, being, and health as a framework for school well-being. The success of these programmes relies heavily on strict nutritional standards and efficient management. Additionally, school feeding programmes not only enhance student well-being but also stimulate economic growth by supporting local farmers through the use of locally sourced food products.

Keywords: free meal school, literature review, school feeding programs, well-being

1 INTRODUCTION

Implementing school feeding programs significantly impacts school well-being by improving nutritional intake and food security. These programs are crucial in addressing child malnutrition and health inequalities. Research in China has shown that school meal programs significantly reduced the urban-rural health gap by 21.6% (Ren *et al.* 2023). Similarly, research in New Zealand shows that government-funded school lunch programs enhance food security, reduce financial pressure on families, and improve well-being (McKelvie-Sebileau 2023). Other research also indicates that the free school meal policy in the UK has shown promising results in reducing food insecurity, hunger, and family stress (Carlisle *et al.* 2023).

School feeding programs have rapidly expanded, reaching 60.1 million children in 41 countries, with most funding coming from domestic sources. This reflects a shift towards national ownership and local procurement (Wineman *et al.* 2022). These programs also play a crucial role in education by increasing school enrolment, attendance, and academic performance, as food-secure Children are more inclined to attend school consistently and achieve better academic performance (Fitawek 2024). School feeding aims for the welfare of students and the success of their development and prosperity (Latifa *et al.* 2023). In addition, the school meal program also contributes to anticipating unexpected health hazards in the future (Rahiem 2024).

School well-being includes various dimensions contributing to a positive and effective learning environment based on educational goals. The objective condition of students in

 DOI: 10.1201/9781003645542-48

modern society is that they are very susceptible to experiencing stress (Mujib and Rena 2020). In the context of school, well-being refers to a conceptual model that consists of the dimensions of belonging, loving, becoming, and health status (Konu and Rimpelä 2002). The dimension of "having" refers to the material and physical resources available to students, such as safe school facilities and adequate learning materials that support the well-being of students in schools (Karan and Erdemir 2023). The "loving" dimension emphasises the importance of positive relationships and emotional support in the school community, including relationships with peers, interactions with teachers, and other elements. The dimension of "being" involves personal growth through perseverance, self-esteem, and optimism, closely related to student happiness. The "health" dimension refers to students' physical and mental well-being as they participate in the learning process at school(Norwich *et al.* 2022). Previous research has mentioned that access to regular and nutritious food can improve students' self-esteem and social interaction (Gordon *et al.* 2007), closely related to school well-being.

In this context, there is an increasing need to strengthen, organise, and synthesise the existing literature on free school meal programs and school well-being about efforts to achieve the SDGs. A systematic review is valuable as it addresses the need to comprehensively examine all existing research related to a specific research question. This method is highly beneficial for academics to identify, analyse, critically evaluate, and summarise knowledge on a specific field in a clear, replicable, and rigorous manner. The significance of establishing free school meal programs for enhancing school well-being and achieving the SDGs is highlighted by the absence of similar studies in the existing research. This gap presents an opportunity to fill this void and serve as a scholarly resource by conducting a systematic literature review on free meal programs.

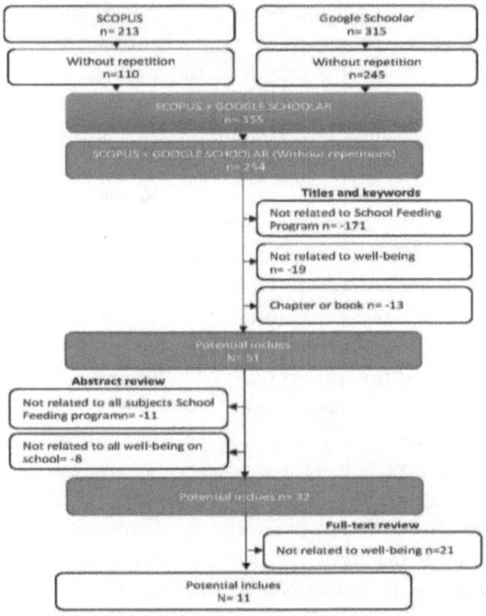

Figure 1. Flow diagram to show the study selection process.

2 METHOD

A prevalent approach to developing research questions in systematic reviews involves using the PICO framework established by the National Institute for Health and Clinical Excellence (Schardt *et al.* 2007). For a systematic review, we examined articles published in academic journals between 2014 and 2024 that used English. Relevant articles were searched using keywords such as "school meals", "school meals programmes", "school meals", "school meals", AND "Well-being". The keywords were also combined to narrow the search in the

Google Scholar and Scopus databases and involved 11 articles for further analysis. The selection of Google Scholar as a database is based on the breadth of its coverage compared to other indexes. Meanwhile, using Scopus as a database is based on the assumption that it is a reputable journal article. Figure 1 shows a flow diagram of the manuscript selection process.

3 RESULTS

In the table below, we have compiled the fundamental ideas from the 11 analysed works, separating on one side the types of interventions or actions proposed for school meal programs in about-being, and on the other side, the main conclusions drawn (Table 1).

Table 1. Research conclusions.

	How is the school feeding program related to school well-being?	What conclusions did they find?
(Aliyar *et al.* 2015)	School feeding programs provide nutritious food that can significantly improve children's nutritional status, which is crucial for their overall health and development. Studies show that school feeding programs can enhance cognitive performance, particularly among disadvantaged students.	Ensuring that schools provide nutritious food is essential for boosting student attendance and retention, improving nutritional status and cognitive growth, addressing poverty and obesity, and acting as a vital social safety net for low-income families.
(Corcoran *et al.* 2014)	School feeding programs can indirectly affect academic performance through improved attendance and behaviour. Increased participation in feeding programs may lead to better classroom engagement and reduced absenteeism, indicative of enhanced student well-being.	The study indicates that while breakfast programs enhance participation, they do not provide short-term gains in academic performance or alleviate concerns about increased obesity. Further research is needed to explore these aspects more comprehensively.
(Norwood 2020)	School breakfast programs can enhance cognitive performance, particularly in memory and attention functions. This effect is especially pronounced in nutritionally deficient children, who show significant improvements in tasks such as math tests and vocabulary. Additionally, school breakfast programs positively influence student behaviour and reduce disruptive conflicts.	Research findings indicate a 16.6% reduction in total disciplinary reports, including a 16.1% decrease in code of conduct violations and a 13.5% decrease in student conflicts. Additionally, the program has reduced school absenteeism and provides direct benefits to improved mental and physical health.
(Cohen 2022)	Free school meal programs help reduce food insecurity, especially among low-income students. This results in improved overall health and well-being of the students. By consuming nutritious meals, students are more likely to develop healthier eating habits that persist into adulthood.	School meal programs have demonstrated a positive link between free meals and academic achievement, especially at lunchtime. Nonetheless, implementing free school meals should be guided by proper nutritional standards.
(Kristjansson 2022)	School meal programs can provide comprehensive meals, including milk, for children aged 5-19, ensuring they receive essential nutrition during school hours. By alleviating food-related stress, school meal programs contribute to a more productive learning environment at school and home, positively impacting students' psychological well-being.	Effect sizes and intervals gauge the magnitude and precision of a program's impact. Larger effect sizes with narrow confidence intervals (CI) are more likely to indicate a genuine effect. Beyond statistical measures, the real-world significance of the findings is assessed by evaluating how observed effects translate into tangible benefits for the children involved.

(*continued*)

Table 1. Continued

	How is the school feeding program related to school well-being?	What conclusions did they find?
(Corpuz and Bantulo 2023)	Food initiatives for education in developing countries aim to achieve the Sustainable Development Goals (SDGs) and reduce hunger. Further research is needed to explore the impact of these programs on students' health and academic outcomes.	School feeding programs are viable even with limited resources as long as there is robust support from stakeholders and the management demands are manageable. Most students from the four surveyed schools attained satisfactory or better academic performance, with only a few reaching exceptional levels.Bagian Atas FormulirBagian Bawah Formulir.
(Garton *et al.* 2023)	School feeding programs have been linked to substantial positive effects on educational attainment in similar initiatives internationally, indicating potential long-term benefits for academic outcomes. Additionally, the program enhances student well-being and reduces financial strain on families. These results align with the program's objectives and suggest it could achieve even more with continued support.	School feeding programs benefit students, families, communities, and the broader food system. Students participating in the program have experienced significant improvements in both physical function and mental well-being, reflecting the positive effects of the program on overall health.
(Kesa 2024)	School feeding programs contribute to academic performance, student attendance, and school participation; there is an improvement in alignment with efforts to enhance nutritional intake and overall school well-being. Improved nutrition helps students concentrate better in class, leading to better academic outcomes.	Stakeholders have observed the beneficial impacts of the school feeding program on school attendance, punctuality, classroom engagement, and academic performance, as well as an overall boost in student well-being.
(Wara *et al.* 2023)	The research found a strong positive correlation between meal tickets and academic performance through cognitive functions. Conversely, poor nutrition can lead to memory impairments and reduced problem-solving abilities.	Meal tickets have been found to positively impact students' academic performance and correlate positively with students' GPA/CGPA. This study recommends that Havilla University continue its meal ticket policy to enhance academic performance and student well-being.
(Hill *et al.* 2023)	The school breakfast program is implemented with a comprehensive approach to student health and well-being. Furthermore, the program aids in children's social-emotional development by addressing their emotional needs and encouraging positive social-emotional learning and relationships, essential for fostering prosocial behaviour, academic engagement, and performance.	The program provides substantial social advantages for students and significantly enhances the overall school environment. The study underscores the potential for additional research into the material, relational, and pedagogical aspects that help reduce negative impacts and deepen understanding of the broader effects.
(Su 2024)	School feeding programs have been shown to increase children's height and reduce the prevalence of obesity. Additionally, they positively impact cognitive skills by decreasing school absenteeism and enhancing learning abilities.	The study contributes significantly to understanding non-cognitive and behavioural outcomes in children, highlighting that previous estimates of the importance of school feeding programs for well-being may have been overlooked.

4 DISCUSSION

The eleven documents reviewed exhibit a range of objectives, scopes, and methodologies. Despite these differences, each study examines the impact of school meal programs on student well-being, whether directly or indirectly. All the studies affirm the essential role of student well-being within these programs. Together, they explore how implementing school feeding programs boosts student well-being and contributes to achieving the SDGs.

Several studies have described several contributions to providing food at school that have implications for school well-being (Table 2).

Table 2. Contribution of school feeding programs to school well-being.

Contribution of school food programs to school well-being			
Having	Loving	Being	Health
Improved physical and mental function Overcoming malnutrition	Improves social relationships Improves social-emotional skills Improves prosocial behaviour Reducing code of conduct violations and conflicts	Increase productivity Improves discipline Improves cognitive abilities Improve problem-solving skills	Overcoming Physical Disorders Reduces memory impairment Reduces obesity

4.1 *Having*

Providing nutritious food can overcome food insecurity and improve student nutrition, which contributes to improving physical and mental function and overcoming student nutritional disorders. Research conducted in South Africa noted that most nutrients provided did not meet 25-30% of the recommended daily food intake (Kesa 2024). School meals should at least include protein, vitamins and minerals for energy. Thus, policymakers must establish research-based nutritional standards and menu composition guidelines to improve the quality of school meals (Aliyar *et al.* 2015).

4.2 *Loving*

The school feeding program can improve social relationships, social-emotional skills, and prosocial behaviour and reduce conflict. School meal programs directly reduce the impact of food insecurity among children from low-income families (Gundersen 2015). This impacts improving social relationships, prosocial behaviour, and social and emotional skills and reduces code of ethics violations and conflict among students (Garton *et al.* 2023). Increasing social relationship status and prosocial behaviour can improve school well-being.

4.3 *Being*

The school feeding program highlights the positive impact of increasing productivity, discipline, cognitive ability, and problem-solving ability. Improving nutritional status indirectly positively impacts cognitive development, with increased concentration during the learning process and a conducive school environment through connections with the socio-ecological model that connects individuals, families, schools and the wider community. In addition, the free school meal program is associated with increased school attendance, decreased dropout rates, and improved memory (Cohen 2022; Corcoran *et al.* 2014; Delbiso *et al.* 2021; Kesa 2024; Su 2024; Wara *et al.* 2023). In addition, this program contributes to creating a more productive learning environment at school and at home (Kristjansson 2022).

4.4 *Health*

The literature review provides evidence that movement reduces the prevalence of anaemia and malnutrition and improves students‘ physical health. Indirectly, school meal programs reduce concerns about increasing obesity (Aliyar *et al.* 2015; Corcoran *et al.* 2014) and overall health problems (Garton *et al.* 2023).

5 CONCLUSION

Research interest in achieving SDGs has increased in response to countries' challenges, particularly improving education quality and food security. This literature review has highlighted the importance of free school lunch programs in enhancing students' well-being. School well-being can be seen from the dimensions of belonging, love, existence, and health status that students perceive. These programs act as a bridge to minimise risks for students from low-income families, providing them with better future opportunities. This study recommends that governments and stakeholders establish strict guidelines on nutritional content, meal timings, and the involvement of local agricultural products. In addition to improving school well-being, these programs can also have a positive impact on improving the economy of local farmers. Further research, based on longitudinal studies and other research approaches and areas, is highly recommended to support this study's findings.

REFERENCES

Aliyar, R., Gelli, A., and Hamdani, S. H. (2015). A review of nutritional guidelines and menu compositions for school feeding programs in 12 countries. In *Frontiers in public health*. frontiersin.org. https://doi.org/10.3389/fpubh.2015.00148

Carlisle, V. R., Jessiman, P. E., Breheny, K., Campbell, R., Jago, R., Leonard, N., Robinson, M., Strong, S., and Kidger, J. (2023). A Mixed Methods, Quasi-Experimental Evaluation Exploring the Impact of a Secondary School Universal Free School Meals Intervention Pilot. *International Journal of Environmental Research and Public Health*, *20*(6), 5216. https://doi.org/10.3390/ijerph20065216

Cohen, J. F. W. (2022). Implementation of Universal School Meals during COVID-19 and beyond: Challenges and Benefits for School Meals Programs in Maine. *Nutrients*, *14*(19). https://doi.org/10.3390/nu14194031

Corcoran, S. P., Elbel, B., and Schwartz, A. E. (2014). The Effect of Breakfast in the Classroom on Obesity and Academic Performance: Evidence from New York City. *Institute for Education and Social Policy*. https://eric.ed.gov/?id=ED559664

Corpuz, S. G., and Bantulo, J. S. (2023). Level of Effectiveness, Sustainability of School-Based Feeding Program, Nutritional Status and Academic Performance of Pupils Amidst Covid-19 Pandemic: Basis for a Proposed Project Busog Talino Program. *European Journal of Education Studies*. http://oapub.org/edu/index.php/ejes/article/view/4880

Delbiso, T. D., Kotecho, M. G., and Asfaw, F. M. (2021). Effects of COVID-19 imposed school closure on school feeding program in Addis Ababa, Ethiopia. *Social Sciences and Humanities Open*, *4*(1), 100185. https://doi.org/10.1016/j.ssaho.2021.100185

Fitawek, W. (2024). Approaches to assessing the impact of education on food security. In *Handbook on Public Policy and Food Security* (pp. 131–141). Edward Elgar Publishing. https://doi.org/10.4337/9781839105449.00018

Garton, K., Riddell, C., McKelvie-Sebileau, P., and ... (2023). Not Just a Free Lunch a logic model and evidence review for the Ka Ora, Ka Ako| Healthy School Lunch programme. *Policy* https://ojs.victoria.ac.nz/pq/article/view/8243

Gordon, A., Crepinsek, M. K., Nogales, R., and Condon, E. (2007). *School Foodservice, School Food Environment, and Meals Offered and Served.*

Gundersen, C. (2015). Food Assistance Programs and Child Health. *The Future of Children*, *25*(1), 91–109. https://doi.org/10.1353/foc.2015.0004

Hill, S. M., Byrne, M. F., Wenden, E., Devine, A., and ... (2023). Models of school breakfast program implementation in Western Australia and the implications for supporting disadvantaged students. In *Children and Youth* .... Elsevier. https://www.sciencedirect.com/science/article/pii/S0190740922004066

Karan, Ş., and Erdemir, E. (2023). Subjective school wellbeing of refugee and local children: Voices from having, loving, and being dimensions. *Early Child Development and Care*, *193*(9–10), 1141–1158. https://doi.org/10.1080/03004430.2023.2234671

Kesa, H. (2024). Menu Evaluation versus Health and Well-being of Children Participating in School Feeding Programmes in South Africa. *International Journal of School Health, 11*(1), 2–13. https://doi.org/10.30476/intjsh.2023.99878.1342

Konu, A., and Rimpelä, M. (2002). Well-being in schools: a conceptual model. *Health Promotion International, 17*(1), 79–87. https://doi.org/10.1093/heapro/17.1.79

Kristjansson, E. (2022). School feeding programs for improving the physical and psychological health of school children experiencing socioeconomic disadvantage. *Cochrane Database of Systematic Reviews, 2022* (8). https://doi.org/10.1002/14651858.CD014794

Latifa, R., Chang, K., Liem, A., Idriyani, N., Roebianto, A., Rahmah, M., and Puri, V. G. S. (2023). Indonesian Version of WHO-5 Well-being Index Amidst COVID-19 Pandemic Settings: Scale Validation and Standardisation. *JP3I (Jurnal Pengukuran Psikologi Dan Pendidikan Indonesia), 12*(2), 145–161. https://doi.org/10.15408/jp3i.v12i2.33964

McKelvie-Sebileau, P. (2023). Health, wellbeing and nutritional impacts after 2 years of free school meals in New Zealand. *Health Promotion International, 38*(4). https://doi.org/10.1093/heapro/daad093

Mujib, A., and Rena, S. (2020). The Moderating Effect of Spirituality on the Relationship Between Academic Life Stressors and Perceived Stress in Medical Undergraduate Students. *Proceedings of the Proceedings of the 1st International Conference on Religion and Mental Health, ICRMH 2019, 18 - 19 September 2019, Jakarta, Indonesia.* https://doi.org/10.4108/eai.18-9-2019.2293361

Norwich, B., Moore, D., Stentiford, L., and Hall, D. (2022). A critical consideration of 'mental health and wellbeing' in education: Thinking about school aims in terms of wellbeing. *British Educational Research Journal, 48*(4), 803–820. https://doi.org/10.1002/berj.3795

Norwood, B. (2020). Breakfast of champions: Universal Free Breakfast and student conflict and test scores in Texas schools. *Available at SSRN 3576085.* https://papers.ssrn.com/sol3/papers.cfm?abstract_id=3576085

Rahiem, M. D. H. (2024). Early childhood education contingencies for sustaining learning during school closures: Lessons from preschool remote education home visits in Indonesia during the COVID-19 pandemic. *Children and Youth Services Review, 166*, 107955. https://doi.org/10.1016/j.childyouth.2024.107955

Ren, J., Zheng, X., Smith, R., and Fang, X. (2023). School feeding program and urban–rural inequality of child health: Evidence from China. *Agribusiness, 39*(S1), 1399–1416. https://doi.org/10.1002/agr.21862

Schardt, C., Adams, M. B., Owens, T., Keitz, S., and Fontelo, P. (2007). Utilization of the PICO framework to improve searching PubMed for clinical questions. *BMC Medical Informatics and Decision Making, 7*, 16. https://doi.org/10.1186/1472-6947-7-16

Su, D. (2024). Nutrition and Child Development: Evidence from the Student Nutrition Improvement Program. *Available at SSRN 4708989.* https://papers.ssrn.com/sol3/papers.cfm?abstract_id=4708989

Wara, S. T., Owan, R. A., Wara, H. U., Ibe, A. E., and ... (2023). Impact of Meal Ticketing on Students' Academic Performance and Well-Being in Havilla University, Nde-Ikom, Cross River State. In *ASRIC JOURNAL ON ....* asric.africa. https://www.asric.africa/documents/Journal/Journal of Social Sciences and Humanities 2023 Vol. 4 Issue 1.pdf#page=297

Wineman, A., Ekwueme, M. C., Bigayimpunzi, L., Martin-Daihirou, A., de Gois V. N. Rodrigues, E. L., Etuge, P., Warner, Y., Kessler, H., and Mitchell, A. (2022). School Meal Programs in Africa: Regional Results From the 2019 Global Survey of School Meal Programs. *Frontiers in Public Health, 10.* https://doi.org/10.3389/fpubh.2022.871866.

Towards Resilient Societies: The Synergy of Religion, Education, Health, Science, and Technology – Rahiem (Ed.)

Bridging the gap: Education, religion, and technology as catalysts for Indonesian decent work

Nida Husna & Reza Anis Maulidya
Universitas Islam Negeri Syarif Hidayatullah Jakarta, Indonesia

Suyany Marth
Cambodian Islamic Center, Cambodia

ABSTRACT: This chapter explores the intersection of education, religion, and technology as key drivers for promoting decent work and sustainable economic growth in Indonesia. The study addresses skill mismatches, encourages ethical leadership, and fosters inclusive economic opportunities by incorporating religious values into educational curricula and utilising technological advancements. Drawing on case studies and scholarly literature, it examines the transformative effects of vocational training programs integrated with corporate social responsibility (CSR) principles, digital platforms that expand market access, and ethical business practices grounded in religious teachings. The findings reveal significant challenges, including the digital divide, rural-urban infrastructure disparities, and resistance to integrating technology within traditional religious contexts. These obstacles underscore the need for targeted policy interventions and collaborative efforts among policymakers, educators, religious leaders, and technology innovators. The study highlights the importance of aligning these collaborative efforts with Indonesia's commitment to Sustainable Development Goal 8 (decent work and economic growth) to improve productivity, equity, and resilience. By overcoming these challenges and fostering partnerships, Indonesia can build a more inclusive and sustainable economy that bridges societal divides and empowers marginalised communities.

Keywords: decent work, educational reforms, ethical leadership, religious values, sustainable economic growth

1 INTRODUCTION

In Indonesia, addressing unemployment is intertwined with sustainable economic growth, which is crucial for socio-economic development. As the largest economy in Southeast Asia, Indonesia faces the formidable challenge of attaining economic growth and excellent employment opportunities that safeguard the fundamental rights of workers (World Bank 2021). The International Labour Organization (ILO) defines decent work as productive employment, offering fair compensation, and facilitating social interaction among stakeholders and workers, among other business-related factors.

Given Indonesia's dedication to the principles of decent employment, several recent trends are troubling. Barriers such as inadequate education, skills misalignment, and an inability to address the technological divide hinder the achievement of the benchmarks established by Sustainable Development Goal 8 related to decent work and economic growth. Despite the Inclusive Economic Growth Strategy within the RPJMN being a commendable endeavour capable of generating numerous employment opportunities, deficiencies in human capital and disparities between regional and educational requirements impede overall advancement

DOI: 10.1201/9781003645542-49

(ADB 2022). The underlying difficulties underscore the theme elements of gender equality, ensuring all individuals have access to good job opportunities (United Nations 2021).

In Indonesia's pursuit of economic sustainability and inclusive growth. Education, religion, and technology, among other elements, are considered pivotal. The premise is that education equips people with skills and promotes innovations essential for economic growth and competitiveness (OECD 2020). The reorganisation of the vocational education system enhances employability by aligning skill development with labour market requirements. The improvements in vocational training facilitate improved access to industry post-graduation by equipping graduates with the skills required by the sector (BPS 2018). Moreover, ethical teachings derived from Indonesia's religious heritage can effectively enhance the education system to cultivate the character of future professionals who will champion justice, honesty, integrity, and social responsibility (Evita *et al.* 2023). These forces enhance the nation's human capital and foster socio-cultural conditions that promote achieving established sustainability goals (United Nations 2021). Religion is undeniably regarded as a significant social institution in Indonesia due to its ability to shape beliefs, values, and conventions, facilitating its integration into business and community growth. The collaboration between religious and educational institutions has successfully integrated ethics into the curriculum, equipping graduates to become ethical leaders (Evita *et al.* 2023). This emphasises ethics over profits and promotes corporate social responsibility initiatives that facilitate inclusive growth. Policies designed for wealth redistribution, based on religious principles of fairness and equity, incorporate the disadvantaged into the economic development agenda, promoting equity-focused economies and sustainable development (United Nations 2021; Weber and al-Ansari 2022).

Innovation has emerged as a significant catalyst for creating quality employment opportunities in Indonesia. The rise of e-commerce enables smaller companies in rural regions to participate in local and international trade (Jakarta Post 2021). Tokopedia and Gojek exemplify platforms enabling small vendors to transcend spatial limitations and access broader markets, promoting economic inclusion. Indonesian adolescents are targeted via digital literacy and online vocational training programs designed to equip them for the digital economy (UNESCO 2021). These projects enhance productivity and efficiency while generating additional employment in emerging industries such as fintech and e-services (ADB 2022). Consequently, technology can assist Indonesia in fully using its demographic dividend to stimulate inclusive growth.

Substantial obstacles persist in integrating education, religion, and technology to promote dignified employment. The digital gap represents a significant obstacle, characterised by considerable inequalities in internet availability between urban and rural regions (INTI Media 2024). This disparity restricts access to economic possibilities and intensifies inequalities in education and employment. A second-order digital divide endures, characterised by individuals possessing physical access to technology yet lacking the requisite skills or motivation for effective utilisation. These discrepancies highlight the need for focused investments in internet infrastructure, literacy initiatives, and training for rural regions.

A further issue is the disparate levels of technological advancement between urban and rural regions. Educators in remote areas lack the necessary assistance and expertise to effectively integrate technology into their teaching. Moreover, a divergence arises between technology and religious education. Integrating technologically sophisticated curricula with conventional religious teachings in various social contexts results in inconsistencies between anticipated learning outcomes and actual learning experiences. These difficulties necessitate governance actions and policies that facilitate comprehension of the socio-economic, cultural, and religious complexities that define Indonesia.

Despite the aforementioned challenges, rural communities connected to the internet through initiatives like "Digital Villages" can access various programs that assist them, transforming their disadvantages into distinctive opportunities (ADB 2022; Ministry of Communication and Information Technology 2019). These projects aid villagers in overcoming poverty by providing access to broader markets and facilitating their integration into the digital economy, thereby fostering resilience in these regions. The integration of education, technology, and religion has demonstrated the capacity to generate opportunities that stimulate growth, with these instances serving as prime examples of that potential.

Indonesia's capacity for sustained economic progress hinges on its comprehension of the interplay among technology, religion, and education. Should this triangle be endorsed, Indonesia can guarantee that all its inhabitants receive quality education and ethical management as the nation advances technologically. This vision constitutes a profound dedication to attaining sustainable development objectives and introduces a novel approach to fostering social, economic, and other forms of cohesiveness, resilience, and dignity. As Indonesia rapidly integrates into global markets, these complements can really enhance communities and establish the groundwork for a more equal and successful society.

In conclusion, Indonesia stands at a pivotal juncture where the convergence of education, religion, and technology offers transformative potential for sustainable economic growth and equitable prosperity. By prioritising investments in education to enhance skills and innovation, integrating ethical teachings from diverse religious traditions, and harnessing digital advancements to democratise economic opportunities, Indonesia is poised to foster inclusive development that benefits all its citizens. This study will pose two research questions: first, how can Indonesia further leverage these synergies to ensure marginalised communities also benefit from economic growth?. Second, what role can international partnerships play in supporting Indonesia's efforts to integrate ethical leadership and technological advancements into its financial strategies? Embracing these synergies aligns with global sustainable development goals and establishes a blueprint for fostering social cohesion, economic resilience, and human dignity. As Indonesia navigates the complexities of a globalised economy, these efforts promise to pave the way for a more just and prosperous future for all Indonesians.

2 METHODS

This chapter synthesises insights from scholarly literature and case studies to explore the synergies among education, religion, and technology in fostering sustainable economic growth and decent work opportunities. The case studies were selected based on their relevance to the research objectives, particularly their alignment with Sustainable Development Goal 8 (SDG 8), emphasising inclusive economic growth and equitable access to decent work. Selection criteria focused on examples demonstrating innovative integration of religious values into education, using technology to empower marginalised communities, and initiatives addressing socio-economic disparities in Indonesia. The methodological approach involved reviewing recent publications and reports from 2018 to 2023, including government policies like the National Medium-Term Development Plan (RPJMN) and analyses by international organisations such as UNESCO and the World Bank. These sources provided comprehensive insights into how educational, religious, and technological initiatives contribute to ethical leadership, corporate responsibility, and inclusivity. The time frame and selection criteria ensured the data reflected current developments and directly supported the chapter's objectives.

3 FINDINGS & DISCUSSIONS

3.1 *Findings*

3.1.1 *Leveraging synergies between education, religion, and technology to ensure marginalised communities benefit from economic growth*

Indonesia has made significant strides in leveraging education, religion, and technology to ensure economic growth benefits all, including marginalised communities. One of the key avenues is through the integration of religious values into education, particularly in business and vocational training programs. For instance, the growing trend towards Islamic business ethics has informed educational reforms encouraging ethical conduct grounded in religious teachings. Islamic principles like justice ('adl) and fairness (istiqamah) promote transparency and social responsibility in business environments, ensuring that economic activities align with societal well-being (Muhammad 2020). Similarly, Christian ethical values, emphasising love and justice, guide business leaders to make decisions that benefit their organisations and society (Chase 2004; Kim *et al.* 2009). By embedding religious principles into the curriculum, Indonesia equips future graduates with the ethical framework to navigate complex business landscapes, fostering sustainable growth that benefits marginalised communities (Triyuwono *et al.* 2021).

Furthermore, these religious values-based educational reforms cultivate a business environment that promotes social justice and inclusivity. Programs emphasising compassion and social responsibility encourage entrepreneurs to create business models addressing poverty alleviation and environmental sustainability, aligning with Indonesia's cultural and religious beliefs (UNESCO 2021; World Bank 2021). Initiatives like Corporate Social Responsibility (CSR) modules and sustainability-focused vocational training equip students with the knowledge to integrate ethical and sustainable practices into their businesses (Darmawan *et al.* 2023; BPS 2018). These programs foster a socially responsible business ecosystem, ensuring that economic benefits are equitably distributed among Indonesia's diverse population (Jakarta Post 2021; OECD 2020).

For instance, educational institutions in Indonesia have implemented training modules incorporating CSR and environmental stewardship into vocational courses, enabling students to understand the importance of ethical leadership and sustainability in business (BPS 2018; Darmawan *et al.* 2023). Research shows that such curriculums improve graduates' ability to operationalise sustainable practices, positioning them to create businesses that drive economic growth and impact society positively (Ramadhani and Rahayu 2021). Ethical leadership training in business schools further empowers students to address societal challenges while maintaining profitability, aligning Indonesia's education system with global sustainable development goals (Jakarta Post 2021; OECD 2020). These educational initiatives demonstrate how integrating religion and technology into the curriculum helps build a resilient and inclusive economy that benefits all Indonesians, including marginalised communities.

3.1.2 *The role of international partnerships in supporting Indonesia's integration of ethical leadership and technological advancements into economic strategies*

International partnerships have proven instrumental in supporting Indonesia's efforts to integrate ethical leadership and technological advancements into its economic strategies, benefiting marginalised communities. Technological innovations, in particular, have democratised access to economic opportunities by creating digital platforms that enable rural and underprivileged individuals to participate in the economy. Platforms like Gojek and Tokopedia have empowered micro-entrepreneurs by providing them with tools to access national and global markets, thereby enhancing their ability to grow their businesses (Dewan and Ramaprasad 2014; Kristyanto and Jamil 2023) The integration of technology into the economy not only generates new income streams but also provides marginalised communities with access to digital financial services, allowing them to participate in the formal economy (ADB 2022; OECD 2020; Sharma and Lijuan 2015).

Moreover, international partnerships have supported Indonesia in using high-tech solutions to promote socio-economic inclusivity. Mobile technology, for example, has been instrumental in extending healthcare and educational resources to remote regions, enhancing their quality of life. Mobile health platforms connect people in rural areas with medical experts, breaking down barriers to healthcare access (Kristyanto and Jamil 2023). Similarly, digital learning programs targeting disadvantaged groups have created opportunities to enter the digital economy, making them more employable and improving their economic prospects (ADB 2022; UNESCO 2021). These technological interventions not only empower marginalised communities but also contribute to poverty reduction and the achievement of Indonesia's sustainable development goals.

Examples of successful initiatives include the Indonesian government's "Digital Villages" program, which aims to connect rural communities to the Internet and provide training in online entrepreneurship (ADB 2022; Fahmi and Mendrofa 2023; Ministry of Communication and Information Technology 2019). Recent data (Figure 1) shows that internet penetration in urban areas is approximately 85%, compared to only 45% in rural regions, highlighting the urgent need for such initiatives to bridge the digital divide (INTI Media 2024). These programs help rural entrepreneurs access broader markets and integrate into the digital economy, reducing poverty and fostering resilience in underserved areas.

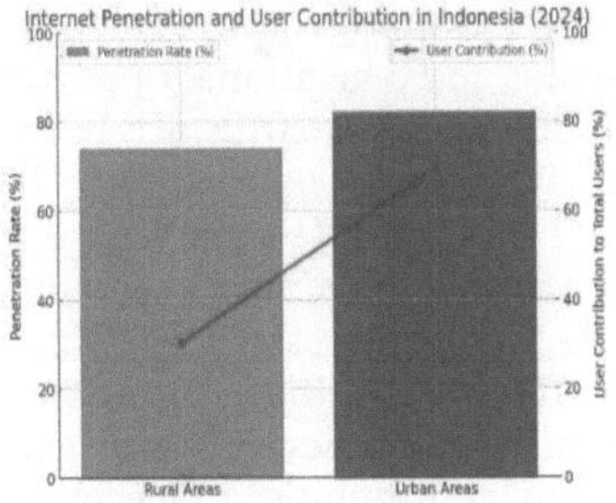

Figure 1. Internet penetration and contribution in rural areas (2024).

Religious teachings also play a critical role in guiding ethical leadership in these international partnerships. Religious principles such as integrity, empathy, and fairness serve as ethical anchors for business leaders, ensuring their decisions benefit stakeholders and promote societal well-being (Chandra 2023; Setiawan and Faisal 2024). For example, Catholic Relief Services (CRS) partnered with Starbucks to promote sustainable agriculture practices in East Timor, helping local farmers increase their yields while adhering to fair trade principles (CRS 2021; Hackett *et al.* 2006; Kropf and Luong 2017). Similarly, the Islamic Development Bank (IsDB) collaborates with member-state enterprises to implement projects to improve social welfare and economic growth, aligning with Islamic teachings on poverty alleviation and community development (Weber and al-Ansari 2022). These partnerships highlight the importance of ethical leadership in creating inclusive economic strategies that benefit all segments of society.

3.2 *Discussion*

The growing trend of incorporating religion into business education curricula in Indonesia to develop ethical leadership that promotes sustainable economic practices is palpable. Indonesian universities and vocational institutions increasingly adopt Islam and Christian teachings in their courses to enable future business leaders to make ethical choices with integrity and fairness (Arif *et al.* 2023; Evita *et al.* 2023). Through this integration, students are equipped to understand what ethics mean when making decisions, which helps them weigh the welfare of stakeholders and profits first. For example, research indicates that Islamic principles of justice ('adl) and fairness (istiqamah) contribute to corporate cultures that value transparency and ethical conduct, while Christian ethics emphasise virtues like love and justice guiding business practices that benefit communities and promote environmental sustainability (Chase 2004; Kim *et al.* 2009; Muhammad 2020). Corporate Social

Responsibility (CSR) and sustainability are also emphasised by educational initiatives in Indonesia, which not only improve the social impact of businesses but also boost financial performance, leading to vibrant business environments aligned with sustainable development goals (OECD 2020; UNESCO 2021; World Bank 2021).

Technological progress in Indonesia has been vital in democratising economic opportunities and promoting inclusive growth. However, significant challenges remain, particularly the digital divide between urban and rural areas, where internet penetration is 85% and 45%, respectively (INTI Media 2024). This disparity highlights infrastructure and access gaps, which hinder rural communities' participation in the digital economy. Platforms like Gojek and Tokopedia have played a crucial role in creating a platform for micro-entrepreneurs from rural areas who can now access larger markets through them, thereby increasing their income generation ability (Dewan and Ramaprasad 2014; Kristyanto and Jamil 2023). This digital integration bridges socio-economic gaps and facilitates access to digital financial services, enhancing financial inclusion throughout the country (ADB 2022; OECD 2020; Sharma and Lijuan 2015). Mobile technology has revolutionised service delivery in remote regions, improving healthcare access and educational opportunities and improving quality of life and societal well-being (MacLeod *et al.* 2012; UNICEF 2021). For this reason, government initiatives such as digital village programs have been established to equip rural communities with essential digital skills and connectivity so that they can effectively engage in online businesses and obtain access to government services while fostering resilience and broader market access (ADB 2022; Ministry of Communication and Information Technology 2019). These innovative technologies promote economic activities in underserved areas and contribute to sustainable development goals, including entrepreneurship and resilience within marginalised communities (Dewan and Ramaprasad 2014; World Bank 2021).

These efforts reflect Indonesia's dual approach, which integrates ethical values into education while leveraging technological advancements for inclusive growth and sustainable development. By doing so, religious teachings are woven into business education to prepare future leaders who will navigate the complexities of right from wrong (Chandra 2023). Consequently, marginalised communities have been empowered by technological innovations that widen economic opportunities while ensuring easy access to essential services, thus upholding Indonesia's economic stamina and commitment to global sustainability goals. These efforts are supported by international collaborations, which supply experts' knowledge and resources for improving the educational reformation process and building technological infrastructures to ensure its transformation into a more just and prosperous society.

To further leverage these synergies in ensuring marginalised communities can benefit from economic growth, Indonesia can expand educational programs that integrate social responsibility and sustainable development teachings, ensuring they reach underserved communities. Initiatives like vocational training courses with CSR modules equip marginalised youth with skills aligned with industry needs, enhancing their employability (BPS 2018; Darmawan *et al.* 2023). However, the digital divide presents significant challenges, particularly in rural areas where limited access to stable internet and digital literacy training restricts participation in the digital economy (INTI Media 2024). Addressing these barriers through targeted policies and infrastructure investments is crucial. Furthermore, leveraging technology to deliver educational content and facilitate digital entrepreneurship in rural areas can empower marginalised populations to participate more actively in the digital economy (ADB 2022; UNESCO 2021). Strengthening partnerships between religious organisations, educational institutions, and businesses could also extend ethical business practices and economic opportunities to marginalised communities, fostering inclusive growth across Indonesia.

Meanwhile, in supporting Indonesia's efforts to integrate ethical leadership and technological advancements into its economic strategies, international partnerships can provide expertise, resources, and best practices to bolster Indonesia's initiatives in ethical leadership and technological integration. Collaborations with global educational institutions can enhance curriculum development in ethical business practices and digital skills training, aligning them with international standards (OECD 2020). Foreign investments and knowledge transfer in technological infrastructure and digital innovation can accelerate Indonesia's digital transformation, creating more opportunities for inclusive economic growth (ADB 2022; World Bank 2021). Additionally, international partnerships focused on sustainable development goals can facilitate exchanges of ideas and solutions, helping Indonesia address complex societal challenges while promoting ethical governance and economic resilience (United Nations 2021). These recommendations underscore the potential for Indonesia to strengthen its educational, technological, and ethical frameworks through targeted initiatives and global collaborations, paving the way for sustainable development and prosperity.

4 CONCLUSION

Integrating education, religion, and technology holds great potential for driving sustainable economic growth in Indonesia. This research has demonstrated that combining these elements fosters inclusivity, resilience, and ethical leadership. Education, especially when aligned with religious teachings, equips individuals with the skills needed for the digital economy while promoting ethical leadership and social responsibility. Vocational training programs incorporating Corporate Social Responsibility (CSR) modules enhance employability and cultivate business practices centred on fairness, transparency, and social responsibility, contributing to a sustainable economic environment. Religion also shapes the ethical dimensions of leadership and business practices, embedding values like justice and integrity into education to prepare future leaders for guiding businesses in ways that benefit society. Meanwhile, technology democratises economic opportunities, empowering micro-entrepreneurs from underserved regions and fostering social equity through digital platforms like e-commerce and fintech. Collaboration among policymakers, educators, religious leaders, and technology innovators is essential. Policymakers must prioritise digital infrastructure and sustainability-focused vocational training, while religious leaders advocate for ethical leadership. International partnerships should fund digital tools and capacity-building programs for rural communities. Technology innovators must create inclusive platforms to bridge the digital divide. Key indicators such as increased rural internet penetration, employment in sustainable industries, and entrepreneurial growth will ensure Indonesia advances towards a resilient, equitable economy aligned with the United Nations Sustainable Development Goals (SDGs).

REFERENCES

Arif, M., Arsad, S., Abdullah, F., and Rusby, F. (2023). Integrating religious teachings in business education in Indonesia. *Journal of Business Ethics Education*, 10(2), 123–139.

Asian Development Bank (ADB). (2022). Inclusive economic growth strategy in Indonesia.

BPS. (2018). Corporate social responsibility and sustainability in Indonesia. *Badan Pusat Statistik Report.*

Catholic Relief Services (CRS). (2021). *Catholic Relief Services: Sustainable agriculture partnerships.* Retrieved from https://www.crs.org

Chandra, R. (2023). Ethical leadership and sustainable business practices in Indonesia. *Business Ethics Quarterly*, 29(1), 67–84.

Chase, J. (2004). Christian ethics and business leadership. In *Christian Ethics and Business Leadership* (pp. 45–61). New York, NY: HarperCollins.

Darmawan, I., Jamil, M., and Rees, M. (2023). Integrating corporate social responsibility in Indonesia's educational system. *Sustainability in Education*, 12(3), 245–258.

Dewan, A., and Ramaprasad, J. (2014). Technology integration in Indonesia's rural economy: A case study of digital platforms. *Journal of Digital Economy*, 8(2), 145–158.
Evita, A., Arsad, S., and Rusby, F. (2023). Integrating Islamic and Christian ethics in business education for sustainable development. *Journal of Religious Education*, 17(1), 37–50.
Fahmi, M., and Mendrofa, E. (2023). Digital villages and rural entrepreneurship in Indonesia. *Ministry of Communication and Information Technology*.
Hackett, L., Luong, M., and Chang, H. (2006). Starbucks and fair trade: Partnerships for sustainable agriculture. *Journal of Agricultural Economics*, 23(4), 345–360.
INTI Media. (2024). Internet penetration rates in urban and rural Indonesia. *INTI Media Report*. Retrieved from https://www.intimedia.com
Jakarta Post. (2021). CSR initiatives and economic growth in Indonesia. *Jakarta Post*. Retrieved from https://www.thejakartapost.com
Kim, Y., Lim, H., and Yoo, D. (2009). Christian ethics and business practices: A comparative analysis. *Journal of Business Ethics*, 55(2), 130–145.
Kristyanto, A., and Jamil, M. (2023). E-commerce and economic inclusion in Indonesia: The role of Gojek and Tokopedia. *Journal of Digital Economics*, 6(1), 75–90.
Kropf, J., and Luong, M. (2017). Fair trade in East Timor: A case study. *Fair Trade Review*, 14(2), 58–72.
MacLeod, D., Smith, R., and Evans, P. (2012). Mobile health platforms in Indonesia: Bridging gaps in healthcare access. *Mobile Health Journal*, 6(4), 219–232.
Ministry of Communication and Information Technology. (2019). *Digital Villages Program: Connecting Rural Communities*. Retrieved from https://www.kominfo.go.id
Muhammad, A. (2020). Islamic business ethics and societal well-being in Indonesia. *Journal of Islamic Business Studies*, 12(1), 95–110.
Organisation for Economic Co-operation and Development (OECD). (2020). *Education and skills for a competitive economy: The case of Indonesia*.
Ramadhani, I., and Rahayu, S. (2021). Ethical leadership in Indonesia's business schools. *Ethics in Business Education*, 9(2), 99–115.
Setiawan, E., and Faisal, R. (2024). Religious ethics in business leadership: A model for Indonesia. *Business Ethics Review*, 32(3), 51–68.
Sharma, P., and Lijuan, C. (2015). Mobile financial services: A path to economic inclusion. *Journal of Financial Technology*, 13(1), 32–49.
Triyuwono, I., Rahayu, S., and Andriani, A. (2021). Integrating Islamic business ethics in vocational education in Indonesia. *Islamic Education Journal*, 8(4), 230–245.
United Nations Educational, Scientific and Cultural Organization (UNESCO). (2021). *Digital Literacy Programs for Youth in Indonesia*. Retrieved from https://www.unesco.org
United Nations Children's Fund (UNICEF). (2021). *Digital Learning and Healthcare Access in Remote Areas of Indonesia*. Retrieved from https://www.unicef.org
United Nations. (2021). The Sustainable Development Goals Report. https://unstats.un.org/sdgs/report/2021/
Weber, M., and al-Ansari, T. (2022). Islamic Development Bank and social welfare projects: A case study. *International Journal of Social Welfare*, 11(2), 144–158.
World Bank. (2021). *Indonesia's Economic Growth and Employment Challenges*. Retrieved from https://www.worldbank.org

Towards Resilient Societies: The Synergy of Religion, Education, Health, Science, and Technology – Rahiem (Ed.)

Development of TPACK-based performance assessment instruments for prospective teachers in elementary education

Nur Luthfi Rizqa Herianingtyas
Universitas Islam Negeri Syarif Hidayatullah Jakarta, Tangerang Selatan, Indonesia
ORCID ID: 0000-0002-3682-8083

Didah Nurhamidah
Universitas Islam Negeri Syarif Hidayatullah Jakarta, Tangerang Selatan, Indonesia
ORCID ID: 0000-0001-7843-7833

Azkia Muharom Albantani
Universitas Islam Negeri Syarif Hidayatullah Jakarta, Tangerang Selatan, Indonesia
ORCID ID: 0000-0001-9080-4149

Wafeeqah Rakiep
Rylands High School, Cape Town, South Africa

ABSTRACT: In the 21st century, many teachers still lack sufficient knowledge to effectively use technology in the classroom. This fact underscores the importance for Educational Institutions and Educators to strive towards producing adaptable graduate outcomes for teachers. The integration of the Sustainable Development Goals (SDGs), especially SDG 4: Quality Education, emphasises the importance of equipping prospective teachers with the skills to foster a quality learning environment that is relevant to the times. Assessment instruments based on the Technological Pedagogical Content Knowledge (TPACK) framework can serve as tools to measure the connection of prospective teachers and teachers' understanding of content, pedagogy, and technology in an integrative manner. This study adopts the Research and Development (R&D) model ADDIE (Analyse, Design, Development, Implementation, and Evaluation). The sample used was 30 students. This instrument is designed to identify and measure the ability of prospective teachers to apply TPACK components, which include Pedagogical Content Knowledge (PCK), Pedagogical Knowledge of Technology (TPK), and Knowledge of Content Technology (TCK), in the planning and implementation of teaching activities. The instrument includes 51 indicators, allowing for a comprehensive assessment of the ability of prospective teachers to integrate knowledge, content, and pedagogical technology in teaching. Instrument Validity Analysis was conducted using Aiken's V. This instrument provides an authentic tool to evaluate the performance of prospective teachers in microteaching and can also be a reference for developing assessment instruments in other pedagogical programs.

Keywords: 21st century learning, TPACK, performance assessment

1 INTRODUCTION

In a broader perspective, global challenges and international agendas require teachers to continuously develop their abilities and professionalism. One way to accomplish the 17 Sustainable Development Goals (SDGs) pertaining to sustainability is through education. Education for Sustainable Development (ESD) is the term used to describe education that aims

DOI: 10.1201/9781003645542-50

to achieve the SDGs (Wals and Kieft 2010). Based on research by Shumba and Kampamba (2013), if learning only relies on Pedagogical Content Knowledge (PCK), it is not enough to help teachers integrate ESD, the ability of teachers to integrate ICT in learning is needed. Learning needs to be future-oriented by utilizing ICT (Arifin *et al.* 2020).

However, several studies indicate that in the 21st century, many teachers still lack sufficient knowledge to use technology effectively in the classroom (Bakir 2016) Some still struggle to utilize technology as a learning medium (Handrianto *et al.* 2021; Harris and Hofer 2011). According to the Global Education Monitoring (GEM) report released by UNESCO, the quality of education in Indonesia ranks 10th out of 14 developing countries, with teacher competency ranking last at 14th place. One significant issue is that many teachers still rely heavily on textbook-based teaching (Okpatrioka, and Sopian 2023). Research from the Ministry of Education and Culture's Center for Data and Information Technology (Pusdatekom Kemdikbud) in 2021 also reported that 77.5% of Indonesian teachers have standard-level Information and Communication Technology (ICT) competency, with only 3.7% considered proficient, while the rest are at an advanced level. These facts underscore the importance for Educational Institutions and Educators to strive towards producing adaptive graduate outcomes for teachers. However, in reality, the TPACK framework paradigm has not yet become the orientation in several Educational Institutions and Educators. Many of these institutions focus more on how to assist prospective teachers in learning extensive teaching theories and subject matter as disciplines (Aydeniz and Kirbulut 2014). Results from document studies conducted between January-June 2023 on Comprehensive Examinations, Microteaching Examinations, and Field Introduction Examinations (PLP) instruments in the Elementary School Teacher Education Program at Tangerang Selatan, it is noted that there is currently no assessment component measuring students' ability to effectively manage learning through the integration of content, pedagogy, and technology.

Assessment instruments based on the Technological Pedagogical Content Knowledge (TPACK) framework can serve as tools to measure the interconnected understanding of prospective teachers and teachers regarding content, pedagogy, and technology integratively (Koehler *et al.* 2014). With this TPACK assessment, it can be a reflection material and feedback for teachers or prospective teachers to improve their capacity in teaching, especially in integrating ICT (Wang *et al.* 2018). Several studies, such as those conducted by So and Kim (2009) and Patahuddin *et al.* (2016), have examined the application of TPACK. Additionally, Pazin Fadzil *et al.* (2022) analysed the ability of prospective teachers to apply TPACK in lesson planning. Meanwhile, Koh *et al.* (2010) and Abbitt (2011) researched the implementation of TPACK in teaching practices. Several studies have shown the success of developing valid and reliable TPACK with performance assessments for teachers in several countries such as (Harris *et al.* 2010) and Harris *et al.*, (2010) at United States, Koh *et al.*, (2010) developed the TPACK rubric at Singapore, Hu and Fyfe (2010) at Australia. Most TPACK studies have utilised self-assessment survey techniques, assessing knowledge domains separately (Abbitt 2011; Graham *et al.* 2009; Landry 2010). There is a lack of research developing comprehensive performance assessment instruments based on TPACK for planning, implementation, and evaluation of learning simultaneously, utilizing detailed and comprehensive assessment models.

Therefore, the researcher is interested in developing a performance assessment instrument to measure the ability of prospective teachers to apply TPACK. The hope is that this study can contribute as a reference and recommendation for assessment instruments in Microteaching and Field Introduction Examinations for Madrasah Ibtidaiyah Teacher Education students. Furthermore, it is anticipated that the results of this development and research can serve as a reflection for lecturers and Madrasah Ibtidaiyah Teacher Education study programs in devising relevant curriculum strategies aligned with field needs and developments.

2 METHOD

This study follows the Research and Development (R&D) model ADDIE (Analyze, Design, Development, Implementation, and Evaluation) (Dick and Carey 1996). The population in this study is all students of the Madrasah Ibtidaiyah Teacher Education study program UIN Syarif Hidayatullah Jakarta. Meanwhile, the sample used is 30 students who are taking Microteaching courses. Madrasah Ibtidaiyah in this case is equivalent to the elementary school level. The data collection techniques used in this study are: Literature review to explore documents and literature related to the research substance of TPACK, Instrument Validation: to assess the content validity of the developed performance assessment instrument. Instrument Validity Analysis is conducted using Aiken's V. This instrument validation technique is effective to use when the number of raters (rater) is more than one for each item. The formula and criteria as shown in (Aiken 1985) Instrument Validity Levels.

3 RESULTS AND DISCUSSION

This study aimed to develop a performance assessment instrument based on Technological Pedagogical Content Knowledge (TPACK) for prospective teachers in Madrasah Ibtidaiyah. The design of the performance assessment instrument followed the ADDIE model stages: (1) Analyze stage, which involved analysing the needs for designing performance instruments by reviewing literature on TPACK components developed by Koehler *et al.* (2014), and examining the 8 domains of TPACK in learning activities developed by Yeh *et al.* (2015). (2) Design stage, where the performance instrument was designed by considering the main components of learning developed by Anderson, L.W and Krathwohl, D.R (2001), covering planning, implementation, and evaluation. This stage also involved developing scales and criteria for each indicator while considering the characteristics of TPACK and 21st-century learning developed by Koehler *et al.* (2014) and Tondeur *et al.* (2012). (3) Development stage, which included instrument validation involving 5 experts, processed using Aiken's V validity scoring technique (1985).

The TPACK Performance Assessment Instrument was developed in the form of a rating scale to identify TPACK components among Madrasah Ibtidaiyah Teacher Education students in the microteaching course. The TPACK components developed included PCK (Pedagogical Content Knowledge), TPK (Technology Pedagogical Knowledge), TCK (Technology Content Knowledge), and TPACK. These components were further detailed into 31 items for planning-based TPACK learning and 20 items for implementation-based TPACK learning (Table 1).

Table 1. Indicators of assessment for TPACK-based lesson planning.

No	Indicator	Code
1	Connection between indicator formulations with basic competencies	PCK 1
2	Integration of indicator formulations based on digital	TPK 1
3	Integration of indicator formulations based on higher order thinking skills	PCK 2
4	Integration of achievement indicator formulations based on 4Cs	PCK 3
5	Connection between learning objectives with achievement indicators	PCK 4
6	Consistency of learning objectives with components Audience, Behaviour, Condition, Degree	PCK 5
7	Integration of learning objectives based on digital	TPK 2
8	Integration of learning objectives based on HOTs (C4, C5, C6)	PCK 6
9	Integration of learning objectives based on 4Cs	PCK 7
10	Accuracy of selection of teaching method with learning objectives	PCK 8
11	Accuracy of selection of teaching method with subject characteristics	PCK 9
12	Accuracy of selection of teaching method with student developmental characteristics	PCK 10
13	Accuracy of selection of teaching method with characteristics of 21st-century learning	PCK 11
14	Accuracy of selection of instructional media with learning objectives	PCK 12
15	Accuracy of selection of instructional media with subject characteristics	PCK 13
16	Accuracy of selection of instructional media with student developmental characteristics	PCK 14
17	Accuracy of selection of IT-based instructional media containing content material	TCK 1
18	Accuracy of selection of IT-based media used by teachers and students in learning	TPK 3
19	Integration of teaching material with HOTs (C4, C5, C6)	PCK 15

(*continued*)

Table 1. Continued

No	Indicator	Code
20	Connection between teaching material with learning objectives	PCK 16
21	Connection between learning steps with learning objectives	PCK 17
22	Connection between learning steps with methods, models, and approaches used	PCK 18
23	Accuracy of learning steps with student developmental characteristics	PCK 19
24	Integration of learning steps based on IT	TPK 4
25	Integration of learning steps based on HOTs (C4, C5, C6)	PCK 20
26	Integration of learning steps based on 4Cs	PCK 21
27	Connection between learning evaluation with learning objectives	PCK 22
28	Connection between learning evaluation with teaching material content	PCK 23
29	Integration of learning evaluation based on HOTs (C4, C5, C6)	PCK 24
30	Integration of learning evaluation based on 4Cs	PCK 25
31	Integration of learning evaluation based on IT	TCK 2

The assessment instrument in this study was developed for the microteaching course by integrating the TPACK components. This research aims to focus on mapping students' proficiency profiles in applying TPACK when designing and implementing learning activities (Table 2). TPACK refers to understanding how to organise learning of a concept or material through technological and pedagogical approaches. In this context, several essential skills include pedagogical capabilities, technological mastery, and subject matter comprehension. Given the demands of the current era, especially in the 21st century and Society 5.0 era, preparing prospective elementary school teachers has become increasingly complex. They need to be equipped with various skills that support the implementation of TPACK to address the complexities of 21st-century learning.

Table 2. Indicators of assessment for TPACK-based lesson implementation.

Indicator	Code
3.1 Begining	
Identifying student readiness for learning	PCK 26
Greeting students and checking attendance	PCK 27
Guiding students in prayer	PCK 28
Conducting preconception in line with learning content and context	PCK 29
Communicating learning objectives	PCK 30
Utilizing IT tools to initiate learning	TPK 5
3.2 Core	
Applying methods, models, and approaches according to lesson plans	PCK 31
Applying learning models according to syntax	PCK 32
Using IT tools to deliver content material	TCK 3
Using IT tools to facilitate students in sharing ideas together	TPK 6
Using IT tools to facilitate student self-learning	TPK 7
Using IT tools to facilitate group learning	TPK 8
Demonstrating mastery of learning material	PCK 33
Demonstrating ability to facilitate class/student-centered learning	PCK 34
Using clear and good oral and written language	PCK 35
Demonstrating appropriate style/gesture	PCK 36
3.3 Conclusion	
Facilitating lesson conclusions	PCK 37
Conducting learning reflection	PCK 38
Conducting comprehensive learning evaluation	PCK 39
Conducting learning evaluation using IT tools	TCK 4

3.1 *Content validity*

Content validity of the questions is an assessment by experts or specialists to examine the accuracy of the test instrument in terms of its relevance to the objectives, construction, and language. Table 3 is a list of experts/specialists who validated the TPACK Assessment Instrument.

Table 3. Expert evaluation of the assessment instrument.

Average Score	ne	CVR	CVRtable	Category
5	5	1,00	0,99	High Validity

The validation results of the Technological Pedagogical Content Knowledge (TPACK)-based performance assessment instrument show that out of 51 indicators developed, all were declared valid after undergoing a validity test using the Aiken's V technique. This validity ensures that each indicator in the instrument can accurately measure the relevant TPACK components, such as Pedagogical Content Knowledge (PCK), Technological Pedagogical Knowledge (TPK), and Technological Content Knowledge (TCK). The instrument is deemed "suitable for use" in measuring the TPACK profile of prospective teachers in the context of microteaching, which involves the planning, implementation, and evaluation of TPACK-based teaching.

Evaluation tools based on the TPACK framework have the potential to serve as indicators to measure the extent to which prospective teachers and educators comprehend content, pedagogy, and technology integratively (Koehler *et al.* 2014). Utilising this framework can guide the creation of effective and creative learning experiences in response to the demands of the 21st century (Henriksen *et al.* 2018). Research findings by Koh *et al.* (2010) indicate that developing TPACK during teacher education significantly impacts their performance when assuming the role of teachers.

4 CONCLUSION

Based on the research findings and discussions, it can be concluded that the TPACK Performance Assessment instrument was developed as a rating scale to identify the TPACK components of Madrasah Ibtidaiyah Teacher Education students in the microteaching course. The TPACK components developed include those essential for 21st-century learning, namely PCK (Pedagogical Content Knowledge), TPK (Technology Pedagogical Knowledge), and TCK (Technology Content Knowledge). The instrument received the rating of "suitable for use" after undergoing the content validity testing phase. There are 51 indicators within the instrument that have been deemed "Valid". Therefore, the developed TPACK-based performance assessment instrument can be utilised to collect research data to measure the TPACK profile of prospective MI teachers.

ACKNOWLEDGEMENT

Thank you to the *Lembaga Penelitian dan Pengabdian kepada Masyarakat* (LPPM) of UIN Syarif Hidayatullah Jakarta for their moral and material support in conducting this research. This study is part of capacity-building research funded by the BLU grant from UIN Syarif Hidayatullah Jakarta in 2023.

REFERENCES

Abbitt, J. T. (2011). Measuring technological pedagogical content knowledge in preservice teacher education: A review of current methods and instruments. *Journal of Research on Technology in Education*, 43(4), 281–300

Aiken, L.R. (1985). Three coefficients for analyzing the reliability and validity of ratings, *Educational and Psychological Measurument*, *45*(1), 131–142. doi: http://doi.org/10.1177/0013164485451012.

Anderson, L. W., and Krathwohl, D. R. (2001). *A taxonomy for learning, teaching, and assessing; a revision of bloom's taxonomy of educational objectives*. New York: Addison Wesley Longman

Arifin, V., Jalaldeen, M. R. M., Kartiwi, M., and Abdullah, L. M. (2020). Investigation of IT Competence and Readiness of IT Students Facing The Industrial Revolution 4.0. *2020 8th International Conference on Cyber and IT Service Management (CITSM)*, 1–6.

Aydeniz, M., and Kirbulut, Z. D. (2014). Exploring challenges of assessing pre-service science teachers' pedagogical content knowledge (PCK). *Asia-Pacific Journal of Teacher Education*, 42(2), 147–166. https://doi.org/10.1080/1359866x.2014.890696

Bakir, N. (2016). Technology and teacher education: A brief glimpse of the research and practice that have shaped the field. *Tech Trends*, 60(1), 21–29.

Dick, W., and Carey, L. (1996). *The Systematic Design of Instruction* (*4th ed.*). New York: Harper Collins College Publishers.

Graham, R. C., Burgoyne, N., Cantrell, P., Smith, L., St Clair, L., and Harris, R. (2009). Measuring the TPACK confidence of inservice science teachers. *TechTrends*, 53(5), 70–79.

Handrianto, C., Jusoh, A. J., Nengsih, Y. K., Alfurqan, A., Muslim, M., and Tannoubi, A. (2021). Effective Pedagogy in Primary Education: A Review of Current Literatures. *Abjadia*, 6(2), 134–143. https://doi.org/10.18860/abj.v6i2.12978.

Harris, J., Grandgenett, N., and Hofer, M. (2010). Testing a TPACK-based technology integration assessment rubric. *Society for Information Technology and Teacher Education International Conference*, 3833–3840.

Harris, J. B., and Hofer, M. J. (2011). Technological Pedagogical Content Knowledge (TPACK) in Action, *Journal of Research on Technology in Education. Journal of Research on Technology in Education*, 43(3), 211–229.

Henriksen, D., Henderson, M., Creely, E., Ceretkova, S., Černochová, M., Sendova, E., Sointu, E. T., and Tienken, C. H. (2018). Creativity and Technology in Education: An International Perspective. *Technology, Knowledge and Learning*, 23 (3), 409–424. https://doi.org/10.1007/s10758-018-9380-1.

Hu, C., and Fyfe, V. (2010). Impact of a new curriculum on pre-service teachers' technical, pedagogical and content knowledge (TPACK). *Proceedings Ascilite Sydney*, 184–189.

Koehler, M. J., Mishra, P., Kereluik, K., Shin, T. S., and Graham, C.R. (2014). *The Technological Pedagogical Content Knowledge Framework*. New York: Springer-Verlag.

Koh, J. H. L., Chai, C. S., and Tsai, C.-C. (2010). Examining The Technological Pedagogical Content Knowledge of Singapore Pre-service Teachers with a Large-Scale Survey. *Journal of Computer Assisted Learning*, 26, 563–573. https://doi.org/10.1111/j.1365-2729.2010.00372.x.

Landry, G. A. (2010). Creating and validating an instrument to measure middle school mathematics teachers' technological pedagogical content knowledge (TPACK).

Okpatrioka, and Sopian. (2023). INCREASING TEACHER COMPETENCE IN THE INDUSTRIAL REVOLUTION INDUSTRI 4.0. *JHSS (Journal of Humanities and Social Studies)*, 07, 380–386.

Patahuddin, S. M., Lowrie, T., and Dalgarno, B. (2016). Analysing mathematics teachers' TPACK through observation of practice. *The Asia-Pacific Education Researcher*, 1–10.

Pazin Fadzil, A. H., Maat, S. M., and Mahmud, M. S. (2022). A Rasch model analysis of the TPACK Instrument in The Creative Teaching of Primary Mathematics Teachers. *Cypriot Journal of Educational Sciences*, 17(11), 4259–4274. https://doi.org/10.18844/cjes.v17i11.7792

Shumba, O., and Kampamba, R. (2013). Mainstreaming ESD into science teacher education courses: A case for ESD pedagogical content knowledge and learning as connection. *Southern African Journal of Environmental Education*, 151–166.

So, H. J., and Kim, B. (2009). Learning About Problem Based Learning: Student Teachers Integrating Technology, Pedagogy and Content Knowledge. *Australasian Journal of Educational Technology*, 25(1), 101–116.

Tondeur, J., van Braak, J., Sang, G., Voogt, J., Fisser, P., and Ottenbreit-Leftwich, A. (2012). Preparing Pre-service Teachers to Integrate Technology in Education: A Synthesis of Qualitative Evidenc, *Computers and Education Journal*, 59 (1), 134–144.

Wals, A. E., and Kieft, G. (2010). *Education for sustainable development: Research overview*.

Wang, W., Schmidt-Crawford, D., and Jin, Y. (2018). Preservice teachers' TPACK development: A review of literature. *Journal of Digital Learning in Teacher Education*, 34(4), 234–258.

Yeh, Y.-F., Chien, S.-P., Wu, H.-K., and Hsu, Y.-S. (2015). Rubrics of TPACK-P for Teaching Science with ICTs. *Development of Science Teachers' TPACK*, 53–70. https://doi.org/10.1007/978-981-287-441-2_4.

Towards Resilient Societies: The Synergy of Religion, Education, Health, Science, and Technology – Rahiem (Ed.)
© 2026 The Author(s), ISBN: 978-1-041-08471-6
Open Access: www.taylorfrancis.com, CC BY-NC-ND 4.0 license

Mapping global trends in design thinking for creative education: A bibliometric analysis

Nur Luthfi Rizqa Herianingtyas
Universitas Islam Negeri Syarif Hidayatullah Jakarta, Tangerang Selatan, Indonesia
ORCID ID: 0000-0002-3682-8083

Rizqy Nur Sholihat
Universitas Islam Negeri Syarif Hidayatullah Jakarta, Tangerang Selatan, Indonesia
ORCID ID: 0000-0003-2937-1481

Afif Faizin
Universitas Islam Negeri Syarif Hidayatullah Jakarta, Tangerang Selatan, Indonesia
ORCID ID: 0009-0009-3781-5557

Habiburrahman Rizapoor
Department of Islamic Studies, Shariah Faculty of Badakhshan University, Afghanistan

ABSTRACT: The 21st-century education system must prioritise developing students' creative thinking skills to foster a generation adaptable to change. Design thinking offers a structured, human-centred process for generating solutions to complex, ill-defined problems, making it a valuable tool for implementing Sustainable Development Goals (SDG 4: Quality Education and SDG 9: Innovation). Initially rooted in industrial and business sectors, design thinking was adapted for education in the 2000s, sparking interest among educational researchers. This study aims to map the research landscape on design thinking in education through bibliometric and content analysis of 2,330 articles published in Scopus-indexed journals between 2014 and 2024. Using VOSviewer and Biblioshiny software, metadata were quantitatively and statistically processed. Findings highlight significant opportunities for further research, as design thinking in education remains underexplored, particularly its connection to creative thinking—a critical 21st-century skill. This study underscores the need for more in-depth investigations into the role of design thinking in enhancing creativity and its broader applications in educational contexts.

Keywords: creative thinking skills, design thinking, research trends, SDGs

1 INTRODUCTION

The world is undergoing rapid transformations, posing increasing challenges in anticipating the future issues and difficulties that upcoming generations will face (Lee 2018). Students will encounter various skills in fields that are currently unknown (World Economic Forum 2020). The education system needs to cultivate relevant skills in students (Lambert 2019), fostering an adaptive attitude towards change (Ryoo and Winkelmann 2021), and preparing students with diverse competencies to be flexible and courageous in embracing unforeseen new developments (Cropley 2015). Creative thinking is considered a higher cognitive level (Anderson *et al.* 2001) as a critical survival skill for the 21st century society and contribute to SDG 9 (innovation) (Oltra-Badenes *et al.* 2023).

Creative thinking involves generating new alternative ideas that have value (Antonio 2022; Sternberg 2003), including value in problem-solving impact (Kaufman and Beghetto 2009), effectiveness that aligns with ethics (Cropley 2015), originality, and uniqueness (Hernández-Torrano and Ibrayeva 2020).

DOI: 10.1201/9781003645542-51

In the context of education, creative thinking is categorised as a trainable skill (Craft 2010). Therefore, learning can serve as a means to cultivate creative thinking skills (Clemente *et al.* 2016; Kampylis 2010), including involving students in problem-solving and project development (Jackson and Shaw 2006).

Design thinking is a thinking process used to solve problems (Calavia *et al.* 2021; Matthews and McBee 2007). The Design Thinking process consists of five steps: empathize, define, ideate, prototype, and test (Plattner 2013). The first two steps (Empathize and define), referred to as the "problem space". The subsequent three steps (Ideate, Prototype, and Test) are known as the "solution space" (Herianingtyas 2024). Design thinking methodology emerged in the 1980s in business, architecture, design, and industry to tackle ill-defined problems (Dorst 2011; Plattner 2013) and wicked problems (Cross 2023; Tsai *et al.* 2023). By applying design thinking in teacher learning, it can encourage the realisation of SDGs (Sustainable Development Goals), especially in improving the quality of education (SDG 4) and innovation (SDG 9) (de Waal and Maritz 2022; Soomro *et al.* 2021). Some researchers also revealed that design thinking can encourage students to think creatively in generating problem-solving ideas when involved in innovation projects (Beckman and Barry 2007; Herianingtyas *et al.* 2023). Teachers can utilise the power of students' creativity through two spaces in the design thinking stage. Students' creativity can be built by the process of empathy in the problem space and the development of ideas in the solution space to encourage sustainable progress. Design thinking can encourage students to provide solutions to environmental problems such as reducing plastic waste (SDG 13) or promoting gender equality (SDG 5), as well as helping students understand how they can contribute to sustainable development solutions in their communities (Gillet and Vonèche-Cardia 2021; Leal Filho *et al.* 2024; Leong *et al.* 2024).

Researchers have since explored the significant benefits of design thinking in the 21st century, leading to its adaptation in the educational realm starting in the 2000s (Brown and Wyatt 2010). Learning in the 21st century is oriented towards thinking processes, therefore, as an innovation in education, it is essential to identify the research trends in design thinking.

2 METHODS

This study employed bibliometric analysis and content analysis (Pritchard 1969). Resources of research articles published in Scopus-indexed journals, including all articles on design thinking in education from Taylor and Francis. This approach aimed to explore the research trends and implementation of design thinking in education from 2014 to 2024. Keywords used for data collection included "design thinking," "education," and "creative thinking skills." Articles were collected from https://www.scopus.com and exported in *csv format. Metadata were subsequently processed quantitatively and statistically using VOSviewer and Biblioshiny software.

3 FINDINGS AND DISCUSSION

3.1 *Findings*

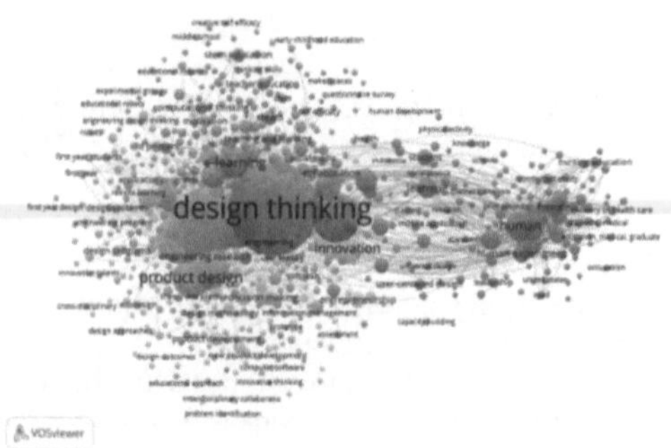

Figure 1. Network visualization of design thinking research.

Research Trends in Design Thinking and Education from 2014 to 2024 This study presented 2,330 Scopus-indexed articles focusing on research topics related to design thinking in education from 2014 to 2024. This research has become a significant trend in the United States. VOSviewer analysis yielded six clusters: dark blue, light blue, red, green, yellow, and orange.

Based on Figure 1, it is also evident that many topics are still faintly coloured, indicating they have been rarely studied despite their close relevance. These include creativity, thinking skills, soft skills, assessment, and others.

3.1.1 *Trends in design thinking and creative thinking skills*

It is known that design thinking and creativity are closely related, yet in the research trends of the last decade (2014-2024), both topics have been rarely studied. Based on Scopus metadata, the number of studies on "design thinking" and "creativity" amounts to 341 documents, whereas "design thinking" and "creative thinking skills" have 7 documents seen in Table 1.

Table 1. Research on design thinking and creative thinking skills (2014-2024).

Authors	Title	Implications
(Koroh 2014)	A development module to teach creative thinking ability based on creative problem solving and design thinking models	The use of the module showed a significant improvement in enhancing students' creative thinking abilities (91.63%) in the fourth semester of the Elementary School Teacher Education program.
(Rumahlatu *et al.* 2021)	Resource based learning design thinking (RBLDT): A model to improve students' creative thinking skills, concept gaining, and digital literacy	The combined syntax of the RBLDT learning model can enhance students' creative thinking, concept acquisition, and digital literacy better than when RBL or DT is applied separately.
(Kuo *et al.* 2021)	The Impact of Design Thinking PBL Robot Course on College Students' Learning Motivation and Creative Thinking	Experimental group students were able to think faster, consider problems from various perspectives, and generate more creative ideas.
(X. Yang *et al.* 2022)	Relationship between creative thinking and experimental design thinking in science education: Independent or related	Design Thinking and Creative Thinking are not entirely separate or distinguishable. Instead, they complement each other, forming an innovative whole of experimental design thinking.
(Arenda and Suwono 2023)	Design thinking and EDS prompt in STEAM learning: The impact on germane cognitive load and creative thinking	Design Thinking and EDS prompt provide enjoyable learning experiences, facilitate students in acquiring knowledge, and encourage students to generate more creative ideas. Students' knowledge and creative thinking are heightened with the inclusion of design thinking and EDS prompt in STEAM education.
(Hua *et al.* 2023)	Developing a Graphic Design Thinking Model to Enhance the Creative Thinking	The GDT model can help students realize graphic creations and significantly enhance their creative thinking.
(Arifin and Siew 2023)	Integration Of Socioscientific Approach and Design Thinking: An Entrepreneurial Creative Thinking Module For Stem Education	There was a significant difference in the post-test scores of the SIA-DT group compared to the control group on ECT and the five ECT constructs.

3.2 *Discussion*

The bibliometric results illustrate an increasing trend in research on design thinking in education each year. Originally contextualised from industry and business into education, design thinking has become a relevant concept facilitating student-centered learning paradigms (Johansson-Sköldberg *et al.* 2013). It aligns with the demands of 21st-century skills (Cain and Henriksen 2022; Pande and Bharathi 2020) and serves as an alternative learning approach. However, based on

Figure 2 creative thinking skills in relation to design thinking have not received sufficient attention despite their intrinsic connection. While numerous studies have explored creativity in relation to design thinking, the development of creative thinking skills within the framework of design thinking remains limited. Creative thinking is considered a trainable skill in educational contexts (Halpern 2003). Design thinking, characterised by iterative thinking centred on human needs and aimed at solving ill-defined problems (Herianingtyas *et al.* 2023), presents an opportunity to cultivate students' creative thinking skills. Students' adept in creative thinking can generate original ideas fluently, view issues flexibly from multiple perspectives, and connect ideas to make them compelling (Kaufman 2016). These characteristics resonate with the iterative nature of design thinking, allowing students to interpret issues flexibly in the problem space and generate their best ideas in the solution space.

Several studies have integrated design thinking into various learning components such as models, methods, approaches, media, and tools to enhance its effectiveness in fostering creative thinking skills. Based on Table 1, the blue cluster in metadata mapping highlights instances where design thinking integrates with STEM approaches in student project development (Chit *et al.* 2023; Ho and Pang 2024; Thomason 2023). Design thinking applied in the development phases of design works effectively enhances student outcomes (Suwankarjank and Rugwongwan 2021; Zhu *et al.* 2024). Moreover, various studies have shown positive contributions of design thinking to learning outcomes, including increased student motivation (Bawaneh and Alnamshan 2023; Tsai *et al.* 2023), collaborative competence (Leem and Lee 2024; Picanço and dos Santos 2022), self-efficacy (Bawaneh and Alnamshan 2023; Voigt *et al.* 2019), computational thinking (Bringardner *et al.* 2023; Chang *et al.* 2023), and student creativity (Von Thienen *et al.* 2017; C.-M. Yang and Hsu 2020). These findings underscore the potential of design thinking not only as a methodological framework but also as a catalyst for developing essential skills in education. This bibliometric study is limited by its reliance on Scopus data from 2014 to 2024, and results may vary with other databases. Future research could expand on these findings by incorporating additional databases or focusing on specific aspects of design thinking and its implications for education.

4 CONCLUSION

Based on the bibliometric analysis of the Scopus database globally, it is evident that despite its origins in industry and business, design thinking has shown a consistent upward trend in research over the past decade. The United States emerged as both the pioneer and the largest contributor to this research, followed by China. The trend of design thinking research in classroom learning often integrates with other components such as learning resources, methods, models, and approaches, demonstrating that design thinking has the potential to enhance the quality of classroom learning. Design thinking can then be implemented in the curriculum by adopting a project-based approach to engage students in real problem-solving.

REFERENCES

Anderson, L. W., Krathwohl Peter W Airasian, D. R., Cruikshank, K. A., Mayer, R. E., Pintrich, P. R., Raths, J., and Wittrock, M. C. (2001). *Taxonomy for_ Assessing a Revision 0F Bl00M'S Tax0N0My 0F Educati0Nal Objectives*. https://www.uky.edu/~rsand1/china2018/texts/Anderson-Krathwohl-A taxonomy for learning teaching and assessing.pdf

Antonio, R. P. (2022). Effectiveness of Blended Instructional Approach in Improving Students' Scientific Learning Outcomes: A Meta-Analysis. In *Journal of Higher Education Theory and Practice* (Vol. 22, Issue 5).

Arenda, T., and Suwono, H. (2023). Design thinking and EDS prompt in STEAM learning: The impact on germane cognitive load and creative thinking. *AIP Conference Proceedings*, *2614*(1).

Arifin, S., and Siew, N. M. (2023). Integration of Socioscientific Approach and Design Thinking: An Entrepreneurial Creative Thinking Module for STEM Education. *Journal of Baltic Science Education*, *22*(5), 767–780.

Bawaneh, A. K., and Alnamshan, M. M. (2023). Design Thinking in Science Education: Enhancing Undergraduate Students' Motivation and Achievement in Learning Biology. *Int J Inf Educ Technol*, *13*(4), 621–633.

Beckman, S. L., and Barry, M. (2007). Innovation as a learning process: Embedding design thinking. *California Management Review*, *50*(1), 25–56.

Bringardner, J., Castroverde, E., Charette, P., Abuelgasim, S. M. S., Yoshinobu, M., Li, R., Bill, V., and Paredes, I. (2023). The Impact of Documenting Design Thinking, the Engineering Design Process Canvas, and Project Communication on Design Self-Efficacy of First-Year Students. *2023 ASEE Annual Conference* and *Exposition*.

Brown, T., and Wyatt, J. (2010). Design thinking for social innovation. *Development Outreach*, *12*(1), 29–43.

Cain, W., and Henriksen, D. (2022). Creativity and Design Thinking: Crucial Mindsets in Instructional Design Education. In *The Instructional Design Trainer's Guide* (pp. 38–47). Routledge.

Calavia, M. B., Blanco, T., and Casas, R. (2021). Fostering creativity as a problem-solving competence through design: Think-Create-Learn, a tool for teachers. *Thinking Skills and Creativity*, *39*, 100761.

Chang, C.-Y., Du, Z., Kuo, H.-C., and Chang, C.-C. (2023). Investigating the Impact of Design Thinking-Based STEAM PBL on Students' Creativity and Computational Thinking. *IEEE Transactions on Education*.

Chit, S. C., Shabli, A. H. M., and Mahmuddin, M. (2023). The Effectiveness of Conducting STEM Projects Using Design Thinking Approach in Rural Schools in Kedah, Malaysia: A Smart Farming Project. *International Conference on Computing and Informatics*, 190–203.

Clemente, V., Vieira, R., and Tschimmel, K. (2016). A learning toolkit to promote creative and critical thinking in product design and development through Design Thinking. *2016 2nd International Conference of the Portuguese Society for Engineering Education (CISPEE)*, 1–6.

Craft, A. (2003). Creative thinking in the early years of education. *Early Years: An International Journal of Research and Development*, *23*(2), 143–154.

Craft, A. (2010). Possibility thinking and wise creativity: *Educational futures in England?*

Cropley, A. (2015). *Creativity in education and learning: A guide for teachers and educators*. Routledge.

Cross, N. (2023). *Design thinking: Understanding how designers think and work*. Bloomsbury Publishing.

de Waal, G. A., and Maritz, A. (2022). A disruptive model for delivering higher education programs within the context of entrepreneurship education. *Education+ Training*, *64*(1), 126–140.

Gillet, D., and Vonèche-Cardia, I. (2021). Promoting Critical and Design Thinking Activities to Tackle Sustainable Development Goals in Higher Education. *2021 World Engineering Education Forum/Global Engineering Deans Council (WEEF/GEDC)*, 224–230.

Halpern, D. F. (2003). *Thinking critically about creative thinking*.

Henriksen, D., Richardson, C., and Mehta, R. (2017). Design thinking: A creative approach to educational problems of practice. *Thinking Skills and Creativity*, *26*, 140–153.

Herianingtyas, N. L. R. I. M. A. R. (2024). *Design thinking based learning: inovasi pembelajaran IPA abad 21*. Rajawali Pers.

Herianingtyas, N. L. R., Marini, A., Zakiah, L., and Chen, Y. (2023). Integration of the Pancasila Student Profile Strengthening Project with Design Thinking in the Merdeka Curriculum: Stimulating Students' Creative Thinking Skills. *TARBIYA: Journal of Education in Muslim Society*, *10*(2), 237–250.

Hernández-Torrano, D., and Ibrayeva, L. (2020). Creativity and education: A bibliometric mapping of the research literature (1975–2019). *Thinking Skills and Creativity*, *35*. https://doi.org/10.1016/j.tsc.2019.100625

Ho, T. E., and Pang, V. (2024). Outcomes of an integrated STEM with design thinking module on preschoolers' engineering practices. *Eurasia Journal of Mathematics, Science and Technology Education*, *20*(4), em2431.

Hua, W. X., Eng, L. S., Tong, W., and Bin, Y. (2023). Developing a Graphic Design Thinking Model to Enhance the Creative Thinking. *Academic Journal of Interdisciplinary Studies*, *12*.

Jackson, N., and Shaw, M. (2006). Developing subject perspectives on creativity in higher education. *Developing Creativity in Higher Education: An Imaginative Curriculum*, 89–108.

Johansson-Sköldberg, U., Woodilla, J., and Çetinkaya, M. (2013). Design thinking: Past, present and possible futures. *Creativity and Innovation Management*, *22*(2), 121–146.

Jussila, J., Raitanen, J., Partanen, A., Tuomela, V., Siipola, V., and Kunnari, I. (2020). Rapid product development in university-industry collaboration: Case study of a smart design project. *Technology Innovation Management Review*, *10*(3).

Kampylis, P. (2010). *Fostering creative thinking: The role of primary teachers* (Issue 115). University of Jyväskylä.

Kaufman, J. C. (2016). *Creativity 101*. Springer publishing company.

Kaufman, J. C., and Beghetto, R. A. (2009). Beyond big and little: The four c model of creativity. *Review of General Psychology*, *13*(1), 1–12.

Koroh, T. R. (2014). A development module to teach creative thinking ability based on creative problem solving and design thinking models. *Journal of Turkish Science Education*, *11*(1), 3–23.

Kuo, H.-C., Yang, Y.-T. C., Chen, J.-S., Hou, T.-W., and Ho, M.-T. (2021). The impact of design thinking PBL robot course on college students' learning motivation and creative thinking. *IEEE Transactions on Education*, *65*(2), 124–131.

Lambert, A. (2019). Developing skills through creative problem-solving. *Childhood Education*, *95*(4), 24–29.

Leal Filho, W., Schmidberger, I., Sharifi, A., Vargas, V. R., Rampasso, I. S., Dibbern, T., Liakh, O., Aina, Y. A., Trevisan, L. V., and Mbah, M. F. (2024). Design thinking for sustainable development: A bibliometric analysis and case study research. *Journal of Cleaner Production*, *455*, 142285.

Lee, D. (2018). Design thinking in the classroom: Easy-to-use teaching tools to foster creativity, encourage innovation and unleash potential in every student. Simon and Schuster.

Leem, S., and Lee, S. W. (2024). Fostering collaboration and interactions: Unveiling the design thinking process in interdisciplinary education. *Thinking Skills and Creativity*, *52*, 101520.

Leong, W. Y., Leong, Y. Z., and San Leong, W. (2024). Engaging SDGs Agenda into a Design Thinking Module. *Educational Innovations and Emerging Technologies*, *4*(2), 1–7.

Matthews, M. S., and McBee, M. T. (2007). School factors and the underachievement of gifted students in a talent search summer program. *Gifted Child Quarterly*, *51*(2), 167–181.

Oltra-Badenes, R., Guerola-Navarro, V., Gil-Gómez, J.-A., and Botella-Carrubi, D. (2023). Design and implementation of teaching–learning activities focused on improving the knowledge, the awareness and the perception of the relationship between the SDGs and the future profession of university students. *Sustainability*, *15*(6), 5324.

Pande, M., and Bharathi, S. V. (2020). Theoretical foundations of design thinking–A constructivism learning approach to design thinking. *Thinking Skills and Creativity*, *36*, 100637.

Picanço, C. T., and dos Santos, S. C. (2022). Promoting Collaboration and Creativity in Process Improvement: A Proposal based on Design Thinking and Gamification. *ICEIS (2)*, 418–429.

Pritchard, A. (1969). Statistical bibliography or bibliometrics. *Journal of documentation*, *25*, 348.

Plattner, H. (2013). An introduction to design thinking. *Iinstitute of Design at Stanford*, 1–15.

Rumahlatu, D., Sangur, K., Berhitu, M. M., Kainama, S. Y., Kakisina, V. V, and Latupeirissa, C. (2021). Resource Based Learning Design Thinking (RBLDT): A Model to Improve Students' Creative Thinking Skills, Concept Gaining, and Digital Literacy. *Cypriot Journal of Educational Sciences*, *16*(1), 288–302.

Ryoo, J., and Winkelmann, K. (2021). Innovative learning environments in STEM higher education: Opportunities, challenges, and looking forward.

Soomro, S. A., Casakin, H., and Georgiev, G. V. (2021). Sustainable design and prototyping using digital fabrication tools for education. *Sustainability*, *13*(3), 1196.

Sternberg, R. J. (2003). Creative thinking in the classroom. *International Journal of Phytoremediation*, *47*(3), 325–338. https://doi.org/10.1080/00313830308595

Suwankarjank, V., and Rugwongwan, Y. (2021). Design Thinking Model in Design Education for Thai Product Design Student. *Review of International Geographical Education Online*, *11*(3).

Thomason, D. (2023). The Effect of STEM Integrated Curriculum on Design Thinking Dispositions in Middle School Students: A Mixed Methods Study. The University of Texas at El Paso.

Tsai, C.-A., Song, M.-Y. W., Lo, Y.-F., and Lo, C.-C. (2023). Design thinking with constructivist learning increases the learning motivation and wicked problem-solving capability—An empirical research in Taiwan. *Thinking Skills and Creativity*, *50*, 101385.

Voigt, C., Unterfrauner, E., Aslan, T., and Hofer, M. (2019). Design thinking with children: The role of empathy, creativity and self-efficacy. In *Proceedings of FabLearn 2019* (pp. 144–147).

Von Thienen, J. P. A., R. A. and M. C. (2017). Design Thinking in Higher Education: How Students Become Dedicated Creative Problem Solvers. 306–328.

Yang, C.-M., and Hsu, T.-F. (2020). Integrating design thinking into a packaging design course to improve students' creative self-efficacy and flow experience. *Sustainability*, *12*(15), 5929.

Yang, X., Zhang, M., Zhao, Y., Wang, Q., and Hong, J.-C. (2022). Relationship between creative thinking and experimental design thinking in science education: Independent or related. *Thinking Skills and Creativity*, *46*, 101183.

Zhu, L., Shu, L., Tian, P., Sun, D., and Luo, M. (2024). Facilitating students' design thinking skills in science class: an exploratory study. *International Journal of Science Education*, 1–22.

Towards Resilient Societies: The Synergy of Religion, Education, Health, Science, and Technology – Rahiem (Ed.)
© 2026 The Author(s), ISBN: 978-1-041-08471-6

Integrating Waqf into Pesantren curriculum: Strengthening ethical values and economic empowerment

Saifudin Asrori
Universitas Islam Negeri Syarif Hidayatullah Jakarta, Indonesia
ORCID ID: 0000-0003-2700-3626

Yazid Hady, Muhammad Isma'il & Ahmad B. Jalhani
Universitas Islam Negeri Syarif Hidayatullah Jakarta, Indonesia

Nurdin Likia
The Patani Center for Organizational and Public Outreach, Thailand

ABSTRACT: This study explores the integration of waqf (Islamic endowment) into the educational curriculum of Islamic boarding schools (pesantren) to enhance ethical values and waqf management. Using a qualitative descriptive approach, the research examines how pesantren implement waqf education within their learning environment. As holistic Islamic institutions, pesantren have the potential to develop waqf systems that not only promote social and economic welfare but also shape students' (santri) character and ethical consciousness. Findings indicate that integrating waqf into the pesantren curriculum strengthens religious and social values, fosters philanthropic awareness, and supports financial independence for educational institutions. However, challenges such as limited understanding of waqf law and resource constraints hinder effective implementation. Addressing these barriers can enhance pesantren's role in fostering ethical leadership and economic sustainability. The study highlights the need for strategic policies and community engagement to optimise waqf's potential in building self-sufficient and socially responsible institutions.

Keywords: education, ethics, pesantren, waqf, welfare

1 INTRODUCTION

The implementation of waqf in Islamic boarding schools is a key component of financial sustainability (Siddiq *et al.* 2021) and ensures the continuity and growth of the institutions (Usman and Rahman, 2022). Through the initiatives involve utilising waqf assets for productive purposes (Amin *et al.* 2024), the income generated from these ventures has enabled schools to offer affordable tuition fees, provide meals for students, offer scholarships for teachers pursuing higher education, and overall improve the quality of education provided (Fanani *et al.* 2021). However, productive waqf management encounters various challenges that affect its successful implementation (Bessais *et al.* 2024). These challenges encompass whit the mismanagement and underutilisation of waqf assets (Maulida 2024), the awareness of waqf risk and Islamic spirituality of Implementing standard operating procedures (Ambrose *et al.* 2018), and issues of legal complexities, the lack of competence and professionalism of waqf administrators, known as nazir Waqf (Suhendi and Sholeh 2021), and even fewer institutional ownership of waqf assets in pesantren as full public leading to reduced community engagement in developing these assets.

DOI: 10.1201/9781003645542-52

Despite the growing recognition of waqf's role in education, existing studies view it as a financing tool rather than a structured component of pesantren curricula. Anshori (2018) highlights its role in funding infrastructure; Azhari *et al.* (2021) focus on managerial aspects, while Suhendi and Sholeh (2021) and Setiawan (2016) discuss governance and asset optimisation but neglect its pedagogical integration. These studies emphasise financial sustainability but overlook waqf's potential to shape ethical and financial literacy among students. Addressing this gap, this research examines the structured integration of waqf education as both a financial strategy and a pedagogical model to foster ethical leadership and economic independence. Unlike prior research, it proposes a curriculum-based approach, embedding waqf literacy into teaching methodologies and character development, aligning with pesantren's holistic educational philosophy. Additionally, it evaluates key challenges in productive waqf implementation, including mismanagement and underutilisation of assets, and explores governance improvements to maximise community benefits. Strengthening responsible waqf stewardship and enhancing community involvement is central to ensuring the long-term sustainability of waqf-based education.

This study examines waqf management in Pondok Pesantren, focusing on Pondok Gontor, and its impact on educational quality. Providing empirical evidence, it informs policies and strategies to strengthen waqf's role in education. By proving its effectiveness, the study advocates for broader implementation and long-term support for integrating ethical values. Additionally, it aligns with SDG 4 (Quality Education) by ensuring sustainable educational access and SDG 8 (Decent Work and Economic Growth) by fostering financial literacy and ethical entrepreneurship for pesantren communities.

2 METHOD

This qualitative study investigates the effective management of waqf assets in achieving institutional goals such as financial sustainability, infrastructure development, scholarships, and community engagement. Pondok Gontor was selected as a case study due to its long-standing waqf-based financial model, which has been widely adopted by other Islamic educational institutions, making its findings relevant beyond a single context. Data were collected through in-depth interviews with teachers, waqf administrators, and randomly selected students, focusing on their experiences and perceptions regarding waqf management (Creswell 2010). Thematic analysis was employed to systematically identify patterns and recurring themes in waqf management practices, ensuring a structured interpretation of qualitative data. To strengthen validity, cross-verifying data from multiple sources applied triangulation, minimizing bias and enhancing reliability (King *et al.* 1994). All research activities adhered to strict ethical protocols, including informed consent, anonymity, and data security, in compliance with institutional review board (IRB) standards for research involving human participants (Flyvbjerg 2006).

3 FUNDING AND DISCUSSION

3.1 *The concept of Pondok Waqf*

Waqf is a voluntary and irrevocable endowment dedicated to religious or charitable causes, ensuring long-term community benefits. As an Islamic trust, it involves transferring asset ownership while ensuring its perpetual use for designated beneficiaries (Morgan 2001). In Islamic law, waqf is a sustainable mechanism for socio-religious development, covering land, intellectual property, and cash waqf (Jahar 2019). At Pondok Gontor, the waqf declaration is essential for sustaining its educational institutions, with assets managed under the Waqf Board to align with socio-religious objectives (Jamal *et al.* 2022; Rofiqo *et al.* 2021). This structure grants pesantren financial independence and institutional autonomy (Ali *et al.* 2023). The Waqf Board institutionalises the nadzir (manager) system, entrusting governance to alumni as custodians (Siddiq *et al.* 2021). The nadzir, or *mutawalli*, plays a key role in supervising assets, managing finances, ensuring legal compliance, and engaging with the community (Awan and Sadiq 2022). This governance model helps maintain the sustainability and integrity of waqf properties (Emha *et al.* 2022).

Despite Pondok Gontor's success, replicating its waqf-based education model in other pesantren presents challenges. Many pesantren lack structured governance, financial autonomy, and adequate regulatory support, leading to inefficiencies in asset management (Hadiyanto 2023). Additionally, variations in legal frameworks and inconsistent enforcement hinder broader adoption (Asni *et al.* 2024). Institutional capacity and administrative expertise are crucial for ensuring the sustainability of waqf-based systems (Faisal *et al.* 2024). Without clear governance mechanisms, smaller pesantren struggle to implement and manage waqf effectively, limiting their ability to sustain educational programs and long-term financial stability.

A structured policy framework is needed to standardise waqf management across pesantren to address these barriers. Collaboration between government agencies, financial institutions, and Islamic education stakeholders can enhance institutional capacity and ensure waqf sustainability. Strengthening regulatory oversight, financial transparency, and capacity-building initiatives will help pesantren implement and sustain this model, fostering a more resilient Islamic education system. A well-governed waqf system secures educational sustainability and enhances pesantren's role in economic empowerment and community development.

3.2 *Self waqf; commitment and sustainability in Pondok Gontor*

Self-waqf at Pondok Gontor is a deep commitment made by prospective cadres to support the institution's sustainability. This practice is firmly grounded in Islamic jurisprudence and the principles of maqasid al-shariah, which aim to maximise benefits and reduce harm (Taufik and Aprillillah 2022). The concept of Self waqf involves more than just financial contributions; it requires the total dedication of the cadres to prioritise the institution's mission over personal interests, strengthening their connection to the values and goals of Pondok Gontor.

Cadres who commit to Self waqf agree to serve without financial compensation, relying instead on institutional support such as housing provided by the pesantren (Asrori and Syauqi 2020). This arrangement allows them to focus entirely on their responsibilities without distractions, and they are expected to be flexible and ready to serve whenever and wherever needed. Formal contracts, witnessed by parents and institutional leaders, help ensure accountability and continuity in managing waqf assets, which are critical to the institution's smooth operation.

By making Self waqf part of its culture, Pondok Gontor fosters a long-term commitment to its waqf assets and educational mission (Kibiya *et al.* 2023). This dedication promotes effective management of resources and strengthens ties between the pesantren and the surrounding community (Yapono 2015). The practice of self-waqf not only reduces dependence on external funding and empowers alumni and community members to contribute actively. This fosters a culture of philanthropy that helps generate sustainable income while addressing the broader needs of society (Ascarya and Sakti 2022; Fanani *et al.* 2021; Siahaan *et al.* 2019).

3.3 *Integrated waqf management and curriculum at Pondok Gontor*

Pondok Modern Darussalam Gontor (PMDG) has developed an integrated model for managing waqf that aligns closely with its educational system, promoting a holistic approach that includes administrative order, work ethics, and fostering independence within a community setting. The well-organised administrative structure ensures efficient financial management, especially for its business units and student cooperatives. This clear governance system defines roles and responsibilities, ensuring that all financial transactions are transparent and accountable while supporting the institution's educational mission (Siddiq *et al.* 2021). Centralised financial management, overseen by the Administration Section, allows for better oversight of funds, ensuring that income and expenditures are well-managed, flexible for daily needs, and conducted transparently (Nurkhin 2023; Rusydiana 2023).

The financial management at Pondok Gontor involves careful budgeting and resource allocation, ensuring funds from waqf, donations, and business activities are properly tracked and utilised for the institution's needs. Strategic planning helps ensure that resources are used effectively to meet the institution's goals, with regular financial reporting for transparency and accountability (Ali *et al.* 2023; Rakhmawan 2023). These financial reports, prepared by the Administration Section, provide essential insights for decision-making and help keep parents, alumni, and other stakeholders informed about the institution's financial health (Omercic 2017). By involving alumni and local stakeholders in financial planning (Ascarya 2024), Pondok Gontor builds trust and fosters community involvement, which enhances long-term financial sustainability (Zulkifli *et al.* 2022).

Work ethics at Pondok Gontor are grounded in values like sincerity, honesty, and responsibility, which are key when selecting administrators (Mu'minah 2015). Business units, jointly managed by students and teachers, foster teamwork, responsibility, and financial independence while upholding moral values and community service. Profits are reinvested into education, minimising reliance on external funding. Pondok Gontor also provides practical training and internships, equipping students with career-ready skills to contribute meaningfully to society (Siahaan *et al.* 2019).

4 CONCLUSION

Education about waqf is very important in equipping pesantren students with waqf asset management knowledge and skills. This supports the sustainable development of pesantren resources and strengthens the role of pesantren in society. Islamic boarding schools are uniquely positioned to integrate waqf in their curriculum, given their tradition of scholarship and leadership in the Muslim community. Teaching about waqf plays a crucial role in fostering economic empowerment within communities. Effective implementation requires strong collaboration between the government, educational institutions, and the pesantren community, supported by clear regulatory and policy frameworks. Integrating waqf into the pesantren curriculum will deepen students' religious understanding and equip them with essential skills in waqf-based economic management, ultimately enhancing societal welfare. This integration strengthens pesantren as key institutions that drive economic and social progress in Muslim communities.

REFERENCES

Ali, S. N. M., Mohd Noor, A. Z., and Osman, A. H. (2023). Exploring the role of fiqh waqie in waqf and zakat collaboration framework for hostel development. *Environment-Behaviour Proceedings Journal*, 8(23), 81–86. https://doi.org/10.21834/ebpj.v8i23.4494

Ambrose, A. H. A. A., Hassan, M. A. G., and Hanafi, H. (2018). A proposed model for waqf financing public goods and mixed public goods in Malaysia. *International Journal of Islamic and Middle Eastern Finance and Management*, 11(3), 395–415. https://doi.org/10.1108/imefm-01-2017-0001

Amin, H., Hassan, M. K., and Shaikh, I. M. (2024). Waqf-based qardhul hassan financing preference in Malaysia: An exploratory study. *International Journal of Ethics and Systems*, 40(3), 651–671. https://doi.org/10.1108/IJOES-02-2023-0025

Anshori, I. (2018). Peran dan manfaat wakaf dalam pengembangan pendidikan Islam: Studi kasus di Madrasah Tsanawiyah, Pesantren al Andalusia Caringin Sukabumi Jawa Barat Indonesia. *Jurnal Tahdzibi: Manajemen Pendidikan Islam*, 3(1), 19–26. https://doi.org/10.24853/tahdzibi.3.1.27-38.

Ascarya, A. (2024). An investigation of waqf-based Islamic micro financial institution models to identify the most effective model for Indonesia. *Journal of Islamic Accounting and Business Research*. https://doi.org/10.1108/JIABR-12-2022-0325

Ascarya, A., and Sakti, A. (2022). Designing micro-fintech models for Islamic micro financial institutions in Indonesia. *International Journal of Islamic and Middle Eastern Finance and Management*, 15(2), 236–254. https://doi.org/10.1108/imefm-05-2020-0233

Asni, F., Mohammed Noor, A., and Hasbullah, M. H. (2024). Management of cash waqf fund generation through the implementation of istibdal in Kedah. *Qualitative Research in Financial Markets*, 16(1), 60–86. https://doi.org/10.1108/QRFM-03-2022-0048

Asrori, S., and Syauqi, A. (2020). Kontribusi pendidikan Islam dalam reproduksi identitas sosial Muslim Indonesia. *Mimbar Agama Budaya*, 13–22.

Awan, M. A., and Sadiq, S. (2022). A waqf empowerment model based on benefit analysis. *Journal of Information Technology and Communication*, 12(1), 1–14. https://doi.org/10.32350/jitc.141.18

Azhari, M., Tanjung, H., and Yono, Y. (2021). Analisis Manajemen Pengelolaan dan Pengembangan Wakaf Produktif di Pondok Pesantren Darussalam Ciomas Bogor. *El-Mal: Jurnal Kajian Ekonomi and Bisnis Islam*, 2(3), 140–148. https://doi.org/10.47467/elmal.v2i3.530

Bessais, S., Zakariyah, H., and Mohd Noor, A. (2024). The possibility of establishing an endowment (waqf) from the Islamic financial institutions' provisioned funds in the United Arab Emirates. *International Journal of Islamic and Middle Eastern Finance and Management*. https://doi.org/10.1108/IMEFM-12-2023-0466

Creswell, J. W. (2010). *Qualitative, Quantitative, and Mixed Approaches*. Yogyakarta.

Emha, Z. I. F., Silviana, A., and Musahadi, M. (2022). Waqf land certification postponement for place of worship due to the obscurity of the toll road expansion project (Re-Overview of Gustav Radbruch's three basic legal values theory). *Al-Ahkam*, 32(1), 17–40. https://doi.org/10.21580/ahkam.2022.32.1.10745

Faisal, F., Mu'in, F., Edy, R. N., Indra, G. L., and Santoso, R. S. (2024). Strengthening the National Waqf Ecosystem through Legal Reform: An Analysis of Waqf Issues in the Provinces of Lampung and South Sumatra. *Al-Istinbath: Jurnal Hukum Islam*, 9(1), 81. https://doi.org/10.29240/jhi.v9i1.9320

Fanani, A., Haq, F., and Wathani, M. Z. (2021). The contribution of waqf on poverty alleviation through digital platforms: A case of Indonesia. *Shirkah Journal of Economics and Business*, 6(2). https://doi.org/10.22515/shirkah.v6i2.386

Flyvbjerg, B. (2006). Five misunderstandings about case-study research. *Qualitative Inquiry*, 12(2), 219–245.

Hadiyanto, R. (2023). The development model of Islamic philosophy to realize the independence of Islamic boarding schools. *Kne Social Sciences*. https://doi.org/10.18502/kss.v8i18.14347

Jahar, A. S. (2019). Bureaucratizing Sharia in modern Indonesia: The case of zakat, waqf and family law. *Studia Islamika*, 26(2), 207–245.

Jamal, M., Haq, F., and Wathani, M. Z. (2022). Waqf land management: Case study on waqf of rice field Gontor. *Journal of International Conference Proceedings*, 5(1). https://doi.org/10.32535/jicp.v5i1.1458

Kibiya, I. U., Usman, M., Kurfi, S. A., and Hamid, K. T. (2023). Awareness and knowledge of Islamic accounting system among accounting students in Nigerian universities. *Journal of Islamic Accounting and Business Research*, 14(7), 1106–1120. https://doi.org/10.1108/JIABR-07-2021-0186

King, G., Keohane, R. O., and Verba, S. (1994). *Designing Social Inquiry: Scientific Inference in Qualitative Research*. Princeton University Press.

Maulida, S. S. (2024). Study of implementation of the eco-pesantren concept at Dayah Terpadu Inshafuddin, Banda Aceh. *IOP Conference Series: Earth and Environmental Science*, 1290(1), 12037. https://doi.org/10.1088/1755-1315/1290/1/012037

Morgan, C. (2001). Islam and civil society: The waqf. *The Good Society*, 10(1), 21–24.

Mu'minah, N. (2015). Character building dalam konsep pendidikan Imam Zarkasyi ditinjau dari filsafat moral Ibnu Miskawaih. *Jurnal Filsafat*, 25(1), 100–133.

Nurkhin, A. (2023). Green-pesantren and environmental knowledge and awareness: Case study at Pondok Pesantren As Salafy Al Asror Semarang. *IOP Conference Series: Earth and Environmental Science*, 1248(1), 12003. https://doi.org/10.1088/1755-1315/1248/1/012003

Omercic, J. (2017). Waqf in Bosnia and Herzegovina in the 20th and 21st century. *ICR Journal*, 8(3), 342–367. https://doi.org/10.52282/icr.v8i3.179

Rakhmawan, Z. (2023). Kurikulum tersembunyi pendidikan karakter dan model parenting di Ma'had Tahfidz Al-Qur'an Bina Madani Putri Grabag. *Bemas Jurnal Bermasyarakat*, 4(1), 148–155. https://doi.org/10.37373/bemas.v4i1.588

Rofiqo, A., Muslih, M., and Sari, D. N. (2021). Reputation, transparency, trust and waqif's perception on nadzir's professional toward intention to act waqf: Empirical study in Pondok Modern Darussalam Gontor (PMDG). *Journal of Islamic Economic Laws*, 4(2), 42–66. https://doi.org/10.23917/jisel.v4i2.14870

Rusydiana, A. S. (2023). Clustering research on sukuk-waqf. *International Journal of Waqf*, 3(1). https://doi.org/10.58968/ijw.v3i1.241

Setiawan, E. (2016). Pengelolaan Wakaf Pondok Pesantren Al-Hikmah Sirampog Kabupaten Brebes. *INFERENSI: Jurnal Penelitian Sosial Keagamaan*, 10(2), 495–516. doi:https://doi.org/10.18326/infsl3.v10i2.495-516

Siahaan, D., Iswati, S., and Fathullah, A. (2019). Social enterprise: The alternative financial support for educational institutions. *International Journal of Economics and Financial*, 9(3).

Siddiq, A., Sulaiman, M., and Baharuddin, A. (2021). Development strategy of Islamic boarding schools-based waqf (Case study: Islamic boarding schools of Daarut Tauhiid, Bandung). *Tadris Jurnal Keguruan dan Ilmu Tarbiyah*, 2(2), 11–22. https://doi.org/10.59202/as.v2i2.625

Suhendi, H., and Sholeh, N. S. M. (2021). Digitizing da'wa movement of Pesantren Baitul Hidayah through productive waqf. *Muharrik Jurnal Dakwah dan Sosial*, 4(2), 365–376. https://doi.org/10.37680/muharrik.v4i02.1068

Taufik, F., and Aprillillah, Y. N. I. (2022). Application of waqf in Modern Islamic Boarding School Darussalam Gontor Ponorogo Regency in the perspective of Sharia Economic Law and Law Number 41 of 2004 concerning Waqf. *Al-Muamalat Jurnal Ekonomi Syariah*, 9(1), 1–6. https://doi.org/10.15575/am.v9i1.14248

Usman, M., and Rahman, A. A. (2022). Funding higher education through waqf: A lesson from Malaysia. *International Journal of Ethics and Systems*, 39(1), 107–125. https://doi.org/10.1108/ijoes-12-2021-0217

Yapono, A. (2015). Filsafat pendidikan dan hidden curriculum dalam perspektif KH. Imam Zarkasyi. *Tsaqafah*, 11(2), 291–312.

Zulkifli, N., Md Nor, N., Ibrahim, N., and Zainal, N. (2022). An empirical investigation on integrated waqf governance in Malaysian waqf institutions. *Environment-Behaviour Proceedings Journal*, 7(22), 3–8. https://doi.org/10.21834/ebpj.v7i22.4146

Towards Resilient Societies: The Synergy of Religion, Education, Health, Science, and Technology – Rahiem (Ed.)

Student teachers' TPACK: Perceptions for effective online instruction

Siti Nurul Azkiyah
Universitas Islam Negeri Syarif Hidayatullah Jakarta, Indonesia
ORCID ID: 0000-0001-5494-1509

Waliyadin & Nur'aeni
Universitas Islam Negeri Syarif Hidayatullah Jakarta, Indonesia
ORCID ID: 0000-0002-0872-0315

Muhammad Rizal Mahfuzo
University of Michigan, USA

ABSTRACT: Amid the rapid advancement of technology and the unprecedented challenges posed by the COVID-19 pandemic, integrating ICT into classrooms has become essential for achieving quality education for all, as outlined in the United Nations' Sustainable Development Goal 4. While prior research has largely focused on the technological competence of in-service teachers, a notable lack of studies examining student teachers raises concerns about whether future educators are adequately prepared with the necessary technological and pedagogical content knowledge. This study aims to address this gap by investigating student teachers' perceptions of their Technological Pedagogical Content Knowledge (TPACK). Using a quantitative survey design, the study involved 350 student teachers from both private and public universities supervised by the Ministry of Religious Affairs (MORA) and the Ministry of Education and Culture (MOEC). The findings revealed that, overall, student teachers perceived themselves as having a good understanding of technology and technological content knowledge. However, they reported only a moderate understanding of content knowledge, technological pedagogical knowledge, pedagogical content knowledge, and the integrated TPACK framework.

Keywords: education, online teaching, student teachers, support for online teaching and learning, teaching profession, technological and pedagogical content knowledge

1 INTRODUCTION

Research across different countries has proven the importance of teachers in influencing students' learning achievement (Azigwe 2016; Azkiyah *et al.* 2018; Chuyun Hu 2020; Stronge 2018). Therefore, it is not surprising to see many studies on teachers and lecturers on many aspects, including their readiness, challenges, and strategies in dealing with the COVID-19 pandemic, which started in early 2020 (Bayrakdar and Guveli 2022; Gamage *et al.* 2020; Muir *et al.* 2024; Rasmitadila *et al.* 2020; Scherer *et al.* 2021). Regarding the

DOI: 10.1201/9781003645542-53

COVID-19 pandemic, these studies indicate similar findings in which schools, colleges, and universities have to switch their face-to-face learning to an online or blended mode of instruction.

This online teaching and learning process requires integrating technology in the classroom, which has taken place before the appearance of the COVID-19 pandemic and has been positively related to better outcomes. In the language field, for instance, some studies have shown that technology has recently played a pivotal role, especially in providing rich and meaningful learning resources and effective delivery of instruction (Rintaningrum 2023) and practical strategies in engaging students (Bal *et al.* 2015). Moreover, technology in education can also enhance autonomous learning (Luckin *et al.* 2016) and increase students' motivation, especially when integrated with gamification (Licorish *et al.* 2018).

Thus, using technology in education to replace traditional face-to-face teaching and learning processes has a robust theoretical framework. Consequently, teachers must integrate technology into their classrooms, which could be measured by Technological and Pedagogical Content Knowledge (TPACK) (Lie *et al.* 2020; Nordin *et al.* 2013; Pokhrel and Chhetri 2021; Scherer *et al.* 2021). In Indonesia, a large-scale study carried out by SMERU Research Institute (Alifia *et al.* 2020) reveals that teachers' responses and abilities towards the pandemic requiring them to integrate technology in the classroom vary deeply, where teachers outside Java have to visit their students due to difficult access to the internet and low level of gadget ownership among parents while those living in Java can utilise various digital application like WhatsApp, YouTube, Google Classroom and Zoom Cloud Meeting. Furthermore, Lie *et al.* (2020) state that teachers in many areas of Indonesia have reported some challenges in delivering online classes due to some factors such as less previous exposure to online learning, unstable internet connection, less adequate technical and pedagogical knowledge, as well as less backing from schools and parents.

According to the studies mentioned above, in-service teachers—especially those who are not Java experts or have not had much experience with technology—have found it challenging to integrate it into the classroom. Therefore, it makes sense to presume that many student teachers—especially those who do not teach in Java—may lack the necessary technological and pedagogical understanding of the subject. Moreover, a case study on the secondary student teachers' technological and pedagogical content knowledge mastery level in New Zealand carried out long before the COVID-19 pandemic began showed a low level of technology knowledge (Nordin *et al.* 2013).

The COVID-19 pandemic has necessitated a rapid transition from traditional face-to-face instruction to online and blended learning environments, prompting numerous studies on the readiness and challenges faced by in-service teachers during this period (Bayrakdar and Guveli 2022; Gamage *et al.* 2020; Muir *et al.* 2024; Rasmitadila *et al.* 2020; Scherer *et al.* 2021). In this post-pandemic era, technology in the classroom is undoubtedly needed, and student teachers must acquire knowledge of technology and be skilful in using it. However, there remains a significant gap in research focusing on student teachers and the future educators tasked with navigating these new teaching landscapes.

This study addresses this gap by examining student teachers' perceptions of their Technological and Pedagogical Content Knowledge (TPACK). While previous research has predominantly centred on in-service teachers, understanding student teachers' preparedness is crucial, especially in the context of the ongoing challenges posed by the pandemic. Integrating Information and Communication Technology (ICT) in education is essential for achieving quality learning outcomes, as highlighted by the United Nations' Sustainable Development Goal 4.

Furthermore, Voogt *et al.* (2013) review many previous studies on TPACK, including how to measure TPACK, which shows the use of a combination of several instruments. One of the most widely used instruments is self-assessment surveys, which report the participants' perception of or confidence in TPACK through a Likert scale questionnaire. This study will use the same questionnaire format to collect the data. The results should help determine if student instructors are prepared to use technology in the classroom and what steps faculty members or study programs need to take.

2 METHODS

This study aims to investigate how student teachers perceive their technological and pedagogical content knowledge. The study, which represented several Indonesian islands, was conducted in Jakarta, Banten, Jambi, and Lombok using a quantitative survey design. The chosen geographical locations—Jakarta, Banten, Jambi, and Lombok—represent a diverse range of socio-economic contexts and access to technology within Indonesia, allowing for a comprehensive understanding of how these factors influence student teachers' perceptions of their Technological and Pedagogical Content Knowledge (TPACK).

A total of 350 students from public and private universities were invited to participate in the study, which was overseen by the Ministries of Religious Affairs (MoRA) and Education and Culture (MoEC). The sample size is significant as it provides a statistically reliable representation, enhancing the validity of the findings and enabling generalisation across the broader population of student teachers in Indonesia.

A 5-Likert-Scale questionnaire examining student teachers' perception of their technological, pedagogical, and content knowledge (TPACK) was developed to collect the data, which were then analysed descriptively. There were 33 main questions with eight technological, pedagogical, and content knowledge components. The instrument was validated using content validity analysis and expert judgment, while the reliability was measured using Cronbach Alpha, which coefficient was very high, at r = .9.

3 FINDINGS AND DISCUSSION

3.1 *Findings*

Table 1. summarises the descriptive findings of the study by showing the mean score of each component in Technology, Pedagogy, and Content Knowledge. In general, student teachers perceived a good understanding of technological knowledge and technological content knowledge. However, they perceived only a moderate understanding of the other six components.

Table 1. Descriptive analysis.

No	Component	Mean	Standard Deviation
1.	Technology knowledge	4,11	0,56
2.	Pedagogical knowledge	3,88	0,67
3.	Content knowledge	3,71	0,68
4.	Technological pedagogical knowledge	3,94	0,75
5.	Pedagogical content knowledge	3,89	0,68
6.	Technological content knowledge	4,10	0,72
7.	Technological, pedagogical, and content knowledge	3,76	0,77

In summary, student teachers generally had a good understanding of technological knowledge and content integration but only a moderate grasp of the other six components.

As Table 1 shows, among the seven components of TPACK, only two, Technology Knowledge and Technological Content Knowledge, scored more than 4, while the remaining five scored less than 4. These findings indicate that student teachers perceived themselves to have a good understanding of Technology Knowledge and Technological Content Knowledge and a moderate understanding of the other components.

Furthermore, it is critical to note that among the other five components that scored less than four, Content Knowledge and Technological, Pedagogical, and Content Knowledge were the two factors with the lowest scores, at 3.71 and 3.76, respectively. These findings indicate that student teachers in this study must continuously improve their knowledge of the subject matter. In addition, they also need to deepen their understanding of how to combine various aspects related to content, technology, and instruction to facilitate effective instruction.

3.2 *Discussion*

Based on the findings, this study aligns with some relevant previous studies. Concerning integrating technology in the classroom, the participants in this study were good at identifying and using some applications to facilitate learning, which supports a previous study by Durriyah and Zuhdi (2018), revealing that English as a Foreign Language (EFL) student teachers were familiar with Facebook, blogs, Skype, and WhatsApp to facilitate and enhance language learning. Similarly, a recent study in Estonia indicated that pre-service teachers feel confident integrating technology into their lessons despite lacking strong pedagogical skills (Luik *et al.* 2018).

Furthermore, the other five components scored less than four, implying the need to pay more attention to those areas, especially Content Knowledge (3,71) and Technological, Pedagogical, and Content Knowledge (3,76), which were less than 3.80. This may suggest that student teachers have less exposure to the experiences of integrating digital technologies into subject areas and subject-specific pedagogies. Content and technology knowledge could be interrelated (Tondeur *et al.* 2020). As discussed in the introduction, technology has been proven to encourage and motivate student learning and positively affect student outcomes (Licorish *et al.* 2018; Li *et al.* 2015). However, Tsui and Tavares (2021) emphasise that there should be a balance between the mastery of digital technology and pedagogy, or the dichotomisation of technology and pedagogy should not exist.

Additionally, the interplay between Technological and Pedagogical Content Knowledge (TPACK) components is essential for fostering pedagogical innovation in teacher education. Each component—Technology Knowledge (TK), Pedagogical Knowledge (PK), and Content Knowledge (CK)—should be understood in relation to one another to create effective teaching strategies that leverage technology. For instance, strong PK enables student teachers to select appropriate technological tools that enhance content delivery, while robust CK ensures these tools are used effectively to meet learning objectives. Teacher training institutions should adapt their curricula to emphasise this interconnectedness, promoting a holistic approach to TPACK that encourages experimentation with innovative pedagogical practices. By integrating TPACK-focused training programs, institutions can better prepare future educators to navigate the complexities of modern classrooms, ultimately leading to improved educational outcomes and a more dynamic learning environment. Therefore, teacher training institutions must promote concrete examples of the TPACK framework teaching practice.

4 CONCLUSIONS

This study concerns student teachers' perception of their combined knowledge of technology, pedagogy, and content, which is expected to enable them to provide quality teaching in the classroom while using technology. Using a questionnaire, the findings reveal that among the seven components measured in the study, student teachers perceived themselves to have a good understanding of technology and technological content knowledge. However, they perceived only a moderate understanding of the remaining factors, especially content knowledge and technological, pedagogical, and content knowledge. Therefore, teacher education programs should revisit their curriculum and include a unit integrating digital technology and pedagogy. To enhance student teachers' perspectives of their technological pedagogical content knowledge, organisations should help them by organising incidental TPACK-based training programs linked to their disciplines, as suggested in previous studies (e.g. Diamah *et al.* 2022).

REFERENCES

Alifia, U., Barasa, A. R., Bima, L., Pramana, R. P., Revina, S., and Tresnatri, F.A. (2020). Learning from home: Portrait of teaching and learning inequalities in times of the Covid-19 pandemic. *Smeru Research Institute*, 1(1), 1–8. https://smeru.or.id/sites/default/files/publication/cp01_covidpjj_eng.pdf

Azigwe, J. B. (2016). Using comparative international studies for modeling educational effectiveness: A secondary analysis of PISA-2009 Study. *International Journal of Education and Practice*, 7(18), 199–209. Microsoft Word - JEP-Vol.7 No.18 2016.docx (ed.gov)

Azkiyah, S. N., Doolaard, S., Creemers, B. P. M., and Van der Werf, M. P. C. (2018). Students' perspective on the impact of English teacher development programs on teaching quality in Indonesia. *Pertanika Journal of Social Science and Humanities*, 26(1), 199–218.

Bal, A. S., Grewal, D., Mills, A., and Ottley, G. (2015). Engaging students with social media. *Journal of Marketing Education*, 37(3), 190–203.

Bayrakdar, S., and Guveli, A. (2022). Inequalities in home learning and schools' remote teaching provision during the COVID-19 school closure in the UK. *Sociology*. https://doi.org/10.1177/00380385221122444

Chuyun Hu, C. (2020). Understanding college students' perceptions of effective teaching. *International Journal of Teaching and Learning in Higher Education*, 32(2), 318–328. https://eric.ed.gov/?id=EJ1286303.

Diamah, A., Rahmawati, Y., Paristiowati, M., Fitriani, E., Irwanto, I., Dobson, S., and Sevilla, D. (2022). Evaluating the effectiveness of technological pedagogical content knowledge-based training program in enhancing pre-service teachers' perceptions of technological pedagogical content knowledge. *Frontiers in Education*, 7. https://doi.org/10.3389/feduc.2022.897447

Durriyah, T. L., and Zuhdi, M. (2018). Digital literacy with EFL student teachers: Exploring Indonesian student teachers' initial perceptions about integrating digital technologies into a teaching unit. *International Journal of Education and Literacy Studies*, 6(3), 53–60. http://dx.doi.org/10.7575/aiac.ijels.v.6n.3p.53

Gamage, K. A. A., Wijesuriya, D. I., Ekanayake, S. Y., Rennie, A. E. W., Lambert, C. G., and Gunawardhana, N. (2020). Online delivery of teaching and laboratory practices: Continuity of university programmes during COVID-19 pandemic. *Education Sciences*, 10(10), 1–9. https://doi.org/10.3390/educsci10100291

Li, L., Worch, E., Zhou, Y., and Aguiton, R. (2015). How and why digital generation teachers use technology in the classroom: An explanatory sequential mixed methods study. *International Journal for the Scholarship of Teaching and Learning*, 9(2), 9. https://doi.org/10.20429/ijsotl.2015.090209

Licorish, S. A., Owen, H. E., Daniel, B., and George, J. L. (2018). Students' perception of Kahoot!'s influence on teaching and learning. *Research and Practice in Technology Enhanced Learning*, 13(1). https://doi.org/10.1186/s41039-018-0078-8

Lie, A., Tamah, S. M., Gozali, I., Triwidayati, K. R., Utami, T. S. D., and Jemadi, F. (2020). Secondary school language teachers' online learning engagement During the Covid-19 pandemic in Indonesia. *Journal of Information Technology Education: Research*, 19, 803–832. https://doi.org/10.28945/4626

Luckin, R., Holmes, W., Griffiths, M., Forcier, L. (2016). *Intelligence unleashed: An argument for AI in education*. London: Pearson Education.

Luik P, Taimalu M, Suviste, R. (2018). Perceptions of technological, pedagogical and content knowledge (TPACK) among pre-service teachers in Estonia. *Education and Information Technologies*, 23(2), 741–755.

Muir, T., Muir, B., Hicks, D., Beasy, K., and Murphy, C. (2024). Learning from home during COVID-19: primary school parents' perceptions of their school's management of the home-learning situation. *Cogent Education*, 11(1). https://doi.org/10.1080/2331186X.2024.2308414

Nordin, H., Davis, N., and Ariffin, T. F. T. (2013). A case study of secondary pre-service teachers' technological pedagogical and content knowledge mastery level. *Procedia - Social and Behavioral Sciences*, 103, 1–9. https://doi.org/10.1016/j.sbspro.2013.10.300

Pokhrel, S., and Chhetri, R. (2021). A literature review on impact of COVID- 19 pandemic on teaching and learning. *Higher Education for the Future*, 8(1), 133–141. https://doi.org/10.1177/2347631120983481

Rasmitadila, R., Aliyyah, R. R., Rachmadtullah, R., Samsudin, A., Syaodih, E., Nurtanto, M., and Tambunan, A. R. S. (2020). The perceptions of primary school teachers of online learning during the COVID-19 pandemic period: A case study in Indonesia. *Journal of Ethnic and Cultural Studies*, 7(2), 90–109. https://www.jstor.org/stable/48710085

Rintaningrum, R. (2023). Technology integration in English language teaching and learning: Benefits and challenges. *Cogent Education*, 10(1). https://doi.org/10.1080/2331186X.2022.2164690

Scherer, R., Howard, S. K., Tondeur, J., and Siddiq, F. (2021). Profiling teachers' readiness for online teaching and learning in higher education: Who's ready? *Computers in Human Behavior*, 118, 106675. https://doi.org/10.1016/j.chb.2020.106675

Stronge, J. H. (2018). *Qualities of effective teachers. Alexandria*, VA: ASCD.

Tondeur, J., Scherer, R., Siddiq, F., and Baran, E. (2020). Enhancing pre-service teachers' technological pedagogical content knowledge (TPACK): A mixed-method study. *Educational Technology Research and Development*, 68(1), 319–343. https://www.jstor.org/stable/48727440

Tsui, A.B.M., and Tavares, N.J. The technology cart and the pedagogy horse in online teaching. (2021). *English Teaching and Learning* 45, 109–118. https://doi.org/10.1007/s42321-020-00073-z

Voogt, J., Fisser, P., Pareja Roblin, N., Tondeur, J., and van Braak, J. (2013). Technological pedagogical content knowledge - A review of the literature. *Journal of Computer Assisted Learning*, 29(2), 109–121. https://doi.org/10.1111/j.1365-2729.2012.00487.

Towards Resilient Societies: The Synergy of Religion, Education, Health, Science, and Technology – Rahiem (Ed.)

Four decades of Islamic environmental teachings: A bibliometric and content analysis

Trisia Megawati Kusuma Dewi
Universitas Islam Negeri Syarif Hidayatullah Jakarta, Indonesia
ORCID ID: 0009-0000-3953-3291

Rena Latifa
Universitas Islam Negeri Syarif Hidayatullah Jakarta, Indonesia
ORCID ID: 0000-0001-7615-7416

ABSTRACT: This research integrates the Sustainable Development Goals (SDGs) of quality education, climate action, and sustainable development by emphasising the role of a strong theological foundation in promoting quality education. Using a combination of bibliometric and content analysis, the study offers a comprehensive assessment of this field's evolution, highlighting major themes, key publications, and emerging trends over the past four decades (1983–2024). A total of 185 articles from the Scopus database were analysed, resulting in the development of bibliometric networks that map the core literature on this topic. The findings reveal those Islamic teachings—centred on stewardship, interconnectedness, and moderation—have significant potential to support global sustainability efforts, especially in Muslim-majority countries. The study underscores the need to integrate religious, ethical, and scientific perspectives to address environmental challenges effectively. It advocates for embedding Islamic principles into sustainable development strategies to promote inclusive and environmentally conscious climate action. Despite limitations, such as reliance on the Scopus database and theme clustering based on bibliographic data, this study lays the groundwork for future research exploring the intersection of ethics, education, and sustainability, contributing to global resilience and equitable progress.

Keywords: bibliometric analysis, content analysis, environmental ethics, Islamic environmentalism, Islamic teaching

1 INTRODUCTION

Muslim environmental stewardship principles have been studied further in the last 40 years. According to Nurhasanah *et al.* (2023), the Quran and Hadith require humans to protect nature as a divine responsibility. This increase in the study has examined how Islamic values might inspire modern environmental practices and address climate change, biodiversity loss, and pollution (Zamhari 2022). Additionally, several Islamic teachings have influenced the process of environmental sustainability, such as Islamic teachings requiring its followers to continue maintaining environmental sustainability. Nevertheless, no research has combined bibliometrics, text mining, and content analysis with Islamic environmental teachings. This study captures the scientific underpinning of research on this topic, identifies significant themes and trends over the past forty years, and creates an inclusive research landscape to benefit environmental policymakers and practitioners. Islamic beliefs on stewardship and trust underpin environmental management (Abdelzaher *et al.* 2019). However, cultural and

DOI: 10.1201/9781003645542-54

geographical differences, fast modernisation and secularisation in some Muslim-majority nations, and political and economic pressures make interpreting and applying these principles difficult (Rahim and Rahiem 2013). Furthermore, there is a lack of systematic approaches to integrating Islamic environmental precepts with contemporary science and sustainability frameworks, resulting in fragmented implementation. Integrating Islamic environmental precepts with current science and sustainability is difficult. Interdisciplinary collaboration between religious scholars, scientists, and policymakers is needed (Subchi *et al.* 2022). Climate change also requires cross-cultural and religious collaboration. For environmental stewardship, Islamic teachings must align with global sustainability efforts (Mathevet *et al.* 2018). The contributions of Islamic scholars to environmental policy are often overlooked.

2 METHODOLOGY

This quantitative and qualitative study examined Islamic environmental teachings. Figure 1 depicts the full study design. Bibliographic and content analysis are used to review Islamic environmental literature. These studies illuminate Islamic ecology's ethical principles, focusing on stewardship, interconnectedness, and moderation and their practical implementations (Setianingrum *et al.* 2024). This methodology shows how Islamic environmental beliefs affect behaviour and outcomes and identifies key principles for Eco-Islam implementation (Abdelzaher *et al.* 2019). As in Jia and Jiang (2018) and Schöggl *et al.* (2020). Articles were clustered using bibliographic coupling analysis. After that, the most influential writings in each cluster were analysed qualitatively to examine Islamic environmental theories. This study demonstrates how content analysis can complement bibliometric findings, bridging quantitative patterns with qualitative depth. This integration provides a more nuanced understanding of the literature, revealing Islamic environmental theory's ethical and practical dimensions and potential contributions to sustainability practice.

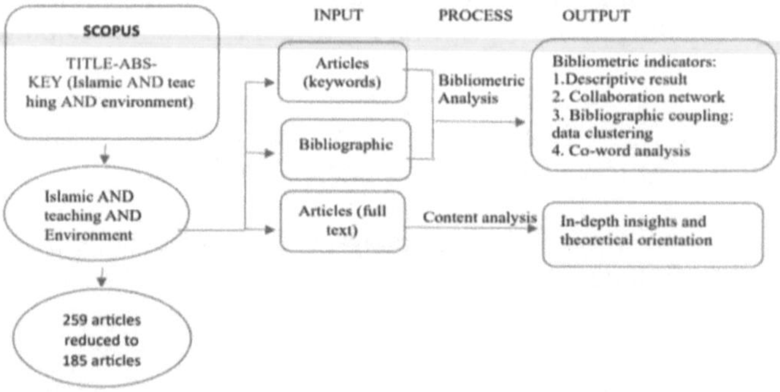

Figure 1. Research framework design.

To ensure comprehensive coverage of research on Islamic environmental teachings, data was gathered from Scopus using the search phrase TITLE-ABS-KEY (Islamic AND teaching AND environment), retrieving 259 papers from 1983 to 2024. Refining the dataset to 50 relevant papers involved applying specific filters related to the topic, which mentioned Islamic environmental teachings. Additionally, this analysis focuses on English and Bahasa Indonesia papers to reduce the risk of ignoring non-English contributions, especially from Muslim-majority regions, when depending solely on English-language publications. Bibliometric analysis quantified literature through publication frequency, citation patterns, and authorship, with final journal articles analysed using methods from Jia and Jiang (2018) and Schöggl *et al.* (2020) for classification and qualitative examination.

3 FINDING AND DISCUSSION

The bibliometric analysis was conducted using VOSviewer version 1.6.20 to examine the development of research on Islamic teachings related to environmental issues. The analysis focused on author keywords to identify research hotspots, performing a co-occurrence analysis with 55 of the 71 retained keywords. The resulting co-occurrence network showed the frequency and connections of keywords, with larger circles indicating more frequent occurrences and thicker lines representing stronger connections. Keywords related to Islamic teaching, education, religion, Islam, and learning environment were the most frequent and interconnected. Revealing three distinct groups: Islamic Teaching perspectives on the environment, Islamic Teaching conceptualisation and implementation, and Islamic Teaching approach to the environment to help identify current research frontiers and attractive investigation areas in Islamic environmental teachings.

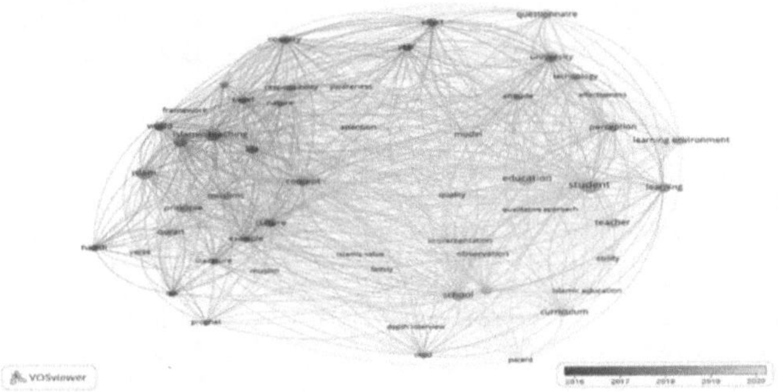

Figure 2. Overlay visualisation of the publication (2016-2020).

The analysis focused on author keywords to identify research hotspots, performing a co-occurrence analysis with 55 of the 71 retained keywords. The resulting co-occurrence network showed the frequency and connections of keywords, with larger circles indicating more frequent occurrences and thicker lines representing stronger connections. Keywords related to Islamic teaching, education, religion, Islam, and learning environment were the most frequent and interconnected. Keywords about Islamic teaching and religion were found to be older, while recent research has focused more on education, models, and learning environments, as seen in Figure 2. Scopus data also highlights the importance of citation counts in identifying influential papers (Merigó *et al.* 2015). Table 1 shows that the study by Abdelzaher *et al.* (2019) on Islamic teaching has the highest citations, totalling 45. Ghafory-Ashtiany (2009) and Mir *et al.* (2015) follow with 36 citations each. The most cited papers focus on Islamic teaching views and implementation (Scopus).

Table 1. Top 10 articles Islamic teaching on environment.

Authors	Title	Year	Source title	Cited by	Author Keywords	Publisher
Abdelzaher D.M.; Kotb A.; Helfaya A.	Eco-Islam: Beyond the Principles of Why and What, and Into the Principles of How	2019	Journal of Business Ethics	45	Eco-Islam; Environment; Individual responsibility; Islam; Islamic environmental responsibility; Sustainability	Springer Netherlands

(continued)

Table 1. Continued

Authors	Title	Year	Source title	Cited by	Author Keywords	Publisher
Ghafory-Ashtiany M.	View of Islam on earthquakes, human vitality and disaster	2009	Disaster Prevention and Management: An International Journal	36	Disasters; Earthquakes; Human physiology; Islam; Risk management	Emerald Publishing
Mir G.; Meer S.; Cottrell D.; McMillan D.; House A.; Kanter J.W.	Adapted behavioural activation for the treatment of depression in Muslims	2015	Journal of Affective Disorders	35	Behavioural activation; Culture Psychotherapy; Depression; Muslim; Religion	Elsevier
Bukhari S.A. A.; Hashim F.; Amran A. B.; Hyder K.	Green Banking and Islam: two sides of the same coin	2020	Journal of Islamic Marketing	33	Environment, Social and Governance (ESG); Green banking; Green Islamic banking; Green Muslims; Islamic corporate social responsibility (ICSR)	Emerald Group Holdings Ltd.
Rana M.H.; Malik M.S.	Impact of human resource (HR) practices on organizational performance: Moderating role of Islamic principles	2017	International Journal of Islamic and Middle Eastern Finance and Management	28	Employee training; HRM practices; Islamic work ethic; Labor compensation; Performance evaluation	Emerald Group Publishing Ltd.
McKay J.E.; Mangunjaya F.M.; Dinata Y.; Harrop S. R.; Khalid F.	Practise what you preach: A faith-based approach to conservation in Indonesia	2014	ORYX	27	Biodiversity; Customary law; Deforestation; Ecosystem services; Indonesia; Islam; REDD; Religion	Cambridge University Press
Farooq S.; Ansari Z.I.	Wastewater reuse in Muslim countries: An Islamic perspective	1983	Environmental Management	24	Islamic law; Religion and environment; Wastewater recycling	Springer-Verlag
Md.Sapir @ Md.Shafik A. S.; Wan Ahmad W.M.	Financial literacy among Malaysian Muslim undergraduates	2020	Journal of Islamic Accounting and Business Research	17	Financial literacy; Financial well-being; Muslims; Undergraduates	Emerald Group Holdings Ltd.
Torabi M.; Noori S.M.	Religious leaders and the environmental crisis using knowledge and social influence to counteract climate change	2019	Ecumenical Review	13	Environment; Islam; Religious events; Religious leaders; Sustainability	World Council of Churches
Zainal D.; Zulkifli N.; Saleh Z.	Corporate social responsibility reporting in Malaysia: A comparison between Shariah and non-Shariah approved firms	2013	Middle East Journal of Scientific Research	13	Comparative study; Corporate social responsibility reporting; Shariah approved firms	International Digital Organization for Scientific Information (IDOSI)

3.1 *Discovering most influence and productive author*

Existing environmental Islamic teaching studies focus on important issues. Social science (25 articles), art and humanities (18 articles), and environmental science (9 articles) are the most common Islamic teaching areas, according to Scopus. According to Table 2, House Allan Oliver is the most influential and industrious author with 18,494 citations from 338 publications.

Table 2. The most influential and productive authors in the Islamic Teaching research towards the environment.

Most influence author			Most productive author		
Rank	Author	Total citation	Rank	Author	Total publication
1	House Allan Oliver	18,494	1	House Allan Oliver	338
2	Kanter Jonathan William	4,040	2	Ramirez Andres Alexis	153
3	Amran Azlan Bin	3,914	3	Amran Azlan Bin	132
4	Cottrell David J	3,255	4	Kanter Jonathan William	107
5	Ghafory Ashtiany, Mohsen Bukhari	1,892	5	Ghafory Ashtiany, Mohsen Bukhari	97
6	Harrop Stuart Ro	1,170	6	Cottrell David J	93
7	Hammid Ali Thaeer	1,250	7	Hammid Ali Thaeer	92
8	Ramirez Andres Alexis	763	8	Abdullah Alias Bin	81
9	Abdullah Alias Bin	713	9	Farooq Shaukat	44
10	Farooq Shaukat	678	10	Harrop Stuart Ro	39

3.2 *Islamic teachings on environmental: content analysis approach*

The content analysis explored "Eco-Islam" through Qur'anic and prophetic teachings, highlighting themes such as ethical duty and stewardship (Abdelzaher *et al.* 2019). Islamic teachings stress humanity's duty in balancing and moderating environmental effects, addressing concerns such as overexploitation and climate change that correspond with the SDGs of sustainable cities and communities (Setianingrum *et al.* 2024). The Qur'an advocates sustainable use and prohibits exploitation, reinforcing a communal responsibility to protect the environment (Kadir *et al.* 2022). Islamic teachings offer valuable insights for aligning with global sustainability goals and underscore the need for community involvement and practical environmental actions (Abdelzaher *et al.* 2019). Integrating Islamic environmental ethics with contemporary environmental policies could enhance their effectiveness. While Islamic teachings provide a robust ethical framework for environmental stewardship, the practical application is often hindered by economic and historical factors (Mangunjaya 2023). Bridging this gap requires leveraging Islamic principles in policymaking and utilising Islamic finance to support climate action. This approach could foster greater alignment with global sustainability efforts and promote more effective environmental conservation practices within Muslim communities (Kadir *et al.* 2022). Furthermore, Islamic texts highlight the importance of water resource management, environmental justice, plant conservation, biodiversity, sustainability, and environmental stewardship, making them a culturally relevant resource for environmental health interventions in Muslim communities, making them a culturally appropriate resource for environmental health interventions in Muslim communities (Aboul-Enein 2018). The Islamic position on human-nature interaction promotes responsible stewardship and the intrinsic value of nature and humans, contrasting anthropocentric ideologies and emphasising Islamic environmental ethics' balance and inclusiveness (Bayat 2023). While the studies give useful insights, it is crucial to emphasise that the bibliometric and content analyses are restricted to certain study areas and may not cover the whole range of Islamic teachings on environmental stewardship, including the possibility of gaps in the existing literature, which necessitates additional research (Helfaya *et al.* 2018; Nurhasanah *et al.* 2023; Mi'raj and Ulev 2024). Moreover, cultural differences make it difficult to apply Islamic environmental teachings in a non-Muslim culture, but these barriers can be overcome via interfaith discussion and shared sustainability values. The findings show that policymakers can integrate Islamic environmental ethics into sustainability policies, and educators can use these teachings to promote sustainability and environmental responsibility in different groups.

4 CONCLUSION

In conclusion, the emphasis on environmental protection in Islam underscores the significance of integrating ecological principles into human actions. Future research opportunities lie in conducting comparative studies with other religious frameworks to broaden our understanding of environmental ethics across diverse belief systems. Expanding the analysis to encompass various regions and languages will also enhance the inclusivity of research findings, revealing how Islamic teachings on environmental stewardship are interpreted and implemented globally. By exploring these avenues, we can foster deeper interdisciplinary collaborations among religious scholars, environmental scientists, and policymakers, ultimately contributing to a more sustainable future aligned with Islamic principles. This multifaceted approach will enrich the discourse on sustainability and highlight the substantial potential contributions of Islamic teachings to global environmental initiatives.

REFERENCES

Abdelzaher, D. M., Kotb, A., and Helfaya, A. (2019). Eco-Islam: Beyond the Principles of Why and What, and Into the Principles of How. *Journal of Business Ethics*, 155(3), 623–643. https://doi.org/10.1007/s10551-017-3518-2

Aboul-Enein, B. H. (2018). "The earth is your mosque": narrative perspectives of environmental health and education in the Holy Quran. *Journal of Environmental Studies and Sciences*, 8(1), 22–31. https://doi.org/10.1007/s13412-017-0444-7

Bayat, M. (2023). A Study of Religious Perspectives on Biodiversity and the Environment. *Journal of Wildlife and Biodiversity*, 7(Special Issue), 74–87. https://doi.org/10.5281/zenodo.10206928

Ghafory-Ashtiany, M. (2009). View of Islam on earthquakes, human vitality and disaster. *Disaster Prevention and Management: An International Journal*, 18(3), 218–232. https://doi.org/10.1108/09653560910965600

Helfaya, A., Kotb, A., and Hanafi, R. (2018). Qur'anic Ethics for Environmental Responsibility: Implications for Business Practice. *Journal of Business Ethics*, 150(4), 1105–1128. https://doi.org/10.1007/s10551-016-3195-6

Jia, F., and Jiang, Y. (2018). Sustainable global sourcing: A systematic literature review and bibliometric analysis. *Sustainability (Switzerland)*, 10(3), 1–26. https://doi.org/10.3390/su10030595

Kadir, F. K. A., Aziz, N. A. A., Rusli, N., Rahman, A. A., Embong, A. H., and Salamun, H. (2022). The Role of Human Beings asCaliphs in Preserving Naturein this Internet ERA. *Journal of Positive School Psychology*, 6(3), 1681.

Mangunjaya, F. M. (2023). How Islamic Tradition Benefits Nature and Climate Change Action. *Journal of the Siam Society*, 111(2), 253–264.

Mathevet, R., Bousquet, F., and Raymond, C. M. (2018). The concept of stewardship in sustainability science and conservation biology. *Biological Conservation*, 217(March 2017), 363–370. https://doi.org/10.1016/j.biocon.2017.10.015

Merigó, J. M., Mas-Tur, A., Roig-Tierno, N., and Ribeiro-Soriano, D. (2015). A bibliometric overview of the Journal of Business Research between 1973 and 2014. *Journal of Business Research*, 68(12), 2645–2653. https://doi.org/10.1016/j.jbusres.2015.04.006

Mi'raj, D. A., and Ulev, S. (2024). A bibliometric review of Islamic economics and finance bibliometric papers: an overview of the future of Islamic economics and finance. *Qualitative Research in Financial Markets*. https://doi.org/10.1108/QRFM-03-2023-0068

Mir, G., Meer, S., Cottrell, D., McMillan, D., House, A., and Kanter, J. W. (2015). Adapted behavioural activation for the treatment of depression in Muslims. *Journal of Affective Disorders*, 180, 190–199. https://doi.org/10.1016/j.jad.2015.03.060

Nurhasanah, N., Rusydiana, A. S., Harahap, D., and Pratomo, W. A. (2023). The Role of Islamic Social Finance on Environmental Issue: A Review. *Review of Economics and Finance*, 21(2020), 2575–2583. https://doi.org/10.55365/1923.x2023.21.275

Rahim, H., and Rahiem, M. D. H. (2013). The Use of Stories as Moral Education for Young Children. In *International Journal of Social Science and Humanity* (pp. 454–458). EJournal Publishing. https://doi.org/10.7763/ijssh.2012.v2.145

Schöggl, J. P., Stumpf, L., and Baumgartner, R. J. (2020). The narrative of sustainability and circular economy - A longitudinal review of two decades of research. *Resources, Conservation and Recycling*, 163(April), 105073. https://doi.org/10.1016/j.resconrec.2020.105073

Setianingrum, D. A., Setiyo, and Dwiyanto, A. (2024). Environmental Education through Islamic Lens: Values and Practices. *E3S Web of Conferences*, 482. https://doi.org/10.1051/e3sconf/202448204014

Subchi, I., Zulkifli, Z., Latifa, R., and Sa'diyah, S. (2022). Religious Moderation in Indonesian Muslims. *Religions* 13: 451. https://doi.org/10.3390/rel13050451

Zamhari, A. (2022). *The Importance of Science and Technology Education in the Muslim worlds*. PIPS. Pakistan.

Towards Resilient Societies: The Synergy of Religion, Education, Health, Science, and Technology – Rahiem (Ed.)

Islamic perspectives on Muslim behaviour and global climate change

Ubaidillah Ubaidillah, Yusuf Rahman, Arif Sumantri & Maswani Maswani
Universitas Islam Negeri Syarif Hidayatullah Jakarta, Indonesia

ABSTRACT: Indonesia, home to one of the world's largest forest areas spanning 933,000 km^2, faces significant environmental challenges due to rapid economic development, resulting in an annual deforestation rate of 10% over the past five years. This study examines the influence of Muslim behaviour within the Islamic ecological paradigm. Employing a qualitative methodology, the research explores the interaction between *dhikr* (remembrance of God) as a foundation for virtuous behaviour (*Ihsan*) in human actions. Through descriptive analysis grounded in a phenomenological framework, the study highlights *Akhlaq* (ethics) principles derived from the Qur'an and environmental *fiqh* (Islamic jurisprudence). Findings reveal that Islamic behavioural principles can be crucial in addressing environmental degradation. Ecological policy interventions guided by Islamic teachings offer a comprehensive approach to environmental governance, aligning with Sustainable Development Goal (SDG) 13 (Climate Action). The study demonstrates how Islamic ecological values and behaviour modification can support climate change mitigation through forest conservation. It also recommends integrating environmental management into broader climate action frameworks and fostering positive behavioural changes in youth to promote health and sustainable environmental conservation.

Keywords: climate, environment, ethics, ecology

1 INTRODUCTION

The widespread forest fires have not only disturbed the global climate balance but have also shown a profound lack of awareness of the implications of climate change among Muslim communities, especially in the areas of waste management and environmental management (Manuel Oyarzún *et al.* 2021). According to data from FAO (2018), Indonesia ranks eighth with an area of 933,000 km^2, with Papua and East Kalimantan having the largest forest reserves in the country. However, this vast expanse of forest is increasingly under pressure from economic development and population growth (Figures 1 and 2).

The impact of deforestation in Indonesia significantly exacerbates the phenomenon of global climate change (Barau 2021). Contemporary forest protection frameworks often show shortcomings in aligning economic activities with forest rehabilitation needs, posing multifaceted challenges to environmental sustainability. An empirical investigation conducted by Duke University (2019), together with data from the Food and Agriculture Organization (FAO), revealed the rate of deforestation in Indonesia, with an estimated loss of around 50 hectares of forest every day since 2007, making Indonesia the country with the highest rate of forest degradation in the world. This accelerating deforestation, driven by industrial progress, poses a major threat to the balance of the local and global environment. The primary function of forests goes beyond carbon sequestration, encompassing a multitude of essential ecosystem services, including water purification, pollution mitigation, climate modulation, and natural disaster prevention (Jasin and Aulia 2022). Seyyed Hossein Nasr

DOI: 10.1201/9781003645542-55

elaborated on this view in his 1992 publication, "Man and Nature: The Spiritual Crisis of Modern Man," which describes nature as sacred and of intrinsic value. Nasr attributes the contemporary ecological crisis to the fading spiritual awareness of humans towards nature (Sayem 2021). This view starkly contrasts the theory of Anthropocentric Ethics, which advocates uncontrolled human domination over the environment, a view that has no theological justification (Nurul Huda *et al.* 2023).

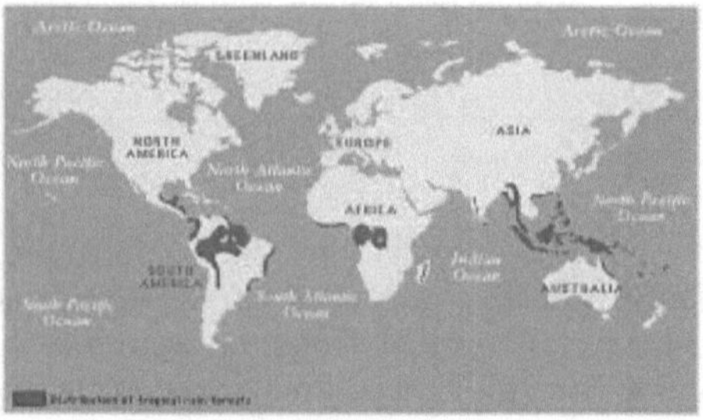

Figure 1. Map of the largest tropical forests coverage.

Figure 2. Illustration of deforestation in Kalimantan.

Social problems or stigma of mental illness often interfere with people with mental illness that can damage the environment that occurs by exploring perceptions and attitudes towards individuals with mental illness among adolescents in Indonesia. This is because some who damage the environment have a high school education background (Subu *et al.* 2024). This scientific research aims to bridge the gap between Islamic doctrine on environmental conservation and pragmatic steps to combat deforestation and climate change in Indonesia. Given the challenges above, this study focuses on three main research questions: a. How can Islamic ecological policy interventions strengthen environmental management in line with the principles inherent in the creation of Allah SWT? b. How important is the media of da'wah communication in fostering local wisdom values related to environmental conservation? c. How can increased productivity, combined with improvements in behavioural and belief systems, produce good results for environmental health?

2 METHOD

Ritzer stated that in research related to Ecology, the dimensions of social reality are described in four quadrants. This study adopts a qualitative methodology and is carried out as field research, where researchers serve as the main instrument for examining natural phenomena (Sorensen *et al.* 2021). The data validation process is done through a systematic methodology with data sources, which involves cross-referencing interviews with religious authorities, community constituents, and environmental specialists. A compares observational findings with documented practices and validates field observations against Islamic textual sources (Husamah *et al.* 2022).

The influence of the rise of Islamic conservatism has not been the main focus of the study of Islamic conservatism in Indonesia as a determining factor in policy. Studies on Islamic conservatism have so far emphasized the national political aspect, which illustrates that there has been a change in the dynamics of Islam towards conservatism, including actions towards environmental preservation so that environmental damage resulting in climate change is still quite large in Indonesia (Zulfadli *et al.* 2024).

Methodological Triangulation combines data from interviews, direct observations, and document review, ensuring consistency between verbal narratives and observed actions.

Furthermore, analytical triangulation includes collaborative analysis by multiple researchers to reduce individual bias and engage in peer review of interpretive conclusions (Effendi and Ahmad Nasir 2020).

This investigation uses a descriptive-analytical methodology (Creswell and Creswell 2018), emphasising the importance of contextual understanding over broad generalisations in qualitative research findings. Qualitative inquiry investigates human experiences and interactions to understand their linguistic expressions and perceptual frameworks. Two key determinants influenced the configuration of this research design: First, the exploration of acculturation requires extensive field data and second, tradition embodies social constructs relevant to this study (Hargreaves *et al.* 2019).

This methodology includes a comprehensive literature synthesis, examining various texts, scholarly articles, and historical records relevant to Islamic doctrine, particularly emphasising principles derived from the Qur'an and Hadith (Durachman *et al.* 2021). The examination interprets and explains the application of these principles in the context of addressing global climate change (Abdellah 2020). Furthermore, this study actualises Muslim behavioural practices aimed at maintaining environmental wisdom and places these behaviours within the framework of Islamic Ecology (Ragozina 2023). The Social Phenomenology Paradigm underlines the importance of *Akhlak* and *Rahmad* in Islamic Ecology, as articulated in the Qur'an, environmental Fiqh, and Islamic intellectual heritage. The data analysis is iterative, including data reduction and coding, thematic identification and categorisation, pattern analysis and interpretation, and validation through triangulation and synthesis of findings in line with Islamic ecological principles (Mohamed 2014).

3 RESULTS AND DISCUSSION

This section investigates the interrelations among Islamic doctrines, environmental stewardship, and productivity within the paradigm of global climate change. To enhance clarity and organisation, the findings are systematically categorised into three thematic domains: Environmental Ethics in Islam, Principles of Islamic Ecology, and Practical Implications, each supplemented with specific instances that exemplify the application of Islamic ecological principles.

3.1 *Environmental ethics in islam*

Islam advocates a profound commitment to environmental ethics, rooted in the Qur'an and Hadith teachings. Several fundamental ethical tenets govern environmental stewardship: the notion of Khalifah (stewardship), which assigns humans the role of custodians of the earth, accountable for its nurture and preservation, rather than as absolute proprietors (Sorensen *et al.* 2021). The principle of *Tauhid* (Oneness of God) positions Allah SWT as the creator and sovereign of the cosmos, thereby instilling a sense of respect and accountability towards His creation. *Ihsan* (excellence or benevolence) prompts Muslims to engage with nature empathetically and to refrain from harmful practices. Lastly, the principle of Mizan (balance) underscores the importance of maintaining harmony between human necessities and environmental sustainability, denouncing excessive exploitation and pollution as manifestations of injustice. These principles collectively furnish a solid ethical framework for the sustainable management of environmental resources.

3.2 *Principles of Islamic ecology*

Islamic ecology aims to comprehend and solve environmental issues by applying Islamic teachings. The interdependence of all species created by Allah SWT and the idea that harm to one ecosystem component affects the entire system are important tenets of this paradigm.

The idea of Amanah (trust) and responsibility highlights the obligation of humans to govern the planet properly, which includes preventing environmental deterioration and using resources wisely (Rakhmat 2022). Recognising the importance of Indigenous knowledge, Islamic ecology also encourages blending Islamic teachings with traditional environmental management techniques and local expertise.

These guidelines provide a thorough foundation for comprehending the interaction between humans and nature and encouraging environmentally friendly behaviours. Specific examples of applying Islamic ecological principles: *Ihsan* in Waste Management. The principle of *Ihsan* can be applied by encouraging efforts to minimise waste generation, practising recycling and composting, and disposing of waste responsibly to prevent pollution and environmental damage. This is a form of proactive approach to environmental awareness to avoid damage. *Amanah* in Resource Consumption. The *Amanah* concept guides the avoidance of wasteful consumption of resources. For example, saving water, using energy efficiently, and consuming food responsibly help humans manage resources sustainably.

Protection of Biodiversity as *Amanah.* Protecting biodiversity can be seen as upholding the *Amanah* to preserve God's creation in various forms. This principle is a conservation effort to protect endangered species and their habitats by recognising the intrinsic value of all living things. Media for Da'wah Communication also plays a very important role. For example, The #HijaukanMasjid campaign has succeeded in increasing environmental awareness among the Muslim community through social media platforms. The campaign leverages digital channels to disseminate information about environmentally friendly practices relevant to everyday life, such as reforestation, energy conservation and tree planting.

3.3 *Practical implications*

The ethical and ecological principles outlined have considerable practical implications for tackling environmental challenges, particularly global climate change. This research reveals several important implications. Da'wah, or the propagation of Islam, is instrumental in enhancing public awareness regarding environmental stewardship. Through diverse media channels, Da'wah can effectively disseminate messages concerning environmental ethics and sustainable practices. Additionally, the Islamic notion of productivity encompasses quantity, quality, efficiency, and job satisfaction (Sayem 2021). Applying Islamic human resource and environmental management principles can promote sustainable work productivity (Karimi *et al.* 2020).

Lastly, character education is grounded in Islamic values, as demonstrated by the teachings of KH. Hasyim Asy'ari, is vital for fostering environmentally responsible behaviours (Awasthi *et al.* 2021; Rassool 2021). These insights indicate that integrating Islamic principles can significantly address environmental issues and advance sustainable development. In addition, the solution to the growth of Islamic religious education in Indonesia is pluralistic. It is important, based on the principles of positive pluralism, that they can hold on to their beliefs while at the same time understanding, accepting, and being involved in handling environmental preservation even though they have different beliefs (Mu'ti 2023).

This study is closely related to SDG 11 (Sustainable Cities and Communities) through an in-depth study of the contribution of Muslim communities in building sustainable cities. By exploring best practices based on Islamic teachings, this study provides a real contribution to achieving the goal of inclusive, safe, resilient, and sustainable urban development with reforestation and environmental management in the Indonesian Capital City (IKN) in Kalimantan. In addition, in line with SDG 4 (Quality Education), this study emphasises the importance of quality education in shaping pro-environmental awareness and behaviours among Muslim individuals. Integrating Islamic values related to environmental conservation into the educational curriculum at Islamic Educational Institutions is hoped to produce a young generation who cares about the environment and can take real action in dealing with climate change.

4 CONCLUSIONS AND RECOMMENDATIONS

The study results indicate that Islamic ecological principles, such as Khalifah and the prohibition of *ifsad fil-ardh*, build a strong ethical foundation for environmental conservation and climate change mitigation. The integration of *Akhlaq* and Rahmah in the behaviours of Muslims, supported by da'wah media, has proven effective in fostering sustainable practices and increasing environmental awareness. Policymakers are urged to integrate Islamic ecological principles into the Environmental Law Update and promote interdisciplinary collaboration. In addition, educators are encouraged to incorporate Islamic environmental principles into educational programs through innovative resources. The implications of this study are in line with Sustainable Development Goal 13 (Climate Action) and also illustrate the significant results of achieving SDG 12 (Responsible Consumption and Production) by consuming responsible products for sustainable lifestyles and reducing waste, as well as achieving Sustainable Development Goal 15 (Life on Land) through biodiversity conservation and forest rehabilitation. This ensures that increasing awareness and practices based on Islamic values contribute significantly to climate change mitigation and achieving sustainable development goals.

REFERENCES

Abdellah, A. (2020). The Islamic Declaration On Global Climate Change; An Ideological Discourse Analysis. *Middle Eastern Journal Of Research In Education And Social Sciences, 1*(2), 77–93. Https://Doi.Org/10.47631/Mejress.V1i2.66

Awasthi, A. K., Cheela, V. R. S., D'adamo, I., Iacovidou, E., Islam, M. R., Johnson, M., Miller, T. R., Parajuly, K., Parchomenko, A., Radhakrishan, L., Zhao, M., Zhang, C., and Li, J. (2021). Zero Waste Approach Towards A Sustainable Waste Management. *Resources, Environment And Sustainability, 3* (January), 100014. Https://Doi.Org/10.1016/J.Resenv.2021.100014

Barau, A. S. (2021). *Environment And Sustainable Development In The Quran And Sustainable Development In The International Institute Of Islamic Thought. July.*

Creswell, J. W., and Creswell, J. D. (2018). Mixed Methods Procedures. In *Research Defign: Qualitative, Quantitative, And Mixed M Ethods Approaches.*

Durachman, Y., Sean Bein, A., Purnama Harahap, E., Ramadhan, T., and Putri Oganda, F. (2021). Technological And Islamic Environments: Selection From Literature Review Resources. *International Journal Of Cyber And It Service Management, 1*(1), 37–47. Https://Doi.Org/10.34306/Ijcitsm.V1i1.12

Effendi, M. R., and Ahmad Nasir, M. (2020). Forest Conservation Base on Religious Values: *'Abqari Journal, 22*(2), 81–94. https://doi.org/10.33102/abqari.vol22no2.324

Hargreaves, S., Rustage, K., Nellums, L. B., Mcalpine, A., Pocock, N., Devakumar, D., Aldridge, R. W., Abubakar, I., Kristensen, K. L., Himmels, J. W., Friedland, J. S., and Zimmerman, C. (2019). Occupational Health Outcomes Among International Migrant Workers: A Systematic Review And Meta-Analysis. *The Lancet Global Health, 7*(7), E872–E882. Https://Doi.Org/10.1016/S2214109x(19)30204-9.

Husamah, H., Suwono, H., Nur, H., and Dharmawan, A. (2022). Innovation integration of Islamic values in learning environmental themes in biology education. *JINoP (Jurnal Inovasi Pembelajaran), 8*(1), 1–19. https://doi.org/10.22219/jinop.v8i1.19565

Jasin, F. M., and Aulia, R. N. (2022). Disaster Mitigation Curriculum in Pesantren in Industry 4.0. *Proceedings of the International Seminar on Language, Education, and Culture (ISoLEC 2021), 612.* https://doi.org/10.2991/assehr.k.211212.019

Karimi, A., Barkhordari, A., Saranjam, B., Abazari, M., and Babaei-Pouya, A. (2020). The Effects Of Implementing An Occupational Health And Safety Management System On Functional Indices: A Five-Year Study In Casting Industry. *Malaysian Journal Of Medicine And Health Sciences, 16*(3), 8–14.

Manuel Oyarzún, G., Fernando Lanas, Z., Marcelo Wolff, R., and Arnoldo Quezada, L. (2021). The Impact Of Climate Change On Health. *Revista Medica De Chile, 149*(5), 738–746. Https://Doi.Org/10.4067/S0034-98872021000500738

Mohamed, N. (2014). Islamic Education, Eco-ethics and Community. *Studies in Philosophy and Education, 33* (3), 315–328. https://doi.org/10.1007/s11217-013-9387-y

Mu'ti, A. (2023). Pluralistic Islamic Religious Education: A Vision For Indonesia. *Review Of Faith And International Affairs*, *21*(2), 121–127. Https://Doi.Org/10.1080/15570274.2023.2200280

Huda, N., Kuswanjono, A., and Utomo, A. H. (2023). Theological Anthropocentrism: An Interpretation Of Scripture And Semitic Theodicy In Overcoming The Environmental Crisis. *Hayula: Indonesian Journal Of Multidisciplinary Islamic Studies*, *7*(2), 161–184. Https://Doi.Org/10.21009/Hayula.007.02.03

Ragozina, S. A. (2023). Environmentalism in Modern Islamic Philosophy. *RUDN Journal of Philosophy*, *27* (2), 233–250. https://doi.org/10.22363/2313-2302-2023-27-2-233-250

Rakhmat, A. (2022). *Islamic Ecotheology: Understanding The Concept Of Khal Ī Fah And The Ethical Sparked A Discussion On Environmental Issues In " The Historical Roots Having An Anthropocentric View That Causes Environmental Destruction* . It. 7170, 1–24.

Rassool, G. H. (2021). Islamic Psychology: Human Behaviour And Experience From An Islamic Perspective. In *Islamic Psychology: Human Behaviour And Experience From An Islamic Perspective* (Issue March). Routledge. Https://Doi.Org/10.4324/9780429354762

Sayem, M. A. (2021). Islam And Environmental Ethics. *Islamic Studies*, *60*(2), 157–172. Https://Doi.Org/10.52541/Isiri.V60i2.1438

Sorensen, G., Dennerlein, J. T., Peters, S. E., Sabbath, E. L., Kelly, E. L., and Wagner, G. R. (2021). The Future Of Research On Work, Safety, Health And Wellbeing: A Guiding Conceptual Framework. *Social Science And Medicine*, *269*, 113593. Https://Doi.Org/10.1016/J.Socscimed.2020.113593

Subu, M. A., Dias, J. M., Mottershead, R., Ahmed, F. R., Narulita, S., Maryuni, M., Zakiyah, Z., Nurbaeti, I., Mohamed Al Marzouqi, A., and Al-Yateem, N. (2024). Exploring Mental Health Stigma Among Indonesian Healthcare Students Towards Individuals With Mental Illnesses: A Qualitative Study. *International Journal Of Qualitative Studies On Health And Well-Being*, *19*(1). Https://Doi.Org/10.1080/17482631.2024.2327103

Zulfadli, Makruf, J., Zulkifli, and Jahroni, J. (2024). From Islamic Modernism To Islamic Conservatism: The Case Of West Sumatra Provinces, Indonesia. *Cogent Social Sciences*, *10*(1). Https://Doi.Org/10.1080/23311886.2024.2406297.

Towards Resilient Societies: The Synergy of Religion, Education, Health, Science, and Technology – Rahiem (Ed.)
© 2026 The Author(s), ISBN: 978-1-041-08471-6
Open Access: www.taylorfrancis.com, CC BY-NC-ND 4.0 license

Digital leadership: Developing digital literacies in schools in Indonesia for empowering education and advancing sustainable development goals (SDGs)

Ummi Kultsum & Didin Nuruddin Hidayat
Universitas Islam Negeri Syarif Hidayatullah Jakarta, Indonesia

Jee Young Lee
University of Canberra, Australia

ABSTRACT: This study examines digital leadership in Indonesia, focusing on how leaders' interests, motivations, and challenges influence technology integration for online learning in schools. Using a mixed-methods approach, the research assesses the validity and reliability of a digital media literacy framework through an online survey of 577 teachers across Indonesia. The findings reveal that 61.5% of teachers enjoy using technology, and 65.5% are open to embracing its challenges. However, 43.2% face difficulties with technology integration due to limited technical skills and inadequate internet access. Only 23% effectively utilise technology in the classroom, and 25.8% feel comfortable using it. The study highlights significant gaps in digital literacy, underscoring the importance of effective digital leadership, improved communication, and targeted resources to address these challenges. This research is particularly relevant as it sheds light on the specific obstacles teachers encounter in adopting technology within Indonesia's educational context and offers strategies to enhance technology implementation and digital literacy in underdeveloped and remote areas. The study concludes that strong educational leadership and comprehensive support are essential to empower teachers to overcome digital challenges and maximise technology for improved teaching practices.

Keywords: digital leadership, digital literacy, Indonesian school

1 INTRODUCTION

In late 2019, the pandemic affected many areas of society, with education being crucially affected. All schools were forced to shut their doors and switch to distance learning. This caused many teachers to have to rethink many of their lessons, as several of them clearly stated how unprepared they were for this new kind of teaching. This transition was a form of remote teaching and learning driven purely by necessity.

The use of the Internet is not entirely unprecedented in Indonesia. By 2020, Indonesia had approximately 81.87 million smartphone users, placing it among the highest globally for smartphone penetration. This high level of internet accessibility corresponds with substantial engagement in social media. According to APJII (2019), internet users frequently access chat applications (24.7%) and social media platforms (18.9%). Research indicates that online learning presents significant challenges in Indonesia, particularly within schools in geographically and culturally marginalised communities, such as Madrasahs. In these settings, disparities in internet infrastructure and educational resources contribute to difficulties in implementing digital technology for instruction. Rural students often struggle to access online learning platforms (Astini 2020; Hermawan *et al.* 2018).

The issue of digital exclusion in these areas is compounded by complex factors, including local customs, traditional practices, and resistance to new ideologies and technologies (Amin

DOI: 10.1201/9781003645542-56

2018). Moreover, as reported by the World Bank (2020), approximately 10% of Indonesians live in poverty, a figure likely exacerbated by the pandemic. These further impact access to online education for both students and teachers, particularly in remote regions.

In response, the Indonesian government has initiated a phased reopening of schools, contingent upon adherence to the requirements outlined in the Joint Decree of the Minister of Education and Culture No. 03/KB/2021, the Minister of Religious Affairs No. 384 of 2021, the Minister of Health No. HK 01.08/Menkes/4242/2021, and the Minister of Home Affairs No. 440-717 of 2021 (Surat Keputusan Bersama Menteri Pendidikan dan Kebudayaan Nomor 03/KB/2021, Menteri Agama Nomor 384 Tahun 2021, Menteri Kesehatan Nomor HK 01.08/Menkes/4242/2021, and Menteri Dalam Negeri Nomor 440-717 Tahun 2021). However, only 39% of schools meet these criteria (UNICEF 2021). Consequently, education in Indonesia continues to depend heavily on online modalities. Furthermore, the reopening of schools necessitates that teachers are equipped to deliver blended or hybrid learning, as face-to-face instruction is limited to two hours per session, with only 25% of students able to attend each class in person.

Despite the importance of digital literacy in the context of the pandemic, research on digital leadership in Indonesia, especially in remote and underprivileged areas, still needs to be explored. While the existing literature on digital media literacy and leadership emphasises the importance of digital skills in teaching (Amin 2018; UNESCO 2020), few studies have specifically addressed how school leaders can support teachers in marginalised communities to develop these skills and integrate technology effectively into their teaching practices. This study aims to fill this gap by focusing on how digital leadership, particularly at the local government and school principal levels, can promote digital inclusion and empower educators to enhance their digital literacy skills in the face of the ongoing crisis. This study aims to fill this gap by focusing on how digital leadership, particularly at the local government and school principal levels, can promote digital inclusion and empower educators to enhance their digital literacy skills in the face of the ongoing crisis.

Leadership, as a multifaceted field, can be analysed from various perspectives. In this context, leaders are pivotal in influencing and motivating organisational members to follow their guidance. Digital leadership primarily focuses on fostering digital inclusion within institutions and supporting members in effectively using technology to enhance workplace efficiency. This study examines explicitly digital leadership in the context of local government and school principals, aiming to provide digital resources and support for teachers during the global pandemic. The research questions are as follows: 1) What aspects of digital media literacy are necessary for teachers to build resilience in response to the COVID-19 crisis in education? 2) How can school leaders facilitate teachers' positive use of digital media literacy to mitigate the impacts of the COVID-19 crisis?

This study contributes to the broader field of digital leadership by exploring the unique challenges teachers face in remote and underserved areas of Indonesia and by providing insights into how leadership can support the development of digital literacy in these contexts.

2 METHOD

This study employs a mixed methods approach to examine digital leadership and media literacy in Indonesian schools. The qualitative phase involves focus group discussions (FGDs) and interviews with randomly selected teachers and principals from various provinces, ensuring a diverse sample. Twenty teachers participated in the FGDs, and five teachers were interviewed individually.

For the quantitative phase, an online survey was distributed to a broader group of teachers across Indonesia, focusing on digital media literacy and attitudes toward technology in teaching. The survey, mobile-friendly and distributed via WhatsApp, includes questions informed by both the qualitative findings and existing literature on digital literacy. It measures various aspects of digital media literacy, socio-demographic factors, and challenges related to technology integration.

To ensure the validity of the digital literacy framework, the study incorporates a content validation process during the survey design. Expert reviews were conducted to assess the framework's alignment with existing theories of digital literacy, and pilot testing was performed to refine the survey. Additionally, reliability will be measured using Cronbach's alpha to evaluate internal consistency. Statistical descriptive analyses were used to explore relationships between digital literacy and other factors, such as experience with technology, providing insights into the digital literacy landscape in Indonesian schools.

3 FINDINGS AND DISCUSSIONS

As stated before, the survey was conducted on 577 teachers in Indonesia. It consisted of questions about the availability of technology devices for teachers, teacher digital literacy, and digital leadership regarding school and personal readiness to facilitate teachers' and students' online learning.

3.1 *The use of technology in the teaching and learning process*

The use of technology in the teaching and learning process has become increasingly critical during the global pandemic (Figure 1). Teachers must be proficient in integrating technological devices into their lessons. The current study involving 577 teachers found that approximately 89.6% (517 participants) use smartphones for teaching and learning purposes. Teachers recognise that smartphones are essential tools for online education due to their simplicity and ease of use for educators and students. This finding is consistent with Kultsum (2022), who asserts that smartphones are fundamental for enhancing lesson delivery. However, Kultsum also observed that some students struggle with internet access due to limited credit and inadequate Wi-Fi facilities. The study further indicated that teachers utilised various media platforms, including YouTube (79.8%), Kahoot (24.8%), and WordPress (18.5%), believing these tools significantly aid students in comprehending lessons during online learning. Prastiyo and Purnawan (2018) corroborate this, showing that digital learning media such as YouTube and Google Classroom-based e-learning platforms considerably support student achievement compared to those lacking digital resources.

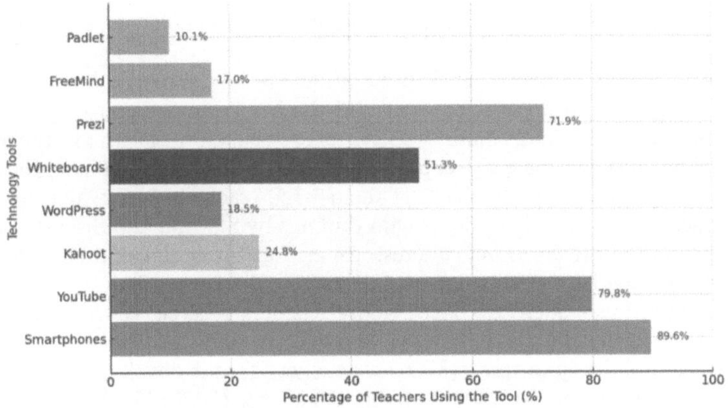

Figure 1. The use of technology in the teaching and learning process.

While many teachers continued to use traditional media like whiteboards (51.3%), they increasingly adopted alternative learning tools to deliver educational content during the pandemic. Some educators explored other applications like Padlet (10.1%), FreeMind (17%), and

Prezi (71.9%), indicating a concerted effort to enhance their ICT skills and find alternative methods to engage with students in an online learning environment. The significant impact of ICT on various educational aspects, including teaching, assessment, mobile learning, and research, highlights its transformative potential in education.

In summary, the pandemic in Indonesia has compelled schools and teachers to seek alternative media and tools to support online teaching and learning. Educators have found that these technologies are effective, simplify the learning process, and sustain it.

3.2 *Digital literacy*

The findings reveal several aspects of teachers‘ digital literacy, such as their interest, motivation, and objectives in employing technology (Figure 2). Many of these aspects align with previous observations that teachers increasingly use alternative media to enhance online learning. Teachers’ high levels of interest and motivation are reflected in their technology use, which may indicate a strong foundation in digital literacy. Despite their growing proficiency, the pandemic has compelled teachers to integrate various media technologies into their teaching, even though they may still experience some challenges in operating these tools effectively.

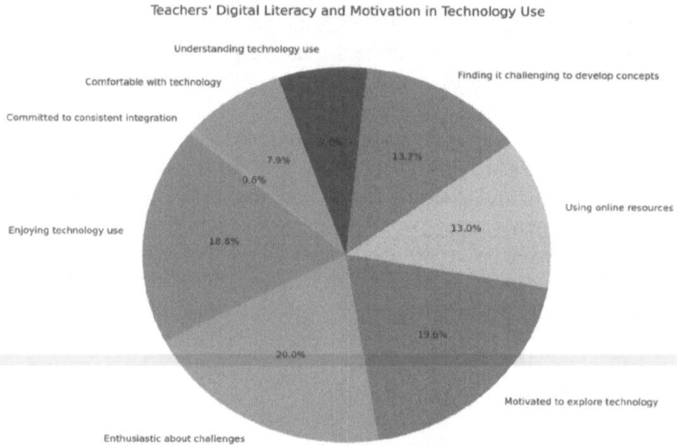

Figure 2. Teachers‘ digital literacy and their motivation to use technology in education.

The data on teachers’ interest and motivation in using technology is encouraging. Approximately 61.5% of the 577 participants, or 355 teachers, reported enjoying using technology in their teaching processes. Additionally, 65.5% expressed enthusiasm for the challenges of operating technology in online classes. Teachers noted that online learning has prompted them to enhance their ICT knowledge, with 64.1% of them—370 teachers—motivated to explore technology use in education further. This aligns with findings by Sharma and Srivastava (2019), who suggest that educators‘ motivation and interest are crucial for optimising ICT use in teaching, potentially improving teachers’ performance and pedagogical expertise.

However, the study also uncovered that many teachers rely heavily on extensive internet access to develop their teaching ideas, with 42.6% (246 teachers) using online resources to prepare their materials. Despite this, 43.2% of teachers found developing learning concepts through the internet and technology challenging. Schleicher (2020) highlights that adapting to new pedagogical approaches and teaching methods, for which many teachers are not formally trained, can be difficult. This is supported by the OECD‘s Teaching and Learning International Survey, which found that 40% of teachers had not received professional development in technology use, and nearly 20% recognised a significant need for additional training.

Furthermore, the study identified that some teachers struggle with their technical skills. Only 23% (138 out of 577) of teachers demonstrated an understanding of how to use technology effectively in the classroom, while just 25.8% (149 teachers) felt comfortable using technology. Additionally, only 1.9% (11 teachers) were committed to integrating technology into their teaching consistently. This data suggests that technology use remains a significant challenge for many teachers, who rely on limited knowledge, making sustained technology integration problematic. These findings are consistent with Kultsum *et al.* (2022), which indicate that issues with technology operation among teachers, students, and parents hinder the effectiveness of online learning.

3.3 *Digital leadership*

Based on the results of the above surveys, solutions are needed to change teachers' perspectives and reach higher fulfilment rates, which is a requirement to drive digitalisation. As a starting point, the most important topics must be addressed.

Therefore, it was defined which leadership aspect should be improved first. This resulted in two main problems which can be focused on. One improvement focus should be to close the digital literacy among teachers because they should have to develop their motivation along with the demand for the use of technology in education. The second improvement focus is on the leadership categories, which could have the highest potential for teachers‘ improvements since they represent the weakest points in digital leadership.

3.4 *Communication as a solution*

As an archipelagic nation, Indonesia experiences significant disparities in the distribution of information, which impacts the mobility and activities of its population. For instance, while technology and information have advanced rapidly in major cities on western islands such as Java and Sumatra—particularly in economic, political, and educational hubs like Jakarta, Bandung, Medan, and Surabaya—access remains limited in the central and eastern islands. This disparity is evident in residents' challenges in these remote areas, particularly in accessing information and technology for educational purposes.

In remote regions, teachers and students often face low levels of digital literacy, hindering their ability to use information technology effectively in teaching and learning. Many in these areas lack essential devices such as laptops, smartphones, or tablets, restricting their engagement with digital educational resources.

Effective communication is crucial for school leaders addressing teachers‘ digital literacy challenges. Principals often struggle to maintain close contact with all staff members due to the demands of managing the madrasah. As a result, principals typically engage directly only with those who are available and willing to meet. Despite this, principals prefer to work in a unified office environment to foster direct interactions with their staff.

Digital communication tools, such as chat rooms, can be utilised to mitigate time constraints. These tools facilitate both internal and external communication. However, communication barriers persist, particularly between principals and teachers. Teachers in urban areas, introduced to technology-based communication earlier, tend to find it easier to interact with their peers. A demographic survey revealed that 47.7% of teachers are located in western Indonesia, with 36% from urban areas. These teachers find that technology platforms like WhatsApp, email, and social media enhance communication and collaboration. Consequently, principals in urban schools face fewer challenges in managing teacher activities during crises.

Nevertheless, the adoption of such tools is not universal. The vast amount of information available complicates selecting and producing valuable content for educators. The survey indicated that 36.6% of teachers—approximately 211 individuals—use the Internet solely for WhatsApp communication. Financial constraints likely contribute to this issue, as teachers may lack the resources for additional tools requiring internet access. The costs associated with comprehensive communication systems and the introduction of new technologies often exceed the budgets of financially constrained schools. Leaders may also prioritise funding for immediate school goals over investing in communication tools (Bolte *et al.* 2018).

Communication is crucial in effectively disseminating expectations. However, inadequate communication may also stem from poor leadership orientation or unclear expectations, which can be traced to the organisation's underlying visions and values (Bolte *et al.* 2018; Eberl and Drews 2021). Visionary digital leaders play a key role in guiding change during crises and conveying their vision to teachers, enabling them to incorporate it into their teaching practices. Leaders should articulate easily understood and motivating values to transfer this vision effectively.

Values inspire school members and can empower individuals to make decisions independently of extensive coordination (Bolte *et al.* 2018). School leaders must foster coordination and ensure that all members embrace values to address digital literacy challenges effectively. Leaders should provide appreciation and support for teachers' efforts to integrate ICT, helping them feel accepted and confident in overcoming their digital literacy obstacles.

4 CONCLUSION

The COVID-19 pandemic has marked a critical juncture in the evolution of digital engagement within the educational sector. Despite expanding media and educational platforms and developing emerging application technologies, teachers still encounter significant challenges in effectively integrating technology into their teaching practices. This study reveals a need for further development of digital literacy to enhance the effectiveness of technology use in online learning environments. School leaders, particularly principals, are crucial in optimising teacher performance in this context. Principals must foster effective communication with all teaching staff, providing a platform for addressing their challenges. Additionally, principals should integrate the vision for digital advancement into their teaching staff's professional development by aligning it with their values and encouraging active participation.

REFERENCES

Amin, M. (2018). ICT for Rural Area Development in Indonesia: A literature review. *Journal of Information Technology and Its Utilization*, 1(2), 32–37.

Asosiasi Penyelenggara Jasa Internet Indonesia (APJII). (2019). *Penetrasi dan profil perilaku pengguna internet Indonesia tahun 2018* (pp. 1–51). APJII. Retrieved from www.apjii.or.id

Astini, N. K. S. (2020). Tantangan Dan Peluang Pemanfaatan Teknologi Informasi Dalam Pembelajaran Online Masa Covid-19. *Cetta: Jurnal Ilmu Pendidikan*, 3(2), 241–255.

Bolte, S., Dehmer, J., and Niemann, J. (2018). Digital Leadership 4.0. Acta Technica Napocensis-Series: *Applied Mathematics, Mechanics, and Engineering*, 61(4).

Eberl, J. K., and Drews, P. (2021, March). Digital Leadership™Mountain or molehill? A literature review. In *International Conference on Wirtschaftsinformatik* (pp. 223–237). Springer, Cham.

Hermawan, H. D., Deswila, N., and Yunita, D. N. (2018). Implementation of ICT in Education in Indonesia during 2004–2017. Paper presented at the 2018 International Symposium on Educational Technology (ISET).

Kultsum, U., Zada, K., Defianty, M., and Roup, M. (2022). Muslim Women Leadership: The Catalytic Style in Developing School Members' Psychological Well-Being Amidst Global Pandemic. *Islamic Guidance and Counseling Journal*, 5(2), 119–134.

Prastiyo, W., and Purnawan, P. (2018). Development of Youtube integrated google classroom based e-learning media for the light-weight vehicle engineering vocational high school. *Jurnal Pendidikan Vokasi*, 8(1), 53–66.

Schleicher, A. (2020). *The Impact of COVID-19 on Education: Insights from" Education at a Glance 2020"*. OECD Publishing.

Sharma, L., and Srivastava, M. (2019). Teachers' motivation to adopt technology in higher education. *Journal The World Bank*. (2020). The World Bank in Indonesia. https://www.worldbank.org/en/country/indonesia/overview, Accessed 17 October, 2020.

UNESCO. (2020) *Education: From disruption to Recovery*, https://en.unesco.org/covid19/educationresponse

UNICEF. (2021). Indonesia: 18 bulan setelah sekolah ditutup, kini waktunya anak-anak kembali ke sekolah dengan aman sesegera mungkin : Murid dan guru membutuhkan berbagai bentuk dukungan untuk mengisi ketertinggalan dalam pembelajaran. (15 September 2021) https://www.unicef.org/indonesia/id/press-releases/indonesia-18-bulan-setelah-sekolah-ditutup-kini-waktunya-anak-anak-kembali-ke

Towards Resilient Societies: The Synergy of Religion, Education, Health, Science, and Technology – Rahiem (Ed.)

Enhancing environmental attitudes through lesson study: A step towards achieving sustainable development goals in education

Yanti Herlanti
Universitas Islam Negeri Syarif Hidayatullah Jakarta, Indonesia
ORCID ID: 0000-0002-8380-2417

Qumilaila Qumilaila
Yayasan Semarak Pendidikan Indonesia, Tangerang Selatan, Indonesia

Shinji Nobira
Aichi University of Education, Kariya, Japan

Yasunobu Kuboki
Indonesia Education Promoting Foundation, Toyama, Japan

ABSTRACT: Environmental issues in South Tangerang City are largely attributed to limited environmental awareness and attitudes. Addressing this challenge requires the implementation of effective environmental education initiatives. This study evaluates the impact of integrating environmental education (EE) into lesson study projects on students' environmental attitudes, supporting Sustainable Development Goals (SDGs), particularly Goal 4 (Quality Education) and Goal 13 (Climate Action). The study involved 193 junior high school students aged 12 to 15, divided into two groups: 116 students were taught by science teachers who had previously participated in lesson study projects. The findings indicate that students from schools engaged in EE lesson studies demonstrated improved environmental attitudes in Factor II (attitudes toward recovery), Factor III (attitudes toward recycling), and Factor IV (environmental consciousness and behaviour). However, they showed weaker performance in Factor I (attitudes toward prevention). Additionally, student participation in environmental projects was significantly higher in schools implementing EE lesson study programs. These results highlight the effectiveness of lesson study-based environmental education in fostering positive environmental attitudes and encouraging active student involvement in sustainability initiatives.

Keywords: attitude, education, environmental, lesson study, SDGs

1 INTRODUCTION

As per the national environmental quality index report released by the Ministry of Environment and Forestry of the Republic of Indonesia in 2019, it has been identified that eight provinces, including Banten, exhibit a poor environmental quality index falling within the range of 40 to 50. The inferior environmental quality in Banten Province is attributed to the water quality, which has been assessed at a value of 43.11. Additionally, Asiantoday in 2020 highlighted that South Tangerang City, an integral part of Banten Province, is grappling with poor water quality stemming from industrial activities and the accumulation of substantial waste materials.

Seeing a river with cloudy water and trash is a common sight for locals. They have not recognised the danger of the decrease in the quality of this river. The South Tangerang City government has made several efforts to improve environmental quality by issuing laws to prevent river pollution and empowering communities through environmentally conscious activities.

DOI: 10.1201/9781003645542-57

Efforts to address environmental issues encompass not only reactive measures but also proactive initiatives. One such proactive step involves imparting environmental education at an early stage. Between 2017 and 2021, the IEPF, with the support of JICA, undertook a project to introduce the subject of 'Environment' in Junior High Schools and disseminate environmental education. Previously, a similar initiative targeting elementary schools in South Tangerang City was implemented from 2014 to 2017. The overarching objective of this project is to cultivate an environmental attitude among students, particularly in elementary education. Students receive educational resources, and environmental learning is seamlessly integrated into the science curriculum. In addition, teachers collaborate through lesson study activities to enhance their competence in delivering environmental education. The projects played a pivotal role in advancing the Sustainable Development Goals (SDGs) in Indonesia. These initiatives mainly addressed the indicators concerning the number of formal education units that foster an environmental culture (climate action). Furthermore, significant strides were made in integrating sustainable development education within the teacher education and training curriculum (quality education).

SDGs 4 (quality education) underscores the importance of adequately trained teachers. Lesson study directly contributes to this by providing a structured platform for continuous professional development. Through reflective collaboration, teachers refine their pedagogical skills, which leads to improved teaching effectiveness and student outcomes. Previous research has indicated that lesson study impacts teachers' capacity to teach and collaborative learning. The implementation of lesson studies by teachers has a direct impact on enhancing teacher competence. Lesson study improved the quality of teacher instruction, particularly in learning that is meaning- and application-oriented (Vermunt *et al.* 2019). Using lesson study increases core instructional practices, especially problem-solving teaching strategies (Huang *et al.* 2017). Lesson study also contributes to developing pedagogy content knowledge for both beginner and experienced teachers (Coenders and Verhoef 2019). Further research indicates the indirect influence of lesson study activities on students. Lesson study has changed students' culture, beliefs, practices, expectations, and learning (Dudley *et al.* 2019). Lesson study also improves student learning outcomes in the natural sciences, although the difference is insignificant compared to other classes where the teachers do not take lesson study (Ayra and Kösterelioğlu 2021). Other studies apply cooperative learning in biology subjects, and teacher applies lesson study, which impacts increasing students' activity and learning outcomes (Hasan *et al.* 2021).

However, lesson study applications specifically to enhance environmental attitudes remain underexplored. According to data from online lesson study activities in 2020-2021, 91 science teachers from 24 public junior high schools actively participated in these efforts (Herlanti *et al.* 2022; Herlanti *et al.* 2024). Environmental education (EE) lesson study enhances teachers' abilities to integrate environmental concerns into the learning process. It aims to ascertain whether teachers' involvement in environmental education projects impacts students' environmental literacy. However, this study still emphasises knowledge acquisition rather than attitude transformation. The EE lesson study explores pedagogical strategies to prioritise fostering pro-environmental attitudes as a key component of achieving the Sustainable Development Goals (SDGs). Nonetheless, more research is needed on whether it similarly affects students' environmental attitudes. This study seeks to elucidate the environmental attitude of students under the instruction of teachers who actively participate in EE lesson study as compared to students under teachers who have not engaged in EE lesson study activities.

2 METHODS

The study used a quasi-experimental approach featuring a post-test-only control group design. The study was conducted in South Tangerang City, with 213 public and private schools divided into seven regions. The research focused on a target population of schools in the Serpong and Ciputat regions. The research sample was chosen based on the engagement of science teacher working groups in EE lesson study activities on August 2023 – February

2024. The selection of schools for the treatment group will be based on the active participation of science teachers in EE lesson study online during the 2020-2021 period and their commitment to continuing the implementation of offline lesson study. Non-treatment schools were selected based on their willingness to permit students aged 12-15 in grades 7-9 to complete the environmental attitudes instrument (Table 1).

Table 1. Sampling.

Experimental treatment	Name of School	Pupils' population	Sum of random sampling
Science teachers have followed the EE lesson study	LS1	526	47
	LS2	548	38
	LS3	218	31
Science teachers haven't yet followed the EE lesson study	NLS1	92	27
	NLS2	84	22
	NLS3	74	28
Sum		1542	193

The research instrument was adopted from high school students' environmental attitudes (Ugulu *et al.* 2013). Following translation and adjustment for the Indonesian setting, the instrument underwent testing with 40 junior high school students, yielding a Cronbach's Alpha coefficient of 0.678. Among the 36 questions administered, seven were identified as valid (items 2, 4, 9, 13, 14, 17, and 19). Subsequently, the instrument's content underwent scrutiny to ensure alignment with the context of students in Indonesia, identifying 25 questions pertinent to environmental attitudes (Table 5). Questions are assessed using a 1-4 Likert scale, ranging from strongly disagree to strongly agree. This data is then segregated into three groups comprising various environmental attitude categories (Table 2).

Table 2. Categories of environmental attitude.

Achievement (Score)	Categories of environmental attitude
25-50	Low
51-74	Moderate
75-100	High

3 FINDINGS & DISCUSSION

Based on the t-test results (Table 3), it is evident that there is no statistically significant difference in the average environmental attitudes of students who are instructed by science teachers participating in the EE lesson study ($t=0.69$, $p > 0.05$). Furthermore, it is worth noting that the average environmental attitude of the students aligns with the moderate category, as detailed in Table 4.

Table 3. t-Test: Two-sample assuming unequal variances

	Variable 1 (NLS)	*Variable 2 (LS)*
Mean	74,03	73,21
Variance	70,21	59,73
Observations	77	116
Hypothesized Mean Difference	0	
df	154	
t Stat	0,69	

While not statistically significant, the results indicate that a higher percentage of students in schools where science teachers did not participate in EE lesson study (NLS) demonstrated excellent environmental attitudes compared to schools where teachers did participate in EE lesson study (LS). Further analysis of Table 5 shows that the elevated environmental attitudes among NLS group students are linked to heightened environmental awareness. Conversely, LS group students exhibited heightened attitudes in other areas, including attitudes toward recovery and recycling also environmental consciousness and behaviour.

Table 4. Environmental attitude of students.

Treatment	Name of School	Low	Moderate	High	Total
Science teachers have followed the EE lesson study (X)	LS1	0	26	21	47
	LS2	0	16	22	38
	LS3	1	22	8	31
		1 (1%)	64 (55%)	51 (44%)	116 (100%)
Science teachers haven't yet followed the EE lesson study	NLS1	0	13	14	27
	NLS3	0	9	13	22
	NLS3	0	14	12	28
	Sum	0 (0%)	38 (49%)	39 (51%)	77 (100%)

Both students from schools that have and have never participated in lesson study lack an understanding of the role of humans in protecting nature. Most students from both groups agree that people can alter nature to fulfil their needs (Table 5, question number 8). However, students from schools that have never participated in lesson study show a better grasp of the fundamental requirement for ecological balance, as demonstrated by the number of students who disagree with the notion that the extinction of insects such as flies is beneficial for the environment (Table 5, number 5). Furthermore, students from schools that participated in lesson study project engagement in environmental activities exceeded that of those that did not. Most students are willing to participate in door-to-door activities for environmental campaigns (Table 5, numbers 16, 23 and 25).

Table 5. Students' environmental attitude.

		Students in Lesson Study Participant School (%)		Students in Non-Lesson Study Participant School (%)	
No.	Questions	Agree	Disagree	Agree	Disagree
A.	**Factor I. Environmental Awareness**				
1	The primary purpose of three planting is so beautifying the environment in terms of aesthetics	71	29	57	43
2	Since the environment can clean itself, human waste does not cause problem.	25	75	29	71
3	Instead of spending money on historical places, it is more advantageous for us to build luxurious roads	28	72	34	66
4	Some species are unnecessary for the environment.	25	75	22	78
5	The extinction of the insects such as files is useful for environment	53	47	44	56
6	Nature renews itself with substances cycle, therefore recycling helps economy only in term time.	39	61	30	70
7	The best way to build houses is to dry up the wetlands and	55	45	52	48

(*continued*)

Table 5. Continued

No.	Questions	Students in Lesson Study Participant School (%)		Students in Non-Lesson Study Participant School (%)	
		Agree	Disagree	Agree	Disagree
	build there.				
8	It is meaningless to buy paper bags instead of nylon bags given for free in the markets.	34	66	30	70
9	People have the right to make changes in nature for meeting their needs.	74	26	74	26
10	Money can save by buying drinks in plastic bottles since drink in glass bottles are expensive.	50	50	49	51
B.	**Factor II. Attitudes towards recovery**				
11	Using rechargeable batteries instead of disposable batteries supports recycling.	85	15	82	18
12	Shopping only as much as needed is an important step in recycling.	91	9	94	6
13	I believe that we should be economical for environment.	96	4	94	6
14	We should use both sides of white papers to support recycling.	82	18	82	18
15	For saving energy, I turn off the light in my house when they are not used.	98	2	94	6
C.	**Factor III. Attitude toward recycling**				
16	I can go from door to door to teach people recycling.	81	19	6	31
17	I separate waste materials in my house for recycling	89	11	86	14
18	I feel sad when I see people throwing away objects than can be recycled.	86	14	86	14
19	It makes me happy when people recycle used bottles, can, and paper.	97	3	94	6
D.	**Factor IV. Environmental consciousness and behaviour**				
20	For a liveable environment, I can work voluntarily for a long time if needed.	85	15	84	16
21	I don't waste water while I am brushing my teeth.	85	15	83	17
22	I prefer environmentally harmless products even if they are more expensive.	80	20	78	22
23	I participate in environmental projects	84	16	77	23
24	My friends know me as sensible person toward environment.	78	22	78	22
25	I talk with people around me on environment matters.	87	13	77	23

A noteworthy finding from this research pertains to the lesson study of environmental education. It reveals increased attitudes toward recovery and recycling, as well as heightened environmental consciousness and behaviour (Table 5). Although the difference is statistically insignificant, it has important implications for continuously advancing environmental education lesson studies. Environmental attitudes significantly correlate with Pro-Environmental Behaviour (PEB) (Miller *et al.* 2022; Razali *et al.* 2023). PEB embodies community participation and concern for the environment (Soeharso *et al.* 2023); students attending schools involved in environmental education lesson studies demonstrate a strong inclination toward engaging in environmental activities.

The study had limitations in finding that teachers' participation in environmental education lesson study had no significant impact on students' environmental attitudes. In this study, environmental education was integrated with science learning. Furthermore, research suggests that teaching environmental education as a monolithic subject was more effective than integrating it with science learning (Istiqomah *et al.* 2020). However, environmental education is intended to enhance environmental attitudes but yields mixed results (Gifford and Sussman 2012; van de Wetering *et al.* 2022). Multiple factors contribute to students' environmental attitudes. One of these factors is the need for more practical, hands-on experience in environmental education (Ramchunder and Ziegler 2021). Engaging students in real-world environmental activities can significantly improve their awareness and commitment to environmental issues (Dale *et al.* 2020; Rosa and Collado 2019). Furthermore, societal and cultural influences beyond the school environment, such as social norms, cultural values, and family practices, also significantly shape students' environmental attitudes (Jaime *et al.* 2023; Mónus 2022).

4 CONCLUSION

Environmental education lesson study improves students' participation in environmental movement; 81% of students will go from door to door to teach people recycling, 87% will talk with people around them on environmental matters, and 84% will participate in environmental projects. However, it still requires improvement because the integrated environmental education lesson study activities in science learning have only improved environmental attitudes in 50% of students. It is essential to continue improving environmental attitudes in environmental education because they correlate with PEB, reflecting students' participation and concern for the environment.

REFERENCES

Ayra, M., and Kösterelioğlu, İ. (2021). Effect of the Lesson Study Practice on Students' Academic Achievements in Life Sciences Course. *Educational Policy Analysis and Strategic Research*, 16(1), 249–270. https://doi.org/10.29329/epasr.2020.334.14

Coenders, F., and Verhoef, N. (2019). Lesson Study: professional development (PD) for beginning and experienced teachers. *Professional Development in Education*, 45(2), 217–230. https://doi.org/10.1080/19415257.2018.1430050

Dale, R. G., Powell, R. B., Stern, M. J., and Garst, B. A. (2020). Influence of the natural setting on environmental education outcomes. *Environmental Education Research*, 26(5), 613–631. https://doi.org/10.1080/13504622.2020.1738346

Dudley, P., Xu, H., Vermunt, J. D., and Lang, J. (2019). Empirical evidence of the impact of lesson study on students' achievement, teachers' professional learning and on institutional and system evolution. *European Journal of Education*, 54(2), 202–217. https://doi.org/https://doi.org/10.1111/ejed.12337

Gifford, R., and Sussman, R. (2012). Environmental Attitudes. In *The Oxford Handbook of Environmental and Conservation Psychology* (pp. 65–80). Oxford University Press. https://doi.org/10.1093/oxfordhb/9780199733026.013.0004

Hasan, R., Lukitasari, M., and Ernawati, Y. (2021). Students' achievement and teachers' perception in the implementation of lesson study-based cooperative learning. *Journal of Physics: Conference Series*, 1731(1), 12004. https://doi.org/10.1088/1742-6596/1731/1/012004

Herlanti, Y., Fadlilah, D. R., and Fakhrotun, N. (2022). Establishing Environmental Education on Students' Environmental Literacy at Junior High School. *Indonesian Journal Of Educational Research and Review*, 5 (3), 598–613. https://doi.org/10.23887/ijerr.v5i3.51756

Herlanti, Y., Nobira, S., Kuboki, Y., and Qumilaila, Q. (2024). Online lesson study design: integrating environmental issues with science learning to enhance students' environmental literacy. *International Journal for Lesson and Learning Studies*, ahead-of-print(ahead-of-print). https://doi.org/10.1108/IJLLS-08-2024-0169

Huang, R., Barlow, A. T., and Haupt, M. E. (2017). Improving core instructional practice in mathematics teaching through lesson study. *International Journal for Lesson and Learning Studies*, 6(4), 365–379. https://doi.org/10.1108/IJLLS-12-2016-0055

Istiqomah, I., Suwondo, S., and Firdaus, L. N. (2020). Environmental Education in Forming Attitudes of Environmental Care for Students. *Journal of Educational Sciences*, 4(1), 200. https://doi.org/10.31258/jes.4.1.p.200-211

Jaime, M., Salazar, C., Alpizar, F., and Carlsson, F. (2023). Can school environmental education programs make children and parents more pro-environmental? *Journal of Development Economics*, 161, 103032. https://doi.org/https://doi.org/10.1016/j.jdeveco.2022.103032

Miller, L. B., Rice, R. E., Gustafson, A., and Goldberg, M. H. (2022). Relationships Among Environmental Attitudes, Environmental Efficacy, and Pro-Environmental Behaviors Across and Within 11 Countries. *Environment and Behavior*, 54(7–8), 1063–1096. https://doi.org/10.1177/00139165221131002

Mónus, F. (2022). Environmental education policy of schools and socioeconomic background affect environmental attitudes and pro-environmental behavior of secondary school students. *Environmental Education Research*, 28(2), 169–196. https://doi.org/10.1080/13504622.2021.2023106

Ramchunder, S. J., and Ziegler, A. D. (2021). Promoting sustainability education through hands-on approaches: a tree carbon sequestration exercise in a Singapore green space. *Sustainability Science*, 16(3), 1045–1059. https://doi.org/10.1007/s11625-020-00897-5

Razali, G., Iskandar, Y., Fatmawati, S., Hidayat, R., Umer, M., and Mujahid, F. (2023). Psychological Factors Influencing Pro-Environmental Behavior in Urban Areas. *In West Science Interdisciplinary Studies* (Vol. 01, Issue 07).

Rosa, C. D., and Collado, S. (2019). Experiences in nature and environmental attitudes and behaviors: Setting the ground for future research. In *Frontiers in Psychology* (Vol. 10, Issue APR). Frontiers Media S.A. https://doi.org/10.3389/fpsyg.2019.00763

Soeharso, S. Y., Ningtyas, L. D., and Sundari, R. (2023). The Role of Environmental Attitude as Mediator Between Perceived Sustainability Policy and Spiritual Well-Being Toward Pro-Environmental Behavior [Peran EA Sebagai Mediator PSP dan SpWB Atas PEB]. *ANIMA Indonesian Psychological Journal*, 38(1), 038109. https://doi.org/10.24123/aipj.v38i1.4906

Ugulu, I., Sahin, M., and Baslar, S. (2013). High School Students' Environmental Attitude: Scale Development and Validation. *International Journal of Educational Sciences*, 5(4), 415–424. https://doi.org/10.1080/09751122.2013.11890103

van de Wetering, J., Leijten, P., Spitzer, J., and Thomaes, S. (2022). Does environmental education benefit environmental outcomes in children and adolescents? A meta-analysis. *Journal of Environmental Psychology*, 81, 101782. https://doi.org/https://doi.org/10.1016/j.jenvp.2022.101782

Vermunt, J. D., Vrikki, M., van Halem, N., Warwick, P., and Mercer, N. (2019). The impact of Lesson Study professional development on the quality of teacher learning. *Teaching and Teacher Education*, 81, 61–73. https://doi.org/https://doi.org/10.1016/j.tate.2019.02.009

Towards Resilient Societies: The Synergy of Religion, Education, Health, Science, and Technology – Rahiem (Ed.)

Integrating local culture into biotechnology education for multicultural character building

Zulfiani Zulfiani
Universitas Islam Negeri Syarif Hidayatullah Jakarta, Indonesia
ORCID ID: 0000-0001-9369-1418

Indah Juwita Sari
Universitas Islam Negeri Syarif Hidayatullah Jakarta, Indonesia

Prasart Nuangchalerm
Mahasarakam University, Thailand

ABSTRACT: Multicultural education is essential for promoting equality among students by respecting cultural diversity, necessitating the reconstruction of an integrative curriculum. This article explores the integration of local cultural wisdom into biotechnology education, specifically in traditional biotechnology topics, by emphasising fermentation-based local food and beverage products. The study employs a qualitative approach using document analysis and literature review methods. The findings present a science learning model grounded in local culture to foster students' multicultural character. This model incorporates Aikenhead's (2015) inquiry-based syntax and sociocultural framework through three stages: (1) guiding students to identify science content aligned with local wisdom (*Enculturation*), (2) encouraging critical thinking by connecting local wisdom with scientific concepts (*Assimilation*), and (3) helping students compile relevant scientific content to enhance the quality of life within local communities (*Acculturation*). This model effectively supports the development of students' multicultural character, aligns with societal needs, and reinforces local culture-based education policies. However, further research is necessary to evaluate its practical implementation in classroom settings.

Keywords: character values, local culture, multicultural, science learning model

1 INTRODUCTION

Multiculturalism is the concept of diversity in a community of nations that accepts differences in culture, race, ethnicity, ethnicity, and religion, among others. This concept has become an interesting discourse in various fields, as well as the desired social principle and state ideology (Sahrasad *et al.* 2024). In the context of education, Rosyada (2014) explained that multicultural education is an educational process that provides equal opportunities for all students regardless of ethnicity, culture, or religion. It respects diversity, recognises equality of rights, strengthens nationalism, and builds national and international harmony. Diversity should be seen as a valuable asset that complements perfection (Sopyan 2020).

Multicultural education can be developed through (1) independent fields of study/subjects, (2) programs, and (3) practices planned by educational institutions (Murdiono 2012). In the form of a multicultural education course, it becomes a scientific study that includes knowledge, theory, and several skills and attitudes. This course already exists at several PTKIN

 DOI: 10.1201/9781003645542-58

universities (such as IAIN Kendari (FTIK IAIN Kendari 2022), IAIN Salatiga, and others) for undergraduate students majoring in PAI with the nomenclature of Multicultural Education courses (Ul-hasanah 2017).

For example, creative program forms can be embedded in the cross-border Real Work Lecture (KKN) program initiated by the UIN Jakarta Community Service Program (PPM). This program fosters multicultural values/character in collaborative activities of Indonesian students at the international level. Apart from developing soft skills, students are involved in private teaching activities, education campaigns, and international book values, one of which is in Turkey (RDK Fidkom UIN Jakarta 2018). Practices planned by educational institutions can be carried out in formal academic or non-formal activities to respond to the needs or aspirations of their community. The various variations of activities above provide evidence that, in its application, multicultural education is flexible and a continuous process.

In practice, educators have the task of implementing multicultural education. In the case above, the task of educators is (1) to diagnose students' needs, (2) to formulate learning objectives, and (3) to determine the learning strategies to be adopted (Majid 2005). So, in lectures, it is essential to analyse multicultural concepts that can be integrated into courses. Next, the teacher identifies learning objectives and selects strategies and assessments.

According to Banks (1993), five dimensions of multicultural education: (1) Content integrations in instructional. Efforts to design multicultural learning by creating integrity in content and teaching materials (accommodating various cultural materials diversely). (2) The knowledge construction process in instructional. The construction of knowledge/teaching materials is operationalised in the learning process in the classroom (3) An equity instructional pedagogy. The teacher's ability to display pleasant learning and a respectful attitude when interacting with students is very important. (4) Training participation in instructional. Student involvement in classroom activities that will meet learning objectives. (5) Prejudice reduction in instruction. Educators play a role in creating class discussions with an open attitude towards forms of diversity present in the class.

However, efforts to integrate local cultures in science education offer a new dimension that has not been widely explored. This approach provides local relevance and creates more authentic learning by connecting science to students' daily lives. Considering the importance of the essence and goals of multicultural education, reconstruction efforts are needed by designing integrative curriculum development. Curriculum models and integrative learning designs are alternatives that can be developed in multicultural education. Conditions or situations are needed so that students can understand the social context. Remember that the specific learning objectives focus on identifying the development and challenges of cultural differences.

Efforts to integrate local culture in science education open up new dimensions that are locally relevant and create authentic learning by connecting science to students' lives. To support multicultural education, reconstruction is needed through the development of an integrative curriculum, including the idea of an Inclusive Curriculum integrating harmony and tolerance in collaboration with educational institutions (Muthoifin *et al.* 2024).

Knowledge of local culture is one component that can be integrated into science learning. Tomi *et al.* (2018) have developed learning tools at the Madrasah Aliyah (MAS) level by integrating local wisdom with biological concepts (biodiversity), which can build cultural character and preserve social and cultural values. A literature review by Rahmatih *et al.* (2020) stated that linking awik-awik or local customary rules with science learning in elementary schools provides authentic experiences for students, which implies a sense of love for local cultural values. The operationalisation is prepared in the learning plan, and according to Sudarmin (2014), wise educators will insert local cultural values into their learning.

The model is novel in its approach by directly incorporating local culture into science, creating more authentic, relevant, and contextual learning than traditional methods. In addition, this integration allows science learning to be cognitive and involves students' affective and psychomotor aspects, which are relevant to the educational needs of the 21st century.

2 METHOD

This study uses a qualitative approach with document analysis and literature review (Creswell and David Creswell 2018). The documents analysed include Semester Learning Plans (SLP), learning tools, and learning reports. This document was chosen because it reflects the integration of local cultures. Research participants include lecturers and students at the Biology Education Study Program of FITK UIN Jakarta for the 2022/2023 Academic Year. Analysis Procedure: (1) Collecting SLP and relevant documents, (2) Identifying elements of multiculturalism, local culture, and inquiry learning, (3) The data were analysed to align with the research objectives, which is to integrate local cultures in science education for the strengthening of multicultural characters. The instruments used include observation sheets, code lists, and field notes to support data validation and in-depth analysis. The analysis of the documents is used because it provides an in-depth picture of multicultural education policies and implementations that are relevant to the research context.

3 DISCUSSION

3.1 *Local culture science learning model*

Research results show that there have been many applications of science learning models that link local culture and local wisdom to foster values or character at every level of education. Below, we will discuss the development of a science learning model with local culture at the tertiary level in the biotechnology course of the Biology Education Department, FITK UIN Jakarta. The development of a science learning model with local culture refers to Aikenhead (2015), namely, enculturation, assimilation, and acculturation (Mulyani and Julianto 2018).

The local culture-based science learning process has stages:

(1) Enculturation is building cultural identity and character by introducing local culture. Students identify science content that is in harmony with local wisdom. Multicultural values, such as respect for diversity, are operationalised through group discussions on traditional foods to find connections between cultural products and scientific principles, thus fostering awareness to appreciate diversity.
(2) Assimilation is the process of combining cultural elements. This stage focuses on an in-depth study of science materials that can develop students' mindsets so that their scientific thinking changes in understanding science phenomena in the context of local wisdom. Multicultural values, such as respect for cultural diversity and strengthening cultural identity, are operationalised through group discussions.
(3) Acculturation is the process of adjusting to a new culture by inventorying science materials relevant to the community's local potential, replacing old ideas, and adding new, more relevant ideas. Cross-cultural group projects are implemented to instil values of tolerance and cooperation, where students practice traditional biotechnology, document the process, and evaluate the contribution of innovation to local communities.

Several stages must be considered in implementing local culture science learning, such as preparing a learning plan with objectives and assessments. Soenaryo (2000) states that the process is the primary key to planning activities. Scrivener (2011) states that planning is a thinking skill, namely imagining teaching before it happens. Learning planning is the organisation of students' learning experiences that teachers deliberately design to achieve learning goals (Majid 2005). Teachers can develop learning plans, including choosing strategies, media, and assessments.

Science learning designs can utilise local culture as context and discover issues that are relevant to the subject material. This way, the science taught is contextual and can preserve local culture. The primary capital of Indonesian citizens is ethnic, religious, and ethnic diversity, which makes it possible for students to get to know each other's local culture. Developing science learning with a local culture to foster character values requires teacher creativity and innovation to achieve learning objectives. Understanding is not limited to subject matter but includes multicultural character values and their implications.

3.2 *Science learning model: inquiry model*

The Inquiry Model emphasises student activities (student-centred) to discover knowledge through a series of scientific processes. Inquiry activities can develop cognitive, psychomotor, and scientific attitude competencies. Hinrichsen and Jarrett (1999) stated four inquiry characteristics: connection, design, investigation, and building knowledge. Joyce *et al.*, (2008) in their book *Models of Teaching* state that inquiry, besides having the main effect on scientific learning, also has an accompanying effect, namely commitment to inquiry and developing aspects of values/character. Many researchers have implemented the inquiry model, namely developing inquiry learning programs in biotechnology courses at the tertiary level (Zulfiani 2022) and general Biology courses (Anggraini 2006). In this way, inquiry learning can be synergised by linking local culture to develop multicultural values.

3.3 *Multicultural character*

The character of the Indonesian nation is genuinely based on the values of Pancasila, the norms of the 1945 Constitution, the principle of Bhinneka Tunggal Ika, and the integrity of the unitary state of the Republic of Indonesia. Thus, the nation's character is unique as a manifestation of awareness, understanding, creativity, feeling, and initiative, which are the collective agreement of citizens. Education has a strategic role in realising the cultivation of character values in its citizens. Thus, a structured, measurable, planned effort is needed to develop it in an active and meaningful learning process and to develop religious and spiritual potential, personality, and noble morals.

According to Afifah *et al.* (2021), character development is carried out in three stages: planning, implementation, and evaluation of results. Character identification is based on philosophical, theoretical, and empirical sources in the planning phase. Philosophical sources are derived from the nation's ideological foundation, operationalised in the National Education System Law. Theoretical aspects include psychology, sociocultural education, and empirical, including experience or best practices (Suryaman *et al.* 2012).

National values can grow by instilling an attitude of national awareness, the formulation of which is outlined in graduate competency standards, namely, minimum criteria regarding the unity of competency in attitudes, skills, and knowledge that shows student achievements (KEMENDIBUD 2023). In the aspect of attitude, relevant multicultural values can be developed.

3.4 *Learning design: development of local culture-based Biotechnology SLP to develop multicultural character values*

Conceptually, a learning model that can accommodate the development of multicultural character is student-centred, one of which is the Inquiry Model. The following are the stages of developing a Semester Learning Plan for the local culture-based biotechnology course.

3.4.1 *Analysis of the Biotechnology course curriculum*

Curriculum analysis related to the demands of Program Learning Outcome (PLO), Learning Outcome (LO), and Course Learning Outcome (CLO) from the Biotechnology course. The PLO component has aspects of knowledge, skills, and attitudes. Teachers can insert multicultural values into attitudes such as respecting diversity, caring for the environment, and love of knowledge. Local cultural values will make someone love culture, preserve it, and foster a spirit of nationalism and patriotism. Science learning also contributes to developing attitudes of curiosity, logical, critical, creative, and innovative thinking, honesty, discipline, independence, self-confidence, and responsibility. Teachers analyse relevant biotechnology content to develop multicultural values.

3.4.2 *Planning a science learning strategy based on local culture*

Educators can specify more operational goals for the indicators based on PLO, LO, and CLO. Indicator descriptions are adapted to inquiry learning models based on local culture. Content that can be developed, such as traditional food biotechnology. The following is the preparation of indicators from CLO in the Biotechnology course.

Course: Biotechnology

LO: Analysing the process of producing traditional and modern food biotechnology products.

CLO: Analysing the process of producing biotechnology products.

Indicators: (1) Identify various products and traditional biotechnology biological processes by relating these products to the local context of the region; (2) Carrying out practical work on making traditional biotechnology products (activities documented in video form to be uploaded on YouTube); (3) Conduct class discussions (offline or online) to reconstruct biotechnology knowledge, the relationship between product history and context.

3.4.3 *Local culture based science learning procedures*

The indicators prepared from the CLO are then described in detail in the learning stages. The following learning stages integrate the Inquiry Model of Hinrichsen and Jarrett (1999) and Aikenhead (2015) (Table 1).

Table 1. Stages of learning integration with the inquiry model.

Inquiry model	Stages of learning	Student activities
Connection	Enculturation	Every student comes from various regions in Indonesia; realizing this, students are directed to work in cooperative groups using an inquiry learning model. Students identify traditional biotechnology products from their region and the biological processes that occur and relate these products to various local contexts in their region.
Design	Enculturation	Students discuss plans for trials/experiments on traditional biotechnology products. Students formulate problem statements and hypotheses by identifying control, independent, and dependent variables.
Investigation	Assimilation	Students carry out practical work on making traditional biotechnology products and make videos to be uploaded on social media such as YouTube.
Building knowledge	Assimilation	Students are asked to get an assessment/review of the video they have made. This stage is also part of the Assessment.
	Acculturation	Conduct class discussions (offline or online) to: reconstruct biotechnology knowledge; linkage of product history with regional context; Respond and show appreciation for differences in local cultural products; Follow-up to preserve local food culture products. Students complete a self-assessment to reflect their attitudes as citizens who accept, respect, and cooperate with ethnic, linguistic, and cultural diversity in their society.

The description above is a learning design that develops multicultural character values with a science learning model based on local culture. Assessment/peer review of the practical videos made, self-assessment to respond and provide an appreciation for the differences in local cultural products, and follow-up to preserve local food culture products.

4 CONCLUSION

Local culture-based science learning design is an effective strategy to foster multicultural character values, such as respect, acceptance, and cooperation in a multi-ethnic society. Analysis of Biotechnology teaching materials is an essential initial step, followed by selecting appropriate learning and assessment strategies, such as using an inquiry model. This model strengthens mastery of concepts and instils a scientific attitude. Policymakers can support implementation by designing flexible curricula integrating local culture into science education. Teacher training to implement local culture-based learning is also needed to ensure the quality of teaching. Further research can evaluate this model at a broader range of educational levels or culturally diverse areas.

REFERENCES

Afifah, N. P. N., Dewi, D. A., and Furnamasari, Y. F. (2021). Pendidikan Multikultural Dalam Upaya Membangun Karakter Anak Bangsa. *Jurnal Pendidikan Tambusai, 5*(3), 7170–7175. https://doi.org/https://doi.org/10.31004/jptam.v5i3.2107

Aikenhead, G. (2015). Acculturation. In R. Gunstone (Ed.), *Encyclopedia of Science Education* (pp. 7–9). Springer Netherlands. https://doi.org/10.1007/978-94-007-2150-0_348

Anggraini, S. (2006). *Pengembangan Program Perkuliahan Biologi Umum berbasis Inkuiri bagi Calon Guru Biologi* [Universitas Pendidikan Indonesia]. http://digilib.upi.edu/digitalview.php?digital_id=1180

Banks, J. A. (1993). Multicultural Education: Historical Development, Dimensions, and Practice. *Review of Research in Education, 19*, 3–49. https://education.uw.edu/sites/default/files/ReviewofResearchAERA.pdf

Creswell, J. W., and David Creswell, J. (2018). *Research Design: Qualitative, Quantitative, and Mixed Methods Approaches.* SAGE Publications.

FTIK IAIN Kendari. (2022). *Mata Kuliah IAIN Kendari.* FTIK IAIN Kendari. https://fatik.iainkendari.ac.id/mata-kuliah-prodi-pai/

Hinrichsen, J., and Jarrett, D. (1999). *Science Inquiry for The Clasroom on Program Report.*

Joyce, B., Calhoun, E., and Hopkins, D. (2008). *Models of Learning, Tools for Teaching* (3rd Editio). McGraw-Hill Education (UK).

KEMENDIBUD. (2023). *Peraturan Menteri Pendidikan, Kebudayaan, Riset, Dan Teknologi Republik Indonesia N0 53/2023.*

Majid, A. (2005). *Perencanaan Pembelajaran: Mengembangkan Standar Kompetensi Guru.* Remaja Rosdakarya.

Mulyani, M., and Julianto, J. (2018). Pembelajaran Sains Berbasis Budaya Lokal Sebagai Bentuk Integratif Pendidikan Karakter. *EduStream: Jurnal Pendidikan Dasar, 2*(1), 35–42. https://doi.org/https://doi.org/10.26740/eds.v2n1.p35-42

Murdiono, M. (2012). Strategi Pembelajaran Pendidikan Multikultural Berbasis Budaya Lokal. *PKn Progresif, 7*(1), 24–33.

Muthoifin, Abuzar, M., Mahmudulhassan, Afiyah, I., Nirwana, A., Nuha, Setianto, G., and Waston. (2024). Fostering Multicultural Community Harmony to Enhance Peace and Sustainable Development Goals (SDG's). *Journal of Lifestyle and SDGs Review, 5*(1 SE-Artigos), e01687. https://doi.org/10.47172/2965-730X.SDGsReview.v5.n01.pe01687

Rahmatih, A., Maulyda, M., and Syazali, M. (2020). Refleksi Nilai Kearifan Lokal (Local Wisdom) dalam Pembelajaran Sains Sekolah Dasar: Literature Review. *Jurnal Pijar Mipa, 15*(2), 151. https://doi.org/10.29303/jpm.v15i2.1663

RDK Fidkom UIN Jakarta. (2018). *Ikuti KKN Internasional Mahasiswa UIN Jakarta Mengajar ke Turki.* RDK Fidkom UIN Jakarta. http://rdk.fidkom.uinjkt.ac.id/index.php/2018/07/05/ikuti-kkn-internasional-mahasiswa-uin-jakarta-mengajar-ke-turki/

Rosyada, D. (2014). Pendidikan Multikultural Di Indonesia Sebuah Pandangan Konsepsional. *SOSIO DIDAKTIKA: Social Science Education Journal*, *1*(1), 1–12. https://doi.org/10.15408/sd.v1i1.1200
Sahrasad, H., Nurdin, M. A., Dar, M., and Baihaky, R. (2024). Multiculturalism, Islamophobia and the Muslim Minority in Australia: A Reflection. *Journal of Al-Tamaddun*, *19*(1 SE-Articles), 335–353. https://doi.org/10.22452/JAT.vol19no1.23
Scrivener, J. (2011). *Learning Teaching. The Essential Guide to English Language Teaching* (J. Scrivener (ed.); Third Edit). Macmillan Education. https://www.ircambridge.com/books/Learning_Teaching.pdf
Soenaryo, E. (2000). *Pengantar Teori Perencanaan Pendidikan Berdasarkan Pendekatan Sistem* (Ed.1, Cet.). Adicita Karya Nusa.
Sopyan, Y. (2020). Urgency of Legal Pluralism Study for Students of Sharia and Law Faculty After Changes of Academic Degree. *Journal de Jure*, *12*(2). https://doi.org/10.18860/j-fsh.v12i2.10186
Sudarmin. (2014). *Pendidikan Karakter, Etnosains dan Kearifan Lokal (Konsep dan Penerapannya dalam Penelitian dan Pembelajaran Sains)*. Fakultas Matematika dan Ilmu Pengetahuan Alam Universitas Negeri Semarang.
Suryaman, M., Wiyatmi, W., Hartono, H., and Efendi, A. (2012). Pengembangan Model Panduan Pendidik Pengajaran Sastra Berbasis Pendidikan Karakter. *Jurnal Kependidikan: Penelitian Inovasi Pembelajaran*, *42*(1).
Tomi, D., Anggereini, E., and Muhaimin, M. (2018). The Development of Biology Learning Toolswith Kerinci's Local Wisdom-Based on Biodiversity Material for Senior High Schools. *Edu-Sains: Jurnal Pendidikan Matematika Dan Ilmu Pengetahuan Alam*, *7*(2), 11–20. https://doi.org/https://doi.org/10.22437/jmpmipa.v7i2.8203
Ul-hasanah, N. (2017). *Kurikulum 2015 PAI FINAL koding*. Dropbox. https://www.dropbox.com/scl/fi/iup-siday7s0u9qko5a0iu/Kurikulum-2015-PAI-FINAL-koding.doc?rlkey=07x272zszrzbu9980janjb5v2&e=2
Zulfiani. (2022). *Inkuiri dan Teknologi: Konsep dan praktik Pembelajaran Sains*. Rajawali Press. https://repository.uinjkt.ac.id/dspace/handle/123456789/66035.

Towards Resilient Societies: The Synergy of Religion, Education, Health, Science, and Technology – Rahiem (Ed.)

The impact of religiosity on mental health: Insights for emerging adults

Pihasniwati, Rizka Amalia, Maswani Maswani & Zidni Immawan Muslimin
Universitas Islam Negeri Sunan Kalijaga Yogyakarta, Indonesia

Erlina Listyanti Widuri
Universitas Ahmad Dahlan, Yogyakarta, Indonesia

ABSTRACT: The emerging adulthood (EA) phase is widely recognised by scholars as a period marked by complex dynamics in religious behaviour, characterised by exploration, openness, and reduced formal participation in religious rituals. During this phase, engagement in religious activities tends to decline while the risk of mental health disorders increases. Although some studies suggest that religiosity positively influences mental health, others indicate that it can also negatively impact mental well-being in emerging adults. This study aims to explore the relationship between religiosity and mental health in emerging adults, identify mediating variables between these two factors, and examine additional influences on this relationship. A Systematic Literature Review was conducted using predefined inclusion and exclusion criteria, with data from Scopus, Taylor and Francis, and JSTOR databases. The findings reveal that religiosity is associated with improved mental health outcomes, such as reduced depressive symptoms and increased life satisfaction, contributing to the Sustainable Development Goals (SDGs), particularly health and well-being. Additionally, self-regulation, life goals, and social support are key mediating variables linking religiosity to positive psychological adjustment. Other influencing factors include gender, intrinsic versus extrinsic religiosity, and the degree of religious internalisation.

Keywords: emerging adults, mental health, religiosity

1 INTRODUCTION

According to Arnett, the emerging adult period is a developmental period between the ages of 18 and 29, in late adolescence to early adulthood, characterised by five main characteristics: identity exploration, instability, self-focus, a sense of being between adolescence and adulthood, and a sense of future possibilities (Arnett *et al.* 2014). This is a critical period characterised by identity exploration, transition to independence, and major decisions about careers, relationships, and outlook on life, often accompanied by uncertainty, anxiety, and stress (Arnett *et al.* 2014).

This situation makes mental health an important issue in this phase. It becomes even more relevant when pressures from social, academic and work environments increase, while psychosocial support is sometimes inadequate. Religion is expected to be a moral and emotional foundation when facing this challenging transition phase. In terms of religiosity, young adulthood is characterised as a time when young people question: (a) the beliefs they were raised with, (b) emphasise the importance of personal spirituality over affiliation with religious institutions, and (c) take on their own the aspects of religion that best suit them (Novianty and Garey 2020). Some EA experience scepticism, agnosticism, strengthening beliefs or switching to a religion that better matches their personal values (Arnett 2014). Relatedly, for some EA, dogmatic religiosity often clashes with personal values and can lead to inner conflict, feelings of guilt, or excessive anxiety (Pihasniwati *et al.* 2023).

DOI: 10.1201/9781003645542-59

Research conducted by Koenig (2009) shows that religious beliefs and practices are believed to reduce helplessness and anxiety in problems. Religious activities such as prayer can provide comfort and life satisfaction and control individuals in the face of psychological illness (Desmond *et al.* 2018; Giannone *et al.* 2019; Kirk and Lewis 2013). However, Utami's research (in Qolbi *et al.* 2020) shows that religiosity does not affect anxiety and subjective well-being because individuals may view religiosity only as a formal religious practice, not as a coping strategy in everyday life. These two research gaps are of interest to investigate so that the research to be conducted aims to understand the relationship between religiosity on mental health and what factors affect the relationship.

Based on the description above, some of the questions raised are as follows:

(a) What is the relationship between religiosity and mental health in *emerging adults*?
(b) What variables mediate the relationship between religiosity and mental health?
(c) What other factors influence the relationship between religiosity and mental health in emerging adults?

2 METHODS

This research employs the Systematic Literature Review method, focusing on religiosity, mental health, and emerging adults. Literature selection followed inclusion and exclusion criteria (Table 1), using Scopus, JSTOR, and Taylor and Francis databases. Keywords used were "religious AND mental health AND emerging adult," with further filtering based on subject areas such as arts and humanities, social sciences, psychology, and religion. Research published between 2014 and 2024 was considered. Initial screenings based on titles and abstracts were followed by a final screening, where the content and methodology were assessed to select the relevant journals for analysis (Figure 1).

Table 1. Inclusion and exclusion criteria.

No	Inclusion Criteria	No	Exclusion Criteria
1	Research discusses religiosity and mental health	1	Does not address the link between religiosity and mental health
2	The subject of the research is emerging adults	2	Study subjects are not emerging adults
3	Documents in the form of journals	3	Not accessible
4	Journals published between 2014 - 2024	4	Not in English
5	Accessible	5	Qualitative research
6	In English		

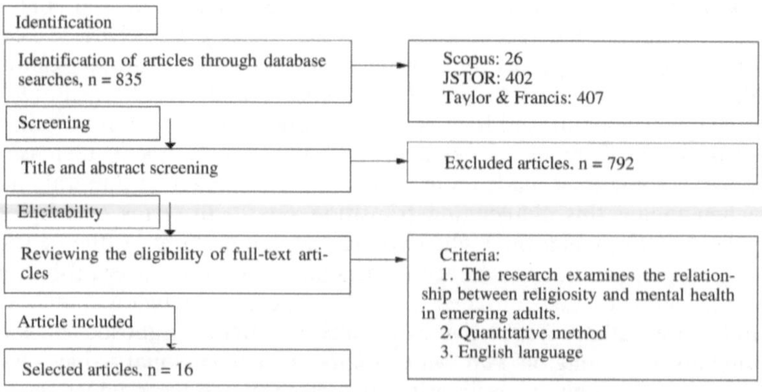

Figure 1. Literature search.

3 RESULTS AND DISCUSSION

Based on the inclusion and exclusion criteria set, the results obtained are 15 journal articles with the following description.

Journal	Results
(1) Adverse childhood experiences, religious/ spiritual struggles, and mental health symptoms: examination of mediation models (2017)	The results of this study showed a significant relationship between adverse childhood experiences (ACEs) and symptoms of post-traumatic stress disorder (PTSD) and depression in young adults. This study also found that *Religious/ Spiritual Struggles* mediated the relationship between ACEs and mental health symptoms in this case PTSD and depression.
(2)"Let the heavens be glad, and the earth rejoice": religion and life satisfaction among emerging adults in the United States	The results of this study indicate that the dimension of religiosity that has the greatest significant value with life satisfaction is private religiosity, while individuals who are spiritual but not religious show a lack of satisfaction in life.
(3) The relationships between religiosity, stress, and mental health for Muslim immigrant youth (2018)	The results of this study show that two components of religiosity, namely religious identity and religious practice, are positively related to adolescents' mental health. Religious identity can be a moderator between acculturation stress and mental health. However, religious identity does not always have a positive impact on mental health, sometimes exacerbating the effects of acculturation stress on mental health.
(4) Influence of religion, religiosity and spirituality on positive mental health of young people (2013)	This study provides results that there is a relationship between religion, religious behaviours, and perceived spirituality on Positive Mental Health (PMH). This study provides results that the religiosity scores of Muslim and Christian youths in Kerala, India are higher than Hindu youth. However, the PMH score of Hindu youth is higher than that of Muslim youth.
(5) Exploring two approaches to an existential function of religiosity in mental health (2019)	The results of this study show that life purpose mediates the relationship between religiosity and mental health. Purpose in life fully mediates the relationship between religiosity and depression but partially mediates anxiety.
(6) Relations of religious beliefs with distress and well-being among Hindu college students (2020)	The results of this study indicate that religious belief has a positive relationship with depression and anxiety in Hindu students. However, other results from this study indicate that individuals with high levels of religious belief will have a high meaning of life.
(7) Religiosity and social support: A mediational model of adjustment in emerging adults (2017)	The results of this study indicate that social support is able to mediate the relationship between religiosity and psychological well-being.
(8) Associations Between Depressive Symptoms and Religiosity in Young Adults (2020)	The results of this study suggest that religiosity beliefs and practices can contribute to better mental health.

(*continued*)

Continued

Journal	Results
(9) Multiple dimensions of religiosity, self-regulation, and psychological adjustment among emerging adults (2023)	This study divides religiosity into three dimensions: religious commitment, religious involvement, and religious engagement. The results in this study show that the dimension of religious commitment can improve self-regulation which in turn is able to have more positive self-adjustment.
(10) Intrinsic Religiousness and Spirituality as Predictors of Mental Health and Positive Psychological Functioning in Latter-Day Saint Adolescents and Young Adults (2015)	The results of this study indicate that intrinsic religiosity and spirituality are significantly associated with mental health and positive psychological functioning in LDS (Letter-Day Saints).
(11) Religion and Life Satisfaction: A Correlational Study of Undergraduate Student (2018)	The result of this study is that there is a significant relationship between religiosity and life satisfaction. However, this effect varies depending on socio-demographic contexts such as gender, religious affiliation, and type of religion practiced.
(12) Religious Integration and Psychological Distress: Different Patterns in Emerging Adult Males and Females (2018)	The results of this study showed that female participants had higher religious integration than males. In addition, there was a negative correlation between religious integration and psychological stress only in female participants. While in men no significant value was found.
(13) The Effects of Religiosity on Psychopathology in Emerging Adults: Intrinsic Versus Extrinsic Religiosity (2014)	The results of this study show that intrinsic religiousness can reduce various types of psychopathologies.
(14) The impact of religious behaviours on the health and well-being of emerging adults (2013)	The results of this study show that religious behaviours such as attendance or participation in religion and prayer/meditation can reduce the use of alcohol and addictive substances. In addition, the frequency of prayer/meditation in this study can be positively related to life satisfaction.
(15) Why are women more religious than men? Testing the explanatory power of personality theory among undergraduate students in Wales (2015)	The results of this study show that Christian or post-Christian women have a higher level of religiosity. However, this can be explained by the fact that religiosity is controlled by the basic personality dimensions of men and women.
(16) The Role of Religiosity to Address the Mental Health Crisis of Students: A Study on Three Parameters (Anxiety, Depression, and Stress) (2021)	The findings of this study indicate that religiosity and its components act as protective factors against depression and anxiety. In other words, a religious lifestyle is closely associated with reduced levels of psychological distress.

*remarks: Quantitative methods were chosen for their objectivity and applicability to large-scale studies, facilitating generalisation to emerging adults (Kandel 2020).

From these results, the discussion of the previous 3 questions is as follows:

3.1 *The relationship of religiosity to mental health in emerging adulthood*

From several articles, researchers analysed 13 out of 16 articles, which revealed that religiosity could positively impact mental health. Religiosity has a positive impact on aspects of mental health such as life satisfaction, general mental health, positive mental health, meaning of life, positive psychological functioning, decreased stress, decreased depression, decreased use of alcohol and addictive substances (Author *et al.* 2018; Desmond *et al.* 2018;

Ganga and Kutty 2013; Gwin *et al.* 2020; Kirk and Lewis 2013; Lace *et al.* 2018; Sanders *et al.* 2015; Misran *et al.* 2021; Stuart and Ward 2018).

The association of religiosity with general mental health functioning and life satisfaction is most prominent from these findings. Research conducted by Kirk and Lewis (2013) suggested that frequency of prayer and/or meditation was a significant predictor of life satisfaction and decreased cannabis use. In addition, attendance at religious services is associated with a lower risk of substance use and sexual behaviours. In addition to reducing substance use and social behaviours, religiosity also acts as a protective factor against NSSI as religious practices and principles are inversely associated with NSSI (Pihasniwati *et al.* 2023). Other research conducted by Desmond *et al.* (2018) also shows that the dimensions of religiosity that have a significant value in life satisfaction are private religiosity, namely personal devotion and religious efficacy. These personal devotions relate to prayer, scripture reading, and spiritual reflection. Meanwhile, religious efficacy relates to belief in divine support, positive experiences, and perceptions of religious influence.

Author *et al.* (2018) found a positive correlation between religiosity and life satisfaction, influenced by socio-demographic contexts such as gender, religious affiliation, and type of religion practised. However, three other studies revealed contrasting findings, indicating a positive relationship between faith and depression. Park and Kamble's (2020) study among Hindu Indian students found a positive correlation between trust and depression levels. Stuart and Ward (2018) discovered that religious identity can exacerbate acculturation stress's impact on mental health. Meanwhile, McCormick *et al.* (2017) identified a positive relationship between adverse childhood experiences, religious struggle, and depression. These findings highlight the importance of considering socio-demographic factors and religious identity when understanding the relationship between religion and mental health.

3.2 *Variables that can mediate the relationship between religiosity and mental health*

Three out of 16 articles discuss variables mediating the relationship between religiosity and emerging adult mental health. Self-regulation plays a key role in this relationship, with individuals with high religious commitment showing better psychological adjustment (Zong and Cheah 2023). Additionally, life purpose fully mediates the relationship between religiosity and depression and partially mediates religiosity and anxiety (Giannone *et al.* 2019). Other research conducted by Milevsky (2017) shows that social support can mediate the relationship between religiosity and psychological well-being. In religious practice, individuals often build strong social networks that can foster feelings of being cared for, thus improving well-being. However, it is important to prioritise the quality over the quantity of religious friends because in the research conducted by Desmond *et al.* (2018) shows that religious friends do not have a high impact on life satisfaction unless they are quality friendships.

3.3 *Related factors that are thought to influence the relationship between religiosity and mental health in emerging adults*

3.3.1 *Gender factor*

Research conducted by Lace *et al.* (2018) shows the results that women provide significant value than men in terms of religious integration with psychological discomfort. Women who have high religious integration have low psychological discomfort. It was not shown in male respondents. There are several factors explained by the researcher in this case, namely, the number of male respondents is not large and religious integration and relationship with God in women have a more significant influence on their emotional and psychological functioning. However, other studies reveal that differences in religiosity do not depend on men and women but on personality dimensions. A study conducted by Penny *et al.* (2015) found that the difference in religiosity levels was more due to the basic personalities of femininity and masculinity. Other research from Mercer and Durham, Smith, and Abu-Ali (in Penny *et al.* 2015) It was also mentioned that high femininity scores were associated with high religiosity levels in Christianity and Islam. Therefore, it is not gender that causes low religiosity levels but basic orientation and personality that can affect it.

3.3.2 *Intrinsic vs extrinsic religiosity*

Research conducted by Kuyel (in Lace *et al.* 2018) revealed that extrinsic religious orientation is related to increased levels of anxiety, depression, and interpersonal hostility. Other research conducted by Bravo and Sander (in Lace *et al.* 2018) explains that intrinsic religious orientation significantly contributes to decreased depression and anxiety as well as protective factors associated with psychological well-being.

Research by Sanders *et al.* (2015) found that intrinsic religiosity and spirituality were linked to positive mental health and psychological functioning. Higher levels of intrinsic religiosity were associated with fewer mental health issues like depression and antisocial personality problems, as shown by Power and Mckinney (2014). However, intrinsic religiosity did not significantly affect anxiety, as it may reduce some anxieties but increase others related to religious expectations. Power and Mckinney (2014) also found no link between extrinsic religiosity and psychopathology, suggesting that these findings may be specific to college students who already have positive social support.

3.3.3 *Internalisation of religiosity*

Specifically, research conducted by Zong and Cheah (2023) stated that young adults with stronger religious commitment reported higher levels of self-regulation and were able to result in positive psychological adjustment. In contrast, young adults with greater religious involvement and religious struggle report lower levels of self-regulation, which in turn, is associated with poor psychological adjustment. Such high self-regulation is influenced by good religious internalisation.

Unexamined and imposed religious beliefs become less meaningful and cannot guide individuals' decision-making and behaviours. Religious involvement that lacks a psychological component can be a picture of routine activities that are not truly internalised. It can take up a lot of one's limited psychological resources for self-regulation and ultimately lead to poor psychological adjustment.

4 CONCLUSION

Research suggests religiosity positively impacts emerging adults' mental health, enhancing life satisfaction, reducing depression, and promoting psychological adjustment (Park and Kamble 2020). Mediating factors like self-regulation, life purpose, and social support strengthen this relationship. Intrinsic religiosity and internalised spiritual practices foster psychological well-being, whereas extrinsic engagement may increase mental health risks. Fostering personal devotion practices (prayer, scripture reading, spiritual reflection) builds resilience and improves life satisfaction. Divine support and positive religious experiences enhance life satisfaction, mitigating depression, anxiety, stress, substance use, and risky behaviours.

Practitioners and religious leaders are encouraged to promote effective self-regulation and integrate religiosity as a coping strategy to address psychological issues. Religious leaders should emphasise the meaningful internalisation of rituals to ensure they contribute to well-being rather than becoming burdensome formalities. This aligns with the Sustainable Development Goals (SDGs), emphasising human well-being as a vital component of sustainable living.

REFERENCES

Arnett, J. J. (2014). Emerging Adulthood. In *Emerging Adulthood.* Oxford University Press New York. https://doi.org/10.1093/acprof:oso/9780199929382.001.0001

Arnett, J. J., Žukauskiene, R., and Sugimura, K. (2014). The new life stage of emerging adulthood at ages 18-29 years: Implications for mental health. In *The Lancet Psychiatry* (Vol. 1, Issue 7, pp. 569–576). Elsevier Ltd. https://doi.org/10.1016/S2215-0366(14)00080-7

Author, T., Habib, D. G., Donald, C., and Hutchinson, G. (2018). Religion and Life Satisfaction: A Correlational Study of Undergraduate Students in. *Source: Journal of Religion and Health*, 57(4), 1567–1580. https://doi.org/10.1007/sl0943-018-0602-6

Desmond, S. A., Kraus, R., and Dugan, B. J. L. (2018). "Let the heavens be glad, and the earth rejoice": religion and life satisfaction among emerging adults in the United States. *Mental Health, Religion and Culture*, 21(3), 304–318. https://doi.org/10.1080/13674676.2018.1478397

Ganga, N. S., and Kutty, V. R. (2013). Influence of religion, religiosity and spirituality on positive mental health of young people. *Mental Health, Religion and Culture*, 16(4), 435–443. https://doi.org/10.1080/13674676.2012.697879

Giannone, D. A., Kaplin, D., and Francis, L. J. (2019). Exploring two approaches to an existential function of religiosity in mental health. *Mental Health, Religion and Culture*, 22(1), 56–72. https://doi.org/10.1080/13674676.2019.1592136

Gwin, S., Branscum, P., Taylor, L., Cheney, M., Maness, S. B., Frey, M., and Zhang, Y. (2020). Associations Between Depressive Symptoms and Religiosity in Young Adults. *In Source: Journal of Religion and Health* (Vol. 59, Issue 6). https://about.jstor.org/terms

Kandel, B. (2020). Qualitative Versus Quantitative Research. *Marsyangdi Journal*, 1, 1–5.

Kirk, C. M., and Lewis, R. K. (2013). The impact of religious behaviours on the health and well-being of emerging adults. *Mental Health, Religion and Culture*, 16(10), 1030–1043. https://doi.org/10.1080/13674676.2012.730037

Koenig, H. (2009). *Research on Religion, Spirituality, and Mental Health: A Review*. 54(5), 283–291.

Lace, J. W., Haeberlein, K. A., and Handal, P. J. (2018). Religious Integration and Psychological Distress: Different Patterns in Emerging Adult. *Source: Journal of Religion and Health*, 57(6), 2378–2388. https://doi.org/10.1007/sl0943-018-0608-0

McCormick, W. H., Carroll, T. D., Sims, B. M., and Currier, J. (2017). Adverse childhood experiences, religious/spiritual struggles, and mental health symptoms: examination of mediation models. *Mental Health, Religion and Culture*, 20(10), 1042–1054. https://doi.org/10.1080/13674676.2018.1440544

Milevsky, A. (2017). Religiosity and social support: A mediational model of adjustment in emerging adults. *Mental Health, Religion and Culture*, 20(5), 502–512. https://doi.org/10.1080/13674676.2017.1359243

Misran, R. N., Khaiyom, J. H. A., and Razali, Z. A. (2021). The Role of Religiosity to Address the Mental Health Crisis of Students: A Study on Three Parameters (Anxiety, Depression, and Stress). *Pertanika Journal of Social Sciences and Humanities*, 29(4), 2833–2851. https://doi.org/10.47836/pjssh.29.4.40

Novianty, A., and Garey, E. (2020). Memahami Makna Religiusitas/Spiritualitas pada Individu Dewasa Muda Melalui Photovoice. *In Jurnal Psikologi Integratif* (Vol. 8, Issue 2).

Park, C. L., and Kamble, S. V. (2020). Relations of religious beliefs with distress and well-being among Hindu college students. *Mental Health, Religion and Culture*, 23(10), 902–911. https://doi.org/10.1080/13674676.2020.1856801

Penny, G., Francis, L. J., and Robbins, M. (2015). Why are women more religious than men? Testing the explanatory power of personality theory among undergraduate students in Wales. *Mental Health, Religion and Culture*, 18(6), 492–502. https://doi.org/10.1080/13674676.2015.1079603

Pihasniwati, P., Subandi, S., Mujib, A., Taqiyyah, R., Sajaroh, W., and Ningrum, D. (2023, June 16). *A Scoping Review: Religiosity/Religious Coping as An Approach for Non-Suicidal Self-Injury* (NSSI). https://doi.org/10.4108/eai.19-10-2022.2332312

Power, L., and Mckinney, C. (2014). The Effects of Religiosity on Psychopathology in Emerging Adults: Intrinsic Versus Extrinsic Religiosity. *Journal of Religion and Health*, 53(5), 1529–1538. https://doi.org/10.1007/s

Qolbi, F. H., Chotidjah, S., and Ariez, M. (2020). Masa Emerging Adulthood pada Mahasiswa: Kecemasan akan Masa Depan, Kesejahteraan Subjektif, dan Religiusitas Islam. *Psikoislamika: Jurnal Psikologi Dan Psikologi Islam*, 17(1), 44. https://doi.org/10.18860/psi.v17i1.8821

Sanders, P. W., Kawika Allen, G. E., Fischer, L., Scott Richards, P., Morgan, D. T., and Potts, R. W. (2015). Intrinsic Religiousness and Spirituality as Predictors of Mental Health and Positive Psychological Functioning in Latter-Day Saint Adolescents and Young Adults I ■ 1 CrossMark (§)' Intrinsic Religiousness and Spirituality as Predictor of Mental Health and Positive Psychological Functio in Latter-Day Saint Adolescents and Young Adults. *Source: Journal of Religion and Health*, 54(3), 871–887. https://doi.org/10.1007/sl0943-015-0043-4

Stuart, J., and Ward, C. (2018). The relationships between religiosity, stress, and mental health for Muslim immigrant youth. *Mental Health, Religion and Culture*, 21(3), 246–261. https://doi.org/10.1080/13674676.2018.1462781

Zong, X., and Cheah, C. S. L. (2023). Multiple dimensions of religiosity, self-regulation, and psychological adjustment among emerging adults. *Current Psychology*, 42(5), 4133–4142.

Towards Resilient Societies: The Synergy of Religion, Education, Health, Science, and Technology – Rahiem (Ed.)

Enhancing foreign language learning in Indonesian universities through ChatGPT: Opportunities and challenges

Kisno Umbar
Universitas Islam Negeri Syarif Hidayatullah Jakarta, Indonesia
ORCID ID: 0000-0002-8896-8117

Azkia Muharom Albantani
Universitas Islam Negeri Syarif Hidayatullah Jakarta, Indonesia
ORCID ID: 0000-0001-9080-4149

Khabib Syaikhu Rohman
Mohammed bin Zayed University for Humanities (MBZUH), Abu Dhabi, United Arab Emirates

ABSTRACT: This study investigates using artificial intelligence (AI), specifically ChatGPT, as a digital language learning assistant in foreign language instruction at higher education institutions in Indonesia. Employing a mixed-methods approach, the research combines quantitative surveys and qualitative interviews to explore users' experiences with ChatGPT. The findings reveal that many students utilise ChatGPT as a language learning tool, contributing to improvements in their language skills. However, several challenges persist, including technical limitations, insufficient lecturer training, and resistance to adopting new teaching methods. The discussion highlights the need for greater support, such as comprehensive training for educators and improved technical infrastructure, to fully harness the benefits of ChatGPT in language education. Despite these challenges, the study concludes that ChatGPT holds significant potential to enhance foreign language teaching in higher education. Additionally, the study recommends strategies for improving lecturer training and integrating AI into the curriculum to optimise its educational impact.

Keywords: artificial intelligence, ChatGPT, digital language learning assistant, Indonesia

1 INTRODUCTION

In the rapidly evolving digital era, artificial intelligence (AI) technology has become one of the key pillars in transforming education worldwide. These technological improvements have opened up new opportunities for more interactive, personalised, and efficient language learning (Cota-Rivera *et al.* 2024). The existence of adaptive learning systems in AI adjusts content and speed based on personal performance, accelerating learning progress and serving diverse educational needs (Jameer Md 2024). Furthermore, the existence of AI also creates immersive learning experiences that improve language retention and practical applications (Jameer Md 2024). This technology allows learners to experience language in diverse cultural and situational contexts, improving linguistic skills and cultural understanding.

Research on AI in Indonesia has shown important progress in many areas, such as education, health, and the economy. Studies indicate that AI helps make teaching more efficient in colleges and creates more adaptable learning methods in schools. While there is plenty of

 DOI: 10.1201/9781003645542-60

research on AI in education, there is little focus on how tools like ChatGPT can assist with learning languages. Most studies look at general technologies rather than the specific advantages of AI language models like ChatGPT. For example, Civit *et al.* (2024) which researched the use of AI-tutoring methods in universities, Oyarzún Yáñez and Rodríguez Rodríguez (2024) which compares students' writing experiences with traditional and utilises Chat GPT, in addition to the research of Bartolome Muñoz De Luna and Martin Gomez (2024) which examines students' ability to utilise AI efficiently. In the context of Indonesia, Hatmanto and Sari (2023) examines the congruence between Chat GPT and well-established theoretical frameworks and best practices in the field of English language education. In addition, ChatGPT has shown great potential in supporting language learning due to its ability to provide contextual and interactive responses in multiple languages (Alam and Asmawi 2024). Several studies have shown the effectiveness of ChatGPT in helping the teaching and learning process in a more personalised and adaptive way than traditional methods (Babu and Adhithya 2023). The potential of Chat GPT in teaching has been tested in fact, but the majority of its research has focused on aspects of learning in general, specifically alluding to the teaching of existing languages (Alam and Asmawi 2024; Hatmanto and Sari 2023). However, despite its great potential, the use of Chat GPT in language teaching in the context of Indonesian language learning is still minimally explored and requires more in-depth study.

Research on AI in education is growing quickly, but there are few studies on using ChatGPT for language learning in Indonesian universities. Most research focuses on AI technology in general or on traditional teaching methods. It often overlooks important challenges, such as internet access and digital literacy, and how these issues affect teaching methods and learning strategies. This study aims to explore the opportunities of ChatGPT as a digital language learning assistant in Indonesian universities, focusing on enhancing language skills, identifying integration challenges for lecturers and students (such as internet accessibility and digital literacy), and assessing its influence on teaching methods and learning strategies.

2 METHOD

This study uses a mixed methods approach with *an explanatory sequential* design, where the research begins with quantitative data collection in the first stage, followed by qualitative exploration in the second stage. This approach was chosen to provide a thorough understanding of the use of ChatGPT as a digital language learning assistant in universities in Indonesia, focusing on the potential use of AI, its challenges, and its impact (Creswell and Clark 2017).

The population of this research consists of 530 university students enrolled in foreign language programs at a university in Jakarta, Indonesia. The sample size is determined using Slovin's formula with a 5% margin of error, resulting in a total of 228 respondents. The sampling process adopts a simple random sampling technique to ensure every student in the population has an equal chance of being selected.

The research is structured in two stages: initially involving quantitative data analysis from a Likert scale-based questionnaire, followed by qualitative data collection through in-depth interviews with 10 purposefully selected students familiar with ChatGPT. The semi-structured interviews aim to explore the challenges and perceptions surrounding ChatGPT's impact on teaching and learning, utilising a thematic analysis approach. This process includes several key steps: familiarisation, coding, theme generation, theme review, theme definition, and naming the themes (Terry *et al.* 2017).

This study also pays attention to the ethical aspects of research by ensuring that all participants give *informed consent*, as well as maintaining the anonymity and confidentiality of the collected data (Bryman 2016).

3 FINDINGS AND DISCUSSION

The study measured students' perceptions of ChatGPT as a foreign language learning assistant through 15 questions. These questions were consolidated into a single composite variable, which reflects the overall perception of ChatGPT's effectiveness and acceptance in language learning. A descriptive analysis of the results is also provided.

Table 1 shows 228 respondents in this study, which shows an adequate sample size to describe student perceptions. The average score is 101.21, which indicates a generally positive perception of using ChatGPT as a language learning assistant. A standard deviation of 19.22 indicates considerable variation among respondents, with a minimum score of 59 and a maximum of 147, reflecting a wide range of perceptions from less to heavily helped by ChatGPT. The percentile also showed a fairly high distribution, with 50% of respondents scoring below 98 and 75% below 116, indicating that most respondents gave a relatively high rating. However, there were significant differences between individuals.

Table 1. Descriptive statistical test results.

Metric	Value
Count	228
Mean	101.2105263
Standard Deviation	19.2243759
Minimum	59
25th Percentile	92
Median	98
75th Percentile	116
Maximum	147

Descriptive analysis shows that Indonesian students generally have a positive perception of using ChatGPT as an assistant in foreign language learning. Most respondents felt that ChatGPT provided clear and relevant answers to the questions, with a consistent average score of 3 to 4 on the Likert scale (3 = neutral, 4 = agree). However, there are indications that although ChatGPT is considered beneficial, the frequency of use and motivation for learning languages using ChatGPT can still be increased.

Furthermore, some respondents admitted that ChatGPT influences the ability to speak, write, and read in foreign languages. Consider the following Table 2 data:

Table 2. ChatGPT and language skills.

Item	Strongly disagree	Disagree	Neutral	Agree	Strongly agree
speaking	26,3%	21,1%	47,4%	5,3%	0%
writing	15,8%	15,8%	42,1%	21,1%	5,3%
reading	10,5%	15,8%	31,6%	36,8%	5,3%

Table 2 indicates that most respondents remained neutral regarding the impact of ChatGPT on foreign language learning, particularly in speaking (47.4%) and writing (42.1%). However, there is a notable increase in positive responses for reading skills, with 36.8% agreeing and 5.3% strongly agreeing. These findings suggest that ChatGPT is more effectively utilised as a digital language learning assistant to support reading and writing skills than speaking.

When asked about their satisfaction with using ChatGPT as a language learning tool, 21.1% of respondents agreed, and 15.8% strongly agreed. In contrast, 36.8% remained neutral, while 26.3% disagreed. Respondents had mixed feelings about whether ChatGPT can help them learn the language: 42.1% agreed, while another 42.1% were neutral. Additionally, 5.3% agreed, and 5.3% strongly disagreed or disagreed.

The data shows that ChatGPT has not been fully helpful in teaching foreign languages. Most respondents chose neutral answers rather than agree or strongly agree. However, this suggests that college students are starting to see some benefits of using ChatGPT as a teaching aid for foreign languages. These findings are relevant to the SDGs 4 (Quality Education) point, which focuses on inclusive and quality education and supports lifelong learning (UNESCO 2017). ChatGPT aims to use technology to enhance learning. It supports the development of important 21st-century skills, such as digital literacy and cross-cultural communication. Additionally, it helps provide a curriculum that is more flexible and relevant for today's digital world (Hidayati *et al.* 2024; Ramli *et al.* 2024). Thus, this technology contributes to efforts to achieve SDG 4 targets in higher education.

3.1 *Technology adaptation and challenges in the use of chatgpt*

Using technology in language learning with ChatGPT shows that students who are good at using technology are often more adaptable and benefit more from this tool. They like the flexibility and interactivity ChatGPT offers, which helps them learn in a more personalised way. On the other hand, students who are not comfortable with technology find it hard to use these tools at first. This situation highlights the need for digital skills to adapt to new technology. Research supports this idea, showing that being comfortable with technology and having good digital skills significantly affect how well students can use technology for learning.

Additionally, the personalisation of learning offered by ChatGPT, where students can customise the material and learning style according to their individual needs, is highly appreciated by tech-savvy students. This personalisation has been shown to improve learning engagement and effectiveness, in line with the findings (Krause *et al.* 2015), which show that technologies that enable personalisation can significantly improve learning outcomes. In the context of education in Indonesia, research by Dwihadiah *et al.* (2024) also supports that the personalisation provided by ChatGPT can enrich students' learning experiences. However, access and digital literacy challenges still need to be considered.

Using ChatGPT for language learning has some challenges. One issue is that the answers it provides can be too generic or irrelevant, often leading users to ask their questions again or ask for clarification. This problem is especially common when questions are unclear, highlighting how AI can struggle to understand specific situations. It can also be tough for users with limited tech skills to ask clear questions. Research supports these points, showing that how easily users find the technology to use and the quality of the responses influence whether they accept it.

In addition, unstable internet connections are also a major obstacle, especially for users who rely on ChatGPT for daily learning. This instability can disrupt the learning curve and reduce the effectiveness of using this technology. Research by Bahrini *et al.* (2023) highlights a similar problem, where inadequate internet infrastructure is a major barrier to adopting AI-based learning technologies. While ChatGPT offers significant benefits, its successful use depends on the user's ability to overcome technological challenges.

3.2 *The effectiveness of ChatGPT compared to traditional methods*

The use of ChatGPT in language learning shows an advantage in speed and efficiency compared to traditional methods. Many participants felt that ChatGPT could provide instant solutions that are difficult to find in textbooks, improving the efficiency of their study time. The interactivity offered by ChatGPT, which allows for direct feedback, is also the reason why some users prefer it over traditional methods. Recent studies, such as those conducted by Davidson (2023), suggest that learning-based AI can accelerate the learning process and improve student engagement.

However, despite ChatGPT's effectiveness, some participants are still more comfortable with traditional methods, especially for more in-depth and structured references. This preference reflects findings by Chang *et al.* (2021) who found that while digital learning technologies offer many benefits, students still often return to traditional methods for materials that require a deeper understanding. This shows that ChatGPT and traditional methods can complement each other in language learning.

3.3 *Motivation to learn with ChatGPT*

The use of ChatGPT in language learning has been shown to increase student's motivation to learn, especially due to its flexibility and accessibility. Most participants reported feeling more motivated to learn because they could access help anytime and anywhere without being bound by rigid schedules. These findings are supported by recent research, such as those conducted by Zaghlool and Khasawneh (2023), which shows that AI technology can improve student engagement and motivation by providing a more personalised and flexible learning experience.

Additionally, many users see ChatGPT as a support tool that complements their primary learning method. They use ChatGPT to overcome specific difficulties that traditional methods cannot overcome, making it an effective complementary tool. Research by Wong (2024) shows that the use of digital learning technology as a supporting tool can improve learning outcomes, especially when integrated with conventional methods. This suggests that ChatGPT is most effective when used with traditional learning methods, offering additional assistance to improve understanding and learning outcomes.

4 CONCLUSION

Students generally have a positive view of using ChatGPT for language learning. They find it helpful for improving their language skills because it is flexible, fast, and interactive. ChatGPT also motivates students to learn and offers a more personalised learning experience. However, students have different opinions, and some still prefer traditional methods for deeper and more structured learning.

The study found several challenges in using ChatGPT in teaching and learning. These include problems like inconsistent internet access, difficulties in asking the right questions, and a lack of training for teachers on how to use AI technology. The research tool used in this study is reliable, ensuring that the results accurately reflect student opinions. Despite the challenges, the study concludes that ChatGPT can significantly improve foreign language teaching in universities, as long as schools provide better infrastructure and training for teachers to make the most of it.

These insights highlight the importance of focused efforts to improve lecturer training programs and incorporate AI technologies such as ChatGPT into the curriculum. By tackling these challenges and bolstering infrastructure support, universities can maximise the potential of ChatGPT to foster a more personalised, adaptive, and effective language learning experience.

REFERENCES

Alam, M. S., and Asmawi, A. (2024). Chat GPT in Constructivist Language Pedagogy: Exploring Its Potential and Challenges in Legal Vocabulary Acquisition. *Applied Research on English Language*, 13(1), 109–126. Scopus. https://doi.org/10.22108/are.2024.139708.2188

Babu, C. V. S., and Adhithya, S. (2023). ChatGPT and other generative AI tools in education: Transformative potential and ethical considerations. In *Creat. AI Tools and Ethical Implic. In Teach. And Learn.* (pp. 135–152). *IGI Global*; Scopus. https://doi.org/10.4018/979-8-3693-0205-7.ch007

Bahrini, A., Khamoshifar, M., Abbasimehr, H., Riggs, R. J., Esmaeili, M., Majdabadkohne, R. M., and Pasehvar, M. (2023). ChatGPT: Applications, Opportunities, and Threats. *2023 Systems and Information Engineering Design Symposium (SIEDS)*, 274–279. https://doi.org/10.1109/SIEDS58326.2023.10137850

Bartolome Muñoz De Luna, A., and Martin Gomez, S. (2024). Advances and challenges of Artificial Intelligence in the university context: An empirical study. *Journal of Infrastructure, Policy and Development*, 8(13), 8388. https://doi.org/10.24294/jipd.v8i13.8388

Bryman, A. (2016). *Social Research Methods* (Fifth edition). Oxford University Press.

Chang, C.-Y., Lee, D.-C., Tang, K.-Y., and Hwang, G.-J. (2021). Effect sizes and research directions of peer assessments: From an integrated perspective of meta-analysis and co-citation network. *Computers and Education*, 164, 104123. https://doi.org/10.1016/j.compedu.2020.104123

Civit, M., Escalona, M. J., Cuadrado, F., and Reyes-de-Cozar, S. (2024). Class integration of ChatGPT and learning analytics for higher education. *Expert Systems*, 41(12), e13703. https://doi.org/10.1111/exsy.13703

Cota-Rivera, E. I., González Correa, M. E., Bernal Marín, L. A., Marquez Montenegro, M. Y., Herrera, A. M., and Martinez Martinez, M. A. A. (2024). Transforming Education With the Power of Artificial Intelligence: Case Studies. In Ş. Ş. Demir and M. Demir (Eds.), *Advances in Higher Education and Professional Development* (pp. 113–140). IGI Global. https://doi.org/10.4018/979-8-3693-1666-5.ch006

Creswell, J. W., and Clark, V. L. P. (2017). *Designing and Conducting Mixed Methods Research.* Sage publications.

Davidson, C. C. (2023). Enhancing outcomes, fostering values, and shaping pedagogy with ChatGPT in english learning courses: AI in two value-oriented english language courses. In *Transform. The Lang. Teach. Exp. In the Age of AI* (pp. 234–255). *IGI Global*; Scopus. https://doi.org/10.4018/978-1-6684-9893-4.ch013

Dwihadiah, D., Gerungan, A., and Purba, H. (2024). Penggunaan ChatGPT di Kalangan Mahasiswa dan Dosen Perguruan Tinggi Indonesia. *CoverAge: Journal of Strategic Communication*, 14(2), 130–145.

Hatmanto, E. D., and Sari, M. I. (2023). Aligning Theory and Practice: Leveraging Chat GPT for Effective English Language Teaching and Learning. *E3S Web of Conferences*, 440, 05001. https://doi.org/10.1051/e3sconf/202344005001

Hidayati, S., Hidayah, N., and Kamarusdiana. (2024). Sustainability challenges of community-based social security program in Indonesia: Effective collaboration and digital technology adoption. In M. D. H. Rahiem, *Religion, Education, Science and Technology towards a More Inclusive and Sustainable Future* (1st ed., pp. 249–253). Routledge. https://doi.org/10.1201/9781003322054-42

Jameer Md, Ms. R. (2024). Augmenting EFL/ ESL Education: The Outcome Of Technology On Contemporary Language Learning. *Educational Administration: Theory and Practice*, 30(5), 2196–2202. https://doi.org/10.53555/kuey.v30i5.3260

Krause, M., Mogalle, M., Pohl, H., and Williams, J. J. (2015). A Playful Game Changer: Fostering Student Retention in Online Education with Social Gamification. *Proceedings of the Second (2015) ACM Conference on Learning @ Scale*, 95–102. https://doi.org/10.1145/2724660.2724665

Oyarzún Yáñez, R., and Rodríguez Rodríguez, M. N. (2024). Writing in University with AI Support: Anxiety and Perceived Usefulness in Using Chat GPT. *European Public and Social Innovation Review*, 9, 1–15. https://doi.org/10.31637/epsir-2024-922

Ramli, M., Fatra, M., Murtadlo, M., Albana, H., Hana Susanti, B., and Aldeia, S. (2024). Navigating the tech-savvy generation; key considerations in developing of an artificial intelligence curriculum. *IAES International Journal of Artificial Intelligence (IJ-AI)*, 13(4), 3942. https://doi.org/10.11591/ijai.v13.i4.pp3942-3950

Terry, G., Hayfield, N., Clarke, V., and Braun, V. (2017). Thematic analysis. *The SAGE Handbook of Qualitative Research in Psychology*, 2(17–37), 25.

UNESCO. (2017). *Education for Sustainable Development Goals: Learning objectives.* UNESCO. https://doi.org/10.54675/CGBA9153

Wong, K. K. (2024). Blended Learning and AI: Enhancing Teaching and Learning in Higher Education. In W. W. K. Ma, C. Li, C. W. Fan, L. H. U, and A. Lu (Eds.), *Blended Learning. Intelligent Computing in Education* (Vol. 14797, pp. 39–61). Springer Nature Singapore. https://doi.org/10.1007/978-981-97-4442-8_4

Zaghlool, Z. D., and Khasawneh, M. A. S. (2023). Incorporating the Impacts and Limitations of AI-Driven Feedback, Evaluation, and Real-Time Conversation Tools in Foreign Language Learning. *Migration Letters*, 20(7), 1071–1083.

Towards Resilient Societies: The Synergy of Religion, Education, Health, Science, and Technology – Rahiem (Ed.)

Understanding scrupulosity in adults: A comprehensive systematic review

Salwa Alfina Siregar, Gusty Syaiim Framita, Aulia Tiara Savitri & Rena Latifa
Universitas Islam Negeri Syarif Hidayatullah Jakarta, Indonesia

ABSTRACT: Scrupulosity is a complex form of obsessive-compulsive disorder (OCD) characterised by excessive concerns about morality, ethics, and spirituality, often deeply intertwined with religious beliefs and predominantly affecting adults. This systematic review, conducted in accordance with PRISMA guidelines, analysed 1,525 publications from Science Direct and PubMed. After applying inclusion criteria focused on quantitative studies from the past five years, 10 relevant articles were selected. The findings reveal that scrupulosity involves an intense preoccupation with sin and moral correctness, with cognitive behavioural therapy (CBT) and selective serotonin reuptake inhibitors (SSRIs) identified as effective treatment options. The review underscores the need for personalised interventions that integrate pharmacological and psychological therapies, aligning with Sustainable Development Goal 3 (Good Health and Well-being). It also highlights the importance of recognising and addressing scrupulosity in clinical practice to enhance mental health outcomes. Furthermore, the study recommends that future research develop more inclusive treatment models and assessment strategies to support better individuals struggling with this condition.

Keywords: adult, CBT, obsessive-compulsive disorder in religion, OCD, scrupulosity

1 INTRODUCTION

The phenomenon of scrupulosity, often described as "religious OCD," presents a unique intersection of obsessive-compulsive disorder (OCD) and deeply held moral or spiritual beliefs. According to the Diagnostic and Statistical Manual of Mental Disorders (DSM) 5, obsessive-compulsive disorder (OCD) is a mental illness characterised by recurrent experiences of unwanted, distressing, and intrusive thoughts (obsessions) and (compulsions) that cannot be controlled. Compulsive behaviours are performed to alleviate the anxiety associated with the obsessions (Van Roessel *et al.* 2023).

In DSM-5, OCD is divided into four parts, one of which is scrupulosity. Scrupulosity is a well-known but rare form of OCD (Abramowitz and Jacoby 2014). Scrupulosity is an excessive fear of contextualising religion or morality (Siev *et al.* 2017). In literature, the terms religious OCD, religious obsessions, and compulsions are often used together with scrupulosity interchangeably in research.

One primary challenge in diagnosing scrupulosity is distinguishing it from genuine religious devotion. The symptoms can often mimic sincere concern for moral behaviours, making it difficult for both patients and healthcare providers to recognise the underlying psychological disorder. Additionally, the subjective nature of religious experiences complicates the assessment process, as individuals may be hesitant to disclose their obsessive thoughts due to fear of judgment or misunderstanding from both mental health professionals and religious leaders.

 DOI: 10.1201/9781003645542-61

Treatment for scrupulosity also faces hurdles. Many individuals with scrupulosity fear that therapeutic interventions may conflict with their religious beliefs or diminish their faith. This concern can create resistance to seeking help or adhering to treatment protocols. Furthermore, there is a lack of standardised assessment tools designed explicitly for scrupulosity, which can lead to inconsistencies in diagnosis and treatment approaches.

This systematic review aims to address these gaps in the literature by synthesising current findings on scrupulosity, focusing on its diagnosis, assessment, and effective treatment strategies. By critically evaluating recent studies and clinical practices over the past five years, this review will provide a comprehensive overview of the challenges associated with scrupulosity and highlight the most effective therapeutic interventions.

2 METHOD

This study employs the Systematic Literature Review approach. This writing technique begins by searching the literature on a specific topic to identify the literature based on inclusion and exclusion criteria. Thus, it can be synthesised following the study's aims (Selcuk 2019). In this SLR review, the researcher followed the Preferred Reporting Items for Systematic Reviews and Meta-Analyses (PRISMA) procedure using the Science Direct and PubMed databases. The PRISMA framework was chosen for this systematic review due to its established guidelines for conducting systematic reviews and meta-analyses. Utilising PRISMA ensures a structured approach that enhances the transparency and reproducibility of the review process. It also facilitates comprehensive reporting of the methods used, which is essential for evaluating the quality and reliability of the findings.

2.1 *Literature identification*

Science Direct and PubMed were the two databases used to search the literature. We used the search term "scrupulosity OR obsessive-compulsive in religion AND adult." The use of these keywords is a literature search method based on Boolean operators, which are search operators that can assist in entering and combining words/concepts throughout the literature search process. OR: to mix alternative terms or synonyms to broaden search results; AND: to combine concepts to get more specific results; NOT: to remove certain terminology from the search.

2.2 *Selection criteria*

The selection criteria for this review were as follows:

Inclusion Criteria (Table 1):

- Quantitative research articles published in English within the last five years.
- Studies focusing explicitly on scrupulosity in adult populations.
- Articles that discuss treatment methods or assessment tools for scrupulosity.

Table 1. Inclusion criteria.

Type	Inclusion
Article	Research Article (Full text)
Publication Year	2019-2024
Sample	Adults (up to 18 years old)
Methods	Quantitative
Language	English
Article Topic	Literature about scrupulosity

Exclusion Criteria:

- Non-English publications to avoid potential biases in interpretation and understanding of the literature. Excluding non-English studies was necessary to mitigate bias related to language interpretation and comprehension. Language barriers can lead to misinterpretation of research findings and cultural nuances that may affect understanding scrupulosity. By focusing solely on English-language publications, we aimed to maintain consistency in terminology and conceptualisation across the studies reviewed, thereby enhancing the validity of our conclusions.
- Qualitative studies, case reports, or opinion pieces that do not provide quantitative data relevant to the objectives of this review.
- Articles not focused specifically on scrupulosity or those that address unrelated psychological conditions.

Following these criteria, a total of 1,525 publications were initially identified. In the screening stage, 202 articles were excluded based on the criteria of publication in English in the last 5 years, articles with quantitative methods, and the topic of scrupulosity in adults. In the eligibility stage, the researchers read the full texts of the 27 articles. The researchers did not find any duplicate articles. In the final stage, researchers concluded that only 10 journals met the criteria and synthesised them according to the research objectives. The Diagram can be seen in Figure 1.

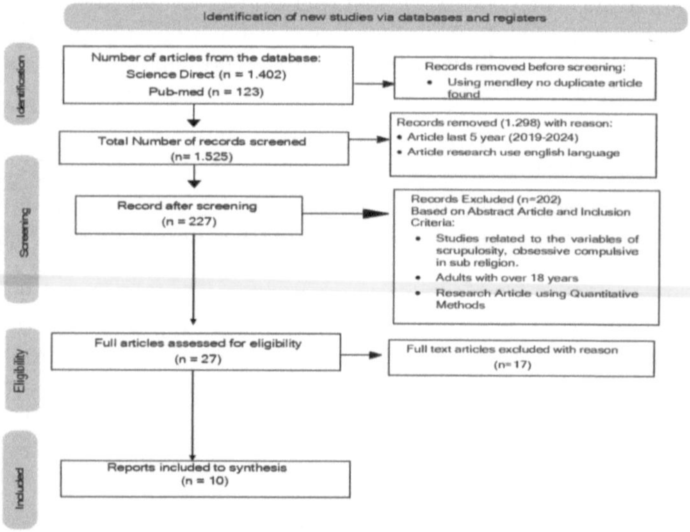

Figure 1. PRISMA diagram: Scrupulosity in adults.

3 FINDINGS AND DISCUSSION

This systematic review aimed to synthesise current findings on scrupulosity, focusing on its diagnosis, assessment, and treatment in adults. The review highlighted several key findings: (1) the definition and characteristics of scrupulosity, (2) the effective treatment of scrupulosity, and (3) assessment tools for the diagnosis and evaluation of scrupulosity.

3.1 *The definition and characteristics of scrupulosity*

Scrupulosity is a subtype of obsessive-compulsive disorder (OCD) characterised by excessive concerns about sin and morality, often with religious implications. This condition can significantly impair the quality of life and mental health, particularly among vulnerable populations such as college students (Mancini *et al.* 2023). Common symptoms include fear of committing sins, intrusive thoughts about offending God, compulsive rituals like repeated prayers, and avoidance of triggering religious texts. The development of scrupulosity can be influenced by traumatic experiences that heighten moral anxieties, as well as religious beliefs where a positive view of God may exacerbate symptoms (Pinciotti and Fisher 2022).

3.2 *Effective treatment of scrupulosity*

Effective treatment for scrupulosity focuses on addressing the guilt and anxiety associated with moral concerns, primarily through Cognitive Behavioural Therapy (CBT). CBT helps individuals recognise and modify harmful thought patterns, with Exposure and Response Prevention (ERP) playing a crucial role by gradually exposing patients to anxiety-provoking situations while guiding them to resist compulsive behaviours like excessive praying (Mancini *et al.* 2023; Senderecka *et al.* 2019). In addition to therapy, medications such as selective serotonin reuptake inhibitors (SSRIs) may be used to alleviate severe symptoms by increasing serotonin levels in the brain. The combination of therapy and medication often leads to better outcomes for those who struggle with scrupulosity (Mancini *et al.* 2023; Myers *et al.* 2021). Collaborative interventions, including group or family therapy and relaxation techniques like meditation and deep breathing, can further support treatment by fostering connection and enhancing emotional resilience (Myers *et al.* 2021). Incorporating guidance from spiritual advisors can also provide a balanced perspective, addressing the spiritual dimensions of scrupulosity (Mancini *et al.* 2023; Senderecka *et al.* 2019).

3.3 *Assessment tools for the diagnosis and evaluation of scrupulosity*

To effectively treat scrupulosity, evaluating its varying states in individuals is essential, which necessitates the development of accurate assessment tools for proper diagnosis and intervention (Myers *et al.* 2021). Instruments such as the Daily Spiritual Experiences Scale (DSES) by Capel *et al.* (2023), Penn Inventory of Scrupulosity (PIOS) by Abramowitz *et al.* (2002), and Yale-Brown Obsessive Compulsive Scale (Y-BOCS) by Rosario-Campos *et al.* (2006) help measure symptom severity and identify specific obsessions and compulsions. Recent advancements include the Scrupulosity Inventory (SI) from Miller *et al.* (2024), which has shown strong predictive capabilities for diagnosing scrupulosity, and the Dimensional Obsessive-Compulsive Scale-Scrupulosity or Religion (DOCS-SR) by Wetterneck *et al.* (2021), which assesses stress and avoidance related to moral or religious issues. These tools are vital for understanding the distinct clinical features of scrupulosity compared to OCD and ensuring that treatment aligns with individuals' religious beliefs.

3.4 *The heterogeneity of the studies*

The review of studies reveals significant heterogeneity in sample size, methodology, and cultural context, complicating the synthesis of findings while enriching the understanding of diverse phenomena. For instance, Pinciotti and Fisher (2022) utilised a large sample, enhancing generalizability, whereas Abdel Hamid *et al.* (2019) had a smaller cohort, potentially limiting robustness. Methodological discrepancies also exist, with Miller *et al.* (2024) employing a mixed methods approach while Wetterneck *et al.* (2021) focusing solely on quantitative measures. Additionally, cultural context plays a crucial role; for example, Myers *et al.* (2021) studied scrupulosity within a Western framework, contrasting with De Jong and Cook (2021) non-Western perspective. These variations underscore the need for more accurate assessment tools and inclusive research designs to understand scrupulosity across different populations better.

3.5 *Clinical application*

Clinicians treating scrupulosity must consider cultural contexts to enhance treatment effectiveness and patient rapport. Practical applications include using culturally adapted assessment tools like the Yale-Brown Obsessive Compulsive Scale (Y-BOCS), which has been translated into various languages for diverse populations; involving religious leaders in therapy, especially in religious cultures, to bridge clinical practices with spiritual beliefs; and adapting cognitive behavioural therapy (CBT) to incorporate culturally relevant techniques that address specific religious fears. Additionally, educating patients and families about scrupulosity can demystify the condition and reduce stigma, while integrating mindfulness practices tailored to the patient's beliefs can help manage anxiety. Ethical considerations are also crucial, as clinicians must respect patients' religious beliefs, avoid pathologising normal religious behaviours, and ensure informed consent when incorporating religious figures into treatment plans (Myers *et al.* 2021; Wetterneck *et al.* 2021).

3.6 *Limitations of the study and recommendation*

This review study highlights several limitations regarding the research on scrupulosity, particularly its focus on Western populations and Christian samples, which may not accurately represent the experiences of individuals from diverse cultural backgrounds, such as Muslims. The reliance on convenience sampling can skew results, as seen in Pittenger *et al.* (2021), whose small, homogeneous sample may limit generalisability. Additionally, researchers' cultural backgrounds and beliefs can introduce biases in interpreting data related to scrupulosity. To address these gaps, future research should conduct thorough reviews to identify emerging trends, formulate specific research questions, employ diverse methodologies, engage interdisciplinary researchers for innovative approaches, and prioritise longitudinal designs better to understand the long-term impacts of scrupulosity across different cultures.

4 CONCLUSION

This literature review describes the psychological conditions, treatment, and measurement of scrupulosity in adults. In conclusion, the heterogeneity observed across the reviewed studies underscores the complexities inherent in synthesising research findings from diverse methodologies, sample sizes, and cultural contexts. While this variability presents challenges for interpretation and generalisation, it also enriches our understanding by highlighting different dimensions of human behaviours and societal phenomena. Future research should prioritise methodological rigour and consider cultural contexts to enhance the validity and applicability of findings across various settings. Addressing scrupulosity within diverse cultural contexts requires clinicians to adopt culturally sensitive approaches that respect patients' beliefs while providing effective treatment strategies. By integrating practical applications tailored to individual cultural needs, clinicians can enhance therapeutic outcomes while navigating ethical considerations and potential biases inherent in existing research.

REFERENCES

Abdel Hamid, A. A. L., Nasreldin, M., Gohar, S. M., Saleh, A. A., and Tarek, M. A. (2019). Sexual and religious obsessions in relation to suicidal ideation in bipolar disorder. *Suicide and Life-Threatening Behavior*, 49(6), 1552–1559. https://doi.org/10.1111/sltb.12540

Abramowitz, J. S., D. Huppert, J., Cohen, A. B., Tolin, D. F., and Cahill, S. P. (2002). Religious obsessions and compulsions in a non-clinical sample: The Penn Inventory of Scrupulosity (PIOS). *Behaviour Research and Therapy*, 40(7), 825–838. https://doi.org/10.1016/S0005-7967(01)00070-5

Abramowitz, J. S., and Jacoby, R. J. (2014). Scrupulosity: A cognitive–behavioral analysis and implications for treatment. *Journal of Obsessive-Compulsive and Related Disorders*, 3(2), 140–149. https://doi.org/10.1016/j.jocrd.2013.12.007

Capel, L. K., Zurita Ona, P., Moller, C., and Twohig, M. P. (2023). An open trial of acceptance and commitment therapy with exposure and response prevention in an intensive outpatient setting for adults with OCD. *Cognitive and Behavioral Practice*, 30(2), 218–228. https://doi.org/10.1016/j.cbpra.2022.01.004

De Jong, D. C., and Cook, C. (2021). Roles of religiosity, obsessive–compulsive symptoms, scrupulosity, and shame in self-perceived pornography addiction: A preregistered study. *Archives of Sexual Behavior*, 50(2), 695–709. https://doi.org/10.1007/s10508-020-01878-6

Mancini, C. J., Quilliam, V., Camilleri, C., and Sammut, S. (2023). Spirituality and negative religious coping, but not positive religious coping, differentially mediate the relationship between scrupulosity and mental health: A cross-sectional study. *Journal of Affective Disorders Reports*, 14, 100680. https://doi.org/10.1016/j.jadr.2023.100680

Miller, C. H., Hedges, D. W., Brown, B., Olsen, J., and Baughan, E. C. (2024). Development of the scrupulosity inventory: A factor analysis and construct validity study. *Journal of Behavior Therapy and Experimental Psychiatry*, 83, 101926. https://doi.org/10.1016/j.jbtep.2023.101926

Myers, S. G., Pirutinsky, S., and Rosmarin, D. H. (2021). Implicit and explicit beliefs about god and scrupulosity symptoms: A prospective study. *Journal of Obsessive-Compulsive and Related Disorders*, 28 (October 2020), 100621. https://doi.org/10.1016/j.jocrd.2021.100621

Pinciotti, C. M., and Fisher, E. K. (2022). Perceived traumatic and stressful etiology of obsessive-compulsive disorder. *Psychiatry Research Communications*, 2(2), 100044. https://doi.org/10.1016/j.psycom.2022.100044

Pittenger, C., Brennan, B. P., Koran, L., Mathews, C. A., Nestadt, G., Pato, M., Phillips, K. A., Rodriguez, C. I., Simpson, H. B., Skapinakis, P., Stein, D. J., and Storch, E. A. (2021). Specialty knowledge and competency standards for pharmacotherapy for adult obsessive-compulsive disorder. *Psychiatry Research*, 300, 113853. https://doi.org/10.1016/j.psychres.2021.113853

Rosario-Campos, M. C., Miguel, E. C., Quatrano, S., Chacon, P., Ferrao, Y., Findley, D., Katsovich, L., Scahill, L., King, R. A., Woody, S. R., Tolin, D., Hollander, E., Kano, Y., and Leckman, J. F. (2006). The dimensional yale–brown obsessive–compulsive scale (DY-BOCS): An instrument for assessing obsessive–compulsive symptom dimensions. *Molecular Psychiatry*, 11(5), 495–504. https://doi.org/10.1038/sj.mp.4001798

Selcuk, A. A. (2019). A guide for systematic reviews: PRISMA. *Turkish Archives of Otorhinolaryngology*, 57 (1), 57–58. https://doi.org/10.5152/tao.2019.4058

Senderecka, M., Kossowska, M., Sekerdej, M., and Szewczyk, J. (2019). Religious fundamentalism is associated with hyperactive performance monitoring: ERP evidence from correct and erroneous responses. *Biological Psychology*, 140, 96–107. https://doi.org/10.1016/j.biopsycho.2018.12.007

Siev, J., Huppert, J. D., and Zuckerman, S. E. (2017). Understanding and treating scrupulosity. In J. S. Abramowitz, D. McKay, and E. A. Storch (Eds.), *The Wiley Handbook of Obsessive Compulsive Disorders* (1st ed., pp. 527–546). Wiley. https://doi.org/10.1002/9781118890233.ch29

Van Roessel, P. J., Grassi, G., Aboujaoude, E. N., Menchón, J. M., Van Ameringen, M., and Rodríguez, C. I. (2023). Treatment-resistant OCD: Pharmacotherapies in adults. *Comprehensive Psychiatry*, 120, 152352. https://doi.org/10.1016/j.comppsych.2022.152352

Wetterneck, C. T., Rouleau, T. M., Williams, M. T., Vallely, A., La Torre, J. T., and Björgvinsson, T. (2021). A new scrupulosity scale for the dimensional obsessive-compulsive scale (DOCS): Validation with clinical and nonclinical samples. *Behavior Therapy*, 52(6), 1449–1463. https://doi.org/10.1016/j.beth.2021.04.001.

Towards Resilient Societies: The Synergy of Religion, Education, Health, Science, and Technology – Rahiem (Ed.)

Balancing flexibility and challenges: Teachers' perspectives on implementing the Merdeka curriculum in science education

Dzikri Rahmat Romadhon
Universitas Islam Negeri Syarif Hidayatullah Jakarta, Indonesia
ORCID ID: 0000-0003-2796-7281

Dede Rosyada, Murodi Murodi, Wahdi Wahdi & Asmu'i Asmu'i
Universitas Islam Negeri Syarif Hidayatullah Jakarta, Indonesia

Solahudeen Mani
The Patani Center for Organizational and Public Outrech, Thailand

ABSTRACT: This study explores the Merdeka Curriculum's implementation in Indonesian science education, focusing on its flexibility and innovative teaching practices, alongside execution challenges. The guiding research question is: What opportunities does the Merdeka Curriculum offer for flexible and innovative teaching in science education, and what specific challenges hinder its effective implementation? Employing a qualitative methodology through a comprehensive literature review and content analysis of both primary and secondary data, the study reveals that while the curriculum enables educators to design learning experiences that are more responsive to local needs and global dynamics, key challenges persist in the form of an increased administrative burden, inadequate training support, and marked infrastructural disparities between regions. These challenges detrimentally affect teacher–student interactions and impede the realisation of the intended pedagogical innovations. Consequently, this study contributes to SDG 4 (Quality Education) by underscoring the need for strategic, collaborative efforts among all stakeholders to overcome these barriers and develop innovative strategies that fully harness the curriculum's potential to enhance overall educational quality.

Keywords: Merdeka curriculum, educational transformation, science education, teaching innovation, administrative burden, infrastructural disparities

1 INTRODUCTION

The introduction of the Merdeka Curriculum marks a significant shift in Indonesian education, promoting teaching flexibility and active teacher-student participation in instructional management. As globalisation and technological advancements accelerate, education systems require methods responsive to local needs (Hunaepi 2024; Jamilah 2024; Kemendikbud 2022). Over 300,000 institutions have adopted this curriculum to address 21st-century educational challenges (Hunaepi 2024; Kemendikbud 2022). However, while the policy empowers teachers to innovate, it also increases administrative demands, requiring major managerial adjustments (Faiz and Kurniawaty 2020; Reza 2023; Sidik 2023).

Despite its promise of flexibility, implementation issues undermine its effectiveness. Studies indicate a discrepancy between policy goals and classroom reality, with 65% of teachers stating that administrative tasks reduce teacher-student interaction time (Nasrulhaq

DOI: 10.1201/9781003645542-62

et al. 2022). Inadequate training and regional disparities further hinder the adoption of student-centered approaches (Hidayat 2019; Kemendikbud 2022; Mardapi 2017). The lack of empirical studies on administrative burdens and teacher support in Southeast Asia highlights the need for further research (Wibowo and Suyanto 2018).

To optimise implementation, deeper analysis is crucial. Initial findings suggest that structural obstacles limit the curriculum's success (Mardapi 2017; Nasrulhaq *et al.* 2022). PISA data reveals that Indonesia lags behind OECD averages in literacy and numeracy, underscoring urgent reforms (Sari 2019; Smith and Larimer 2017). Therefore, comprehensive research is essential to refine implementation strategies, enhance educational quality, and contribute to discussions on curricular innovation in Southeast Asia (Creswell 2014; Hargreaves and Fullan 2012; Sowell 2015).

2 METHODS

This qualitative study used semi-structured interviews to explore teachers' perspectives on Merdeka Curriculum implementation in science education (Creswell 2014). Data were collected from 11 science educators across junior high schools (SMP), senior high schools (SMA), and Islamic madrasahs (MA) in Jakarta and South Tangerang. The predominantly female sample, with two male participants, had diverse academic backgrounds—mainly in Physics, Chemistry, Biology, and Mathematics—and teaching experience ranging from less than five years to over a decade. The sample size was deemed sufficient based on data saturation principles (Creswell 2014). Interviews focused on curriculum benefits, challenges, and strategic implementation approaches, particularly its impact on instructional methods and administrative workload. Data were recorded, transcribed, and thematically coded to identify emerging themes. Inter-coder reliability checks and triangulation methods, including member checking and peer debriefing, ensured credibility. Ethical protocols were strictly followed, with informed consent obtained and data protection measures enforced. Figure 1 presents a flowchart summarising data collection and analysis. This methodology provides key insights into curriculum implementation and challenges faced by educators in realising its potential.

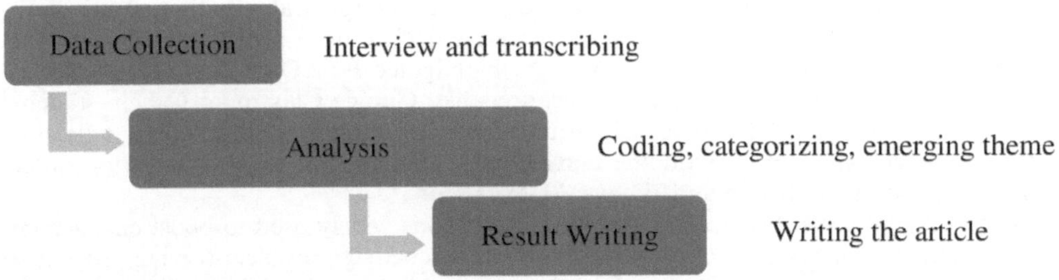

Figure 1. Flowchart summarising the research process.

3 RESULTS AND DISCUSSIONS

3.1 *Advantages and challenges in implementing the Merdeka Curriculum in Science Education*

The Merdeka Curriculum in science education offers unprecedented flexibility, allowing teachers to adapt methods to student needs. Many educators welcome this autonomy in selecting content and pedagogy. Ms E.S. stated, "The main benefit of the Merdeka Curriculum is the flexibility it offers teachers. We can adjust the material according to the students' conditions and the class's needs." Similarly, Ms Y.Y. noted, "We are freer to decide on teaching methods that best meet our students' needs." This aligns with research showing teacher autonomy enhances student engagement (Saa 2024).

The curriculum also fosters student independence in learning. Ms S.R. observed, "What I have noticed is that children are more active and eager to seek out additional sources," while Ms H.T. added, "Students are more independent in finding extra materials due to the restrictions of the standard textbooks." This supports research on active learning improving higher-order cognitive skills (Hamka *et al.* 2024; Tekkol and Demirel 2018). However, challenges such as limited technological access and unequal infrastructure hinder equitable implementation (Astuti *et al.* 2024).

Despite these benefits, significant challenges remain. Many educators report insufficient training during the transition. Ms P.D. stated, "Initially, it was challenging because with the Merdeka Curriculum, which operates in phases, I had to learn a great deal." Ms I.Y. added, "We received very brief curriculum socialisation, so many teachers remain confused about its implementation." This highlights a lack of support systems (Astuti *et al.* 2024; Hamka *et al.* 2024). Additionally, student adaptation varies. Mr S.M. noted, "The main challenge is that because the curriculum is 'free', students are no longer under the same pressure as before when there was a National Examination. However, as a consequence, some students have become too relaxed and less motivated." Ms N.A.L. remarked, "Not all students can immediately adjust to the more independent learning methods." These concerns reflect structural and inclusivity challenges (Saa 2024).

Addressing these issues requires enhanced teacher training and support systems. Canada's mentorship models foster professional development (Field *et al.* 2023), while Singapore integrates technology to bridge resource gaps (Zhang *et al.* 2023). Pakistan's problem-based learning promotes self-directed exploration (Bhuttah *et al.* 2024), mirroring Indonesia's STEM initiatives, which emphasize collaborative problem-solving despite infrastructure limitations (Baiduri 2018). The GOAL framework, structuring self-directed learning, has shown success in K-12 settings (Yang *et al.* 2024).

3.2 *Teacher strategies for overcoming learning challenges in the Merdeka Curriculum*

To implement the Merdeka Curriculum, educators must adapt strategies that ensure learning effectiveness. One key approach is integrating technology, providing teachers with diverse tools for engaging instruction. Ms I.Y. highlighted how Quizizz enhances lessons, while Mr B.H. stated, "We integrate applications like Google Classroom to help students access material more easily." These insights underscore technology's role in creating flexible, self-paced learning, aligning with the curriculum's emphasis on digital tools for differentiated instruction (Safrudin and Wijaya 2024).

Beyond digital tools, Project-Based Learning (PBL) is widely used to boost engagement and comprehension. Ms Z.M. explained, "We more often assign problem-solving projects so that students become more interested in the material." Ms S.M. added, "We employ a project-based learning model to encourage greater student participation." This method transforms students into active problem-solvers, fostering critical thinking, collaboration, and practical application, as emphasised in studies on innovative pedagogical strategies (Asmahasanah 2023; Hazlinda and Settiawan 2024; Safrudin and Wijaya 2024).

Other emerging methods, such as differentiated and integrated learning, also promote student autonomy (Witraguna 2024). Like PBL, these approaches emphasise real-world applications, reinforcing a flexible, student-centered learning model in Indonesian education (Afgany 2023; Mutmainah 2024; Trisna *et al.* 2024).

The shift toward student-centered teaching requires educators to act as facilitators rather than sole knowledge providers. Ms I.Y. observed, "Now, teachers function more as facilitators, providing space for students to explore the material on their own." However, Ms L.

warned, "While the curriculum encourages student independence, the challenge lies in ensuring that they remain on the correct path." Mr B.H. added, "Students who are already familiar with technology may adjust more quickly, but those who struggle require extra guidance from the teacher." Similarly, Ms Y.Y. noted, "We assign challenging projects to foster creativity in understanding concepts, yet the challenge lies in guiding them to ensure that the projects remain focused." These challenges highlight the need for sustained teacher support and careful planning to maximise the benefits of both technology-based and project-based learning (Hazlinda and Settiawan 2024; Ndari *et al.* 2023; Safrudin and Wijaya 2024).

3.3 *Administrative burden and its impact on teachers in the implementation of the Merdeka Curriculum*

A major concern among teachers regarding Merdeka Curriculum implementation is the increased administrative workload. While the curriculum offers greater teaching flexibility, many feel this is offset by bureaucratic demands. Ms N.A.L. noted, "The administrative burden under the Merdeka Curriculum feels considerably heavier compared to previous curricula," while Ms P.D. stated, "The volume of administration required of teachers is much greater than before." These concerns highlight how, despite efforts to enhance pedagogy, teachers remain burdened with paperwork (Ndari *et al.* 2023; Sholeh *et al.* 2024).

The rise in reporting and documentation requirements is particularly challenging. Previously, teachers mainly prepared lesson plans and assessments, but now they must complete teaching modules, student progress reports, and project activity records. Ms S.R. explained, "We now have to fill in many more student progress reports, which consequently reduces the time available for teaching." Although designed to support personalised learning, this added complexity reduces direct instructional time (Elsa *et al.* 2024). Studies confirm that teachers struggle with adapting lesson plans and new documentation systems, central to the curriculum (Elsa *et al.* 2024; Sholeh *et al.* 2024).

The heavy administrative workload also affects teaching quality. Ms I.Y. stated, "The administration required under the Merdeka Curriculum is far more extensive than under previous curricula," while Ms S.M. added, "We find ourselves dealing with more documentation than engaging directly with our students." Insufficient training on new formats forces teachers into self-directed learning. Ms L. admitted, "I have had to learn on my own how to complete the teaching modules and student progress reports because there wasn't sufficient training provided." Research confirms limited professional development worsens adaptation challenges (Milla *et al.* 2024; Ndari *et al.* 2023; Sholeh *et al.* 2024).

Beyond pedagogy, administrative demands impact teachers' well-being. Ms Y.Y. noted, "I sometimes have to take my administrative work home because there isn't enough time at school." This highlights how growing workloads disrupt work-life balance, lowering motivation and job satisfaction (Milla *et al.* 2024). Rural and under-resourced schools face greater stress due to limited infrastructure and time constraints (Milla *et al.* 2024; Ndari *et al.* 2023).

Despite these challenges, teachers seek solutions. Mr B.H. shared, "We try to allocate our time more efficiently, but it remains challenging as the volume of administration continues to grow." Ms S.R. emphasised collaboration, stating, "We often work together on preparing reports so that the burden does not fall on one individual alone." These efforts, including task-sharing, digital tools, and peer mentoring, aim to streamline workflows and sustain teaching quality (Elsa *et al.* 2024; Ndari *et al.* 2023).

4 CONCLUSION

This study concludes that the Merdeka Curriculum offers greater flexibility and teaching innovation, yet its implementation faces major challenges, including increased administrative workload, inadequate training, and regional infrastructure gaps. A strategic, collaborative approach is essential to maximise its potential in improving education quality. Key recommendations include reducing administrative burdens through streamlined reporting and professional development programs, while school administrators should foster student-centered teaching environments. Expanding teacher collaboration and mentoring can further ease transitions. These efforts will enhance curriculum implementation and support SDG 4 (Quality Education) by promoting inclusive learning, while addressing regional disparities in line with SDG 10 (Reduced Inequalities).

REFERENCES

Afgany, D. (2023). The differences between the 2013 curriculum and the merdeka curriculum: teacher's understanding. *English Education and Applied Linguistics Journal (Eeal Journal)*, 6(3), 164–172. https://doi.org/10.31980/eeal.v6i3.78

Asmahasanah, S. (2023). Navigating merdeka curriculum in first grade: teacher challenges and strategies. *Journal of Integrated Elementary Education*, 3(2), 137–149. https://doi.org/10.21580/jieed.v3i2.17592

Astuti, D. F., Samanhudi, U., and Pratiwi, W. R. (2024). The Challenges Teachers and Students Face in Implementing the Merdeka Belajar Curriculum for Teaching and Studying English at SMP Negeri 5 Sangatta Utara. *EDUKASIA: Jurnal Pendidikan dan Pembelajaran*, 5(1). https://doi.org/10.62775/edukasia.v5i1.1066

Baiduri, B. (2018). Analysis Of Students' Critical Thinking and Independence of Learning Through Problem-Based Learning. Mathematics Education Journal. https://www.academia.edu/65584347/Analysis_Of_Students_Critical_Thinking_and_Independence_of_Learning_Through_Problem_Based_Learning

Bhuttah, T. M., Xusheng, Q., Abid, M. N., and Sharma, S. (2024). Enhancing student critical thinking and learning outcomes through innovative pedagogical approaches in higher education: The mediating role of inclusive leadership. *Scientific Reports*, 14(1), 24362. https://doi.org/10.1038/s41598-024-75379-0

Creswell, J. W. (2014). *Research Design: Qualitative, Quantitative, and Mixed Methods Approaches* (4th ed.). SAGE Publications.

Elsa, E., Rahim, T. R., and Tohamba, C. P. P. (2024). From Theory to Practice: Teachers' Perspectives and Challenges in Implementing Merdeka Curriculum for Lesson Planning. *Journal of English Language and Education*, 9(6), 88–96.

Faiz, A., and Kurniawaty, I. (2020). Konsep merdeka belajar pendidikan Indonesia dalam perspektif filsafat progresivisme. *Konstruktivisme: Jurnal Pendidikan dan Pembelajaran*, 12(2), 155–164. https://doi.org/10.35457/konstruk.v12i2.973

Field, E., Spiropoulos, G., Nguyen, A., and Grewal, R. (2023). Climate change education within canada's regional curricula: a systematic review of gaps and opportunities. *Canadian Journal of Educational Administration and Policy*, (202), 155–184. https://doi.org/10.7202/1099989ar

Hamka, J., Asrifan, A., and Sadapotto, A. (2024). Teachers' and students' perspective on the implementation of "Kurikulum Merdeka" in English language learning at Upt SD Negeri 2 Sereang. *MACCA Journal*. https://journal.adityarifqisam.org/index.php/macca/article/view/94

Hargreaves, A., and Fullan, M. (2012). *Professional Capital: Transforming Teaching in Every School*. Teachers College Press.

Hazlinda, N., and Settiawan, D. (2024). Exploring English Teachers' Pedagogical Strategy in Implementing Merdeka Curriculum. *Journal of English and Arabic Language Teaching*, 15(1), 54–61.

Hidayat, R. (2019). *Pengantar Sosiologi Kurikulum*. Rajawali Pers.

Hunaepi, H. (2024). Transforming education in indonesia: the impact and challenges of the merdeka belajar curriculum. *Path of Science*, 10(6), 5026–5039. https://doi.org/10.22178/pos.105-31

Jamilah, J. (2024). The integration of local cultural arts in the context of teaching materials on the implementation of the merdeka belajar curriculum. *Journal of Education Research and Evaluation*, 8(2), 404–413. https://doi.org/10.23887/jere.v8i2.78359

Kemendikbud. (2022). Kurikulum Merdeka. *Kementerian Pendidikan dan Kebudayaan*. Retrieved from https://kurikulum.kemdikbud.go.id

Mardapi, D. (2017). Analisis perbandingan Kurikulum 2013 dan Kurikulum Merdeka dalam konteks implementasi kurikulum nasional. *Jurnal Pendidikan Nasional*, 1(1), 1–10.

Milla, H., Nasral, N., Masri, M., Suharmi, S., and Efendi, R. (2024). Evaluation of the Challenges of Merdeka Curriculum in Education in Indonesia. *International Journal of Multicultural and Multireligious Understanding*, 11(11), Article 11. https://doi.org/10.18415/ijmmu.v11i11.6349

Mutmainah, A. (2024). Unveiling success: exploring strategies of pre-service teachers in implementing merdeka curriculum in indonesia. *Asian Journal of Applied Education (Ajae)*, 3(1), 23–42. https://doi.org/10.55927/ajae.v3i1.7833

Nasrulhaq, N., Harakan, A., Syukri, S., Arfah, S. R., Khaerah, N., Tahir, N., and Wahid, N. (2022). Diseminasi kebijakan pendidikan melalui program Merdeka Belajar – Kampus Merdeka pada Perguruan Tinggi Muhammadiyah di Makassar, Indonesia. *Cetta: Jurnal Ilmu Pendidikan*, 1(1), 102–109.

Ndari, W., Suyatno, S., Sukirman, S., and Mahmudah, F. N. (2023). Implementation of the Merdeka Curriculum and Its Challenges. *European Journal of Education and Pedagogy*, 4(3), 111–116. https://doi.org/10.24018/ejedu.2023.4.3.648

Reza, F. (2023). Challenges in implementing kurikulum merdeka for efl teachers. *Jeels (Journal of English Education and Linguistics Studies)*, 10(2), 439–469. https://doi.org/10.30762/jeels.v10i2.1899

Saa, S. (2024). Merdeka Curriculum: Adaptation of Indonesian Education Policy in the Digital Era and Global Challenges. *Revista de Gestão Social e Ambiental*, 18(3), e07323. https://doi.org/10.24857/rgsa.v18n3-168

Safrudin, S., and Wijaya, E. (2024). Innovative Learning Strategies (Differentiated Learning Perspective on Merdeka Curriculum). *Jurnal Ilmiah Mandala Education*, 10(1), 25. https://doi.org/10.58258/jime.v10i1.6467

Sari, R. M. (2019). Peningkatan mutu pendidikan di era global: Analisis data PISA Indonesia. *Jurnal Pendidikan Global*, 3(2), 45–60.

Sholeh, M. I., Wahrudin, B., and Muzakki, H. (2024). Management Challenges in Implementing the Merdeka Curriculum. *Al-Hayat: Journal of Islamic Education*, 8(3), 1051–1071.

Sidik, D. (2023). Empowering teachers: "merdeka belajar" implementation insights from education conference – a qualitative study. *Ijess*, 1(2), 75–87. https://doi.org/10.33650/ijess.v1i2.7216

Smith, K. B., and Larimer, C. W. (2017). *The Public Policy Theory Primer* (3rd ed.). Routledge.

Sowell, E. J. (2015). *Curriculum: An Integrative Introduction* (4th ed.). Pearson Education, Inc.

Tekkol, I. A., and Demirel, M. (2018). An investigation of self-directed learning skills of undergraduate students. *Frontiers in Psychology*. https://www.frontiersin.org/journals/psychology/articles/10.3389/fpsyg.2018.02324/full

Trisna, N., Arafat, Y., and Furkan, N. (2024). Liberating learning: insights into the merdeka curriculum implementation. *Jurnal Ilmiah Ilmu Administrasi Publik*, 14(1), 313. https://doi.org/10.26858/jiap.v14i1.64713

Wibowo, U. B., and Suyanto, S. (2018). *Evaluasi Kurikulum Dan Pembelajaran*. UNY Press.

Witraguna, K. (2024). Learning in the merdeka curriculum: elementary school teachers' understanding of differentiated learning. *International Journal of Elementary Education*, 8(1), 47–56. https://doi.org/10.23887/ijee.v8i1.69779

Yang, Y., Li, H., Majumdar, R., and Ogata, H. (2024). GOAL system for online self-direction practice: Exploring students' behavioral patterns and the impact on academic achievement in the high school EFL context. *Journal of Computers in Education*, 11(2), 595–614. https://doi.org/10.1007/s40692-023-00272-0

Zhang, S., Sulong, R., and Hassan, N. (2023). Perceived double reduction policy, perception of educational involution: exploring the mediating role of parental educational anxiety and the moderating influence of gender—insights from a survey conducted in a city in china. *European Journal of Education*, 59(2). https://doi.org/10.1111/ejed.12604

Towards Resilient Societies: The Synergy of Religion, Education, Health, Science, and Technology – Rahiem (Ed.)

The promise and challenges of generative AI in science education: Teachers' perspectives and policy considerations

Dzikri Rahmat Romadhon
Universitas Islam Negeri Syarif Hidayatullah Jakarta, Indonesia
ORCID ID: 0000-0003-2796-7281

Amany Burhanudin Lubis, Nurbaini Futuhat Wulansari & Yudi Yudi
Universitas Islam Negeri Syarif Hidayatullah Jakarta, Indonesia

Amirfajar Masae
The Patani Center for Organizational and Public Outrech, Thailand

ABSTRACT: This study explored generative AI's role in science education from teachers' perspectives, examining its integration into learning and the challenges it presents. A qualitative descriptive approach was chosen over quantitative or mixed methods to provide an in-depth analysis of teachers' experiences. Semi-structured interviews were conducted with 12 science teachers across Greater Jakarta (Jabodetabek), and thematic coding was used to categorise data into key areas, including lesson planning, interactive teaching materials, assessment development, and technical and infrastructural challenges. Findings show AI enhances teaching by providing personalised, adaptive learning materials, yet barriers persist, such as limited digital infrastructure and insufficient teacher training. The study also addresses AI ethics and offers policy recommendations aligned with Sustainable Development Goal (SDG) 4 (Quality Education) and SDG 9 (Industry, Innovation, and Infrastructure). It highlights the need for a comprehensive, interdisciplinary AI integration strategy to support educators while fostering students' critical and creative skills in the digital era.

Keywords: generative AI, science education, adaptive learning, teacher training, digital infrastructure

1 INTRODUCTION

Education is experiencing a paradigmatic shift with generative AI, which autonomously creates text, images, and interactive simulations. This technology enables personalised, adaptive learning (H. Nguyen and Hayward 2024; Park 2023). AI literacy is increasingly vital, particularly in K-12 education (H. Zhang *et al.* 2022). Studies show AI enhances learning strategies (Łodzikowski *et al.* 2023), improves classroom interactions by accelerating assessments and feedback (Krause *et al.* 2025), and varies in its educational content generation (Yazdani Motlagh *et al.* 2023). AI also boosts operational efficiency and fosters a responsive learning environment, making it a key driver of digital education (Chen *et al.* 2020).

However, challenges remain, particularly at the secondary level, where most research focuses on higher education, leaving teachers with little pedagogical guidance (Li 2024; Salas-Pilco and Yang 2022; Zawacki-Richter *et al.* 2019). Infrastructure gaps, inadequate teacher training, and digital unpreparedness hinder AI adoption, especially in resource-limited areas (Jafari and Keykha 2023; Swiecki *et al.* 2022). The rapid pace of AI innovation often outpaces policy development, creating uncertainty for educators (J. Kim *et al.* 2022).

 DOI: 10.1201/9781003645542-63

Addressing these barriers requires interdisciplinary research to balance operational efficiency, critical thinking, and structural challenges.

This study seeks to bridge AI integration knowledge gaps in science education by proposing a strategic, adaptive framework. Integrating AI ethics into science education is crucial for responsible use (Al-Qerem *et al.* 2023). Sustainable AI integration also requires alignment with national policies, such as those emerging in Indonesia (Tong *et al.* 2024).

2 METHODS

This study used a descriptive qualitative approach to examine generative AI use in science education from teachers' perspectives (Creswell and Poth 2018). Semi-structured interviews were conducted in November 2024 with 12 science teachers from lower secondary (SMP), upper secondary (SMA), and Islamic schools (madrasah) across Greater Jakarta (Jabodetabek). Participants included four males and eight females, mostly holding bachelor's degrees in physics education, with some specialising in biology or other sciences. Their teaching experience ranged from early-career educators to those with over a decade in both public schools and madrasahs. Interviews were conducted online and in person.

Table 1. Demographic overview of science teachers participating in the study.

Initials	Education Level (B.Ed)	Institution	Level of High School	Region
E.S	Physics Education	MTs Pembangunan	Junior	Jakarta
A.S	Physics Education	MTs Pembangunan	Junior	Jakarta
A.N	Physics Education	SMA Alfath Cirendeu	Senior	South Tangerang
Z.F	Physics Education	SMA Darussalam	Senior	South Tangerang
S.A	Physics Education	SMA Yapim	Senior	Subang
F.A	Biology Education	SMP Bunda Kandung	Junior	Jakarta Selatan
D.P	Physics Education	SMA Al-Hasra	Senior	Depok
Y.T	Physics Education	SMP Negeri 17	Junior	South Tangerang
A.H	Physics Education	MAN 4 Jakarta	Senior	South Jakarta
M.J	Physics Education	SMP and SMA Kharisma	Junior and Senior	South Tangerang
M.R	Chemistry Education	SMP Negeri 21	Junior	Bogor
Z.H	Science Education	SMP Negeri 9	Junior	Bekasi

Participants were purposively selected based on their science teaching experience and prior AI integration, identified through screening questionnaires and institutional records. This strategy ensured representation from urban and suburban areas. Though the sample size is small, it was sufficient for a qualitative study, achieving data saturation. A summary of participant demographics is provided in Table 1.

Figure 1. Flowchart summarising the research process.

Data analysis followed thematic coding (Creswell and Poth 2018), involving iterative reading, open coding, and refinement of categories to identify key themes on AI understanding, teaching strategies, integration challenges, and required support. To enhance transparency and reduce bias, reflexive journaling and peer debriefing were used throughout the interpretation process. Figure 1 presents a flowchart summarising the research process. While findings focus on secondary education and are not universally generalisable, they provide valuable insights into AI adoption and its challenges. Data validity was ensured through cross-verification with supporting documents and follow-up discussions with selected participants.

3 RESULTS AND DISCUSSION

3.1 *AI as a tool for enhancing learning effectiveness*

3.1.1 *AI as a tool for lesson planning*

Generative AI acts as a digital assistant, streamlining lesson planning, material creation, and project-based curricula, freeing time for interactive learning. Unlike traditional methods requiring manual effort, AI tools like PangoGPT generate lesson plans in minutes, adapt content to learning styles, and automate tasks like quiz creation (Pango.education 2025). One teacher noted, "I use AI to design my lessons, particularly to summarise learning materials." (E.S). Another added, "AI is extremely helpful in preparing lesson plans and teaching materials." (M.J). AI also enhances project-based learning, as one teacher stated, "I rely on AI to develop more engaging project-based learning curricula." (S.M).

Beyond planning, AI aids in developing interactive materials. One teacher emphasised, "AI helps me develop interactive teaching materials for students who struggle with abstract concepts." (F.K). By simplifying complex topics, AI improves student comprehension. Tools like Copilot and Midjourney create adaptive content, enhancing engagement, a trend noted in AI education reviews (Fu *et al.* 2024).

Traditional tools like textbooks and static multimedia lack AI's adaptability (Holmes *et al.* 2019; Luckin *et al.* 2016). While methods like flipped classrooms (Bishop and Verleger 2013) and blended learning (Garrison and Kanuka 2004) boost engagement, they rely on fixed content. In contrast, AI provides real-time, personalised adaptation that traditional methods cannot (Zawacki-Richter *et al.* 2019).

3.1.2 *AI in question development and learning assessment*

Generative AI helps create practice questions and assessments with diverse formats and adjustable difficulty levels (Lye and Lim 2024). One teacher noted, "AI helps me create practice questions more quickly." (A.N), a practice supported by studies showing AI reduces workload while maintaining pedagogical rigour (Celik *et al.* 2022). AI tools can save teachers up to 30% of assessment preparation time, significantly improving efficiency (Hashem *et al.* 2023). Another teacher added, "I use AI to provide alternative perspectives for students in understanding physics concepts." (S.A). AI also increases question variety, as one teacher stated, "I use AI to create more diverse and creative question examples." (V.P).

Beyond practice exercises, AI aids in exam question development. One interviewee remarked, "AI assists me in designing exam questions with a broader range of variations and difficulty levels." (L.S), a process validated by studies showing AI aligns assessments with learning objectives while minimising bias (Archibald *et al.* 2023). AI thus streamlines question generation while ensuring diverse and fair assessments.

3.1.3 *AI as a medium for visualising and simulating scientific concepts*

Teaching science often involves explaining abstract concepts that students find challenging. Generative AI helps by creating interactive simulations, scientific illustrations, and concept maps, making theories easier to grasp. One teacher shared, "I use AI to create simple simulations that help students understand concepts more easily." (L.S). AI also enhances presentations, as another teacher noted, "AI helps me develop more interactive and engaging presentations for students." (W.T).

Additionally, AI expands teaching materials and references. One teacher explained, "With AI, I can access a wider range of teaching references and innovative ideas." (H.N). These tools go beyond content delivery, serving as catalysts for modernising teaching methods, with studies highlighting their role in boosting student engagement and conceptual mastery (Celik *et al.* 2022; Fu *et al.* 2024).

3.2 *Challenges in using AI Ethically and responsibly*

3.2.1 *Student dependence on AI for completing assignments*

Teachers worry that AI's easy access may hinder critical thinking by encouraging cognitive offloading, where students rely on technology instead of developing independent reasoning (Minh 2024; Zhong *et al.* 2024). Research shows frequent AI use for problem-solving lowers critical thinking scores, especially in younger learners (S. Zhang *et al.* 2024). Many copy AI-generated answers without understanding concepts. One teacher noted, "Students often use AI to find instant answers without grasping the concepts." (M.A). Another added, "Some students have started to overly rely on AI to complete their assignments without engaging in critical thinking." (L.A). This dependency challenges teachers in fostering deep understanding. Another participant stressed, "The key issue is ensuring that information from AI is validated with other sources." (S.T). AI should support learning, not replace critical engagement.

To address this, some teachers guide students in AI use and information evaluation. One teacher shared, "I always teach my students not to take AI-generated information at face value and to compare it with other sources." (U.K). This approach positions AI as a learning tool, enhancing critical thinking. Structured AI literacy programs can also help reduce dependency by teaching critical evaluation skills (Akram and Mahmood 2021; Liu and Wang 2024). Additionally, collaborative learning environments that integrate AI encourage active engagement and critical thinking (Lane *et al.* 2024; Larsson 2019).

3.2.2 *Validity and accuracy of information from generative AI*

Generative AI is not always accurate and can produce misleading information that does not align with the national curriculum. Teachers stress the need for critical evaluation of AI-generated content. One teacher expressed concern, "I worry about the validity of AI-generated information because it is not always accurate." (N.R). Another noted, "AI often provides answers without explaining the reasoning behind them, making it difficult for students to develop a deeper understanding of concepts." (J.L). This reflects critiques of AI's "black box" nature, which lacks transparency in how outputs are generated.

In science education, conceptual understanding is crucial. AI misinformation can create misconceptions, as it often oversimplifies complex topics like climate change or evolutionary biology, omitting key interdisciplinary connections (Fu *et al.* 2024). To counter this, teachers stress verifying AI content with textbooks or peer-reviewed journals. One teacher explained, "I prefer to use AI as a supplementary reference while still relying on validated textbooks." (B.A). Research supports using AI as a supplementary tool, ensuring its speed and efficiency are balanced with content accuracy and depth (Monib *et al.* 2024).

3.2.3 *Technical barriers and school infrastructure limitations*

Beyond validity concerns, teachers face technical barriers in AI integration. Some schools lack stable internet, adequate hardware, and institutional support. One teacher noted, "Although our school has Wi-Fi, not all classrooms have stable access to use AI effectively." (H.S). Another added, "My school has limited technological resources, which hampers the optimal use of AI." (M.R).

Financial constraints also hinder AI adoption, as advanced tools often require paid subscriptions. One teacher highlighted, "The main issue is the lack of school resources to provide stable AI access." (B.A). Many hope for greater government or school support. As one teacher suggested, "We need government support to provide premium AI licenses for schools." (Z.H). Expanding access to high-quality AI tools would help teachers enhance learning effectiveness.

3.3 *Lack of training and support for teachers in using AI*

3.3.1 *Insufficient teacher training in AI utilisation*

Generative AI is advancing rapidly, but many teachers lack the knowledge to use it effectively. While they recognise its potential, integration remains challenging. One teacher stressed, “Training for teachers is essential so that we know how to use AI optimally in the classroom.” (Z.F). Another added, “More training is needed to help teachers understand how to integrate AI effectively.” (D.W). This issue highlights disparities in preparedness, as some teachers experiment with AI while others hesitate due to limited technical skills. A participant noted, “Some teachers are still reluctant to use AI due to a lack of technical knowledge.” (A.I). Training opportunities are scarce, and one teacher suggested, “AI training should be a fundamental part of teacher capacity-building programmes.” (K.P).

Collaboration with higher education institutions and EdTech providers is crucial. Research shows AI literacy enhances teachers‘ ability to integrate technology effectively (Blancia 2024; Güneyli 2024; Omar *et al.* 2024; Zhao *et al.* 2022). Ongoing professional development tailored to AI tools improves teachers’ confidence and competence (Arvin 2023; Davis 2024; Mbambo 2024). Thus, integrating AI training into professional development is essential for maximising AI’s impact in education.

3.3.2 *Limited facilities and access to AI in Schools*

Beyond training challenges, teachers face technological barriers. Many schools lack stable internet, adequate computers, and AI platforms. One teacher noted, “My school has limited technological resources, making AI implementation less effective.” (M.R). Another added, “Even though Wi-Fi is available at my school, not all classrooms have stable access to use AI in real time.” (H.S), reflecting broader infrastructure issues that limit equitable AI access (Monib *et al.* 2024; Pardosi 2024). However, targeted infrastructure investments have enabled effective AI integration in low-resource settings (S. Kim 2023; G. Nguyen *et al.* 2022).

High costs of AI subscriptions further restrict adoption, especially in budget-constrained schools. One teacher explained, “The main issue is the lack of school resources to provide stable AI access.” (B.A). Even well-equipped schools struggle to obtain premium AI licenses. As one teacher stated, “We need government support to provide premium AI licenses for schools.” (Z.H). Research shows government initiatives and tech partnerships can help overcome these financial barriers, as seen in successful AI implementations (Wolff *et al.* 2021; Yim 2024).

3.3.3 *Teachers’ expectations for support from government and educational institutions*

Teachers emphasised that AI integration requires structural support from governments, universities, and educational institutions. Clear policies would enable better training, technology access, and ethical guidelines. Some proposed government-developed AI platforms for education. One teacher suggested, “If the government could create a free AI system integrated with the national curriculum, it would greatly benefit both teachers and students.” (T.C).

Teachers also stressed the need for regulatory oversight. One participant stated, “The government should regulate AI usage in education to prevent misuse by students.” (N.R). Another added, “AI must be properly monitored, particularly to ensure that students use it for learning rather than just retrieving instant answers.” (L.A). Stronger regulations, better infrastructure, and widespread training would ensure AI enhances education while promoting critical thinking and intellectual exploration.

4 CONCLUSION

This study finds that generative AI can transform education by providing personalised, interactive, and adaptive learning materials, improving teaching and assessment efficiency. However, infrastructure limitations and insufficient teacher training hinder its implementation. A research-based AI integration strategy is essential to maximise benefits while maintaining educators 'roles and fostering students' critical and creative skills. To address these issues, schools, education ministries, and AI developers must collaborate on targeted training, infrastructure upgrades, and clear regulations for responsible AI use. Policies should establish sustainable AI integration guidelines and promote digital and AI literacy to reduce student over-reliance on AI. These measures will enhance educational outcomes and support Sustainable Development Goals related to quality education and reduced inequalities. Improving AI literacy among educators is urgent.

REFERENCES

Akram, M., and Mahmood, A. (2021). A correlation study of pakistani efl teachers' professional discerning and their critical thinking. *Arab World English Journal*, (2), 378–387. https://doi.org/10.24093/awej/mec2.27

Al-Qerem, W., Eberhardt, J., Jarab, A., Al Bawab, A. Q., Hammad, A., Alasmari, F., Alazab, B. a., Husein, D. A., Alazab, J., and Al-Beool, S. (2023). Exploring knowledge, attitudes, and practices towards artificial intelligence among health professions' students in Jordan. *BMC Medical Informatics and Decision Making*, 23(1). https://doi.org/10.1186/s12911-023-02403-0

Archibald, A., Hudson, C., Heap, T., Thompson, R. R., Lin, L., DeMeritt, J., and Lucke, H. (2023). A Validation of AI-Enabled Discussion Platform Metrics and Relationships to Student Efforts. *TechTrends*, 67(2), 285–293. https://doi.org/10.1007/s11528-022-00825-7

Arvin, N. (2023). Teacher experiences with ai-based educational tools. *aitechbesosci*, 1(2), 26–32. https://doi.org/10.61838/kman.aitech.1.2.5

Bishop, J. L., and Verleger, M. A. (2013). The flipped classroom: A survey of the research. *In ASEE National Conference Proceedings*, Atlanta, GA.

Black, N., George, S., Eguchi, A., Dempsey, J., Langran, E., Fraga, L., … and Howard, N. (2024). A framework for approaching ai education in educator preparation programs. *Proceedings of the Aaai Conference on Artificial Intelligence*, 38(21), 23069–23077. https://doi.org/10.1609/aaai.v38i21.30351

Blancia, G. (2024). The mediating effects of artificial intelligence literacy on the association between computational thinking skills and organizational agility among secondary school teachers. *Problems of Education in the 21st Century*, 82(5), 616–629. https://doi.org/10.33225/pec/24.82.616

Celik, I., Dindar, M., Muukkonen, H., and Järvelä, S. (2022). The promises and challenges of artificial intelligence for teachers: A systematic review of research. *TechTrends*, 66(4), 616–630. https://doi.org/10.1007/s11528-022-00715-y

Chen, L., Chen, P., and Lin, Z. (2020). Artificial intelligence in education: A review. *IEEE Access*, 8, 75264–75278. https://doi.org/10.1109/ACCESS.2020.2988510

Creswell, J. W., and Poth, C. N. (2018). *Qualitative inquiry and research design: Choosing among five approaches* (4th ed.). SAGE Publications.

Davis, R. (2024). Korean in-service teachers' perceptions of implementing artificial intelligence (ai) education for teaching in schools and their ai teacher training programs. *International Journal of Information and Education Technology*, 14(2), 214–219. https://doi.org/10.18178/ijiet.2024.14.2.2042

Fu, Y., Weng, Z., and Wang, J. (2024). Examining AI Use in Educational Contexts: A Scoping Meta-Review and Bibliometric Analysis. *International Journal of Artificial Intelligence in Education*. https://doi.org/10.1007/s40593-024-00442-w

Garrison, D. R., and Kanuka, H. (2004). Blended learning: Uncovering its transformative potential in higher education. *The Internet and Higher Education*, 7(2), 95–105.

Güneyli, A. (2024). Exploring teacher awareness of artificial intelligence in education: a case study from northern cyprus. *European Journal of Investigation in Health Psychology and Education*, 14(8), 2358–2376. https://doi.org/10.3390/ejihpe14080156

Hashem, R., Ali, N., Zein, F., Fidalgo, P., and Khurma, O. (2023). Ai to the rescue: exploring the potential of chatgpt as a teacher ally for workload relief and burnout prevention. *Research and Practice in Technology Enhanced Learning*, 19, 023. https://doi.org/10.58459/rptel.2024.19023

Holmes, W., Bialik, M., and Fadel, C. (2019). Artificial intelligence in education: Promises and implications for teaching and learning. *Center for Curriculum Redesign.*

Jafari, F., and Keykha, A. (2023). Identifying the opportunities and challenges of artificial intelligence in higher education: a qualitative study. *Journal of Applied Research in Higher Education*, 16(4), 1228–1245. https://doi.org/10.1108/jarhe-09-2023-0426

Kim, J., Lee, H., and Cho, Y. H. (2022). Learning design to support student-AI collaboration: Perspectives of leading teachers for AI in education. *Education and Information Technologies*, 27(5). https://doi.org/10.1007/s10639-021-10831-6

Kim, S. (2023). Change in attitude toward artificial intelligence through experiential learning in artificial intelligence education. *International Journal on Advanced Science Engineering and Information Technology*, 13(5), 1953–1959. https://doi.org/10.18517/ijaseit.13.5.19039

Krause, S., Panchal, B. H., and Ubhe, N. (2025). The Evolution of Learning: Assessing the Transformative Impact of Generative AI on Higher Education. *Lecture Notes on Data Engineering and Communications Technologies*, 356–371. https://doi.org/10.1007/978-981-97-9255-9_24

Lane, S., Haley, T., and Brackney, D. (2024). Tool or tyrant: guiding and guarding generative artificial intelligence use in nursing education. *Creative Nursing*, 30(2), 125–132. https://doi.org/10.1177/10784535241247094

Larsson, K. (2019). Using essay responses as a basis for teaching critical thinking – a variation theory approach. *Scandinavian Journal of Educational Research*, 65(1), 21–35. https://doi.org/10.1080/00313831.2019.1650824

Li, W. (2024). An explanatory study of factors influencing engagement in ai education at the k-12 level: an extension of the classic tam model. *Scientific Reports*, 14(1). https://doi.org/10.1038/s41598-024-64363-3

Liu, W., and Wang, Y. (2024). The effects of using ai tools on critical thinking in english literature classes among efl learners: an intervention study. *European Journal of Education*, 59(4). https://doi.org/10.1111/ejed.12804

Łodzikowski, K., Foltz, P. W., and Behrens, J. T. (2023). Generative AI and its educational implications. *arXiv*. https://arxiv.org/abs/2401.08659

Luckin, R., Holmes, W., Griffiths, M., and Forcier, L. B. (2016). *Intelligence unleashed: An argument for AI in education*. Pearson Education.

Lye, C. Y., and Lim, L. (2024). Generative Artificial Intelligence in Tertiary Education: Assessment Redesign Principles and Considerations. *Education Sciences*, 14(6), 569. https://doi.org/10.3390/educsci14060569

Mbambo, G. (2024). Impact of artificial intelligence on teacher training in open distance and electronic learning. *International Journal of Learning Teaching and Educational Research*, 23(5), 370–386. https://doi.org/10.26803/ijlter.23.5.19

Minh, A. N. (2024). Leveraging chatgpt for enhancing english writing skills and critical thinking in university freshmen. *Journal of Knowledge Learning and Science Technology* ISSN: 2959–6386 (Online), 3(2), 51–62. https://doi.org/10.60087/jklst.vol3.n2.p62

Monib, W. K., Qazi, A., Apong, R. A., Azizan, M. T., De Silva, L., and Yassin, H. (2024). Generative AI and future education: a review, theoretical validation, and authors‘ perspective on challenges and solutions. *PeerJ Computer Science*, 10, e2105. https://doi.org/10.7717/peerj-cs.2105

Nguyen, G., Nguyen, N., and Giang, N. (2022). Situation and proposals for implementing artificial intelligence-based instructional technology in vietnamese secondary schools. *International Journal of Emerging Technologies in Learning (Ijet)*, 17(18), 53–75. https://doi.org/10.3991/ijet.v17i18.31503

Nguyen, H., and Hayward, J. (2024). Applying Generative Artificial Intelligence to Critiquing Science Assessments. *Journal of Science Education and Technology*. https://doi.org/10.1007/s10956-024-10177-x

Omar, Z., Harun, M., Ishar, N., Mustapha, N., and Ismail, Z. (2024). Leveraging artificial intelligence for enhancing postgraduate teaching: a framework to engaging professional learners. *IJELHE*, 19(2), 177–198. https://doi.org/10.24191/ijelhe.v19n2.19212

Pango.education. (2025, January 5). AI vs traditional lesson planning: Which is better? Pango Blog. https://blog.pango.education/ai-vs-traditional-lesson-planning

Pardosi, V. (2024). Implementation of an artificial intelligence based learning management system for adaptive learning. *Al-Fikrah Jurnal Manajemen Pendidikan*, 12(1), 149. https://doi.org/10.31958/jaf.v12i1.12548

Park, J. (2023). Integrating artificial intelligence into science lessons: teachers' experiences and views. *International Journal of Stem Education*, 10(1). https://doi.org/10.1186/s40594-023-00454-3

Salas-Pilco, S. Z., and Yang, Y. (2022). Artificial intelligence applications in Latin American higher education: A systematic review. *International Journal of Educational Technology in Higher Education*, 19(1). https://doi.org/10.1186/s41239-022-00326-w

Swiecki, Z., Khosravi, H., Chen, G., Martinez-Maldonado, R., Lodge, J. M., Milligan, S., Selwyn, N., and Gašević, D. (2022). Assessment in the age of artificial intelligence. *Computers and Education: Artificial Intelligence*, 3(August 2021). https://doi.org/10.1016/j.caeai.2022.100075

Tong, W., Zhang, X., Zeng, H., Pan, J., Gong, C., and Zhang, H. (2024). Reforming china's secondary vocational medical education: adapting to the challenges and opportunities of the ai era. *Jmir Medical Education*, 10, e48594–e48594. https://doi.org/10.2196/48594

Wolff, J., Pauling, J., Keck, A., and Baumbach, J. (2021). Success factors of artificial intelligence implementation in healthcare. *Frontiers in Digital Health*, 3. https://doi.org/10.3389/fdgth.2021.594971

Yazdani Motlagh, N., Khajavi, M., Sharifi, A., and Ahmadi, M. (2023). The impact of artificial intelligence on the evolution of digital education: A comparative study of OpenAI text generation tools including ChatGPT, Bing Chat, Bard, and Ernie. arXiv. https://arxiv.org/abs/2309.02029

Yim, I. (2024). Teachers' perceptions, attitudes, and acceptance of artificial intelligence (ai) educational learning tools: an exploratory study on ai literacy for young students. *Future in Educational Research*, 2(4), 318–345. https://doi.org/10.1002/fer3.65

Zawacki-Richter, O., Marín, V. I., Bond, M., and Gouverneur, F. (2019). Systematic review of research on artificial intelligence applications in higher education – Where are the educators? *International Journal of Educational Technology in Higher Education*, 16(1). https://doi.org/10.1186/s41239-019-0171-0

Zhang, H., Lee, I., Ali, S., DiPaola, D., Cheng, Y., and Breazeal, C. (2022). Integrating ethics and career futures with technical learning to promote ai literacy for middle school students: an exploratory study. *International Journal of Artificial Intelligence in Education*, 33(2), 290–324. https://doi.org/10.1007/s40593-022-00293-3

Zhang, S., Zhao, X., Zhou, T., and Kim, J. H. (2024). Do you have AI dependency? The roles of academic self-efficacy, academic stress, and performance expectations on problematic AI usage behavior. *International Journal of Educational Technology in Higher Education*, 21, 34. https://doi.org/10.1186/s41239-024-00467-0

Zhao, L., Wu, X., and Luo, H. (2022). Developing ai literacy for primary and middle school teachers in china: based on a structural equation modeling analysis. *Sustainability*, 14(21), 14549. https://doi.org/10.3390/su142114549

Zhong, W., Luo, J., and Lyu, Y. (2024). How do personal attributes shape AI dependency in Chinese higher education context? Insights from needs frustration perspective. *PLOS ONE*, 19(11), e0313314. https://doi.org/10.1371/journal.pone.0313314.

Towards Resilient Societies: The Synergy of Religion, Education, Health, Science, and Technology – Rahiem (Ed.)

Benefits and challenges of environmental education in early childhood education in Indonesia: A systematic literature review

Ratna Faeruz
Universitas Islam Negeri Syarif Hidayatullah Jakarta, Indonesia
ORCID ID: 0000-0001-7782-222X

Nasirin Roset
Institute An-Nikmah Al-Islamiyah Phnom Penh, Cambodia

Dzikri Rahmat Romadhon
Universitas Islam Negeri Syarif Hidayatullah Jakarta, Indonesia
ORCID ID: 0000-0003-2796-7281

Evi Muliyah
Universitas Islam Negeri Syarif Hidayatullah Jakarta, Indonesia
ORCID ID: 0000-0002-1223-1581

Maila Dinia Husni Rahiem
Universitas Islam Negeri Syarif Hidayatullah Jakarta, Indonesia
ORCID ID: 0000-0002-5618-2486

Ali Munhanif
Universitas Islam Negeri Syarif Hidayatullah Jakarta, Indonesia

Supriyanto Suharman
Dharmaila Pradita Foundation, Indonesia

ABSTRACT: Environmental education (EE) in early childhood education (ECE) is crucial for shaping a generation that values and actively engages in environmental conservation. This systematic literature review explores the benefits and challenges of integrating EE into ECE. Utilising a systematic literature review and thematic analysis, this review synthesises insights from peer-reviewed studies published in English and Indonesian academic journals between 2013 and 2023. The findings reveal significant benefits, including the development of environmental responsibility, enhanced cognitive skills, and improved emotional well-being. Additionally, this study highlights the contribution of EE to achieving the Sustainable Development Goals (SDGs), particularly SDG 4 (Quality Education) and SDG 13 (Climate Action), which can provide long-term positive impacts. However, challenges such as limited resources, inadequate teacher training, and curriculum constraints are also identified. This review emphasises the need for comprehensive strategies to effectively integrate EE into ECE, as well as policy recommendations to support better implementation, to foster sustainable practices from an early age.

Keywords: environmental education, early childhood education, sustainability, systematic literature review, thematic analysis

DOI: 10.1201/9781003645542-64

1 INTRODUCTION

Environmental challenges such as climate change, pollution, and biodiversity loss demand urgent action from all sectors (Arora *et al.* 2018; Seddon *et al.* 2021). These issues are particularly critical in Indonesia due to its rich biodiversity and environmental vulnerability (Mukti and Sobirov 2023). Education, especially from an early age, is key to fostering environmental awareness and action (Ardoin *et al.* 2020, Lamanauskas 2023 especially from an early age Ardoin and Bowers 2020). Early Childhood Education (ECE) provides an ideal stage to introduce Environmental Education (EE) and instil sustainable values (Guevara-Herrero *et al.* 2024, Goldman *et al.* 2017, Hebe 2017, Rahmania 2024).

In a global context, early childhood education plays a crucial role in achieving the Sustainable Development Goals (SDGs), particularly SDG 4 (Quality Education), which emphasises the importance of inclusive and quality education and promoting lifelong learning opportunities for all, and SDG 13 (Climate Action), which advocates for action to combat climate change and its impacts. EE in ECE contributes to achieving both goals by fostering cognitive and social skills that support quality education and shaping behaviours that are mindful of climate change from an early age (Hayati *et al.* 2023).

Indonesia faces deforestation, pollution, and climate change impacts (Adyasari *et al.* 2021, Sulistiawati 2024, Shivanna 2022). EE in ECE can help build an environmentally conscious generation equipped to tackle these issues. Early exposure to nature shapes children's worldviews and behaviours (Ernst and Burcak 2019, Feriver *et al.* 2016, Green *et al.* 2016, Molinario *et al.* 2020). However, EE implementation in Indonesia faces barriers like limited resources, inadequate teacher training, and curriculum constraints (Paananen *et al.* 2023, Nirmala *et al.* 2024, Yeşilpınar Uyar and Eti 2023, 2024). Addressing these challenges is vital for effective integration.

This paper systematically reviews EE in Indonesian ECE, analysing its benefits, challenges, and best practices. It contributes to a broader dissertation that seeks to develop a framework for EE integration, considering Indonesia's environmental, cultural, and educational context. This study lays the groundwork for future empirical studies and policy recommendations by examining existing research.

2 METHODS

This study employs a systematic literature review to gather, analyse, and synthesise research on the importance of Environmental Education (EE) in Early Childhood Education (ECE). Systematic literature reviews are rigorous methods for summarising existing research, ensuring a comprehensive and unbiased overview of a particular topic (Gentles *et al.* 2016, Simsek *et al.* 2023, Shaheen *et al.* 2023).

A comprehensive search of academic databases such as ERIC, JSTOR, Scopus, and Google Scholar was conducted using keywords like "environmental education," "early childhood education," "sustainability," and "systematic literature review." The search was limited to peer-reviewed articles published between 2013 and 2023 to capture recent developments, challenges, and advancements in EE in ECE, including recent policy changes and educational practices in Indonesia and globally.

Inclusion criteria focused on peer-reviewed studies in English and Bahasa Indonesia, examining EE's impacts, challenges, and practices in ECE. Both empirical and conceptual papers were included. Exclusion criteria excluded studies not specifically on EE in ECE and non-peer-reviewed articles, theses, or dissertations.

Data extraction involved identifying key themes, findings, and methodologies, followed by a thematic analysis to identify patterns and insights related to EE in ECE. The themes were synthesised for a comprehensive understanding of the topic. Figure 1 summarising the study selection process is provided below.

Figure 1. Flowchart of the literature selection process.

3 RESULTS AND DISCUSSIONS

The systematic literature review identified several key benefits and challenges of integrating EE into ECE.

3.1 *Benefits*

3.1.1 *EE in ECE fosters a sense of responsibility and care for the environment among young children*

Early exposure to nature promotes pro-environmental attitudes and behaviours that can persist into adulthood (Collado and Evans 2019). Children who frequently interact with natural environments—such as playing in forests, visiting beaches, or observing wildlife—develop greater environmental awareness than those with limited exposure. These experiences promote habits like recycling, energy conservation, and waste reduction. More importantly, they lead to behaviour changes that cultivate sustainable practices in children, such as reducing plastic use, conserving water, and adopting eco-friendly consumption patterns. As children learn about the environment, they are more likely to turn off lights, pick up litter, or plant trees, which contribute to their environmental stewardship.

The impact of these experiences depends on outcome expectancy or a child's belief in their ability to influence the environment. Children with low outcome expectancy benefit more from direct nature contact, increasing their self-efficacy and confidence in making a difference. Therefore, EE programs should provide nature-based experiences and reinforce that small actions, like turning off taps or planting trees, contribute to sustainability. When children understand the positive impact of their actions, they are more likely to adopt pro-environmental behaviours.

Studies show that direct exposure to nature strengthens children's emotional connection to the environment, encouraging pro-environmental behaviours (Cheng and Monroe 2012). Activities like playing outdoors and observing wildlife enhance environmental awareness, foster a sense of belonging and build an understanding of the human impact on ecosystems. This emotional connection is linked to lasting behaviour changes, including increased environmental advocacy, reduced wastefulness, and greater participation in conservation activities.

Nature-based education is a powerful tool for enhancing environmental awareness and fostering sustainable behaviours from an early age. EE plays a key role in shaping future generations' commitment to sustainability through emotional and behavioural transformation.

3.1.2 *Engaging with EE activities can improve children's cognitive abilities, including critical thinking, problem-solving, and decision-making skills*

Natural activities like nature walks, gardening, and recycling projects offer hands-on learning that enhances cognitive development (Kiewra and Veselack 2016). Research shows children in outdoor classrooms improve in critical thinking, problem-solving, and decision-making. Observing trees, insects, and weather helps them identify patterns and understand ecological relationships. Gardening teaches plant growth, observation, hypothesis formation, and cause-effect reasoning while recycling projects foster innovation and problem-solving as children repurpose materials creatively.

Williams and Dixon (2013) highlight that school gardening enhances science and nutrition understanding, boosting engagement and motivation. Through gardening, children learn about plant life cycles, ecosystems, and sustainability, sharpening critical thinking and problem-solving skills (Williams and Dixon 2013). Haimovitz and Dweck (2017) emphasise that recycling projects stimulate creativity and cognitive growth, encouraging a growth mindset by framing challenges as learning opportunities (Haimovitz and Dweck 2017). Thus, nature-based activities strengthen knowledge and essential cognitive skills for academic and social success.

3.1.3 *Interaction with natural environments has been shown to reduce stress and anxiety in young children*

EE activities enhance emotional well-being by promoting physical activity, creativity, and social interaction (Shanahan *et al.* 2016). Research shows that natural environments reduce stress and anxiety, as green spaces offer psychological restorative effects. Children in parks or forests experience greater stress hormone reduction than those in artificial settings. Nature exposure also improves emotional regulation, helping children manage stress and focus on learning.

Beyond stress reduction, EE fosters creativity and social skills. Nature-based activities like forest walks, gardening, and environmental exploration encourage physical activity, collaboration, and imaginative play. Children with frequent nature exposure show higher self-confidence and resilience than those in enclosed spaces. Integrating EE into ECE curricula provides long-term benefits for emotional well-being, creativity, and social competence.

3.1.4 *EE encourages children to work collaboratively and develop social skills*

EE fosters collaboration, social skills, and ethical values. Participation in recycling, energy conservation, and reforestation teaches children to work together toward common goals, strengthening teamwork and responsibility(Cooper 2015).

Research shows EE enhances empathy toward living beings, laying the foundation for compassion, respect, and responsibility (Williams and Dixon 2013). Children make decisions that benefit their communities and the planet by understanding their environmental impact. Well-integrated EE also improves communication, negotiation, and conflict resolution skills, crucial for future personal and professional growth (Juvonen *et al.* 2019). Thus, EE builds environmental awareness and develops essential social and ethical skills for life.

3.2 *Challenges*

3.2.1 *Many ECE institutions face challenges in implementing EE due to limited financial and material resources*

Many early childhood education (ECE) institutions struggle implementing Environmental Education (EE) due to limited outdoor spaces, materials, and funding (Cooper 2015). Although the benefits of nature-based learning are well-documented, many schools lack green spaces, gardens, or educational farms. Urban areas face a scarcity of open spaces, while rural areas suffer from inadequate infrastructure. Funding constraints further limit access to educational tools and environmental programs (Suryana and Suryani 2019).

Another major barrier is the lack of teacher training in EE. Most teachers receive insufficient training in nature-based education, making it difficult to design effective learning strategies (Hidayati and Wulandari 2018). EE is often treated as a secondary activity rather than being integrated into national education policies. Additionally, the lack of regulations and dedicated funding weakens EE implementation. Stronger policies are needed, including allocating funding for EE, curriculum integration, and continuous teacher training to ensure effective and widespread adoption (Hidayati and Wulandari 2018).

Economic barriers are significant in EE implementation. Many institutions depend on government or NGO funding to overcome financial constraints. In Finland, the government has established a robust funding mechanism for nature-based education programs, contributing to EE's success (Juvonen *et al.* 2019). Similarly, initiatives in Canada and Australia, such as the *EcoAction Community Fund* (Government of Canada 2020) and the *Sustainability Education Grants* program (Australian Government Department of Education 2021), offer financial support for schools that integrate environmental education. These models highlight how partnerships between governments, NGOs, and the private sector can mitigate financial limitations and improve EE program quality in ECE institutions.

These initiatives demonstrate that while financial barriers persist, public-private partnerships can overcome resource constraints. Sustainable funding models and comprehensive EE integration into national policies are essential for long-term support in early childhood education.

3.2.2 *Effective EE requires teachers who are well-trained in environmental concepts and pedagogical strategies*

The success of Environmental Education (EE) depends largely on teachers who are well-trained in both environmental concepts and pedagogical strategies. However, many teachers lack professional development opportunities, hindering EE integration into early childhood education curricula (Juvonen *et al.* 2019). Without proper training, teachers struggle to design nature-based learning, manage outdoor activities, and instil sustainability values. Moreover, limited institutional support and education policies that prioritise academic achievement over environmental awareness make EE a secondary concern (Kullberg and Myhr 2020).

While teacher training is crucial, practical recommendations for improving EE training are still lacking. Studies suggest that a more pragmatic approach, including practice-based training, workshops, and collaborative learning, is needed (Hedefalk *et al.* 2016, Stevenson 2007). EE training must integrate sustainability into the ECE curriculum and provide facilities for interactive outdoor activities. In addition to training gaps, EE faces challenges such as limited resources, insufficient green spaces, and low prioritisation within the education system (Tan and Pedretti 2010). A conflict exists between EE's sustainability goals and the focus of traditional education on existing social structures (Stevenson 2007). Therefore, a proactive approach is needed, emphasising continuous teacher training and the development of infrastructure that supports EE.

Teachers should engage in sustainable, practice-based training to ensure effective EE implementation. Hands-on workshops and field-based training will improve their ability to manage nature-based education activities. According to Gagnon and DeLuca (2022), incorporating environmental education into curricula can strengthen future generations' environmental awareness. Supported by adequate policies, this training will create a generation more aware of sustainability and better equipped to address future environmental challenges (Gagnon and DeLuca 2022).

3.2.3 *The standard ECE curriculum often does not prioritise EE, making it difficult for teachers to allocate time and resources to environmental topics*

The pressure to meet academic standards limits EE integration in ECE. Current curricula prioritise literacy and numeracy, while sustainability and environmental awareness receive less focus. Without explicit policies, many schools lack guidelines and materials for effective EE implementation. Rigid, target-driven curricula restrict teachers' flexibility in using interactive methods like nature exploration, gardening, and recycling (Kioupi and Voulvoulis 2019). Teachers often prioritise assessed core subjects, treating EE as supplementary. The study suggests systematically integrating EE from an early age to build environmental awareness without compromising academic achievement.

3.2.4 *Cultural attitudes towards nature and the environment can influence the implementation of EE*

Cultural values influence EE implementation, with some societies resisting it due to differing priorities (Kopnina 2018). Education in developing countries or resource-dependent economies often prioritises economic growth and workforce skills, making EE a lower priority. For example, in mining and plantation areas, curricula focus on job preparation with less emphasis on environmental sustainability.

EE is often taught with a human-centred (anthropocentric) approach, focusing on resource use rather than ecosystem preservation. Education for Sustainable Development (ESD) highlights social and economic aspects of environmental protection. Kopnina (2018) argues that true sustainability requires balancing economic, social, and ecological concerns, teaching human benefits from nature and responsibility for its preservation.

4 CONCLUSION

Integrating environmental education (EE) into early childhood education (ECE) offers numerous benefits, including the development of environmental stewardship, enhanced cognitive and emotional well-being, and improved social and ethical skills. However, significant challenges such as lack of resources, insufficient teacher training, and curriculum constraints must be addressed to realise these benefits fully. To overcome these challenges, a multifaceted approach is essential. This includes increasing investment in EE resources, such as outdoor learning spaces and environmental programs, and providing comprehensive teacher training programs to equip educators with the necessary knowledge and skills for effective EE implementation. Advocating for the inclusion of EE in the standard ECE curriculum is also vital to ensure it becomes a core component of educational policies.

Furthermore, policy interventions should play a crucial role in improving EE implementation. Stronger policies are needed to promote EE integration within ECE curricula, with dedicated funding mechanisms to support these initiatives. Continuous professional development for teachers focusing on EE is another important step in ensuring the long-term success of these programs. Additionally, future research is needed to explore the best models for integrating EE into national curriculums and assess early environmental education's long-term effects on young learners. By addressing these barriers, we can lay the groundwork for a sustainable future, starting with the education of our youngest learners.

REFERENCES

Adyasari, D., Pratama, M. A., Teguh, N. A., Sabdaningsih, A., Kusumaningtyas, M. A., and Dimova, N. (2021). Anthropogenic impact on Indonesian coastal water and ecosystems: Current status and future opportunities. *Marine Pollution Bulletin*, 171, 1–14. https://doi.org/10.1016/j.marpolbul.2021.112689

Ardoin, N. M., and Bowers, A. W. (2020). Early childhood environmental education: A systematic review of the research literature. *Educational Research Review*, 31, 100353. https://doi.org/10.1016/j.edurev.2020.100353

Ardoin, N. M., Bowers, A. W., and Gaillard, E. (2020). Environmental education outcomes for conservation: A systematic review. *Biological Conservation*, 241, 108224. https://doi.org/10.1016/j.biocon.2019.108224

Arora, N. K., Fatima, T., Mishra, I., Verma, M., Mishra, J., and Mishra, V. (2018). Environmental sustainability: challenges and viable solutions. *Environmental Sustainability*, 1, 309–340. https://doi.org/10.1007/s42398-018-00038-w

Canada, G. of. (2020). *EcoAction Community Fund.* https://www.canada.ca/en/services/environmental-grants/ecoaction-fund.html

Collado, S., and Evans, G. W. (2019). Outcome expectancy: A key factor to understanding childhood exposure to nature and children's pro-environmental behavior. *Journal of Environmental Psychology*, 61, 30–36. https://doi.org/10.1016/j.jenvp.2018.12.001

Cooper, A. (2015). Nature and the outdoor learning environment: The forgotten resource in early childhood education. *International Journal of Early Childhood Environmental Education*, 3(1), 85–97. https://eric.ed.gov/?id=EJ1108430

Education, A. G. D. of. (2021). *Sustainability Education Grants*. https://www.education.gov.au/sustainability-education-grants

Ernst, J., and Burcak, F. (2019). Young children's contributions to sustainability: The influence of nature play on curiosity, executive function skills, creative thinking, and resilience. *Sustainability*, 11(15), 1–22. https://doi.org/10.3390/su11154212

Feriver, Ş., Teksöz, G., Olgan, R., and Reid, A. (2016). Training early childhood teachers for sustainability: Towards a 'learning experience of a different kind.' *Environmental Education Research*, 22(5), 717–746. https://doi.org/10.1080/13504622.2015.1027883

Gagnon, A., and DeLuca, C. (2022). Building environmental literacy in early childhood education: A critical review of current strategies. *Journal of Environmental Education*, 53(4), 303–315. https://doi.org/10.1080/00958964.2022.2033129

Gentles, S. J., Charles, C., Nicholas, D. B., Ploeg, J., and McKibbon, K. A. (2016). Reviewing the research methods literature: principles and strategies illustrated by a systematic overview of sampling in qualitative research. *Systematic Reviews*, 5(172), 1–11. https://doi.org/10.1186/s13643-016-0343-0

Goldman, D., Pe'er, S., and Yavetz, B. (2017). Environmental literacy of youth movement members–is environmentalism a component of their social activism? *Environmental Education Research*, 23(4), 486–514. https://doi.org/10.1080/13504622.2015.1108390

Green, C., Kalvaitis, D., and Worster, A. (2016). Recontextualizing psychosocial development in young children: A model of environmental identity development. *Environmental Education Research*, 22(7), 1025–1048. https://doi.org/10.1080/13504622.2015.1072136

Guevara-Herrero, I., Bravo-Torija, B., and Pérez-Martín, J. M. (2024). Educational practice in education for environmental justice: A systematic review of the literature. *Sustainability*, 16(7), 1–17. https://doi.org/10.3390/su16072805

Haimovitz, K., and Dweck, C. S. (2017). The origins of children's growth and fixed mindsets: New research and a new proposal. *Child Development*, 88(6), 1849–1859. https://doi.org/10.1111/cdev.12955

Hayati, M., Rahiem, M. D. H., Romadhon, D. R., Faeruz, R., Rahmatullah, R., Subchi, I., and Hadi, Y. (2023). Science in Islamic kindergarten in Indonesia: Content and methods. *AIP Conference Proceedings*, 2595(1), 040030-1-040030–040038. https://doi.org/10.1063/5.0123803

Hebe, H. N. (2017). Towards a theory-driven integration of environmental education: The application of Piaget and Vygotsky in Grade R. *International Journal of Environmental and Science Education*, 12(6), 1525–1545.

Hedefalk, M., Almqvist, J., and Häggström, M. (2016). A comparison of the environmental knowledge and attitudes of children in different countries: A review of studies. *Environmental Education Research*, 22(1), 79–92. https://doi.org/10.1080/13504622.2015.1045209

Hidayati, K., and Wulandari, S. (2018). Kendala guru dalam implementasi pendidikan lingkungan hidup di PAUD. *Jurnal Pendidikan Anak*, 7(1), 1–10. https://doi.org/10.21831/jpa.v7i1.18199

Juvonen, J., Lessard, L. M., Rastogi, R., Schacter, H. L., and Smith, D. S. (2019). Promoting social inclusion in educational settings: Challenges and opportunities. *Educational Psychologist*, 54(4), 250–270. https://doi.org/10.1080/00461520.2019.1655645

Kiewra, C., and Veselack, E. (2016). Playing with Nature: Supporting Preschoolers' Creativity in Natural Outdoor Classrooms. *International Journal of Early Childhood Environmental Education*, 4(1), 70–95. https://eric.ed.gov/?id=EJ1120194

Kioupi, V., and Voulvoulis, N. (2019). Education for sustainable development: A systemic framework for connecting the SDGs to educational outcomes. *Sustainability*, 11(21), 1–18. https://doi.org/10.3390/su11216104

Kopnina, H. (2018). Education for sustainable development (ESD): The turn away from 'environment'in environmental education? In *Environmental and sustainability education policy* (pp. 135–153). Routledge. https://doi.org/10.1080/13504622.2012.658028

Kullberg, A., and Myhr, A. (2020). Sustainability education in early childhood: Teachers' perspectives and practices. *Environmental Education Research*, 26(6), 895–911. https://doi.org/10.1080/13504622.2020.1776923

Lamanauskas, V. (2023). The importance of environmental education at an early age. *Journal of Baltic Science Education*, 22(4), 564–567. https://doi.org/10.33225/jbse/23.22.564

Molinario, E., Lorenzi, C., Bartoccioni, F., Perucchini, P., Bobeth, S., Colléony, A., Diniz, R., Eklund, A., Jaeger, C., and Kibbe, A. (2020). From childhood nature experiences to adult pro-environmental behaviors: An explanatory model of sustainable food consumption. *Environmental Education Research*, 26 (8), 1137–1163. https://doi.org/10.1080/13504622.2020.1784851

Mukti, H., and Sobirov, B. B. (2023). Environmental Justice at the Environmental Regulation in Indonesia and Uzbekistan. *Journal of Human Rights, Culture and Legal System*, 3(3), 476–512. https://doi.org/10.53955/jhcls.v3i3.171

Nirmala, B., Solihatin, E., and Sukardjo, M. (2024). Augmented Reality in Early Childhood Education: Trends, Practices, and Insights from a Literature Review. *International Journal of Interactive Mobile Technologies*, 18(22), 50–67. https://doi.org/10.3991/ijim.v18i22.50337

Paananen, M., Centeno, V. G., Fjällström, S., Bhattarai, D. P., Jovanović, G., Ferreira, J. M., Millei, Z., Ribeiro, C. R., Paakkari, A., and Sajjadieh, N. S. (2023). Imaginaries of ECEC: A comparison of policies of access in early childhood education in Brazil, Finland, Hungary, Indonesia, Iran, Nepal, and Serbia. *Journal of Early Childhood Education Research*, 12(3), 145–180. https://doi.org/10.58955/jecer.122962

Rahiem, M. D. H. (2024). Early childhood education contingencies for sustaining learning during school closures: Lessons from preschool remote education home visits in Indonesia during the COVID-19 pandemic. *Children and Youth Services Review*, 166, 107955. https://doi.org/10.1016/j.childyouth.2024.107955

Rahmania, T. (2024). Exploring school environmental psychology in children and adolescents: The influence of environmental and psychosocial factors on sustainable behavior in Indonesia. *Heliyon*, 10(18), 1–16. https://doi.org/10.1016/j.heliyon.2024.e37881

Seddon, N., Smith, A., Smith, P., Key, I., Chausson, A., Girardin, C., House, J., Srivastava, S., and Turner, B. (2021). Getting the message right on nature-based solutions to climate change. *Global Change Biology*, 27 (8), 1518–1546. https://doi.org/10.1111/gcb.15513

Shaheen, N., Shaheen, A., Ramadan, A., Hefnawy, M. T., Ramadan, A., Ibrahim, I. A., Hassanein, M. E., Ashour, M. E., and Flouty, O. (2023). Appraising systematic reviews: a comprehensive guide to ensuring validity and reliability. *Frontiers in Research Metrics and Analytics*, 8, 1–9. https://doi.org/10.3389/frma.2023.1268045

Shanahan, D. F., Franco, L., Lin, B. B., Gaston, K. J., and Fuller, R. A. (2016). The benefits of natural environments for physical activity. *Sports Medicine*, 46(7), 989–995. https://doi.org/10.1007/s40279-016-0502-4

Shivanna, K. R. (2022). Climate change and its impact on biodiversity and human welfare. *Proceedings of the Indian National Science Academy*, 88(2), 160–171. https://doi.org/10.1007/s43538-022-00073-6

Simsek, Z., Fox, B., and Heavey, C. (2023). Systematicity in organizational research literature reviews: A framework and assessment. *Organizational Research Methods*, 26(2), 292–321. https://doi.org/10.1177/10944281211008652

Stevenson, R. B. (2007). Schooling and environmental education: Contradictions in purpose and practice. *Environmental Education Research*, 13(2), 139–153. https://doi.org/10.1080/13504620701295726

Sulistiawati, L. Y. (2024). Climate change related litigation in Indonesia. *Communications Earth and Environment*, 5(1), 522–544. https://doi.org/10.1038/s43247-024-01684-1

Tan, M., and Pedretti, E. (2010). Negotiating the complexities of environmental education: A study of Ontario teachers. *Canadian Journal of Science, Mathematics and Technology Education*, 10, 61–78. https://doi.org/10.1080/14926150903574320

Williams, D. R., and Dixon, P. S. (2013). Impact of garden-based learning on academic outcomes in schools: Synthesis of research between 1990 and 2010. *Review of Educational Research*, 83(2), 211–235. https://doi.org/10.3102/00346543134758

Yeşilpınar Uyar, M., and Eti, İ. (2023). Curriculum fidelity of Turkish preschool teachers in school-based practices. Education 1-13, 51(5), 820–835. https://doi.org/10.1080/03004279.2021.2024867.

Towards Resilient Societies: The Synergy of Religion, Education, Health, Science, and Technology – Rahiem (Ed.)

The challenge to reduce plastic-based waste: A case study of snack packages vs. green campus movement

Eny S. Rosyidatun
Universitas Islam Negeri Syarif Hidayatullah Jakarta, Indonesia

Nasry Voeut
An-Nikmah Al Islamiyah Phnom Penh, Cambodia

ABSTRACT: Providing snacks for meeting attendees is a universal practice. In the Indonesian context, snacks are often served in individual packages, allowing participants to consume them as needed. While practical, this packaging generates various types of waste, particularly plastics. This study examines the typical snack packaging used in official meetings and its implications for the green campus movement at State Islamic University (UIN) Syarif Hidayatullah Jakarta. Employing a descriptive qualitative analysis, the research utilises 2024 snack consumption data from the University Business Centre. The findings reveal that snacks are presented in individual packages comprising plastic-based materials alongside organic materials, paper-based wrappers, and aluminium cups. These snack packages generate 34.07 kilograms of plastic waste per month. This study highlights the urgent need for the university to implement specific policies on snack packaging to reduce plastic waste, aligning with the principles of the green campus movement and contributing to sustainable waste management practices.

Keywords: green campus, snacks, waste

1 INTRODUCTION

Snacking is a universal practice in human daily lives around the world. Its implementation is not only in the informal events but also in the formal ones. Snacking takes place and covers a range of situations, from family gatherings to campus meetings. People accept the importance of snacking as it serves as "fuel" during a meeting or gathering (Hess and Slavin 2018; Lamport and Saunders 2020). In Indonesia, the cultural practice of snacking during meetings has become deeply ingrained, reflecting both hospitality traditions and practical considerations in professional settings. This habitual practice involves serving various snacks, typically individually packaged and sealed, using materials such as plastic and aluminium (UNDP Indonesia 2020). Each snack, whether crispy *keripik*, savoury crackers, or sweet pastries, is meticulously prepared and presented to accommodate varying tastes and preferences among attendees. However, the convenience and customisation of these individually wrapped snacks come at a significant environmental cost.

The proliferation of single-use plastic and aluminium packaging not only adds to the volume of waste generated but also exacerbates challenges in waste management infrastructure across the country (Suzuki 2019). These gatherings occur frequently in offices, community centres, and social events nationwide, so the cumulative impact on Indonesia's waste footprint is substantial. According to recent research published in the Journal of Environmental Management, Indonesia faces significant challenges in managing its waste effectively, with inadequate infrastructure and limited recycling facilities compounding the issue (Sugiyama 2020).

On the other hand, green campus movements are running on its pathways. Sustainability awareness mandates that university stakeholders implement the concept in the campus atmosphere. Furthermore, as UIN Jakarta is an Islamic institution, the message from the Holy Quran surely

 DOI: 10.1201/9781003645542-65

encourages the enforcement of sustainability programs. It is stated in Chapter 30 *Surah ArRum* (the Romans) Verse 41: *"Corruption has spread on the land and sea as a result of what people's hands have done, so that He may cause them to taste (the consequences of) some of their deeds, and perhaps they might return (to the Right path)*". This verse implicitly states that humans are the main actors in environmental quality deterioration. Thus, Syarif Hidayatullah State Islamic University has declared its commitment to sustainability as action. It is reflected in the university motto of 2024: Innovation, Green, Humanity. The tag "Green" refers to the commitment of all civitas of UIN Jakarta to have an environment-friendly mindset as well as habits in daily routines. This way, humans might return to the Right path, maintaining the balance of nature that could support lives for the current generation and future generations. This way, a campus might contribute to making sustainability come true in its proper quality.

According to UI Green Metric (2024), a global ranking system for green campuses, there are six criteria for a sustainable campus to look at, namely 1) Setting and Infrastructure, (2) Energy and Climate Change, (3) Waste, (4) Water, (5) Transportation, and (6) Education and Research. One crucial aspect of the green campus program is to deal with the waste problem. Several activities on campus raise waste. Snacking in meetings could be directly intervened to reduce generated waste from campus activities.

2 METHODS

This study implemented descriptive qualitative analysis as the method. The data was collected using a time series approach. This allows to predict the trend in the future (Shumway and Stoffer 2011). In UIN Syarif Hidayatullah Jakarta, the snacks supply is managed by a unit called The University Business Centre. Since the snack supply is centralised, all the snack data for the whole university units, which consist of 12 faculties and 3 centres/units, are controlled in this centre. The data covers the distribution and number of snacks from January to December 2024. Descriptive analysis is applied on the data.

3 FINDINGS AND DISCUSSION

3.1 *Typical snack packages in campus meeting*

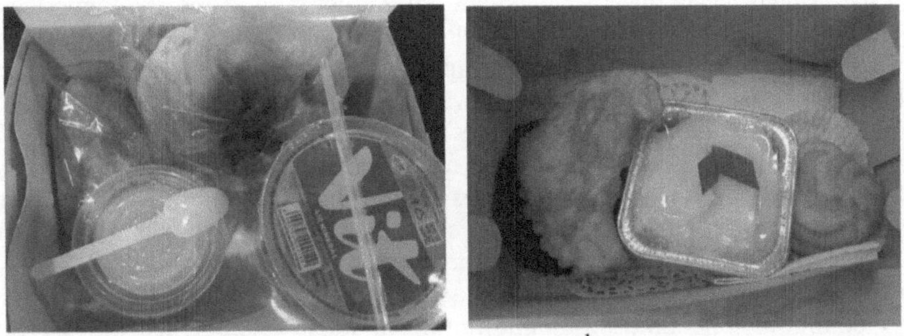

a b

Figure 1. Typical current snack package samples (a. no aluminium cup, b. with aluminium cup).

Snack packages are available in various campus events, such as seminars, workshops, and regular management meetings. Figure 1 shows the presentation of personal package samples.

Since the function of snacks is to maintain the attendees' energy during the meeting, they are practically served in a personal package. The attendees take a package before the meeting to be consumed when needed. It is a good way, considering that some people have not had a proper breakfast, so the snack will help to keep them energised in the meeting. In the Indonesian context, it is common that a package of snacks includes a carbo-based dish, one or two side dishes that can vary from savoury taste to sweet, and a glass of water. Figure 1a shows a snack box containing a glass of water, a plastic cup of pudding, a piece of sweet bread, and a savoury pastry. Figure 1b shows a snack box containing an aluminium cup of rice pudding, a piece of sweet cake, and a savoury pastry.

3.2 *The snacks' waste*

Table 1. Snack package and its waste.

Part/Type	Material	Weight (gr/unit)
Personal box	paper	2.63
Water (Glass)	plastic	3.92
Straw	plastic	0.19
Pastry sheet	paper	1.54
Pastry wrapper	plastic	0.78
Pudding cup	plastic	1.04
Dessert cup	aluminium	1.56
Spoon	plastic	0.55

Figure 1 clearly shows that the personal snack packages potentially contribute to adding a number of waste to the environment. It varies from the organic materials to the inorganic ones, especially plastics. Table 1 below shows the generated waste from a package of snacks.

It can be derived from Table 1 that one personal package of snacks would contribute not less than 6.48 grams of plastic-based waste.

Snack wrappers are predominantly made from various types of plastic, such as polyethylene (PE) or polypropylene (PP). These materials are derived from fossil fuels and require substantial energy and resources during extraction, refining, and manufacturing processes (Geyer *et al.* 2017). The other material is aluminium, produced from bauxite ore through energy-intensive mining and refining processes, contributing to resource depletion and environmental impacts (Norgate 2010). Due to their non-biodegradable nature, plastic snack wrappers persist in the environment for long periods, accumulating in landfills and exacerbating waste management challenges (Thompson *et al.* 2004). Despite efforts to promote recycling, snack wrappers are often disposed of as residual waste due to inadequate infrastructure and the economic feasibility of separating mixed-material packaging (Hopewell *et al.* 2009). Thus, encouraging using biodegradable or compostable materials for snack wrappers can reduce environmental impact, as these materials break down more easily in natural environments (Muthu 2015).

Plastic pollution, particularly, has become emblematic of Indonesia's waste crisis. The country is ranked second globally regarding marine plastic pollution, trailing only behind China (Jambeck *et al.* 2015). Annually, Indonesia contributes an estimated 10% of the world's marine plastic debris, threatening marine ecosystems and biodiversity and posing serious risks to public health and livelihoods of coastal communities (Jambeck *et al.* 2015). The sight of plastic waste clogging waterways, littering beaches, and entangling marine life has become distressingly common, reflecting a systemic failure in waste management infrastructure and practices. According to the World Bank, global waste generation in Indonesia is projected to increase from 2.01 billion tonnes in 2016 to 3.40 billion tonnes by 2050, underscoring the escalating pace at which societies produce waste (World Bank 2018).

Table 2 below describes the monthly waste generation data, highlighting the fluctuation in waste quantities throughout the year of 2024. The highest number of snack supplies was in January 2024 (12,805 packages of snacks), while the least number was in March-April 2024 (1,365 and 1,420 packages of snacks, consecutively). It reflects that in 2024, January was the densest calendar for academic agendas such as seminars, conferences, workshops, etc. Faculties' big events, such as international conferences, clinical examinations, and welcoming events for new students are several of the most contributed events to snack providing (Pusat Bisnis UIN Jakarta 2024). The number of the snacks in March and April 2024 were the lowest, below 1,500 packages. It could be explained that the mid of March to the mid of April 2024 coincided with Ramadhan, the month of fasting for Muslims; so that the number of snacks were the lowest.

Table 2. Monthly snacks number and the waste volume in 2024.

Month	Number of packages	Plastic-based Waste (kg)
January	12.805,00	82,98
February	5.587,00	36,20
March	1.365,00	8,85
April	1.420,00	9,20
May	3.140,00	20,35
June	6.238,00	40,42
July	3.379,00	21,90
August	6.527,00	42,29
September	4.046,00	26,22
October	8.867,00	57,46
November	7.133,00	46,22
December	2.596,00	16,82
TOTAL	63.103,00	408,91

Table 2 also states the volume of plastic-based waste generated from snack packages. It reached 408,91 kilograms throughout the year of 2024. The average plastic-based waste from snack consumption of 2024 was 34.07 kilograms per month.

Considering the significant volume of plastic-based waste generated from snack packages, it is important to work more on waste management. "Prevention is better than cure" should be implemented. Addressing this issue requires a multifaceted approach that promotes sustainable alternatives to packaging, encourages recycling initiatives, and cultivates awareness among all university stakeholders about the environmental consequences of single-use materials (Ellen MacArthur Foundation 2021). By re-evaluating consumption habits and adopting more eco-friendly practices in snack packaging and distribution, the university could support Indonesia in reducing its contribution to the global waste crisis while preserving its rich cultural traditions of hospitality and communal dining.

Figure 2. Traditional snacks presentation.

The richness of Indonesian food heritage would provide many alternatives to more eco-friendly snacks. Some snacks do not need a synthetic wrapper. Figure 2 represents a traditional presentation of snacks with organic waste only. The package consists of steamed sweet corn, peanut, sweet potato, and purple potato. Steamed banana (with local varieties: *kepok, uli, nangka, tanduk*, etc.), various kinds of peas (soybean, edamame, *kacang Bogor)*, and various kinds of corms (cassava, taro, black potato, etc.) are the other alternatives which often served as traditional snacks. In contrast to plastic-based waste, organic material is not harmful to the environment. Since it can be decomposed, even it would add useful biochemical elements for the environment. The consequences of inadequate waste management extend beyond environmental degradation. In urban centres like Jakarta, Surabaya, and Bandung, overflowing landfills and open burning of waste contribute to air and water pollution, exacerbating respiratory illnesses and other health hazards among residents (United Nations Environment Programme 2020). Moreover, the economic toll of ineffective waste management is substantial, with direct costs of clean-up and remediation activities burdening local governments as well as university management and hindering sustainable development efforts.

4 CONCLUSIONS

This study provides empirical data on waste generation linked to snacking habits in a campus environment. The findings highlight the significant volume of plastic-based waste generated from snack packages. Sustainability is in need of holistic approaches to waste management. Addressing the environmental impact of snacking requires collaborative efforts from all campus stakeholders to adopt more sustainable practices and reduce the environmental footprint associated with snack packaging. Policies on operational budgeting, including snack providing, must accommodate sustainable concepts. In conclusion, although snack wrappers made from materials like plastic and aluminium are known for their durability and ability to preserve food freshness, they pose significant environmental challenges. They do serve a practical purpose in packaging and preserving food products, but their environmental impact is substantial due to production processes, disposal challenges, and contribution to pollution and waste accumulation. Furthermore, enhancing waste management infrastructure and campus communities-based recycling initiatives are crucial to achieving sustainable waste reduction goals.

REFERENCES

Ellen MacArthur Foundation. (2021). *The Circular Economy in Cities: Evolving the Model for a Sustainable Urban Future*.

Geyer, R., Jambeck, J. R., and Law, K. L. (2017). Production, use, and fate of all plastics ever made. *Science Advances*, 3(7). https://doi.org/10.1126/sciadv.1700782

Hess, J. M., and Slavin, J. L. (2018). Snacking: A review of the literature. *Journal of the Academy of Nutrition and Dietetics*, 118(5), 833-842.e3. doi: 10.1016/j.jand.2017.12.015

Hopewell, J., Dvorak, R., and Kosior, E. (2009). Plastics recycling: challenges and opportunities. *Philosophical Transactions of the Royal Society B: Biological Sciences*, 364(1526), 2115–2126. https://doi.org/10.1098/rstb.2008.0311

Jambeck, J. R., Geyer, R., Wilcox, C., Siegler, T. R., Perryman, M., Andrady, A., Narayan, R., and Law, K. L. (2015). Plastic waste inputs from land into the ocean. *Science*, 347(6223), 768–771. https://doi.org/10.1126/science.1260352

Lamport, D. J., and Saunders, C. (2020). Snacking and cognitive function. *Journal of Nutrition and Metabolism*, 25, 1–9. doi: 10.1186/s12986-020-00438-4

Muthu, S. S. (2015). Environmental implications of plastic consumption in global scenario. *Journal of Cleaner Production*, 103, 467–477.

Norgate, T. (2010). Aluminium in sustainable engineering. *Journal of Cleaner Production*, 18(7), 637–646

Pusat Bisnis UIN Jakarta. (2024). *Internal Report of Snack Distribution in 2024*.

Quran.com. the Holy Quran and the translation available at https://quran.com

Shumway, R. H., and Stoffer, D. S. (2011). *Time Series Analysis and Its Applications*. Springer.

Sugiyama, T. (2020). Solid waste management in Indonesia: Current status and future projections. *Journal of Environmental Management*, 266, 110597.

Suzuki, S. (2019). Challenges and opportunities in plastic pollution management in Indonesia. *Marine Pollution Bulletin*, 149, 110510.

Thompson, R. C., Olsen, Y., Mitchell, R. P., Davis, A., Rowland, S. J., John, A. W. G., McGonigle, D., and Russell, A. E. (2004). Lost at Sea: Where Is All the Plastic? *Science*, 304(5672), 838–838. https://doi.org/10.1126/science.1094559

UI Green Metric. (2024). Available at https://greenmetric.ui.ac.id

United Nations Development Programme (UNDP) Indonesia. (2020). Sustainable Consumption and Production: Plastic Waste in Indonesia.

United Nations Environment Programme. (2020). *Global Environment Outlook - GEO-6: Healthy Planet, Healthy People*.

World Bank. (2018). *What a Waste 2.0: A Global Snapshot of Solid Waste Management to 2050*.

Towards Resilient Societies: The Synergy of Religion, Education, Health, Science, and Technology – Rahiem (Ed.)

Integrating the Muamalah Curriculum for anti-corruption education: The role of Islamic law, Gender, and SDG 16

Yayat Sujatna & Yulianti Muthmainnah
Institut Teknologi dan Bisnis Ahmad Dahlan Jakarta
ORCID ID: 0000-0002-0168-3128

Ilham Mundzir
Universitas Muhammadiyah Prof. Dr. Hamka Jakarta

Dzikri Rahmat Romadhon
Universitas Islam Negeri Syarif Hidayatullah Jakarta
ORCID ID: 0000-0003-2796-7281

Ahmad Qisa'i
Universitas Paramadina Jakarta

Fadliya Kadir
Institute An-Nikmah Al-Islamiyah Phnom Penh, Cambodia

ABSTRACT: Integrating religious values, ethics, and clean leadership behaviour remains a fundamental principle continuously reinforced in various institutions and organisations across Indonesia. This commitment is exemplified by the AKHLAK values (Amanah-Trustworthy, Kompeten-Competent, Harmonis-Harmonious, Loyal-Loyal, Adaptif-Adaptive, Kolaboratif-Collaborative), implemented by a government ministry to promote corruption eradication and ethical leadership, aligning with Sustainable Development Goal 16. This study explores the Muamalah curriculum, specifically the Fiqh of Anti-Corruption module, at Institut Teknologi dan Bisnis Ahmad Dahlan (ITB-AD) Jakarta, using a qualitative approach that includes literature review, student assignment analysis, and alumni interviews. The findings reveal four key insights. First, ITB-AD has incorporated the Muamalah curriculum as a mandatory General Education Course to instil anti-corruption values. Second, the curriculum integrates theological principles, Islamic law, moral values, gender, legal culture, and ethical-worship perspectives, reinforcing the fight against corruption. Third, students' understanding of anti-corruption values is reflected in their final assignments, which include videos or opinion narratives focused on corruption-related themes. Fourth, the integration of religious and moral values within the Muamalah curriculum and SDG 16 effectively fosters ethical leadership and an anti-corruption culture among graduates. Alumni, whether working in governmental institutions or as entrepreneurs, demonstrate a strong commitment to ethical conduct, refusing to justify unethical means to achieve their goals.

Keywords: anti-corruption education, ethical leadership, Islamic law, muamalah, SDGs 16

DOI: 10.1201/9781003645542-66

1 INTRODUCTION

Corruption is a global crisis requiring urgent action. According to the United Nations Convention Against Corruption (UNCAC), it is a transnational issue affecting societies and economies, necessitating international cooperation (UNCAC 2024). In a 2003 UN address, Kofi Annan emphasized its harm to economies, the impoverished, and development. Indonesia combats corruption through Law No. 31 of 1999, though enforcement remains controversial. The Harvey Moeis case, involving IDR 300 trillion in losses but only a 6.5-year sentence and IDR 210 billion fine, triggered public outrage and protests over judicial failures. (Mahendra 2024; CNN Indonesia 2024; Nuranisa 2024). Including a report from Transparency International's 2022 Corruption Perception Index showing a steep decline from 38 to 34, the highly credible Corruption Eradication Commission (KPK) recently lost its identity after its leader Lili Pintauli resigned over allegations of graft. Firli Bahuri was named a suspect in a Ministry of Agriculture corruption case (ICW 2023).

Education is key to anti-corruption efforts. Indonesia's SDG 16 Roadmap stresses instilling anti-corruption values from early education (Bappenas 2021). Muhammadiyah-'Aisyiyah Higher Education Institutions (PTMA) contribute by producing ethically grounded graduates. Institut Teknologi dan Bisnis Ahmad Dahlan (ITB-AD) Jakarta, a PTMA-affiliated university, aligns with Muhammadiyah's Progressive Islam, fostering a just, democratic, and ethical society (Nashir 2022). This vision is upheld through five core principles: tauhid, deep Qur'anic and Sunnah understanding, institutionalized good deeds, future orientation, and moderation (Biyanto 2022). ITB-AD mandates the Muamalah course, which includes the Fiqh of Anti-Corruption module in the fourth session, with complementary courses such as Islamic Economics and Business Ethics.

Research confirms higher education's role in anti-corruption efforts (Dikyanmas KPK *et al.* 2018; Kemenristekdikti 2018; KPK 2016, 2019; Parellangi 2020). Studies show university curricula, student organisations, and character education effectively instill integrity (Burhanudin 2022; Candra 2023; Dewi & Dedi 2022; Hasan 2015; Suhandi & Agustin 2023). Social norms also shape behaviours, as outlined by U4's step-by-step anti-corruption strategies (Bicchieri 2016; U4 2020).

This study expands on prior research by examining how anti-corruption values are internalised through literature review, student assignments (videos, essays), and alumni interviews. It assesses how ITB-AD's Muamalah curriculum translates into professional ethics. Using Imam Al-Syatibi's Maqasid al-Shari'ah framework, particularly hifz al-din (preserving religion), this study evaluates anti-corruption implementation across three levels: (1) daruriyyat—Qur'anic, Hadith, and prophetic references to corruption; (2) hajiyyat—Muhammadiyah's legal interpretations and ITB-AD policies; (3) tahsiniyyat—student projects demonstrating applied learning.

A significant difference between the ITB-AD Muamalah-Anti Corruption curriculum and other secular anti-corruption curricula is that ITB-AD integrates stronger religious values. Referring to Al-Islam and Kemuhammadiyahan (AIK) into PTMA, AIK is integrated in the PTMA curriculum effectively touching all aspects of human personality with three key aspects including knowing (cognitive), loving (affective), and acting (psychomotor) based on divinity, morals, and law (Mundzir 2024). This research examines why students prioritize anti-corruption themes in Muamalah and how alumni apply these principles in their careers, providing insights into the long-term impact of anti-corruption education at ITB-AD.

2 METHODOLOGY

This study employs a qualitative research approach, combining literature review, student final project analysis, and alumni interviews. The literature review focuses on the Muamalah curriculum, which includes Qur'anic verses, Hadith, prophetic traditions, Muhammadiyah's

Tarjih and *Tajdid* legal decisions, and institutional policies on anti-corruption. Additionally, the study examines student final projects, particularly those selecting anti-corruption themes in videos or opinion essays, as part of the 2022–2023 academic period. Given that Muamalah covers 14 topics. When students chose or preference for anti-corruption as a final project subject provides valuable insights. The way to analyse their assignments is whether the materials in the Muamalah curriculum given during lectures are applied in their projects such as Islamic law, SDG 16, gender, anti-corruption laws that apply in Indonesia, including personal reflections after attending lectures. If applied then their assignments are selected in this paper. Finally, alumni interviews assess how graduates apply anti-corruption values in their professional careers, offering a practical perspective on the curriculum's long-term impact. The selection of alumni was based on an invitation to be interviewed. There were 10 alumni selected based on their occupations, namely bank employees, entrepreneurship, working in the tax office. Of these 10 people, three alumni who work in government banks and philanthropic organisations that handle money on a daily basis agreed to be interviewed. While the other seven were not willing due to personal reasons.

3 FINDINGS AND DISCUSSION

3.1 *Islam's firm stance against corruption*

Islam emphasises not only individual rights and freedoms but also family and societal responsibilities. Wealth ownership, for instance, is not absolute in Islam; it entails obligations such as family sustenance, inheritance distribution, and zakat payments upon reaching specified thresholds. The importance of fulfilling others' rights is exemplified by Caliph Abu Bakr As-Siddiq's firm stance against those who refused to pay zakat (Shobirin 2014). Islam also strictly prohibits fraud, dishonesty (*khianat*), deceptive trade practices, and any form of rights infringement. The punishment for theft includes hand amputation, with harsher penalties for violent robbery, such as cross-limb amputation (Fikih Sunnah, Sayyid Sabiq). This strict stance against corruption is reflected in Muhammadiyah-'Aisyiyah Higher Education Institutions (PTMA) through their Catur Dharma, particularly in Al-Islam and Kemuhammadiyahan or Muhammadiyah Studies (AIK), which distinguishes PTMA from other institutions that adhere only to Tri Dharma (teaching, research, and community service).

The AIK curriculum is incorporated into Muamalah courses, particularly the Fiqh of Anti-Corruption module, which covers (1) the definition of corruption, (2) its causes, (3) forms and methods, (4) its consequences, (5) Muhammadiyah's stance, and (6) strategies for prevention and eradication from educational, socio-economic, legal, and political perspectives. Given that Muamalah carries 3 credits (150 minutes per session), the curriculum has been expanded to include corruption definitions from legal, Islamic, and international perspectives, its historical context, forms under Islamic, Indonesian, and international law, and elements of corruption crimes. The course also explores causative factors, such as weak faith, consumerism, greed, and corrupt environments, alongside the operational methods of corruption and its impact, including poverty, gender inequality, deteriorating infrastructure, and potential loss of life. Furthermore, the course delves into Muhammadiyah's stance on corruption and anti-corruption strategies, emphasizing monotheistic values (tauhid), moral integrity, honesty in social and economic interactions, and legal accountability. Lastly, case studies of women in anti-corruption efforts are included to highlight gender dimensions in corruption.

During class, when the issue of anti-corruption was discussed from a gender perspective, the classroom atmosphere became emotional and some cried. This was because the two cases discussed came from true stories. Firstly, the story of a pregnant woman who died due to being taken to the hospital too late in X village: there was no hospital in the village, only a first-level clinic with a general practitioner instead of an obstetrician, the pregnant woman was carried on a stretcher, the road was impassable by car because the road was muddy, badly damaged, and there had been no road repairs for the last 10 years, and there was no ambulance car. The second example is the true story of a women chaiperson official in village Y who succeeded in developing her village into a village with zero tolerance for corruption and saved many women from death due to childbirth during her 5-year tenure. During this discussion, three students cried, because they remembered the true stories of their mothers, neighbours, or friends who died because they were not helped. Broken roads, no doctors, no ambulances, no health budget for women to detect early pregnancy complications are examples of corruption that have an impact on women's deaths. So the death of a mother in labour is not martyrdom as in hadith of the Prophet Muhammad, but due to systemic murder through corruption. Mothers who die are martyrs as per the hadith of the Prophet Muhammad are those who have received health care during pregnancy, not because of limited facilities due to corruption. So this material is relevant to the issues of Islam, gender, SDG 16 and 5.

3.2 *Muamalah syllabus – fiqh of anti-corruption*

The Muamalah syllabus integrates Qur'anic injunctions, Hadith, and historical examples from Prophet Muhammad (PBUH) and his companions, categorised under *daruriyyat* (primary principles). Corruption, referred to as *ghulul* (betrayal of trust) in Qur'an 3:161 and al-suht in Qur'an 5:42, 62, 63, is strictly condemned in Islamic law. Hadith further defines corruption as bribery (*risywah*), acceptance of gifts by officials, and embezzlement. The Prophet exemplified anti-corruption measures by refusing to lead funeral prayers for corrupt individuals, emphasising that even minor corruption leads to hellfire, divine condemnation, and the rejection of charitable acts derived from illicit gains (HR. Muslim).

Muhammadiyah's stance on corruption, outlined in "Fikih Anti Korupsi: Perspektif Ulama Muhammadiyah" (2006), reflects *hajiyyat* (secondary principles) and was developed in collaboration with Majelis *Tarjih* dan *Tajdid* and the Centre for the Study of Religion and Civilization (PSAP). The book discusses Islamic legal perspectives, corruption typologies, historical cases, and misconceptions about sin absolution—a particularly relevant topic given recent proposals for sentence reductions for corrupt officials by President Prabowo.

At ITB-AD Jakarta, anti-corruption commitment is institutionalised in the Integrity Pact, which mandates that university leaders pledge to actively combat corruption, collusion, nepotism, and extremism while maintaining ethical conduct. Additionally, ITB-AD's "Unqualified Opinion" audit rating from a Public Accounting Firm affirms its corruption-free financial management.

A distinguishing feature of ITB-AD's Muamalah syllabus is its focus on corruption's impact on women and children. It examines cases where corruption obstructs maternal healthcare access, resulting in preventable deaths—deemed "systemic murder" rather than martyrdom. The course also highlights women's roles in anti-corruption efforts and profiles global female leaders in corruption eradication. Furthermore, corruption is analysed from three legal perspectives: Islamic Law, International Law, and National Law, offering a comprehensive framework for understanding its broader implications.

3.3 *Internalisation of anti-corruption values in students and alumni*

The tertiary (*tahsiniyyat*) aspect of students' final projects in the Muamalah course reflects their understanding of anti-corruption principles gained through coursework.

3.3.1 *Student video projects on youtube – Rumah Muthmainnah*

As part of their final assignments in the Muamalah course, students created video projects exploring various aspects of corruption from an Islamic perspective. These videos analyse corruption's moral, legal, and societal implications, emphasizing the role of individuals and communities in preventing corrupt practices.

Table 1. Student video projects on anti-corruption.

Student Name and Sex (F/M)	Project Title	YouTube Link
Selvi Fazela (F)	Corruption: When Families Ignore the Source of Their Wealth, Prevent Corruption from Home!	https://www.youtube.com/watch?v=ox2zvB97XPw
Herlambang Agung Pangestu (M)	Corruption in Fiqh: Moral Violations and Religious Perspectives	https://www.youtube.com/watch?v=9WWpoIYnPV0&t=7s
Firly Zendian Fernando (M)	Fiqh of Corruption in Islam	https://www.youtube.com/watch?v=i-PKmqJePHI
Dzulkahfi Zaki Adilah (M)	Fiqh of Corruption	https://www.youtube.com/watch?v=rM2wf-veBm4
R. Haryo Dewanto GS (M)	Fiqh of Corruption	https://www.youtube.com/watch?v=Y6oDW_sslpM
Luluk Nur Septianawati (F)	Fiqh of Corruption	https://www.youtube.com/watch?v=kpIPshm1IeE
Yulinda Arisandy Sunardi (F)	Corruption	https://www.youtube.com/watch?v=jxq-TTLvZyc
Larasati (F)	Islamic Perspectives on Corruption	https://www.youtube.com/watch?v=nBqTx6u7_HY
Puspa Dwi Destiana (F)	Corruption from an Islamic Perspective	https://www.youtube.com/watch?v=XiQensx-R6w

These student-led projects serve as educational tools that promote anti-corruption awareness, integrating Islamic teachings, moral values, and legal principles to combat corruption effectively.

3.3.2 *Opinion articles on Nun Institute.co*

Students also contributed opinion pieces on Nun Institute.co, exploring various aspects of corruption from an Islamic perspective. These articles examine moral, legal, and societal implications, emphasising the role of ethics, law, and governance in eradicating corruption. these articles reflect a comprehensive analysis of corruption, providing critical insights into its causes, consequences, and solutions from an islamic perspective. based on the data above, the number of male students who chose anti-corruption issues is greater than female students, as shown below:

Table 2. Anti-corruption opinion articles published on Nun Institute.co.

Author and Sex	Article Title	Article Link
Reihansyah (M)	The material 'Fiqh of Anti-Corruption in Muamalah Perspective' was very Impressive	https://nun-institute.co/materi-fiqih-korupsi-dalam-perspektif-muamalah-sangat-berkesan/
Rehan Dhyka (M)	Fiqh of Anti-Corruption Effectively Combats Corruption	https://nun-institute.co/fiqih-korupsi-efektif-perangi-tindak-pidana-korupsi/
A'ang Mustapa (M)	When Habit Becomes Justification (Village Fund Corruption)	https://nun-institute.co/ketika-kebiasaan-dijadikan-pembenaran-korupsi-dana-desa/
Muhammad Al Ilal Hamdi (M)	Fighting Corruption Through Strengthening Morality and Integrity	https://nun-institute.co/korupsi/

(*continued*)

Table 2. Continued

Author and Sex	Article Title	Article Link
Desta Anggre (F)	Fiqh of Corruption: Transforming Anti-Corruption Efforts by Integrating Islamic Principles	https://nun-institute.co/fiqih-korupsi-transformasi-anti-korupsi-mengintegrasikan-prinsip-prinsip-islam-dalam-upaya-pemberantasan-korupsi/
Zaki Adilah (M)	Islamic Jurisprudence on Corruption in Indonesia	https://nun-institute.co/hukum-fikih-korupsi-islam-di-indonesia/
Herlambang Agung Pangestu (M)	Corruption in Fiqh: Moral Violations and Religious Perspectives	https://nun-institute.co/korupsi-dalam-fikih-pelanggaran-moral-dan-perspektif-agama/
Firly Zendian Fernando (M)	Fiqh of Corruption in Islam & Five Strategies to Combat Corruption in Indonesia and Appropriate Punishments for Corruptors	https://nun-institute.co/fikih-korupsi-dalam-islam-5-cara-memerangi-korupsi-di-indonesia-serta-hukuman-yang-pantas-bagi-koruptor/
Aryo Dewanto (M)	Fiqh of Corruption	https://nun-institute.co/fiqih-korupsi/
Zahwa Zia Asy-Syifa (F)	Corruption and Its Impact on Development	https://nun-institute.co/korupsi-dan-dampaknya-terhadap-pembangunan/
Wildan Hadi Purnama (M)	Corruption and Poverty	https://nun-institute.co/korupsi-dan-kemiskinan/
Subhan Hidayatullah (M)	Understanding Corruption in Islam and Four Ways Islam Fights Corruption	https://nun-institute.co/pengertian-korupsi-dalam-islam-dan-4-cara-islam-memerangi-korupsi/
Alfarezi Khulafa Haq (M)	Leaders Who Betray Trust for Wealth & Use Corrupt Gains for Charity	https://nun-institute.co/pemimpin-yang-mengkhianati-amanah-karena-harta-menggunakan-hartanya-untuk-bersedekah/
Puspa Dwi (F)	Is the Death Penalty for Corruptors Permissible in Islam?	https://nun-institute.co/apakah-pidana-hukum-mati-bagi-koruptor-dalam-islam-diperbolehkan/
Fidella Helga (F)	Islamic Perspectives on the Death Penalty for Corruptors	https://nun-institute.co/bagaimana-perspektif-islam-terhadap-hukuman-mati-bagi-koruptor/
Fitriyani TI (F)	Can Charity for Orphans Be Given Using Corrupt Money?	https://nun-institute.co/bagaimana-hukum-memberi-santunan-kepada-anak-yatim-dengan-uang-korupsi-dalam-islam/
Yulinda Arisandy Sunardi (F)	Corrupt Practices in Islam	https://nun-institute.co/tindak-koropsi-dalam-islam/
Tasya Annisa Salsabila (F)	Three Types of Minor Corruption: University and School Students as Perpetrators from an Islamic Perspective	https://kumparan.com/ttaassaa1995/3-jenis-korupsi-kecil-mahasiswa-dan-pelajar-pelaku-sampai-pandangan-islam-1zSXmcNJQ09/1
Ita Yosita Ratna (F)	Corruption in Indonesia from Islamic and Positive Legal Perspectives	https://tanwir.id/korupsi-di-indonesia-dalam-perspektif-hukum-islam-dan-positif/

Youtube Rumah Muthmainnah = 9		Nun-Institute.co = 19	
Female = 5	Male = 4	Female = 8	Male = 11

3.3.3 *Alumni perspectives on anti-corruption values*

To assess the impact of the Muamalah course, three alumni were randomly selected to share their views and experiences regarding the application of Islamic anti-corruption principles in their professional lives.

Table 3. Alumni reflections on the impact of the muamalah course on anti-corruption ethics.

Alumni Name	Workplace	Key Reflections
Dendi Raihan	Bank BUMN	Gained deeper understanding of Islamic values in economic transactions and social interactions. Prioritizes honesty, fairness, and professionalism in workplace decision-making.
Tatan Mulyana	Bank BNI	Finds Islamic anti-corruption teachings highly beneficial in daily life and professional career. Emphasizes honesty, integrity, and confidentiality in banking operations.
Adirosadi	LAZISMU Pusat	Values the Muamalah course for clarifying corrupt practices. Increased vigilance and adherence to ethical and legal standards in professional duties.

Although the topic is covered in only a single lecture session, integrating religious and moral values within the Muamalah curriculum effectively fosters ethical leadership and an anti-corruption culture among graduates. This is evident in their continued adherence to the values instilled during their academic journey, which they actively apply in their professional environments, starting with small, impactful actions. These responses indicate that anti-corruption principles have been effectively internalized by students and continue to shape their professional ethics and leadership behaviour.

4 CONCLUSION

Incorporating anti-corruption education into university curricula is highly strategic. ITB-AD Jakarta offers a unique approach by addressing corruption from the perspectives of Islamic Law, International Law, and National Law, as well as corruption in the gender perspective its impact on women, the gendered effects of corruption have been extensively explored. This Muamalah-Fiqh Anti-Corruption curriculum is very easy to adopt. ITB-AD already has a learning module and syllabus, which is equipped with methods and procedures for providing material in class so that it can be applied on religious campuses and secular campuses. As a recommendation, in future research, follow-up studies should be conducted to track alumni's long-term ethical behaviour.

ACKNOWLEDGEMENTS

We extend our gratitude to the students and alumni of ITB-AD Jakarta for their willingness to participate in interviews and allow the publication of their work.

REFERENCES

Bappenas. (2021). *Peta jalan SDGs Indonesia menuju 2030.* https://sdgs.bappenas.go.id/website/wp-content/uploads/2021/02/Roadmap_Bahasa-Indonesia_File-Upload.pdf

Bicchieri, C. (2016). *Norms in the wild: How to diagnose, measure, and change social norms.* Oxford University Press.

Biyanto. (2022). Merawat nilai-nilai utama untuk Muhammadiyah berkemajuan. *Universitas Muhammadiyah Surabaya.* https://data.um-surabaya.ac.id/Materi_3_Presentasi%20UM%20Surabaya_Biyanto,%209-3-2022.pdf

Burhanudin, A. A. (2022). Peran perguruan tinggi dalam penanaman dan penguatan pendidikan karakter dan anti korupsi. *Salimiya: Jurnal Studi Ilmu Keagamaan Islam,* 3(4). https://ejournal.iaifa.ac.id/index.php/salimiya/article/view/872

Candra, B. Y. (2023). Kepemimpinan dan kontrol kebijakan: Pembentukan budaya anti korupsi di perguruan tinggi. *Studia Manageria: Jurnal Manajemen Pendidikan Islam,* 5(2). https://alumni.radenfatah.ac.id/index.php/studiamanageria/article/download/19707/6716/

CNN Indonesia. (2024, December 31). MA respons kemarahan publik soal vonis Harvey Moeis. https://www.cnnindonesia.com/nasional/20241231081857-12-1182485/ma-respons-kemarahan-publik-soal-vonis-harvey-moeis

Dewi, N. A., & Dedi. (2022). Pendidikan anti korupsi di perguruan tinggi sebagai upaya preventif pencegahan. *Justice Journal,* 1(1). https://justices.pdfaii.org/index.php/i/article/view/3/4

Dikyanmas KPK, Universitas Paramadina, & USAID Cegah. (2018). *Modul Pendidikan Antikorupsi Untuk Mahasiswa.* https://aclc.kpk.go.id/materi-pembelajaran/pendidikan/buku/modul-pendidikan-antikorupsi-untuk-mahasiswa

Hasan, R. (2015). Penerapan pendidikan anti korupsi di kampus sebagai bagian integral dari pendidikan karakter. *Edukasi: Jurnal Pendidikan,* 13(2). https://ejournal.unkhair.ac.id/index.php/edu/article/download/51/143

Institut Teknologi dan Bisnis Ahmad Dahlan (ITB-AD). (2017). *Visi misi tujuan ITB-AD Jakarta.* https://www.itb-ad.ac.id/tentang-kami/tujuan-kami/

Indonesia Corruption Watch (ICW). (2023). *Laporan Akhir Tahun ICW 2023; Bertumbuh dalam Kolaborasi.* https://antikorupsi.org/id/laporan-akhir-tahun-icw-2023

Kementerian Riset, Teknologi, dan Pendidikan Tinggi. (2018). *Buku pendidikan anti korupsi untuk perguruan tinggi.* https://aclc.kpk.go.id/materi-pembelajaran/pendidikan/buku/buku-pendidikan-antikorupsi-untuk-perguruan-tinggi

Komisi Pemberantasan Korupsi (KPK). (2016). *Modul Siap Berantas Korupsi.* https://aclc.kpk.go.id/materi-pembelajaran/pendidikan/buku/modul-siap-berantas-korupsi

Komisi Pemberantasan Korupsi (KPK). (2019). *Modul Materi Tindak Pidana Korupsi.* https://aclc.kpk.go.id/materi-pembelajaran/pendidikan/buku/modul-materi-tindak-pidana-korupsi

Mahendra, K. (2024, December 31). *Harvey Moeis: Vonis ringan yang mengecewakan dan tanpa keadilan.* Tempo.co. https://www.tempo.co/hukum/harvey-moeis-vonis-ringan-yang-mengecewakan-dan-tanpa-keadilan-1188217

Mundzir, Ilham. (2024). Kebijakan Pengembangan Religiositas Mahasiswa di Perguruan Tinggi Muhammadiyah-'Aisyiyah: Studi Kasus UHAMKA dan UNIMUS. Dissertation of Doctoral Program at UIN Syarif Hidayatullah Jakarta.

Nashir, H. (2022, October 29). *Islam Berkemajuan.* Koran Republika. https://muhammadiyah.or.id/2022/10/islam-berkemajuan/

Nuranisa, A. (2024, December 28). 6 Cuitan netizen "Beri kami 300 T, kami siap dipenjara 6 tahun," bentuk protes. Liputan6. https://www.liputan6.com/hot/read/5854701/6-cuitan-netizen-beri-kami-300-t-kami-siap-dipenjara-6-tahun-bentuk-protes

Parellangi, A. L., Gasma, A., Lestari, T. J., Kusmiyati, Y., & Pascoal, M. E. (2020). Modul pendidikan budaya antikorupsi di politeknik kesehatan Kementerian Kesehatan RI. https://repositori-ditjen-nakes.kemkes.go.id/468/1/PBAK_final.pdf

Shobirin. (2014). Pemikiran Abu Bakar Ash-Shiddiq tentang memerangi orang yang membangkang. *Jurnal Ziswaf,* 1(1). https://journal.iainkudus.ac.id/index.php/Ziswaf/article/download/1532/1403

Suhandi, M. F., & Agustin, S. (2023). Pendidikan anti korupsi pada jenjang perguruan tinggi. *Jurnal Saskara Pendidikan dan Pengajaran,* 1(1), 19–27. https://sj.eastasouth-institute.com/index.php/spp/article/view/75

U4. (2020). *Anti-corruption through a social norms lens.* https://www.u4.no/publications/anti-corruption-through-a-social-norms-lens/fullversion

United Nations Convention Against Corruption (UNCAC). (2024). United Nations Convention Against Corruption. https://www.unodc.org/documents/brussels/UN_Convention_Against_Corruption.pdf.

Towards Resilient Societies: The Synergy of Religion, Education, Health, Science, and Technology – Rahiem (Ed.)

Assessing artificial intelligence literacy among prospective chemistry teachers: Promoting quality education (SDG 4) in the era of technological advancement

Munasprianto Ramli
Universitas Islam Negeri Syarif Hidayatullah Jakarta, Indonesia
ORCID ID: 0000-0002-5618-2486

Ira Apriliani
Universitas Islam Negeri Syarif Hidayatullah Jakarta, Indonesia

Media Putri Yohana
Abak Academy Jakarta, Indonesia

Murnee Masae
The Patani Center for Organizational and Public Outreach, Thailand

ABSTRACT: Although AI literacy among students has been widely studied, research focusing on prospective teachers, especially in chemistry, remains limited. This study addresses this gap by examining the level of artificial intelligence literacy among prospective chemistry teachers. It also investigates differences in AI literacy based on gender, school of origin, and parental income, contributing to SDG 4 (Quality Education) by promoting equitable access to AI knowledge. Using a mixed-methods approach, 310 students were selected through probability sampling. The AILQ (AI Literacy Questionnaire) and interviews were employed for data collection. Descriptive statistics analysed questionnaire data, while thematic analysis examined interview responses. Findings reveal that 67.10% of prospective chemistry teachers exhibit high AI literacy. These results underscore AI literacy's inclusivity in teacher education and its potential to support equitable access to technological competence, aligning with SDG 4.

Keywords: artificial Intelligence literacy, prospective chemistry teachers, quality education, sdg4, technological advancement

1 INTRODUCTION

In the 21st century, emerging technologies are expected to impact social life significantly. Education is neither immune to nor detached from this transformation. Many implications of new information technologies for the teaching profession have emerged, driven by rapid technological development and the interconnectedness of advancements, giving rise to unprecedented phenomena in the Fourth Industrial Revolution—particularly artificial intelligence (AI) (Supriyadi and Asih 2021).

In today's digital world, citizens need to engage positively, critically, and competently, effectively communicating and collaborating through the responsible use of technology (Fleaca and Stanciu 2019). AI applications are increasingly pervasive across industries such as business, science, arts, and education, aiming to enhance user experience and efficiency. Despite widespread awareness of AI services and devices, many people do not understand the underlying concepts and ethical concerns associated with AI use (Ghallab 2019).

DOI: 10.1201/9781003645542-67

AI plays a crucial role in daily life, making it vital for future generations to develop AI literacy to navigate and leverage technology effectively (Wong *et al.* 2020). AI literacy, defined as the ability to understand, evaluate, and use AI technologies ethically, is a critical skill for thriving in the digital age (Long and Magerko 2020). AI also offers transformative potential in education, enabling personalised learning, optimising teaching strategies, and improving educational outcomes, thus contributing to SDG 4 (Quality Education) by enhancing equitable access to quality learning experiences (Mambu and Saputra 2023).

Given the importance of AI literacy, particularly for prospective teachers tasked with preparing future generations, this study investigates the AI literacy levels of prospective chemistry teachers. It also explores differences based on gender, school of origin, and parental income, filling the gap in research and supporting efforts to integrate AI into education for sustainable development.

2 METHOD

This study employs a mixed-methods research approach. Mixed methods research combines elements of quantitative and qualitative approaches to produce broader and deeper data (Creswell and Clark 2018). In this study, quantitative and qualitative methods are combined sequentially. In the first phase, quantitative methods are used to collect measurable data, which can be descriptive, comparative, or associative. In the second phase, qualitative methods validate, deepen, expand, or refute the quantitative findings. The research design follows the Sequential Explanatory Model, which involves collecting and analysing quantitative data in the first stage, followed by qualitative data collection and analysis in the second stage. Subsequently, all data is analysed comprehensively to conclude. After quantitative methods, qualitative methods aim to better understand and explain the quantitative results obtained earlier.

The population of this study consists of all chemistry education students at a university in the southern part of Jakarta. The sample comprises 310 students selected randomly. For qualitative data, interviews were conducted with 9 students, with 3 selected from each high, medium, and low AI literacy category. Quantitative data was collected using a questionnaire adapted from the instrument developed by Ng *et al.* (2024) in their AI Literacy Questionnaire: The Affective, Behavioural, Cognitive, and Ethical Approach. The results were then analysed using descriptive statistics. Meanwhile, qualitative data was obtained through semi-structured interviews and analysed using a thematic analysis approach.

3 RESULTS AND DISCUSSION

The quantitative analysis results were obtained from a 51-item questionnaire completed by 310 chemistry education students from cohorts 2020, 2021, 2022, and 2023. The questionnaire used was adapted from the instrument developed by Ng *et al.* (2024), the AILQ (AI Literacy Questionnaire). This questionnaire covered 9 indicators: intrinsic motivation, self-efficacy, belief, career interest, commitment behaviour, collaboration, knowing and understanding, applying, evaluating, and creating, and AI ethics. The qualitative analysis was obtained through semi-structured interviews. Nine students were interviewed and selected based on the results of the quantitative analysis to represent each category: high, medium, and low.

Overall, AI literacy indicators showed that 208 chemistry education students had high AI literacy, representing 67.10%, 101 students had medium AI literacy, representing 32.58%, and 1 student had low AI literacy, representing 0.32%. Similar results were found in a previous study by Wang *et al.* (2023), where the average overall AI literacy for middle school students was 77.78 (out of a total of 100 points). Of the total, 43.9% and 22.0% scored "very good" and "good," respectively, while 20.0% and 14.0% scored "medium" and "poor." Based on the research findings from chemistry education students, they provided varied answers across the indicators of AI literacy. This article discusses 4 indicators: belief, knowing and understanding, applying, evaluating, and creating, and AI ethics.

3.1 *Belief in AI*

The study reveals that 69.68% of students believe highly in Artificial Intelligence (AI) literacy. This result is consistent with previous research, such as the study by Zawieska and Duffy (2015), which showed that public understanding and self-confidence regarding AI ethics are generally higher than AI development. This suggests that students, as part of society, tend to have strong confidence in AI literacy. Students‘ confidence is important, as it demonstrates their readiness to engage with AI technology and its potential applications.

Student statements further support this belief in AI literacy during interviews. One student remarked, "What makes me confident in AI is that it is a very important technology in our time. AI uses computer technology with specialised programs to think, process information, and solve human problems efficiently. (student A interview) " Another student added, "I am confident that I can handle tasks related to artificial intelligence well because I have been trained with various AI concepts and applications throughout my learning process. Moreover, I keep up with the latest developments in AI and am always ready to learn and adapt to evolving technology. (student C interview) " These statements reflect a strong understanding of AI's significance and capabilities.

The students' high confidence in AI literacy aligns with the findings of Janssen *et al.* (2022), which indicate that high AI literacy helps individuals make informed decisions, solve problems, and collaborate effectively with AI technology. This is crucial for SDG 4 (Quality Education), as promoting AI literacy among students contributes to building an inclusive and sustainable education system. Empowering students with AI literacy fosters their ability to engage with the rapidly evolving technological landscape, enhancing both their personal development and their potential to contribute meaningfully to society.

3.2 *Knowing and understanding*

The research data shows that the largest percentage for the "knowing and understanding" indicator falls under the medium category, with 51.29% of students (159 students) displaying this level of AI literacy. According to Burgsteiner and Kandlhofer (2016), AI literacy is defined as the ability to understand the techniques and fundamental concepts behind AI in various products and services. Many students are familiar with AI-powered services but are not well-versed in foundational concepts such as computational thinking and data structures. A solid understanding of AI and computer science principles is crucial for their future careers (Burgsteiner and Kandlhofer 2016). This highlights the importance of building a strong AI foundation in education to prepare students for the increasingly technology-driven job market.

Previous research has defined AI literacy as understanding basic AI techniques and concepts embedded in various products and services (Estevez *et al.* 2019; Williams *et al.* 2023). A solid understanding of AI and machine learning principles is essential for students' future careers. Furthermore, several studies have linked AI literacy to self-perception, confidence, and students‘ readiness to learn about AI. However, a lack of confidence in AI's potential, especially among teachers, has also been documented (Vazhayil *et al.* 2019). This gap in confidence and understanding reflects the need for improved AI literacy to ensure that both students and educators are prepared for the evolving technological landscape.

During interviews, students GA and IY expressed the growing importance of AI, noting that understanding AI basics is essential to using it wisely: "*AI is rapidly developing and touching almost every aspect of our lives. We need to understand the basics of AI to use it responsibly*" (interview with student G). Another student stated, "*By understanding AI, we can use it optimally and minimise the risks that may arise from this technological development*" (interview with student I. To foster AI literacy in students, researchers have designed curricula and activities that promote AI literacy, focusing on how teachers learn and understand AI concepts (Cope *et al.* 2021). Thus, understanding and knowing AI forms the foundational AI literacy, where educators must grasp the basic concepts, knowledge, and guidelines for using AI in teaching. This aligns with SDG 4 (Quality Education), emphasising the need for inclusive and equitable education that prepares students for future challenges through AI literacy.

3.3 *Applying, evaluating, and creating AI*

The largest percentage for the "applying, evaluating, and creating" indicator falls under the medium category, with 47.10% of students (146 students) displaying this level of AI literacy. It is clear that educating teachers on how to apply AI concepts in various contexts is crucial (Cope *et al.* 2021). Rodrigues and Araújo (2012) evaluated machine learning (ML) models to educate the public on understanding AI applications and their impact on our lives, as well as to raise awareness of ethical issues related to AI technology. How *et al.* (2020) utilised AI thinking by performing data analysis through computation and interpreting new findings from hidden patterns in machine-learned data.

A student shared in an interview: "I see examples from other teachers who might now shift to technology and use AI, which could be displayed on a projector screen, making learning more engaging and interactive. *This is a small example, but it represents a breakthrough in education, especially as we can use technology, not just books and handwriting.*" (Interview with student E).

Han *et al.* (2019) enhanced students' scientific and technological knowledge, which was then applied in inquiry-based learning to solve practical problems. This Study involved the public in creating AI facilities in public spaces to expand public AI literacy and experiences. Considering the research by How et al, data analysis through computation allows teachers to discover new patterns hidden in data through machine learning, which is a practical application of AI-based thinking (Sarker 2022). For teachers, AI application literacy is about knowing when and how to use AI correctly and appropriately in the classroom. This directly ties into SDG 4 (Quality Education), emphasising the importance of providing inclusive, equitable, and quality education and promoting lifelong learning opportunities. AI literacy empowers educators to adopt innovative teaching methods, ensuring that all students have access to modern, technology-driven education.

3.4 *AI ethics*

Based on the research data, it is evident that the largest percentage, 84.84%, for the AI ethics indicator falls under the high category, with 263 students demonstrating strong awareness of AI ethics., with 263 students demonstrating high AI literacy in this area. AI plays a critical role in everyday decision-making, and if misused or poorly designed, it can cause irreversible harm to individuals and society. Gong *et al.* (2020) found that students often pay insufficient attention to ethical issues such as AI bias, legal responsibilities (8%), and intellectual property (9%). These findings emphasise the importance of fostering a deeper understanding of AI ethics, especially as AI becomes increasingly integrated into daily life.

The significance of AI ethics is underscored by student perspectives shared during interviews. One student mentioned, "*In my opinion, AI ethics is very important because AI has a huge impact on human life, so we must use it wisely to avoid negative consequences. The moral principles related to AI use are privacy, because AI often collects personal data, so privacy must be protected to ensure that someone's personal data is not misused by others.*" (Interview with student F). This highlights the students' awareness of the ethical challenges associated with AI, particularly the protection of personal data and privacy, which are key concerns in today's digital landscape.

Another student further emphasised the importance of ethical principles in AI, stating, "*All knowledge, whatever it is, should be accompanied by ethics so that it does not deviate from its original purpose. The moral principle is that AI should be accessible to everyone, anyone can use or learn AI technology without discrimination, ensuring user privacy, and, most importantly, benefiting humanity.*" (Interview with student G). This statement reflects a growing awareness among students of the ethical implications of AI and the need for inclusive, non-discriminatory access to technology that prioritises privacy and human well-being.

These student insights align with the research by Boden *et al.* (2017), which highlight the high ethical awareness among students and the importance of preparing the younger generation to have ethical competence in AI development. As AI continues to advance, ensuring that students are equipped with the knowledge to navigate its ethical challenges is essential. Ethical literacy in AI is necessary to avoid harmful consequences and promote a fair and just society where technology serves the collective good.

This focus on AI ethics is directly linked to SDG 4 (Quality Education), which emphasises providing inclusive and equitable quality education and promoting lifelong learning opportunities for all. Therefore, the ethical aspect of AI must be given greater attention and integrated into AI education and curriculum development. A curriculum designed with principles of active learning, integrated ethics, and easy access can help students become responsible and informed users and creators of AI (Akgun and Greenhow 2022; Ramli *et al.* 2024; Williams *et al.* 2023). By fostering ethical AI literacy, we empower students to contribute to a sustainable and just digital future, ensuring that technological advancements are used to benefit society while safeguarding human rights and privacy.

4 CONCLUSION

This study highlights the growing importance of AI literacy among prospective chemistry teachers, particularly in AI ethics, where most students showed high awareness. With AI playing an increasingly important role in everyday decision-making, students must understand its ethical implications, particularly concerning issues such as privacy and fairness. The findings suggest that fostering AI ethics education is essential for ensuring responsible and ethical use of technology.

Furthermore, the study revealed that while many students demonstrated moderate understanding and application of AI concepts, a solid foundation in AI literacy is necessary for preparing students for future careers. Educating students in the fundamentals of AI and its ethical considerations aligns with SDG 4 (Quality Education), which calls for inclusive, equitable, and quality education that equips students with the skills to succeed in the digital age. Integrating AI literacy and ethics into curricula will help empower students to navigate technological advancements responsibly and contribute to a more sustainable, just society.

Given that this study was limited to a single university, it is recommended that future research expands to include multiple institutions across different regions to gain a more diverse and comprehensive understanding of AI literacy levels among students. By including a broader sample, researchers can examine variations in AI literacy based on different cultural, socioeconomic, and educational contexts, providing a clearer picture of how AI literacy is developing globally. Additionally, research could explore how AI literacy influences students' ability to integrate ethical considerations into their technological practices. This will align with SDG 4 (Quality Education), which emphasises the need for inclusive, equitable, and quality education that prepares students to be responsible and ethical users of emerging technologies. Further studies could also examine the effectiveness of specific curricula or teaching methods in enhancing AI literacy and ethics in diverse educational settings.

REFERENCES

Akgun, S., and Greenhow, C. (2022). Artificial intelligence in education: Addressing ethical challenges in K-12 settings. *AI and Ethics*, 2(3), 431–440. https://doi.org/10.1007/s43681-021-00096-7

Boden, M., Bryson, J., Caldwell, D., Dautenhahn, K., Edwards, L., Kember, S., Newman, P., Parry, V., Pegman, G., Rodden, T., Sorrell, T., Wallis, M., Whitby, B., and Winfield, A. (2017). Principles of robotics: regulating robots in the real world. *Connection Science*, 29(2), 124–129. https://doi.org/10.1080/09540091.2016.1271400

Cope, B., Kalantzis, M., and Searsmith, D. (2021). Artificial intelligence for education: Knowledge and its assessment in AI-enabled learning ecologies. *Educational Philosophy and Theory*, 53(12), 1229–1245. https://doi.org/10.1080/00131857.2020.1728732

Creswell, J. W., and Clark, J. D. (2018). Research Design Qualitative, quantitative and mixed methods approach. *In Introducing English Language*. https://doi.org/10.4324/9781315707181-60

Estevez, J., Garate, G., and Grana, M. (2019). Gentle Introduction to Artificial Intelligence for High-School Students Using Scratch. *IEEE Access*, 7, 179027–179036. https://doi.org/10.1109/access.2019.2956136

Fleaca, E., and Stanciu, R. D. (2019). Digital-age Learning and Business Engineering Education- a Pilot Study on Students' E-skills. *Procedia Manufacturing*, 32, 1051–1057. https://doi.org/10.1016/j.promfg.2019.02.320

Ghallab, M. (2019). Responsible AI: requirements and challenges. *AI Perspectives*, 1(1), 1–7. https://doi.org/10.1186/s42467-019-0003-z

Gong, X., Tang, Y., Liu, X., Jing, S., Cui, W., Liang, J., and Wang, F. Y. (2020). K-9 artificial intelligence education in Qingdao: Issues, challenges and suggestions. *In 2020 IEEE International Conference on Networking, Sensing and Control (ICNSC)* (pp. 1–6). IEEE.

Han, E.-R., Yeo, S., Kim, M.-J., Lee, Y.-H., Park, K.-H., and Roh, H. (2019). Medical education trends for future physicians in the era of advanced technology and artificial intelligence: an integrative review. *BMC Medical Education*, 19(1). https://doi.org/10.1186/s12909-019-1891-5

How, M.-L., Cheah, S.-M., Khor, A. C., and Chan, Y. J. (2020). Artificial Intelligence-Enhanced Predictive Insights for Advancing Financial Inclusion: A Human-Centric AI-Thinking Approach. *Big Data and Cognitive Computing*, 4(2), 8. https://doi.org/10.3390/bdcc4020008

Janssen, M., Hartog, M., Matheus, R., Yi Ding, A., and Kuk, G. (2022). Will Algorithms Blind People? The Effect of Explainable AI and Decision-Makers' Experience on AI-supported Decision-Making in Government. *Social Science Computer Review*, 40(2), 478–493. https://doi.org/10.1177/0894439320980118

Long, D., and Magerko, B. (2020). What is AI Literacy? Competencies and Design Considerations | *Proceedings of the 2020 CHI Conference on Human Factors in Computing Systems*. 1–16.

Mambu, J. G. Z., and Saputra, A. (2023). Pemanfaatan Teknologi Artificial Intelligence (AI) Dalam Menghadapi Tantangan Mengajar Guru di Era Digital. *Journal on Education*, 06(01), 2689–2698.

Ng, D. T. K., Wu, W., Leung, J. K. L., Chiu, T. K. F., and Chu, S. K. W. (2024). Design and validation of the AI literacy questionnaire: The affective, behavioural, cognitive and ethical approach. *British Journal of Educational Technology*, 55(3), 1082–1104. https://doi.org/10.1111/bjet.13411

Ramli, M., Fatra, M., Murtadlo, M., Albana, H., Hana Susanti, B., and Aldeia, S. (2024). Navigating the tech-savvy generation; key considerations in developing of an artificial intelligence curriculum. *IAES International Journal of Artificial Intelligence (IJ-AI)*, 13(4), 3942. https://doi.org/10.11591/ijai.v13.i4.pp3942-3950

Rodrigues, F., and Araújo, L. (2012). Automatic assessment of short free text answers. *CSEDU 2012 - Proceedings of the 4th International Conference on Computer Supported Education*, 2, 50–57. https://doi.org/10.5220/0003920800500057

Sarker, I. H. (2022). AI-Based Modeling: Techniques, Applications and Research Issues Towards Automation, Intelligent and Smart Systems. *SN Computer Science*, 3(2). https://doi.org/10.1007/s42979-022-01043-x

Supriyadi, E. I., and Asih, D. B. (2021). Implementasi Artificial Intelligence (Ai) Di Bidang Administrasi Publik Pada Era Revolusi Industri 4.0. *Jurnal RASI*, 2(2), 12–22. https://doi.org/10.52496/rasi.v2i2.62

Vazhayil, A., Shetty, R., Bhavani, R. R., and Akshay, N. (2019). Focusing on Teacher Education to Introduce AI in Schools: Perspectives and Illustrative Findings. *2019 IEEE Tenth International Conference on Technology for Education (T4E)*, 71–77. https://doi.org/10.1109/t4e.2019.00021

Wang, X., Li, X., and Huang, J. (2023). Junior High School Artificial Intelligence Literacy: Connotation, Evaluation and Promotion Strategy. *Open Journal of Social Sciences*, 11(05), 33–49. https://doi.org/10.4236/jss.2023.115004

Williams, R., Ali, S., Devasia, N., DiPaola, D., Hong, J., Kaputsos, S. P., Jordan, B., and Breazeal, C. (2023). AI + Ethics Curricula for Middle School Youth: Lessons Learned from Three Project-Based Curricula. *International Journal of Artificial Intelligence in Education*, 33(2), 325–383. https://doi.org/10.1007/s40593-022-00298-y

Wong, G. K. W., Ma, X., Dillenbourg, P., and Huan, J. (2020). Broadening artificial intelligence education in K-12: where to start? *ACM Inroads*, 11(1), 20–29. https://doi.org/10.1145/3381884

Zawieska, K., and Duffy, B. R. (2015). The Social Construction of Creativity in Educational Robotics. *Advances in Intelligent Systems and Computing*, 329–338. https://doi.org/10.1007/978-3-319-15847-1_32.

Towards Resilient Societies: The Synergy of Religion, Education, Health, Science, and Technology – Rahiem (Ed.)

The role of vocational training in addressing income disparities in Indonesia

Dharendra Wardhana
Ministry of National Development Planning (Bappenas)

ABSTRACT: This study aims to examine the role of vocational training (VT) in shaping the socio-economic disparities among workers in 2021. Utilising data from the National Labor Force Survey (Sakernas), the research identifies significant differences in average earnings across various worker quintiles. The analysis employs Equiplots—a dot plot used to visualise and compare inequalities within and between subgroups—offering a comprehensive overview of income disparities and accessibility to VT across regions. This study is among the first to use Equiplots to analyse labour income inequality. Disaggregated data on selected socio-economic indicators reveal clear wage advantages for males, urban residents, and individuals on the Java mainland. The analysis of equitable access to vocational training is intended to enrich discussions on achieving employment-related targets outlined in Sustainable Development Goal (SDG) 8 on decent work and in Indonesia's RPJPN 2025–2045 agenda. The findings of this study are expected to inform policymaking aimed at reducing income disparities through targeted skills development initiatives and improved access to vocational training.

Keywords: earnings, equiplot, vocational training

1 INTRODUCTION

In the context of Technical and Vocational Education and Training (TVET), most economic theories propose or imply that vocational training (VT) enhances employment outcomes in a relatively shorter term compared to vocational education (VE). According to Indrawati and Kuncoro (2021), improving TVET is crucial for national development, particularly in accompanying reform efforts.

Indonesia has set ambitious goals as stipulated in their national targets for the Sustainable Development Goals (SDGs) and, more recently, in the long-term national development plan (*Rencana Pembangunan Jangka Panjang Nasional* or RPJPN) 2025-2045. However, at least presently, some problems might hinder achieving the goals. Premature deindustrialisation needs to be addressed through effective policies on supply-side advancement (Araujo *et al.* 2021; Glawe and Wagner 2020; Ikhsan *et al.* 2021). This will involve developing infrastructure and human resources. Therefore, reindustrialisation and its accompanying policies need to serve as the backdrop.

The economic crises following the COVID-19 pandemic also cause "scarring effects" (see Pritadrajati *et al.* 2021; Shasta Pratomo and Manning 2022) that eventually required a major overhauls in labour policies and employment creation. The low capacity of workers raises concerns about the country's ability to develop its human resources to reform its economic structure and escape the middle-income trap. In this study, VT is assumed to play a pivotal role in breaking the middle-income trap by providing the workforce with industry-relevant skills. VT aligns educational outcomes with labour market needs, facilitating a smoother transition from education to employment and supporting economic development.

DOI: 10.1201/9781003645542-68

The connection between VT, income disparities, and SDGs goal number 8, which focuses on promoting sustained, inclusive, and sustainable economic growth, full and productive employment, and decent work for all, is multifaceted. In this vein, TVET is crucial in addressing income disparities through skills that enhance employability and productivity. However, there is a lacuna in studies of VT impact in the context of Indonesia. In addition, the most recent policy like Presidential Regulation No. 68 Year 2022 on TVET Revitalisation, invites further research and feedback.

This study has two main purposes. First, assess VT's impact and identify factors influencing labour premia in 2021. Second, this study's novelty lies in investigating the role of VT in providing the advantage and competitiveness of the Indonesian labour force. To meet both objectives, this study implements equiplot graphs for the analysis.

The paper is structured as follows: the following section provides a brief literature review, followed by the methodology section. The results section outlines the study's findings, the discussion section interprets these findings, and the conclusion provides recommendations and proposes directions for future research.

1.1 *Relevant literature*

A handful of studies have sought to examine the effects of vocational education (VE) in Indonesia. First, Astuti and Setyonaluri (2022) demonstrate that the return on investment in VE in 2020 was 16.82% higher than that of general high school education. Second, utilising Indonesian Family Life Survey (IFLS) data sets, Pritadrajati (2022) examined the impact of enrolling in vocational senior-secondary schools on labour market outcomes. The analysis reveals that public VE yields better female labour outcomes than public general schools. Third, a pair of studies by Purnastuti *et al.* (2015, 2013) unravel important findings on Indonesia's declining returns to education compared to other Asian peer countries.

Recent research on labour income disparity in Indonesia rarely touches upon the nexus of human capital and skills development. For instance, a study by Husain *et al.* (2020) scrutinises the dynamic impact of financial development on income inequality. Another study by Aginta *et al.* (2021) aims to re-examine the regional convergence hypothesis on income in Indonesia over the 2000–2017 period. The result shows the significance of five convergence clubs in Indonesian districts' income dynamics, implying the persistence of income disparity problems across districts after decentralisation policy implementation. Handayani *et al.* (2022) investigate Indonesia's potential macroeconomic determinants of income inequality, particularly energy consumption, corruption, foreign direct investment, and other supporting determinants. Though it seemingly encompasses wider dimensions of inequality, the paper does not explicitly mention labour competitiveness.

2 METHOD

This study mainly uses comparative analysis aided by equiplots i.e. a type of dot plot used to visualise and compare inequalities within and between subgroups. This graph is particularly useful for depicting and highlighting gaps in estimates among different groups commonly used in epidemiology, public health, and nutrition studies.

There are at least three key features of the equiplot: visualisation of gaps, data disaggregation, and common axis, which help interpret gaps. Arguably, equiplot is the developed version of the previous technique, like biplot graphic display and statistical properties such as component analysis, as mentioned in Gabriel (1971). Furthermore, equiplots emphasise effect sizes, providing a more comprehensive view of data analysis. They facilitate a shift from dichotomous thinking to quantitative reasoning, crucial for more nuanced data interpretation (Ho *et al.* 2019).

There is an increasing number of widely-cited articles displaying equiplots mostly in the subject of public health like Hounton *et al.* (2015); Jiwani and Antiporta (2020); de Castro *et al.*, (2018); McClarty *et al.* (2021); Kang *et al.* (2022). Meanwhile, the use of equiplot in the Indonesian context can be found in a pivotal article published in the Lancet journal titled "Universal health coverage in Indonesia: concept, progress, and challenges" (Agustina *et al.* 2019). An equiplot in this article is mentioned as "Dartanto plot", which tells about the equity gap in insurance coverage.

Equiplot allows one to elucidate the relations between VT and other variables connected with income levels. It provides a more nuanced understanding of socioeconomic data by integrating various dimensions of endowments and income inequality. In addition, by incorporating geographic variability, the equiplot can highlight spatial patterns of income disparities, offering insights into regional differences and helping to identify areas that require targeted interventions (Gluhih and Portnov 2004).

Having said that, interpreting equiplots can be fraught with biases and challenges that affect the accuracy of their insights. These biases stem from both the plots' visual elements and the viewers' cognitive biases (Filipowicz *et al.* 2023). Interpreting visualised data involves complex perceptual and cognitive processes, which can be influenced by both the visualisation design and the viewer's cognitive biases (Xiong *et al.* 2023).

3 METHODS

The analysis relies on Indonesian Labor Force Survey (*Survei Angkatan Kerja Nasional* or Sakernas) data sets, regularly published by Statistics Indonesia (*Badan Pusat Statistik* or BPS) twice a year. For this study, and taking advantage of a larger sample size (50,000 household samples in February and 200,000 in August), the data set for August 2021 is selected to build the equiplots. However, capturing only a period of data set will provide a single snapshot dismissing the dynamics or trends. Caveats need to be considered for the usage of data sets in periods where anomalous events occurred; in this case, the COVID-19 pandemic might affect the quality and responses of Sakernas respondents.

4 FINDINGS AND DISCUSSION

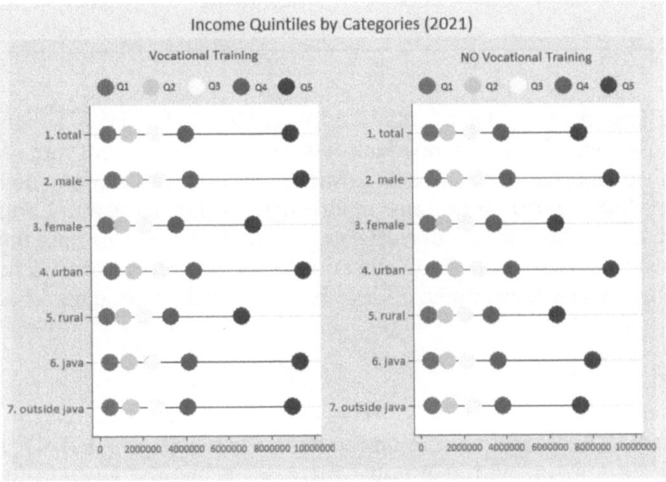

Figure 1. Income quintiles by categories for workers with and without VT, year 2021.

Analysis based on equiplot observations shows that categorical variations in the role of VT present a rich narrative. Figure 1 presents a comparative analysis of income distribution across various demographic categories for individuals with and without VT in 2021. The left panel in each year depicts the income distribution (in quintile groups) of workers who participated in VT while the right panel describes the quintiles for non-trained workers. The relationship between SDGs Goal Number 8 and labour income disparity is multifaceted, involving productivity, labour rights, and economic policies. Yet, these elements do not guarantee equitable income distribution. The disparity in labor income is influenced by various factors, including the socio-economic context of different countries in which the equiplot findings below attempt to explain.

4.1 *Income analysis*

In 2021, workers who undertook VT have higher incomes. The left panel (workers attending VT) shows a significant spread across quintiles, with Q5 (top quintile) displaying the highest income. While the right panel (workers without VT), the income distribution is more compressed, especially in Q1 to Q4. In summary, VT does provide a wage premium, but it is more conspicuous, particularly to the wealthiest groups (those in the top quintile or Q5). This fact provides a hint of inequality that might occur or correlate with the access on VT facilities.

Table 1. Income quintiles by categories for workers with and without VT, year 2021.

Categories	Income-q1	Income-q2	Income-q3	Income-q4	Income-q5
Vocational training					
Male	708,443	1,735,765	2,781,657	4,287,615	9,459,099
Female	398,757	1,121,846	2,167,114	3,602,091	7,204,318
Urban	638,294	1,649,149	2,889,159	4,428,796	9,530,531
Rural	399,585	1,139,239	2,095,188	3,313,908	6,646,443
Java	505,439	1,385,578	2,434,776	4,173,798	9,348,282
Outside Java	498,354	1,487,504	2,637,181	4,113,633	8,982,102
No vocational training					
Male	639,955	1,649,284	2,668,165	4,129,601	8,933,321
Female	411,975	1,140,815	2,163,497	3,464,207	6,349,995
Urban	632,845	1,635,510	2,747,752	4,257,042	8,863,436
Rural	407,719	1,143,281	2,072,942	3,323,242	6,384,192
Java	459,002	1,253,099	2,170,291	3,621,832	7,994,776
Outside Java	502,966	1,328,663	2,471,044	3,816,002	7,445,486

For reference, a study in India by Bhatt *et al.* (2024) finds that VT significantly boosts employability and income. Trained individuals have a 70% employment rate compared to 50% for untrained individuals, and they earn a 29% wage premium, with median monthly incomes of INR 18,000 versus INR 14,000 for untrained individuals. Vandenberg and Laranjo (2021) show that TVET graduates in the Philippines earn significantly higher wages than those who entered the job market with a secondary school education or below. However, individuals trained in TVET and pursued tertiary education earn less than those with secondary school or below.

4.2 *Gender analysis*

Regarding the gender differences, male workers have higher income levels than females, both with and without VT. Analysis focusing on the female group shows that VT appears to have a more pronounced impact, with a notable increase in income levels compared to those female workers without VT. It is important to know why VT impacts female workers differently, which entails further investigations. At this moment, cursory analysis indicates that gender gaps in employment still persist and affect labour policies' implementations (let alone the impact). Perhaps unsurprisingly, barriers in labour markets and TVET facilities explain lower income distribution levels for females attending and not attending VT.

4.3 *Regional analysis*

For the workers in urban areas, VT significantly boosts incomes, with a sharp rise in Q5. Meanwhile, in rural areas, the role of VT is still positive but less pronounced than in urban areas. Workers in rural areas without VT show a more compressed income distribution. For Java workers, VT's impact is substantial, especially in the higher quintiles. Meanwhile, for those living outside Java, the income distribution is more balanced across quintiles, with VT still providing a notable increase, particularly in Q5. This fact aligns with the context of the regional development gap and its correlation with either access to quality VT or labour market that seemingly is favourable towards certain groups with privileged access to local infrastructure and resources. Although there are seemingly no discernible differences in the impact of VT on labour market premia, the significant gap between Java and outside Java cannot be denied. The fact that Q5 incomes are higher in Java and urban areas suggests that regional inequality persists. A higher level of income implies higher access and quality of VT.

5 CONCLUSION

The depictions of the above equiplots reveal privilege for certain groups, with the highest income groups, benefiting the most. Moreover, the second group that benefited the most from VT was workers in the fourth quintile group. Again, this indicates that the bottom group and some middle-level quintile groups (Q2 to Q3) are the ones that are 'left behind' from policy outcomes and implementation of VT.

The analysis in this study underscores the positive impact of VT on income levels across various demographic categories. However, it also highlights existing disparities based on gender, urban-rural divide, and regional differences. Undeniably, VT significantly boosts incomes, especially in higher quintiles, and increases income inequality within the trained workforce groups. These insights can guide policymakers in targeting VT programs to maximise benefits and address income disparities effectively.

For future research, the exploration of other relevant variables is important. In response to the problems posed in this study, it is also expected that policies will be directed to achieve strategies as stipulated in SDGs and RPJPN 2025-2045. With the context set as a background of this study, policy orientation needs to address the sources and causes of inequality that range from gender, geography, and other socio-economic characteristics. This ultimately leads to the policy designs of VT to be more inclusive, particularly towards vulnerable groups such as persons with disability, unemployed youth, refugees, families with young children, ethnic minorities, elderly widows, etc. On a related note, the discourse on whether policy approaches should be needs-based or rights-based will be worth exploring.

Increasing the scope and coverage of VT (and other policies in general) will require massive additional resources, which obviously will burden the limited fiscal capacity. Therefore, relevant to this context is the private sector's involvement and role, especially in industries with an interest in cultivating human capital. Collaboration in many forms can support the expansion of VT at many levels.

REFERENCES

Aginta, H., Gunawan, A. B., and Mendez, C. (2021). Regional income disparities and convergence clubs in Indonesia: new district-level evidence. *Journal of the Asia Pacific Economy*, *28*(1), 101–132. https://doi.org/10.1080/13547860.2020.1868107

Agustina, R., Dartanto, T., Sitompul, R., Susiloretni, K. A., Suparmi, S., Achadi, E. L., Taher, A., ... Khusun, H. (2019). Universal health coverage in Indonesia: concept, progress, and challenges. *The Lancet*, *393*(10166), 75–102. https://doi.org/10.1016/s0140-6736(18)31647-7

Araujo, E., Araújo, E., Peres, S. C., and Punzo, L. F. (2021). An investigation into shapes and determinants of deindustrialization processes: Theory and evidence for developed and developing countries (1970–2017). *EconomiA*, *22*(2), 129–143. https://doi.org/10.1016/j.econ.2021.03.001

Astuti, N. F. V., and Setyonaluri, D. (2022). Labor Market Outcomes of Vocational High Schools (SMK) and General High Schools (SMA) during the COVID-19 Pandemic. *The Journal of Indonesia Sustainable Development Planning*, *3*(3), 278–293. https://doi.org/10.46456/jisdep.v3i3.328

Bhatt, V., Yadav, R., and Khan, M. (2024). An Empirical Study on the Impact of Integrating Vocational Education and Skill Development Programs on the Labor Market. *Research and Advances in Education*, *3*(6), 15–25. https://doi.org/10.56397/rae.2024.06.02

de Castro, D. B., de Seixas Maciel, E. M. G., Sadahiro, M., Pinto, R. C., de Albuquerque, B. C., and Braga, J. U. (2018). Tuberculosis incidence inequalities and its social determinants in Manaus from 2007 to 2016. *International Journal for Equity in Health*, *17*(1), 187. https://doi.org/10.1186/s12939-018-0900-3

Filipowicz, A., Carter, S., Bravo, N., Iliev, R., Hakimi, S., Shamma, D. A., Lyons, K., Hogan, C., and Wu, C. (2023). Visual Elements and Cognitive Biases Influence Interpretations of Trends in Scatter Plots. ArXiv. https://arxiv.org/abs/2310.15406

Gabriel, K. R. (1971). The Biplot Graphic Display of Matrices with Application to Principal Component Analysis. *Biometrika*, *58*(3), 453–467. https://doi.org/10.2307/2334381

Glawe, L., and Wagner, H. (2020). The Middle-Income Trap 2.0: The Increasing Role of Human Capital in the Age of Automation and Implications for Developing Asia*. *Asian Economic Papers*, *19*(3), 40–58. https://doi.org/10.1162/asep_a_00783

Gluhih, R., and Portnov, B. A. (2004). Visualization of the spatial patterns of inter-urban income disparities using coordinate transformations. *International Journal of Geographical Information Science*, 18(3), 281–297. https://doi.org/10.1080/13658810310001620951

Handayani, B. D., Yanto, H., Pujiati, A., Ridzuan, A. R., Keshminder, J. S., and Shaari, M. S. (2022). The Implication of Energy Consumption, Corruption, and Foreign Investment for Sustainability of Income Distribution in Indonesia. *Sustainability*, *14*(23), 15915. https://doi.org/10.3390/su142315915

Ho, J., Tumkaya, T., Aryal, S., Choi, H., and Claridge-Chang, A. (2019). Moving beyond P values: data analysis with estimation graphics. *Nature Methods*, *16*(7), 565–566. https://doi.org/10.1038/s41592-019-0470-3

Hounton, S., Winfrey, W., Barros, A. J. D., and Askew, I. (2015). Patterns and trends of postpartum family planning in Ethiopia, Malawi, and Nigeria: evidence of missed opportunities for integration. *Global Health Action*, *8*(1), 29738. https://doi.org/10.3402/gha.v8.29738

Husain, S., Sohag, K., Hasan, R., and Shams, S. M. R. (2020). Interest rate and income disparity: Evidence from Indonesia. *Strategic Change*, *29*(6), 665–672. https://doi.org/10.1002/jsc.2381

Ikhsan, M., Indrawati, S. M., Virananda, I. G. S., Abdi, Z., and Can, C. K. (2021). The Productivity and Future Growth Potential of Indonesia. *Economics and Finance in Indonesia*, *67*(2), 235–256. https://doi.org/10.47291/efi.v67i2.996

Indrawati, S. M., and Kuncoro, A. (2021). Improving Competitiveness Through Vocational and Higher Education: Indonesia's Vision For Human Capital Development In 2019–2024. *Bulletin of Indonesian Economic Studies*, *57* (1), 29–59. https://doi.org/10.1080/00074918.2021.1909692

Jiwani, S. S., and Antiporta, D. A. (2020). Inequalities in access to water and soap matter for the COVID-19 response in sub-Saharan Africa. *International Journal for Equity in Health*, *19*(1), 82. https://doi.org/10.1186/s12939-020-01199-z

Kang, Y., Park, C., Young, A. M. P., and Kim, J. (2022). Socio-economic disparity in food consumption among young children in eight South Asian and Southeast Asian countries. *Nutrition Research and Practice*, *16*(4), 489–504. https://doi.org/10.4162/nrp.2022.16.4.489

McClarty, L. M., Blanchard, J. F., and Becker, M. L. (2021). Leaving no one behind? An equity analysis of the HIV care cascade among a cohort of people living with HIV in Manitoba, Canada. *BMC Public Health*, *21*(1), 281. https://doi.org/10.1186/s12889-021-10225-w

Pritadrajati, D. S. (2022). From School to Work: Does Vocational Education Improve Labour Market Outcomes? An Empirical Analysis Of Indonesia. *Buletin Ekonomi Moneter Dan Perbankan*, *25*(3), 471–492. https://doi.org/10.21098/bemp.v25i3.1315

Pritadrajati, D. S., Kusuma, A. C. M., and Saxena, S. C. (2021). Scarred for life: Lasting consequences of unemployment and informal self-employment: An empirical evidence from Indonesia. *Economic Analysis and Policy*, *70*(2), 206–219. https://doi.org/10.1016/j.eap.2021.02.009

Purnastuti, L., Miller, P. W., and Salim, R. (2013). Declining rates of return to education: Evidence for Indonesia. *Bulletin of Indonesian Economic Studies*, 49(2), 213–236. https://doi.org/10.1080/00074918.2013.809842

Purnastuti, L., Salim, R., and Joarder, M. A. M. (2015). The returns to education in Indonesia: Post reform estimates. *The Journal of Developing Areas*, 49(3), 183–204. https://doi.org/10.1353/jda.2015.0174

Shasta Pratomo, D., and Manning, C. (2022). Structural Change and Formal Sector Employment Growth in Indonesia. *Journal of Southeast Asian Economies*, *39*(1), 1–20. https://doi.org/10.1355/ae39-1a

Vandenberg, P., and Laranjo, J. (2021). Vocational training and labor market outcomes in the Philippines. *International Journal of Educational Development*, 87, 102501. https://doi.org/10.1016/j.ijedudev.2021.102501

Xiong, C., Stokes, C., Kim, Y. S., and Franconeri, S. (2023). Seeing What You Believe or Believing What You See? Belief Biases Correlation Estimation. *IEEE transactions on visualization and computer graphics*, *29*(1), 493–503. https://doi.org/10.1109/TVCG.2022.3209405.

Towards Resilient Societies: The Synergy of Religion, Education, Health, Science, and Technology – Rahiem (Ed.)
© 2026 The Author(s), ISBN: 978-1-041-08471-6

Well-being based on *Tazkiyatun Nafs* At The *Darut Tasbih Ar-Rafi*, Tangerang Regency

Abqorina Abqorina, Sururin Sururin & Erba Rozalina Yulianti
Universitas Islam Negeri Syarif Hidayatullah Jakarta

Karoma Karoma
Universitas Islam Negeri Raden Fatah Palembang

Yuslaini Yuslaini
Universitas Muhammadiyah Palembang

ABSTRACT: This study examines adolescent well-being based on *tazkiyatun nafs* at Pondok Jiwa Pesantren Darut Tasbih, Tangerang Regency. Initial interviews revealed that more than 35 adolescents at the Islamic boarding school experienced mental disorders from various factors, including anxieties that led to the teachings of drugs and alcohol, trauma from family, school, and community problems, and the practice of black magic. The Darut Tasbih Islamic Boarding School implements the *tazkiyatun nafs* method as a 24-hour healing approach. This study aims to understand better how *tazkiyatun nafs* contributes to students' mental well-being at Darut Tasbih.

Keywords: mental health, well-being, tazkiyatun nafs, religious, and boarding school

1 INTRODUCTION

Humans are creatures that are given the potential to actualise everything they have. He has various facilities supporting his survival (Taylor 2009). Humans in psychological development at the age of 18-29 years are called early adulthood (Negovan 2010; Smet 2004). According to Santrock, early adult development is one of the transitions from adolescence to adulthood characterised by changes such as cognitive, physical, psycho-social and emotional changes (Diana Mutiah 2015; Dierendonck 2008; Keyes and Ryff 2002; Ryff and Singer 2008). This stage can be called the stage of self-exploration with all the greater challenges, such as facing a quarter-life crisis, which is a factor of change in adolescents characterised by frustration, confusion, anxiety, panic, career, relationships, and also education (Diener 2003; Krekel *et al.* 2019). In line with the above opinion, according to Pollard and Lee's research, mental health in early adulthood is both negative and positive, stimulated by self-esteem measures, depressive symptoms, and fatigue. Exhaustion is caused by environmental changes so that transformations occur in their bodies, emotions, thoughts and relationships with others (Buchanan and Bowen 2008; Ryan and Deci 2001).

According to WHO (Global Health Estimates), death cases in Indonesia in 2020 reached 2.4 per 100,000 people, with an estimated 1,800 cases per year. The death rate is one of which is caused by suicidal behaviour. The largest contributor to suicide deaths is mostly young people of productive age, 46% aged 25-49 years and 75% aged 15-64 years with various suicide motives. The worst impact of impaired mental health on individuals is suicide. The act is one of the impacts of mental health problems that is important to pay attention to today. The number of deaths due to suicide is increasing every year, with a total of 800,000 people, or the same thing as 1 death every 40 seconds.

DOI: 10.1201/9781003645542-69

In religious life, mental well-being is characterised by one's adherence to religious teachings. Religion is very important in its application in life (Ariati 2010). Therefore, religion, especially Islam, is a belief that has a close relationship with mental well-being which begins with a process of purifying the soul (*Tazkiyatun Nafs*) (Anas Ahmad Karzon 2016; Ibnu Qayyim Al-Jauziyah 2001; Said Hawwa 1999). *Tazkiyatun nafs* has two meanings, *tazkiyatun* is purification and cleansing, while nafs means soul or lust. Purification of the soul to avoid the soul tainted by shirk, revenge, envy, jealousy and other diseases (Noordin and Rahman 2017; Moh. Toriquddin 2008; Muhammad Zuhdi 2022). Human mentality is formed based on fitrah through tawhid, faith, Islam and *Ihsan* so that the human soul is pure so that the formation of self-potential has a spiritual quality, and the practice of Islamic teachings (Masyhuri 2012; Mustafa Zahri 1995; Nur Heni 2017).

The concept of *tazkiyatun nafs* (purification of the soul) has a very close relationship with Sustainable Development Goal (SDG) number 3, namely "Good Health and Well-being". SDG 3 focuses on ensuring healthy lives and promoting well-being for all ages. *Tazkiyatun nafs*, emphasising cleansing and developing the soul, make an important contribution to achieving this goal, especially regarding mental health and preventing risky behaviour. *Tazkiyatun nafs* has a significant role in supporting the achievement of SDG 3, especially in the aspects of mental health and prevention of substance abuse. By cleansing the soul from negative and developing positive traits, individuals can achieve better mental and emotional well-being and avoid risky behaviours that threaten health. Therefore, integrating the values of *tazkiyatun nafs* into health and development programs can strengthen efforts to achieve SDG 3 and create a healthier and more prosperous society.

2 RESEARCH METHOD

This study uses a qualitative approach with a case study type of research (Moh. Nasir 2017). The qualitative approach was chosen because this study aims to deeply understand the phenomenon of well-being associated with *tazkiyatun nafs* in the specific context of the Darut Tasbih Islamic Boarding School. The case study was chosen to explore in depth and detail how *tazkiyatun nafs* is implemented and how it impacts the well-being of students in the Islamic boarding school environment (Sugiyono 2018).

3 RESULTS AND DISCUSSION

3.1 *Stages of Tazkiyatun Nafs at Darut Tasbih Islamic Boarding School in Tangerang Regency*

3.1.1 *Self-control*

3.1.1.1 *Repentance bath* Patients are asked to take a midnight bath early at 01:00 WIB. The midnight bath aims to straighten and soften the nerves in the heads of the patients.

3.1.1.2 *Instilling the values of tauhid* The implementation of tawhid was carried out by KH Rafiuddin by bathing and cleaning patients, then gathering patients in a special hall by providing wejangan, advice and input to restore patient awareness to return to fitrah by remembering and worshipping Allah SWT. Slowly, patients will realise that only Allah is the Almighty and the Most Merciful. Kyai is an intermediary given by Allah; whether they understand or not, it is clear that Allah provides understanding and guidance to understand tawhid to them.

3.1.1.3 *Ruqyah* The application of *ruqyah* is different for each scholar. In Pondok Pesantren Darut Tasbih Tangerang Regency, it is used with shariyyah therapy, namely using the asmas of Allah to obtain mercy from Allah. Ruqyah is done every Friday night and Sunday morning directly by KH. Rafiuddin.

For patients who are sick, ruqyah is done every Monday night, and Tuesday ba'da maghrib is handled directly by KH. Rafiudin. Patients who are sick are the same as healthy people carried out in mass but one by one with different times. The bathed patients are then gathered in the patient hall, and the administrators sit behind them. Kyai is communicated to instill tawhid by saying shahada, reading shalawat, istighfar then reading Ya Allah Ya Latif, Subhanallah Wabihamdihi Subhanallah Wabihamdihi Subhanallah Wabihamdihi Subhanallahiladzim, one by one touched to the patient with KH. Rafiuddin's hand until it is finished, then closed with Al-Fatihah and wejangan.

3.1.2 *Self-development*

3.1.2.1 *Dhikr* Zikr is a way to restore the patient's awareness of what they have done, both small mistakes and big mistakes. This remembrance asks for forgiveness from Allah for all deeds and words done both consciously and unconsciously. Zikr is done every time, and anytime and anywhere; whether the patient is sick or not, everything is the same as reciting zikr. KH. Rafiuddin leads the dhikr. Patients are seriously ill; they listen and sit in an orderly manner in the patient hall. Meanwhile, patients who have just recovered from illness follow the dhikr.

3.1.2.2 *Ritual worship (Fardhu prayer, repentance prayer and tasbih prayer)* Prayer is one of the methods used in Pondok Pesantren Darut Tasbih Tangerang Regency to overcome the problems that exist in patients. This prayer gives an understanding always to remember Allah, who owns the entire universe, including the patient himself. Praying here the patient will be directed and guided by the administrators. Before carrying out the prayer, the patient is taught starting from the procedure for ablution so that he memorises outside the head in the practice of ablution.

The purpose of this method is to eliminate dependence on creatures. When humans get into trouble, many ask for help with something other than Allah SWT. This tawhid can emphasise that Allah has the will for everything in life, and nothing can intervene from any creature.

3.1.2.3 *Ruqyah* Patients who are sick are ruqyah done every Monday night and Tuesday ba'da maghrib handled directly by KH. Rafiudin. Patients who are sick are the same as those who are healthy, done en masse but one by one at different times. The bathed patients are then gathered in the patient hall and the administrators sit behind them. Kyai is communicated to instill tawhid by saying shahada, reading shalawat, istighfar then reading Ya Allah Ya Latif, Subhanallah Wabihamdihi Subhanallah Wabihamdihi Subhanallah Wabihamdihi Subhanallahiladzim, one by one touched to the patient with KH. Rafiuddin's hand until it is finished then closed with Al-Fatihah and wajengan.

3.1.2.4 *Taubah prayer* Before performing the taubah prayer, the patients are asked to take a midnight bath early in the morning at 01:00 WIB. The midnight bath aims to straighten and soften the nerves in the head of the patients, after bathing the patients are asked to perform taubah prayers as much as possible with the number of each two rak'ahs one time greeting. After performing the taubah prayer, they are asked to recite istghfar as much as possible. Although these activities are carried out together, but every three patients are directly guided by one ustad or administrator, so that each patient who will carry out these activities is controlled one by one.

3.1.2.5 *Tasbih prayer* Tasbih prayers are performed every Thursday at 19.00 WIB until completion and Sunday at 07.00 WIB until completion. This prayer activity began with the establishment of Darut Tasbih Islamic Boarding School in Tangerang Regency in 2000 until now. Before carrying out prayer tasbih, usually KH. Rafiuddin gives tausiah and prays first. Tasbih prayer aims to get closer to Allah, support and motivate to carry out activities so that patients can return to their activities.

3.1.2.6 *Prayer* Prayer is one of the ways to overcome the problems that occur in the patient, aiming to provide awareness that he will not be able to do something without His will. Prayer is a form of asking for forgiveness for every mistake made. Prayers are held during istighosah and after prayer. This way of praying can teach patients to depend on God. The content of prayer is the purpose of the wishes and hopes of these patients.

3.1.3 *Self-strengthening*

3.1.3.1 *Fasting* These patients are taught to fast. Fasting aims to control the emotions and lust that exist in these patients. Although they do not fast fully, they can understand that fasting can help treat their mental health. When entering the month of Ramadan, the administrators tell the importance of fasting because fasting during Ramadan is an obligation.

3.1.3.2 *Reading the Qur'an* The patients are taught iqro' by the boarding school administrators, while the normal students also read al-qur`an so that the boarding school is in the shade of blessings and finds grace. The administrators teach reading iqro' which is one form of endeavor in healing the patients. Iqro' is what stimulates the nerves of the patient's back again. The goal is that the patients' tongues are not stiff and know the hijaiyah letters even though they stammer in their pronunciation. So that when they and if they die, their tongues have pronounced the hijaiyah letters. This habit of reading the Qur'an can bring grace and blessings so that they can read it.

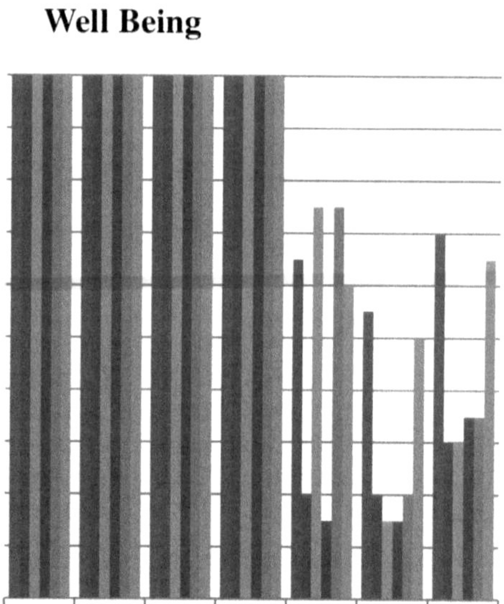

	A	B	F	JK	MI	D	L
■ Self Acceptence	100	100	100	100	65	55	70
■ Goals	100	100	100	100	20	20	30
■ Realitionship	100	100	100	100	75	15	30
■ Otonomy	100	100	100	100	15	15	35
■ Environmental Control	100	100	100	100	75	20	35
■ Self Development	100	100	100	100	60	50	65

Figure 1. Psychological condition of well being of patients at Darut Tasbih Islamic Boarding School, Tangerang Regency.

3.1.3.3 *Practicing dhikr and salawat* After providing understanding to patients to do dhikr, the patients have reflex to say the words of dhikr. The patients who do dhikr a day as much as 50 dhikr after each prayer and coupled with salawat.

3.1.3.4 *Intensive communication* The administrators and patients conduct dialog to stimulate reasoning and responsiveness. The existence of this intensive communication for patients is so that they do not feel sad and do not get stuck in the problems they experience.

3.1.3.5 *Following the study* After implementing the congregational prayer, the patients are taught not to rush from the prayer seat immediately but to listen to Tausiah and advice directly from KH. Rafiuddin is the Darut Tasbih Islamic Boarding School leader in Tangerang Regency.

Based on the Figure 1, after the researcher observes a change in the condition of well-being after undergoing tazkiyatun nafs-based treatment, almost all patients can return according to their respective nature. In terms of well-being, each respondent has different changes and responses. This is due to the patient's background, which underlies the ease of healing. After reviewing the researcher's observations, three patients experienced difficulties in the healing process, so the patients needed a longer period than other patients.

4 CONCLUSION

According to the description and explanation of well-being based on tazkiyatun nafs (case study at the Darut Tasbih Islamic Boarding School, Tangerang Regency), it can be concluded that in providing treatment to survivors of mental disorders at the Darut Tasbih Islamic Boarding School, there are several stages taken to determine the level of mental development. These stages include self-control, self-development and self-strengthening. Meanwhile, in each stage, forms of Islamic treatment are applied, such as ritual worship, dhikr and ruqyah.

From the stages of tazkiyatun nafs, namely the stages of self-control, self-development and self-strengthening, there are important aspects at each stage. The self-control stages are carried out to control, focus and straighten three things: thought patterns, feelings and behaviour. The development stage focuses on the changes that occur in each individual with the hope that each patient can control emotions, maintain mental stability and be able to overcome the stresses of life so that they can build their ability to be more confident, responsible, productive, independent and create a spiritual bond with the Almighty. Finally, the self-strengthening stage aims to ensure that patients have boundary-setting, toughness or resilience and can respond to all problems to experience growth in their mindset and view life more positively.

REFERENCES

Anas Ahmad Karzon. (2016). *Tazkiyatun nafs*. Jakarta, Indonesia: Media Akbar.

Ariati. (2010). Subjective well being (kesejahteraan subjektif) dan kepuasan kerja pada staff pengajar (dosen) di lingkungan Fakultas Psikologi Universitas Diponogoro. *Jurnal Psikologi Undip*, 8(2).

Buchanan, R. L., and Bowen, G. L. (2008). In the Context of Adult Support: The Influence of Peer Support on the Psychological Well-Being of Middle-School Students. *Child and Adolescent Social Work Journal*, 25(5), 397–407. https://doi.org/10.1007/s10560-008-0143-z

Diana Mutiah. (2015). *Peningkatan Keterampilan Pengasuhan Remaja Menuju Kesejahteraan Psikologis (psychology well being)*. Jakarta, Indonesia.

Diener. (2003). *Subjective Well Being is Desirable, but Not the Summum Bonum* [Handout for University of Minnesota Interdisciplinary Workshop].

Dierendonck. (2008). Ryff's six factor model of psychological well being: A span is exploration. *Social Indicators Research, 1*(87).

Ibnu Qayyim Al–Jauziyah. (2001). *Tazkiyatun nafs*. Solo, Indonesia: Arafah Press.
Keyes, K., and Ryff, R. (2002). Optimizing well being: The empirical encounter of two traditions. Journal of Personality and Social Psychology.
Krekel, C., Ward, G., and De Neve, J.-E. (2019). Employee Wellbeing, Productivity, and Firm Performance. *SSRN Electronic Journal*. https://doi.org/10.2139/ssrn.3356581
Masyhuri. (2012). *Prinsip-Prinsip Tazkiyah Al-Nafs Dalam Islam dan Hubungannya Dengan Kesehatan Mental*. Depok: Raja Grafindo.
Moh. Nasir. (2017). *Metode Penelitian*. Jakarta, Indonesia: Ghalia Indonesia.
Moh. Toriquddin. (2008). *Sekularitas Tasawuf: Membumikan Tasawuf Dalam Dunia Modern*. Malang, Indonesia: UIN Press.
Muhammad Zuhdi. (2022). *Konsep Dan Implementasi Well-being Mahasiswa di Perguruan Tinggi Dalam Perspektif Pendidikan Islam dan Sekuler*. Jakarta, Indonesia: LP2M UIN Syarif Hidayatullah Jakarta.
Mustafa Zahri. (1995). *Kunci Memahami Ilmu Tasawuf*. Surabaya, Indonesia: Bina Ilmu.
Negovan, V. (2010). Dimensions of students' psychosocial well-being and their measurement: Validation of a students' Psychosocial Well Being Inventory. *Europe's Journal of Psychology*, *6*(2). https://doi.org/10.5964/ejop.v6i2.186
Noordin, N., and Rahman, R. (2017). Perbandingan proses tazkiyah al–nafs menurut Imam al–Ghazali dan Ibnu Qayyim. *Al–Turath Journal of Al–Qur'an and Sunnah*, *2*(1).
Nur Heni. (2017). Kesehatan mental perspektif Zakiah Daradjat.
Ryan, R. M., and Deci, E. L. (2001). On happiness and human potentials: A review of research on hedonic and eudaimonic well-being. *Annual review of psychology*, 52(1), 141–166.
Ryff, C. D., and Singer, B. H. (2008). Know Thyself and Become What You Are: A Eudaimonic Approach to Psychological Well-Being. *Journal of Happiness Studies*, 9(1), 13–39. https://doi.org/10.1007/s10902-006-9019-0
Said Hawwa. (1999). Almustakhlash fii tazkiyatul anfus, mensucikan jiwa: Konsep tazkiyatun nafs terpadu (A. R. Shaleh, Tahmid, Trans.). Jakarta, Indonesia: Rabbani Press.
Smet. (2004). *Psikologi Kesehatan*. Jakarta: PT. Gramedia Widiasarana Indonesia.
Sugiyono. (2018). *Metode Penelitian Kuantitatif*. Bandung, Indonesia: Alfabeta.
Taylor. (2009). *Health Psychology*. Los Angeles, CA: University of California.

Towards Resilient Societies: The Synergy of Religion, Education, Health, Science, and Technology – Rahiem (Ed.)

Enhancing environmental attitudes through project-based learning on mangrove ecosystems

Irsyaad Hafizd
Universitas Islam Negeri Syarif Hidayatullah Jakarta, Indonesia

Evi Muliyah
Universitas Islam Negeri Syarif Hidayatullah Jakarta, Indonesia
ORCID ID: 0000-0002-1223-1581

Yanti Herlanti
Universitas Islam Negeri Syarif Hidayatullah Jakarta, Indonesia
ORCID ID: 0000-0002-8380-2417

Fonny Dameaty Hutagalung
Universiti Malaya, Malaysia

ABSTRACT: Jakarta's mangrove ecosystem is rapidly degrading due to human activities, underscoring the urgent need for targeted educational interventions aligned with the Sustainable Development Goals (SDGs). This study develops a project-based learning workbook for seventh-grade students in North Jakarta, addressing SDG 4 (Quality Education), SDG 13 (Climate Action), SDG 14 (Life Below Water), and SDG 15 (Life on Land). Designed using the ADDIE instructional model, the workbook integrates ecological content with structured learning to enhance students' understanding of mangrove conservation while fostering critical thinking. Given mangroves' role as carbon sinks and coastal protectors, the study supports SDG 13 by emphasising climate resilience. It also reinforces SDG 14 through marine ecosystem conservation and SDG 15 by promoting sustainable land management. Findings indicate that structured ecological education significantly improves students' environmental awareness and attitudes. Expert validation confirms the workbook's effectiveness in cultivating proactive environmental stewardship, demonstrating education's pivotal role in driving sustainability.

Keywords: education, environmental attitude, mangrove, project, worksheets

1 INTRODUCTION

Mangrove forest ecosystems are critical to natural preservation and support various biotic components. These functions can be assessed from physical, biological, economic, chemical, and other perspectives (Supriadi *et al.* 2020). However, human activities that fail to protect these areas threaten the delicate balance required for environmental protection. For instance, mangrove forest areas in Indonesia have undergone substantial changes in use, being transformed into residential areas, ports, waste disposal sites, ponds, and mining areas. In 1982, the mangrove area covered 4.25 million hectares, but by 1993, it had drastically reduced to 2.50 million hectares (Elminsyah 2020). The degradation of mangrove ecosystems poses a significant threat to coastal biodiversity, climate resilience, and

DOI: 10.1201/9781003645542-70

the livelihood of communities dependent on these critical habitats (Asari *et al.* 2021). In response to these challenges, environmental education has increasingly focused on fostering a sense of responsibility and proactive conservation attitudes among students (Begum *et al.* 2021). Human activities in and around mangrove areas have led to a significant reduction in the coverage of this crucial ecosystem. It is imperative to undertake preventive measures and educate the younger generation about the importance of preserving mangrove areas. Educational platforms can serve as effective channels for instilling knowledge and skills in students, thereby fostering environmental attitudes. The education process encompasses more than merely transmitting information; it also develops and enhances attitudes, values, knowledge, and skills (Ardoin *et al.* 2020). Therefore, education should persist in developing efficacious strategies to cultivate positive environmental attitudes (Ma *et al.* 2023)

Educational platforms can serve as effective channels for instilling knowledge and skills in students, thereby fostering environmental attitudes. As educational institutions, schools are responsible for imparting moral values related to the natural environment, encompassing the ethical treatment and preservation of nature (Hasnawati 2020). An impactful approach involves the utilisation of student worksheets within the framework of project-based learning (PBL), catalysing and enhancing students' involvement in project-based learning activities. Project-based student worksheets are proven to improve student learning outcomes significantly and are effectively used in learning (Murni and Yasin 2021). Through project-based learning, students reported significant improvement in various skills, including the 4C skills (Critical thinking, Communication, Collaboration, Creativity), problem-solving, and higher-order thinking (Romadhon *et al.* 2023).

Studies have shown that project-based learning can significantly enhance students' attitudes and knowledge regarding environmental conservation. For example, a research study demonstrated that project-based worksheets on mangrove ecosystems improved students' understanding and positive attitudes toward conservation (Sigit *et al.* 2019). The effectiveness of project-based worksheets is not only limited to knowledge acquisition but also extends to developing environmental problem-solving skills. Research has highlighted the role of such educational tools in equipping students with the skills needed to tackle environmental issues (Rianti *et al.* 2024).

The study promotes environmental literacy and fosters attitudes conducive to sustainable ecosystem management by employing project-based learning. Mangroves are critical for coastal protection, biodiversity conservation, and carbon sequestration, making them pivotal in climate change mitigation and marine ecosystem health. Integrating mangrove conservation into education empowers communities, especially youth, to engage in sustainable practices, thereby supporting a holistic approach to achieving environmental sustainability (United Nations 2015). By working on projects addressing real environmental challenges, students can better understand the complexities of ecosystem conservation and the importance of collaborative efforts in finding solutions to preserving their natural surroundings (Novitha and Suhartini 2023). This article aims to delve deeper into the potential of project-based worksheets to improve environmental attitudes towards mangrove ecosystems. Ultimately, the goal is to highlight the transformative potential of project-based learning in shaping the environmental attitudes of future generations and ensuring the sustainability of vital ecosystems like mangroves.

2 METHODS

The research is being carried out using the ADDIE (Analysis, Design, Development, Implementation, and Evaluation) model for the developmental research method. It encompasses the following stages in Table 1.

Table 1. Steps and activities in ADDIE developmental research.

Stages	Activity
Analysis	The needs analysis was conducted through interviews and document analysis. The interview involved junior high school science teachers. Two documents were analyzed: the National Curriculum of Science released by the Indonesia Ministry of Education and the Mangrove Curriculum released by the Peat and Mangrove Restoration Agency.
Design	Formulating learning objectives and project activities, organizing the worksheet layout, and creating the project-based worksheet prototype.
Development	This phase comprises two key activities: evaluating the project-based worksheet prototype and developing an instrument to gauge environmental attitudes. The project-based worksheet prototype underwent assessment by four experts in language, graphics, project learning activity, and mangrove material content. The instrument was derived from an article titled High School Students' Environmental Attitude: Scale Development and Validation (Ugulu *et al.* 2013) and was subsequently adapted to focus on the mangrove ecosystem..
Implementation	The project-based worksheet was implemented at North Jakarta Junior High School, involving 189 7th-grade students. The activities spanned six stages: (1) Administering a pre-test on environmental attitudes to the students. (2) Delivering focused instructional material on mangroves to augment the students' understanding. (3) Encouraging students to engage in local mangrove environment observations to foster their environmental consciousness. (4) Facilitating the creation of posters by the students to articulate their comprehension and experiences. (5) Arranging for the students to present their posters to their teacher and peers. (6) Conducting post-tests on environmental attitudes to evaluate the students' learning.
Evaluation	The examination of environmental attitude accomplishments involves processing data using both descriptive and inferential approaches. Descriptively, environmental attitude results are categorized into two groups: • Positive environmental attitude: scores ranging from 90 to 120 • Environmental attitudes reflecting less concern: scores below 90 Inferential statistical analysis of environmental attitudes employs the Wilcoxon signed-rank test to assess attitudes before and after the utilization of project-based worksheets.

3 FINDING & DISCUSSION

3.1 *Analysis*

Environmental issues are essential for higher education to instil attitudes and values in students to protect the environment (Hayat *et al.* 2021). Based on the analysis of the curriculum used in the North Jakarta junior high school with the mangrove area, it is possible to implement project-based learning because the school uses the new Indonesian curriculum based on a project to strengthen character values. The school provides one month for students to work on the project. The curriculum's theme is sustainable lifestyle projects that can be integrated with project learning in mangrove ecosystems. Based on the school needs assessment, project learning in the Mangrove area can be done by doing a basic introduction to Mangroves, visiting the Mangrove ecosystem area, acting in the form of making posters that show the preservation of the Mangrove area and presenting the poster results, which the teacher will evaluate.

3.2 *Design*

The design and structure of the learner worksheets contained 19 parts, which include the following:

(1) A cover page containing the title and theme of the project activity.
(2) The preface delivered by the author is an introduction to learning activities.
(3) The purpose of writing the worksheet is to contain the aims and objectives of researchers in developing the worksheet.
(4) Instructions for use contain a flowchart of the worksheet's use.
(5) The table of contents contains page information on the content contained in the worksheet.
(6) The learner identity page contains a section for learners to write their identities using group names, member names, and learner classes.
(7) Learning outcomes contain the learning outcomes to be achieved by students.
(8) The introduction stage contains illustrations and images that lead learners to understand the material better.
(9) Introductory material contains Mangrove material that students will study
(10) The first and second learning activities contain theoretical questions about Mangrove.
(11) The contextualisation stage contains illustrations and pictures that lead learners to see the actual conditions of the Mangrove Area.
(12) Mangrove Story contains a story for learners to start learning.
(13) Pantai Indah Kapuk (PIK) Mangrove review contains a slideshow on the condition of the Mangrove area in the North Jakarta area.
(14) The learning activity contains observation instruments while students are in the Mangrove Area.
(15) The action stage contains illustrations and pictures introducing the activities to be carried out.
(16) The learning activity contains an introduction for students to make a poster about mangroves.
(17) The reflection stage contains illustrations and pictures introducing reflection activities.
(18) Learning Activity contains an introduction for learners to present their posters.
(19) The bibliography contains references used to prepare the learner worksheet.

3.3 *Development*

Some experts from the Indonesia Peat and Mangrove Restoration Agency, School Teachers, and Environmental Lectures assess the project-based worksheets designed. The results of the expert's assessment can be seen in Table 2.

Table 2. Assessing from experts.

Expert	Assessment aspects	Rating result (%)	Category
I	Language, graphics, and learning models	94,66	Very valid
II	Language, graphics, and learning models	100	Very valid
III	Mangrove Material Content	82	Very valid
IV	Environmental care attitude	93,33	Very valid
Percentage (%) average		92,49	Very valid

The instrument of the environmental attitude was adopted from a journal written by Ilker Ugulu (2013). The results found in developing the instrument are appropriate for assessing students' attitudes toward environmental awareness. The instrument used in measuring environmental care attitudes was tested to determine the instrument's validity with subjects who were not samples in the study. The validity test results obtained 30 instrument items from the previous 40 instrument items. In the process of validity and reliability of instruments involving 90 respondents, the results are in Tables 3 and 4.

Table 3. Item selection in the environmental stewardship survey.

Environmental care factor	Question item
Factor I (Environmental awareness)	1, 2*, 3*, 4, 5*, 6*, 7, 8*, 9, 10*, 11, 12*, 13*, 14*, 15*
Factor II (Attitude toward recovery)	16, 17*, 18*, 19*, 20*, 21*, 22*, 23*
Factor III (Attitude toward recycling)	24*, 25*, 26, 27*, 28*, 29*, 30*
Factor IV (Environmental awareness and behavior)	31*, 32*, 33*, 34, 35, 36*, 37*, 38*, 39*, 40*

*Valid

Table 4. The results of the reliability process of the environmental care attitude instrument.

		A
B	Pearson Correlation	,546**
	Sig. (2-tailed)	0,000
	N	90

**. Correlation is significant at the 0.01 level (2-tailed).

3.4 *Implementation*

The project activities are oriented to four syntaxes: orientation, contextualisation, action, and reflection (Table 5). Before and after doing the project, students filled out the questionnaire on environmental attitude.

Table 5. The project activities.

Stage		Students' activity
Pre-test		Students filled out an environmental attitude questionnaire.
Orientation		Students were given the knowledge of the mangrove ecosystem. The worksheet provided some information about ecosystem mangrove information, types of mangrove plants, roles, functions, conditions, and efforts to conserve mangroves.
Contextualisation		Students visit the Mangrove Ecosystem Area. The worksheet has observation points that will guide students during the observation activities.
Action		Students made posters to encourage other students to participate in conserving the mangrove ecosystem. The worksheet provides some guidelines for making a poster.
Reflexion		Students presented the poster and had a discussion and reflection.
Post-test		Students filled out an environmental attitude questionnaire.

3.5 *Evaluating*

There was an increase in students' environmental attitudes after and before the project activities. Table 6 shows ten students' changes in environmental attitudes. However, the average environmental care attitude still needs to improve, and the increase after project activities could be more significant (Table 7).

Table 6. Descriptive analytics.

	Criteria	Result 1 (Before project learning)	Result 2 (After project learning)
Maximum score	120	110	112
Minimum score	30	60	65
Average	–	87 (Bad environmental attitude)	89 (Bad environmental attitude)
Good environmental attitude	90 - 120	58 students (64%)	68 students (76%)
Bad environmental attitude	<90	32 students (36%)	22 students (24%)

Table 7a. Wilcoxon test questionnaire data processing results (ranks value).

		N	Mean Rank	Sum of Ranks
After-before Ranking	Negatif Rank	106[a]	85,75	9090,00
	Positif Rank	80[b]	103,76	8301,00
	Ties	3[c]		
	Total	189		

Description:

[a]After < Before
[b]After > Before
[c]After = Before

Table 7b. Wilcoxon test questionnaire data processing results (Test Statistics[a]).

	Rank of Pre-Post
Z	–,537[b]
Asymp. Sig (2-tailed)	,592

The formation of environmental attitudes requires a long time. Project learning is carried out for one month (one weekly meeting) and cannot quickly change students' environmental attitudes. Similar research also gave the same results. The environmental care attitude of students who studied with the Mangrove Encyclopedia showed an insignificant increase (Ratminingsih and Subiantoro 2023). The other studies also demonstrated that students' attitudes toward mathematics after using project-based learning do not differ significantly (Alenezi 2023).

4 CONCLUSION

The research that has been done shows that the development of student worksheets based on project-based learning shows results that are feasible to be implemented based on validity tests by experts where the average score validity reaches 92.49% with a very valid category.

Regarding developing students' environmental attitudes, the improvement category was less significant, in which the increase in students only obtained a proportion of 1.26%.

ACKNOWLEDGEMENTS

We thank the BRGM Republic of Indonesia, especially Ms. Novi and Mr. Febri, who collaborated with the Biology Education Study Program to develop an environmental education curriculum for mangrove and peat ecosystems.

REFERENCES

Alenezi, A. (2023). Using Project-Based Learning Through the Madrasati Platform for Mathematics Teaching in Secondary Schools. *International Journal of Information and Communication Technology Education, 19* (1). https://doi.org/10.4018/IJICTE.332372

Ardoin, N. M., Bowers, A. W., and Gaillard, E. (2020). Environmental education outcomes for conservation: A systematic review. *Biological Conservation, 241.* https://doi.org/10.1016/j.biocon.2019.108224

Asari, N., Suratman, M. N., Mohd Ayob, N. A., and Abdul Hamid, N. H. (2021). Mangrove as a Natural Barrier to Environmental Risks and Coastal Protection. In *Mangroves: Ecology, Biodiversity and Management* (pp. 305–322). Springer Nature. https://doi.org/10.1007/978-981-16-2494-0_13

Begum, A., Jingwei, L., Haider, M., Ajmal, M. M., Khan, S., and Han, H. (2021). Impact of environmental moral education on pro-environmental behaviour: Do psychological empowerment and islamic religiosity matter? *International Journal of Environmental Research and Public Health, 18*(4), 1–19. https://doi.org/10.3390/ijerph18041604

Elminsyah, E. (2020). *Skeneario Berkelanjutan Pengelolaan Hutan Mangrove.* CV. Nas Media Pustaka.

Hasnawati. (2020). Akhlak Kepada Lingkungan. *Jurnal Pendais, 2*(2), 208.

Hayat, B., Hidayat, R., and Putra, M.D.K. (2021). Exploring the Factor Structure of Environmental Attitudes Measure in a Sample of Indonesian College Students. *Rev. CES Psico*, 14(1), 112–129.

Ma, L., Shahbaz, P., Haq, S. ul, and Boz, I. (2023). Exploring the Moderating Role of Environmental Education in Promoting a Clean Environment. *Sustainability (Switzerland), 15*(10). https://doi.org/10.3390/su15108127

Murni, A.W. and Yasin, F.N. (2021). Pengembangan Lembar Kerja Peserta Didik (LKPD) Berbasis Proyek pada Materi Siklus Air Kelas V Sekolah Dasar. *JURNALBASICEDU, 5*(6), 6208.

Novitha, and Suhartini. (2023). Development of Problem-Based Learning LKPD Based Local Potential of Baros Mangroves in Biology Subjects Environmental Pollution Material. *Jurnal Penelitian Pendidikan IPA, 9*(12), 10538–10545. https://doi.org/10.29303/jppipa.v9i12.5695

Rianti, N. S., Utaya, S., Purwanto, P., and Shrestha, R. P. (2024). Uncovering The Effectiveness of the Project-Based Learning Model and Ecological Intelligence: Impact on Environmental Problem-Solving Ability in Senior High School. *JAMBURA GEO EDUCATION JOURNAL, 5*(1), 33–43. https://doi.org/10.37905/jgej.v5i1.24461

Romadhon, D. R., Rahiem, M. D. H., Faeruz, R., Muliyah, E., Asmui, and Wulansari, N. F. (2023). Enhancing student engagement in computational physics through project-based learning: An exploration of VBA physics simulation. *2023 11th International Conference on Cyber and IT Service Management (CITSM)*, Makassar, Indonesia, 1–6. https://doi.org/10.1109/CITSM60085.2023.10455497

Sigit, D. V., Miarsyah, M., Komala, R., Suryanda, A., Fadrikal, R., and Ichsan, I. Z. (2019). Improvement of knowledge and attitude in conservation of mangrove and coral reefs through environmental education community network model. *Journal of Physics: Conference Series, 1317*(1). https://doi.org/10.1088/1742-6596/1317/1/012201

Supriadi, D., Widayaka, R., and Khan, A. (2020). *Memaksimalkan Potensi Sumber Daya Perikanan dan Kelautan di Indonesia.* Lekeisha.

Ugulu, I., Sahin, M., and Baslar, S. (2013). High School Students' Environmental Attitude: Scale Development and Validation. *International Journal of Educational Sciences, 5*(4), 415–424. https://doi.org/10.1080/09751122.2013.11890103

United Nations. (2015). Transforming our world: The 2030 Agenda for Sustainable Development. Retrieved from https://sdgs.un.org/2030agenda.

Towards Resilient Societies: The Synergy of Religion, Education, Health, Science, and Technology – Rahiem (Ed.)
© 2026 The Author(s), ISBN: 978-1-041-08471-6
Open Access: www.taylorfrancis.com, CC BY-NC-ND 4.0 license

Innovative interventions during the COVID-19 pandemic: A sustainable model for health and education

Baequni Boerman, Fajar Ariyanti, Narila Mutia Nasir & Shafira Salsabila Samara
Public Health Department, Universitas Islam Negeri Syarif Hidayatullah Jakarta, Indonesia

ABSTRACT: This study examines the collaboration between government and community efforts in South Tangerang Regency to combat COVID-19, focusing on factors influencing adherence to health protocols: predisposing (knowledge), enabling (accessibility), and reinforcing (law enforcement) elements. A methodologically rigorous cross-sectional survey using a stratified sampling approach was conducted in 2020 and 2021 across 13 Public Health Service areas to assess changes in compliance over time. The survey engaged 106 public health students and reached 5,326 households online in 2020 and 1,301 households in 2021. The findings revealed significant correlations between these factors and adherence to health protocols, with a marked decline in handwashing and mask-wearing practices from 2020 to 2021. Identified challenges included limited health literacy, cultural barriers, misinformation, and insufficient government oversight. The study underscores the need for culturally sensitive communication, targeted educational campaigns, and trust-building strategies to strengthen positive health behaviours. Integrating health and education initiatives is essential for supporting Sustainable Development Goals (SDGs) 3 (Good Health and Well-being) and 4 (Quality Education), fostering sustainable community impact and resilience during health crises.

Keywords: community collaboration, community empowerment, COVID-19, education initiatives, enabling factors, health protocol obedience, predisposing factors, public health education, reinforcing factors, stratified sampling

1 INTRODUCTION

The COVID-19 pandemic has profoundly impacted urban and rural regions, exposing significant disparities in response and resilience. Despite these challenges, community initiatives have emerged as critical in combating the virus and navigating crises (Noor *et al.* 2022). Resilient communities, defined by their ability to adapt, respond, and recover from adversity, have played a pivotal role during this global health emergency. The pandemic has revealed vulnerabilities and inequalities across societal sectors, emphasising the need to foster community resilience through collective action and innovation (Schmidt-Sane *et al.* 2021).

Digital platforms have become indispensable in fostering communication, resource sharing, and a sense of belonging among community members, particularly for marginalised groups who often face heightened challenges during crises (Noor *et al.* 2022). Urban areas with higher awareness of COVID-19 due to dense populations and elevated virus prevalence have mobilised grassroots movements to enforce health protocols (Mekonen *et al.* 2021). In lower-income urban neighbourhoods identified as hotspots, local resources have been

 DOI: 10.1201/9781003645542-71

organised to support affected families and improve access to healthcare (Juhn *et al.* 2021). The economic challenges posed by the pandemic, particularly income reductions in urban and semi-urban areas, have spurred community-led initiatives to provide economic support and sustain livelihoods (Syafiq *et al.* 2022).

The pandemic has also reshaped community education and public health strategies, driving innovative approaches to maintain engagement and awareness. Technology-based education has enhanced community knowledge and practices while addressing mental health issues through targeted health campaigns. Educators have adapted to virtual modalities, emphasising effective learning standards and community collaboration to combat misinformation and improve vaccination understanding (Listiowati *et al.* 2023; Weygandt *et al.* 2021). In Indonesia, community-driven initiatives have bolstered the healthcare system, addressing the mental health impacts on healthcare workers and supporting home-quarantined patients through tailored responses (Mahendradhata *et al.* 2021; Setiawati *et al.* 2021; Wibowo *et al.* 2022).

Though sometimes slow, government efforts have been supplemented by community actions that mitigated economic impacts and supported key sectors like tourism and trade (Nasution *et al.* 2020). The pandemic underscores the importance of international collaboration, local adaptations, and public health education in managing crises, offering opportunities for students to design and implement interventions that bridge theory with practice (Himmel and Frey 2022).

In Tangerang, initiatives such as Large-Scale Social Restrictions (PSBB) significantly reduced COVID-19 cases (Furqon and Mulyadi 2020). This research investigates predisposing, enabling, and reinforcing factors influencing adherence to health protocols, providing insights for sustainable public health practices. Data collected during two critical periods of active COVID-19 cases in Indonesia offer a unique perspective on pandemic dynamics, emphasising the importance of integrating local and international strategies for long-term resilience.

2 METHODS

This study employed quantitative techniques to conduct a comprehensive and multi-dimensional analysis. In November 2020 and 2021, a methodologically rigorous cross-sectional survey was conducted using a stratified sampling approach in South Tangerang. The target population included residents of South Tangerang, with one questionnaire allocated per household. If the head of the household was unavailable, the spouse or an adult child (aged 17 or older) residing in the household could complete the questionnaire.

The sample framework was developed using Slovin's formula, incorporating a 5% margin of error, a 95% confidence level, a z-score of 1.96, and a standard deviation of 0.5. Based on these parameters, the minimum required sample size was 385 respondents. The study surveyed 5,326 households in 2020 and 1,301 households in 2021.

Data was collected by 106 public health students, who administered online and offline questionnaires across 13 Public Health Service areas. Quantitative data were collected using Google Forms and analysed using statistical software. A systematic analysis was performed using univariate and bivariate methods, supplemented by Chi-Square tests, with a significance level set at $p < 0.05$ and a 95% confidence interval.

The chosen methodology aligns with the research objectives by providing reliable insights into health protocol adherence. Using stratified sampling ensures representative data, while the large sample size enhances reliability. Rigorous statistical analyses, including Chi-Square tests, allow for a robust evaluation of key factors, supporting the development of resilient public health strategies.

3 RESULT AND DISCUSSION

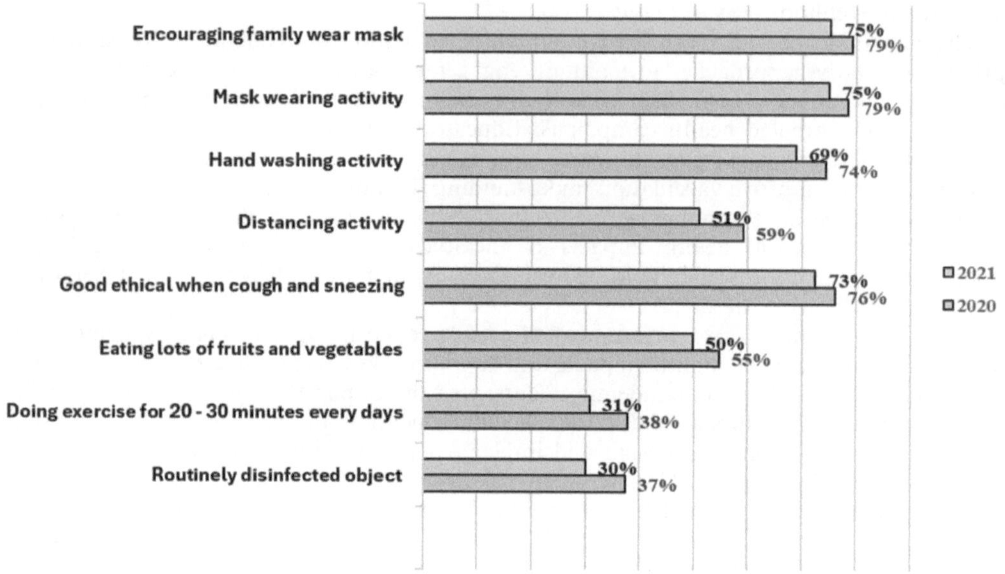

Figure 1. Comparison of the adherence percentage to health protocols in South Tangerang, Banten.

The data in Figure 1, "Comparison of the Percentage of Obedience to Health Protocols in South Tangerang, Banten," reveals a decline in adherence to health protocols in Indonesia from 2020 to 2021. Key behaviours such as cleaning objects, daily exercise, consuming fruits and vegetables, proper cough etiquette, maintaining distance, handwashing, and mask-wearing all saw reductions. For example, handwashing compliance dropped from 74% to 69%, while mask-wearing fell from 79% to 75%. This downward trend underscores the need for stronger efforts to sustain public discipline in health protocols. Despite extensive government-led information and education campaigns, compliance remains low. Research indicates this may stem from differences in risk perception. Those perceiving a higher risk are more likely to follow preventive measures (Nurhidayah 2024). In contrast, groups like younger individuals and males, who often underestimate the threat, show lower adherence. Factors such as personal beliefs, cultural influences, and misinformation further exacerbate this issue, undermining public health campaigns' effectiveness (Dörfler 2024).

Moreover, the effectiveness of health communication plays a pivotal role in shaping public behaviour during the pandemic. For instance, culturally tailored communication strategies have been shown to enhance compliance among Indigenous and Culturally and Linguistically Diverse (CALD) populations, who may face additional barriers such as stigma and mistrust towards health services (Kayrouz *et al.* 2021). Implementing mobile education models during health emergencies has also been advocated to address public perceptions and improve health behaviours (Handayani *et al.* 2023). Such strategies are essential in fostering a better understanding of health protocols and their importance in mitigating the spread of COVID-19.

Table 1 demonstrates the strong link between knowledge, attitudes, enabling, and reinforcing factors and adherence to health protocols in 2020 and 2021. Individuals with poor knowledge showed an increase in non-compliance from 21.8% to 23.6%, highlighting the critical need for improved health literacy. Enabling factors exhibited a strong connection to adherence, with poor enabling conditions showing slight improvements and favourable conditions driving significant compliance. Likewise, reinforcing factors played a vital role, as supportive environments strengthened adherence and reduced non-compliance (Koh *et al.* 2018; Mitchell *et al.* 2021; Mohammadali *et al.* 2022).

Table 1. Impact of predisposing, enabling, and reinforcing factors on health protocol behaviours among respondents (2020-2021), including p-values and odds ratios (or) with 95% confidence intervals (ci).

Predisposing (Knowledge & Attitude), Enabling and Reinforcing	Behaviour Toward Health Protocols (Percents)				P value OR (95% CI)	
	Not Good		Good			
	2020	2021	2020	2021	2020	2021
Respondent's knowledge*						
Poor knowledge	21.8	23.6	23.1	19.3	0.01 2.008 (1.796-2.245)	0.001 1.988 (1.591-2.485)
Good knowledge	17.6	21.8	37.5	35.4		
Respondent's attitude*						
Poor attitudes	27.2	32.7	17.1	19.5	0.01	0.001
Good attitudes	20.2	12.6	35.1	35.1	2.219 (1.988-2.478)	4.674 (3.689-5.921)
Respondent's enabling**						
Poor enabling	26.5	25.7	18.4	22.8	0.01	0.001
Good enabling	21,7	19.6	33.4	31.8	2.684 (2.401-3.000)	1.831 (1.468-2.284)
Respondent's reinforcing***						
Poor reinforcing	31.5	30.5	13.4	19.4	0.01 2.917 (2.603-3.269)	0.001 3.724 (2.957-4.689)
Good reinforcing	24.6	14.8	30.5	35.2		

* **Respondent's knowledge** and **attitude**: "The questions about knowledge on predisposing factors, numbered from 20 to 34, consist of the following: (20) What are the symptoms or signs of a person suffering from COVID-19?; (21) Can a clean and healthy lifestyle prevent the transmission of the COVID-19 virus?; (22) How is the COVID-19 virus transmitted from one patient to another?; (23) When should we wash our hands with soap to prevent COVID-19?; (24) Can wearing a mask prevent the transmission of COVID-19?; (25) When should we wear masks to avoid COVID-19?; (26) How should the mask be worn correctly?; (27) What is a safe distance for interaction with others outside the home during the pandemic?; (28) If you have flu or cold symptoms, should you self-isolate and avoid contact with others?; (29) How should hands be washed effectively?; (30) Can hand sanitiser kill viruses?; (31) Do you engage in at least 20 minutes of physical activity every day?; (32) What is the recommended dietary pattern for adults?; (33) If not wearing a mask, should you cover your mouth and nose with a handkerchief, tissue, or arm when sneezing or coughing?; (34) Is regular disinfection necessary?"

** Respondent's **enabling:** "Enabling questions begin at number 45 and are as follows: (45) Are you provided with a public handwashing facility? (46) Where is the handwashing facility located in your neighbourhood? Have you received information on preventing COVID-19 transmission in your area? (48) Through what media was this information provided? (49) Has disinfection spraying been conducted in your area?"

*** **Respondent's reinforcing:** "Questions regarding reinforcement measures are as follows: (53) Are there any social sanctions or fines imposed on individuals who do not wear masks outside their homes? (54) Have there been any instances of the police or authorities enforcing mask-wearing for COVID-19 by seizing masks? (57-61) These questions pertain to the support for health protocols for families and others."

Law enforcement has also played a vital role in ensuring compliance. Police efforts, including monitoring, fines, and public education, have been pivotal in promoting adherence, particularly in crowded areas or during public gatherings (Sanjaya and Rakhmawan 2020; Wijaya and Barthos 2022). Enforcement measures have supplemented voluntary compliance, which may be insufficient to achieve public health goals.

The analysis of odds ratios (ORs) from 2020 to 2021 revealed shifts in the impact of attitudes and reinforcing conditions on adherence. Attitudes shaped by perceived disease threats, quarantine benefits, trust in government actions, and individual well-being strongly influence health behaviours (Guillon and Kergall 2020). These findings underscore the importance of designing interventions that address attitudes and reinforce positive behaviours.

Health literacy, trust in healthcare systems, and social determinants of health are key modifiable factors influencing compliance (Rucinski 2023). Effective communication strategies are essential to foster understanding and encourage adherence, particularly in managing infectious diseases like COVID-19 (Indrayathi *et al.* 2021; Mendoza 2023). Moreover, addressing psychological factors, such as anxiety and mood disorders, is crucial, as they can hinder public compliance during health crises (Dilas *et al.* 2023). Sustained motivation, support, and health education reinforce compliance, enabling individuals to make informed decisions and adopt health protocols that prevent disease transmission (Pasquale *et al.* 2016; Setyorini 2023).

4 CONCLUSION

The decline in adherence to health protocols from 2020 to 2021 underscores the urgent need for multifaceted strategies. Key factors, including risk perception, health literacy, cultural barriers, and misinformation, must be addressed through culturally tailored communication, scalable mobile education models, and targeted misinformation campaigns. Reinforcing positive behaviours requires improved enabling conditions, psychological support, and trust in healthcare systems, while law enforcement should combine monitoring and education with enforcement. Future research should focus on scalable models integrating health and education to advance SDG 3 and SDG 4, fostering sustainable community resilience. This study has undergone ethical review by the Department of Public Health, Islamic State University Syarif Hidayatullah Jakarta.

REFERENCES

Dilas, D., Flores, R., Morales-García, W. C., Calizaya-Milla, Y. E., Morales-García, M., Sairitupa-Sanchez, L. Z., and Saintila, J. (2023). Social Support, Quality of Care, and Patient Adherence to Tuberculosis Treatment in Peru: The Mediating Role of Nurse Health Education. *Patient Preference and Adherence, Volume 17*, 175–186. https://doi.org/10.2147/ppa.s391930

Dörfler, V. (2024). *Dubito Ergo Sum: Exploring AI Ethics*. https://doi.org/10.24251/hicss.2023.671

Furqon, E., and Mulyadi, E. (2020). The Harmonization of the Central and Local Governments Authority: Handling Public Health Emergencies on Coronavirus Disease 2019. *UNIFIKASI: Jurnal Ilmu Hukum, 7* (2). https://doi.org/10.25134/unifikasi.v7i2.3569

Guillon, M., and Kergall, P. (2020). Attitudes and Opinions on Quarantine and Support for a Contact-Tracing Application in France During the COVID-19 Outbreak. *Public Health, 188*, 21–31. https://doi.org/10.1016/j.puhe.2020.08.026

Handayani, O. W. K., Zainafree, I., and Farida, E. (2023). *Analysis of Using the Mobile Education Model in the Community in Health Emergency Conditions*. https://doi.org/10.4108/eai.29-6-2022.2326080

Himmel, M., and Frey, S. (2022). SARS-CoV-2: International Investigation Under the WHO or BWC. *Frontiers in Public Health, 9*. https://doi.org/10.3389/fpubh.2021.636679

Indrayathi, P. A., Januraga, P. P., Pradnyani, P. E., Gesesew, H. A., and Ward, P. (2021). Perceived Social Norms as Determinants of Adherence to Public Health Measures Related to COVID-19 in Bali, Indonesia. *Frontiers in Public Health, 9*. https://doi.org/10.3389/fpubh.2021.646764

Juhn, Y. J., Wheeler, P., Wi, C.-I., Bublitz, J., Ryu, E., Ristagno, E. H., and Patten, C. (2021). Role of Geographic Risk Factors in COVID-19 Epidemiology: Longitudinal Geospatial Analysis. *Mayo Clinic Proceedings: Innovations, Quality and Outcomes, 5*(5), 916–927. https://doi.org/10.1016/J.MAYOCPIQO.2021.06.011

Kayrouz, R., Schofield, C. B. S., Nielssen, O., Karin, E., Staples, L. G., and Titov, N. (2021). A Review and Clinical Practice Guideline for Health Professionals Working with Indigenous and Culturally and Linguistically Diverse (CALD) Populations During COVID-19. *Frontiers in Public Health, 9*. https://doi.org/10.3389/fpubh.2021.584000

Koh, J. J. K., Cheng, R. X., Yap, Y., Haldane, V., Tan, Y. G., Teo, K. W. Q., Srivastava, A., Ong, P. S., Perel, P., and Legido-Quigley, H. (2018). Access and Adherence to Medications for the Primary and Secondary Prevention of Atherosclerotic Cardiovascular Disease in Singapore: A Qualitative Study. *Patient Preference and Adherence*, *Volume 12*, 2481–2498. https://doi.org/10.2147/ppa.s176256

Listiowati, E., Samsudin, A., Iqbal Nurmansyah, M., Nurjuman, H., Wulandari, Y., Asmawati, W., and Suryo Soularto, D. (2023). Training for COVID-19 vaccination educator to counter vaccination misinformation in 10 cities in Indonesia. *Journal of Community Service and Empowerment*, *4*(1). https://doi.org/10.22219/jcse.v4i1.24905

Mahendradhata, Y., Andayani, N. L. P. E., Hasri, E. T., Arifi, M. D., Siahaan, R. G. M., Solikha, D. A., and Ali, P. B. (2021). The Capacity of the Indonesian Healthcare System to Respond to COVID-19. *Frontiers in Public Health*, *9*. https://doi.org/10.3389/fpubh.2021.649819

Mekonen, E. G., Workneh, B. S., Ali, M. S., and Muluneh, N. Y. (2021). The psychological impact of COVID-19 pandemic on graduating class students at the university of Gondar, northwest Ethiopia. *Psychology Research and Behavior Management*, *14*, 109–122. https://doi.org/10.2147/PRBM.S300262

Mendoza, M. (2023). Health Protocol Adherence to Emerging and Re-Emerging Diseases: Basis for Community Engagement. *International Journal of Research Studies in Psychology*, *9*(1). https://doi.org/10.5861/ijrsp.2023.2005

Mitchell, E., Hakim, A., Nosi, S., Kupul, M., Boli-Neo, R., Aeno, H., Redman-MacLaren, M., Ase, S., Amos, A., Hou, P., Narokobi, R., Willie, B., Vallely, A., Kaldor, J., Badman, S. G., and Kelly-Hanku, A. (2021). A Socio-Ecological Analysis of Factors Influencing HIV Treatment Initiation and Adherence Among Key Populations in Papua New Guinea. *BMC Public Health*, *21*(1). https://doi.org/10.1186/s12889-021-12077-w

Mohammadali, R., Solaimanizadeh, L., and Mohammadinia, N. (2022). Effect of Educational Intervention Based on PRECEDE Model on Improving Menstrual Health Behaviors Among Female Students. *Shiraz E-Medical Journal*, *23*(12). https://doi.org/10.5812/semj-123461

Nasution, D. A. D., Erlina, E., and Muda, I. (2020). Dampak Pandemi COVID-19 terhadap Perekonomian Indonesia. *Jurnal Benefita*, *5*(2). https://doi.org/10.22216/jbe.v5i2.5313

Noor, S. M., Mustafa, H., Rahman, N. A. A., Jaafar, M., and Hou, E. N. C. (2022). *Transcending Pandemics: Conceptualising the Centrality of Communication in Fostering Community Resilience*. https://doi.org/10.15405/epsbs.2022.01.02.4

Nurhidayah, L. (2024). *The Effect of Risk Perception and Self-Control on Preventive Behaviour After Covid-19 Vaccination*. https://doi.org/10.4108/eai.18-7-2023.2343403

Pasquale, C. D., Veroux, M., Fornaro, M., Sinagra, N., Basile, G., Gozzo, C., Santini, R., Costa, A., and Pistorio, M. L. (2016). Psychological Perspective of Medication Adherence in Transplantation. *World Journal of Transplantation*, *6*(4), 736. https://doi.org/10.5500/wjt.v6.i4.736

Rucinski, K. (2023). Patient Adherence Following Knee Surgery: Evidence-Based Practices to Equip Patients for Success. *The Journal of Knee Surgery*, *36*(14), 1405–1412. https://doi.org/10.1055/a-2154-9065

Sanjaya, S. A., and Rakhmawan, S. A. (2020). *Face Mask Detection Using MobileNetV2 in the Era of COVID-19 Pandemic*. 1–5. https://doi.org/10.1109/icdabi51230.2020.9325631

Schmidt-Sane, M., Hrynick, T., and Niederberger, E. (2021). *Community Resilience: Key Concepts and Their Applications to Epidemic Shocks*. https://doi.org/10.19088/sshap.2021.003

Setiawati, Y., Wahyuhadi, J., Joestandari, F., Maramis, M. M., and Atika, A. (2021). Anxiety and resilience of healthcare workers during COVID-19 pandemic in Indonesia. *Journal of Multidisciplinary Healthcare*, *14*. https://doi.org/10.2147/JMDH.S276655

Setyorini, A. (2023). Covid-19 Health Education for Compliance in Implementing Health Protocols. *Caring Indonesian Journal of Nursing Science*, *5*(1), 54–59. https://doi.org/10.32734/ijns.v5i1.12076

Syafiq, A., Fikawati, S., and Gemily, S. C. (2022). Household food security during the COVID-19 pandemic in urban and semi-urban areas in Indonesia. *Journal of Health, Population and Nutrition*, *41*(1). https://doi.org/10.1186/S41043-022-00285-Y

Weygandt, P. L., Jordan, J., Caretta-Weyer, H., Osborne, A., and Moore, K. G. (2021). Impact of the COVID-19 Pandemic on Emergency Medicine Education: Insights from Faculty and Residents. *Aem Education and Training*, *5*(3). https://doi.org/10.1002/aet2.10603

Wibowo, Y. I., Irawati, S., Setiawan, E., Halim, S. V., Sunderland, B., and Setiadi, A. P. (2022). Potential roles of community-based pharmacists in assisting home-quarantined patients with COVID-19 in a resource-challenged setting. *International Journal of Pharmacy Practice*, *30*(6). https://doi.org/10.1093/ijpp/riac053

Wijaya, S., and Barthos, M. (2022). *The Role of the Police in Law Enforcement During the Covid-19 Pandemic*. https://doi.org/10.4108/eai.30-10-2021.2315621

Towards Resilient Societies: The Synergy of Religion, Education, Health, Science, and Technology – Rahiem (Ed.)
© 2026 The Author(s), ISBN: 978-1-041-08471-6

Integrating Thurats in Brunei's curriculum: Strengthening Islamic education

Muhammad Thohri & Fahrurrozi Fahrurrozi
Universitas Islam Negeri Mataram, Indonesia

Lalu Fauzi Haryadi
Institut Agama Islam Hamzanwadi NW Lombok Timur, Indonesia

Ulyan Nasri
Institut Agama Islam Hamzanwadi NW Lombok Timur, Indonesia
ORCID ID: 0009-0003-5146-2440

ABSTRACT: The integration of Islamic intellectual heritage (*thurats*) into Brunei Darussalam's curriculum has become a key strategy in strengthening Islamic education and identity. This study investigates the processes and impacts of incorporating *thurat* into the curriculum, highlighting its role in fostering a comprehensive understanding of Islamic values and history. Using a qualitative case study design, data were collected through in-depth interviews, observations, and curriculum document analysis. Thematic analysis was applied to identify key themes in the integration process. The findings reveal that *thurats* significantly enrich the curriculum by deepening students' Islamic identity and fostering critical engagement with Islamic intellectual traditions. This integration supports Sustainable Development Goal (SDG) 4 by enhancing the quality of education through the preservation and modernisation of cultural heritage and SDG 16 by promoting inclusive, peaceful, and ethically grounded educational practices. The study concludes that embedding *thurats* into the curriculum is crucial for maintaining the relevance of Islamic education in addressing contemporary challenges. Strengthening these efforts is essential to ensure a resilient and culturally rooted Islamic education system in Brunei Darussalam.

Keywords: Thurats, curriculum integration, Islamic education, Brunei Darussalam, sustainable development goals

1 INTRODUCTION

Integrating Islamic intellectual heritage, known as *Thurats*, into educational curricula is essential in preserving and promoting Islamic identity in today's rapidly evolving societies (Abu-'Uksa 2015; 'Aziizah and Rofiq 2023; Seitakhmetova *et al.* 2022; Zuhri *et al.* 2024). Brunei Darussalam, a nation firmly rooted in Islamic traditions, offers a compelling case study examining how *Thurats* is embedded within its educational framework (Gleason and Manca 2019). Despite the profound historical and intellectual contributions that *Thurats* represents, there is a notable gap in understanding how these elements are systematically incorporated into modern educational practices and their impact on students' holistic development (Alwaqaa 2024; Hidayat 2018; Ruswandi *et al.* 2022; Viersen 2024). This research seeks to fill that void by exploring the strategies employed in integrating *Thurats* into Brunei's curriculum and assessing its influence on students' moral, intellectual, and social growth.

 DOI: 10.1201/9781003645542-72

Brunei's approach to integrating *Thurats* highlights its dual commitment to preserving Islamic heritage while meeting the demands of a globalised educational landscape. By aligning *Thurats* integration with universal educational goals—such as fostering critical thinking, ethical awareness, and cultural preservation—Brunei offers a model that harmonises Islamic values with contemporary pedagogical practices. This balance positions the nation as a significant reference point for other Muslim-majority countries striving to uphold their religious heritage amidst modernisation. Furthermore, this study examines the depth of *Thurats* integration in Brunei and aims to provide practical insights and recommendations for policymakers and educators, contributing to a global discourse on the sustainable development of Islamic education.

2 METHODS

This study employs a qualitative approach with a case study design to explore the integration of *thurats* into Brunei Darussalam's educational curriculum, offering a holistic understanding of its implementation within the cultural, religious, and educational context. Participants selected through purposive sampling include educators, curriculum developers, policymakers, and students, ensuring diverse perspectives and comprehensive insights (Cissé and Rasmussen 2022; Elizabeth *et al.* 2023; Mainwaring and Aujla 2023; Mey 2023; Tshabangu *et al.* 2022). Data collection involves semi-structured interviews, classroom observations, and analysis of curriculum materials, focusing on teaching methods, student engagement, and the formal integration of *thurats* (Barrett and Twycross 2018; Dursun 2023; Davison *et al.* 2023; Gill *et al.* 2008; Li 2024). Thematic analysis is used to identify patterns and themes interpreted within existing literature and theoretical frameworks. Reliability and validity are enhanced through data triangulation, member checking, and detailed contextual descriptions, while ethical considerations such as informed consent and confidentiality are upheld (Lerigo-Sampson 2022; Mihas 2023). By adopting this rigorous methodology, the study provides a nuanced understanding of *thurats* integration and its broader implications for Islamic education.

3 FINDING AND DISCUSSION

3.1 *Findings*

3.1.1 *Integration of Thurats into the curriculum*

Thurats is integrated into Brunei Darussalam's educational curriculum systematically across different stages of education. At the primary level, simplified narratives and stories of Islamic history introduce students to key figures and events, laying the foundation for understanding Islamic values. In secondary education, students engage more deeply with the intellectual contributions of Islamic scholars and the evolution of Islamic civilisation through primary sources and classical texts. Tertiary education offers comprehensive exploration through theology, philosophy, jurisprudence, and sciences courses. Interactive teaching methods, digital resources, and specialised teacher training further enhance the delivery and understanding of *Thurats* (Abdelrahman *et al.* 2006; Arini *et al.* 2021; Drake and Reid 2022; Fraser and Paraha 2017).

3.1.2 *Impact on students*

The inclusion of *Thurats* strengthens students' Islamic identity and fosters pride in their heritage by deepening their understanding of Islamic values and connecting them with their cultural and religious roots. It also enhances critical thinking as students engage with complex texts and historical contexts, developing analytical skills and intellectual growth. Furthermore, students gain a greater appreciation for the contributions of Islamic civilisation, inspiring them to continue the legacy of intellectual and ethical excellence.

3.1.3 *Challenges in integration*

The integration of *Thurats* faces challenges in balancing traditional content with modern educational needs, requiring innovative approaches to ensure relevance without compromising authenticity. High-quality resources remain limited, highlighting the need for localised and engaging content. Continuous adaptation, professional development for educators, and incorporating feedback from stakeholders are essential to address these challenges and sustain the effective integration of *Thurats* (Arini *et al.* 2021; Hidayat 2018).

3.1.4 *Policy implications*

Strategic planning and inclusive collaboration among educators, curriculum developers, policymakers, and the community are critical for the successful integration of *Thurats.* Providing ongoing support for educators through professional development and accessible resources is necessary for effective teaching. Feedback mechanisms and continuous research are vital to refining the curriculum and ensuring its relevance and impact. Recommendations include curriculum updates, diversification of content, enhanced teacher training, and leveraging digital tools to support the integration of *Thurats* in the education system.

3.2 *Discussion*

This section examines the research findings, analysing how integrating *Thurats* into Brunei Darussalam's curriculum has shaped educational practices and influenced student outcomes. By incorporating Islamic intellectual heritage into the educational framework, the study highlights strengths and challenges while uncovering its broader pedagogical impacts on teaching methods and student learning experiences.

3.2.1 *Pedagogical techniques and impact*

One of the notable findings is the use of specific teaching techniques tailored to make *Thurats* more engaging and applicable to contemporary learners. Techniques such as storytelling, Socratic questioning, and case-based learning connect historical Islamic knowledge with modern-day issues. For example, educators often use stories from Islamic history to illustrate moral dilemmas and foster critical discussions among students. Socratic questioning encourages deeper analysis, challenging students to think critically about the relevance of *Thurats* principles in addressing current societal challenges. These methods not only enhance understanding but also promote critical thinking and moral reasoning among students (Hoyert and O'Dell 2019; Silva *et al.* 2024; Tymoshenko *et al.* 2024).

Moreover, integrating *Thurats* contributes to students' moral development by embedding values such as justice, compassion, and humility into the curriculum. Practical activities like role-playing and collaborative projects allow students to apply these values in real-life contexts, bridging the gap between theoretical knowledge and practical application. This holistic approach to teaching ensures that students are academically equipped and morally grounded, preparing them for future leadership roles.

3.2.2 *Challenges in integration*

Despite the evident benefits, integrating *Thurats* into the curriculum presents several challenges. One significant issue is balancing traditional *Thurats* content with the requirements of modern education. For instance, aligning classical Islamic texts with contemporary pedagogical practices can be challenging, especially when catering to diverse student needs (Ansari *et al.* 2024; McKeown 2023). Another hurdle is the limited availability of localised teaching resources that effectively convey the depth and relevance of *Thurats* within the Bruneian context. While efforts to digitise content have progressed, developing culturally resonant materials remains an area for improvement.

3.2.3 *Policy implications*

The research findings suggest that policy adjustments are essential to enhance the integration of *Thurats* into the educational framework. Regular updates to the curriculum are necessary to ensure its relevance in addressing contemporary challenges and integrating advancements in educational technology. Additionally, ongoing professional development programs for educators are crucial. These programs should focus on equipping teachers with innovative teaching strategies and fostering a deeper understanding of *Thurats*, enabling them to engage students better and facilitate meaningful learning experiences (Bray 1988; Van Der Velden and Allen 2011).

3.2.4 *Future research*

This study paves the way for future research to explore the long-term impacts of *Thurats* integration on students' personal and professional lives. Investigating how exposure to *Thurats* influences students' career choices, ethical decision-making, and leadership abilities could provide valuable insights. Furthermore, comparative studies with other Islamic countries could shed light on best practices and innovative approaches, enriching the global discourse on integrating Islamic intellectual heritage into education.

Through this comprehensive analysis, the study underscores the importance of integrating *Thurats* into educational curricula to cultivate well-rounded individuals with intellectual and moral competencies. This endeavour strengthens Islamic identity and prepares students to contribute positively to society in an increasingly globalised world.

4 CONCLUSION

Integrating Thurats into Brunei Darussalam's educational curriculum has significantly impacted students' understanding of Islamic history and values, strengthened their Islamic identity, and enhanced their critical thinking and moral reasoning skills. The curriculum fosters a deeper connection between classical Islamic knowledge and contemporary issues by incorporating specific pedagogical techniques such as storytelling, Socratic questioning, and practical activities like role-playing. Despite challenges such as balancing traditional content with modern educational needs and lacking localised resources, the overall impact remains positive. These findings underscore the importance of preserving and promoting Islamic intellectual heritage in education while ensuring its relevance and accessibility to modern learners. Continued efforts in curriculum development, professional teacher training, and resource enhancement are crucial for sustaining and expanding the positive outcomes of this integration.

4.1 *Novelty*

This research makes several novel contributions to the field of Islamic education. Firstly, it provides a detailed examination of how Thurats is integrated into the curriculum in the specific and unique context of Brunei Darussalam, addressing a relatively underexplored area in the academic literature. Secondly, the study employs a comprehensive qualitative methodology that includes in-depth interviews, classroom observations, and curriculum analysis, offering a nuanced understanding of the integration process. It also uncovers innovative pedagogical strategies—such as modern teaching techniques and digital resources—that enhance student engagement and learning outcomes. Additionally, the study explores the dual impact of Thurats on fostering critical thinking and moral development, bridging theoretical knowledge with practical application. These contributions address critical gaps in existing research and provide actionable insights and recommendations for improving the integration of Islamic intellectual heritage into educational systems worldwide.

ACKNOWLEDGMENTS

We would like to express our sincere gratitude to all individuals and institutions that supported this research. Our deepest thanks go to the educators, students, and policymakers in Brunei Darussalam who participated in the study and provided invaluable insights. We also thank our academic colleagues and mentors for their guidance and feedback throughout the research process. Special thanks to Universitas Islam Negeri Mataram for their financial support and resources, which were crucial for the successful completion of this study. Your contributions have significantly enriched our work and made this research possible.

REFERENCES

Abdelrahman, M. A., Elsawy, A., and Canfield, S. (2006). Design of TTUbot: A Modular Learning Platform for Integration into Engineering Curriculum. *2006 Proceeding of the Thrity-Eighth Southeastern Symposium on System Theory*, 235–239. https://doi.org/10.1109/SSST.2006.1619079

Abu-'Uksa, W. (2015). Liberal Renewal of the Turath: Constructing the Egyptian Past in Sayyid al-Qimni's Works. In M. Hatina and C. Schumann (Eds.), *Arab Liberal Thought after 1967* (pp. 101–117). Palgrave Macmillan US. https://doi.org/10.1057/9781137551412_6

Alwaqaa, M. A. M. (2024). Adept adaptation and intelligent employment of ancient Arabic heritage: A reading in the poetry of Muhammad al-Thubayti, a modern Saudi poet. *Digest of Middle East Studies*, 33 (4), 476–496. https://doi.org/10.1111/dome.12342

Ansari, M., Waris, S., and Zara, C. (2024). Barriers to Educational Technology Adoption: Navigating Challenges in Integration. *Qlantic Journal of Social Sciences*, 5(3), 240–247. https://doi.org/10.55737/qjss.123732538

Arini, S. D., Mudjito, M., and Hariyati, N. (2021). Curriculum Integration: Optimizing Multiple Intelligence of Children. *JMKSP (Jurnal Manajemen, Kepemimpinan, Dan Supervisi Pendidikan)*, 6(2), 259. https://doi.org/10.31851/jmksp.v6i2.5084

'Aziizah,' A. U., and Rofiq, M. (2023). The Amalgamation of Traditional and Contemporary Elements in Islamic Education. *Edukasia: Jurnal Penelitian Pendidikan Islam*, 18(1), 165. https://doi.org/10.21043/edukasia.v18i1.23264

Barrett, D., and Twycross, A. (2018). Data collection in qualitative research. *Evidence Based Nursing*, 21(3), 63–64. https://doi.org/10.1136/eb-2018-102939

Bray, M. (1988). Policy Implications and Conclusions. In *Community Financing of Education* (pp. 203–218). Elsevier. https://doi.org/10.1016/B978-0-08-035858-1.50027-7

Cissé, A., and Rasmussen, A. (2022). Qualitative Methods. In *Comprehensive Clinical Psychology* (pp. 91–103). Elsevier. https://doi.org/10.1016/B978-0-12-818697-8.00216-8

Drake, S. M., and Reid, J. L. (2022). Integrated Curriculum. In *Integrated Curriculum. Routledge*. https://doi.org/10.4324/9781138609877-REE135-1

Dursun, B. (2023). A qualitative research technique: Interview. *Disiplinlerarası Eğitim Araştırmaları Dergisi*, 7 (14), 100–113. https://doi.org/10.57135/jier.1245193

Elizabeth M. Pope, Catarina Brandão, Kyle Bower, and Sonia Verdugo-Castro. (2023). Qualitative Research for Interdisciplinary Studies: Multiple Methodologies for Multiple Disciplines. *New Trends in Qualitative Research*, 16, e825. https://doi.org/10.36367/ntqr.16.2023.e825

Fraser, D., and Paraha, H. (2017). Curriculum integration as treaty praxis. *Waikato Journal of Education*, 8 (1). https://doi.org/10.15663/wje.v8i1.444

Gill, P., Stewart, K., Treasure, E., and Chadwick, B. (2008). Methods of data collection in qualitative research: Interviews and focus groups. *British Dental Journal*, 204(6), 291–295. https://doi.org/10.1038/bdj.2008.192

Gleason, B., and Manca, S. (2019). Curriculum and instruction: Pedagogical approaches to teaching and learning with Twitter in higher education. *On the Horizon*, 28(1), 1–8. https://doi.org/10.1108/OTH-03-2019-0014

Hidayat, S. (2018). Integrating Indonesian Muslim Intellectual Discourse into the Islamic World. *Journal of Indonesian Social Sciences and Humanities*, 6(1), 81–84. https://doi.org/10.14203/jissh.v6i1.61

Hoyert, M., and O'Dell, C. (2019). Developing Faculty Communities of Practice to Expand the Use of Effective Pedagogical Techniques. *Journal of the Scholarship of Teaching and Learning*, 19(1). https://doi.org/10.14434/josotl.v19i1.26775

Lerigo-Sampson, M. (2022). Qualitative data analysis. In S. Childe and A. Soares (Eds.), *Handbook of Research Methods for Supply Chain Management* (pp. 187–201). *Edward Elgar Publishing*. https://doi.org/10.4337/9781788975865.00016

Li, Y. (2024). The Methodology and Methods of the Research. In Y. Li, *Internal Migration and Health in China* (pp. 47–59). Springer Nature Singapore. https://doi.org/10.1007/978-981-99-8624-8_3

M. Davison, R., Tan, B., H.M. Wong, L., and Ng, E. (2023). Teaching qualitative research methods. In R. M. Davison (Ed.), *Handbook of Qualitative Research Methods for Information Systems* (pp. 10–29). Edward Elgar Publishing. https://doi.org/10.4337/9781802205398.00007

Mainwaring, L., and Aujla, I. (2023). Qualitative Methodologies. In L. Mainwaring and I. Aujla, *Research Methods in the Dance Sciences* (pp. 180–196). University Press of Florida. https://doi.org/10.5744/florida/9780813069548.003.0013

McKeown, A. (2023). Ethical challenges and principles in integrated care. *British Medical Bulletin*, 146(1), 4–18. https://doi.org/10.1093/bmb/ldac030

Mey, G. (2023). Qualitative Methodology. In J. Zumbach, D. A. Bernstein, S. Narciss, and G. Marsico (Eds.), *International Handbook of Psychology Learning and Teaching* (pp. 453–478). Springer International Publishing. https://doi.org/10.1007/978-3-030-28745-0_22

Mihas, P. (2023). Qualitative research methods: Approaches to qualitative data analysis. In *International Encyclopedia of Education*(Fourth Edition) (pp. 302–313). Elsevier. https://doi.org/10.1016/B978-0-12-818630-5.11029-2

Ruswandi, A., Natsir, N. F., and Haryanti, E. (2022). The Concept of Integration of Religion and Science in The Context of Islamic Education. *Ta Dib Jurnal Pendidikan Islam*, 11(2), 181–192. https://doi.org/10.29313/tjpi.v11i1.9315

Seitakhmetova, N., Kaldybekov, N., Bektenova, M., Toktarbekova, L., and Turganbayeva, Z. (2022). The Role of Intellectual Islamic Heritage in the Discourse of Modern Islamic Philosophy. *WISDOM*, 24(4), 172–183. https://doi.org/10.24234/wisdom.v24i4.803

Silva, V. G. K. D., Pícoli, A., Frauches, E. M. F., Barcelo, K. B., Santos, S. M. A. V., and Secchin, V. D. (2024). A Pedagogia E O Caminho Tecnológico: Perspectivas E Impactos. *Revista Ibero-Americana de Humanidades, Ciências e Educação*, 10(10), 5868–5881. https://doi.org/10.51891/rease.v10i10.16563

Tshabangu, I., Ba', S., and Madondo, S. M. (2022). Qualitative Methods in Research: Alternative Approaches and Navigating Complexities. In I. R. Management Association (Ed.), *Research Anthology on Innovative Research Methodologies and Utilization Across Multiple Disciplines* (pp. 587–603). IGI Global. https://doi.org/10.4018/978-1-6684-3881-7.ch030

Tymoshenko, N., Velyka, A., Yatsiv, O., Rovnyi, V., and Tovtyn, Y. (2024). Evaluation of pedagogical approaches, instructional techniques, and their influence on student progress and growth. *Multidisciplinary Reviews*, 7, 2024spe010. https://doi.org/10.31893/multirev.2024spe010

Van Der Velden, R., and Allen, J. (2011). Conclusions and Policy Implications. In J. Allen and R. Van Der Velden (Eds.), *The Flexible Professional in the Knowledge Society* (Vol. 35, pp. 241–253). Springer Netherlands. https://doi.org/10.1007/978-94-007-1353-6_9

Viersen, H. (2024). *The Time of Turāth: Authenticity and Temporality in Contemporary Arab Thought*. De Gruyter. https://doi.org/10.1515/9783110984286

Zuhri, S., Ningsih, T., Kurniawan, H., and Nurbaiti, A. (2024). Moral and Intellectual Integration in Islamic College. *International Journal of Social Science and Human Research*, 7(07). https://doi.org/10.47191/ijsshr/v7-i07-99

Towards Resilient Societies: The Synergy of Religion, Education, Health, Science, and Technology – Rahiem (Ed.)

Developing character education values in the *Perang Topat* tradition of Kemaliq Lingsar Lombok, Indonesia

Muhammad Zainul Majdi
Institut Agama Islam Hamzanwadi (IAIH) NWDI, Pancor, Indonesia

Abdul Fattah & Murdianto
Universitas Islam Negeri Mataram, Indonesia

Heri Hadi Saputra
Universitas Mataram, Indonesia

ABSTRACT: This article describes the phenomenon of social cohesion in the Perang Topat procession at Kemaliq Lingsar, Lombok, as well as the development of character education values in this tradition. This ex post facto study applies a descriptive qualitative approach, using observation, documentation, and interviews for data collection. Perspectives from Talcott Parsons' structural-functional model, Thomas Lickona's theory of character education, and the Indonesian Ministry of Education and Culture's model are employed. The results highlight three main findings. First, the Perang Topat was carried out in Kemaliq Lingsar-West Lombok by Sasak Muslims on the Sasak calendar during Kembangwaru rarak, in three stages: (1) preparation, (2) preliminary ceremony, and (3) the peak of the Perang Topat ceremony. This commemoration honours the Islamic spreader, Kiyai Haji Abdul Malik. Second, the values of character education embedded in the tradition include (1) religious, (2) tolerance, (3) social concern, and (4) nationalism.

Keywords: social cohesion, character education value, perang topat, Kemaliq Lingsar, Sasak Muslims

1 INTRODUCTION

Culturally, Indonesia is a heterogeneous country with various ethnicities, languages, and religions as its treasures. Ideally, differences should be a colour for harmony instead of being the reason for disputes and hostilities that impact the growth and development of this nation's civilisation and culture compared to others (Nata 2016).

On the other hand, Indonesian people are not superior in character education compared to other nations. This implies that in terms of character education, Indonesia's position is still at an alarming level compared to other countries. The decline in character education is caused by various factors, one of which is education that has forgotten its main purpose; the benchmark is determined by academic grades or abilities (IQ) and is only prepared to enter college. This is supported by lessons more oriented towards mathematical and abstract logic (Nata 2016). Meanwhile, little attention is paid to students‘ affective and motor dimensions that create the generation of characters.

Supratiningrum and Agustin (2015) stated that character education is necessary to create a moral generation because moral degradation almost destroys Indonesian society. Student fights, gangs among students, drug abuse, sexual harassment, promiscuity, and other crimes are increasingly spreading upsetting parents. Various cases are rampant among youth due to the lack of instilling moral character, such as the case of intimidation that led to death at SDN Longkewang, Cicantayan District, Sukabumi Regency, West Java, as reported by Harian Merdeka.com. The above case is one of the many cases that befall young people due to moral character disorders.

DOI: 10.1201/9781003645542-73

The previous description revealed that character education in Indonesia is no longer empowered to create students with good character. For this reason, in addition to the role of family and school, it is also necessary to revitalise the role of the community. This can be realised through a religious, cultured, and civilised environment (cooperation of all agencies and community institutions is needed) (Nata 2016). Indonesia, which is famous for its rich culture and traditions, can be used as the primary learning medium for supporting characters. As done by the Sasak people on the island of Lombok, to instil religious characters, honesty, tolerance and social concern they use the culture of *Perang Topat* followed by two religions (Islam and Hinduism). History proves that most wars are not only triggered by political interests or ideological differences but also because of differences in religious beliefs. Uniquely, in this tradition, two religions (Islam and Muslims) that have different beliefs, history, holy books, and adherents united in the *Perang Topat* which was carried out in Kemaliq Lingsar, West Lombok Regency (West Lombok Voice News Rises 2015).

The Perang Topat is an annual ceremony held at Lingsar Temple. It is a symbol of peace between Muslims and Hindus in Lingsar. The ceremony is usually carried out in the afternoon, when the Waru *Flower* falls or the Waru Tree falls. Envy and greed are evil desires that must be destroyed in the *ceremony of the Perang Topat*. This belief has been preserved for many years, starting in 1759 AD until today (Suparman Taufik 2019). This can be used as a learning medium that supports character education. Therefore, it is important to know the formation of character in tradition (Mustari 2011). For this reason, this research is focused on *the Perang Topat.* There are three focuses of research: (1) the *Perang Topat* procession in Kemaliq Lingsar, West Lombok, (2) the values embedded in each procession, and (3) the process of character formation in *the Perang Topat*.

2 METHODS

The approach used in this study is a descriptive qualitative approach with a sampling method, namely purposive sampling (Sugiyono 2012). This study uses an *ex post facto* paradigm using the perspective of Talcott Parsons' structural functional theory. In short, *ex post facto research* is a systematic empirical discovery (Sukardi 2012). This research is expected to comprehensively find the process, values, and character formation in the tradition of *the Perang Topat* in Kemaliq Lingsar, Lombok Barat. List of research informants regarding the tradition of the *Perang Topat* shown in the Table 1.

Table 1. List of names of research informants regarding the *Perang Topat* in Kemaliq Lingsar.

No	Number	Position	Place	Date
1	Suparman Taufik	Sasak Traditional Leaders	Lingsar	August 20, 2019 August 21, 2019 15 September 2019 20 September 2019 23 September 2019 25 September, 2019
2	Gede Rai	*Mangku* Hindu	Lingsar	21 September, 2019
3	Gede Reme	*Mangku* Hindu	Lingsar	22 September 2019 24 September, 2019

3 RESULT AND DISCUSSION

Suparman Taufik as the *Pemangku* (traditional leader/imam) of Sasak in Kemaliq Lingsar outlined the stages of *the Perang Topat* ritual as follows (Suparman Taufik 2019):

First, before the implementation of the *Perang Topat*, the community carried out *gotong royong* (mutual assistance) to clean the place and equipment needed during the event from the beginning to the end, including, *Momot* and *Gedah* (a type of bottle, glass, or jar). The container consists of a brass tray, a tray, and *a tabag*. Kemaliq decoration consists of *lelingsir, lelangsa, lamak, leluhur*, and *bukus teken*. Cups, plates, and lastly, large umbrellas, and spears were also cleaned. Everything is stored in a place called "*storage bale*" or (storage room) located north of *Pemangku*'s house.in addition to the people working together to clean their tools, they also clean the temple with the division of duties, Gaduh Temple is cleaned by Hindus, while close relatives of Pemangku Sasak clean the Kemaliq Temple area.

The next step is the installation of the *harness. A harness* or *layer lelamak* is a cushion as an offering base. *Lelamak* consists of three layers of fabric, namely (1) white, (2) yellow, and (3) brown, yellow which are installed on the *Kemaliq background. The installation* is carried out after the installation *of the harness. Penaek Gawe* is the beginning of the *Perang Topat ritual* which is carried out the day before the *Perang Topat ritual*. There are a number of activities carried out in *penaek gawe*, including making *Kebon Odek* (a small garden that symbolises the earth), *lekes* (*kinangan* and cigarettes), and *sekar* (flowers are placed on trays, various colours of flowers such as Mussaenda, ylang-ylang, marigold, etc.) (Suparman Taufik 2019).

Second, the preliminary/*mendak* (parade) ceremony from Lingsar Temple to the east to Mount Rinjani, and west to Mount Agung which is carried out at 4-5 pm. The order of the parade is the *batek baris* dance, followed by those who carry offerings, flanked by umbrellas, spears, and banner bearers. The last line is music art. Then *the nam butang* (welcome ceremony) was held at the T-junction to Lingsar Temple. After *climbing*, it is *climbing Kebon* (choose *Kebon Ode*) from *bale* to Kemaliq. Before entering Kemaliq, going around Kemaliq three times (*ngilahan*) is done. After *the infatuation, Kebon Ode'* was taken to Kemaliq Temple and placed in the courtyard. In closing, all participants held a joint prayer led by the Pemangku.

Third, the peak of the *Perang Topat* event. Starting from the morning they carry out various activities such as *Nampah Kaoq* (slaughtering buffaloes), *Miaq Presenter* (making food as many as nine *trays*), *Nyerahang Tupat* (handing over the tupat), *Mendak Sesaji* (taking food that has been prepared in the *storage bale* at 2.00 p.m. before *raraq kembang waru*), *Serving, Ngatur Serving* (Meal Placement Arrangements/*the presenter* led by the pengangku), *the Perang Topat* (after the Hindus on the upper terrace have finished performing pujawali worship and those on the upper terrace are ready and so are those on the lower terrace (sasak people) who are waiting for the ketupat to be thrown from Kemaliq. The war will begin after hearing the sound of *kul-kul (kentongan*/drums made of bamboo or wood) which continues to be sounded for approximately an hour, even though there is throwing at each other, but no one objects but feels that the form of kinship and brotherhood is proven by those who laugh and ask for another ketupat to be thrown. After completing the *Perang Topat*, the community continues to carry out the *Lalang* ritual (a period of rest to prepare one more ceremony filled at night with various kinds of entertainment such as dance, and other entertainment). After *Lalang*, it was continued with the *Beteteh* ceremony. The *Beteth ceremony* is a period where Hindus carry out the *Ngelukar* ceremony, which is the closing ceremony of the entire series of *Pujawali* ceremonies at Lingsar Temple and the Sasak people who carry out *the beteteh* ceremony hold a joint prayer ceremony in Kemaliq, led by *the Regent* of Kemaliq. With the Beteteh ceremony's end, the *Perang Topat* events in Kemaliq Lingsar ended.

Based on the analysis conducted by the researcher on the results of observations, interviews, and documentation at the research location of the origin and purpose of the implementation *of the Perang Topat* tradition, it can be explained that the implementation of the tradition contains moral values/characters. Anis Matta (2006) explained that broadly speaking, two factors affect a person's character: internal and external factors. Internal factors are all the elements of personality that continue to influence human behaviour, which includes biological instincts, psychological needs, and thinking needs. While external factors are factors that come from outside humans but can affect human behaviour, either directly or indirectly.

If using the Anis Matta classification above, the character values contained in *Perang Topat* traditions, all of which are external factors for the cultivation and strengthening of the character of the participants, especially for the younger generation. Furthermore, efforts to strengthen the character triggered by the Government of the Republic of Indonesia are realised by developing 18 national cultural characters. These characters are (1) religious, (2) honest, (3) tolerance, (4) discipline, (5) hard work, (6) creative, (7) independent, (8) democratic, (9) curiosity, (10) national spirit, (11) patriotism/nationalism, (12) respect for achievements, (13) friendly or communicative, (14) peace, (15) love of reading, (16) environmental care, (17) social concern, and (18) responsibility (Ministry of Education and Culture) Indonesia 2017).

Character does not appear just because it is not something that is innate in nature. Character is formed throughout life, especially during childhood and adolescence (Berkowitz and Bier 2005). According to Thomas Lickona (2015). All of these environments (family, school and community environments) must support each other and work together in shaping character to avoid confusion in children and adolescents about character behaviour or activities. For Lickona (2015), the deliberate effort of all dimensions of social life is to help the formation of character optimally. Thus, according to Thomas Lickona (2015), children will be able to respect themselves, respect others, and appreciate the environment and life more broadly. After analysis, there are 4-character values that can be seen from the implementation of the *Perang Topat* in Kemaliq Lingsar, Narmada, West Lombok. The description of the 4-character value can be explained as follows:

First, religious values are obedient attitudes and behaviours in carrying out the religious teachings they hold, being tolerant of the implementation of other religions, and living in harmony with adherents of other religions. Religious values in implementing *the Perang Topat* tradition in Kemaliq Lingsar can be seen starting from the purpose of this tradition being held. When viewed from an Islamic perspective, this event was held as a form of gratitude to Allah SWT for bringing a saint who taught Islamic teachings, as well as to commemorate the spread of Islam in the region, namely K.H. Abdul Malik (Suparman Taufik 2019). *The Perang Topat* is the same as the ritual of throwing hajj in the hajj ritual. Throwing pilgrimages as a symbol to fight the devil that exists in humans who always invite mistakes or do damage to the earth's surface. This devil is always present in humans which is a test given by God so that he is vigilant and subdued. Of course, the religious values possessed by every citizen make this celebration even more interesting where every believer applies what they believe. As stated by Fahrurrozi (Indonesia and The Malay World 2018).

Second, the value of tolerance. The value of tolerance is attitudes and actions that respect the differences in religion, ethnicity, ethnicity, opinions, attitudes, and actions of others who are different from them (Centre for Curriculum Research and Development 2010). In this tradition, it is the tolerance of diverse communities that is most visible, which is reflected in the attitudes and behaviors of both groups (Muslims and Hindus). This is reflected in the preparation process, the two groups met, conveying their respective opinions. This attitude reflects tolerance caused by the belief that everyone has their own rights.

Third, the values of social care. The value of social concern is the character of attitude and action that always wants to assist others and the community in need. One of the values of social concern in this tradition is reflected in the distribution of buffalo meat. This buffalo meat is divided by the Sasak and Hindus according to their capacity. No one feels the most entitled to take part more than both sides. In addition, this value is also seen in the viewing program, which is an event to throw Kebon Ode' and its derivatives in the river. The goal is for all Lombok people to participate in the blessing *of the Perang Topat* Ceremony.

Fourth, the value of nationalism. The value of the spirit of nationalism is a way of thinking, acting and having insight that puts the interests of the nation and state above the interests of the self and the interests of the group (Central Research and Development Agency for Curriculum 2010), and the state is above the interests of individuals or individuals and groups. In this case, the spirit of nationalism is urgently needed by citizens, because Indonesia is a country consisting of thousands of islands and inhabited by hundreds of races, ethnicities, and languages. The Indonesian people have been built on a history of diversity in equality (Nudji 2015).

The value of the spirit of nationality in the *Perang Topat* is reflected in the function of this tradition, namely as a social media to instil values and develop the spirit of togetherness and solidarity between believers with slogans such as: 'different but not separate', 'different but united', 'beliefs may be different, but togetherness must be maintained for the sake of survival in the future' are the main moral messages that are always echoed in this event (Gede Reme 2019).

All character values contained in *the Perang Topat* as mentioned above can be the initial foundation in character formation, especially for students/young people. The character formation in the *Perang Topat* tradition in *Kemaliq* Lingsar is carried out through several stages 1). Introducing the *tradition of Perang Topat* to the younger generation, 2). Explaining the meaning of symbols in the tradition *of Perang Topat* to the younger generation, 3). Conducting the *tradition of Perang Topat* routinely every year, 4). Maintaining, preserving, and inheriting traditions to the younger generation (Mulyana 2004).

4 CONCLUSION

The tradition of *the Perang Topat* is one of Lombok's most valuable cultural assets. In the tradition held every year, in addition to implying the size symbol of the Sasak Muslims and Balinese Hindus, it also implies very interesting character values. Furthermore, Talcott Parsons' structural-functional model, Thomas Lickona's theory of character education, and the Indonesian Ministry of Education and Culture model were used to express the character values contained in the *Perang Topat* tradition. There are 4-character values in "war activities" that can be used as lessons, especially in forming the character of all parties/social classes, both upper and lower classes, especially for the younger generation. Comprehensive character values, such as those in Thomas Lickona's perspective, include oneself, others, and the general environment or life. These character values use the perspective of the Ministry of Education and Culture of the Republic of Indonesia. Communicative/friendly, caring for the environment and caring for society. The overall value of the character refers to two values, namely, the value *of Ilahiyyah* (divinity) and the value *of Insaniyah* (humanity); in principle, everything is based on religious values. *The Perang Topat* is one of the domains of informal education that can be a reference for formal education practitioners to develop and shape students' character. According to the researchers, the character formation and development strategy in this tradition will be very helpful if applied in formal education such as schools and universities, especially Islamic-based ones.

REFERENCES

Berkowitz, M. W., and Bier, M. C. (2005). *What works in character education: A research-based guide for educators.* University of Missouri.

Lickona, T. (2015). *Character matters: Character issues, how to help children develop good appraisal, integrity, and other important virtues.* PT Bumi Aksara.

Matta, M. A. (2006). *Shaping the character of the Islamic way.* Al-I'tishom Cahaya Umat.

Ministry of Education and Culture. (2017). *Konsep dan pedoman penguatan pendidikan karakter tingkat sekolah dasar dan sekolah menengah [Concept and guidelines for strengthening character education at elementary and middle school level].* Ministry of Education and Culture Republic of Indonesia.

Mulyana, R. (2004). *Articulating value education.* Alfabeta.

Mustari, M. (2011). *Character value: Reflections on character education.* Laksbang Pressindo.

Nata, A. (2016). *Islamic education innovation.* Salemba Diniyah.

Nudji. (2015). Efforts to increase the sense of nationalism of high school students through Pancasila education and citizenship (PPKN). *International Academic Research*, 6(1), 405.

Sugiyono. (2012). *Understanding qualitative research.* Alfabeta.

Sukardi. (2012). *Educational research methods.* Bumi Aksara.

Supratiningrum, S., and Agustin, A. (2015). Building student character through school culture in primary schools. *Journal of Character Education*, 5(2), 215.

West Lombok Voice News Rises. (2015, November). Perspectives of the Topat fight, a symbol of Muslim-Hindu peace [News article]. (pp. 3–5).

Session 3: Transforming economies and legal frameworks for a sustainable future

Towards Resilient Societies: The Synergy of Religion, Education, Health, Science, and Technology – Rahiem (Ed.)

Sharia and digital finance: Pathways to Indonesia's financial inclusion and resilience

Firman El Amny Azra
Universitas Islam Negeri Syarif Hidayatullah Jakarta, Indonesia
ORCID ID: 0000-0003-3333-4739

Murdiyah Hayati & Indo Yama Nasarudin
Universitas Islam Negeri Syarif Hidayatullah Jakarta, Indonesia

ABSTRACT: The influence of faith on financial behaviours remains a significant issue amidst the rapid growth of Indonesia's digital finance sector. This research explores how Sharia principles are integrated into financial technology and how this integration fosters financial inclusion and resilience, contributing to Sustainable Development Goals (SDGs) 8 (Decent Work and Economic Growth) and 9 (Industry, Innovation, and Infrastructure). This study employs a multiple case study analysis of two faith-based fintech companies. These companies effectively merge Sharia principles with financial technology by embedding Islamic values into their core operations, marketing strategies, and services. This integration addresses barriers faced by underserved populations, particularly in rural and remote areas, while enhancing financial literacy and empowering users to make informed financial decisions aligned with Islamic teachings. Furthermore, their peer-to-peer lending models play a crucial role in supporting micro and small businesses, driving local economic development. This study underscores the potential of combining Sharia principles with financial technology to enhance financial inclusion and resilience in Indonesia. It recommends that policymakers and practitioners develop regulatory frameworks that encourage growth while strictly adhering to Sharia principles.

Keywords: digital finance, financial technology, sharia, sharia business, sustainable development goals

1 INTRODUCTION

Indonesia is cited as a country with the largest digital economy in the ASEAN region, with high digital transformation opportunities (Alfarizi *et al.* 2024). This digitalization wave reshapes the wider financial services sector where fintech motivates banks to work more efficiently, improving financial literacy and even improving small business productivity (Junarsin *et al.* 2023). Indonesia stands out as the country with the most Muslims in the world, with unique geographic conditions creating both problems and opportunities in making finances more inclusive (Iman 2018). Even though the economy has grown a lot, many Indonesians still don't use banks much or at all in country areas and far-off places. The merging of digital tech and finance has sparked a new crop of fintech firms. Many of these companies blend Sharia approach and digital finance to serve the country's Muslim people. This mix of faith-based principles, tech, and financial service business opens up a promising path to boost capital access and build strong communities throughout Indonesia (Hassan *et al.* 2020).

Existing literature has highlighted the importance of the integration of Sharia principles with technology. For example, a study by Usman *et al.* (2022) has shown that compliance with Sharia principles has a significant effect in shaping customer satisfaction. In another study, Muryanto

DOI: 10.1201/9781003645542-74

(2023) has argued that Sharia compliance serves as a basic element for Islamic Fintech growth with its role in driving trust and protecting all relevant stakeholders. However, the integration of Sharia principles into modern business remains a challenge with hard requirements of deep understanding of Islamic law combined with the application of innovation in products and ethical governance (Ramadhani and Budiarti 2024). This challenge presents a gap in existing literature where most scholars more focused only on compliance instead of governance and operations of the companies (Mujiatun *et al.* 2022; Muryanto 2023; Usman *et al.* 2022). Furthermore, the contextualizing of the religious aspect of Fintech development in the Islamic world remains an underexplored realm in the existing literature (Pandey *et al.* 2023).

This paper aims to evaluate the extent of integration between faith-based principles and financial technology business, as well as its role in financial inclusion and resilience in underserved communities in Indonesia, which is crucial for contributing to sustainable development goals (SDGs). In order to achieve that aim, this study focuses on answering the following questions: how does an Indonesian fintech company integrate Sharia principles with financial technology business? How to integrate Sharia principles with financial technology business to promote inclusion and community resilience? By examining innovative models of faith-based fintech, analysing their impact, and identifying best practices, this paper will contribute to the ongoing dialogue on building inclusive and sustainable financial systems.

As Indonesia keeps moving towards becoming one of the biggest digital economies in the world, it's become more important to understand how religion, education, and tech work together in finance. This study will also provide useful ideas for policy makers, fintech practitioners, and religious groups who are working to make finance more inclusive and resilient in the future.

2 METHODOLOGY

This research utilises a qualitative approach that uses multiple case studies to investigate how people combine their religious faith with digital financial technologies. In-depth analysis of complex phenomena can best be done using qualitative methodology with real-life context. This case study approach provides an opportunity for understanding events as they happen naturally and thereby giving profound explanations to the research question.

This research's multiple case study design involves two prominent Islamic fintech institutions in Indonesia: ALAMI and Danasyariah.id. The criteria for choosing these two cases were based on the assessment of company prominence, company information transparency, and a strong track record that proved through ALAMI achieving Excellence in the Islamic Fintech Category by Global Islamic Finance Award and Dana Syariah as Best Fintech Award by Investor Daily.

Thematic analysis will be used in data analysis to identify, analyse, and report patterns within the information given. Some of the activities involved in this process include getting familiar with the data through reading documents and observation notes several times; coming up with initial codes; looking for themes; reviewing and refining themes; defining themes as well as naming them;

After analysing individual cases, a cross-case analysis will be conducted to enhance understanding and identify similarities or differences. Each case will be analysed individually and then will be evaluated in order to find out their similarities, differences, and emerging patterns to acquire substantive, generalisable findings concerning the incorporation of religious principles in digital finance.

3 RESULTS AND DISCUSSION

Integration of sharia principles into an emerging business such as financial technology requires an in-depth understanding of Islamic law to ensure that all aspects of business comply with the sharia law (Ramadhani and Budiarti 2024). Based on the case study analysis, it is indicated that the integration of Sharia principles into the financial technology business is mainly reflected in the company's core values aspect, strategic aspect, marketing aspect, and operational aspect, which data structure summarised in Figure 1.

The first case study will focus on ALAMI, which was incorporated in 2019 and obtained a license from the Indonesia Financial Service Authority or Otoritas Jasa Keuangan (OJK) in 2020. OJK is one of the biggest Sharia fintech companies in Indonesia. ALAMI is a peer-to-peer lending financial technology which is a business model that provides a platform to bridge investors and borrowers through the internet. As of July 2024, ALAMI has disbursed loans in the amount of Rp5.46 trillion, as well as serving 11,478 active investors and 1671 beneficiaries since its foundation.

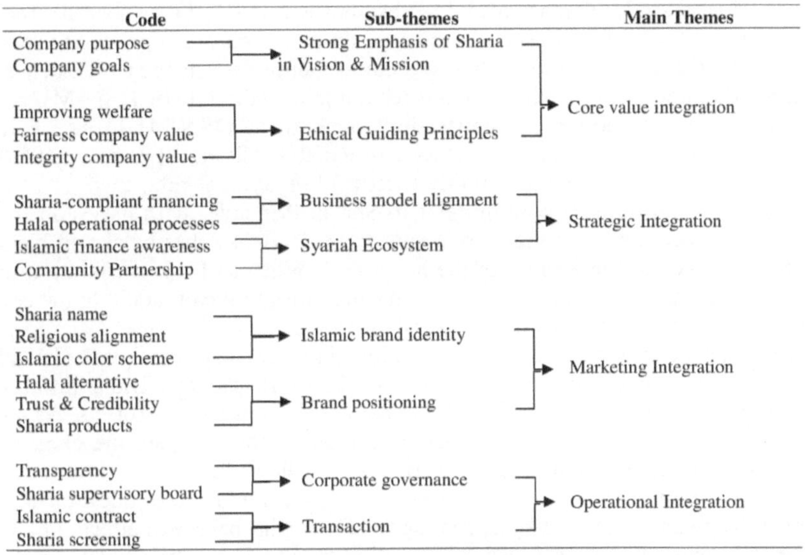

Figure 1. Thematic analysis results (source: researcher data 2024).

ALAMI was founded with a vision to "*realise a more prosperous society evenly, using technology based on sharia principles*". This vision statement reflects ALAMI's core purpose to make Sharia finance, which still lags behind conventional finance, more widely adopted by utilising digital technology. Additionally, ALAMI's stated mission is to "*build sharia ecosystem that empower Muslim society to realise life that adheres to sustainable Islamic values*". ALAMI's stated mission in this case indicates a core value of social responsibility and community development, which strongly aligns with the Islamic principle of mutual cooperation (ta'awun). Strong emphasis on sharia principles in both the vision and mission of ALAMI means that ALAMI integrates Islamic principles on a philosophical level and adopts them as important core values.

ALAMI integration with faith-based principles is also reflected on its branding strategy. First, the official legal name of ALAMI includes the word "Sharia" which is PT ALAMI Fintek Sharia. ALAMI brand name itself is an abbreviation of Alif-Lam-Mim, which is derived from the first verse of Surah Al-Baqarah, the longest Surah of the Quran. Additionally, ALAMI's branding strategy implements Islamic Branding by compliance, which is supported by its commitment to obtain a registration mark from the National Sharia Board/*Dewan Syariah National.*

It should be noted that a true sharia company should not only use the name 'Sharia' for marketing or gimmick purposes, but its operation should also comply with Sharia principles. In ALAMI case, strong integration of Sharia principles is shown by its compliance with Fatwa DSN MUI No. 117/DSN-MUI/IX/2018 (DSN MUI No.117/2018). DSN MUI No.117/2018 specifically sets out that the covenant that the parties use in the implementation of information technology-based financing shall be in the form of covenants that are appropriate with the financing service characteristic, such as *al-bai, ijarah, mudharabah, musyarakah, wakalah bil al ujrah,* and *qardh.* ALAMI, in this case shows good compliance based on observation of 20 projects at ALAMI based on appropriate covenant: 13 *musyarakah* projects, *six wakalah bil al ujrah* projects and one *mudharabah* project. Furthermore, the most important aspect of Sharia principles in financing is ensuring the project is free from usury, *gharar*, *maysir*, *tadlis*, *dharar*, *zhulm*, and haram. Upon examination of the project and covenant used in ALAMI, it is evident that ALAMI aligns with these Sharia principles. This finding supports (Alshater *et al.* 2022) the study that argues that lending prudence and Sharia compliance are important for Islamic fintech sustainability.

In the second case study, Danasyariah.id is established as PT. Dana Syariah Indonesia in 2017. Danasyariah.id is a peer-to-peer sharia-based financing platform provider which licensed and supervised by OJK. In addition, Danasyariah.id mentioned on their official website that they have a Sharia Supervisory Board which is recommended directly by DSN-MUI. However, Danasyariah.id does not possess a registration mark from DSN-MUI. As of July 2024, Danasyariah.id has disbursed financing in excess of Rp3,04 trillion, supported by 95.000 active funders, and claims to have a 99.85% 90-day successful repayment rate.

Danasyariah.id vision reflects commitment to Sharia Principles with its vision to "inviting people to carry out economic activities following Islamic law, so that halal and blessed sustenance can be obtained for the welfare of the hereafter". Whereas Danasyariah.id mission is to "to become a place and centre for sharia economic activities that can facilitate the community, to carry out economic activities according to Islamic law."

From the strategy perspective, Dana Syariah Indonesia is based on principles of Islamic finance that refrains from usury, uncertain activities, and speculative transactions. Danasyariah specialises as a company that provides financing based on *murabahah* in property projects. As most of the property financing market in Indonesia is based on interest, these models are expected to be a suitable option for financing homes or properties that follow Sharia law principles. Through organising its service offerings around these tenets, Danasyariah.id carves out a niche within the highly competitive financial market by appealing to clients who believe in ethical finance.

Danasyariah.id marketing aspect also reflects adherence to Sharia principles. First, the company highlights its compliance with Sharia law in their promotional materials. Additionally, Danasyariah.id includes branding elements that align with Islamic values such as Islamic public figure endorsement and halal certifications. Danasyariah.id marketing campaign notably focused on the religious value, ethical, and community benefit of their product. This approach reflects their focus on appealing to customers, which includes religious value in choosing financial products. By doing so, Dana Syariah Indonesia not only enhances its brand credibility but also builds trust within the Muslim community. This finding supports Nur'aini *et al.*, (2022) argument that MSME actors in Indonesia have a strong desire to improve halalpreneurship practices.

Operationally, Danasyariah.id integrates Sharia compliance into every aspect of its business processes. This includes the establishment of an SSB whose appointment is based on DSN MUI recommendation, who comprised of Islamic scholars who oversee all financial transactions and ensure they comply with Islamic law. Furthermore, Dana Syariah Indonesia incorporates ethical guidelines in its investment decisions, avoiding industries and activities prohibited in Islam, such as alcohol, gambling, and non-halal food production. This reflects the finding of Arif *et al.*, (2023) a study that highlights that by adhering to Sharia principles financial service company can improve its performance and market share in Indonesia.

ALAMI and Danasyariah.id have leveraged digital technology to extend financial services to previously underserved segments of the population. By operating online platforms, these companies have overcome geographical barriers that often limit access to traditional banking services, especially in rural and remote areas of Indonesia (Dwijayanti *et al.* 2022). This is reflected by ALAMI's significant contribution in financing the fisheries sector, which of December 2021, ALAMI has disbursed more than Rp109.5 billion to more than 682 fish farmers (Alami 2022).

Both companies integrate educational components into their platforms, helping improve their users' financial literacy. By offering Sharia-compliant products and explaining the principles behind them, ALAMI and Danasyariah.id provide financial services and educate users about ethical financial practices aligned with Islamic principles (Saifurrahman and Kassim 2021). For example, Danasyariah.id has organised training and workshops for students, small and medium enterprise owners. This educational strategy supports Menne *et al.*, (2023) argument that improving the financial literacy of customers is needed to further develop the sharia economy.

The peer-to-peer lending models of both ALAMI and Danasyariah.id are particularly well-suited for financing micro and small enterprises, which traditional banks often underserve (Yuneline 2022). By connecting these businesses with a pool of investors, these platforms help bridge the financing gap for small entrepreneurs, promoting economic growth and resilience at the grassroots level (Barata 2019). This aligns with the objectives of SDG 8, which aims to promote inclusive and sustainable economic growth, employment, and decent work for all. For instance, ALAMI provides financing to small-scale aquaculture producers, broiler breeders, and other small and medium enterprises.

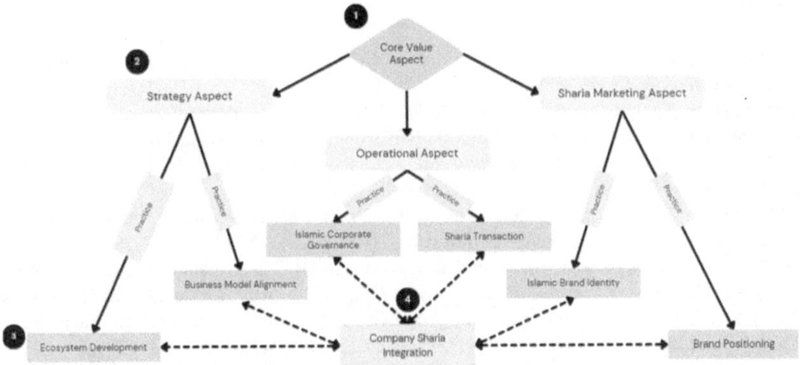

Figure 2. Sharia integration model (source: researcher data 2024).

ALAMI and Danasyariah.id case study reflects a strong integration of Sharia principles in their company which supports the company's growth and prominence. This is because the sharia principles are not only for marketing purposes but it is also practiced in operational aspects, strategy aspects, as well as adopted as a core value of the company, as shown in Figure 2. Notably, the company's contribution to the development of Sharia ecosystems as part of their strategy indicates the crucial role of Sharia financial inclusion in the expansion of an inclusive and equitable halal value chain (Qizam *et al.* 2024). This initiative ties in with the goals of SDG 9, which seeks to develop resilient infrastructure and foster inclusive and sustainable industrialisation and innovation.

4 CONCLUSION

Integrating Sharia principles with digital finance technology presents a promising avenue for promoting financial inclusion and building resilience in Indonesia's underserved communities. ALAMI and Danasyariah's case studies demonstrate how faith-based fintech companies successfully combine Islamic financial principles with modern technology to create innovative and ethical financial solutions.

ALAMI and Danasyariah.id have shown a strong commitment to integrating Sharia principles into their business models, from their vision and mission statements to their operational practices. These companies are leveraging technology to bridge the gap between traditional Islamic finance and the growing digital economy in Indonesia, catering to the country's large Muslim population. The integration of faith-based principles is evident in various aspects of these companies, including their branding strategies, product offerings, and compliance with Islamic financial regulations.

In conclusion, integrating Sharia principles with digital finance technology shows great potential for enhancing financial inclusion and resilience in Indonesia. As the country continues its trajectory towards becoming one of the world's largest digital economies, the synergy between religion, technology, and finance will likely play a crucial role in shaping an inclusive and ethical financial future. The study's implications for policymakers and practitioners reflect the need to encourage the merger of religious values and technology by building regulatory frameworks that promote growth while adhering strictly to Sharia principles. However, this study was limited to two cases of prominent companies in Indonesia; therefore, future research should be conducted in a wider scope, such as including multiple foreign companies in the study as a comparison.

REFERENCES

Alami. (2022). *Alami Impact Report 2021 Grow Sustainably* (Impact Report) [Impact Report]. https://alamisharia.co.id/impact-report/impact-report-2021/

Alfarizi, M., Widiastuti, T., and Ngatindriatun. (2024). Exploration of Technological Challenges and Public Economic Trends Phenomenon in the Sustainable Performance of Indonesian Digital MSMEs on Industrial Era 4.0. *Journal of Industrial Integration and Management*, *09*(01), 65–96. https://doi.org/10.1142/S2424862223500045

Alshater, M. M., Saba, I., Supriani, I., and Rabbani, M. R. (2022). Fintech in islamic finance literature: A review. *Heliyon*, *8*(9). https://www.cell.com/heliyon/fulltext/S2405-8440(22)01673-5

Arif, M. N. R. A., Ihsan, D. N., Zulpawati, and Fatah, D. A. (2023). The impact of conversion on market share in Indonesian Islamic banks. *Banks and Bank Systems*, *18*(2), 1–12. https://doi.org/10.21511/bbs.18(2).2023.01

Barata, A. (2019). Strengthening national economic growth and equitable income through sharia digital economy in Indonesia. *Journal of Islamic Monetary Economics and Finance*, *5*(1), 145–168.

Dwijayanti, N., Iqbal, M., and Zulfikar, M. (2022). The role of Islamic FinTech P2PL in increasing inclusion and financial literacy of MSMEs. *Journal of Islamic Finance*, *11*(1), 94–101.

Hassan, M. K., Rabbani, M. R., and Ali, M. A. M. (2020). Challenges for the Islamic Finance and banking in post COVID era and the role of Fintech. *Journal of Economic Cooperation and Development*, *41*(3), 93–116.

Iman, N. (2018). Is mobile payment still relevant in the fintech era? *Electronic Commerce Research and Applications*, *30*, 72–82.

Junarsin, E., Pelawi, R. Y., Kristanto, J., Marcelin, I., and Pelawi, J. B. (2023). Does fintech lending expansion disturb financial system stability? Evidence from Indonesia. *Heliyon*, *9*(9). https://www.cell.com/heliyon/fulltext/S2405-8440(23)05592-5

Menne, F., Mardjuni, S., Yusuf, M., Ruslan, M., Arifuddin, A., and Iskandar, I. (2023). Sharia Economy, Islamic Financial Performance and Factors That Influence It—Evidence from Indonesia. *Economies*, *11*(4), 111.

Mujiatun, S., Effendi, I., and Badawi, A. (2022). Sharia Financial Technology (Fintech) Management Model in Indonesia. *Cuadernos de Economía*, *45*(128), 145–156.

Muryanto, Y. T. (2023). The urgency of sharia compliance regulations for Islamic Fintechs: A comparative study of Indonesia, Malaysia and the United Kingdom. *Journal of Financial Crime*, *30*(5), 1264–1278.

Nur'aini, D., Kristiana, R., and Al, M. N. R. (2022). The Relationship Between Human Capital, Halalpreneurs' desire, and Halal Awareness: Empirical Evidence from Indonesia's Micro-Entrepreneurs. *Asian Economic and Financial Review*, *12*(7), 450–458.

Pandey, D. K., Hassan, M. K., Kumari, V., Zaied, Y. B., and Rai, V. K. (2023). Mapping the landscape of FinTech in banking and finance: A bibliometric review. *Research in International Business and Finance*, 102116.

Qizam, I., Berakon, I., and Ali, H. (2024). The role of halal value chain, Sharia financial inclusion, and digital economy in socio-economic transformation: A study of Islamic boarding schools in Indonesia. *Journal of Islamic Marketing*. https://www.emerald.com/insight/content/doi/10.1108/jima-03-2024-0108/full/html

Ramadhani, Y. C., and Budiarti, A. (2024). Innovative Strategy in Integrating Sharia Principles with Modern Business Practices. *International Journal of Sharia Business Management*, *3*(1), 9–14.

Saifurrahman, A., and Kassim, S. (2021). Islamic financial literacy for Indonesian MSMEs during COVID-19 Pandemic: Issues and importance. *Journal of Islamic Finance*, *10*, 045–060.

Usman, H., Projo, N. W. K., Chairy, C., and Haque, M. G. (2022). The exploration role of Sharia compliance in technology acceptance model for e-banking (case: Islamic bank in Indonesia). *Journal of Islamic Marketing*, *13*(5), 1089–1110.

Yuneline, M. H. (2022). Implications of Shariah Financial Technology in Increasing Financial Inclusion to Micro, Small, and Medium Enterprises. *Islamiyyat*, *44*(1), 89–99.

Towards Resilient Societies: The Synergy of Religion, Education, Health, Science, and Technology – Rahiem (Ed.)

Sustainable South Korea Tourism through BTS in the economic and social sector (2020-2022)

Agus Nilmada Azmi & Cut Syifa Azzahra Chairullah
Universitas Islam Negeri Syarif Hidayatullah Jakarta, Indonesia

ABSTRACT: This research explores South Korea's sustainable tourism strategies through nation branding, focusing on the role of the idol group BTS in the economic and social sectors from 2020 to 2022. Through collaboration between the government and BTS in tourism promotion campaigns, South Korea has successfully enhanced its national image, generating positive impacts across economic, cultural, and social dimensions. This qualitative study employs a descriptive approach and relies on secondary data obtained through library research, including official government and organisational websites, books, journals, official reports, and news sources. The findings reveal three key nation branding initiatives involving BTS during the 2020–2022 period: collaboration between the Korea Tourism Organization (KTO) and BTS in the 'Feel the Rhythm of Korea' campaign, BTS's role as tourism ambassadors for Seoul City, and their appointment as honorary ambassadors for the Busan Expo 2030. BTS significantly contributes to South Korea's economy as cultural representatives and global musical icons. This research offers valuable insights into sustainable tourism development through pop culture and suggests potential comparisons with similar strategies in other countries, such as Japan with Anime or India with Bollywood.

Keywords: BTS, nation branding, South Korea, sustainable tourism, tourism

1 INTRODUCTION

Today, the tourism industry has become an important element for a country. This sector plays an important role in improving the country's economy, as well as shaping the country's positive image in the international arena. Every country is competing to improve the quality of tourism for the long term, known as sustainable tourism. Sustainable tourism is defined as tourism that takes full account of current and future economic, social, and environmental impacts, meeting the needs of visitors, industry, the environment, and host communities. In ensuring long-term sustainability, it is necessary to refer to the development and balance of tourism in environmental, economic, and socio-cultural aspects (UNEP 2005).

Sustainable tourism practices promoted by a country can strengthen the country's nation branding in the eyes of the world. When a country is recognised as a tourist destination that is environmentally friendly, has social responsibility, and preserves local culture, the public perception of the country will be formed positively, and attract foreign tourists to visit these destinations (Yoeti 2013).

South Korea is one of the countries rich in tourism resources. The strong traditional culture combined with modern values, or traditional elements in modern locations, provides an interesting experience for tourists in tourism. Especially since the emergence of Korean popular culture (Hallyu), such as K-Pop, K-drama, and cosmetics have become new elements and characteristics in South Korean tourism destinations, which is called Hallyu Tourism (Bae *et al.* 2017).

One of the hallyu products that successfully occupied the world stage is the idol group BangtanSonyeondan (BTS). This group received various international awards, thus successfully bringing the good name of South Korea. It is even rumoured that this group is the most successful manifestation of the Hallyu phenomenon (Oh and Lee 2014). BTS became a famous group with various messages contained in its music that became a lot of concern among today's youth, especially about social issues. Starting from mental health issues, wrong social standards, political issues, to bullying. Not only that, but BTS also highlight their culture and daily life as South Koreans in their songs and other entertainment content. This is what makes fans feel connected to their music and lyrics and become interested in following the various aspects highlighted by the group (Jin 2021).

BTS has significantly impacted South Korea, bolstered by the unwavering support of their dedicated fan base, known as ARMY (Adorable Representative M.C. for Youth). In 2018, the Hyundai Research Institute (HRI) stated that BTS contributed four to five billion US dollars per year to the South Korean economy in tourism, and consumer goods such as fashion, cosmetics, and food (Huang 2022).

In 2020, the worldwide tourism sector experienced a drastic decline. This was due to the spread of the COVID-19 virus that has attacked countries around the world. This event certainly has a significant impact on the country's tourism sector (Lee 2021). In overcoming this, South Korea has carried out various branding strategies to restore the country's tourism. One of them is by conducting a South Korean tourism promotion campaign through an online media platform by cooperating with the idol group BTS as its brand ambassador (Jarlis and Djumala 2022).

Through their music, public presentation, and positive image, BTS has turned the group into K-pop artists with high value and great potential for South Korea. One of the South Korean government's efforts to use BTS on the country's agenda was to appoint the group as South Korea's Cultural Ambassadors in 2018 and as the President's official envoy to become South Korea's diplomatic actor in 2021. This makes BTS an ideal messenger to carry out sustainable tourism promotion as one of South Korea's nation branding strategies (Le 2021).

Therefore, this paper will answer the research questions: how are South Korea's strategies on sustainable tourism through BTS in the economic and social sectors?

2 METHODS

This research uses a qualitative method descriptive approach. This method is used to explain what phenomena are experienced by research subjects, such as behaviour, perceptions, and actions that are poured in descriptions in the form of language and words (Moleong 2017). The descriptive approach is used to make descriptions and images systematically, actual and accurate regarding the facts, properties, and relationships between the phenomena being investigated. In this case, researchers analysed how the BTS group's role in the South Korean government's agenda in carrying out its national branding to support South Korea's sustainable tourism. Data obtained through secondary data sources such as official government and organisation websites, articles, journals, books, and other digital traces are developed and processed with a descriptive analysis approach. Where the author describes the data in detail and according to the facts to find out the description and state of the existing phenomenon to produce an accurate analysis.

3 FINDING AND DISCUSSION

3.1 *South Korea's public diplomacy through the idol group BTS*

BTS's existence in the music industry on the world stage has its interest. Based on the results of a survey conducted by The Korean Business Research Institute, BTS became the K-pop group with the first-ranked brand reputation for 31 consecutive months with a value of

13,693,956 in January 2020 (Cha 2020). The term brand reputation refers to the image and public perception of a group, and how well-liked or recognised they are among the foreign public (Pagunsan 2018).

Seeing the positive image that BTS has in the international scene, the South Korean government involves BTS in various public diplomacy agendas of the country. The influence that this group has on foreign communities is large enough to have a positive impact on South Korea. Starting from the spread of culture that they pour through their music, videos, and uploaded content to the economic improvement they provide through their achievements (Han 2017).

According to Cull, public diplomacy is a country's engagement and communication with foreign publics in communicating certain narratives and images of its country in order to achieve its goals in fulfilling national interests (Cull 2009). Through BTS, South Korea can communicate narratives and images of their country to foreign publics and indirectly capture the hearts and interests of the global public to get to know and learn more about South Korea. This will build a positive perception of the global community, making it easier for South Korea to promote state agendas to the public in order to fulfill national interests (Lee 2022).

So that in supporting its nation branding, there are several South Korean public diplomacy agendas through BTS including:

3.1.1 *KTO's collaboration with BTS on the feel of the rhythm of Korea*

The Korean Ministry of Culture, Sports, and Tourism and the Korea Tourism Organization (KTO) are actively engaging BTS in promoting Korean tourism. 'Hallyu Tourism', a destination theme focusing on hallyu tourism, features recommendations for locations the idol group has visited in music videos and other entertainment content. In short, almost all locations that the group has visited have become tourist destinations in their own right among ARMYs.

This phenomenon certainly opens up opportunities for the South Korean government to develop the country's tourism promotion more broadly. KTO collaborated with BTS in the 'Feel The Rhythm of Korea' campaign to carry out various Korean tourism promotion agendas as a whole. In 2022, there are two Korean tourism promotion agendas implemented through the campaign collaboration.

In April 2022, BTS held the "Permission To Dance (PTD) On Stage" concert for 4 days with a total of about 624 thousand fans watching live, online, and off-site. According to Big Hit Music, there were about 200 thousand fans who watched the concert live at the stadium. The online concert broadcast on the last day of the concert attracted 402 thousand viewers from 182 countries around the world. As well as an off-site concert held in real time through Live-Play at the MGM Grand Garden Arena, Las Vegas, bringing 22 thousand viewers during the 4 days of the concert (Hicap 2022). This momentum was utilised by the KTO to conduct offline and online campaigns to promote South Korean tourism and culture and invite fans to visit Korea.

In October of the same year, KTO released a promotional video titled '2022: Feel the Rhythm of Korea with BTS' with two BTS members, Suga and Jimin, showcasing the beautiful charm of 4 Korean cities, Busan, Daejeon, Pohang, and Jeju, accompanied by Korean classical music and K-pop. HS Ad reported that the promotional video uploaded through the YouTube account "Imagine Your Korea" received a lot of attention and reached more than 230 million views in the first 3 weeks of release (Yonhap 2022).

Through this collaboration, the South Korean government can promote tourism and culture more optimally. Seeing that the group has loyal fans who are enthusiastic about their idols, this momentum is utilized by the country to develop its national branding in the international arena. Through booths and a collection of videos showing South Korean tourism and culture, the KTO can introduce and invite foreign fans to visit the ginseng country (Hur and Kim 2020).

3.1.2 *BTS as Seoul City's ambassador*

Since 2017, the Seoul City Government has engaged BTS as a tourism promotion ambassador. The Seoul Tourism Organization (STO) seeks to promote the city's tourism charm through 'Visit Seoul' campaigns and promotional content with the group uploads on various social media platforms such as YouTube, Instagram, and Facebook (Parc and Kim 2020).

The Director of STO, Hong Jae-Sun, said that by engaging BTS in the promotion of Seoul tourism, the message can be more easily transmitted to fans around the world and invite them to visit Seoul to see how much influence the global artist has (Alam *et al.* 2022). By using BTS, Seoul City tourism has gained a lot of attention from foreigners. The natural beauty, tourist destinations, and cultural heritage contained in the Visit Seoul promotional video were successfully communicated and widely spread throughout the world. So that public opinion and positive perceptions of the global public towards Seoul tourism are formed.

In 2020, the COVID-19 pandemic hampered all activities of the world community. Almost all countries implemented lockdown policies and stopped the flow in and out of the country to avoid increasing the spread of the virus, including South Korea. This certainly has an impact on the tourism sector. The number of foreign tourists visiting the ginseng country decreased by 98% from 1.23 million tourists in May 2019 to only 6,111 foreign tourists visiting Korea in May 2020.

The government's decision to involve BTS in Seoul's tourism promotion agenda was the right decision, given the idol group's popularity among the global community. The national branding carried out by South Korea through collaboration with BTS has not only succeeded in increasing economic growth but has also indirectly formed a positive image and perception from the world public.

3.1.3 *BTS as honorary ambassador of Busan Expo 2030*

In July 2022, the Busan City government officially appointed BTS as the Honorary Ambassador of the World Expo 2030 Busan, Korea. This decision was made based on the South Korean government making every effort to win the tender between its competitors, namely Riyadh, Saudi Arabia, and Rome, Italy (Busan.go.kr).

As ambassadors, BTS participated in various promotional agendas. One of them is the 'Yet To Come in Busan' concert in October 2022. This agenda was created as an effort to promote Busan as a candidate for the 2030 World Expo and to win the hearts and support of the Korean people and abroad. The concert was held for free at the Busan Asiad Main Stadium and brought in around 52 thousand spectators. Not only the physical concert, but the concert was also broadcast online around the world through Korea's JTBC TV channel and Weverse platform free of charge. The online broadcast through Weverse also successfully penetrated more than 49 million views (Frater 2022).

The election of BTS as the honorary ambassador of Busan Expo 2030 is one of South Korea's public diplomacy efforts. The decision to appoint BTS as the honorary ambassador of Busan Expo 2030 is a form of hope that the idol group will take an important role and increase public interest and support in Busan to host the World Expo 2030 (Busan.go.kr).

Through the 'Yet To Come in Busan' concert, which was held on a large scale, many fans from around the world came to attend the concert in person. This was BTS's first concert as a group after the hiatus announcement in June 2022. The concert attracted about 523 million US Dollars and 100 thousand domestic and international visitors to Busan (Nassar 2022).

The success of BTS's concert held in support of Busan's winning bid to host the World Expo shaped the favorable perception of the foreign public. South Korea's victory as the host is very important for the country because the Expo helps to improve the country's image and national prestige on an international scale. In addition, hosting the 2030 World Expo will bring huge economic benefits, especially in creating jobs and boosting tourism. South Korea estimates that it will generate about 61 trillion won, create 500 thousand jobs (Hae 2022), and bring about 34.8 million foreign tourists to Busan during the exhibition period of approximately six months (Lee 2023).

3.2 *Sustainable tourism South Korea: nation branding South Korea through the BTS idol group in the tourism sector for the period 2020-2022*

In practice, a country needs to have a capable and good nation branding. Nation branding itself means the process in which a country's image is formed, managed, and communicated proactively to enhance the country's reputation by targeting international audiences (Fan 2010). Every country needs to develop and manage its national branding as well as possible, because a positive image and value of a country can determine the country's position in the international arena (Dannita and Deniar 2021).

In the context of South Korea, the efforts made in nation branding in the tourism sector are by cooperating with the idol group BTS in various tourism promotion agendas of the country. BTS, as one of the biggest idol groups in South Korea, not only has an influence in the field of music, but also in other fields. The group's worldwide popularity has made the global public increasingly interested in South Korea as the group's country of origin. It is for this reason that the country involves BTS' existence in nation branding to fulfill its national interests (Dubey 2019).

The image and value of the idol group have contributed to shaping the country's image in an international context, especially in the tourism sector. Through the sophistication of technology and the internet, tourism promotion is much easier to do and can reach a wider scope. Seoul's tourism promotion video, by cooperating with BTS as an advertising model, succeeded in attracting hundreds of millions of viewers (Yonhap 2020). Similarly, a promotional video was run through a collaboration between KTO and BTS in the 'Feel the Rhythm of Korea' campaign (Yonhap 2022).

Until the end of 2022, BTS maintained its brand reputation at number one with a value of 12,276,480 in December 2022 (Kim 2022). The data proves that the positive image and reputation of BTS can increase the interest of foreign fans to visit South Korea. This is supported by a report from the Hyundai Research Institute (HRI), BTS managed to bring more than 800 thousand foreign tourists to South Korea per year (Lee 2023). Especially when organizing the 'Yet To Come in Busan' concert, the number of domestic and international tourists attending Busan reached up to 100 thousand people in the fourth quarter of 2022 (Nassar 2022). Through BTS, South Korea managed to convey positive values of the country and get a good image from the foreign public, thus forming the trust of the global community to learn and visit the country so that it can have a positive impact on South Korea's sustainable tourism (Lyu 2024).

BTS holds a vital position in the South Korean economy as both a cultural ambassador and a musical phenomenon. Their influence is evident not only in the volume of CD sales but also in various economic dimensions, including corporate collaborations and brand partnerships. An article by Seth Abramovitch in the Hollywood Reporter indicates that this boy band generated a revenue of 465 billion dollars for South Korea's GDP, placing them in competition with major corporations like Samsung and Hyundai (Abramovitch 2023).

According to NH Investment & Securities, BTS, as the most lucrative K-pop performance ensemble, generated approximately 880 billion won (around 615 million U. S. dollars) in sales during the year 2021, representing nearly 70% of the agency's total revenue. In 2020, the contribution of BTS was even more pronounced, with figures approaching 92% (Business World Online 2022). This remarkable statistic underscores the substantial economic influence exerted by BTS. According to the Statistics Research Department, as illustrated in Figure 1, the direct economic effect of the concerts amounts to 330. 7 billion South Korean won, equivalent to roughly 250 million U. S. dollars. Notably, 23,000 out of a total audience of 130,000, approximately 18%, consisted of international fans.

Besides that, BTS is recognised for its exceptional album sales. The group initiated its album sales in 2013 and gained significant popularity around 2017, attributed to their remarkable album sales. As shown by Chartmasters (see Figure 2), even though BTS was the last to debut among the artists listed, they outperformed numerous global icons such as Rihanna, Beyoncé, Justin Bieber, and Adele, among others (ChartMasters.org 2019).

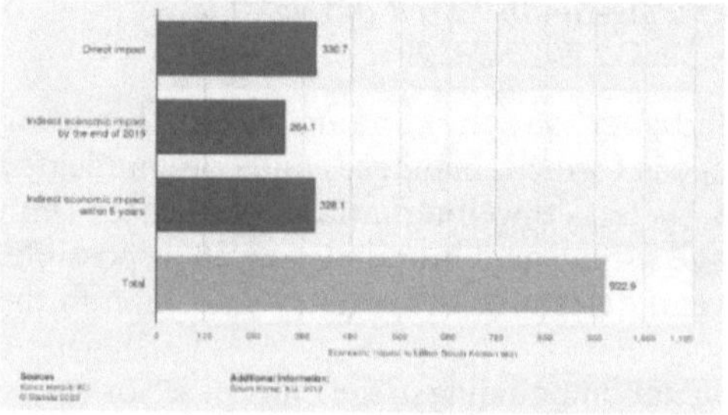

Figure 1. Estimated economic impact of BTS concerts held in Seoul in October 2019.

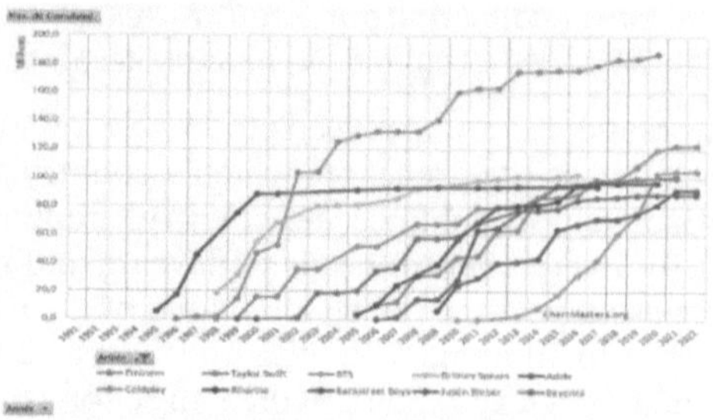

Figure 2. Album sales from 1995 to 2022.

4 CONCLUSION

As an idol group that has a positive image on an international scale, BTS' existence is trusted by the South Korean government in carrying out national interests, especially in the country's nation branding agenda. There are three South Korean nation branding efforts through BTS in the tourism sector for the 2020-2022 period. First, the Korea Tourism Organization (KTO) collaborated with BTS on the 'Feel The Rhythm of Korea' campaign. Second, BTS is the ambassador of Seoul City. Third, the selection of BTS as the honorary ambassador of the World Expo 2030 in Busan.

The three agendas above are in line with South Korea's sustainable tourism. Through collaboration between the government and BTS in the country's tourism promotion agenda, South Korea can improve the image of the country. Moreover, the cultural elements contained in these agendas can be an important point for Korea in showing the characteristics of the country. So that it can attract greater attention and positive perceptions from the international community. BTS has made important contributions to South Korea's sustainable economic growth, such as holding concerts, album sales, and so on. Thus, South Korea's decision to use BTS in the sustainable tourism agenda is the right decision in improving the image and positive perception of the foreign public towards the country, and is a good step in fulfilling its national interests, especially nation branding.

REFERENCES

Abramovitch, S. (2019). BTS is back: Music's billion-dollar boy band takes the next step. The Hollywood Reporter. Retrieved from https://www.hollywoodreporter.com/movies/movie-features/bts-is-back-musics-billion-dollar-boy-band-takes-next-step-1244580/

Alam, G., Affandi, R., Dermawan, W., Azmi, F., and Mahyudin, E. (2022). Seoul's paradiplomacy to the global public: The involvement of BangtanSonyeondan (BTS) in the tourism industry. *Journal of Governance*, 7(4).

Bae, E., Chang, M., Park, E., and Kim, D. (2017). The effect of Hallyu on tourism in Korea. *Journal of Open Innovation: Technology, Market, and Complexity*, 3(22).

Busan.go.kr. (2022). *BTS* officially appointed as honorary ambassadors for World Expo 2030 Busan, Korea: News & Notices. Retrieved from https://www.busan.go.kr/eng/bsnews01/1535513/

Business World. (2022). Shares in HYBE up after BTS clears uncertainty over military service. Retrieved from https://www.bworldonline.com/world/2022/10/18/481396/shares-in-hybe-up-after-bts-clears-uncertainty-over-military-service/

ChartMasters.org. (2022). The bestselling artists to debut in the last 30 years. Retrieved from https://twitter.com/chartmastersorg/status/1553809613371457536

Cha, E. (2020). January idol group brand reputation rankings announced. Soompi. Retrieved from https://www.soompi.com/article/1377884wpp/January-Idol-Group-Brand-Reputation-Rankings-Announced

Cull, N. J. (2009). Public diplomacy before Gullion: The evolution of a phrase. In *Routledge handbook of public diplomacy* (pp. 11). Routledge.
Dannita, S., and Deniar. (2021). The chaebol's contribution on South Korea nation branding through Korean wave. *Satwika Kajian IlmuBudaya Dan Perubahan Sosial*, 5(2), 318.
Dubey, H. (2019). The concept of nation branding and its adoption in South Korea. *International Journal of Social Science and Economic Research*, 4(4), 3032.
Fan, Y. (2010). Branding the nation: Towards a better understanding. *Place Branding and Public Diplomacy*, 6 (2), 101.
Frater, P. (2022). BTS: Concert in Busan viewed by 50 million worldwide. Variety. Retrieved from https://variety.com/2022/global/asia/bts-busan-concert-worldwide-views-1235404813/
Hae, K. (2022). Why is hosting the 2030 World Expo so important to South Korea? The Diplomat. Retrieved from https://thediplomat.com/2022/09/why-is-hosting-the-2030-world-expo-so-important-to-south-korea/
Han, B. (2017). K-Pop in Latin America: Transcultural fandom and digital mediation. *International Journal of Communication*, 11, 2252.
Hicap, J. (2022). BTS' Las Vegas concert draws 624,000 fans; 4 million watch 'Permission to Dance' tour series. Manila Bulletin.
Huang, Y. (2022). Analysis on South Korean soft power—Taking BTS, Parasite and Squid Game as examples. Proceedings of the 2022 International Conference on Comprehensive Art and Cultural Communication (CACC 2022), 663.
Hur, J., and Kim, T. (2020). Understanding contraflow pop-culture tourism: The case of transnational fandom of South Korean pop-culture and the "Hallyu" tourism. *Journal of Tourism Insights*, 10(1), 5.
Jarlis, A., Djumala, D., &Rezasyah, T. (2022). Feel the rhythm of Korea: South Korea nation branding in promoting tourism towards Indonesia. *JurnalSosioteknologi*, 21(3), 231.
Jin, D. (2021). The BTS sphere: Adorable representative M.C. for youth's transnational cyber-nationalism on social media. *Communication and the Public*, 6(1), 5.
Kim, C. (2022). Top 50 Kpop boy group brand reputation rankings in December 2022. Kpop Post. Retrieved from https://www.kpoppost.com/top-50-kpop-boy-group-brand-reputation-rankings-in-december-2022/
Le, N. (2021). No BTS: From K-Pop idols to diplomats: The right formula for effective public diplomacy? USC Center on Public Diplomacy. Retrieved from https://uscpublicdiplomacy.org/blog/bts-k-pop-idols-diplomats-right-formula-effective-public-diplomacy
Lee, S., and Kim, H. (2021). Nation branding in the COVID-19 era: South Korea's pandemic public diplomacy. *Place Branding and Public Diplomacy*, 17(3).
Lee, S. (2022). Film as cultural diplomacy: South Korea's nation branding through Parasite. *Place Branding and Public Diplomacy*, 18, 96.
Lee, M. (2023). BTS 10th anniversary celebration to supercharge domestic economy. The Korea Times. Retrieved from https://www.koreatimes.co.kr/www/biz/2023/06/602_353041.html
Lyu, P. (2024). The multifaceted impact of BTS: Driving South Korea's economy, soft power, and cultural exchange. *Transactions on Social Science, Education, and Humanities Research*, 11, 5.
Moleong, L. (2017). *Metodepenelitiankualitatif* (17th ed.). Bandung: PT. RemajaRosdakarya.
Nassar, A. (2022). BTS yet to come in Busan. South Lakes Sentinel. Retrieved from https://www.southlakessentinel.com/news/2022/10/18/bts-yet-to-come-in-busan/
Oh, I., and Lee, H. (2014). K-Pop in Korea: How the pop music industry is changing a post-developmental society. *Cross-Currents: East Asian History and Culture Review*, 3(3), 73.
Pagunsan, S. (2018). What does "brand reputation" of group in K-Pop mean? How do they distribute them in their places? Quora. Retrieved from https://qr.ae/py5dxc
Parc, J., and Kim, Y. (2020). Analyzing the reasons for the global popularity of BTS: A new approach from a business perspective. *Journal of International Business and Economy*, 21(1), 17.
UNEP. (2005). UN tourism: Making tourism more sustainable—A guide for policy makers. Pp. 11–12.
Yoeti, O., and Gunadi, I. (2013). Sustainable tourism sebagaiinstrumenstrategisdalamperencanaanpembangunan. *Journal of Tourism Destination and Attraction*, 1(1), 37.
Yonhap. (2020). Seoul city's promotional video featuring BTS tops 100 million views. Yonhap News Agency. Retrieved from https://en.yna.co.kr/view/aen20200923011500315
Yonhap. (2022). Feel the rhythm of Korea with BTS' surpasses 230 million views. K-Odyssey. Retrieved from https://m.k-odyssey.com/news/newsview.php?ncode=1065579412401619.

Towards Resilient Societies: The Synergy of Religion, Education, Health, Science, and Technology – Rahiem (Ed.)

Islamic philanthropy regulation and poverty alleviation in Indonesia: A systematic review

Evy Flamboyan Minanda, Hamdani Hamdani & Usep Abdul Matin
Universitas Islam Negeri Syarif Hidayatullah Jakarta, Indonesia

ABSTRACT: Cases of alleged misappropriation of aid funds or donations by individuals or philanthropic institutions often undermine the spirit of generosity among Indonesians. Previous studies have explored the role of zakat in economic empowerment, the regulation of Islamic philanthropy amid opportunities and challenges, and its contribution to improving welfare. In the context of philanthropy, Law 9/1961 and Law 23/2011 share a similar legal standing; however, Law 23/2011 addresses explicitly the Muslim community. Given the evolving landscape of global philanthropy, Law 9/1961 requires amendments to align with current legal and practical needs. Harmonising these two laws is expected to support the achievement of Sustainable Development Goals (SDGs), particularly in eradicating poverty, promoting justice, and building strong institutions. Strengthening blended funding mechanisms is essential to optimise the use of philanthropic funds, reduce financial risks, and encourage greater participation from non-public sectors in national development. This harmonisation aims to create a more transparent, accountable, and effective philanthropic ecosystem that contributes meaningfully to sustainable national growth.

Keywords: harmonisation, Islamic philanthropy, philanthropy, regulation

1 INTRODUCTION

The word "philanthropy" is a new term in Islam; however, recently, a number of Arabic terms have been used as its equivalent. Philanthropy is sometimes equated with al-ata' al-ijtima'i which means social giving, al-takaful al-insani which means human solidarity, 'ata' khayri which means giving for good, or sadaqah which means alms (Ibrahim 2011). The colonial government did not interfere in Islamic philanthropy, so the practice of philanthropy became a private matter. Ulama and Islamic boarding schools became the centres of philanthropy, and many people gave them zakat, alms, and waqf. Since the 1990s, Islamic philanthropy in Indonesia has grown rapidly. Institutions for collecting zakat, infaq, alms, and waqf are increasing, such as Lazismu, Lazis NU, or Dompet Dhuafa. Philanthropic institutions quite well-known to the public are Aksi Cepat Tanggap and Kita Bisa.com.

Conceptually, philanthropy differs from the tradition of giving in Islam, such as zakat, infak and shadaqah. Philanthropy is more oriented towards love for humanity and moral motivation. Meanwhile, Islam's philosophical basis is Allah's obligation to realise social justice. However, recently, these terms have been popularly used to identify the practice of religious-based generosity, including among Muslims. However, the author has not found a comprehensive legal analysis that compares the regulation of philanthropy, Islamic philanthropy, and poverty regulation, which is seen in how the harmonisation and implementation are especially in community empowerment programs. Related to the distribution of zakat, there have been cases of misappropriation and abuse in the distribution of zakat in poverty programs in several countries. In Malaysia, for example, some zakat funds are used for

 DOI: 10.1201/9781003645542-76

programs other than poverty alleviation programs. In several other countries, most of the allocation is not on target for other purposes, such as those related to political and terrorist agendas. The suboptimal achievement of zakat targets has caused the assistance given to zakat recipients to be less than adequate, so it has not followed the expected goals. In addition to these problems, the large operational costs of managing zakat are suspected to be caused by the high salaries of zakat collectors/managers, such as in Sudan and Malaysia. This is also a similar issue in Indonesia.

With the increasing cases of misappropriation of philanthropic funds carried out by institutions and individuals and the overlapping authority of licensing the implementation of non-religious-based philanthropy with religious-based philanthropy, it is hoped that it will strengthen and provide alternative management of philanthropy as input for amendments to Law 9/1961 to improve justice, social welfare, and contribute to overcoming poverty. There are several gaps in the current regulation of philanthropy, including not yet regulating online implementation activities, not yet regulating transparency, reporting and accountability mechanisms, and light sanctions. Islamic philanthropy is guided by the Qur'an and hadith: zakat, infak, shadaqoh, and waqf. Meanwhile, global philanthropy is carried out by collecting money/goods to help with problems of poverty, education, water crisis, health, illiteracy, or climate change. This gap presents an opportunity to fill this void and serves as a scientific source by conducting systematic literature reviews on alternative management of philanthropy with religious philanthropy. Dissect cases of misappropriation of philanthropic funds carried out by institutions and individuals, provisions of laws and regulations violated, and the formulation of appropriate regulations between Law 9/1961 and Law 23/2011. Systematic reviews are interesting because they meet the need to broadly view all existing research related to a particular research question (Gaur 2018). This method is very helpful for academics to identify, analyse, critically evaluate, and summarise knowledge about a particular field of knowledge in a clear, replicable and rigorous manner (D. Gough 2017) . This provides an opportunity to fill the gap in this theme as a scientific reference in studies based on a systematic review of the literature related to harmonising philanthropic management with religious philanthropy.

2 METHOD

Based on the identification of the problem as described above, the author uses a qualitative research design approach with the systematic literature review method, *hereinafter* abbreviated as SLR. The SLR method has been used in several studies, one of which was used by Triandini (Triandini 2019). It is explained in their publication that SLR is used to identify, review, evaluate, and interpret research in phenomenal topic areas and certain relevant research questions. The SLR method can be used to review and identify journal articles systematically, each process of which follows steps and protocols. Meanwhile, in a different publication article, Hariyati (Hariyati 2010) in her nursing study, explains in her publication that SLR is a method that uses reviews, reviews, structured evaluations, classifications, and categorisations of *previously produced evidence-based evidence.* The limitation of the SLR method in social and multidisciplinary research is that it can limit researchers from obtaining other, more relevant research.

This paper uses the SLR method to reveal the regulation of philanthropy in Indonesia after the case of misappropriation of aid carried out by philanthropic institutions or individuals, then tries to see the regulation of philanthropy based on the provisions of laws and regulations that have an Islamic character, as well as the impact of philanthropic funding on poverty reduction. Several authors argue that this SLR methodology is unsuccessful in social and multidisciplinary research. The data collection rules used in this study are to create a special search by focusing on two databases: the Google Scholar web page and Semantic Scholar. Google Scholar is a search engine provided by Google which mostly consists of academic journals, theses, dissertations, or research results. Google Scholar allows users to search for scientific literature based on keywords, article titles, author names, and research topics. Unlike Google Scholar, Semantic Scholar is designed to highlight a paper's most important and influential elements. Based on the search for the phrase "Regulation of Islamic Philanthropy in Handling Poverty in Indonesia" in Google Scholar, 1610 articles were found, while 65 articles were found in Semantic Scholar. If using the phrase "Jurid analysis is the regulation of philanthropy" in Google Scholar, 689 articles were found, and in Semantic Scholar, 1600 articles were found.

3 RESULTS

The data collection rules used in this study are to create a special search by focusing on two databases: the Google Scholar web page and Semantic Scholar. Google Scholar mostly consists of academic journals, theses, dissertations, and research results, and Semantic Scholar is a system that includes various publications from all fields of science. Semantic Scholar uses natural language processing and machine learning models to drive its search engine, thus providing more relevant search results. The first search was conducted by writing the background, namely "Islamic philanthropy regulation in poverty management in Indonesia" in Google Scholar, 1610 results were found and 65 results were found in Semantic Scholar. In the phrase "Legal analysis of philanthropy regulation" in Google Scholar, 689 results were found, and in Semantic Scholar, 1600 results were found. If the author looks at the regulatory aspect in the Google Scholar search engine with the keyword "philanthropy regulation in Indonesia" with a limitation of August 5, 2014, to August 5, 2024, 37 results were found. In the Semantic Scholar search engine with the same keywords, the time limit is from 2014 to 2024 with a limitation of legal study articles of 565 articles.

Philanthropy regulations are regulated in Law Number 9 of 1961 concerning the Collection of Money or Goods (Law 9/1961). However, when using the nomenclature of philanthropy, most searches do not display Law 9/1961. For this reason, the author tried to search the semantic scholar search engine using the keyword "Law Number 9 of 1961 concerning the Collection of Money or Goods" and found 27 results. Meanwhile, no research results were found in Google Scholar. The next stage is to conduct a selection of relevant study results, then select quality study results, and then write the study results in a document. Related to the research theme, the author obtained 9 related studies that showed regulatory gaps in research with the theme of philanthropy based on Law 9/1961 and more use of Law 23/2011. The challenge going forward is how the regulation of Law 9/1961 and Law 23/2011 can synergise and be harmonious in reducing poverty, creating justice, and increasing welfare by strengthening institutions through utilisation of funds, which is a catalyst in national development.

4 DISCUSSION

4.1 *Philanthropy regulation in Indonesia*

Based on discussion above, related with arrangement philanthropy in Indonesia is divided into 2 (two) types, namely philanthropy in a social or humanitarian context with philanthropy based on religion, namely zakat, infak, and shodaqoh. Philanthropy socially and humanitarianly based on the provisions of Law 9/1961 in conjunction with PP 29/1980 and Permensos 8/2021. While philanthropy based on religion, namely zakat, infak, and shodaqoh based on Law 23/2011 in conjunction with Government Regulation Number 14 of 2014 concerning the Implementation of Law 23/2011 (PP 14/2014). Law 9/1961 states that what is meant by collecting money or goods in this law is every effort to obtain money or goods for development in social welfare, mental/religious/spiritual, physical and cultural fields. According to the author, the contents of Law 9/1961 can no longer be a qualified legal basis for philanthropic activities today. This is because the development of philanthropy is progressing very rapidly while the legal basis used is still the old legal basis.

Zakiyuddin (Baidhawy 2007) in the book "Islam Against Capitalism; Concepts of Justice in Islam" states that the Islamic economic system and ethics can be an alternative solution to answer the problems of economic injustice that plague world civilisation today. For Zakiyuddin, the problem of injustice occurs because the world currently adheres to an economic system based on capitalism. Frances (2008), in her book "The End of Charity: Time for Social Enterprise", states that we should reformulate the idea of the fate of philanthropy and charity movements with various reasons and arguments to overcome the problems of poverty, injustice, and inequality issues in life.

In the context of the ratification of Law 23/2011, according to Asmawi, it is a step backwards for the national zakat world because it failed to carry out its main mission of optimising the potential of large Islamic philanthropic funds and its strategic role in poverty alleviation and improving people's welfare. In the context of a democratic Indonesian civil society, Law 23/2011 should strengthen the role of the state in providing protection for citizens who pay zakat (muzakki), maintaining public order by preventing misuse of zakat funds, facilitating the Islamic philanthropic sector for social change and providing incentives for the development of the national zakat world (Asmawi 2017).

Article 1 of Law 9/1961 states that every business in the field of social welfare, mental/religious/spiritual, physical and cultural fields, meaning that the scope of regulation of Law 9/1961 is not only in the field of social welfare but also includes business activities in the field of mental/religious/spiritual, physical and cultural fields. This is interesting to discuss because after the ratification of Law 23/2011, there are regulations regulating money and goods in the field of religion, namely for Muslims. In the context of institutions that collect zakat, infaq, and shodaqoh as a religious activity, the institution that organises the collection is called a zakat collection institution (LAZ) whose licensing is the authority of the Ministry of Religion. However, when the context touches on social and humanitarian matters, the licensing is based on the provisions of Law 9/1961 in conjunction with PP 29/1980 in conjunction with Permensos 8/2021 through the Ministry of Social Affairs or social agencies according to their authority.

The amendment to Law 9/1961 to update the legal basis for philanthropy in Indonesia is actually to adjust to the current conditions of society about philanthropic activities. The renewal of the legal basis for philanthropy is expected to be a fairer legal basis for stakeholders related to philanthropy. This is because it covers the rights and obligations of related parties and other provisions that are more accommodating to philanthropic activities today. Law 9/1961 no longer follows the development of philanthropic activities. Several provisions in Law 9/1961 must be changed, updated, or added to Indonesia's development goals. In addition, the phrase regulating religious philanthropy in Law 23/2011 must be removed/removed from the regulation of philanthropy. Lawrence M. Friedman, in his book American Law: An Introduction, put forward the theory of the Legal System. According to him, a legal system in actual operation is a complex organism in which structure, substance, and culture interact (Friedman 2001). According to Friedman, the legal system must touch three legal components, namely: legal structure, legal substance, and legal culture.

The 'inherent' problem of the rule of law is regulatory disharmony caused by too many regulations being formed or what is known as *hyper regulations,* which are then popularly referred to as legal obesity. Regulatory disharmony, marked by the existence of norm conflicts, occurs in Indonesia and other countries. Many things cause norm conflicts to occur frequently, including the existence of laws and regulations which are required to always be dynamic in following the development of community needs, laws and regulations are divided into levels arranged in a hierarchy, and the legal substance covers aspects of community life that are so complex. In the context of philanthropy, Law 9/1961 and Law 23/2011 have the same position, but Law 23/2011 regulates a special object, namely Muslims, so the material contained in Article 1 of Law 9/1961 no longer applies to Muslims. They should no longer use the provisions in Law 9/1961 because the principle of lex specialis derogat legi generali applies. However, it is necessary to discuss philanthropic activities outside the religious context that are not included in the regime of Law 9/1961.

Bagir Manan, in his book "Positive Law of Indonesia" as quoted by AA Oka Mahendra in his article entitled "Harmonisation of Legislation" states that there are several things that can be used as guidelines in applying the principle of lex specialis derogat legi generali, namely: 1. provisions found in general legal rules remain valid, except those specifically regulated in the special legal rules; 2. provisions of lex specialis must be equal to provisions of lex generalis (for example, laws with laws); and 3. provisions of lex specialis must be in the same legal environment (regime) as lex generalis (Manan 2004). As a country whose population has been proven to be generous, Indonesia must be given certainty and confidence that the funds that have been disbursed are managed with good accountability principles. As the legislator and the Government, the DPR, which then establishes derivative regulations, must begin to prioritise regulatory changes related to philanthropy, especially in synergy and harmony with Islamic philanthropy. By harmonising Law 9/1961 and Law 23/2011, it can provide contributions to institutions, fund management, and data utilisation in community empowerment and improving welfare to eradicate poverty.

4.2 *Systematic literature review (SLR)*

Several previous studies have discussed the role of zakat in economic empowerment, the regulation of Islamic philanthropy between opportunities and challenges, Islamic philanthropy in improving welfare or other themes. However, the author has not found a comprehensive legal analysis that compares the regulation of philanthropy, the regulation of Islamic philanthropy, and the regulation of poverty, which is then seen how the harmonisation and implementation are especially in community empowerment programs. As for the dissection of cases of misappropriation of philanthropic funds carried out by institutions or individuals, provisions of the laws and regulations that are violated, and the appropriate formulation of regulations between Law 9/1961 and Law 23/2011.

Based on the legal regulations, namely Law 9/1961 (collection of money or goods/philanthropy), Law 23/2011 (zakat, infak, shodaqoh), and a comparison of Law 9/1961 and Law 23/2011, it is known that the discussion related to the regulation of philanthropy uses the regulation based on Law 23/2011 (zakat, infak, shodaqoh) as many as 2 studies, based on Law 9/1961 (collection of money or goods/philanthropy) as many as 7 studies, and research that conducts a comparative study of the two regulations is 1 study. Meanwhile, Law 11/2009 (social welfare) and Law 13/2011 (handling the poor) were not found in the discussion of philanthropy based on the SLR search conducted by the author. From the results of the SLR search above, there is a gap in the regulation of laws and regulations regarding philanthropy when linking Islamic philanthropy but not discussing the regulation of philanthropy as regulated in Law 9/1961, each study stands alone as if they are not interconnected and there is no harmonisation of regulations or in the author's language, it is not comprehensive.

Other studies also found that research based on the approach to philanthropic material in Law 9/1961 no longer follows the legal needs and developments in philanthropy, starting from criminal sanctions that are no longer appropriate, governance that is not in accordance with the times, and other problems. Most research methods use normative legal research/normative law, while the rest use sociological legal research and case studies. So, based on the research that has been done, there is a gap in research from the regulatory side related to the impact of philanthropy on poverty using *a mixed* method (Creswell 2018).

5 CONCLUSION

From the results of the SLR search above, there is a gap in the regulation of laws and regulations regarding philanthropy when linking Islamic philanthropy, but the regulation of philanthropy is not discussed as regulated in Law 9/1961. Each study stands alone as if it is not interconnected and not comprehensive. The same thing when looking at research related to the implications of Islamic philanthropy, most of which use the legal basis of Law 23/2011 on poverty, only looking at the sociological and historical aspects but forgetting the aspect of harmonisation with the provisions of other laws and regulations, namely Law 13/2011 and Law 11/2009, so that it is necessary to make changes to Law 9/1961 and Law 23/2011 with do strengthening mix funding through utilisation funds which intended reduce financial risks for non-public sector participation in national development for justice, improving welfare, and reducing poverty.

REFERENCES

Asmawi. (2017). Contextualization law zakat in Indonesia (Study about legislation and constitutionality act management zakat in perspective yes, it is valid Syar'iyyah). *Journal of Islamic Law/Journal Law Islam*, 16 (1).

Baidhawy, Z. (2007). *Islam Against Capitalism*. Yogyakarta: Resist Book.

Creswell, J. W. (2018). *Designing and Conducting Mixed Methods Research*. Sage Publications.

Frances, N. (2008). *The End of Charity: Time for Social Enterprise*. Crows Nest: Allen & Unwin.

Friedman, L. M. (2001). *American Law: An Introduction (W. Basuki, Trans.)*. Jakarta: Tatanusa.

Gaur. (2018). A systematic approach to conducting review studies: An assessment of content analysis in 25 years of IB research. *Journal of World Business*, 53(2), 280–289. https://doi.org/10.1016/j.jwb.2017.11.003

Gough, D. (2017). *An Introduction to Systematic Reviews* (2nd ed.). London, UK: SAGE Publications Ltd.

Hariyati. (2010). Understanding systematic review theory and case studies. *Indonesian Nursing Journal*, 13(2). Retrieved from http://jki.ui.ac.id/index.php/jki/article/view/242

Ibrahim, J. (2011). *Normative Legal Research Theory & Approach*. Malang: Bayu Media Publishing.

Manan, B. (2004). *Positive Law of Indonesia*. Yogyakarta: FH UII.

Triandini. (2019). Systematic literature review method for identification of platforms and methods of information system development in Indonesia. *Indonesian Journal of Information Systems (IJIS)*, 1(2).

Towards Resilient Societies: The Synergy of Religion, Education, Health, Science, and Technology – Rahiem (Ed.)

The political economy of media in Indonesia: Case study of commodification in Tempo Magazine

Gun Gun Heryanto & Nurlaillah Sari Amallah
Universitas Islam Negeri Syarif Hidayatullah Jakarta, Indonesia

ABSTRACT: This study examines the commodification of content and workers in Tempo Magazine using Vincent Mosco's critical paradigm and a qualitative approach. The research focuses on the political economy of media in Tempo Magazine, employing an instrumental case study method to gain an in-depth understanding of specific issues within a real-life context. The findings reveal that Tempo Magazine adopts a unique approach by implementing the "story behind the news" concept in its content. This is reflected in the choice of diction, including terms such as "kusut," "kubu Banteng," "partai merah," "wong cilik," "bancakan," "upeti," "cokok," "rusuh," "lancung," and "digelontor," particularly in the Main Report Rubric, which achieves an average circulation of 120,000 copies, highlighting content commodification. Regarding worker commodification, Tempo Magazine journalists face irregular working hours, insufficient wages, and multiple overlapping responsibilities. The study concludes that Tempo Magazine's editorial team, under the guise of noble journalism, operates within a political economy framework, strategically highlighting social assistance corruption cases as central themes in the Main Report and Opinion Column.

Keywords: commodification of content, commodification of workers, journalist, main report rubric, Tempo magazine

1 INTRODUCTION

Political economy states that media markets are part of the capitalist economic system (Fourie 2007). Viewed through a political economy perspective, Vincent Mosco highlights the influence of power and social relations in shaping the production, distribution, and consumption of resources, including communication resources, is a crucial consideration (Mosco 2009). Mosco's framework for understanding the political economy of communication includes three main concepts: commodification, spatialisation, and structuration. Commodification involves the transformation of goods and services, including communication, into commodities that can be bought and sold. Commodification is the process of changing use value into exchange value (Mosco 2009). There are several forms of commodification, such as content commodification, which is the process of transforming raw data into material ideas that have economic value (Heryanto 2018). Then commodification of workers is a process where labour is considered as goods or services that can be bought and sold. This research is very important because communication studies often overlook workers who are a crucial part of the media industry.

In the social assistance (bansos) corruption case reported by *Tempo Magazine*, in each of its contents, *Tempo Magazine* has its unique way of covering this issue. This can be seen from diction such as "kusut," "kubu Banteng," "partai merah," "*wong cilik*," "bancakan," "upeti," "cokok," "rusuh," "lancung," and "digelontor," which shows that content commodification has occurred. *Tempo Magazine* is now expanding into digital form. This is done to make it easier for readers to access news updates that are similar to the printed version of that. This

 DOI: 10.1201/9781003645542-77

proves that the rapid development of digitalisation and media convergence requires media industries such as magazines to adapt to new business models (Malmelin *et al.* 2021).

Based on research data and findings, various forms of exploitation occur against the journalists at Tempo Magazine who work flexible hours, often exceeding standard working hours, without overtime compensation and are also assigned double duties. They adhere to the ideology of false consciousness, which prevents them from recognising themselves as workers and instead idealises journalism as a noble profession. This perspective is shaped by long working hours, low wages, lack of overtime compensation, and the need to undertake diverse marketing promotions for various Tempo products. This research aims to reveal the political economy of media in Indonesia.

Previous collective research conducted by Muslikhin, Mulyana, Hidayat, and Utari showed that content from social media was converted into news published by traditional media or other news outlets (Muslikhin *et al.* 2021). According to previous research conducted by Wahyudi Marhaen Pratopo, the commodification of workers is evidenced by increased working hours, low wages, multiple tasks, and obligations for marketing promotion (Pratopo 2018). Then research from Claudia Mellado and Alfred Hermida shows that journalists try to increase their market value by building and marketing their brands on social media (Mellado and Hermida 2022). Besides that, journalists are forced to work harder with inflexible time arrangements and more assignments as part of the process of commodification of their labour (Rohman 2023). On the other hand, a journalist, whose role is to seek and verify facts, feels frustrated as they have to compete with the speed and headlines of the digital era (Heryanto *et al.* 2024). In the end, some media neglected journalistic standards, such as clarifying issues, to serve certain political interests (Neyazi *et al.* 2022)

Despite the research conducted in the field, there are still gaps that have not been fully explored. Specifically, researchers do not comprehensively examine the concept of commodification within political economy, especially in relation to social assistance corruption issues. To fill this research gap, we comprehensively analyse both the commodification of content and workers in *Tempo Magazine.* This study is guided by the following research questions: (1) what is the political economy of media in Indonesia?, (2) what does the commodification of content and workers look like at *Tempo Magazine*?

The structure of this paper is as follows: This section reviews the existing literature on political economy and commodification. It is followed by a methodology that integrates paradigm, approach, and method to provide a comprehensive framework for the study. The findings and discussion present the research results in detail and interpret them within the existing body of knowledge. Finally, the conclusion advocates ethical journalistic practices and suggests future research.

2 METHODS

Guba explains that the critical paradigm seeing that the object or social reality observed is an expression of false consciousness that humans have, not an objective reality or a reality that is by the true essence (Guba 1991). Researchers seek to examine how power is exercised, who has access to it, and how it is used to influence and control individuals and groups in society. Thus, this study employs the critical political economy framework as conceptualised by Vincent Mosco (2009). Agger's critical social theory helps us understand that structures of domination in society are not only maintained through direct power but also through ideologies and false consciousness that are internalized by society (Agger 1998). According to political economy, the media market is an integral part of the capitalist economic system and is closely connected to a country's political system (Fourie 2007). Looking through a political economy lens, Mosco emphasises the significant role that power and social relations play in shaping how society produces, distributes, and consumes resources, including communication resources (Mosco 2009).

In this context, the researcher utilises a critical paradigm to uncover the political economy and uses a qualitative approach because it uses various methods such as field notes, interviews, conversations, photos, recordings, and personal memos to record and describe what they encounter in the field (Denzin and Lincoln 2018). The positioning of theory in qualitative research serves as a tool for verification and analysis during fieldwork (Creswell 2002). In this study, a qualitative approach was adopted, as the data were derived from interview activities; researchers also interviewed the Editor-in-Chief of *Tempo Magazine, Tempo* journalists, and six former *Tempo Magazine* journalists as key informants to see different sides of *Tempo*. The literature review focused on academic databases like Taylor & Francis Online, Google Scholar, and searched for peer-reviewed articles published in the last ten years using keywords such as "commodification," commodification of content," and "commodification of workers." This research is more suitable for using the case study method because researchers can develop new theories or test and improve existing theories (Denzin and Lincoln 2018).

In this research, the case study used is an instrumental case study to gain an in-depth understanding of certain issues in a real-life context (Stake 2006). This research is more suitable for using instrumental case study, namely the method used in qualitative research to obtain an in-depth understanding of certain issues in real-life contexts.

3 FINDINGS AND DISCUSSION

Tempo Magazine uses investigative journalism and engaging storytelling to maintain readership, but its journalists face demanding workloads and low wages, highlighting the commodification of content production and labour exploitation.

3.1 *Commodification of content*

The analysis revealed that to increase traffic, media compete to provide interesting news content so that readers remain loyal to their online media (Widiastuti 2021). Because the basis of *Tempo Magazine* is investigative journalism, the easiest thing to uncover is corruption in Indonesia. In this case, each media type competes to attract readers' attention and time. With its unique diction, *Tempo Magazine* uses the concept of the story behind the news to convey more complete information and present in-depth analysis so that it is comfortable for readers to read. *Tempo Magazine* made this case the main menu in the Main Report and Opinion Rubric and published it in three special editions. The first is entitled "Kusut Bantuan Sosial." Second, "Korupsi Bansos Kubu Banteng." Third, "Bancakan Bansos Banteng."

In each of its contents, *Tempo Magazine* is unique in covering the issue of corruption in social assistance. This can be seen from the diction such as "kusut," "kubu Banteng," "partai merah," "*wong cilik*," "bancakan," "upeti," "cokok," "rusuh," "lancung," and "digelontor" which shows that content commodification has occurred. Based on supply, there is a lot of supply to investigate, one of which is corruption and our demand includes investigations. We have conducted internal research, so the news readers want the most is about corruption involving big figures (Setri Yasra, Editor in Chief of *Tempo Magazine*, in-depth interview, December 2, 2022).

Tempo Magazine is now expanding into digital form. This is done to make it easier for readers to access news updates like the printed version. This proves that the rapid development of digitalisation and media convergence requires media industries such as magazines to adapt to new business models (Malmelin *et al.* 2021). Data obtained by researchers, the *Tempo Magazine* circulation team in three special editions of social assistance corruption had an average circulation of 120,000 copies. Additionally, pageviews on their digital platform *(Tempo.co)* increased. Notably, "Upeti Bansos untuk Tim Banteng" garnered 130,390 pageviews, "Basi Setelah Sehari" received 45,540 pageviews, and "Paket Mewah Partai Merah" had 31,105 pageviews.

3.2 *Commodification of workers*

The case studies highlighted several best practices, including there is no specific working time pattern for *Tempo Magazine* journalists, The Editor-in-Chief of *Tempo Magazine* also said the same thing that journalism is a profession that demands a 24-hour commitment. He further emphasised that journalism is a noble profession, where obligations are not confined by working hours. For him, the achievement of securing a source is a matter of professional pride and remains a memorable accomplishment (Setri Yasra, Editor in Chief of *Tempo Magazine*, in-depth interview, August 8, 2023). In Marxist ideology, this is what is called false consciousness (Littlejohn and Foss 2009). This refers to the failure of workers who do not understand that the exploitation they experience is part of the capitalist system or the larger social structures affecting them (Lewy 1982). This phenomenon is referred to as the mystification process, wherein attributes such as independence, bravery, and credibility are equated with the status of *Tempo* journalists, who are often captivated by the prestigious image associated with their profession (Amallah *et al.* 2024).

Additionally, the researchers sought to examine *Tempo* from an alternative perspective. Researchers tried to interview the informant as a former *Tempo Magazine* journalist. He said that journalists frequently must adapt to the demands of their profession, requiring them to remain on standby and work extended hours and their own rights are neglected (Adi Prasetya, Former *Tempo Magazine* Journalist, in-depth interview, August 6, 2023). Even though journalists work more than normal working hours, they do not receive overtime pay or other compensation. Taking into account their working hours, *Tempo* journalists do not receive adequate wages and are relatively low compared to other media. This is inherent in the finding that lower-income workers tend to be more vulnerable to the impacts of the pandemic. It also highlights the dominant injustice within the social and economic system (Soderborg and Muhtadi 2020). In a capitalist system, commodities are not just physical goods that can be used or exchanged. They are also given symbolic and economic value that obscures the social relations and human labour involved in their production (Tsogas 2018). Thus, civil society is a place where the value of commodities is shaped to align with the capitalist way of thinking, particularly its focus on profit (Bakti 2018).

This process is referred to as commodity fetishism. In the context of *Tempo Magazine*, consumers buy magazines or subscribe to get access to information. However, they are often unaware of the complex social relationships and hard work involved in the production of such news. Marx explained that in the capitalist system, workers are not paid according to the true value of their contribution to production. Because workers' wages only cover the minimum cost of living, while the actual value produced by workers is greater, exploitation occurs by capital owners (Holborow 2018). As Fuchs points out, the unequal distribution of ownership results in capitalists being able to exploit labour for their own profit. The value created by workers through their contributions to the production process frequently exceeds the wages they receive in compensation (Fuchs 2021). In this particular context, a journalist is tasked with managing a diverse array of responsibilities simultaneously. These duties encompass conducting interviews, capturing photographs and videos, and gathering data.

This dual task is also carried out by the Chief Editor of *Tempo Magazine*, Setri Yasra. According to the individual, *Tempo* seniors are expected to execute their responsibilities to the best of their abilities, even to the point of great sacrifice. Despite holding the esteemed position of editor-in-chief, he continues to identify himself as a reporter. Researchers want to show that although journalists consider their work to be a noble profession, they actually experience the commodification of workers, namely exploitative working conditions.

4 CONCLUSION

Tempo Magazine, in the political-economic context, turns news about corruption involving big figures into a product that is traded to increase circulation. They have exploitative power over journalists by creating injustice in the work system. Then the absence of any form of compensation can create an economic structure that affects workers' welfare. In addition, the ideology of false consciousness makes them accept non-ideal working conditions as part of their moral and ethical responsibilities as journalists. Overall, the researcher explains that the editorial team of *Tempo Magazine* is carrying out political economy through the commodification and the need for better working conditions and protection of rights in the media industry. Future research should explore spatialisation within political economy to provide a broader and more comprehensive understanding of the subject.

REFERENCES

Agger, B. (1998). *Critical Social Theories An Introduction*. Westview Press.

Amallah, N. S., Heryanto, G. G., and Adeni, A. (2024). Commodification of Workers in Tempo Magazine. *International Journal of Social Science and Human Research*, 07(03). https://doi.org/10.47191/ijsshr/v7-i03-54

Bakti, A.F. (2018). Media and Religion: Rodja TV's Involvement in The Civil Society Discourse for Community Development. *Jurnal Komunikasi, Malaysian Journal of Communication*, 34(3), 226–244. https://doi.org/10.17576/JKMJC-2018-3403-13

Creswell, J. W. (2002). *Research Design: Qualitative and Quantitative Approaches*. SAGE Publications.

Denzin, N. K., and Lincoln, Y. S. (2018). *The SAGE Handbook of Qualitative Research* (5th ed.). SAGE Publications.

Fourie, P. J. (2007). *Approaches to the Study of Mass Communication In Media Studies: Media History, Media and Society* (2nd ed.). Juta and Company.

Guba, E. G. (1991). *The Paradigm Dialog* (2nd ed.). Sage.

Heryanto, G. G. (2018). *Media Komunikasi Politik: Relasi Kuasa Media di Panggung Politik*. IRCiSoD.

Heryanto, G. G., Zamroni, M., and Astuti, Y. D. (2024). Disinformation Unveiled: Tracking Media Hoaxes to Build Public Literacy for Indonesia's 2024 Elections. *Studies in Media and Communication*, 12(4), 28. https://doi.org/10.11114/smc.v12i4.6931

Holborow, M. (2018). Language, commodification and labour: the relevance of Marx. *Language Sciences*, 70, 58–67. https://doi.org/10.1016/j.langsci.2018.02.002

Lewy, G. (1982). *False Consciousness: An Essay on Mystification*.

Littlejohn, S. W., and Foss, K. A. (2009). *Encyclopedia of Communication Theory (Issue v. 1)*. SAGE Publications.

Malmelin, N., Virta, S., and Kuismin, A. (2021). Transforming Media: Reviewing the Issues and Contexts of Change in Media Management Research. *Journal of Media Business Studies*, 19(3), 203–224. https://doi.org/10.1080/16522354.2021.1960619

Mellado, C., and Hermida, A. (2022). A Conceptual Framework for Journalistic Identity on Social Media: How the Personal and Professional Contribute to Power and Profit. *Digital Journalism*, 10(2), 284–299. https://doi.org/10.1080/21670811.2021.1907203

Mosco, V. (2009). *The Political Economy of Communication* (2nd ed.). SAGE Publication.

Muslikhin, M., Mulyana, D., Hidayat, D. R., and Utari, P. (2021). The Commodification, Spatialization, and Structuration of Social Media in the Indonesian Cyber Media News. *Media and Communication*, 9(2), 110.

Neyazi, T. A., Yi Kai Ng, A., Kuru, O., and Muhtadi, B. (2022). Who Gets Exposed to Political Misinformation in a Hybrid Media Environment? The Case of the 2019 Indonesian Election. *Social Media + Society*, 8(3). https://doi.org/10.1177/20563051221122792

Rohman, M. N. (2023). It is Hard for Journalists to Pray on Time: Commodification of Media Workers in the Digital Age. *Kalijaga Journal of Communication*, 5(1), 41–56. https://doi.org/10.14421/kjc.51.03.2023

Soderborg, S., and Muhtadi, B. (2020). Policy, Partisanship, and Pay: Diverging COVID-19 Responses in Indonesia. *SSRN Electronic Journal*. https://doi.org/10.2139/ssrn.3636486

Stake, R. E. (2006). *Multiple Case Study Analysis*. Guilford Publications. https://books.google.co.id/books?id=rQWT5aDHiZYC

Tsogas, G. (2018). Transnational labor regulation, reification, and commodification: A critical review. *Journal of Labor and Society*, 21(4), 517–532. https://doi.org/10.1111/wusa.12362

Widiastuti, T. (2021). The commodification of virtual community content in increasing media traffic. *Media and Communication*, 9(2), 98–109. https://doi.org/10.17645/mac.v9i2.3737

Towards Resilient Societies: The Synergy of Religion, Education, Health, Science, and Technology – Rahiem (Ed.)

Applying the capability approach to sustainable economic development in Pesantren

Habibi Zaman Riawan Ahmad, Hamdani & Desmadi Saharuddin
Graduate School of Universitas Islam Negeri Syarif Hidayatullah Jakarta, Indonesia

ABSTRACT: This study explores how pesantren (Islamic boarding schools in Indonesia) develop sustainable economic initiatives aligned with their educational and social missions using a capability approach. A qualitative research methodology, combining library and field research, was employed. The findings reveal that three dimensions of the capability approach—freedom, justice, and functioning—explain the rapid economic development of Pesantren Walisongo Ngabar. First, economic activities initiated and managed by the pesantren, such as establishing sustainable businesses, creating employment opportunities, and providing skills training, contribute significantly to the well-being of both individuals and the broader pesantren community. Second, pesantren equip students with access to resources and capacity-building opportunities, empowering them to make autonomous economic choices and achieve their financial goals. Third, justice, rooted in Islamic ethical principles, fosters social harmony, reduces inequalities, and ensures equal access to educational and economic opportunities for all members of the pesantren community. Additionally, collaboration with surrounding communities strengthens sustainable economic development efforts. The capability approach offers a comprehensive framework for assessing and advancing sustainable economic initiatives in pesantren, emphasising equity, empowerment, and long-term impact.

Keywords: capability approach, economic development, pesantren, sustainable

1 INTRODUCTION

Pesantren (Islamic boarding school) has become one of Indonesia's education centres. It also has significant potential to contribute to the economic field. According to data from the Minister of Religion (2024), there are at least 41,599 Pesantren in Indonesia, with 3,143,555 students in 2024; among them, they have economic potency. Based on official data from the Coordinating Minister for Economic Affairs, at least 40% of pesantren have economic potential spread across various fields, such as agriculture, farming, fisheries, and micro and small businesses (Tempo 2023).

Despite this potency, the economic contributions of Pesantren are often underdeveloped because of the focus primarily on their educational mission. That is why pesantren needs to explore sustainable economic development models that can enhance their financial independence and have an important role in contributing to the well-being of the surrounding communities (Fazlurrahman *et al.* 2022). Therefore, to realise the effective role of Pesantren in enhancing economic development, whether for its group or its contribution to the larger society, a practical approach is needed to analyse some key factors of best achievement.

One of the appropriate approaches to analysing economic development is the capability approach. The capability approach offers an alternative perspective on sustainable economic development, focusing on human well-being and societal needs while safeguarding future generations' abilities to meet their needs without compromise (Sheehy and Farneti 2021). That well-being can be measured by the extent of freedom individuals or communities have in determining their preferences by selecting

DOI: 10.1201/9781003645542-78

from various meaningful 'functionings' (Zitcer *et al.* 2016). It is in line with the high potential of Pesantren, whether the number of human resources, various educational programs, or the social trust, which creates the dynamic of how the capability level of Pesantren is measured. All those complex potencies could generate great change in economic development in some good ways, which generally impact the development of local and global economies.

The study on the development of the pesantren economy has been growing, yet within a partial and non-comprehensive spectrum. Some studies have not addressed the institutional aspects of pesantren economics and the procedures through which pesantrens enhance economic activities effectively. Several studies on Pesantren's economics limited their work to only view the entrepreneurial side, Pesantren's financial management (Badrudin *et al.* 2021; Murdayanti and Puruwita 2017; Sarwenda *et al.* 2024). This study will propose recommendations for stakeholders to apply the capability approach in the context. The research questions guiding this study are: (1) what are the main challenges in pesantren economic development? (2) how does the capability approach work to develop a sustainable economy in Pesantren?

This approach helps to identify and explore the economic potency of pesantren to acquire their skills, resources, and potency to enhance their economy and improve their living standards. This approach provides a solid foundation to design holistic, sustainable, and empowerment-oriented economic development strategies for pesantren. It requires active participation from both pesantren and the community in the economic development, enhancing their role as local economic development agents.

2 METHODS

The research uses a qualitative investigation applying a capability approach to analyse sustainable economic development in Pesantren Walisongo, Ngabar. Library and field research are conducted to collect the data (Cresswell and Poth 2016). Library data related to this topic was taken from 2015 to 2024. At the same time, field data is obtained through documentation, observations, and in-depth interviews, which involve five informants. They are Kyai, foundation administrators, unit managers, senior students, and employees in Pesantren Ngabar. The data is analysed using Miles and Hubarman's model, including data reduction, data presentation, conclusion drawing, and triangulation.

The reduction of data in developing pesantren economic capabilities begins with selecting and sorting data from both field and the literature review, followed by a simplification process to focus and direct the data for subsequent presentation. The data is classified into three categories: first, the issues in pesantren economic development; second, applying the capability approach in pesantren economic development through agency theory, resource utilisation, and achievement of pesantren economic development. This categorisation considers the alignment between research objectives and issues to present the research data (Akinyode and Khan 2018). After being reduced, data is presented as descriptive analysis by applying a capability approach, which is inferred to draw a holistic conclusion.

3 FINDINGS AND DISCUSSION

This study aims to analyse the potency of sustainable economic development in Pesantren by identifying an appropriate developmental framework through the capabilities approach, drawing from literature discussing the economic development of these institutions.

3.1 *Main challenges to sustainable economic development in Pesantren*

As a religious educational institution deeply rooted in society, Pesantren has a long history of contributing to education and social services. They have demonstrated remarkable adaptability to keeping pace with contemporary developments and challenges and have some mechanisms for strategic renewal, enabling them to contextualise their role as original religious and educational institutions in Indonesia, centres for training and developing skilled human resources to community empowerment (Falikul Isbah 2020; Hudaefi and Heryani 2019).

A strong pesantren economy supports these goals. Many Pesantren still face challenges in economic development, particularly in sustaining economic activities. Key challenges include limitations in skilled resources, including financial management, restricted access to capital, technology, market access, and business unit management (Adnan and Mulyany 2023; Isbah and Sakhiyya 2023). The challenges of each pesantren are closely aligned with their unique characteristics, which each have various potency including their available resources, leadership styles, educational focus, community engagement, and economic activities based on their different culture such as traditional or modern, and geographic location. These issues above impact the quality and productivity of Pesantren's economic initiatives. However, human capital is crucial for economic development in any institution, including Pesantren.

3.2 *Applying the capability approach to develop the sustainable economy of Pesantren*

The capability approach does not assess the success of economic development merely by asset ownership but also considers the achievement of capabilities in reaching desired goals. It emphasises qualitative aspects as essential components of economic development (Stikkers 2019). It comprises three components, namely Freedom, Justice, and functioning. Freedom refers to the opportunity for individuals or communities to decide, choose, and live with their values without any pressure. Justice ensures equitable access to opportunities and resources, addresses systemic barriers, and promotes fairness to help individuals and communities recognise and actualise their full potential. Then, all valued achievements attained success by individuals and communities, which are all related to freedom and justice, is called functioning.

Based on Kabeer's empowerment theory, the capability approach is operationalised through three important aspects: agency, resource, and achievement (Huq 2023). It means that freedom, justice, and functioning work within the agency. While agency theory emphasises the role of individuals and groups in shaping their life trajectories, it is influenced by social structures and capital, including economic, cultural, social, and symbolic capital (Delva *et al.* 2021; Hayes 2020).

In the context of economic development at Pesantren Wali Songo Ngabar Ponorogo, the agency's role encompasses the Kyai, foundation administrators, unit managers, senior students, and employees. The Kyai and foundation administrators hold the highest hierarchical positions and possess significant symbolic capital, which attracts community trust. This trust directly impacts the increasing number of students and the human resources managed by the pesantren to cultivate superior capabilities (Observation). These capabilities are future investments for individuals, the pesantren, and society. In formulating economic policies, the Kyai and administrators exercise *freedom* by making policies that meet the pesantren's potential while upholding the value of *justice*, exemplified by consensus-based decision-making processes in designing economic development programs (Interview). The implementation pattern of the capability approach at the Ngabar Islamic Boarding School is illustrated in more detail in Figure 1 below.

Pesantren Ngabar, with 3,257 students and 543 teachers (as of December 2024), manages 20 economic units involving 17 managers and 69 employees. The appointment of managers is not based on social hierarchy but on individual capabilities. For instance, Mr Oki Iskandar, a senior student with expertise in agriculture, leads the Ngabar Farm sector. His expertise was acquired through hands-on experience, knowledge exchange with professionals, and participation in agricultural and livestock training programs. In managing enterprises such as melon, shallot, and maise cultivation, the managers collaborate with foundation administrators to analyse internal conditions and market challenges (Interview). The managers' performance is profit-oriented and focuses on meeting the students' essential and secondary needs, such as barbershops, distros, minimarkets, and cafeterias. This demonstrates a management approach grounded in *freedom* and *justice* (Observation). Figure 1 illustrates Ngabar Pesantren's business activities' contribution to economic development, closely linked to the capability approach.

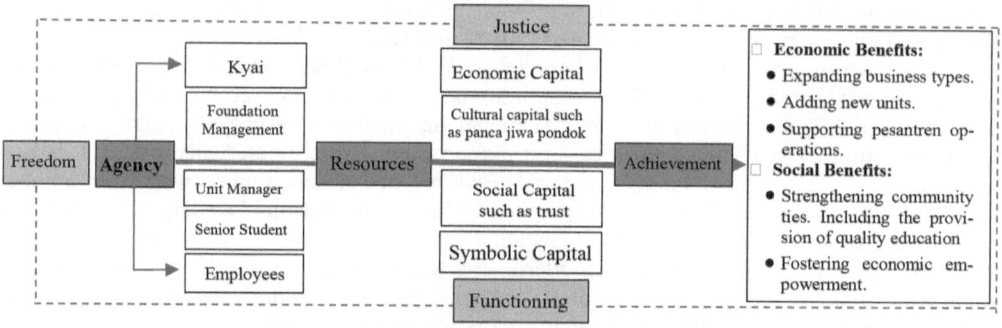

Figure 1. The application pattern of the capability approach in developing the pesantren economy (Source: Researchers processed from various sources).

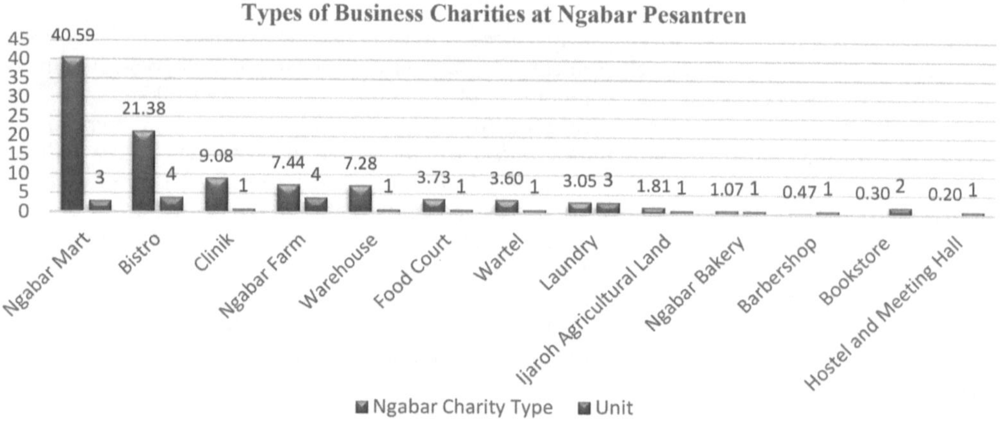

Figure 2. The economic benefits of implementing pesantren economic capability (in percent) (Source: Ngabar Pesantren 2024).

Pesantren Ngabar's openness in managing business units and adapting to the digital era reflects the application of freedom (Figure 2). The principle of *justice* is evident in how resources are managed without discrimination against employees, whether internal or external (Interview with employee). These values extend to partnership programs, granting all stakeholders the liberty to maximise their capabilities for collective economic advancement. This policy reflects not only respect for individual freedom but also social justice. By providing opportunities for the community to actualise their abilities, the pesantren contribute to the independence, economic resilience, and sustainable economic development of the institution and its surrounding society (Observation). In response to the digital era, the payment systems in Pesantren's economic units have transitioned to a card-based, digitally integrated method. This transformation enables parents to monitor student fund usage in real time. At the same time, for the pesantren, this system supports realising the principle of freedom through a digital financial information governance policy monitored by stakeholders (Observation). Furthermore, transparency in financial management reflects *justice*, ensuring fairness in financial reporting. This digital integration enhances *functioning* by improving efficiency, enabling the pesantren to focus more effectively on sustainably developing its economic and social capacities.

The importance of freedom, justice, and functioning in developing the sustainable economy of Pesantren as the pillar of a resilient community's economy is practically shown in some studies conducted in another Pesantren. Alhifni *et al.*, (2021) reveal that Pesantren Gontor enhances economic development by training students and communities with skills and facilitating their potential through business units. Similarly, Sunan Drajat and Sidogiri pesantren foster alumni collaborations, creating sustainable partnerships that promote economic independence and strengthen the pesantren's role in community development (Bawono *et al.* 2020; Fazlurrahman *et al.* 2022). In this context, there is potential for halal products owned by Pesantren to be collaboratively developed through the existing Pesantren network (Qizam *et al.* 2024).

Another Pesantren who implicitly works with freedom, justice, and functioning values is Pesantren Rakyat Sumberpucung. Economic freedom within pesantren involves pursuing economic activities aligned with religious values without external pressures. Even in some cases, determining freedom in development is done through the istikharah prayer, which Muslims believe guides them in seeking the best choice. This freedom supports independent economic management. Justice in this context ensures fair access to economic opportunities and resources, regardless of social status, and promotes the well-being of all pesantren community members. The practices at Pesantren Rakyat Sumberpucung exemplify how pesantren-driven economic activities, grounded in Islamic moral values like *ta'awun* (mutual assistance) and *maslahat* (the common good), contribute to equitable development (Musnandar and Alzitawi 2024).

The capability approach in community economic development is evident in case studies of several countries. Crick and Crick (2024) found that this approach helps Canadian university students create startup business models aligned with their aspirations. Arafat *et al.* (2020) highlighted that women's entrepreneurial involvement in India is strongly influenced by their capabilities. Similarly, countries like Japan, Thailand, and Nepal have successfully implemented the One Village One Product (OVOP) model in rural areas, driven by capabilities, training, and skills support (Gupta *et al.* 2015). Supported by competent resources, the capability approach significantly contributes to economic success in various contexts (Ahamad *et al.* 2020).

4 CONCLUSION

As an approach that offers sustainable development, the capability approach offers a more comprehensive perspective on the dynamics of Pesantren's economic development in achieving significant economic development. Using the capability approach, with its three elements, proves that Pesantren Walisongo Ngabar possesses an excellent agency comprising professional agents such as Kyai, business unit managers, employees, and stakeholders. In developing The Economy of Pesantren, various policies or regulations related to the economic development reveal their commitment to realising freedom and justice, then appreciate individual's capability by functioning, that is, giving them chances to actualise individual or group capability through some facilities provided by Pesantren. Maximalizing Pesantren's resources with the open-minded agency's management enables Pesantren Walisongo Ngabar to achieve the best in its economy.

REFERENCES

Adnan, M. I., and Mulyany, R. (2023). Accountability of Islamic Boarding School: A Comparative Insights from Modern and Traditional Pesantren in Aceh, Indonesia. *2023 International Conference on Sustainable Islamic Business and Finance*, SIBF 2023, 246–250.

Ahamad, S., Al-jaifi, H. A., and Mostafiz, M. I. (2020). Conceptualizing recourses as antecedents to the economic performance of family-based microenterprise – the moderating role of competencies. *Journal of Family Business Management*, 0–25. https://doi.org/10.1108/JFBM-04-2019-0027

Akinyode, B. F., and Khan, T. H. (2018). Step by Step Approach for Qualitative Data Analysis. *International Journal of Built Environment and Sustainability*, 5(3), 163–174.

Alhifni, A., Ahwarumi, B., and Kurniasari, W. (2021). Application of Entrepreneurial Values Rahmatan Lil'alamin Through the Islamic Boarding School Incubator and Supporting the Independence of Islamic Boarding Schools (Case Study Islamic Boarding School Sunan Drajat). *American Journal of Humanities and Social Sciences Research*, 5(3), 434–441. www.ajhssr.com

Arafat, M. Y., Ali, J., Dwivedi, A. K., and Saleem, I. (2020). Social and Cognitive Aspects of Women Entrepreneurs: Evidence from India. *Vikalpa*, 45(4), 223–239. https://doi.org/10.1177/02560909211015457

Badrudin, Satori, D., Komariah, A., and Kurniady, D. A. (2021). The Implementation of Pesantren Financing Based on Agribusiness Social Entrepreneurs. *Jurnal Ilmiah Peuradeun*, 9(1), 17–38. https://doi.org/10.26811/peuradeun.v9i1.504

Bawono, A., Saerozi, M., Irkhami, N., Nabila, R., Anwar, S., and rof. (2020). The Determinants of Creative Economy in Developing the Independence of Pesantren. *International Journal of Innovation, Creativity and Change*, 13(7), 1540–1558.

Cresswell, J. W., and Poth, C. N. (2016). *Qualitative Inquiry and Research Design Choosing Among Five Approaches (Fourth Edi)*. SAGE Publications Inc. https://books.google.co.id

Crick, J. M., and Crick, D. (2024). Student entrepreneurs with triple bottom line objectives: capabilities in creating viable business models. *Education and Training*, 66(9), 1230–1245. https://doi.org/10.1108/ET-04-2024-0175

Delva, J., Forrier, A., and De Cuyper, N. (2021). Integrating Agency and Structure in Employability: Bourdieu's Theory of Practice. *Journal of Vocational Behavior*, 127. https://doi.org/https://doi.org/10.1016/j.jvb.2021.103579

Falikul Isbah, M. (2020). Pesantren in the changing indonesian context: History and current developments. *Qudus International Journal of Islamic Studies*, 8(1), 65–106.

Fazlurrahman, H., Abqari, L. S., and Wardoyo, D. T. W. (2022). Pesantren (Islamic Boarding Schools): The Largest Form of Social Entrepreneurship in Indonesia. In V. Ratten (Ed.), *Driving Entrepreneurship in Southeast Asia* (1st Editio, pp. 54–61). Routledge.

Gupta, S., Yadav, G., Choudhary, A., and Kaur, A. (2015). Emergent Trends in Sustainable Technologies in Thailand: Developing OTOP-Based Manufacturing Capabilities in Rural India. In *Promoting Socio-Economic Development through Business Integration* (pp. 48–59). IGI Global. https://doi.org/10.4018/978-1-4666-8259-7.ch004

Hayes, A. S. (2020). The behavioral economics of pierre bourdieu. *Sociological Theory*, 38(1), 16–35. https://doi.org/10.1177/0735275120902170

Hudaefi, F. A., and Heryani, N. (2019). The Practice of Local Economic Development and Maqāṣid al-Sharī'ah. *International Journal of Islamic and Middle Eastern Finance and Management*, 12(5), 625–642. https://doi.org/10.1108/IMEFM-08-2018-0279

Huq, A. (2023). Naila Kabeer: Deconstructing Empowerment of Poor Women Entrepreneurs in Postcolonial Bangladesh. *In Postcolonial Feminism in Management and Organization Studies: Critical Perspectives from India, Pakistan, and Bangladesh* (pp. 17–35). Taylor and Francis. https://doi.org/10.4324/9781003197270-3

Isbah, M. F., and Sakhiyya, Z. (2023). *Pesantren in Contemporary Indonesia: Negotiating Between Equity and the Market* (Z. Sakhiyya and T. Wijaya Mulya (eds.); pp. 137–152). Springer Nature Singapore. https://doi.org/10.1007/978-981-99-1878-2_8

Muhith, A., Umiarso, U., Mardiana, D., Rafik, A., and Fatmawati, E. (2021). Quality Development of Pesantren Education In East Java. *In Proceedings of the 6th Batusangkar International Conference, BIC 2021, 11-12*. https://doi.org/10.4108/eai.11-10-2021.2319578

Murdayanti, Y., and Puruwita, D. (2017). Transparency and Accountability of Financial Management in Pesantren (Islamic Education Institution). *Advanced Science Letters*, 23(11), 10721–10725. https://doi.org/10.1166/asl.2017.10138

Musnandar, A., and Alzitawi, D. U. D. M. (2024). Religious-Sociocultural Networks and Social Capital Enhancement in Pesantren. *Jurnal Pendidikan Islam*, 10(1), 137–148. https://doi.org/10.15575/jpi.v10i1.19997

Qizam, I., Berakon, I., and Ali, H. (2024). The role of halal value chain, Sharia financial inclusion, and digital economy in socio-economic transformation: a study of Islamic boarding schools in Indonesia. *Journal of Islamic Marketing*. https://doi.org/10.1108/JIMA-03-2024-0108

Sarwenda, S., Rahim, H., Rosyada, D., Zamhari, A., and Salim, A. (2024). Development of Digital Entrepreneurship Programs in Pesantren in Indonesia. *Religion, Education, Science and Technology towards a More Inclusive and Sustainable Future: Proceedings of the 5th ICIIS 2022, Lombok, Indonesia, 19-20 October 2022, 2021*, 273–278. https://doi.org/10.1201/9781003322054-46

Sheehy, B., and Farneti, F. (2021). Corporate Social Responsibility, Sustainability, Sustainable Development and Corporate Sustainability: What is the Difference, and Does it Matter? *Sustainability (Switzerland)*, 13(11). https://doi.org/10.3390/su13115965

Stikkers, K. W. (2019). Growth', Economic and Human: Reconstructing Economics through Pragmatism and the Capabilities Approach. *Contemporary Pragmatism*, 16(2–3), 286–306.

Tempo. (2023, April). *Potensi Ekonomi Pesantren*. https://www.tempo.co (Accessed: 27/8/2024)

Zitcer, A., Hawkins, J., and Vakharia, N. (2016). A capabilities approach to arts and culture? Theorizing community development in West Philadelphia. *Planning Theory and Practice*, 17(1), 35–51. https://doi.org/10.1080/14649357.2015.1105284

Towards Resilient Societies: The Synergy of Religion, Education, Health, Science, and Technology – Rahiem (Ed.)

Bibliometric insights into poverty alleviation strategies in Indonesia

Aldi Munandar
Syarif Hidayatullah State Islamic University Jakarta, Indonesia

Ali Rama
Syarif Hidayatullah State Islamic University Jakarta, Indonesia
ORCID ID: 0000-0002-6335-1890

Nurul Ichsan
Syarif Hidayatullah State Islamic University Jakarta, Indonesia

ABSTRACT: This study examines the evolving landscape of poverty alleviation research in Indonesia using Scopus metadata, comprising 351 entries. A bibliometric analysis conducted via Biblioshiny reveals an 8.83% annual growth rate in poverty research from 1991 to 2023, with a sharp increase from 2015 onward, peaking at 45 publications in 2023. The thematic map identifies eight key clusters: welfare and economic growth intersect as core themes alongside microfinance; impact evaluation and microfinance emerge as driving themes; social protection and livelihoods are categorized as niche themes; while education and social forestry appear in the declining or emerging themes quadrant. Each cluster is labeled with representative research areas, offering a structured overview of Indonesia's poverty research landscape. This study contributes to Sustainable Development Goal 1 (No Poverty) by identifying research gaps and providing evidence-based insights for policymakers, academics, and practitioners to develop targeted, sustainable poverty reduction strategies within Indonesia's policy framework.

Keywords: Poverty alleviation, Indonesia, bibliometric, microfinance, research

1 INTRODUCTION

Despite the abundance of research on poverty alleviation in Indonesia, there are still a notable gap of studies using bibliometric analysis approaches. Globally, Yu and Huang (2021) have investigated poverty alleviation research. The study reveals a significant surge in publications on this topic in the 21st century, especially post-2015, underscoring disparities in both quantity and quality of research between China and Kenya. The article advocates for enhancing interdisciplinary integration, emphasizing contributions, addressing intergenerational poverty, and exploring diverse research dimensions to guide future poverty alleviation research efforts.

In a more specific national context, bibliometric analysis research has been conducted in China by Li *et al.* (2021). The study explores poverty alleviation efforts during China's 13th Five-Year Plan, employing statistics and bibliometric analysis from the Web of Science database. The research indicates sustainable development and multidimensional poverty as emerging and critical topics in China's poverty alleviation initiatives. Focusing on sustainability and addressing poverty comprehensively across all dimensions are key elements of China's strategy to combat poverty. Similarly, in Latin America, bibliometric studies by Amegah and Agyei-Mensah (2017) have outlined within the broad framework of the international policy instruments, measures for addressing urban air pollution and its associated health impact in targeted underserved populations.

DOI: 10.1201/9781003645542-79

Conducting a bibliometric analysis on poverty alleviation in Indonesia is crucial for understanding the current research landscape and guiding future efforts. By mapping out existing studies, such an analysis identifies knowledge gaps and tracks the evolution of research in this field. It also evaluates the impact of specific works, helping to highlight influential studies and emerging thought leaders. This information is valuable for aligning research efforts with Indonesia's national poverty alleviation goals, ensuring that both academic and practical approaches contribute effectively to addressing the country's unique challenges. Moreover, a bibliometric analysis informs policymaking and resource allocation by providing evidence of what research has been effective in guiding poverty reduction strategies. It helps funding agencies prioritize underexplored but critical areas and enables academic collaboration by identifying key researchers and institutions. Ultimately, it supports the development of more informed, targeted, and impactful poverty alleviation initiatives in Indonesia.

2 METHODOLOGY

2.1 *Data extraction*

The data extraction process follows a protocol akin to PRISMA guidelines for systematic reviews (Page *et al.*, 2021). This involves several key steps: identification, screening, and inclusion. In this research, identification begins with focusing on themes such as poverty alleviation and sustainable development goals in Indonesia. The Scopus database is utilised for comprehensive searches across titles, abstracts, and keywords. The search terms are adapted from Hassan *et al.* (2023), encompassing phrases like "poverty* alleviation," "reduce* poverty," "poverty reduction," "poverty elimination," "poverty mitigate*," and "poverty eradicate*." Including the keyword "Indonesia" ensures alignment with the primary research goal of understanding poverty alleviation research, specifically within Indonesia.

In the final stage of the search on the Scopus database, the research query was formulated to target specific fields, such as titles and keywords, using the TITLE-ABS-KEY field. The query included terms like "poverty* alleviation," "reduce* poverty," "poverty reduction," "poverty elimination," "poverty mitigate*," and "poverty eradicate*," alongside the term "Indonesia." Exclusions were applied for regions outside Indonesia, including "Southeast Asia," "Asia," "Eurasia," and specific countries such as "Vietnam," "China," "India," "South Asia," "Malaysia," "Nepal," "Bangladesh," "South America," "Africa," "Uganda," "Thailand," "Sub-Saharan Africa," "Philippines," "eastern hemisphere," "East Africa," and "Colombia." The search focused on articles published between 1989 and 2023, with document type restricted to "article," publication stage set to "final," and language specified as English. During the screening phase, applying these filters reduced the metadata to 351 articles, which were prepared for further analysis.

2.2 *Analysis technique*

We employ a bibliometric analysis, guided by contemporary procedures as outlined by Donthu *et al.* (2021) and Öztürk *et al.* (2024). Biblioshiny software will focus on extracting main information and annual publication data from the metadata. For knowledge mapping, the conceptual structure technique will be utilised, leveraging Biblioshiny (Aria and Cuccurullo, 2017). Specifically, a thematic map will be used in the conceptual structure analysis, providing a comprehensive overview of the evolving concepts in poverty alleviation research in Indonesia.

3 FINDINGS AND ANALYSIS

3.1 *Main information about data*

The metadata on poverty alleviation research in Indonesia covers the period from 1991 to 2023 (Table 1). This dataset includes 246 journals or sources, encompassing 351 journal articles. The research in this field has a growth rate of 8.83%, with the average age of articles being approximately 7 years. Each article receives an average of 15 citations. The total number of references across all articles in the metadata is 14,156. A key focus of this analysis will be the 977 author keywords. The 351 articles have been authored by a total of 1,014 individuals, with an international collaboration rate of 24.5%.

Table 1. Summary of research publications (1991-2023): growth, authorship, and citation metrics.

Description	Results
MAIN INFORMATION ABOUT DATA	
Timespan	1991:2023
Sources (Journals, Books, etc)	246
Documents	351
Annual Growth Rate %	8.83
Document Average Age	6.82
Average citations per doc	15.48
References	14156
DOCUMENT CONTENTS	
Keywords Plus (ID)	706
Author's Keywords (DE)	977
AUTHORS	
Authors	1014
Authors of single-authored docs	62
AUTHORS COLLABORATION	
Single-authored docs	63
Co-Authors per Doc	3.19
International co-authorships %	24.5
DOCUMENT TYPES	
article	351

3.2 *Annual publication*

Research on poverty alleviation in Indonesia began in 1991, with 3 journal articles published that year. In 1993, the number of publications remained the same, with 3 articles. Initially, research in this field was relatively sparse. However, there was a noticeable increase in 2014-2015, with 30 publications over these two years. Since then, research on poverty alleviation in Indonesia has gained significant momentum. The increased research on poverty alleviation in Indonesia can be largely attributed to the Sustainable Development Goals (SDGs), particularly Goal 1 (No Poverty). The adoption of the SDGs in 2015 provided a global framework and clear targets for eradicating poverty, significantly highlighting the focus on poverty-related issues. This global agenda has increased funding, attention, and collaboration among researchers, policymakers, and NGOs. The peak of poverty alleviation research in Indonesia was reached in 2023, with 45 publications.

3.3 *Thematic map*

The thematic map analyses the topics discussed in poverty alleviation research in Indonesia, categorising clusters into four quadrants (Figure 1). The first quadrant, basic themes, includes topics with high impact but limited development, such as welfare, economic growth, and an intersection with microfinance. The second quadrant, motor themes, consists of clusters that drive research on poverty alleviation in Indonesia, including microfinance and impact evaluation. The third quadrant, niche themes, encompasses specialised topics that are well-developed in the context of poverty alleviation in Indonesia, such as livelihoods and social protection. The final quadrant, declining or emerging themes, includes clusters that are either developing or being phased out, such as education and social forestry.

Cluster 1 encompasses chronic poverty, empowerment, food security, government policy, productive zakat, quality of life, resilience, urban poverty, and welfare. This cluster emphasises the critical role of community empowerment and resilience in addressing chronic poverty, especially in urban areas. Ensuring food security and enhancing quality of life are key priorities for promoting overall well-being. Effective government policies and productive zakat programs are essential for supporting social and economic welfare. A comprehensive approach that involves multiple stakeholders is necessary for sustainably combating poverty.

Cluster 2 covers development, livelihoods, Papua, and small-scale fisheries. Sustainable development and improving local livelihoods are key to addressing poverty in Indonesia. Small-scale fisheries are significant for providing livelihoods and enhancing the well-being of Indonesian communities. A holistic and inclusive approach is crucial for achieving long-term sustainability and prosperity.

Cluster 3 encompasses agriculture, ARDL, COVID-19, efficiency, gender, income inequality, investment, microfinance, unemployment, and zakat. The primary focus is on the agricultural sector and its effects on poverty, particularly during the COVID-19 pandemic, which impacted efficiency and food security. Income inequality and unemployment are also significant issues, with investment and microfinance identified as potential solutions to enhance well-being. Additionally, zakat and other social assistance programs are vital for poverty reduction, while gender issues underscore the need for inclusive policies to empower women. A comprehensive approach that integrates these various elements is crucial for achieving sustainable poverty alleviation.

Cluster 4 encompasses economic development, economic growth, financial inclusion, fiscal decentralisation, inequality, Islamic microfinance, social capital, and sustainability. The focus is primarily on economic development and growth, crucial for sustainable poverty reduction. Financial inclusion is vital in expanding access to formal financial services and fostering inclusive economic growth. Fiscal decentralisation is also important, as it allows local governments to allocate resources more effectively and implement policies tailored to local needs. Income inequality remains a significant challenge, and Islamic microfinance offers a Sharia-compliant solution to address this issue. Social capital helps strengthen community networks and boost participation in economic activities, thereby supporting sustainable development. Integrating these elements into a comprehensive approach is essential for achieving lasting poverty alleviation in Indonesia.

Cluster 5 encompasses impact evaluation, Kalimantan, multidimensional poverty, oil palm, rural development, and smallholders. The primary focus is assessing the impact of various programs and policies to reduce multidimensional poverty in Kalimantan. The development of oil palm plantations is a central strategy for rural development, involving smallholders as key participants. This approach seeks to improve rural welfare by boosting income and access to economic resources. However, it is essential to address the environmental and social impacts of expanding oil palm plantations and ensure that economic benefits are equitably shared among all community members. A comprehensive approach that combines impact evaluation, rural development, and smallholder empowerment is crucial for achieving sustainable poverty alleviation.

Cluster 6 focuses on education and income, examining how enhancing access to and quality of education can boost community income levels. Improved education gives individuals the skills needed in the job market, increasing their chances of obtaining better-paying and more stable employment. Additionally, targeted educational programs for low-income groups can help bridge income gaps and promote social mobility. Therefore, investing in education is a critical strategy for effective poverty alleviation in Indonesia.

Cluster 7 includes topics such as the panel vector error correction model (PVECM), rural poverty, and social protection. This cluster primarily focuses on how PVECM can be used to understand rural poverty dynamics and evaluate social protection programs' effectiveness. PVECM helps analyse long-term and short-term relationships between economic variables affecting poverty. Since rural poverty is often more severe than urban poverty, it requires targeted policy interventions. Social protection programs, such as cash transfers and subsidies, are designed to mitigate the impact of economic shocks on rural communities and improve their well-being. Integrating PVECM analysis focusing on rural poverty and social protection provides valuable insights and more effective solutions for poverty alleviation in Indonesia.

Cluster 8 focuses on social forestry, a policy designed to provide local communities with legal access to manage and sustainably use forest resources. This initiative aims to improve the well-being of impoverished communities living in and around forests by granting them rights to forest management. These communities can engage in economic activities such as agroforestry, ecotourism, and the harvesting of non-timber forest products through social forestry. This approach not only helps to alleviate poverty but also contributes to forest conservation. Thus, social forestry is a vital strategy for poverty alleviation in Indonesia, especially in rural areas that rely on forest resources.

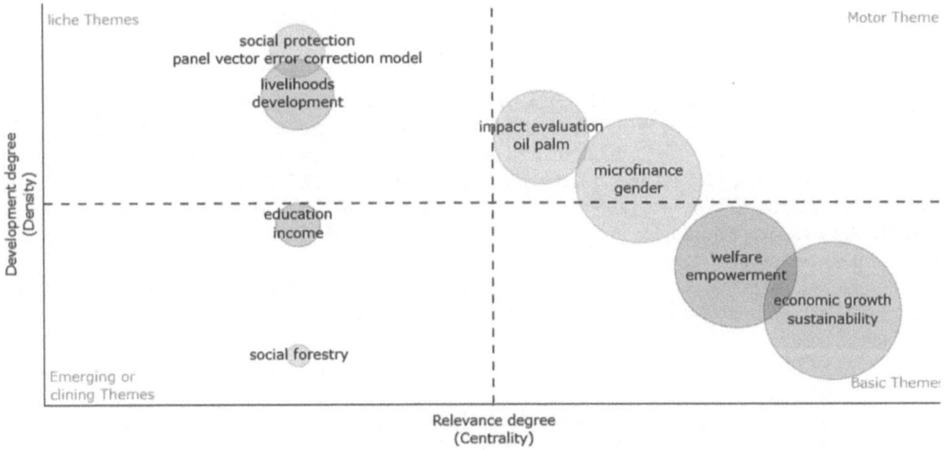

Figure 1. Thematic map of research topics: relevance degree vs. development degree.

Parameter Top 50 Words, Min Cluster Frequency (Per Thousand Docs) 10, Label 2, StopWords 'poverty, Indonesia, Poverty Reduction, Poverty Alleviation', Synonym (example 'social welfare' merge to 'welfare').

We assign labels to represent the topics within each cluster on the thematic map. The thematic map analysis includes a document option detailing each cluster's articles. These articles are identified and extracted from their full texts by the author. The author then provides examples of research for each cluster, supported by evidence from the thematic map analysis. For future research, it is advisable to focus on clusters in the motor themes and basic themes quadrants, as the topics in these areas significantly impact poverty alleviation in Indonesia.

4 CONCLUSION

This bibliometric analysis offers a comprehensive and holistic view of poverty alleviation research in Indonesia from 1991 to 2023. Research in this field has grown by 8.83%, peaking in 2023, which saw 45 publications. The thematic map analysis identifies eight key clusters: welfare, livelihoods, microfinance, economic growth, impact evaluation, education, social protection, and social forestry. Future research should focus on the clusters of economic growth, welfare, microfinance, and impact evaluation, particularly in the context of their alignment with Sustainable Development Goals (SDGs).

A limitation of this research is its dependence on a single database, Scopus. To enhance the quality and comprehensiveness of the metadata, it would be beneficial to include additional databases, such as Web of Science and Google Scholar. Incorporating a broader range of sources could provide a more robust and detailed analysis. Future studies should prioritise exploring the impact of innovative financial tools, such as digital Islamic microfinance and fintech solutions, on poverty reduction (Rama and Yaman, 2024). Moreover, longitudinal studies employing mixed-method approaches could investigate the effectiveness of integrated poverty alleviation strategies across diverse regions in Indonesia. Researchers should also delve deeper into the role of social forestry and sustainable rural development in addressing multidimensional poverty while mitigating environmental degradation. Finally, conducting comparative analyses of Indonesia's poverty alleviation programs with those of other countries could offer valuable insights and policy recommendations.

REFERENCES

Amegah, A. K., and Agyei-Mensah, S. (2017). Urban air pollution in Sub-Saharan Africa: Time for action. *Environmental Pollution*, 220, 738–743. https://doi.org/https://doi.org/10.1016/j.envpol.2016.09.042

Aria, M., and Cuccurullo, C. (2017). bibliometrix: An R-tool for comprehensive science mapping analysis. *Journal of Informetrics*, 11(4), 959–975. https://doi.org/10.1016/j.joi.2017.08.007

Donthu, N., Kumar, S., Mukherjee, D., Pandey, N., and Lim, W. M. (2021). How to conduct a bibliometric analysis: An overview and guidelines. *Journal of Business Research*, 133, 285–296. https://doi.org/https://doi.org/10.1016/j.jbusres.2021.04.070

Hassan, M. K., Alshater, M. M., Banna, H., and Alam, M. R. (2023). A bibliometric analysis on poverty alleviation. *International Journal of Ethics and Systems*, 39(3), 507–531.

Li, H., Chen, J., and Li, J. (2021). A bibliometric analysis of poverty alleviation researches in China during the 13th Five-year Plan. *Journal of Physics: Conference Series*, 1827(1), 12205. https://doi.org/10.1088/1742-6596/1827/1/012205

Öztürk, O., Kocaman, R., and Kanbach, D. K. (2024). How to design bibliometric research: an overview and a framework proposal. *Review of Managerial Science*, 1–29.

Page, M. J., McKenzie, J. E., Bossuyt, P. M., Boutron, I., Hoffmann, T. C., Mulrow, C. D., Shamseer, L., Tetzlaff, J. M., Akl, E. A., Brennan, S. E., Chou, R., Glanville, J., Grimshaw, J. M., Hróbjartsson, A., Lalu, M. M., Li, T., Loder, E. W., Mayo-Wilson, E., McDonald, S. and Moher, D. (2021). The PRISMA 2020 statement: An updated guideline for reporting systematic reviews. *The BMJ*, 372. https://doi.org/10.1136/bmj.n71

Rama, A., and Yaman, B. (2024). Islamic Crowdfunding and Social Entrepreneurship: The Effect of Organizational Orientation Rhetoric on Crowdfunding Success. *Journal of Social Entrepreneurship*, 1–26. https://doi.org/10.1080/19420676.2024.2326128

Yu, Y., and Huang, J. (2021). Poverty Reduction of Sustainable Development Goals in the 21st Century: A Bibliometric Analysis. *Frontiers in Communication*, 6(October), 1–15. https://doi.org/10.3389/fcomm.2021.754181

Towards Resilient Societies: The Synergy of Religion, Education, Health, Science, and Technology – Rahiem (Ed.)
© 2026 The Author(s), ISBN: 978-1-041-08471-6
Open Access: www.taylorfrancis.com, CC BY-NC-ND 4.0 license

Integrating blockchain technology into waqf governance: Enhancing transparency and accountability in Sharia finance for achieving SDGs

Kaharuddin Kaharuddin
Pamulang University, Indonesia

Asep Saepudin Jahar
Universitas Islam Negeri Syarif Hidayatullah Jakarta, Indonesia

Arta Amaliah Nur Afifah
Raden Fatah State Islamic University Palembang, Indonesia

ABSTRACT: Blockchain technology introduces an innovative approach to waqf governance by incorporating smart contracts based on distributed ledger technology and encryption security. This advancement enhances administrative services and strengthens public trust in waqf management. This research uses a qualitative descriptive approach based on a literature review to explore the benefits of applying blockchain technology in waqf governance. The findings reveal that blockchain technology significantly improves the previously unstructured waqf governance system. By enabling encrypted and decentralised transaction recording, blockchain ensures transparent management of waqf funds, facilitates efficient tracking of fund status, and prevents data manipulation. Additionally, blockchain enhances the efficiency and security of transactions, paving the way for an innovative and digital waqf ecosystem. The technology also promotes financial inclusion by expanding public access to participate in waqf initiatives. This broader participation contributes to poverty alleviation (SDG 1) and reduces economic disparities (SDG 10) through waqf funds' transparent and accountable distribution. Integrating blockchain technology into waqf governance offers a transformative solution for advancing transparency, efficiency, and inclusivity in Sharia financial systems.

Keywords: blockchain, governance system, waqf

1 INTRODUCTION

The Society 5.0 era has accelerated the development of information and communication technology, including in the world of cryptography. One of its key innovations is blockchain technology, which offers an encrypted transaction recording system with high security and transparency without involving third parties (Muchlis Gazali and Che Ismail 2019). The Indonesian Waqf Agency (BWI) sees the potential of this technology to modernize the waqf management system that has been conventional. Waqf as one of the instruments of Islamic philanthropy plays an important role in overcoming poverty (SDG 1) and economic inequality (SDG 10). With the development of cash waqf, the potential of waqf is increasingly dynamic in empowering the people's economy and distributing its benefits more effectively to people in need (Afifah 2021). The application of blockchain in waqf management is expected to increase the security, efficiency, and transparency of fund management. This technology provides innovative solutions to key challenges in philanthropic governance, such as lack of transparency, data manipulation risks, and distribution efficiency (Mohamed and Ali 2022).

DOI: 10.1201/9781003645542-80

Recent research shows that blockchain not only improves the quality and accountability of Islamic philanthropic fund management but also facilitates more effective tracking and distribution of funds (Al-Saudi 2023; Huang and Nordin 2021; Ibrahim 2023). Blockchain also strengthens the principles of Islamic finance by creating a fairer and more inclusive system and provides opportunities for broader socio-economic integration (Zaman *et al.* 2023). Through the application of blockchain in the waqf governance system, the potential of this technology can be aligned with several other SDG goals, such as financial inclusion (SDG 8) and economic infrastructure innovation (SDG 9). By providing easier and more affordable access to finance, this technology allows more individuals to participate in waqf schemes. In addition, a more accurate and transparent distribution of waqf funds can help reach marginalised groups often marginalised in the traditional economic system (Abd Razak *et al.* 2020). showing its benefits in improving the operational efficiency and security of financial transactions. Given this potential, there is an urgency for the government and related stakeholders to promote and implement blockchain technology in the waqf governance system. This step is expected to increase public understanding, optimise the management of waqf funds by nazhir, and improve the community's economic welfare more effectively and sustainably (Elasrag 2019).

This research aims to complement existing shortcomings by analysing blockchain technology is effectiveness in the waqf governance system. For this reason, this paper aims to answer three important questions: (a) how effective is blockchain technology in the waqf governance system? (b) what are the characteristics of blockchain applied to the waqf governance system? (c) what are the implications of using blockchain technology in the waqf governance system? These three statements can later be a way for business actors to explain how effective blockchain technology is in the waqf governance system. This paper is based on the argument that blockchain technology is possibly one that has a major impact in transforming waqf governance to digital (Mohamed and Lahsasna 2020). With the integration of smart contracts, this technology can reduce the risk of fraud, reduce administrative costs, and improve overall operational efficiency. In this context, blockchain serves as a technological tool and an innovative solution to address socio-economic challenges relevant to global development goals. Therefore, the waqf blockchain is integral to the digital paradigm shift. From securing financial transactions to creating solutions for more efficient supply chains, blockchain has opened the door to innovations that have significantly impacted a wide range of industries, including waqf (Meskini and Aboulaich 2024). The challenge of shaping a safer and more efficient digital future through blockchain technology is exciting and promising.

2 METHOD

This research adopts a descriptive qualitative approach, examining social conditions through the lens of blockchain phenomena such as those in the cryptocurrency domain, which are widely used by millennials as storage systems that utilise cryptography. This study seeks to analyse and describe the influence of blockchain technology on the waqf governance system. Furthermore, an interactive and constructive paradigm is applied, emphasising detailed exploration and synthesis from various sources (Sugiyono 2018). The researcher will analyse and describe the influence of blockchain technology on the waqf governance system. In addition, the researcher uses an interactive and constructive paradigm that looks at detailed explanations from various existing sources.

To obtain data, this study uses two types of data sources: primary and secondary. Primary data sources are from information media such as newspapers, magazines, academic articles, journals, and other social media platforms relevant to the blockchain theme. Secondary data sources are collected through library research, including books and scientific works on blockchain governance and waqf issues (Hadi 2004). The literature review employs a structured approach to enhance transparency and replicability, using keywords like "blockchain technology," "waqf governance," and "smart contracts" to identify relevant studies. Inclusion criteria focus on recent, peer-reviewed works discussing blockchain and waqf governance, while exclusions cover irrelevant or non-empirical sources. Thematic coding is applied to analyse recurring themes, such as transparency, efficiency, and accountability in blockchain-based waqf systems. This systematic process synthesises insights into blockchain's effectiveness and growing adoption in modern waqf governance.

3 RESULTS AND DISCUSSION

3.1 *Blockchain technology in the context of islamic economics*

Blockchain is an innovative technology that serves as a distributed ledger in a peer-to-peer system without a central authority. The system consists of interconnected blocks of digital information, forming a chain stored in a public database (Mbaidin *et al.* 2024). The main characteristic of blockchain is its permanent and transparent nature, where the data that has been stored cannot be changed without the consent of the majority of network users. The application of this technology gave birth to the concept of smart contracts, which are contracts converted into programming code format and integrated into the blockchain network. The advantage of blockchain lies in its ability to store data consistently and chronologically and prevent unauthorised transaction entries, thus creating a secure and reliable system for tracking different types of transactions (Ghezal and Lahsasna 2020).

The application of blockchain technology in the Islamic economic sector has significant potential to increase the efficiency and effectiveness of Islamic financial institutions. This system ensures transaction fees' security, transparency, and efficiency while minimising the risk of corruption and fraud. A World Bank study (2019) shows that blockchain has the potential to trigger an industrial revolution that will change economic and business models, especially in the Islamic finance industry. Blockchain implementation is expected to encourage increased productivity and public trust in the Islamic economic sector (Aysan and Al-Saudi 2023). However, applying this technology requires a deep understanding from scholars regarding its concept and its conformity with the principles of Maqashid al-Sharia. Thus, blockchain is seen as a potential tool to develop and strengthen the position of the Islamic economic sector in the future (Makarim and Hamzah 2024).

3.2 *The use of blockchain technology in the waqf governance system*

The merger of waqf with blockchain technology is a significant innovation in developing the sharia economy, especially in waqf. This technology offers a more efficient, transparent, and secure governance system for managing waqf assets. Blockchain allows the recording of transactions in a decentralised, open, and non-manipulateable manner, thereby increasing trust between wakif, nazhir, and mauquf alaih (Huang and Nordin 2021). This system makes the waqf process simpler, with each transaction recorded in interconnected blocks and equipped with a timestamp to prevent duplication. The application of blockchain in waqf also facilitates the digital collection and investment of waqf assets, as well as the distribution of its benefits to people in need (Mohaiyadin and Aman 2021). The advantage of this technology lies in its ability to ensure data security, transparency in management, and efficiency in the distribution of waqf benefits so that it has the potential to increase community participation in waqf and optimise the use of waqf assets for the welfare of the community (Musah *et al.* 2023).

Finterra, a pioneering company in Singapore, has implemented blockchain technology in managing waqf, creating an innovative ecosystem of platforms to facilitate user participation in token trading with various payment methods (Mohamed and Ali 2022). This initiative aims to develop a blockchain-based crowdfunding platform that uses smart contracts for waqf projects, offering a more effective and efficient governance system in the collection, management, development, and utilisation of waqf assets (حازم فضل الله ساسي, 2019). Implementing this blockchain is expected to increase public trust and encourage the productivity of waqf assets through transparent project proposals (Kunhibava *et al.* 2024). Finterra develops a dedicated network for specific assets, allowing contributors to evaluate and select projects before participating. Each project will have a digital token based on the ERC-777 standard, allocated into a crowdfunding smart contract and only activated after meeting the set requirements, ensuring integrity and accountability in managing waqf funds. Participants can claim tokens after reaching the fundraising goal due to the project (Naoual 2022). The process as described above can then be further illustrated in the form of the following Figure 1:

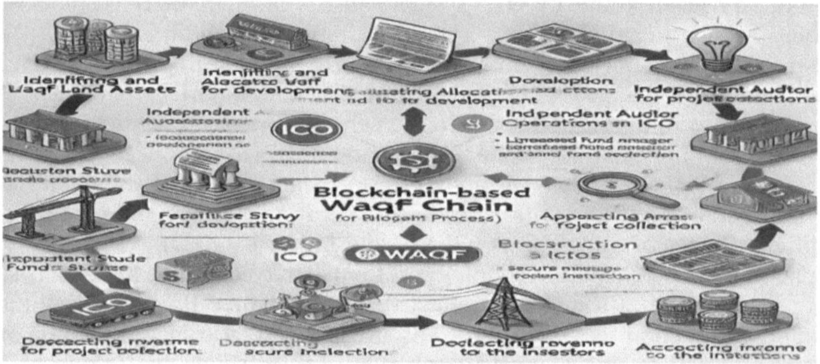

Figure 1. The Finterra chain process based on blockchain technology.

According to Salsabila *et al.*, (2023) the explanation of the Finterra chain process based on blockchain technology is as follows:

(1) Waqf institutions identify and allocate land assets for development purposes.
(2) The development project plan includes land status, feasibility study, building architecture, construction project plan, project budget, profit and loss projection, recommended funding instruments
(3) Independent auditors conduct thorough evaluation and validation of development project plans.
(4) Licensed fund managers initiate ICOs (Initial Coin Offerings) for specific waqf development projects, facilitating the sale of crypto tokens to qualified global investors.
(5) The Waqf Chain, built on the Finterra blockchain, operates as follows: tokens are launched to raise the required capital, global investors who have qualified purchase tokens, funds are safely stored in designated wallets, designated fund managers collect the funds raised, investors can choose from the available financial instruments, Waqf Chain offers four instruments: cash waqf, loans in accordance with sharia, Mudarabah, and sukuk
(6) After the successful capital acquisition, the fund manager contracts a construction company to start the construction process.
(7) Once construction is complete, the fund manager appoints an asset manager to oversee and manage the property.
(8) Fund managers collect the income or income generated by assets.
(9) The accumulated income or income is further distributed to investors based on the chosen investment instrument and the terms and conditions underlying the investment.

Blockchain offers significant benefits in waqf management, such as transparency, security, efficiency, and increased public participation. It surpasses centralised database systems and fintech crowdfunding applications through smart contract features that comply with sharia principles. However, challenges like low digital literacy, immature regulations, and resistance to conventional methods persist. Legal uncertainties and nazhir adaptation require collaboration among scholars, technology practitioners, and regulators to ensure adoption aligns with maqashid sharia. This approach allows blockchain to optimise waqf management and build public trust through a more inclusive and efficient system.

3.3 *Implications of the application of blockchain technology in the waqf governance system*

Applying blockchain technology in the waqf governance system offers various significant advantages. Umam *et al.* (2020) identified several key reasons for the effectiveness of blockchain in waqf management, including the flexibility and permanence of philanthropic asset listing and higher data security through decentralised storage systems. This system allows accurate tracking from waqif to end users, overcoming the complexity of recording the use of waqf assets (Dahdal *et al.* 2022). Blockchain also expands access to previously unreachable waqf, encouraging waqf innovation that has been stagnant. Other advantages include cost efficiency, better governance, greater transparency, and increased public trust. This implementation has the potential to connect waqif with business units that need funds, as well as encourage the development of a digital-based sharia economic community. Further implications include increasing the credibility of waqf institutions and the Indonesian Waqf Council through better disclosure of waqf asset management (Kanwal *et al.* 2023).

4 CONCLUSION

This research reveals the significant potential of blockchain technology in revitalising the waqf governance system. Through encrypted and decentralised transaction recording, blockchain offers a solution to overcome the stagnation of waqf asset management. The effectiveness of the implementation of this technology is reflected in increased transparency, data security, and operational efficiency. Important implications include the ease of tracking waqf funds, the prevention of data manipulation, and increasing the credibility of waqf management institutions such as the Indonesian Waqf Agency. This study emphasises the urgency of adapting the waqf governance system to technological developments, opening up opportunities for the dynamism of waqf assets in the digital era. The contribution of this research lies in expanding the insight into the integration of modern technology in sharia economic practices, especially waqf, encouraging stakeholders to improve the waqf management system according to the demands of the times. This has the potential to accelerate the development of a digital-based sharia economy and increase public understanding of the evolution of waqf management from conventional methods to a more effective and efficient technology-based system.

REFERENCES

Abd Razak, N. I., Ag Omar, P. M. F. F., Samsulbahri, M. N., Gazali, H. M., and Ishak, N. (2020). Review of Waqf Blockchain for Islamic Finance Education: A Case of Universiti Malaysia Sabah. *Journal of Critical Reviews*.

Afifah, A. A. N. (2021). Analisis Prinsip Good Coorporate Governance Dalam Pengelolaan Wakaf Produktif Di Yayasan Sekolah Islam Terpadu (Sit) Al-Furqon. *I-Philantrhropy: A Research Journal On Management Of Zakat and Waqf*, *1*(1), 1–16. https://doi.org/10.19109/iphi.v1i1.9625

Al-Saudi, H. A. (2023). Waqf Fund Management and the Blockchain Horizon. *In Eurasian Studies in Business and Economics*. https://doi.org/10.1007/978-3-031-30061-5_17

Aysan, A. F., and Al-Saudi, H. (2023). Blockchain As A Waqf Disruptor: Lessons Learned From Finterra's Waqf Chain. *Sustainability Practices: Cases from Businesses and Charities*, 144–160. https://doi.org/10.53478/tuba.978-625-8352-51-1.ch07

Dahdal, A., Truby, J., and Ismailov, O. (2022). The Role and Potential of Blockchain Technology in Islamic Finance. *European Business Law Review*. https://doi.org/10.54648/eulr2022005

Elasrag, H. (2019). Blockchains for Islamic finance: Obstacles Challenges. *MPRA Paper: Munich Personal RePEc Archive*, *92676*, 1–33.

Ghezal, M., and Lahsasna, A. (2020). Blockchain Waqf: Enabling Access to Social Islamic Finance. In *EasyChair Preprint*.

Hadi, S. (2004). *Metodologi Research*. Andi Offset.

Huang, N. S. M., and Nordin, S. (2021). A Comparative Study of Blockchain Framework for Waqf Management System. *International Journal of Academic Research in Business and Social Sciences*. https://doi.org/10.6007/ijarbss/v11-i12/11979

Ibrahim, S. S. (2023). Potential Blockchain Applications in Waqf for Sustainability: A Middle East and Asia Perspective. *Islamiyyat*. https://doi.org/10.17576/islamiyyat-2023-4502-04

Kanwal, A., Tayyab, M., and Idrees, S. (2023). Exploring the Nexus of Financial Technologies, Financial Inclusion, and Blockchain in Islamic Finance within Digital Transformation. *Pakistan Journal of Humanities and Social Sciences*. https://doi.org/10.52131/pjhss.2023.1104.0675

Kunhibava, S., Muneeza, A., Mustapha, Z., Khalid, M., and Kiran, G. (2024). BLOCKCHAIN USE CASE IN ISLAMIC SOCIAL FINANCE. *ISRA International Journal of Islamic Finance (IJIF)*, *16*(1), 93–110. https://doi.org/10.55188/ijif.v16i1

Makarim, D. F., and Hamzah, M. Z. (2024). Peran dan Potensi Digitalisasi Manajemen Zakat: Sebuah Systematic Literature Review. *Jurnal Ilmiah Ekonomi Islam*, *10*(1), 463. https://doi.org/10.29040/jiei.v10i1.12406

Mbaidin, H. O., Sbaee, N. Q., Almubydeen, I. O., and Alomari, K. M. (2024). Key success drivers for implementation blockchain technology in UAE Islamic banking. *Uncertain Supply Chain Management*. https://doi.org/10.5267/j.uscm.2023.11.016

Meskini, F. Z., and Aboulaich, R. (2024). Insurance Based on Waqf and Blockchain Technology: A Strong Social Impact and Efficiency. *WSEAS Transactions on Business and Economics*, *21*, 741–752. https://doi.org/10.37394/23207.2024.21.62

Mohaiyadin, N. M. H. J., and Aman, A. (2021). Understanding the Issues of Waqf at Public University: Preliminary Findings. *International Journal of Islamic Thought*, *20*, 95–108. https://doi.org/10.24035/ijit.20.2021.214

Mohamed, G., and Lahsasna, A. (2020). Blockchain Waqf: Enabling Access to Social Islamic Finance. *EasyChair*.

Mohamed, H., and Ali, H. (2022). Blockchain, fintech, and islamic finance: Building the future in the new islamic digital economy. In *Blockchain, Fintech, and Islamic Finance: Building the Future in the New Islamic Digital Economy*. https://doi.org/10.1515/9783110745016

Muchlis Gazali, H., and Che Ismail, C. M. H. (2019). A Conceptual Framework for Cash Waqf with Blockchain in Financing Education for the Islamic Religious School in Malaysia. *Journal of Islamic Sciences and Comparative Studies*.

Musah, M. K., Alhaj, Z. M., Hasan, A., and Aziz, D. A. (2023). Beyond Management Challenges: A Demise of a Cash Waqf Foundation. *Journal of Islamic Social Finance*, *1*(2), 1–16.

Naoual, B. (2022). Using Blockchain Technology to Revolutionize Waqf: The Finterra Waqf Chain Model. *Journal of Economics and Sustainable Development*, *5*(1), 136–148.

Sugiyono. (2018). *Metode Penelitian Kuantitatif, Kualitatif dan R&D*. Alfabeta.

Zaman, A., Tlemsani, I., Matthews, R., and Mohamed Hashim, M. A. (2023). Assessing the potential of blockchain technology for Islamic crypto assets. *Competitiveness Review*. https://doi.org/10.1108/CR-05-2023-0100

حازم فضل الله ساسي,. (2019). Use of Blockchain Applicationsin Developing Waqf: A Platform by Finterra Companies as a Modal Study. *Journal of Islam in Asia*, *16*(3), 141–160.

Towards Resilient Societies: The Synergy of Religion, Education, Health, Science, and Technology – Rahiem (Ed.)

The urgency for regional policy reform post-2019 marriage law amendment: Enhancing child marriage prevention in Indonesia

Loveria Sekarrini
Universitas Indonesia/Universitas MH Thamrin, Indonesia

Agustin Kusumayati & Sabarinah Prasetyo
Universitas Indonesia, Indonesia

ABSTRACT: Exploring regional policies to prevent child marriage is essential, especially given the 97% approval rate for marriage dispensations in religious courts. Urgent measures are required to achieve Sustainable Development Goal (SDG) 5.3 and eliminate child marriage by 2030. This practice violates children's rights, exacerbates poverty (SDG 1), limits educational and economic opportunities (SDG 4), contributes to nutrition issues (SDG 2), and leads to poor health outcomes (SDG 3). Strengthening legal frameworks is crucial to protect children and support SDG 16. This policy analysis examines regional prevention programs addressing child marriage dispensations following the 2019 amendments to the marriage law. A literature review of 36 articles from *peraturan.bpk.go.id* and grey literature, using keywords such as "regional regulations," "regional action plan," and "child/early marriage" from 2020–2024 reveals that only 10 provinces and 26 districts/cities have revised their child marriage prevention policies. Most policies emphasise education for children and parents on the impacts of child marriage, child supervision, premarital counselling, and institutional support. However, only seven regional regulations mandate health examinations before marriage to assess reproductive and mental health risks associated with dispensations in religious courts. In conclusion, a more collaborative and holistic approach is urgently needed to strengthen regional efforts in preventing child marriage dispensations in Indonesia.

Keywords: child marriage, child marriage prevention, child rights, early marriage, policy analysis

1 INTRODUCTION

Global data estimates that 650 million girls and women were married before age 18, with 12 million new cases annually and 28 girls marrying every minute. South Asia accounts for 45% of these cases (UNICEF 2023; United Nation Human Rights 2023). Although the global percentage of women married before 18 has decreased from 25% to 21% over the past decade, efforts must accelerate 12 times faster to achieve "zero child marriage by 2030". Indonesia ranks 8th globally and 2nd in ASEAN for child marriage, with an estimated 25.5 million cases among 46 million adolescents aged 10-19. The prevalence of child marriage in Indonesia decreased from 14.67% to 6.92% between 2008 and 2023 (Badan Pusat Statistik 2023). This practice undermines national development goals, perpetuates poverty and gender inequality, and is linked to issues like stunting in children. It violates children's rights and limits girls' access to education and economic opportunities (SDG 1 and 4),

DOI: 10.1201/9781003645542-81

leading to lower literacy rates and job prospects. Additionally, it contributes to nutrition issues (SDG 2) and poor sexual and reproductive health outcomes, increasing the risk of complications during childbirth (SDG 3). Strengthening legal frameworks is essential to protect children and support their rights, ensuring that girls can reach their full potential (Bappenas *et al.* 2021; BPS *et al.* 2020; Duadji and Tresiana 2022; Hendra and Hayyuning 2022; Schaffnit and Lawson 2021).

Indonesia has implemented various strategies to prevent child marriage, including raising the minimum marriage age from 16 to 19 through Law No. 16 of 2019, alongside the Child Protection Law, the Child-Friendly District/City initiative, and the National Strategy for the Prevention of Child Marriage. Despite these efforts, child marriage dispensations increased 2-3 times in 2019 and 2020, even with guidelines requiring urgent reasons and supporting evidence (Subchi *et al.* 2021). Legal dilemmas persist, as conflicts between progressive and conservative ideas in family law lead to varied local interpretations (Grijns and Horii 2018; Nawawi *et al.* 2022). A study of 14 religious courts revealed a non-permissive attitude, with judges often rejecting marriage dispensation applications and lacking supplementary documents despite their significant authority in these decisions (Mustofa 2023). In 2018, only 3% of child marriage dispensation cases involved the child being present in court, with judges approving 99% of cases that reached them (AIPJ2 2019; Sumner 2020). These findings underscore the need for enhanced efforts to address the complexities of child marriage and ensure effective implementation of preventive measures, prompting a policy analysis of regional regulations addressing child marriage dispensations following the 2019 marriage law amendments.

2 METHOD

This study identifies 58 domestic laws related to child marriage prevention at the provincial and city/district levels, sourced from peraturan.bpk.go.id and supplemented by grey literature using keywords like "regional regulations" and "child/early marriage." Focusing on laws enacted between 2020 and 2024 to align with the national marriage law, the review screened for duplicates, resulting in 54 documents for further analysis after excluding one duplicate, two irrelevant documents, and two inaccessible ones. Ultimately, 36 articles were selected for in-depth analysis, following the five stages of the national strategy for preventing child marriage with a socio-ecological approach: mapping regional policies, identifying stakeholders, and outlining prevention plans. Planning and evaluation stages were not analysed in depth due to similar information across regulations and the inclusion of other national programs, such as the Child-Friendly City/District Program and premarital health screenings.

3 RESULT

This study reviewed 36 regional policies on child marriage prevention, including 10 at the provincial level, 23 at the regency level, and three at the city level. Currently, 10 of 38 provinces have provincial-level policies, with West Java leading in revisions across eight districts and cities. However, many regions, particularly remote areas like Papua, have not updated their policies, highlighting significant implementation gaps nationwide.

3.1 *Creating child friendly in city/district to prevent child marriage*

The child-friendly Cities initiative prioritises children's rights in UN member countries and encourages local governments to make decisions in the best interest of children, including efforts to prevent child marriage (Bende 2023; Hale Özel 2024). In Indonesia, Presidential

Regulation Number 25 of 2021 on the Child-Friendly City/District (KLA) Policy, introduced by the Ministry of Women Empowerment and Children in 2005-2006, seeks to ensure children's rights in every city and district (Zerlina and Sulaiman 2020). While most prevention policies reference the KLA program, not all effectively link it to child marriage prevention. Although the KLA task force's involvement is noted, the focus remains primarily on implementing the KLA at the district and city levels, strengthening KLA institutions, and conducting evaluations, with limited emphasis on specific child marriage prevention strategies.

3.2 *Premarital physical, reproductive health, and psychological examination*

Premarital health screening is crucial for improving healthcare practices, upholding marital rights, and protecting public health by preventing the spread of life-threatening diseases like HIV and hepatitis. Despite 26 policies at the district and city levels, only 14 focus on premarital physical examinations, reproductive health, and psychological assessments required for marriage dispensation applications to religious courts. These policies mandate screenings in hospitals or primary health care (Puskesmas), ensuring comprehensive evaluations, including pregnancy tests and blood screenings. However, collaboration between the Office of Religious Affairs (KUA) and health facilities is often inadequate, resulting in high rates of unexamined health indicators (Ibrahim *et al.* 2024). Psychological evaluations by trained professionals are essential for assessing readiness for marriage, and effective counselling can significantly prevent child marriages. However, barriers such as insufficient counsellor competency and inadequate facilities limit their effectiveness (Wahid 2024). A successful counselling program in Parigi District improved premarital guidance through structured procedures and collaboration with relevant agencies (Nasir *et al.* 2021). Implementing premarital health screenings can influence marriage decisions and family dynamics, reflecting cultural values about health and well-being. However, accessibility, affordability, and cultural resistance must be addressed, as stigma and discrimination complicate acceptance (Ibrahim *et al.* 2024). Collaboration among stakeholders ensures access to premarital counselling and reduces stigma. Additionally, considerations of maturity and capability and fostering healthy relationships are essential for ensuring family stability in child marriage (Aprilianto *et al.* 2024).

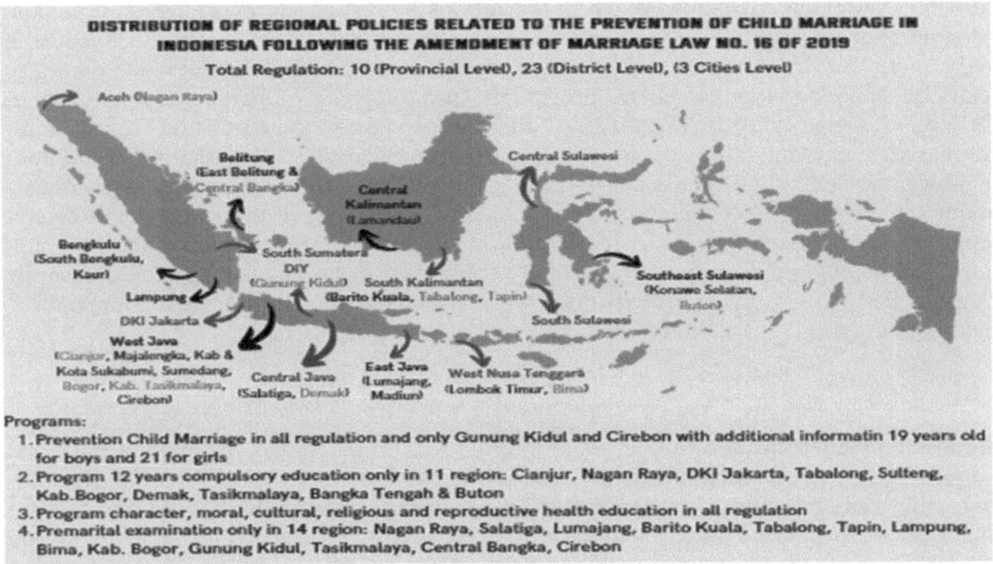

Figure 1. Distribution of regional policies related to the prevention of child marriage in Indonesia following the amendment of marriage law no.16 of 2019.

3.3 *Efforts to prevent child marriage*

All policies emphasise the need for local governments, parents, children, communities, and stakeholders to participate in preventing child marriage through various strategies. These include strengthening parental roles, conducting awareness campaigns, integrating gender mainstreaming into planning and budgeting, regulating and monitoring prevention efforts, enhancing institutional frameworks, engaging stakeholders, establishing Child-friendly areas, increasing educational opportunities, and providing reproductive health education. This collaborative approach focuses on community involvement, education, and supportive policies. However, not all provinces or districts incorporate these strategies in their regulations; one district has developed a comprehensive strategy with approximately 20 outlined approaches, most typically ranging from 4 to 10 strategies for preventing child marriage.

3.4 *Stakeholders involvement in child marriage prevention*

All regulations explain stakeholders' involvement in preventing child marriage. The institutional strengthening for preventing child marriage involves various stakeholders, including local government bodies (district, sub-district, and village levels), family and community organisations (Family Learning Center (Puspaga) and (Regional Technical Implementation Unit for the Protection of Women and Children (UPTD PPA), children's forums, PIK Remaja), social workers, counsellors, psychologists, educational institutions (schools and teachers), health institutions (medical personnel and reproductive health services), law enforcement (legal authorities and KUA officers), and non-governmental organisations (NGOs and community groups). These stakeholders collaborate to provide education, support, and enforce regulations, ensuring a comprehensive approach to preventing child marriage.

3.5 *Parental role strengthening and the role of community*

19 districts have policies requiring parents to prevent child marriage by ensuring 12 years of compulsory education and teaching character, moral, religious, and cultural values (Peraturan Bupati Tabalong No. 46 Tahun 2020). However, only 11 regions clearly define parents' roles in this education. While 18 regulations address character, moral, cultural, and religious education, only 10 specify parents' responsibilities in teaching reproductive health and non-violent relationships, notably in Nagan Raya and Cianjur. Several regulations also focus on child development, care, protection, and supervision, particularly in Cianjur, Demak, Gunung Kidul, Belitung Timur, and Cirebon. Only one regulation from Cianjur emphasises improving family communication (Central Sulawesi). Most regional regulations highlight the community's role in preventing child marriage, including raising awareness about child-related laws and reproductive health, contributing to policy development, reporting forced marriages, and participating in the rehabilitation of young married children. Community engagement is facilitated through organisations like the Child-Friendly District Task Force, which collaborates with local governments on prevention efforts, focusing on the child's best interests, family values, and local wisdom.

3.6 *Enhancing education and economic stability to prevent child marriage*

Socialisation and education campaigns primarily focus on the impact of child marriage and prevention strategies. However, only a few districts address topics like premarital sex, violence, and substance abuse, as seen in regulations from Lampung, Bima City, and Central Sulawesi. Most policies target children and parents, with child-focused initiatives raising awareness about the dangers of child marriage, while parent-focused measures guide and supervise their children. Improving economic stability is also crucial for preventing child marriage. Scholarships for underprivileged students and economic empowerment programs for low-income parents can reduce the pressure to marry off children early. Only East Java Province, Sumedang Regency, and Majalengka Regency have such policies. By tackling education and economic challenges, these efforts can lead to long-term benefits and help eliminate child marriage.

4 DISCUSSION

4.1 *Practice-research gap*

Indonesia's commitment to the United Nations Convention on the Rights of the Child emphasises the importance of ensuring every child can grow up healthy, receive an education, and be treated fairly. The convention outlines four key principles: the right to life and development, non-discrimination, prioritising the best interests of the child, and respecting children's views (KemenPPPA *et al.* 2019). Despite having a national strategy to prevent child marriage, Indonesia lacks a robust legal framework for implementation, leading to inconsistent commitment among provinces and districts. This inconsistency hampers effective policy enforcement and reveals challenges such as limited government capacity, difficulties in integrating Sustainable Development Goals (SDGs), and unclear policies regarding roles, responsibilities, and budget allocations, resulting in reactive rather than strategic initiatives (Duadji and Tresiana 2022; Guner 2023; Musawwamah *et al.* 2023). The role of mothers during critical developmental phases is vital for preventing stunting and promoting healthy growth, with effective parenting linked to preventing child marriage (Saleh *et al.* 2021). Community support and training programs focusing on nutrition, child rights, and mental health are essential for empowering parents (Hale Özel 2024; Surbakti *et al.* 2022). Comparative studies reveal significant regional disparities in Indonesia regarding implementing child-friendly initiatives mandated by Presidential Regulation Number 25 of 2021. Limited support for compulsory and reproductive health education and insufficient stakeholder involvement creates gaps in awareness and enforcement. In contrast, countries like Bangladesh have progressed through strict legal frameworks and community engagement, while India shows mixed results due to socio-cultural factors. Nigeria faces high child marriage rates despite government policies, and Ethiopia has made strides through legal reforms but still contends with traditional practices. These examples highlight the need for tailored, multifaceted approaches to effectively address child marriage, considering each country's unique cultural contexts (Duadji and Tresiana 2022; Retnowulandari *et al.* 2024)

4.2 *Recommended implementation strategies*

To enhance child marriage prevention policies, all regions must adopt and implement Presidential Regulation No. 25 of 2021, creating child-friendly environments. This includes defining parental roles in ensuring 12 years of compulsory education and integrating comprehensive reproductive health education into regional policies. Mandatory standardised premarital health assessments—covering physical, reproductive, and psychological evaluations—should be conducted in government hospitals or primary health care (Puskesmas) to ensure children's readiness for marriage. Expanding campaigns to address premarital sex, violence, and substance abuse will target both children and parents, emphasising the benefits of delaying marriage for education. Strengthening collaboration among local governments, community organisations, educational institutions, health institutions, law enforcement, and NGOs is essential. Engaging stakeholders in the policy-making process ensures relevance and acceptance. Increased advocacy and awareness efforts, along with implementing sanctions based on new regulations, are crucial for reducing child marriage rates and improving outcomes for children in Indonesia.

4.3 *Limitation*

This literature review has several limitations. First, it relied on the Peraturan BPK website for national, provincial, and district regulations, supplemented by hand searches, without exploring individual provincial or city websites due to time constraints. Second, the limited number of available regulations and the diversity of study types led to the inclusion of all studies from 2020 to 2024 to ensure comprehensive results despite challenges in assessing study quality.

5 CONCLUSION

The review of policies in Indonesia highlights significant gaps in preventing child marriage, despite initiatives like Presidential Regulation Number 25 of 2021 for child-friendly cities. There is limited support for 12 years of compulsory education and reproductive health education, with only 14 policies emphasising premarital health examinations. Socialisation campaigns often overlook critical issues such as premarital sex and violence, indicating a need for broader outreach to community leaders and law enforcement. To enhance prevention strategies, stakeholders should improve collaboration, develop training programs, and expand awareness campaigns. Strengthening policy frameworks with clear roles, budget allocations, and effective monitoring is essential to address these gaps and support the achievement of Sustainable Development Goals (SDGs) by 2030, particularly in gender equality and children's rights.

REFERENCES

AIPJ2. (2019). *Decision Analysis of Marriage Dispensation Cases in Indonesia.*

Aprilianto, D., Na'imah, F. U., Fauzi, A., and Maknuun, L. Il. (2024). The Controversy of Child Marriage Culture in The Perspective of Maqāṣid al-Usrah: A Case Study of The Authority of Lebe' in Brebes. *Al-Manahij: Jurnal Kajian Hukum Islam*, 18(2), 199–218. https://doi.org/10.24090/mnh.v18i2.11554

Badan Pusat Statistik. (2023). *Proporsi perempuan umur 20-24 tahun yang berstatus kawin atau berstatus hidup bersama sebelum umur 18 tahun menurut provinsi.*

Bappenas, Australian Government, MAMPU, The University of Melbourne, and Melbourne Insitute. (2021). *Consequences of child marriage in Indonesia.* In MAMPU.

Bende, A. P. (2023). Key Characteristics of Family Cities: A Review. *Bucharest*, 18(1), 74–78.

BPS, BAPPENAS, UNICEF, and PUSKAPA. (2020). *Pencegahan Perkawinan Anak. In UNICEF*: Vol. XVI (Issue 01).

Duadji, N., and Tresiana, N. (2022). Analysis of child marriage and related policies in Indonesia: Sustainable development issue. *Problemy Ekorozwoju*, 17(1), 101–113. https://doi.org/10.35784/pe.2022.1.10

Grijns, M., and Horii, H. (2018). Child Marriage in a Village in West Java (Indonesia): Compromises between Legal Obligations and Religious Concerns. *Asian Journal of Law and Society*, 5(2), 453–466. https://doi.org/10.1017/als.2018.9

Guner, H. G. (2023). Factors Affecting Child-Friendly Policy Making Process in Metropolitan Municipalities: Two Case Studies in Turkey. *In Nucl. Phys.*

Hale Özel, Ç. (2024). Parents' View of Child-Friendly Cities Initiative: The Case of Mersin. *Athens Journal of Tourism*, 11(2), 109–124. https://doi.org/10.30958/ajt.11-2-1

Hendra, T., and Hayyuning, J. (2022). *Seksualitas dan Negara : Permasalahan Dispensasi Perkawinan Anak di Indonesia Sexuality and the State: Dispensation of Child Marriage in Indonesia Pendahuluan.* 13(2).

Ibrahim, A. M., Okesanya, O. J., Ahmed, M. M., Ukoaka, B. M., and Lucero-Prisno III, D. E. (2024). Enhancing public health through the Kano pre-marital health screening bill: analysis of implications, challenges, and opportunities. *PAMJ - One Health*, 14. https://doi.org/10.11604/pamj-oh.2024.14.9.44265

KemenPPPA, PPN/Bappenas, K., and UNICEF. (2019). Modul Pelatihan Sistem Perlindungan Anak: "Modul 1: Hak-Hak Anak, *Perlindungan Anak dan Sistem Perlindungan Anak.* In Kemenpppa.Go.Id.

Musawwamah, S., Taufiq, M., Haryanto, E., Supraptiningsih, U., and Maimun. (2023). Resistance to Child Marriage Prevention in Indonesia and Malaysia. *Ahkam: Jurnal Ilmu Syariah*, 23(1), 259–280. https://doi.org/10.15408/ajis.v23i1.32014

Mustofa, K. N. (2023). ADJUDICATING UNDERAGE MARRIAGES AT RELIGIOUS COURTS A Progressive Collaboration between Judiciary and Health Authority. *Al-Ahwal*, 16(2), 342–363. https://doi.org/10.14421/ahwal.2023.16208

Nasir, M., Abidin, Z., and S. Thahir, L. (2021). Pre-Marital Counseling Effectiveness In Building Happy Muslim Families. *International Journal of Contemporary Islamic Law and Society*, 3(1), 55–68. https://doi.org/10.24239/ijcils.vol3.iss1.27

Nawawi, M. A., Sulastri, Edi, R. N., and Setiawan, A. (2022). Harmonization of Islam and human rights: judges' legal arguments in rejecting child marriage dispensation in Sukadana, Indonesia. *Ijtihad: Jurnal Wacana Hukum Islam Dan Kemanusiaan*, 22(1), 117–133. https://doi.org/10.18326/ijtihad.v22i1.117-133

Retnowulandari, W., Wangga, M. S. E., Notoprayitno, M. I., and Ahmad, N. (2024). The Prevalence Of Child Marriage: Comparitive Study Of Indonesia And Other South Asian States. *Jambura Law Review*, 6 (2), 339–366. https://doi.org/10.33756/jlr.v6i2.24257

Saleh, A., Syahrul, S., Hadju, V., Andriani, I., and Restika, I. (2021). Role of Maternal in Preventing Stunting: a Systematic Review. *Gaceta Sanitaria*, 35, S576–S582. https://doi.org/10.1016/j.gaceta.2021.10.087

Schaffnit, S. B., and Lawson, D. W. (2021). Married too young? The behavioral ecology of 'child marriage.' *Social Sciences*, 10(5). https://doi.org/10.3390/socsci10050161

Subchi, I., Arsadani, Q., Helmi, M. I., and Faridho, E. (2021). Legal Policy of Child Marriage In the Covid-19 Pandemic Period. *Samarah*, 5(2), 846–866. https://doi.org/10.22373/sjhk.v5i2.10570

Sumner, C. (2020). *POLICY PAPER 19 Ending Child Marriage in Indonesia: The Role of the Courts.*

Surbakti, E., Hasibuan, Y., and Pasaribu, R. (2022). The Effectiveness of Local-Leader-Based Social Interventions on Knowledge and Attitudes of Early Marriage among Mother Who Have Teenage Children. *Open Access Macedonian Journal of Medical Sciences*, 10(E), 1215–1220. https://doi.org/10.3889/oamjms.2022.8544

UNICEF. (2023). *A Profile of Child Marriage South Asia.*

United Nation Human Rights. (2023). *Child and Force Marriage Including in Humanitarian Settings.*

Wahid, L. A. (2024). Implications of Madrasa Guidance and Counseling Services To Prevent Child Marriage in Central Lombok Regency. *Al-Tazkiah*, 13(1), 57–76. https://doi.org/10.20414/altazkiah.v13i1.10041

Zerlina, D., and Sulaiman, C. C. (2020). Towards the innovative planning for child-friendly neighbourhood in Jakarta. *IOP Conference Series. Earth and Environmental Science; Bristol*, 592, 1–18. https://doi.org/10.1088/1755-1315/592/1/012023.

Towards Resilient Societies: The Synergy of Religion, Education, Health, Science, and Technology – Rahiem (Ed.)

ZIS distribution and economic growth: Advancing Indonesia's Islamic human development index

Nofrianto
Universitas Islam Negeri Syarif Hidayatullah Jakarta, Indonesia
ORCID ID: 0000-0002-3806-2894

Pandu Adjie Pamungkas
Universitas Islam Negeri Syarif Hidayatullah Jakarta, Indonesia

ABSTRACT: This study investigates Indonesia's Islamic-Human Development Index (I-HDI) from 2017 to 2022 by analysing the influence of poverty, economic growth, government expenditures in the education and health sectors, and the distribution of Zakat, Infaq, and Alms (ZIS). The research utilises secondary data in the form of time series and cross-sections, applying the Panel Data Regression Method with the Fixed Effect Model, which was selected as the most suitable estimation approach. The findings reveal that ZIS distribution, poverty, economic growth, and government spending on education and health significantly impact the I-HDI in Indonesia. Specifically, from 2017 to 2022, these factors have positively and substantially improved Indonesia's I-HDI, highlighting the critical role of economic and social welfare policies in enhancing human development outcomes.

Keywords: Islamic-Human Development Index (I-HDI), ZIS distribution, economic growth, government expenditure, panel data regression

1 INTRODUCTION

Economic development aims to create social welfare. From an Islamic perspective, economic growth places humans at the centre of economic growth by placing them as the subject and focal point of progress. The human development process will develop into a strategy to support various aspects of human life, so human resources are an essential component in the development of a nation. The concept of development in Islam is more comprehensive, encompassing moral, spiritual, social, political, and economic dimensions (Herianingrum *et al.* 2019). The Human Development Index (HDI), developed by the United Nations Development Programme (UNDP), is the most compelling statistic for assessing a country's well-being through human development. According to Kovacevic, HDI never claims to be a more comprehensive measure of human welfare than other economic indicators (Lind 2019)

Islamic economics aims for holistic prosperity (*Al-Falah*) and adheres to *Maqasid al-Shariah*. The intellectual foundations of Islamic development, including Tauhid, *Rububiyyah, Khilafah*, and *Tazkiyah*, which cover moral, spiritual, and material aspects. In response, the Islamic Human Development Index (I-HDI) was established to conduct a complete evaluation of human development in alignment with the principles of *Maqasid al-Shariah*. The I-HDI, which measures economic and spiritual progress, reveals concerning trends across Indonesia's provinces. The Figure 1 below illustrates that by 2022, no province reached the "very high" development level (above 80%), indicating that the system isn't yet achieving its full potential in promoting human welfare according to Islamic principles

DOI: 10.1201/9781003645542-82

(*Maqasid al-Shariah*). The consistent downward trend and stagnation in I-HDI scores suggest that Indonesians have not yet achieved the level of prosperity these religious and social development frameworks envisioned.

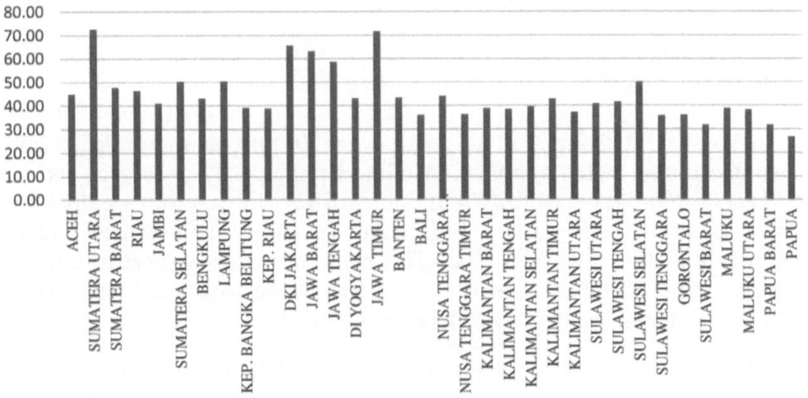

Figure 1. Islamic-Human Development Index, Source: Central Statistics Agency (processed data).

Zakat is a highly effective instrument for poverty reduction, as it is designed to benefit the impoverished and emphasises self-improvement. It is integral to the government's endeavours to assist indigent individuals in breaking free from the cycle of poverty (Zulkifli *et al.* 2021). Zakat has set a pace for abundance, as in Goal 10 of the SDGs, which centres on reductions in well-being, health, education, and other assets (Noor and Pickup 2017). Poverty is primarily caused by deficiencies in infrastructure and technology, often stemming from the opportunistic behaviour of political and bureaucratic elites seeking to fulfil their interests (Atkinson 2021). Additionally, disparities in economic growth across sectors and regions, along with changes in a country's economic structure, significantly influence poverty levels.

In a modern economy, human resources are crucial for economic output, making government efforts to enhance human quality vital for boosting growth (Efthalitsidou *et al.* 2021). Monetary development can be considered an increase in a country's original Gross Domestic Economic development, reflected in GDP growth, signals rising per capita income and improved living standards. Population growth also impacts living standards and is a key policy concern. Economic growth, essential for raising living standards, is primarily measured by GDP growth rates (Simionescu *et al.* 2017). Part of the progress or easing of poverty cannot be achieved without the definition of human resources; the accumulation of human resources relies heavily on teaching and acquiring skills (Shi and Qamruzzaman 2022). Investments in education spending and increases in total social security spending are key factors contributing to developing the Human Development Index (Haque and Khan 2019). Public investment provided by public authorities in the field of welfare will open doors and welfare administration so that it can influence the work of the nature of human resources (Khan *et al.* 2020).

The 1945 Constitution (amended in 2009) mandates a minimum 20% allocation of APBN and APBD to education (Article 31, paragraph 4), while healthcare must receive 5% according to Law Number 20 of 2003. These requirements align with Islamic principles - education spending supports intellectual development (*hifdh al-aql*), while healthcare funding protects life (*hifdh al-nafs*). By combining government funding with Islamic charitable giving (ZIS), Indonesia creates a stronger system for human development.

This research differs from other research because previously, I-HDI research only discussed comparisons with HDI, and I-HDI was used as an independent variable. Another distinction is the analysis method employed in IHDI research. This study is likely to expand the range of references. This research also aims to prove that the ZIS (Zakat, Infaq, and Shadaqah) distribution variable previously contained differences from the results of several similar studies. Apart from that, the I-HDI parameters used are a combination of several previous studies because the use of I-HDI parameters has not been stable until now, and the parameters used have also been adjusted to location conditions.

2 RESEARCH METHODS

This study analysed data from 2017 to 2022, when Indonesia significantly improved its Islamic charitable giving (ZIS) system and related policies. The research used published statistics and economic data from all 34 Indonesian provinces, from wealthy regions like Jakarta to developing areas in eastern Indonesia. The study used Panel Data Regression analysis with a Fixed Effect Model (FEM). This method was chosen because it accounts for the unique characteristics of each province that don't change over time, such as geography, culture, and historical development. Substructural Model 1:

$$I - HDIit = \alpha + \beta 1LN\_ZISit + \beta 2LN\_Povertyit + \beta 3LN\_PEit + \beta 4LN\_PPBPit + \beta 5LN\_PPBKit + \mu it \quad (1)$$

I-HDIit:*Islamic-human development index*(I-HDI) i in period t
LN:Natural Logarithm
ZISit: Distribution of ZIS I in period t
Poverty: Poverty i in period t
PEit: Economic growth in province i period t
PPBPit: Government Expenditures in the Education Sector i in period t
PPBKit: Government Expenditures in the Health Sector i in period t
α: Coefficient Intercept
μit: Nuisance or unobservable factors

3 RESULTS AND DISCUSSION

3.1 *Selection of estimation model*

Based on the obtained data processing findings, it can be observed that the cross-sectional probability F value is 0.0000, which falls below the predetermined alpha significance level of 0.05 (Table 1). Consequently, the decision to employ the Fixed Effect Model (FEM) is justified.

Table 1. Chow test result.

Effects Test	Statistics	df	Prob.
Cross-section F	44.693180	(33,165)	0.0000
Chi-square cross-section	468.471678	33	0.0000

The Fixed Effect Model (FEM) is chosen for use in this study over the Random Effect Model (REM) because the Hausman test (Table 2) processing results show that the Random Cross-section Probability value obtained is 0.0001, which is smaller than the alpha significance level (0.05).

Table 2. Hausman test.

Test Summary	Chi-Sq. Statistics	Chi-Sq. df	Prob.
Random cross-section	105.038631	5	0.0000

3.2 *Statistical test*

After knowing the best approach, the next step will be conducting a panel data regression test.

Table 3. F test (Simultaneous).

F-statistic	138.2345
Prob(F-statistic)	0.000000

The simultaneous significance test is designed to evaluate the simultaneous impact of all independent variables on the dependent variable. The F-stat result is 138.2345, and the prob (F-stat) value is 0.000000, which is less than the alpha 5% threshold ($0.000000 < 0.05$), as revealed by the data processing results (Table 3). The dependent variable, the IHDI, is then significantly impacted by the independent variables ZIS distribution (X1), poverty (X2), economic growth (X3), health sector expenditures by the government (X5), and educational sector expenditures by the government (X4). As a result, the calculated regression model can be considered suitable.

Table 4. Coefficient of determination test (R2).

R-squared	0.969545
Adjusted R-squared	0.962532

The R-squared value is 0.969545, with an adjusted R-squared value of 0.962532 based on the data processing results (Table 4). Since there are more than two variables in this study, the Adjusted R-squared, which is 0.962532, is employed to conduct the test. These findings show the independent variables of ZIS distribution, poverty, economic growth, government health sector expenditures, and government educational sector expenditures (X4). It can account for 96.25% of the influence on the dependent variable, the IHDI. Meanwhile, 3.75% accounted for the impact of variables not covered by the scope of this study.

Table 5. T test (Partial).

Variables	Coefficient	Std. Error	t-Statistics	Prob.
C	55.91177	5.155374	10.84534	0.0000
LN_ZIS	−0.952308	0.217271	−4.383041	0.0000
POVERTY	−0.012318	0.002367	−5.204843	0.0000
P.E	0.106856	0.048564	2.200325	0.0292
LN_EDUCATION	1.304143	0.593419	2.197674	0.0294
LN_HEALTH	−1.294353	0.526351	−2.459107	0.0150

3.2.1 *Zakat, infaq and alms (ZIS)*

Based on the research results, it show that there is a significant negative influence between the distribution of Zakat Infaq Sedekah on the Islamic-Human Development Index (IHDI) in Indonesia. Table 5 showed the t-calculated value of the ZIS variable is more significant than the t-table (-4.383041> 1.971777), and the probability level of the ZIS Distribution variable is 0.0000, indicating that this value is less than the significance level of α= 5% (0.0000 < 0.05). The data support this result. This research has a negative impact in the sense of being less than optimal because the ZIS distribution data is still incomplete, where researchers only get data based on the total number. The strategy and mechanism for distributing zakat must be given more attention so that it can be carried out strategically and on target. Recipients of ZIS funds can change the Mustahik salary but do not yet have the option to change the training entry permit. Their lives are not yet ideal due to the zakat funds given. Addressing humanitarian issues requires transforming a consumer-driven society into a more functional one. Leveraging ZIS assets within organisations play a key role by providing capital, infrastructure, and resources to Mustahik, enabling them to improve the community's overall quality.

3.2.2 *Poverty*

According to preliminary research findings, the I-HDI is adversely affected by Indonesia's poverty variable. The Poverty variable's t-calculated value is higher than the t-table (−5.204843 > −1.971777), and the probability value of 0.0000 is less than the significance level α = 5% (0.0000 < 0.05). Poverty will hurt development, starting with backwardness and lack of capital, which results in low productivity and low income (Nurnaningsih *et al.* 2019). This low value reduces the quality of what is being developed, resulting in a high imbalance. The ongoing cycle of poverty hampers regional development and obstructs the government's ability to assist its citizens effectively.

3.2.3 *Economic growth*

These data show that Indonesia's Islamic-Human Development Index (I-HDI) correlates significantly positively with economic development. The t-calculated value of the Economic Growth variable is more significant than the t-table (2.200325 > 1.971777), and the probability value of the Economic Growth variable is 0.0292, which indicates that this value is less than the significance level α= 5% (0.0292 < 0.05). This result is conclusive. The findings of this study provide credence to Todaro's theory, which holds that rapid economic growth will cause individuals' buying habits to shift to match their purchasing power. As a result, it is directly related to HDI, as purchasing power is one of the HDI indicators, namely payments. This is consistent with Ambya's (2021) findings, which show that GDP Growth has a more or less significant influence on the HDI. Wafa (2022) then applied the T statistical test to the HDI and GRDP. This test has a positive and significant impact when adequately interpreted.

3.2.4 *Educational expenditures by the government*

According to the findings, government spending on education considerably impacts Indonesia's Islamic-Human Development Index (I-HDI). The t-count value is more significant than the t-table (2.197674 > 1.971777), corroborating this assertion. The probability value of 0.0294 indicates that this value is less than the significance level α = 5% (0.0294 < 0.05). Government investments in education are crucial for developing human resources. Funding in each Indonesian region supports necessary training and primary education, enabling individuals to advance through educational levels with the right materials and innovation (Ambya 2021). Such investment enhances knowledge and skills, boosts community productivity, and encourages school attendance, ultimately improving human quality and expanding social assistance.

3.2.5 *Health sector expenditures by the government*

The studies found that health sector expenditures by the government had a significant negative impact on Indonesia's Islamic-Human Development Index (I-HDI). The regression results show that the t-count value for government expenditures in the health sector is −2.459107, whereas the t-table value is 1.971777. The t-count value is higher than the t-table value (−2.459107 > 1.971777). The probability value 0.0150 is lower than the 5% significance level (0.0150 < 0.05). The findings of this research affirm the importance of government expenditure on health, which is essential for providing essential health services and facilities to improve the productivity of communities. The health budget is necessary to develop both central and regional governments. The financial plans that the government established have not been implemented as intended, resulting in a negative impact on health spending. Specifically, the expenditure is based on remedial (recovery) rather than preventive (prevention) measures (Ambya 2021). The government is responsible for formulating welfare financial strategies, including how the public health administration office is organised. According to Adegboye *et al.* (2022), foundation spending promotes human activity and leverages access to monetary activity to boost individual purchasing power and the nation's per capita income.

4 CONCLUSION

From 2017-2022, several factors affected human development across Indonesia's 34 provinces. While economic growth and education spending improved the Islamic-Human Development Index (IHDI), three factors decreased poverty, health spending, and the distribution of religious charitable giving (ZIS). To make ZIS more effective, the government should create a digital system to coordinate ZIS collection and distribution between LAZ and BAZNAS, measure how well ZIS programs work, require quarterly reports on how funds are used, and develop formulas to distribute funds based on each region's needs. The government must also maintain its required spending levels (20% on education and 5% on health) and better monitor how these funds improve people's lives.

REFERENCES

Adegboye, A., Erin, O., and Asongu, S. (2022). Taxing Africa for inclusive human development: the mediating role of governance quality. *Journal of Economic and Administrative Sciences*.

Ambya, A. (2021). Human development index (HDI) in Lampung Province period 2013-2018. *Sriwijaya International Journal of Dynamic Economics and Business*, 4(2), 119–128.

Atkinson, C. L. (2021). Hydropower, development, and poverty reduction in Laos: promises realized or broken? *Asian Journal of Political Science*, 29(1), 67–87.

Efthalitsidou, K., Zafeiriou, E., Spinthiropoulos, K., Betsas, I., and Sariannidis, N. (2021). GDP and Public Expenditure in Education, Health, and Defense. *Empirical Research for Greece. Mathematics*, 9(18), 2319.

Haque, MI, and Khan, MR (2019). Role of oil production and government expenditure in improving human development index: Evidence from Saudi Arabia. *International Journal of Energy Economics and Policy*, 9(2), 251.

Herianingrum, S., Ryandono, MNH, Fauzy, RMQ, AFIFAH, FU, and Laila, N. (2019). The Effect of Government Expenditure on Islamic Human Development Index. *Opción*, 35(88), 685–703.

Khan, SAR, Zhang, Y., Kumar, A., Zavadskas, E., and Streimikiene, D. (2020). Measuring the impact of renewable energy, public health expenditure, logistics, and environmental performance on sustainable economic growth. *Sustainable Development*, 28(4), 833–843.

Lind, N. (2019). Development of the human development index. *Social Indicators Research*, 146(3), 409–423.

Noor, Z., and Pickup, F. (2017). *The Role of Zakat in Supporting Sustainable Development Goals. UNDP Brief*, United Nations Development Programme, New York.

Nurnaningsih, L., Riyanto, WH, and Susilowati, D. (2019). The Influence of Population, Poverty, and Unemployment on the Human Development Index of West Nusa Tenggara Regency/City. *Journal of Economic Sciences*, 3(4), 505–516.

Shi, Z., and Qamruzzaman, M. (2022). Re-visiting the role of education on poverty through the channel of financial inclusion: Evidence from lower-income and lower-middle-income countries. *Frontiers in Environmental Science*, 10, 873652.

Simionescu, M., Lazányi, K., Sopková, G., Dobeš, K., and Balcerzak, A.P. (2017). Determinants of economic growth in V4 countries and Romania. *Journal of Competitiveness*.

Wafa, NRT (2022). The Effect of Local Government Expenditures in the Education Sector and Regional GDP on HDI in Indonesia 2015-2018. *International Journal of Current Economics & Business Ventures*, 2(1).

Zulkifli, MF, Taha, R., Mohd Nor, MN, and Ali, A. (2021). Combating poverty in Malaysia: The role of zakat. *The Journal of Asian Finance, Economics and Business*, 8(5), 505–513.

Towards Resilient Societies: The Synergy of Religion, Education, Health, Science, and Technology – Rahiem (Ed.)

Islamic banking spin-off policy: Transition from mandate to choice

Nur Hidayah
Universitas Islam Negeri Syarif Hidayatullah Jakarta, Indonesia
ORCID ID: 0000-0002-1276-6623

Haryatih
Universitas Islam Negeri Syarif Hidayatullah Jakarta, Indonesia
ORCID ID: 0009-0006-1968-378X

Tira Mutiara
Universitas Islam Negeri Syarif Hidayatullah Jakarta, Indonesia
ORCID ID: 0000-0002-1741-0001

ABSTRACT: In December 2023, the market share of Islamic banking in Indonesia reached 7.38%. To support the growth of Islamic banking, Islamic Banking Law No. 21/2008 required conventional banks with Islamic windows to undergo a spin-off after 15 years of the law's enactment (2023) or when their assets reached 50% of their parent banks' total assets. This mandatory spin-off sparked debate, as many Islamic windows opposed it due to concerns that it would result in numerous small, fully-fledged Islamic banks with limited capital and a shortage of skilled human resources. This paper assesses the spin-off policy using interpretive policy analysis, focusing on arguments from key stakeholders, particularly regulators and industry players. The findings reveal differing perspectives: regulators prioritised quantitative growth, while industry players emphasised quality. This debate was resolved by amendment of the spin-off requirement in Law No. 4/2023 on the Development and Strengthening of the Financial System (P2SK Law), changing the spin-off from a mandatory obligation to an optional strategy. Islamic windows are now only required to spin off if their assets reach 50% of their parent banks' assets. This policy adjustment reflects a more flexible approach, accommodating the dynamic growth of the Islamic banking sector and promoting greater efficiency in supporting Sustainable Development Goal 8: Decent Work and Economic Growth.

Keywords: full-fledged Islamic banks, Islamic bank windows, P2SK law, spin-off

1 INTRODUCTION

The development of the Islamic banking industry urgently needs regulatory support. In Indonesia, Law No. 7/1992 concerning Banking supports the birth of the first full-fledged Islamic Bank, Bank Muamalat (Fathoni *et al.*, 2024; Hidayah *et al.*, 2024). Then, Law No. 10/1998, regarding amendments to Law 7/1992, allowed conventional banks to have Islamic windows(Hidayah, 2019; Hidayah and Haryatih, 2024). This policy is reinforced by Law number 23/1999 concerning the Central Bank of Indonesia, which emphasises the responsibility of the Central Bank for the regulation and supervision of the national banking system, including Islamic banks and Islamic rural banks(Hidayah, 2022). In 2001, the Islamic Banking Bureau was formed, and in 2004, its status was upgraded to the Islamic Banking Directorate. This reflects the existence of Islamic banks in Indonesia's dual banking system.

DOI: 10.1201/9781003645542-83

Regulatory support is getting stronger with the issuance of Law No. 21/2008 concerning Islamic Banking, promoting the growth of the national Islamic banking industry. The article 68 paragraph (1) mandates that Islamic windows whose asset value has reached at least 50% of the total value of its parent assets (Conventional Commercial Banks) or 15 years since the enactment of this Law are obliged to spin off to become a full-fledged Islamic Banks. In 2023, Islamic bank windows are required to undertake a spin-off. As confirmed in Bank Indonesia Regulation No. 11/10/PBI/2009, "Conventional Banks that do not carry out spin-offs as stipulated will be subject to revocation of their Islamic window license", spin-offs will become mandatory in 2023. The banks that successfully conducted the spin-off before the deadline included BRI Syariah, Bukopin Syariah Bank, Bank Jabar Banten Syariah, BNI Syariah, and BTPN Syariah.

The change in bank status to full-fledged Islamic Banks has made Islamic business management more focused and independent. Islamic general banks are considered more resilient than conventional banks when facing crises (Hidayah *et al.*, 2021). Since the issuance of Law No. 21/2008, the market share of Islamic banking has reached 7.38% (until December 2023), with 14 full-fledged Islamic Banks and 19 Islamic windows. This growth can be attributed to the expansion of various financing products, including profit and loss sharing, such as mudharabah and musyarakah financing, as well as non-profit and loss sharing financing, namely murabahah financing (Hidayah *et al.*, 2023). Murabaḥah contract has become one of the dominant financing schemes in Islamic banking (Hidayah *et al.*, 2022). The increasing number of Islamic banks is forcing the Indonesian banking sector to be more competitive, which impacts better performance and higher efficiency in the banking industry (Yuspin and Wardiono, 2017). Spin-offs improve their banks' financial performance (Pambuko *et al.*, 2024), the ratio of operating costs to operating incomes or BOPO (Rokhmawati *et al.*, 2022).

Meanwhile, in other countries, efforts have been made to reduce the establishment of Islamic windows, such as in Malaysia, Qatar, and Nigeria. Previously, Malaysia provided a successful Islamic window model where Islamic bank windows of several conventional banks were later converted into fully licensed Islamic banking subsidiaries. By setting up an Islamic subsidiary with a dedicated marketing department and trained officers to explain the details when dealing with potential clients interested in these products, the demand for these products and services has increased. Then, Qatar was the first and only major player to rescind the right to operate windows. In February 2011, the Qatar Central Bank (QCB) issued a directive to its conventional banks declaring that they would no longer be allowed to open any new branches or accept deposits or offer further Islamic financing. All windows were expected to close by the end of 2011. It was felt that removing windows would pave the way for a fairer and more stable growth of the Islamic banking industry. Likewise, Nigeria is limiting the opening of Islamic windows until the independent Islamic bank (Jaiz Bank) has time to establish itself. In sharp contrast, the National Bank of Ethiopia issued a framework limiting the licensing and establishment of standalone and allowing only the window model. The rationale for this could probably be related to the perceived strength or weakness of a new standalone Islamic bank, as a collapse could destabilise the financial sector (Mohamad *et al.*, 2017).

In its implementation, the Islamic bank windows spin-off encountered several obstacles in several aspects. Other studies state that the spin-off does not impact the performance of Islamic banking. The spin-off criterion, namely reaching 50% of the parent bank's assets, is impossible for Islamic bank windows to achieve (Arif *et al.*, 2017). The spin-off full-fledged Islamic Banks only lasted four years after the spin-off, not even producing a better performance after four years (Trinugroho *et al.*, 2021). Spin-off systems calculated by several financial performance ratios negatively influenced profitability (Jubaedah *et al.*, 2020) and market share (Arif, 2017).

The spin-off policy is not the only strategy to improve the performance of the Islamic banking industry (Arif *et al.*, 2018; Arif *et al.*, 2019). Another strategy that can be done is a merger. Mergers can smooth the competition process (Ahamed, 2012). On the other side, Trinugroho *et al.*, (2021), Arif *et al.*, (2023), and Rahmawati *et al.*, (2024) find that a conversion strategy results in better outcomes, particularly for market share and assets, than a spin-off strategy. One of the converted banks, namely Bank Aceh Syariah, succeeded in achieving a perfect efficiency value after the conversion, which shows effective resource utilisation and optimal operational performance (Wildan Hakim *et al.*, 2024). Naseri *et al.*, (2019) state that size is essential to get better results for Islamic banking. When a bank becomes too large, excessive expansion can and will cause efficiency to drop; and too large can potentially cause systemic risk (too large also fails, which leads to moral hazard). The Islamic bank windows spin-off process is not solely implemented with consideration of regulatory obligations alone, but also based on the readiness of itself to be able to grow sustainably and encourage national economic growth.

Such previous studies indicate contradictory results. Results state that spin-offs have a positive influence on Islamic bank windows. Still, other results state that spin-offs have no influence whatsoever on Islamic bank windows and even reduce bank performance. It is important to know spin-offs from the perspective of regulators (policy makers) and operators (industry/banking practitioners) to enrich information and data from related parties. This study aims to analyse the spin-off policy from the perspective of regulators and industry or Islamic banking practitioners and discuss the policies that mediate the spin-off debate. This study can provide information, data, and facts related to the spin-off policy from various perspectives that form the basis for policy recommendations.

2 METHOD

This study aims to analyse spin-off policies using an interpretive approach by discussing the discourse on spin-off obligations in 2023 from the stakeholders' perspective. The spin-off policy will be reviewed more deeply with a cost-benefit analysis based on the perspectives of regulators and Islamic banking practitioners (Yanow, 2007). Discussions of interpretive policy analysis are then perforce intertwined with issues in interpretive (or interpretive qualitative) research methods. The interpretative method is deemed appropriate for argumentative, deliberative, and other approaches to policy research because researchers do not need to convert words into numbers. This method is word-based, from data collection instruments to data analysis. Data collection methods in interpretive policy analysis include observation, interviews, and reading (Rabin, 2007). This research uses secondary data collected through observation and reading from available sources (such as books, journals, news articles, and presentation materials). Observations were made to observe, feel, and understand the spin-off policy from various stakeholder perspectives to obtain the information needed in the research. The data that has been collected is analysed using an explanatory qualitative method because it is not intended to test a particular hypothesis but rather to explain a phenomenon in more depth, in this case, the policy of spin-off.

3 DISCUSSION

The spin-off policy in Law No. 21/2008 aims to create a higher quality Islamic banking system in terms of management and compliance with Islamic law and promote Islamic banking operations to be more optimal and closer to ideal Islamic principles. This policy is one of the government's efforts to boost the development of national Islamic banking and increase market share so that national banking can compete independently. The dynamic development of national legislation has changed the banking world, particularly in laws related to banking practices, including Islamic banking. The existence of dozens of rules that have been in effect since 1998 shows a need to adjust the Law on Islamic banking to remain harmonious and in line with the dynamics of existing legislation, one of which is the spin-off policy. Teguh Supangkat, as Deputy Commissioner for Banking Supervision I OJK

(Otoritas Jasa Keuangan/Indonesia's Authority of Financal Services), sees that Islamic bank windows' current development and performance are encouraging by recording higher asset growth than full-fledged Islamic Banks. In the last few years, OJK has been in intense discussion to see the prospect of Islamic bank windows' spin-off obligations in the future, both from the Islamic banking industry and other state ministries and institutions.

The OJK received proposals from various stakeholders in the Islamic banking industry to re-consider a spin-off obligation policy in 2023. Then, Wimboh Santoso, as Commissioners of OJK, together with Irwan Trinugroho, Rakianto Irawanto, and Putra Pamungkas (2021), conducted a study related to the spin-off policy and found that the performance and efficiency of full-fledged Islamic Banks were lower than that of Islamic bank windows, while financing risk increases after the spin-off. The low performance of Islamic banks that carried out the spin-off persisted for four years after the spin-off. Likewise, the spin-off policy's effect is that full-fledged Islamic Banks is still constrained by capital and business expansion because they require high investment. To reduce the negative impact of the spin-off policy, OJK made a complementary policy, namely POJK No. 28/POJK.03/2019 concerning Banking Synergy in One Ownership for the Development of Islamic Banking. This POJK was created to support Islamic banking products and services so that they are not less competitive with conventional banking. Implementing POJK's Banking synergy allows full-fledged Islamic Banks to use Conventional Commercial Bank infrastructure in one ownership to provide better customer services by expanding service access, improving product and service quality, and improving operational efficiency. In addition, consolidation among the new Islamic banks can help them achieve economies of scale and become more competitive.

From the practitioner's perspective, ASBISINDO (Asosiasi Bank Syariah Indonesia/ Indonesian Sharia Banking Association) argues that the spin-off policy needs to be reviewed because of the heavy burden that Islamic bank windows must bear in carrying out their activities, which are no longer borne by their parent banks. The structural problems faced by full-fledged Islamic Banks are classic problems experienced by small banks, namely the difficulty competing with conventional commercial banks. The spin-off currently being carried out is counterproductive to its main target, namely the growth and strengthening of Islamic banking. The spin-off will not only harm customers and the Islamic financial industry. Still, it will also cause problems with the increasing number of small full-fledged Islamic Banks with limited capital and professional human resources. The level of service to the community will decline because Islamic banks that are already fully operational due to the spin-off have been unable to provide infrastructure and experts equal to their parent banks. In the future, the spin-off will also have an impact on the growth and quality of assets and the quality of services due to limited capital. Best practices in other countries with Sharia banking market share above double digits allow Conventional Commercial Bank to continue to run its Islamic bank windows so that the performance and market share of the Islamic finance industry can continue to grow healthily and rapidly. With a spin-off, this option will disappear, and investors will turn to a different industry, resulting in an opportunity lost (ASBISINDO, 2022).

Director of CIMB Niaga Syariah, Pandji P. Djajanegara, said that the spearhead of the spin-off was to enlarge the Sharia portfolio, not to increase the number of banks. He said that after implementing the Dual Banking Leveraging Model (DBLM) or Islamic bank windows, CIMB Niaga's portfolio and sharia portion grew more rapidly. Until 2021, the share of book assets grew by 13.8%, total assets grew by 35.9% CAGR, book financing grew by 14.3%, and book deposits grew by 11.3%. In the last five years, the Islamic bank's performance has contributed more to the share of its parent bank. The average contribution of the top five Islamic bank windows assets to the parent bank's share is 14%. Thus, Islamic bank windows are considered reliable in accelerating the achievement of the target of Islamic banking assets of 20% by 2024 by the entire banking sector (Djajanegara, 2022). Director Permata Bank Sharia, Herwin Bustaman, also hoped the spin-off policy could be changed. He said that this spin-off should be left to the business decisions of each bank, whether or not it is a spin-off (Andika and Aldi, 2020). Twin Zakaria, President Director of Maybank, said that his party welcomes the suggestion that spin-offs are no longer mandatory because they have proven quite effective as a structure and provide much efficiency for their conventional parent (Hutauruk and Mahadi, 2021).

Some researchers have found that the impact of spin-off policies on Islamic bank windows which has a large size (such as Bank BNI Syariah, BRI Syariah, Bukopin Syariah, BJB Syariah, Bank Syariah Mandiri, and Mega Syariah) does not affect bank performance. Spin-off systems calculated by several financial performance ratios negatively and significantly influenced profitability and market share (Jubaedah *et al.*, 2020; Arif, 2017). The Islamic bank windows owned by the regional development bank, which is small, Haribowo (2017) found that no Sharia unit could reach 50% of its parent's stock assets. This means that the small banks are not ready to do a spin-off. The requirements for the spin-off of Islamic bank windows into full-fledged must go through a strict process; for example, in terms of adequate capital, the higher the capital, the better the average performance (Prasetyo *et al.*, 2019).

The difference in perspective between regulators and practitioners of Islamic banking in responding to the spin-off policy stems from the different perspectives on the development strategy of Islamic banks between those emphasising the quantity and quality sides. Efforts to improve regulations must be carried out continuously to harmonise Islamic banking operations with conditions that continue to change in the macroeconomy. The regulator has proposed that the Islamic bank windows spin-off will no longer be mandatory in the Development and Strengthening of the Financial Sector Law (P2SK Law). The crucial article of the P2SK Law Part Three concerning Islamic Banking, the Sharia Banking Law Article 68 paragraph (1) is changed to, "If a Conventional Commercial Bank has an Islamic bank window, after fulfilling certain requirements set by the Financial Services Authority, the Bank The Conventional General referred to is obliged to carry out the Separation into a Sharia Commercial Bank", and paragraph (2) was changed to "In addition to the requirements as intended in paragraph (1), the Financial Services Authority may request the Separation of Islamic bank windows into a full-fledged Islamic Banks in the context of consolidating Sharia Banking."

The requirements set by the Financial Services Authority are listed in OJK Regulation No. 12/2023 concerning the Sharia Business Unit, including Article 59 paragraph (1) "Conventional Commercial Bank that has Islamic bank windows with asset value having reached 50% (fifty percent) of the total asset value of its parent Conventional Commercial Bank and/or the amount of Islamic bank windows assets is at least IDR 50,000,000,000,000.00 (fifty trillion rupiahs) is obliged to separate Islamic bank windows with certain stages that take into account the efficient, healthy and sustainable performance of the financial services industry"; Article 60 "Conventional Commercial Bank that has a Islamic bank window can separate the Islamic window before the conditions as intended in Article 59 paragraph (1) are fulfilled"; Article 61 paragraph (1) "OJK may request the separation of Islamic bank window in the context of sharia banking consolidation for the development and strengthening of sharia banking.", paragraph (2) "The OJK request as intended in paragraph (1) is carried out by considering: a. Islamic bank window did not experience significant growth; b. Conventional commercial banks with an Islamic bank window are considered unable to develop Islamic banking windows; c. there is a need for a sharia banking development strategy; and/or d. other factors. OJK Regulation No. 12/2023 removes the sentence "or 15 (fifteen) years since the enactment of this Law" contained in the Sharia Banking Law. This means that Islamic Bank Windows is not required to conduct a spin-off in 2023. Thus, the spin-off is only a choice.

The P2SK Law and OJK Regulation concerning the Sharia Business Unit is a middle ground for the debate among stakeholders regarding the spin-off obligation policy and will be the solution for the current development of the Islamic bank window. The born regulations certainly aim to promote the growth of the Islamic banking sector in Indonesia. The dynamics of Islamic banking regulations can be briefly seen in Figure 1. as follows:

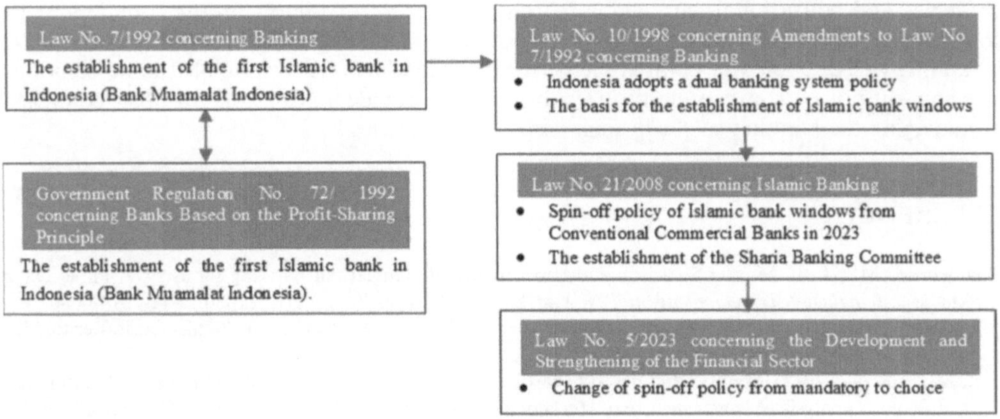

Figure 1. The dynamics of Indonesian Islamic banking regulations.

The new Law has reflected a relaxation policy to accommodate the dynamic development of the Sharia banking industry and is expected to create better efficiency at Islamic bank windows in supporting the 8th SDGs target, namely Decent Work and Economic Growth. In the last five years, Islamic bank windows' performance has proven to contribute more to its full-fledged Islamic Banks. The average contribution of Top five Islamic bank windows assets to its full-fledged Islamic Banks share reached 14%. This means that by maintaining the Islamic bank windows business model, it has the potential to accelerate the achievement of the target of 20% of national banking assets by 2024. However, this policy has risks, namely creating uncompetitive competition and a lack of Islamic bank windows initiative to develop long-term strategies in developing the business. In addition, it has the potential to create ambiguity regarding the level of Sharia compliance, because conventional banking and Islamic banking businesses operate under one management. Islamic banking uses religious values as the basis and indicators for formulating company goals and policies, so Islamic financial institutions must align with Islamic law, ethics, and business (Triasari and de Zwart, 2021; Maksum and Hidayah, 2023; Hidayah and Azis, 2023). Although there is a change in the spin-off policy, Islamic bank windows ares expected to continue formulating a strategy to be independent.

4 CONCLUSION

Islamic banking practitioners, especially in Islamic windows, consider that the rationale of the spin-off is to enlarge the Sharia portfolio, which can be approached by many strategies, not only by increasing the number of banks. The difference in perspective between regulators and practitioners of Islamic banking in responding to the spin-off policy had been attributed to the different perspectives in the development strategy of Islamic banks, between those emphasising the quantity and quality sides. The P2SK Law and OJK Regulation concerning the Sharia Business Unit is a middle ground for the debate among stakeholders regarding the spin-off obligation policy and will be the fair solution for the current development of Islamic bank windows; the spin-off is a choice. Islamic bank windows must immediately evaluate whether to remain as units or spin off into full-fledged Islamic Banks. Key considerations include financial feasibility analysis and investment needs if choosing a spin-off. Furthermore, Islamic bank windows must focus on optimising operations, increasing efficiency, orienting towards product development, improving service quality, leveraging synergies with the parent banks to strengthen the market position and considering marketing strategies and educational campaigns to raise awareness of Sharia products and prepare contingency plans for future policy changes. Regulators must monitor this policy to ensure it is running well and remains compatible with increasing the market share of Islamic banking. The spin-off policy from mandatory to choice might provide an example for other countries in managing the banking sector in their respective countries.

REFERENCES

Ahamed, M. M. (2012). Market Structure and Performance of Bangladesh Banking Industry: A Panel Data Analysis. *Bangladesh Development Studies*, *XXXV*(3), 1–18. https://doi.org/10.1108/jfra-11-2014-0086

Andika, and Aldi. (2020). *Spin-Off Menghantui Pelaku Industri Perbankan*. Https://Knks.Go.Id/Berita/327/Spin-off-Menghantui-Pelaku-Industri-Perbankan?Category=1.

Arif, M. N. R. Al. (2017). Spin-off and market share in the indonesian islamic banking industry: A difference in difference analysis. *Management and Marketing*, *12*(4), 540–551. https://doi.org/10.1515/mmcks-2017-0032

Arif, M. N. R. Al, Haribowo, I., and Suherlan, A. (2018). Spin-off policy and efficiency in the Indonesian islamic banking industry. *Banks and Bank Systems*, *13*(1), 1–10. https://doi.org/10.21511/bbs.13(1).2018.01

Arif, M. N. R. Al, Ihsan, D. N., Zulpawati, and Fatah, D. A. (2023). The impact of conversion on market share in Indonesian Islamic banks. *Banks and Bank Systems*, *18*(2), 1–12. https://doi.org/10.21511/bbs.18(2).2023.01

Arif, M. N. R. Al, Mufraini, M. A., and Prabowo, M. A. (2019). Market Structure, Spin-Off, and Efficiency: Evidence from Indonesian Islamic Banking Industry. *Emerging Markets Finance and Trade*, *56*(2), 329–337. https://doi.org/10.1080/1540496X.2018.1553162

ASBISINDO. (2022). *Kontribusi UUS Dalam Perbankan Syariah Indonesia*.

Djajanegara, P. P. (2022). *Kebijakan dan Strategi Pengembangan Perbankan Syariah* (Issue May).

Fathoni, M. A., Rodoni, A., Al Arif, M. N. R., and Hidayah, N. (2024). Intention to participate in Islamic banking in Indonesia: Does socio-political identity matter? *Journal of Islamic Marketing*. https://doi.org/10.1108/JIMA-12-2023-0397

Haribowo, I. (2017). The Indonesian Islamic Bank's Spin-off: A Study in Regional Development Banks. *Al-Iqtishad: Journal of Islamic Economics*, *9*(1), 53–68. https://doi.org/10.15408/aiq.v9i1.4308

Hidayah, N. (2019). *Fatwa-fatwa Dewan Syariah Nasional: Kajian terhadap Aspek Hukum Islam Perbankan Syariah di Indonesia*. https://repository.uinjkt.ac.id/dspace/bitstream/123456789/50564/1/Buku Fatwa-fatwa DSN Kajian Terhadap Aspek Hukum Islam Perbankan Syariah di Indonesia.pdf

Hidayah, N. (2022). *Ekonomi Syariah di Indonesia: Tinjauan Aspek Hukum*. Dee Publish.

Hidayah, N., and Azis, A. (2023). IMPLEMENTATION OF PROGRESSIVE LAW IN SHARIA BANKING DISPUTE SETTLEMENT: Case Study of Religious Court Decisions in Indonesia. *Ulumuna*, *27*(1), 227–257. https://doi.org/10.20414/ujis.v27i1.652

Hidayah, N., Azis, A., and Muslim, M. B. (2022). Complying with Sharia While Exempting from Value-Added Tax: Murābaḥah in Indonesian Islamic Banks. *Ahkam: Jurnal Ilmu Syariah*, *22*(1), 59–82. https://doi.org/10.15408/ajis.v22i1.22833

Hidayah, N., Azis, A., Mutiara, T., and Larasati, D. (2023). Sharia Banking Disputes Settlement: Analysis of Religious Court Decisions in Indonesia. *Al-Risalah: Forum Kajian Hukum Dan Sosial Kemasyarakatan*, *23*(1), 75–92. https://doi.org/10.30631/alrisalah.v23i1.1347

Hidayah, N., and Haryatih. (2024). *Perbankan Syariah di Indonesia: Perspektif Regulasi dan Perilaku Muslim* (K. Kamarusdiana (ed.)). UIN Jakarta Press.

Hidayah, N., Haryatih, H., and Kamilah, R. A. (2024). *Islam and the State: The Evolution of Sharia Banking Regulation in Indonesia Universitas Islam Negeri Syarif Hidayatullah Jakarta Abstract The interplay of social , economic , and political dynamics influences the legal development of Sharia banking regu* (Vol. 23, Issue 2).

Hidayah, N., Zamilah, A., Rizal, S., and Jaharuddin, J. (2021). The Resilience of Sharia and Conventional Banks in Indonesia during the Covid-19 Pandemic Crisis. *Al-Iqtishad: Jurnal Ilmu Ekonomi Syariah*, *13*(2). https://doi.org/10.15408/aiq.v13i2.23964

Hutauruk, D. M., and Mahadi, T. (2021). *Bank akan pilih pertahankan status UUS jika spin off tak lagi mandatory*. Https://Keuangan.Kontan.Co.Id/News/Bank-Akan-Pilih-Pertahankan-Status-Uus-Jika-Spin-off-Tak-Lagi-Mandatory.

Jubaedah, D., Yusup, D. K., Sobana, D. H., and Suhaeni, I. (2020). The Effect of Spin-Off System and The Third Party Funds On Return On Equity Ratio Of Islamic Bank In Indonesia. *Journal of Critical Reviews*, *7*(13), 2743–2752.

Maksum, M., and Hidayah, N. (2023). The Mechanism of Avoiding Riba in Islamic Financial Institutions: Experiences of Indonesia and Malaysia. *Juris: Jurnal Ilmiah Syariah*, *22*(2), 235–244. https://doi.org/10.31958/juris.v22i2.6952

Mohamad, S., Mahomed, Z., and Kamil, N. (2017). Islamic Banking Business of Conventional Banks: Transition from Windows to Islamic Subsidiaries. In M. Ariff (Ed.), *Islamic Finance in Malaysia: Growth and Development* (pp. 86–114). Pearson Malaysia Sdn. Bhd. Background.

Muhammad Budi Prasetyo, Luxianto, R., Baskoro, R. A., Adawiyah, W., and Putri, N. I. S. (2019). The Role of Capital on Islamic Bank Spin-Offs in Indonesia. *The South East Asian Journal of Management*, *13*(2), 119–139. https://doi.org/10.21002/seam.v13i2.11179

Naseri, M., Bacha, O. I., and Masih, M. (2019). Too Small to Succeed versus Too Big to Fail: How Much Does Size Matter in Banking? *Emerging Markets Finance and Trade*, *0*(0), 1–24. https://doi.org/10.1080/1540496X.2019.1612359

Nur Rianto Al Arif, M., Nachrowi, N. D., Nasution, M. E., and Zakir Mahmud, T. M. (2017). The Islamic banking spin-off: Lessons from Indonesian Islamic banking experiences. *Journal of King Abdulaziz University, Islamic Economics*, *30*(2), 117–133. https://doi.org/10.4197/Islec.30-2.11

Pambuko, Z. B., Medias, F., Dewi, V. S., and Karunia, S. D. (2024). Restructuring Strategy: A Performance Review of Spin-off Islamic Banks in Indonesia. *E3S Web of Conferences*, 1–8. https://doi.org/10.1051/e3sconf/202450005008

Rabin, J. (2007). *Handbook of Public Policy Analysis*. Taylor and Francis Group.

Rahmawati, Y., Ihsan, D. N., Amelia, E., and Al Arif, M. N. R. (2024). Conversion Into Full-Fledge Islamic Banks and It'S Impact on Asset: Evidence From Indonesian Islamic Banking Industry. *International Journal of Professional Business Review*, *9*(4), e04621. https://doi.org/10.26668/businessreview/2024.v9i4.4621

Rokhmawati, A., Desy, and Halim, E. H. (2022). Does Spin-off Obligation of Sharia Business Units from its Parent Commercial Banks Enhance Their Performance? *Iranian Economic Review*, *26*(4), 819–832. https://doi.org/10.22059/ier.2022.90658

Triasari, D., and de Zwart, F. (2021). The Legal Reform Policy on the Shariah Supervisory Board Role's in Indonesian Shariah Banks. *Bestuur*, *9*(2), 113–125. https://doi.org/10.20961/bestuur.v9i2.55173

Trinugroho, I., Santoso, W., Irawanto, R., and Pamungkas, P. (2021). Is Spin-off Policy an Effective way to Improve Performance of Islamic Banks? Evidence from Indonesia. *Research in International Business and Finance*, *56*, 1–30. https://doi.org/10.1016/j.ribaf.2020.101352

Wildan Hakim, F., M. Aziz, R., and Hidayah, N. (2024). Efficiency of Bank Aceh Syariah and Bank NTB Syariah After Conversion. *1st International Conference on Islamic Economics, Business Development and Studies*, *2024*, 645–661. https://doi.org/10.18502/kss.v9i16.16279

Yanow, D. (2007). Interpretation in Policy Analysis: On Methods and Practice. *Routledge Taylor and Francis Group*, *1*(1), 110–122. https://doi.org/10.1080/19460171.2007.9518511

Yuspin, W., and Wardiono, K. (2017). Islamic banking trajectories of Indonesia: Dealing with the present and shaping the future. *International Journal of Economic Research*, *14*(16), 29–42. https://doi.org/https://www.researchgate.net/publication/324795606_Islamic_banking_trajectories_of_Indonesia_Dealing_with_the_present_and_shaping_the_future

Towards Resilient Societies: The Synergy of Religion, Education, Health, Science, and Technology – Rahiem (Ed.)

Price discounts and brands: Their role in urban and rural consumer loyalty in online shopping

Sofyan Rizal & Nurul Apriani
Universitas Islam Negeri Syarif Hidayatullah Jakarta, Indonesia

ABSTRACT: This study aims to analyse the impact of customer loyalty towards brands and discounts in online shopping through marketplaces, comparing urban and rural communities. A quantitative approach was employed, involving a sample of 200 participants—100 from urban areas and 100 from rural areas—determined using the Slovin formula. Data were collected via questionnaires and analysed using IBM SPSS_25 software with multiple linear regression. The analysis included validity and reliability tests, classical assumption tests (normality, heteroscedasticity, and multicollinearity), and hypothesis testing, considering the coefficient of determination (R^2). The t-test and F-test were used for data interpretation. The results reveal that the t-value of 1.987 in urban communities indicates a positive effect of discount pricing and brand on customer loyalty in online shopping through marketplaces. Similarly, in rural communities, the t-value of 1.986 also demonstrates a positive effect of discount pricing and brand on customer loyalty. These findings suggest that both urban and rural customers are influenced by discounts and brand loyalty when shopping online through marketplaces.

Keywords: brand and consumer loyalty, price discount

1 INTRODUCTION

Technological advancements in everyday life persistently evolve, spanning from the agricultural, industrial, and information technology eras to the contemporary era of communication and information technology. These advancements begin with elementary processes and evolve to achieve a level of satisfaction for both individuals and societal entities. The lives of communities, nations, and countries are influenced by these transformations, and everyone wants to use and utilise these changes. With the rapid development of technology, many parties use this progress to take advantage of it. One of them is the trade sector, where almost all goods are sold via the internet. Items that previously could only be bought in traditional markets can now be bought on the internet at a cheaper price. The trend of online shopping has changed people's behaviour, especially in cities and rural areas. Many people now prefer to shop online instead of going directly to the nearest market or mall, which shows this change in habits. With online shopping applications, people can search and buy goods over the internet without having to leave their homes or go to the store (Sharma and Gupta 2022).

Many online marketplaces in Indonesia are already popular. Therefore, online marketplaces use various promotional strategies to survive the increasingly fierce competition and gain loyal customers. Sukesi (2013) states that loyalty can be defined as the level of likelihood that customers will continue to be associated with the organisation. Loyal consumers will buy goods in the same place repeatedly Mu'ah, (2021). Discounts are a common strategy to attract loyal customers. A discount represents a temporary reduction in the product's price, thereby making it lower than the standard price for a specified duration. In this

 DOI: 10.1201/9781003645542-84

context, the price of the product will be reduced from the normal price over a period of time, thereby increasing sales volume. Periodic, promotional, early purchase and final discounts comprise four types of discounts, according to Schindler (2012).

A study conducted by Dewi *et al.* in 2020 examined the effect of discounts on consumer purchasing decisions and online business customer satisfaction. The study shows that factors such as discounts and brands have a significant influence on customers' decisions to buy goods and their level of satisfaction. However, the study has not explained in detail whether customers will return to the same online marketplace when they want to buy goods after the price is discounted. The purpose of this study is to find out how discounts and brands impact customer loyalty in online shopping in urban and rural communities. Research basically measures natural or social phenomena, so good measurement tools are needed. Quantitative research often uses questionnaires as the main way to collect data because they can relate directly to the subject or respondent. To get accurate research data, questionnaires are very important. In quantitative research, a questionnaire or questionnaire is a tool used when the problem is not clear at first. Once the problem is clearly researched, the tool can be developed. Researchers use questionnaires that are made according to local needs to get accurate data (Pradita and Pratiwi 2023).

Consumer behaviour in online marketplaces is significantly influenced by the environment in which they live, whether urban or rural. In urban environments, consumers are typically more open to new products and brands due to greater access to information, technology and infrastructure that supports online transactions. They are also more influenced by modern lifestyles and global consumption trends, which encourage them to shop online more often and try innovative new products. In contrast, consumers in rural areas tend to be more practical and conservative in their purchasing decisions. More limited internet access, as well as stronger social influence from local communities, may influence their preference for products that are more recognisable and relevant to their daily needs. In addition, lower income levels in some rural areas make consumers more cautious in their product choices, preferring items that are more affordable and functional. Therefore, differences in infrastructure, information, and prioritisation of needs create stark differences in consumer behaviour between urban and rural environments.

2 METHODS

This research uses quantitative methodology. Researchers used purposive sampling and nonprobability sampling strategies to select samples (Jannah 2018; Sugiyono 2012). The specific criteria used in sampling in this study are Urban Communities and Rural Communities. To determine the validity and reliability of the questionnaire, the researcher conducted tests on each questionnaire item, this can be tested using methods such as Cronbach's Alpha to measure internal consistency, where values above 0.7 are considered reliable. In addition, to ensure cultural relevance, the questionnaire should be translated and adapted to the cultural context of the respondents, for example through a back-translation process or pilot testing on a small sample of the target population. This process ensures that the questionnaire is not only valid and reliable, but also relevant for the cultural context of the study. According to Nugroho (2021), a survey is a method for collecting information about a population by asking questions to a sample of people. The coefficient of determination (R2) test, simple linear regression test, and F test (ANOVA) were used in this study to determine whether variable X has an effect on variable Y or not. All of these tests were conducted using the SPSS (Ver, 25) program, because this software provides a variety of statistical analysis features suitable for understanding consumer behaviour in depth equipped with the ability to perform descriptive analysis, correlation, regression, and hypothesis testing which is the basis for analysing the relationship between variables in consumer behaviour research. In addition, version 25 offers a user-friendly interface and support for the latest features, such as the ability to process large data with greater efficiency, allowing researchers to process data quickly and accurately. With its reliability in processing quantitative data, SPSS 25 ensures valid and reliable results, making it an appropriate tool for identifying consumer behaviour patterns, testing hypotheses, and providing data-driven recommendations in the context of this study. Consumer loyalty in shopping at the market was compared between urban and rural communities, with a sample size of 1,266 people from urban communities and 2,816 people from rural communities. The sample was calculated using the Slovin formula with a 10% precision level. The reason for choosing the Slovin formula with a precision level of 10% is often done because it provides a balance between the accuracy of the research results and the efficiency of data collection. The Slovin formula is used to determine the sample size of a particular population, especially when the population size is large and it is not possible to conduct a comprehensive survey. A 10% precision level is chosen because it is considered representative enough to get an overview of the population, especially in exploratory research or initial studies that do not require high accuracy such as causal research. In addition, the use of a 10% precision level helps to reduce the time, cost and resources required, without significantly compromising the validity of the data. Therefore, this level of precision provides optimal flexibility and efficiency, especially in studies involving large populations with budget or time constraints.

3 FINDING & DISCUSSION

3.1 *Analysis of findings*

The respondents of this study consisted of people from both urban and rural communities. Rural people, who use other online commerce platforms, accounted for 97 people or 97% of the respondents, and urban people accounted for 93 people or 93%.

Table 1. F test results (city community).

	Model	Sum of Squares	Df	Mean Square	F	Sig.
1	Regression	34,045	2	17,022	2,457	,091^{b}
	Residual	632,568	90	6,929		
	Total	657,613	92			

Table 1 presents the calculated F value of 2.457 with a significance level of 0.091. The F value from the F-distribution table is denoted as Ftabel = F (k, n-k), resulting in F = (2, 91). Consequently, the results from Table 4.13 indicate that Fcount (2.457) exceeds Ftable (3.097), and the significance level is less than 0.05 (5%). This outcome supports the acceptance of H1 and the rejection of H0, thereby concluding that customer loyalty to brand offers and price discounts are interrelated. However, the F-test results for the community data village are as follows:

Table 2. F test results (village community).

	Model	Sum of Squares	Df	Mean Square	F	Sig.
1	Regression	20,325	2	10,163	1,247	,292^{b}
	Residual	766,005	94	8,149		
	Total	786,330	96			

Table 2 presents the calculated F-value of 1.247 with a significance level of 0.292. The F-value from the F-distribution table is denoted as Ftabel = F(k, n-k), where k = 2 and n-k = 95, resulting in F(2, 95). The F-distribution table value is 3.092, as shown in Table 4.13. Since F_count (1.247) exceeds F_table (3.092) and the significance level is less than 0.05, we can conclude that discounts and brands significantly impact customer loyalty in online marketplace purchases.

Table 3. Test results of the coefficient of determination r2 (city society) model summary.

	R	R Square	Adjusted R Square	Std. Error of the Estimate
1	,228[a]	,052	,031	2,632

[a]Predictors: (Constant), price discount, brand.

Table 4. Test results of the coefficient of determination r2 (village community) model summary.

	R	R Square	Adjusted R Square	Std. Error of the Estimate
1	,161[a]	,026	,005	2,855

[a]Predictors: (Constant), price discount, brand.

4 DISCUSSION OF RESEARCH RESULTS

It is known that to compare and illustrate the pattern of consumer loyalty between urban and rural communities, we can create a graph or chart that illustrates the factors that influence consumer loyalty, as well as the differences in loyalty patterns that exist in the two regions. The following are elements that can be considered in such a graph or chart based on the general findings on the differences in consumer loyalty in urban and rural areas:

Factors	Urban (Scale 1-10)	Rural (Scale 1-10)
Price and Discount	7	6
Product Quality	6	8
Familiar Brand	5	7
Shopping Experience	8	5

Purchase Channel	Urban (%)	Rural (%)
Online	60%	20%
Physical Store	30%	70%
Others	10%	10%

4.1 *Analysis of disscusion*

4.1.1 *Effect of price discount, brand on consumer loyalty (city people)*

Compared to the Ftable value of 3.097, the Fcount value of 2.457 indicates a significant influence of the brand on customer loyalty in online shopping through marketplaces in both urban and rural communities. This is attributed to the benefits of discounts that customers can derive. Humans exhibit economic behaviour, utilising minimal effort to achieve optimal results. The ability of an item to achieve a specific goal or performance is crucial. The results demonstrate that the brand's perspective significantly impacts customer loyalty. This is because brands are intricately linked to consumer perceptions. Customers often prefer the dominant brand to create a positive image of the company if they have a favourable attitude towards a particular brand. Furthermore, consumers are less likely to switch to other products due to the satisfaction derived from a superior brand.

4.1.2 *Effect of price discount, brand on consumer loyalty (village community)*

In both urban and rural markets, discounts and brands significantly influence customer loyalty when shopping online, with an Fcount of 2.457 compared to an Ftable of 1.247. The survey results show that rice discounts, customer loyalty, and brand influence customer loyalty. Brand is perspective because it relates to what the customer imagines. If customers have a positive impression of a particular brand, they tend to buy the dominant brand to create a positive company image. Since a better brand will satisfy customers, they will not easily switch to other products.

5 CONCLUSION

This study demonstrates that discounts significantly influence customer loyalty in both urban and rural settings. Discounts motivate customers to make purchasing decisions and maintain loyalty to lower-priced items. Conversely, the impact of brand influence on customer loyalty varies. In urban communities, the influence of the brand on customer loyalty is negligible when shopping online, possibly because consumers prioritise price over brand. Similarly, in rural communities, brand effect on customer loyalty is also insignificant. However, when both brand and discount are combined, they substantially influence customer loyalty in urban and rural settings. An effective strategy to enhance customer loyalty involves offering attractive discounts and cultivating a strong brand reputation.

To increase consumer loyalty in the online marketplace, businesses should focus on personalising the shopping experience, offering attractive loyalty programs, and improving responsive customer service. In addition, simplifying the purchase and return process, strengthening interaction through social media, as well as guaranteeing product quality and fast and safe delivery services will make consumers feel valued and secure. Flexible and transparent pricing strategies, as well as enhanced transaction data security, are also important for creating a deep sense of trust. All these elements complement each other in building long-term relationships with consumers, increasing satisfaction and encouraging them to remain loyal to the brand.

ACKNOWLEDGMENTS

I am grateful to all Lecturers of FEB UIN Jakarta, especially to the Head of the Study Program Mr. Sofyan Rizal who has collaborated in developing a paper on the comparison between urban communities and rural communities in shopping online through the marketplace.

REFERENCES

Alam, A. A. R. N., Purnama Alam, A. R., and Anggi, A. (2019). Pengaruh Kemasan Produk Susu Bear Brand Terhadap Keputusan Pembelian Konsumen Pada PT. Oze Power Switch Di Makassar. *Movere Journal*, 1(2), 111–120. https://doi.org/10.53654/mv.v1i2.63

Anggraini, V. A., and Hananto, A. (2020). The Role of Social Media Marketing Activities on Customer Equity Drivers and Customer Loyalty. *AFEBI Management and Business Review*, 5(1), 1. https://doi.org/10.47312/ambr.v5i1.299

Aprihatiningrum Hidayati, M. M., Sifatu, W. O., Maddinsyah, A., Sunarsi, D., MM, C., and Jasmani, S. E. (2021). Loyalitas dan Kepuasan Konsumen: Tinjauan Teoritik. *Cipta Media Nusantara (CMN).*

Aryatiningrum, S. W., and Insyirah, A. I. (2020). Pengaruh Price Discount terhadap Loyalitas Konsumen pada Pengguna Online Marketplace. *Jurnal Ilmu Manajemen*, 8(3), 946. https://doi.org/10.26740/jim.v8n3.p946–953

Bili, Y. (2020). Analisis Faktor Trust dan Price Discount Bagi Pengguna Traveloka Dampaknya terhadap Kepuasan Konsumen dan Loyalitas Konsumen. *Jurnal Syntax Transformation*, 1(10), 777–781. https://doi.org/10.46799/jst.v1i10.166

Cahyono, N. E., and Fahmi, S. (2022). Pengaruh Kualitas Produk Dan Kepercayaan Terhadap Kepuasan Konsumen Dalam Membentuk Loyalitas Konsumen (Pengguna Platform Aplikasi Marketplace di Kota Malang). *Jemba: Jurnal Ekonomi, Manajemen, Bisnis Dan Akuntansi*, 1(4), 503–516.

Cardia, D. I. N. R., Santika, I. W., and Respati, N. N. R. (2019). Pengaruh Kualitas Produk, Harga, dan Promosi Terhadap Loyalitas Pelanggan. *E-Jurnal Manajemen Universitas Udayana*, 8(11), 6762. https://doi.org/10.24843/ejmunud.2019.v08.i11.p19

Danuri, M. (2019). Perkembangan dan transformasi teknologi digital. *Jurnal Ilmiah Infokam*, 15(2).

Daniel Hermawan. (1970). The Importance of Digital Trust in E-Commerce: Between Brand Image and Customer Loyalty. *International Journal of Applied Research in Management and Economics*, 2(3), 18–30. https://doi.org/10.33422/ijarme.v2i3.268

Jannah, N. (2018). *Quantitative research methods and their applications in social research.* Prenadamedia Group.

Khairawati, S. (2019). Effect of customer loyalty program on customer satisfaction and its impact on customer loyalty. *International Journal of Research in Business and Social Science (2147–4478*, 9(1), 15–23. https://doi.org/10.20525/ijrbs.v9i1.603

Kristanto, F. H., Wimanda Rahma, H., and Nahrowi, M. (2022). Factors Affecting E-Commerce Customer Loyalty In Indonesia. *Jurnal Syntax Transformation*, 3(09), 1150–1164. https://doi.org/10.46799/jst.v3i09.613

Mofokeng, T. E. (2021). The impact of online shopping attributes on customer satisfaction and loyalty: Moderating effects of e-commerce experience. *Cogent Business and Management*, 8(1). https://doi.org/10.1080/23311975.2021.1968206

Mu'ah, S. E. (2021). Loyalitas pelanggan: Tinjauan aspek kualitas pelayanan dan biaya peralihan. *Zifatama Jawara.*

Nugroho, N. E. (2021). Strategi meningkatkan loyalitas konsumen dipengaruhi terpaan iklan, pemberian potongan harga, dan citra merek pada marketplace Shopee dalam kondisi pandemi Covid-19. *Jurnal EKSEKUTIV*, 18(1), 5.

Pradita, S. D., and Pratiwi, N. M. I. (2023). Strategi pemberdayaan mahasiswa magang kampus merdeka dalam upaya mengoptimalkan program PUSPAGA di Balai RW 09 Kelurahan Menur Pumpungan melalui website SSW ALFA dan SIAP PPAK. *Gemah Ripah: Jurnal Bisnis*, 3(4), 77–81.

Rahman, S. H., Rohaeni, H., and Dewi, S. W. K. (2020). Pengaruh price discount dan bonus pack terhadap purchase intention. *Jurnal Sain Manajemen Universitas ARS*, 2(1).

Rosalina, M., Qomariah, N., and Sari, M. I. (2019). Dampak promosi, harga, dan kualitas produk terhadap loyalitas konsumen Oppo Smartphone. *Jurnal Penelitian IPTEKS Universitas Muhammadiyah Jember*, 4 (2), 6.

Sanabria, I. M., Kim, J., and Chan-Olmsted, S. (2022). Online shopping behavior: The role of technological innovations and digital platforms. *Qualitative Market Research: An International Journal*, 25(1), 124–142.

Schindler, R. M. (2012). *Pricing strategies: A marketing approach.* Pearson Education.

Sharma, S., and Gupta, R. (2022). Impact of online shopping applications on consumer behavior and retail market trends. *Journal of E-commerce and Retail Studies*, 14(3), 105–120.

Sugiyono. (2012). *Quantitative, qualitative, and R&D research methods.* Alfabeta.

Sukesi, S. (2013). *Loyalitas Pelanggan Produk Terasi Rebon, Ditinjau Dari Keunggulan Bahan Baku Dan Kualitas Pelayanan 4P.*

Suman, S. K., Srivastava, P., and Vadera, S. (2019). Exploring the behaviour of Indian consumers towards online discounts. *International Journal of Electronic Marketing and Retailing*, 10(1), 78. https://doi.org/10.1504/ijemr.2019.096629.

Towards Resilient Societies: The Synergy of Religion, Education, Health, Science, and Technology – Rahiem (Ed.)
© 2026 The Author(s), ISBN: 978-1-041-08471-6
Open Access: www.taylorfrancis.com, CC BY-NC-ND 4.0 license

Transforming Batam free trade zone and special economic zone: Policy and governance strategies for sustainable investment growth

Rani Murweng Pramesti, Hanifah Marsha Mudita, Muhammad Refardian Imaduddin, Aland Edbert Manurung & Revian Emir Minanda
Directorate of Regional I, Ministry of National Development Planning/Bappenas

Ahmad Rodli Putra Hendrawan
Faculty of Economics and Business, University of Indonesia, Indonesia

ABSTRACT: Special Economic Zones (SEZ) and the Free Trade Zone (FTZ) in Batam face governance and investment challenges that hinder progress toward development goals. This study investigates governance inefficiencies in accelerating investment, aligning with Sustainable Development Goals (SDGs) No. 9 (Industry, Innovation, and Infrastructure) and No. 16 (Peace, Justice, and Strong Institutions). Using a systematic literature review, in-depth interviews (IDI), and focus group discussions (FGD), the study identifies key issues such as overlapping regional development policies, weak leadership competence, governance limitations, insufficient local ownership, and unclear land pricing regulations. The study draws insights from the successful SEZ model in Shenzhen to address these challenges. Recommendations include reforming land and lease regulations, providing competitive tax incentives, investing in infrastructure, and developing integrated policies. Establishing a joint planning committee and implementing a comprehensive infrastructure plan can resolve policy overlaps. Additionally, granting greater autonomy to the SEZ, revising leadership policies, and strengthening local ownership through joint ventures and public-private partnerships are essential strategies for accelerating investment and fostering sustainable economic growth in Batam.

Keywords: economic development, investment, policy implementation, FTZ, SEZ

1 INTRODUCTION

Batam City, located in the Riau Islands of Indonesia, is designated as a SEZ due to its proximity to Singapore. Despite its effort to attract foreign investment, Free Trade and Free Port Zone (KPBPB) Batam achieved just 55.2% of its investment target from 2020 to 2024. Harrison (2017) notes that SEZ success relies on effective management and a stable investment climate, and conflicting leadership can create uncertainty and deter investors. Presidential Decree No. 41 of 1973 established the Batam Authority to accelerate development. In 2007, Batam was designated as KPBPB, and Batam Authority was transformed into BP Batam. Government Regulation No. 62 of 2019 mandated that BP Batam's head is the Mayor of Batam City. Regulations No. 41 of 2021 and No. 67 and 68 of 2021 aimed to boost trade and investment and establish new economic zones.

Batam has faced challenges despite its manufacturing, shipbuilding, and electronics growth. Understanding investment shortfalls is crucial for developing targeted strategies and achieving long-term economic goals (Hutchinson, 2017; Negara and Hutchinson, 2018; Tadjoeddin *et al.*, 2017). Comparisons with successful SEZs like Shenzhen highlight

 DOI: 10.1201/9781003645542-85

bureaucratic inefficiencies and regulatory ambiguities (Zeng, 2016; Hutchinson, 2017; Negara and Hutchinson, 2018; Aritenang, 2017). Shenzhen was chosen as an example for transforming from a small fishing village into a global economic hub, driven by effective governance, strategic policy implementation, and robust infrastructure development. By analysing Shenzhen's model, this study identifies adaptable strategies to address Batam's unique challenges, including overlapping policies, weak leadership, and limited local ownership. Previous studies were conducted to review current investment policies within Batam SEZ, focusing on regulatory concerns and comparison to available practices. Still, none captured the sentiment and commentaries from the investor side. Thus, this study is conducted to find the middle ground between current investment policies and investor's pain points.

This study addresses: (1) shortcomings of Batam's economic zones compared to Shenzhen's; (2) how overlapping regional development policies affect investment; (3) factors contributing to negative investor sentiment towards BP Batam; and (4) legal ambiguities in Batam's SEZs. It offers insights for policymakers by comparing Batam's FTZ with Shenzhen's SEZ, addressing investment challenges, investor sentiment, and legal frameworks to guide policy development and economic growth, and aims to support SDGs. This study includes a literature review, methodology, results, discussion, and recommendations.

2 METHODS

This study combines a literature review and primary data. It aims to provide detailed insights into these regulations' functioning and identify improvement areas (Patton, 2015). Data collection for this study involved conducting FGD and IDI. Six dimensions of investment parameters were reviewed to compare Batam FTZ and Shenzhen SEZ, including natural resources, human resources, policy, infrastructure, market access, and investment. Studies by the World Bank (2020) and UNCTAD (2019) highlight the importance of proximity to natural resources and trade routes for SEZs to enhance market access. Lu (2022) emphasised human capital's role in boosting regional productivity, while Frick and Rodriguez-Pose (2023) identified market access as the primary driver of foreign investments, followed by business environment, policy support, labour productivity, and infrastructure.

Sources interviewed in this study came from various institutions, such as Chamber of Cominding & Discussion merce Industry (KADIN), business actors, regional government, and BP Batam. Respondent from KADIN was selected as a representative of the city-level business association, particularly regarding feedback, aspirations, and complaints about investment growth and underperformance of Batam as FTZ. Commissioner of PT. Energi Listrik Batam, as a business actor, shared experiences regarding challenges in adhering to investment regulations and procedures. Riau Island Regional Government provided perspectives on the development direction, overall policy formulation, and coordination with the Central Government. BP Batam's role in facilitating licensing, investment promotion, and infrastructure development was also explored. Using a multi-perspective approach, the study integrates stakeholder insights to explain differences between SEZ and FTZ while identifying success and failure factors across the six parameters (Baloch *et al.*, 2021; Sa'dia and Fitrady, 2023).

3 FINDINGS AND DISCUSSION

SEZs are designated areas with specific security measures, unified management, and benefits from geographical advantages, featuring FTZ and streamlined procedures. Here is a comparison of SEZs in Shenzhen and Batam (Table 1):

Table 1. Parameter comparison of Shenzhen's SEZ and Batam's FTZ.

No	Parameters	Shenzhen	Batam
1	Natural resources	1,997.47 km^2, 10 districts. Located in Southern China (international trade route).	5 major island clusters. Located on Eastern Sumatra, near Singapore, international trade route.
2	Human resources	Manufacturing leads labor force contribution. COVID-19 reduced migrant workers down 290,000 in 2022.	Batam: 1,390 industries, 169,000 workers. As of 2023, Batam had 460 migrant workers.
3	Policy	Shenzhen City Council manages policies and regulations. Minimum wage: $325.024,16/month in 2023.	Governed by BP Batam and Batam City Government. Minimum wage: $276.19/month in 2023.
4	Infrastructure	Transit network: 229 km rail, 420 km bus lanes. Has one international airport, six train stations, one metro system, and one major port (handles 30 million TEUs).	BRT serves only Batam Island. Lacks rail due to archipelago geography. Has one international airport and five ports (one major port)
5	Market access	China's collaboration with Japan (8.5% investment) and the U.S. benefits. Shenzhen's trade includes 6.3% U.S. investment. Total import-export volume: 3.5 billion Yuan (US$ 528.3 million).	Top five export destinations: Singapore, USA, China, India, and Japan. Cumulative export volume in 2020 was US$6.92 billion.
6	Investment	27th wealthiest city, led by private entrepreneurs. FDI reached $11 billion in 2022. Utilized $8.70 million by 2023. Manufacturing growth: new energy vehicles 173.9%, robots 60.5%, smartphones 40.9%, 3D printing 21.2%.	Total investment IDR 15.6 trillion: IDR 6.8 trillion domestic, IDR 8.8 trillion foreign in 2023. Investment allocation was 76.02% to the tertiary sector (Hotels, Services, etc), 16.59% to secondary sector (Industry), and 7.40% to primary sector (Forestry, Mining, Fisheries).

Shenzhen has maximised its location advantage, outperforming Batam with a strong focus on manufacturing that drives its robust industrial sector. In contrast, Batam remains reliant on the tertiary sector, with Batu Ampar Port handling only 350,000 TEUs annually and suboptimal road infrastructure affecting efficiency. Furthermore, based on the results of the FGD and IDI, several issues were identified, which are detailed as follows:

3.1 *Overlapping regional development planning policies*

In the FGD held on May 13th of 2024, KADIN Batam highlighted the lack of coordination between BP Batam and the Batam City Government, which has resulted in significant management gaps, especially in planning. Suharyo *et al.* (2020) stated that the disjointed nature of planning between BP Batam and the local government has led to inefficiencies and conflicting priorities. This has hampered cohesive development strategies in the region. The lack of integrated planning results in conflicting priorities, where BP Batam focuses on economic growth while local authorities emphasise social and environmental concerns, leading to inconsistencies in policy implementation. This lack of coordination has negatively impacted regional development, resulting in fragmentation and inefficiency. Consequently, overlapping policies and misaligned objectives hinder effective governance and sustainable development. Moreover, the differing development priorities between BP Batam and Batam City Government can lead to misalignment in government spending at the central and regional levels. This situation contradicts the mandate of the Law on Financial Relations between the Central Government and Regional Governments (UU HKPD), which emphasises the importance of fiscal policy alignment and the improvement of regional spending quality for the welfare of the people.

Unclear infrastructure planning heightens investor uncertainty. Kurniawati and Sutanto (2018) emphasise that clear and detailed infrastructure planning is crucial for attracting and retaining investors, as it provides a clear roadmap for future developments and assurances about the region's capacity to support industrial activities.

3.2 *The weakness of leadership competence in investment management*

Appointing ex-officio leaders at BP Batam raises concerns over leadership focus and competence. Paramita *et al.* (2020) argued that such appointments result in less focused leadership, while the Chairman of KADIN Batam (July 29th, 2024) emphasised the need for leaders with expertise in driving investments. However, overlapping duties under ex-officio policy hinder leadership effectiveness. Raharjo (2021) highlighted that competent and consistent leadership is essential to sustain investor confidence and implement development strategies.

Leadership gaps have directly impacted Foreign Direct Investment (FDI) in Batam. Policy uncertainty, stemming from overlapping regulations and unclear roles under Government Regulation No. 62/2019, creates instability, deterring foreign investors. Bureaucratic inefficiencies, including slow licensing and poor inter-agency coordination, further discourage investment, making Batam less competitive than Shenzhen's SEZ. In the FGD held on May 13th of 2024, dissatisfaction with BP Batam's inability to address investor needs was highlighted, intensifying negative sentiment. Additionally, weak leadership has resulted in poor infrastructure development due to the lack of long-term strategies and modernisation. Batam also lags in adapting to global investment trends like green technology.

3.3 *Limitations of authority and governance issues in the Batam Special Economic Zone*

SEZ management in Indonesia, including Batam, faces challenges due to the need for government approval, resulting in limited authority and governance issues. According to academic literature, SEZ governance in Indonesia involves a hierarchical structure where decisions must be approved at various levels. The National Council, Regional Council, and SEZ Administrator all play roles in management; however, the need for approval and recommendations from higher government authorities often results in delays and a lack of autonomy for local SEZ administrators. Law No. 39/2009 concerning SEZs stipulates that SEZs must adhere to provisions set by the central government, including obtaining approval from the Coordinating Ministry for Economic Affairs and other relevant ministries.

In the FGD conducted on May 13th of 2024, it was found that in recent years, the coordinating minister for Economic Affairs has not engaged in the necessary oversight or problem resolution as required. The complexity of complicated and inefficient licensing procedures has become a significant barrier for investors, reducing their interest in the region. Gunawan *et al.* (2019) highlight that excessive bureaucracy in licensing processes reduces the attractiveness of investment regions. KADIN of Batam, in the FGD, also stated that Article 22 of Government Regulation No. 41/2022 stipulates that the authority for FDI licensing has been taken over by the central government, complicating the licensing process. Furthermore, implementing the Online Single Submission (OSS) system requires complex procedures and regulations, prompting entrepreneurs to seek easier alternatives, thereby opening opportunities for fraud by licensing officials. The limited authority of SEZ managers forces entrepreneurs to engage with the relevant ministries proactively.

3.4 *Insufficient local ownership and vague land pricing regulations*

Foreign investors in Batam have expressed concerns about the investment ecosystem due to the lack of local ownership compared to foreign investment. This situation has led to questions regarding their investments' long-term sustainability and stability. Significant foreign ownership has outpaced, especially by investors from Singapore and other countries. KADIN of Batam revealed additional issues exacerbating these concerns. From the beginning, the management authority of Batam was granted to Batam Authority Agency, which included responsibilities as a land manager. However, these responsibilities were removed since the transformation into BP Batam, raising issues regarding land management certainty in Batam. In practice, BP Batam currently provides recommendations to the National Land Agency (BPN) before land certificates are issued. This practice is not following the mandate of Law No. 23 of 2014 on Regional Government and Law No. 36 of 2000 concerning KPBPB, as well as the enactment of Government Regulation in Lieu of Law No. 1 of 2022 on KPBPB into law. This overlap has led to confusion and inefficiencies in land management. Additionally, BP Batam often allocates land to investors/developers but does not issue the necessary land management rights certificates, resulting in conflicts between investors/developers and the local community.

3.5 *Policy options*

Shenzhen's success cannot be fully replicated in Batam due to governance, resources, and strategy differences. Shenzhen benefits from centralised governance, efficient decision-making, high-tech manufacturing, skilled labour, and integrated infrastructure, which attract global investors. In contrast, Batam faces overlapping authorities, fragmented planning, reliance on the tertiary sector, bureaucratic inefficiencies, unclear land pricing, and inadequate workforce development, all hindering its investment appeal. To address these challenges, Batam must adopt Shenzhen's strategies; firstly, several policies can be applied: (1) reforming land and lease regulations; (2) offering attractive tax incentives; (3) investing in productive infrastructure; and (4) developing an integrated investment policy framework. Secondly, addressing the issue of overlapping policies requires focused attention. Several steps should be undertaken: (1) establish a joint planning committee to align priorities and develop a comprehensive infrastructure plan, and (2) initiate investment acceleration programs. Thirdly, to improve leadership commitment in investment management, ex-officio policy must be promptly evaluated to assess its effectiveness and impact. Fourthly, granting autonomy to enhance SEZ area governance should be considered a strategic solution. Fifthly, to enhance local ownership, the following measures are needed: (1) providing incentives for joint ventures; (2) reforming land pricing regulations; and (3) integrating investment policies and promoting public-private partnerships.

4 CONCLUSION

This study has revealed several critical issues impacting the achievement of investment targets in SEZ and FTZ in Batam. Key findings indicate that overlapping regional development planning policies, weak leadership competence, and bureaucratic hurdles in licensing processes are major obstacles. The lack of coordinated planning between BP Batam and local government has resulted in inefficiencies and conflicting priorities. Additionally, the appointment of ex-officio leaders and insufficient alignment between workforce development programs and industry needs have exacerbated these challenges. Investor uncertainty is also heightened by unclear infrastructure planning and vague land pricing regulations. These problems can be solved by implementing policies within these steps: (1) short-term policy within five years should address overlapping policies by establishing a joint planning committee with the Batam City Government, launching investment acceleration programs, and evaluating the ex-officio policy to enhance investment effectiveness. Greater SEZ autonomy could improve oversight, streamline procedures, and boost investor confidence; (2) Long-term policy, spanning up to 20 years, should focus on increasing local ownership, providing incentives for joint ventures, reforming land pricing, and fostering public-private partnerships. Additionally, adopting successful practices from Shenzhen SEZ—such as land regulation reforms, tax incentives, and infrastructure investment—can significantly enhance Batam's SEZ performance. By implementing these steps, it is believed that both domestic and international investments can be effectively established, increased, and sustained. Future research should assess the scalability of interventions, policy reforms, leadership models, and the long-term impact of improved workforce alignment and infrastructure on investment outcomes.

REFERENCES

Aritenang, A. F. (2017). Special economic zone at the crossroads: The case of Batam. *Jurnal Ilmu Sosial dan Ilmu Politik, 21*(2), 132–146.

Baloch, Z. A., Tan, Q., Kamran, H. W., Nawaz, M. A., Albashar, G., and Hameed, J. (2021). A multi-perspective assessment approach of renewable energy production: policy perspective analysis. *Environment, Development and Sustainability*, 1–29.

Frick, S.A. and Rodríguez-Pose, A. (2023) What draws investment to special economic zones? Lessons from developing countries. *Regional Studies*, 57(11), 2136–2147.

Gunawan, S., Irawan, H., and Santoso, T. (2019). Administrative barriers and their impact on investment climate: Evidence from Indonesia. *Asian Journal of Political Science*, *27*(3), 321.

Harrison, L. (2017). *The Role of Special Economic Zones in Developing Economies: A Review of Policy and Practice*. Routledge.

Hutchinson, F. E. (2017). *Rowing against the tide?: Batam's economic fortunes in today's In-donesia*. ISEAS Publishing.

Kurniawati, R., and Sutanto, T. (2018). Infrastructure planning and investor confidence in developing regions. *International Journal of Urban and Regional Research*, *42*(4), 665–682.

Lu, Z. (2022). Human capital spillovers from special economic zones: evidence from yangtze delta in china. *Economies*, 10(5), 99.

Negara, S. D., and Hutchinson, F. (2018). Batam: Life after the FTZ. *Bulletin of Indonesian Economic Studies*, *56*(1), 87–125.

Patton, M. Q. (2015). *Qualitative Research & Evaluation Methods: Integrating Theory and Practice*. Sage Publications.

Paramita, S., *et al*. (2020). Challenges in leadership competence within economic zone management. *Journal of Management Studies*, *57*(5), 1234–1256.

Raharjo, Y. (2021). Leadership and competence in investment management: A comparative study. *International Journal of Business Administration*, *12*(2), 45–62.

Sa'dia, N. H., and Fitrady, A. (2023). Evaluation of Special Economic Zone (SEZ) Impact on Economic Growth: Evidence of Simalungun Regency, North Sumatera, Indonesia. *International Review for Spatial Planning and Sustainable Development*, 11(4), 113–130.

Suharyo, S., *et al*. (2020). The impact of fragmented regional planning on development outcomes. *Urban Studies*, *57*(6), 1290–1308.

Tadjoeddin, M. Z., Auwalin, I., and Chowdhury, A. (2017). Revitalising Indonesia's manufacturing: The productivity conundrum. *European Journal of East Asian Studies*, *16*(1), 124–153.

UNCTAD. (2019) *Location and performance of economic zones*. United Nations Conference on Trade and Development.

World Bank Group. (2020) *Infrastructure development and SEZ success*. Washington, DC: World Bank.

Zeng, D. Z. (2016). Global experience of special economic zones: Implications for developing countries. World Bank.

Towards Resilient Societies: The Synergy of Religion, Education, Health, Science, and Technology – Rahiem (Ed.)

Empowering economic independence through Pesantren initiatives

Suryani
Universitas Islam Negeri Syarif Hidayatullah Jakarta, Indonesia

Rizma Ilfi
Universitas Darunnajah Jakarta, Indonesia

ABSTRACT: Islamic boarding schools (*Pesantren*) have historically been integral to Indonesian society, primarily offering Islamic education and religious instruction. Over time, their influence has expanded into social, cultural, and economic spheres. Traditionally, *Pesantren* operated as self-sufficient communities, sustained by agricultural activities and small-scale enterprises managed by students, laying the foundation for economic independence. Today, *Pesantren* have evolved by integrating vocational training into their curricula, providing students with practical skills in agriculture, craftsmanship, and entrepreneurship. This dual approach equips graduates to contribute economically to their communities, reduce unemployment, and strengthen economic resilience, aligning with Sustainable Development Goal (SDG) 8: Decent Work and Economic Growth. This study aims to identify strategies implemented by *Pesantren* to enhance community economic development, analyse factors influencing their success, and offer policy recommendations to address existing challenges. Using a literature review, the study collects, evaluates, and interprets data from various sources, including journals, books, documentation, online platforms, and libraries. The findings underscore the importance of empowering small and medium-sized enterprises (SMEs) within *Pesantren* through initiatives such as establishing microfinance institutions, offering business training programmes, and forming cooperatives to support local SMEs and create employment opportunities.

Keywords: economic independence, empowering, pesantren

1 INTRODUCTION

Islamic boarding schools (Pesantren) have historically played a pivotal role in Indonesian society, primarily as centres of Islamic education and religious instruction. However, their influence has extended beyond spiritual to social, cultural, and economic spheres. In recent decades, there has been a growing recognition of the potential of Pesantren to contribute to community economic development. They have been integral to Indonesian society for centuries, providing education rooted in Islamic values and teachings. Traditionally, they have functioned as self-sufficient communities, often supported by agricultural activities and small-scale enterprises managed by students. This model of self-sufficiency has laid the groundwork for their role in promoting economic independence.

Nowadays, Pesantren has undergone many changes. While Pesantren continues to play a significant role in Islamic education (tafaqquh fiddin), it has not made significant strides in developing the economy and entrepreneurship. However, many Pesantren have successfully fulfilled their roles and functions in community empowerment and economic development (Dzikri 2019). The role in economic development can strengthen three functions: education, preaching, and community empowerment (Kemenag 2022).

DOI: 10.1201/9781003645542-86

Many pesantren have incorporated vocational training into their curricula, equipping students with practical agriculture, craftsmanship, and entrepreneurship skills. This dual approach of combining religious education with vocational training has empowered graduates to contribute economically to their communities. By instilling entrepreneurial skills, Pesantren contributes to reducing unemployment, fosters economic resilience, and frequently participates in various community development projects, ranging from microfinance initiatives to cooperative enterprises. These projects not only provide economic opportunities for the local population but also encourage a spirit of collaboration and mutual support. Through these activities, Pesantren catalyses economic growth and stability within their regions (Asrol *et al.* 2023).

The existing literature on Pesantren provides a lot of information about its role in the community's economic independence. Several studies have shown that the role of Pesantren is crucial for improving the economy of the community, especially the community around it (Junaidi 2022; Syahputra *et al.* 2022; Ni'am 2015). However, there remains a gap in our understanding of Pesantren's efforts to equalise the community's economy, as follows: 1). Limited access to capital funds is often the primary obstacle. Many Pesantren do not have sufficient funds to develop economic programs or new businesses. 2). Human resources: The lack of competent experts and managers in the fields of economics and business can hinder the development of businesses managed by Pesantren.

The role of pesantren in building an economically independent society has attracted the attention of many parties. However, studies on strategies and economic empowerment models implemented by Islamic boarding schools still require more in-depth studies. Findings on the success of Pesantren in integrating religious education with vocational training can encourage the Ministry of Religious Affairs to formulate policies that require local-based education in Pesantren. The government or Islamic financial institutions can create a special revolving fund program for pesantren to support small business capital, cooperatives, and Islamic boarding school-based agriculture.

Pesantren has implemented various strategies to improve the community's economy, and this study aims to identify and analyse the factors that influence their success. Additionally, this study will identify Pesantren's challenges in empowering the economy and provide policy recommendations to overcome them. These findings can encourage Islamic organisations or NGOs to develop training programs that focus on improving business management skills for pesantren managers and adopt digital technologies such as e-commerce platforms, social media, and fintech applications to support their economic efforts.

2 METHODS

This study employs a literature review, a research method to identify, evaluate, and interpret all relevant research results related to a particular research question, topic, or phenomenon of concern (Kitchenham and Brereton 2013). Individual study is a primary study, while literature review is a secondary study. The literature review will be very useful for synthesising relevant research results so that the facts presented to policymakers are more comprehensive and balanced.

A literature review study collects data or synthesises sources related to the research topic from various sources, including journals, books, documentation, the internet, and libraries. The literature study method is a series of activities related to collecting library data, reading and recording, and managing writing materials (Nursalam 2017).

Research methodology generally includes quantitative and qualitative methods. In the literature review, quantitative and qualitative methods are also used. The quantitative literature review synthesises research results with a quantitative approach. For example, Randomised Control Trials (RCT), Cohort Study, Case-Control Study, or prevalence studies. The statistical approach in synthesising quantitative research results is called meta-analysis, which aggregates data to obtain systematic strength in obtaining causal relationships between risk factors or treatments with effects/outcomes (Perry and Hammond 2002).

The qualitative approach in literature review is used to synthesise (summarise) descriptive qualitative research results. The method of synthesising (summarising) the results of qualitative research is called meta-synthesis, a technique for integrating data to obtain new theories or concepts or a deeper and more comprehensive level of understanding (Perry and Hammond 2002).

Selection of literature is carried out using eligibility criteria. Eligibility criteria consist of inclusion and exclusion criteria (Andriani and Hakam 2022). Inclusion criteria include: 1). Scientific articles written in English or Indonesian; 2). Literature in the form of scientific articles published in journals or proceedings; 3). Articles published in 2020-2024; and 4) Discussion of scientific articles on the role of Pesantren. Exclusion criteria include: 1). Scientific articles cannot be accessed in full text, and 2). Articles in the form of literature reviews. Scientific articles not meeting the criteria are excluded and not used in this study.

3 FINDINGS AND DISCUSSIONS

3.1 *Improving the economy of Pesantren through empowering small and medium enterprises (SMEs)*

Empowering small and medium enterprises (SMEs) has become increasingly important in recent years to strengthen Pesantren's economy. Pesantren were once primarily places for religious instruction, but these days they also play a part in economic growth. Pesantren offers access to vital resources and integrates entrepreneurial programs to promote local SMEs, generate employment, and boost economic growth (Najihah 2022).

Within the Pesantren ecosystem, boosting SMEs involves several strategic initiatives. These include starting microfinance institutions, providing business training courses, and organising cooperatives. Through educating students and community people in entrepreneurship and offering financial support, Pesantren contributes to developing an independent economic environment. Studies show that these programs improve economic resilience and foster an entrepreneurial culture, which is essential for long-term development.

The active involvement and cooperation of numerous parties, including governmental bodies, nonprofits, and the commercial sector, are essential to the success of these initiatives. Pesantren can increase their influence on regional economies by utilising these alliances to obtain more resources and knowledge. Furthermore, combining economic empowerment with religious education guarantees broad advantages, fostering a more inclusive and equitable economic environment. According to recent studies, these coordinated efforts are essential to turning Pesantren into significant economic centres that propel regional growth (Kholis and Rezqia 2020).

3.2 *Community-based economic empowerment*

Improving the role of pesantren in community-based economic empowerment involves leveraging their unique position within society to foster economic growth and development. Pesantren, as traditional Islamic educational institutions, have evolved to become pivotal in community empowerment. Recent studies highlight that pesantren can enhance community welfare by implementing effective economic programs. For example, the Pesantren Darussalam in Rajapolah, Tasikmalaya, significantly improved local incomes and spiritual well-being through its DN Laundry business unit. This model demonstrates the potential of pesantren to act as catalysts for economic upliftment by creating business opportunities and supporting community welfare through both material and spiritual dimensions (Milah *et al.* 2024).

The strategic role of pesantren in economic empowerment is further underscored by their capacity to adapt and innovate within their local contexts. Pesantren Pondok Dalwa, for instance, employs a summative CIPP evaluation model to assess and enhance its economic empowerment initiatives. This approach has allowed the pesantren to successfully run various business units, from retail to hospitality, thereby providing significant economic benefits to the surrounding community. By adopting comprehensive evaluation frameworks, pesantren can ensure that their economic programs are effective and sustainable, thereby contributing to long-term community development (Laili and Zaki 2020).

Furthermore, significant government assistance has resulted from national programs that acknowledge the significance of pesantren in economic empowerment. Pesantren students have access to financial resources through programs like the People's Business Credit (KUR) facility, which enables them to start profitable businesses. Furthermore, a number of initiatives seek to improve pesantren's economic integration and competence by promoting business cooperation among them through online marketplaces and business matching. These initiatives are part of a larger plan to use pesantren's enormous economic potential to establish Indonesia as a global centre for halal products. These programs demonstrate pesantren's vital role in promoting economic growth and empowerment on a local and national scale (Antara News 2020).

3.3 *Building character and social groups*

Building character and fostering social cohesion within the pesantren and the local community is a multifaceted process grounded in Islamic values and educational practices. Pesantren such as Darunnajah use a comprehensive approach to character education that combines religious teachings with practical life skills. This approach focuses not only on students' intellectual growth but also emphasises the development of moral character, preparing them to become responsible and ethical members of society.

Additionally, pesantren uses a variety of outreach initiatives to broaden its influence within the neighbourhood. Through their involvement in social, educational, and economic endeavours, pesantren cultivate a mutually beneficial relationship with the communities around them. The local community gains from this interaction and the student's sense of civic participation and social responsibility are also strengthened. Ultimately, these programs contribute to students' holistic development and society's general well-being by creating a cohesive and encouraging community (Purwanto *et al.* 2021).

Pesantren also act as social institutions that significantly influence their local communities. They engage in various community service activities, such as education and religious guidance, strengthening community bonds. The social structure of pesantren, with its emphasis on communal living and collective activities, fosters a sense of belonging and mutual support among students and community members. This environment helps cultivate social skills and empathy, preparing students to be active and responsible citizens (Bata *et al.* 2022).

Integrating pesantren within local communities also facilitates cultural preservation and social stability. Pesantren respects and incorporates local cultural practices, which helps maintain cultural heritage while promoting Islamic values. This religious and cultural education blend helps students appreciate diversity and develop a balanced worldview. Moreover, by addressing social issues and contributing to community welfare, pesantren enhance their role as vital pillars of social development and cultural continuity in Indonesia (Maula *et al.* 2022).

4 CONCLUSION

The conclusion of the study emphasises the significant role of Pesantren in promoting community economic independence through various strategies. It highlights the importance of empowering small and medium enterprises (SMEs) within the Pesantren ecosystem, which can lead to economic growth and resilience. The study also points out the challenges faced by Pesantren, such as limited access to capital and a lack of competent experts in economics and business. It calls for further research into the strategies and models of economic empowerment implemented by Pesantren. It suggests that active involvement from various stakeholders, including government and private sectors, is essential for the success of these initiatives. Overall, Pesantren are positioned as vital pillars in fostering economic development and social cohesion within their communities.

REFERENCES

Andriani, R., and Hakam, F. (2022). Peran Digital Health untuk Manajemen Penanganan Pandemi COVID-19 di Indonesia: Systematic Literature Review. *Indonesian of Health Information Management Journal (INOHIM)*, *10*(1), 27–40. https://doi.org/10.47007/inohim.v10i1.398

Antara News. (2020). *Pesantren Emerge as Lynchpin of Economic Empowerment*. https://en.antaranews.com/news/162092/pesantren-emerge-as-lynchpin-of-economic-empowerment

Asrol, S., Hesthria, N., and Rizki, O. S. (2023). Role of Pesantren in Improving Sociopreneurs of the Community Around Pesantren in Palembang City. *Peradaban Journal of Economic and Business*, *2*(1), 43–56. https://doi.org/10.59001/pjeb.v2i1.56

Bata, Y., Tahir, L. S., and Mashuri, S. (2022). Modernization Pesantren Education System as the Effort to Face the Challenges in the Globalization Era (a Case Study At Pondok Pesantren Alkhairaat Madinatul Ilmi - Dolo). *The 1st International Conference on Islamic and Interdisciplinary Studies (ICIIS) 2022*, 105–112.

Dzikri, A. D. (2019). Pesantren Dan Perubahan Sosial: Studi Terhadap Peran Pesantren Al-Ishlah, Sidamulya Cirebon. *TAZKIR: Jurnal Penelitian Ilmu-Ilmu Sosial Dan Keislaman*, *5*(1), 59–80. https://doi.org/10.24952/tazkir.v5i1.961

Junaidi, M. (2022). Penyelenggaraan Pesantren Dalam Rangka Memberdayakan Masyarakat. *Jurnal Usm Law Review*, *5*(2), 455. https://doi.org/10.26623/julr.v5i2.5487

Kemenag. (2022). *Visi Kemandirian Pesantren*. https://kemenag.go.id/opini/visi-kemandirian-pesantren-la881e

Kholis, N., and Rezqia, I. (2020). Ekonomi Berjamaah: Relasi Pondok Pesantren dengan Masyarakat dalam Peningkatan Perekonomian (Studi pada Komplek 3 Sunan Pandanaran). *JISA: Jurnal Ilmiah Sosioologi Agama*, *3*(1), 63–79.

Kitchenham, B., and Brereton, P. (2013). A systematic review of systematic review process research in software engineering. *Information and Software Technology*, *55*(12), 2049–2075. https://doi.org/10.1016/j.infsof.2013.07.010

Laili, Y. N., and Zaki, I. (2020). Pemberdayaan Ekonomi Masyarakat Di Pondok Pesantren Dalwa Berdasarkan Model Evaluasi Sumatif Cipp. *Jurnal Ekonomi Syariah Teori Dan Terapan*, *7*(7), 1214. https://doi.org/10.20473/vol7iss20207pp1214-1230

Maula, N., Jazilurrahman, J., and Suhermanto, S. (2022). Strengthening Inclusive Islamic Education: Building Islamic Boarding School-Based Foundations. *Jurnal Islam Nusantara*, *6*(2), 80. https://doi.org/10.33852/jurnalnu.v6i2.368

Milah, J., Nasrulloh, A. A., and Friantoro, D. (2024). Economic Empowerment of Pesantren and Its Impact on Community Welfare in Rajapolah, Tasikmalaya: A Cibest Model Approach. *Al-Infaq: Jurnal Ekonomi Islam*, *15*(1), 162–173.

Najihah, W. 'Ainun. (2022). Peningkatan Ekonomi Pesantren melalui Pemberdayaan Usaha Kecil Menengah (UMKM) Toko Kitab (Studi Pondok Pesantren Sunan Darajat Banjarwati-Paciran-Lamongan). *Al-Maqashid: Jurnal Ekonomi Dan Bisnis Islam*, *2*(2), 10–16.

Ni'am, S. (2015). Pesantren: The miniature of Moderate Islam in Indonesia. *Indonesian Journal of Islam and Muslim Societies*, *5*(1), 111–134. https://doi.org/10.18326/ijims.v5i1.111-134

Nursalam. (2017). *Metodologi Penelitian Ilmu Keperawatan*. Salemba Medika.

Perry, A., and Hammond, N. (2002). Systematic Reviews: The Experiences of a PhD Student. *Psychology Learning and Teaching*, *2*(1), 32–35. https://doi.org/10.2304/plat.2002.2.1.32

Purwanto, E., El Yana, K., Oktarina, S., and Zainal, A. G. (2021). The Role of Islamic Boarding Schools and Character Building of Santri through Increasing Social Capital and Human Capital for Social Changes in the Community. *International Journal of Progressive Sciences and Technologies*, *26*(1).

Syahputra, A., Ismaulina, I., Khairina, K., Zulfikar, Z., and Rofizar, H. (2022). Pendekatan Ekonomi Syariah Bagi Pemberdayaan Ekonomi Pesantren. *Dimasejati: Jurnal Pengabdian Kepada Masyarakat*, *4*(1), 116–130. https://doi.org/10.24235/dimasejati.v4i1.10823

Towards Resilient Societies: The Synergy of Religion, Education, Health, Science, and Technology – Rahiem (Ed.)

Effectiveness of agronative farming cooperatives in empowering Mustahik through productive zakat: A case study of Dompet Dhuafa Farming Villages

Erika Amelia & Shindy Sheptianingsih
Universitas Islam Negeri Syarif Hidayatullah Jakarta

ABSTRACT: The Desa Tani productive zakat programme by Dompet Dhuafa aims to stimulate agricultural development and enhance the economic independence of mustahik. Farming cooperatives are crucial in redistributing economic assets, enabling marginalised communities to achieve self-sufficiency. This study evaluates the effectiveness of agronative farming cooperatives in empowering mustahik within Dompet Dhuafa Farming Villages. Using purposive sampling, data were collected through in-depth interviews with 10 respondents. A mixed-method approach was employed, integrating Data Envelopment Analysis (DEA) for quantitative assessment and SWOT Analysis for qualitative evaluation. The findings indicate that agronative farming cooperatives effectively contribute to mustahik empowerment by improving economic resilience and sustainability. This research highlights the strategic role of cooperatives in optimising productive zakat, reinforcing their impact on poverty alleviation and economic equity.

Keywords: agriculture, cooperatives, empowerment, mustahik, zakat

1 INTRODUCTION

Economic inequality often leads to poverty and unemployment, affecting national economic growth (Hutabarat *et al.* 2023; Smith and Lee 2023; World Bank 2021). As an agricultural nation, Indonesia has abundant natural resources; however, its agricultural productivity remains suboptimal due to weak farmer institutions and limited investment (Isbah 2016). Farmers often lack bargaining power, facing capital constraints, pest attacks, and market limitations (Managanta *et al.* 2019).

Strengthening agricultural institutions via agronative cooperatives enhances farmer competence, technology adoption, and business independence. Islam emphasises economic empowerment through zakat, promoting wealth distribution and reducing poverty (Amsari 2019). Dompet Dhuafa's "Farmer Village" program empowers farmers by providing financial and institutional support (OJK 2013).

2 METHOD

This study investigates the effectiveness of the Agronative Cooperative within the Dompet Dhuafa Farmer Village in Bandung, employing a mixed-method approach to provide a comprehensive assessment (Creswell and Plano Clark 2017). Purposive sampling was utilised to select cooperative members as respondents, ensuring that the study focused on individuals directly involved in the initiative (Campbell *et al.* 2020). Data were collected through

DOI: 10.1201/9781003645542-87

in-depth interviews, observations, and document analysis, facilitating a robust understanding of the cooperative's impact (Creswell and Poth 2018).

Data Envelopment Analysis (DEA) was employed for the quantitative analysis to evaluate the efficiency of resource utilisation and output generation among farmers (Kuang *et al.* 2020). This method objectively measured how effectively inputs were converted into outputs. Meanwhile, for the qualitative assessment, a SWOT analysis was conducted to identify the cooperative's strengths, weaknesses, opportunities, and threats, providing a strategic perspective on its sustainability and growth potential (Gerlach 2018).

The study examined key variables across three dimensions. Inputs included land, seeds, fertilisers, labour, and operational costs, representing the fundamental resources required for agricultural productivity. Outputs comprised total production and income, reflecting the tangible results of the cooperative's farming activities. Lastly, empowerment indicators—such as institutional strengthening, business development, and mustahik welfare—were assessed to determine the broader socio-economic impact of the initiative. This integrated approach ensures a holistic evaluation of the Agronative Cooperative's role in fostering economic resilience and self-sufficiency among mustahik farmers.

3 RESULTS AND DISCUSSION

3.1 *Research overview*

The Dompet Dhuafa Farming Village in West Java is a strategic initiative to empower rural communities through agriculture. This programme seeks to enhance rural populations' economic and social well-being by providing education, training, and essential agricultural resources. By addressing key challenges smallholder farmers face, the initiative aims to foster economic independence and sustainable livelihoods.

According to Ade Rukmana, Head of the Dompet Dhuafa Farmer Village Agronative Cooperative, the Farmer Village programme operates on a community empowerment model where capital and land are provided, allowing farmers to focus solely on land management. In addition to land and capital, farmers receive agricultural inputs, fertilisers, and equipment, ensuring they have the necessary tools for productive farming. The programme consists of two branches, one specialising in greenhouse farming, allowing controlled-environment agriculture to enhance crop yields and quality.

The Dompet Dhuafa Zakat Institution (LAZ DD) is guided by the vision of "The realisation of a just and prosperous society", which serves as the foundation for its commitment to empowering communities towards self-sufficiency and prosperity. Within the Farmer Village programme, two primary missions drive its objectives: first, to establish agricultural enterprises spanning the entire supply chain, from production to distribution, and second, to develop independent farmers who can sustain their livelihoods without external assistance.

To further support this mission, the Agronative Cooperative was established to improve farmers' welfare through collective resource management. This cooperative provides financial assistance, technical training, and market access, enabling farmers to scale their businesses and increase profitability.

The Desa Tani initiative focuses on cultivating high-value horticultural crops using modern agricultural techniques and smart farming methods. Key products include Kenyan baby beans, beef tomatoes, beetroot, and various types of lettuce, which are marketed both online and in retail markets, with some products even reaching international export markets. To maintain high production standards, Desa Tani has implemented a structured work system that ensures consistency in product quality, quantity, and continuity, positioning the initiative as a model for sustainable agricultural development.

3.2 *Results of the level of effectiveness of quantitative methods with data envelopment analysis*

To assess the effectiveness of an institution, a criterion measure serves as a benchmark for evaluating and enhancing efficiency. This measure is established through Data Envelopment Analysis (DEA), a widely recognised method for assessing performance based on input-output relationships (Podinovski and Chambers 2023).

Decision-Making Units (DMUs) in DEA are classified based on their efficiency scores. Efficient DMUs, those achieving a score of 1 or 100%, are deemed fully effective, indicating that they are operating at optimal capacity and cannot increase their output with the current input level. Conversely, inefficient DMUs, with efficiency scores below 1 or 100%, demonstrate potential for improvement, suggesting that they could enhance their output by using available resources better. This classification provides valuable insights into areas where efficiency gains can be realised, guiding strategic improvements in institutional performance.

Table 1. Agricultural input, production, and revenue data for farmers.

DMU	Land (M^2)	Seeds (Kg)	Fertilizer (Kg)	Labor (Hours)	Operational Cost (Rp)	Total Production (Kg)	Revenue (Rp)
Farmer 1	1650	2	50	200	Rp 30.000	22	Rp 500.000
Farmer 2	1000	1	50	200	Rp 50.000	10	Rp 300.000
Farmer 3	1000	7.5	20	250	Rp 30.000	50	Rp 847.000
Farmer 4	2000	5	50	250	Rp 50.000	13	Rp 600.000
Farmer 5	1250	1.5	50	200	Rp 30.000	13	Rp 600.000
Farmer 6	2000	2	50	200	Rp 30.000	13	Rp 600.000
Farmer 7	2500	2.5	50	200	Rp 50.000	7	Rp 300.000
Farmer 8	1000	4	50	200	Rp 30.000	22	Rp 600.000
Farmer 9	2500	10	50	200	Rp 30.000	20	Rp 400.000
Farmer 10	1250	1	50	250	Rp 30.000	20	Rp 200.000

Table 1 shows the input variables: land cultivated with units of square meters, Seeds planted with units of Kg, Fertilizer given, labour used, and operational costs during Farming. The output variables are the total production produced and the income received.

3.3 *Results of the level of effectiveness from the swot analysis method*

Dompet Dhuafa Farmer Village in West Java empowers rural communities through agriculture. The Agronative Cooperative provides capital, land, fertilisers, and training. Smart farming techniques enhance horticultural production for domestic and export markets.

3.3.1 *DEA analysis results*

The Data Envelopment Analysis (DEA) revealed varying levels of efficiency among the participating farmers. Farmers 1, 2, 3, 6, 8, and 10 demonstrated 100% efficiency, indicating optimal utilisation of inputs to achieve maximum output. Conversely, Farmers 4, 5, 7, and 9 exhibited lower efficiency scores, suggesting that improvements in resource allocation or farming techniques could enhance their productivity. Notably, Farmer 7 recorded the lowest efficiency at 47%, highlighting significant inefficiencies that require targeted interventions. These findings align with studies that emphasise the importance of efficient input management and the role of DEA in identifying inefficiencies in agricultural systems

(Bournaris *et al.* 2019; Zhou *et al.* 2018). For instance, DEA has been widely applied to assess eco-efficiency in farming, where inefficiencies in resource use often correlate with environmental pressures such as greenhouse gas emissions (Czyżewski and Kryszak 2025; Picazo-Tadeo *et al.* 2011). Additionally, integrating sustainability dimensions in DEA models, such as economic, environmental, and social factors, provides a holistic framework for improving agricultural productivity (Hoang and Alauddin 2012; Zhou *et al.* 2018).

3.3.2 *SWOT analysis results*

A SWOT analysis provided a strategic overview of the cooperative's strengths, weaknesses, opportunities, and threats. Strengths include robust institutional support, access to financing, and a skilled workforce, all of which contribute to the cooperative's operational stability. However, dependence on external funding and limited innovation emerged as key weaknesses, potentially restricting long-term sustainability. Among the opportunities, the expansion of zakat-based programmes and adopting modern agricultural technologies present avenues for growth and increased efficiency (Mohamed and Shafiai 2021). Nonetheless, the cooperative faces threats from market competition and regulatory changes, which could impact its financial and operational viability.

3.3.3 *Strategic recommendations*

Positioned in Quadrant I of the Grand Strategy Matrix, which signifies high market growth and a strong competitive position, the cooperative is well-placed to capitalise on its strengths and opportunities. To further enhance its impact, it is recommended that the cooperative expands its zakat-based programmes, ensuring greater financial and technical support for farmers (Salim *et al.* 2024). Additionally, increasing mustahik participation can foster greater community engagement and economic inclusion (UNDP 2017). Enhancing training programmes will improve farmers' competencies, particularly in smart farming and sustainable agricultural practices. Finally, establishing strategic alliances with key stakeholders, such as government agencies and private sector partners, will facilitate market expansion, technological adoption, and long-term financial sustainability.

4 CONCLUSION

Agronative Cooperatives effectively empower mustahik, as indicated by high DEA scores among most farmers. SWOT analysis highlights strengths and opportunities for growth. Future strategies should focus on product innovation, market expansion, and capacity building to sustain cooperative effectiveness.

REFERENCES

Amsari, S. (2019). Analysis of the Effectiveness of Productive Zakat Utilization in Mustahik Empowerment (Case Study of LAZISMu Center). *Aghniya Journal of Islamic Economics*.

Bournaris, T., Vlontzos, G., and Moulogianni, C. (2019). Efficiency of vegetables produced in glasshouses: The impact of Data Envelopment Analysis (DEA) in land management decision-making. *Land*, 8(1), 17. https://doi.org/10.3390/land801001713

Campbell, S., Greenwood, M., Prior, S., Shearer, T., Walkem, K., Young, S., Bywaters, D., and Walker, K. (2020). Purposive sampling: Complex or simple? Research case examples. *Journal of Research in Nursing*, *25*(8), 652–661. https://doi.org/10.1177/1744987120927206

Creswell, J. W., and Plano Clark, V. L. (2017). *Designing and Conducting Mixed Methods Research* (3rd ed.). SAGE Publications.

Creswell, J. W., and Poth, C. N. (2018). *Qualitative Inquiry and Research Design: Choosing Among Five Approaches* (4th ed.). SAGE Publications.

Czyżewski, B., and Kryszak, Ł. (2025). Assessing potential reductions of agricultural GHG in countries with different land productivities: Long-term integrated efficiency in DEA hybrid meta-frontier model. *PLoS ONE*, 20(2), e0315571. https://doi.org/10.1371/journal.pone.031557115

Gerlach, R. (2018, December 20). *The sustainability SWOT analysis: A tool for strategic sustainability management and innovation*. Threebility. Retrieved [Jan 7, 2025], from https://www.threebility.com/post/the-sustainability-swot-analysis

Hoang, V. N., and Alauddin, M. (2012). Input-orientated Data Envelopment Analysis framework for measuring and decomposing economic, environmental, and ecological efficiency: An application to OECD agriculture. *Environmental and Resource Economics*, 51(4), 431–452. https://doi.org/10.1007/s10640-011-9506-611

Hutabarat, W., Syahnur, S., and Dawood, T. C. (2023). How population, economic, inequality and unemployment contribute affect indonesian's poverty. *International Journal of Advances in Social Sciences and Humanities*, 2(1), 8–14. https://doi.org/10.56225/ijassh.v2i1.92

Isbah, U. (2016). Analysis of the Role of the Agricultural Sector in the Economy and Employment Opportunities in Riau Province. *Journal of Socio-Economic Development*.

Kuang, B., Lu, X. H., Zhou, M., and Chen, D. L. (2020). Provincial cultivated land use efficiency in China: Empirical analysis based on the SBM-DEA model with carbon emissions considered. *Technological Forecasting and Social Change*, *151*, 119874. https://doi.org/10.1016/j.techfore.2019.119874

Managanta, A. A., Sumardjo, S., Sadono, D., and Tjitropranoto, P. (2019). Institutional support and role in increasing the interdependence of cocoa farmers in central sulawesi province. *Jurnal Tanaman Industri dan Penyegar*, 6(2), 51. https://doi.org/10.21082/jtidp.v6n2.2019.p51-60

Mohamed, M. I., and Shafiai, M. H. M. (2021). Islamic agricultural economic financing based on zakat, infaq, alms and waqf in empowering the farming community. *Iqtishaduna: Jurnal Ilmiah Ekonomi Kita*, 10(1), 144–161. https://doi.org/10.46367/iqtishaduna.v10i1.334

OJK. (2013). Law Number 1 of 2013. *Financial Services Authority*.

Picazo-Tadeo, A. J., Gómez-Limón, J. A., and Reig-Martínez, E. (2011). Assessing farming eco-efficiency: A Data Envelopment Analysis approach. *Journal of Environmental Management*, 92(4), 1154–1164. https://doi.org/10.1016/j.jenvman.2010.11.0253

Podinovski, V. V., and Chambers, R. G. (2023). Data envelopment analysis with embedded inputs and outputs. *Annals of Operations Research*. https://doi.org/10.1007/s10479-023-05426-y

Salim, K., Hamid, B. A., Mahomed, Z., and Hassan, W. (2024). Zakat-based urban farming: A tool for poverty alleviation, community empowerment, financial inclusion, and food security. *Review of Islamic Social Finance and Entrepreneurship*, 3(1), 1–19. https://doi.org/10.20885/RISFE.vol3.iss1.art1

Smith, J., and Lee, K. (2023). The impact of economic inequality on poverty and unemployment. *Journal of Economic Development*, 45(2), 123–145.

UNDP. (2017). *The role of zakat in supporting the sustainable development goals*. United Nations Development Programme. https://www.undp.org/sites/g/files/zskgke326/files/migration/id/INS-Zakat-English.pdf

World Bank. (2021). Poverty, unemployment, and inequality: Pathways to inclusive growth (Report No. 6789).

Zhou, P., Ang, B. W., and Poh, K. L. (2018). A survey of Data Envelopment Analysis in energy and environmental studies. *European Journal of Operational Research*, 189(1), 1–18. https://doi.org/10.1016/j.ejor.2007.04.0421.

Towards Resilient Societies: The Synergy of Religion, Education, Health, Science, and Technology – Rahiem (Ed.)

Evaluating social responsibility in Islamic banking: Insights from Indonesia

Yoghi Citra Pratama
Universitas Islam Negeri Syarif Hidayatullah Jakarta, Indonesia
ORCID ID: 0000-0003-3192-2561

ABSTRACT: This study evaluates the social outcomes of Islamic banks, focusing on integrating religious and social values into the corporate principles of the Islamic banking sector. The research applies the Reactive Defensive Accommodative Proactive (RDAP) scale and employs content analysis to assess the social performance of Islamic banking practices. The analysis is based on the 2021 annual reports of 12 Islamic banking institutions. Findings reveal that 60% of the sampled Islamic banks demonstrated strong social performance, particularly among banks operating independently of conventional banking systems, such as Bank Muamalat, Bank Aceh Syariah, and Bank NTB. In contrast, four out of eight Islamic banks affiliated with or operating as subsidiaries of conventional banks showed weaker social outcomes. This disparity suggests that operational affiliations influence the effectiveness of social engagement within Islamic banks. By prioritising ethical practices, financial inclusion, and social responsibility, Islamic banks have the potential to significantly contribute to the creation of decent work opportunities and support the achievement of Sustainable Development Goals (SDGs), positioning themselves as key drivers of sustainable development.

Keywords: Islamic banking, RDAP approach, social outcomes

1 INTRODUCTION

Islamic banking is founded on Islamic values, particularly the *Maqasid al-Shariah*, encompassing commercial and social objectives. The Islamic Banking Law No. 21 of 2008 underscores its role in providing social benefits to the community, extending beyond just economic gains. Islamic banking serves commercial and social responsibilities, fostering economic growth while promoting ethical practices. Its implementation varies by country, but it consistently aims to balance profit with community welfare, contributing to socio-economic goals. Adapting to a changing global landscape is essential for its future relevance and societal impact.

From the Islamic perspective, the social responsibilities engendered by the Quran and the teachings of the Prophet Muhammad (*the Sunah*) extend to corporate conduct. Islamic business practices are guided by these tenets to fulfil divine will, recognising that work constitutes a form of worship. While the pursuit of profit remains a legitimate business objective, it must align with the principles outlined by Sharia. Organisations asserting compliance with Sharia are expected to transparently communicate their communal responsibilities, as Sharia delineates the norms governing human behaviour and the interactions of businesses with society at large (Asutay and Harningtyas 2015; Haniffa and Hudaib 2007). As noted by Kamel (1997), the objectives of Islamic entities should encompass economic development, value generation, job creation, and the enhancement of community well-being.

 DOI: 10.1201/9781003645542-88

According to the Islamic context, the social responsibilities of individuals that arise from the Quran and Muhammad's deeds and sayings (the Sunah) also pertain to firms (Mirakhor and Askari 2010). Firms that claim their conduct complies with Sharia should be clear about their responsibilities to the community, given that Sharia defines norms of human behaviour and how businesses interact with the outside world. In this regard, Kamel (1997) conveyed that Islamic objectives should encompass economic advancement, value generation, employment opportunities, and the enhancement of community quality (Hamidi and Worthington 2020).

In Indonesia, a noteworthy dynamic exists, where 80% of Islamic banks operate as subsidiaries of or are affiliated with conventional banks. This duality has substantial implications for Islamic banks' strategic direction and operational conduct. Understanding the extent of social commitments within these institutions, particularly those designating themselves as Islamic, is paramount.

At present, there exists a notable gap in the evaluation of the social outcomes of Islamic banks. This study aims to provide a comprehensive measurement framework to assess the social dimensions associated with Islamic banking practices effectively. Implementing such a framework can enhance our understanding of how these institutions contribute to societal welfare and fulfil their ethical commitments.

2 METHODS

This study employed a content analysis methodology to assess the social outcomes of Islamic banking. This study has identified and categorised six distinct areas using a modified version of Hamidi and Worthington's (2020) work that aligns with the social objectives of Islamic banks. In creating the indicators, we utilised existing literature and integrated the Kinder, Lydenberg, Domini Research and Analytics index of corporate social performance alongside the United Nations' 17 Sustainable Development Goals and the five Environment Social Governance Scorecards developed by Oikocredit, a global cooperative and social investor group. All indicators in these references align with Islamic values, reflected in the annual reports of Islamic commercial banks. These reports highlight the organisation's yearly progress and serve as a responsible communication tool for stakeholders while promoting the company's performance.

The first aspect we'll consider is religiosity, closely tied to Islamic banks. Since these banks operate under the banner of religion, they must uphold religious values in all their activities. One way they do this is by managing and contributing to funds for *Zakah* (a religious levy), *Sadaqa* (charity), and *Qard al hasan* (benevolent loans). *Zakah* is one of the five pillars of Islam, and the Qur'an specifies how the proceeds should be spent and who should benefit from them. Unlike traditional banking based on interest (*riba*), Islamic banks are strictly prohibited from participating in such transactions as interest goes against Islamic teachings. Additionally, they avoid speculative transactions (*Maisyir*) and excessive risk (*Gharar*), as detailed in both the Qur'an and hadith, which contains four indicators: 1) Zakah, 2) Qardh al hasan, 3) unlawful transaction (Riba, Gharar, Maisyir), and 4) encouraging spirituality.

The second category is environmental, which we classify into three indicators: 1)preserving and contributing to the environment, 2) managing a green office, and 3) managing green financing. Islam prohibits the destruction of or damage to the physical environment if it is determined to be harmful to an individual or community (Maali *et al.* 2006). Banking is unlikely to directly harm the environment as, for instance, extractive industries might, but Islamic banks should not finance activities that harm the environment, as such projects would harm society. Therefore, Islamic banks should disclose the nature and amount of any donations or environmental activities they have undertaken and any projects they have financed that may negatively affect the environment.

The third category is social community engagement, which consists of five indicators. Islam views public duties as part of its general meritorious and ethical tendencies (Maali *et al.* 2006). It is important for the Islamic bank first to satisfy the needs of the community within which it operates. This social community involvement consists of five indicators: 1) community education and Training, 2) community quality improvement, 3) community economic empowerment, 4) community financial literacy, and 5) assisting micro and small enterprises.

The fourth governance category comprises four indicators: 1) vision and mission, 2) preventing corruption and fraud, 3) handling customer complaints, and 4) ethics and staff compliance. A delicate corporate governance system and extreme risk-taking exposures exacerbate banking sector instability and economic losses (Aslam and Haron 2020). For this reason, effective corporate governance practices are essential. It has been suggested by Ur Rehman *et al.* (2020) that a board that demands good governance can signal a positive message to the market, as well as provide excellent advice to management, which will significantly contribute to the sustainable performance of the banking industry through strategic decision-making.

The fifth dimension concerns how Islamic banks manage their employees and consists of three indicators: 1) rewards, 2) equal work opportunities, and 3) training and education for employees. It is essential to give due attention to the welfare of the business's employees, as they are its most valuable asset. In addition, education and training are important as Islam encourages the search for knowledge. In accordance with Haniffa (2002), this research sets disclosure benchmarks pertaining to wage and bonus payments, employee education and training, equal opportunity, and workplace policies. Such disclosures would help users assess whether the bank violates Islamic principles regarding its dealings with its employees. Employers are responsible for ensuring the well-being of their employees through fair wages, skill development, and equal opportunity.

The sixth dimension evaluates Islamic banking's customer relations, emphasising service quality and SDG alignment. It also measures contributions to women's empowerment and assesses inclusiveness for those lacking access to banking services.

A third indicator is the discount for customers. Clients may not pay amounts due to Islamic banks when they utilise mark-up financing arrangements such as Murabaha and Ijara. The Sharia requires that the lender delay or forego collection if the debtor is experiencing strained circumstances that result in insolvency:

> *"Give him time until it is easy for him to repay the debt." But if you remit it by charity, that is best for you if you only know' (Qur'an, 2:280).*

We, therefore, consider that the Islamic bank should disclose (a) the bank's policy in dealing with insolvent clients, (b) the amounts charged as late penalties, if any, (c) the opinion of the Sharia Supervisory Board regarding whether charging penalties is permissible, and (d) how the bank deals with such penalties (allocation to charity or revenue).

(1) Having diversified financial products, particularly for women and the disadvantages
(2) Giving discounts to customer
(3) Serving unserved areas

Thus, there are six dimensions and 22 indicators in total. Despite the well-established nature of content analysis, it is not immune to challenges. These challenges stem from potential reliability issues due to vague definitions or meanings, as well as validity issues related to the consistency of results with existing theories. Three coders meticulously score the research constructs to address these issues to improve reliability.

To enhance the validity of our analysis, we utilise the categories and definitions from both corporate social performance and Islamic banking. We employ a dichotomous and quantitative scoring method to assess constructs. Our evaluation of the social outcomes of each bank is based on an equal-weighted index that takes into account the following factors:

$$\frac{\sum_{t=1}^{nj} Xij}{nj}$$

When nj is the number of items or constructs disclosed by the jth bank, with $nj = 20$ and where Xij = 2, if Ith construct, is both qualitative and quantitative, Xij = 1 if the construct is only qualitative, and Xij = 0 if neither, consider construct (1) preserving and contributing to the environment as an example. This research first searches the annual report using the key environmental. We then code this construct as follows. If a bank does not address environmental issues, it scores zero; if a bank addresses environmental issues as a conceptual matter, it scores one; and if a bank addresses environmental issues as a conceptual matter as well as does some relevant program and it is recorded in an annual report in detail, it scores two. If a bank does not address environmental issues, it scores zero; if a bank addresses environmental issues as a conceptual matter, it scores one; and if a bank addresses environmental issues as a conceptual matter as well as does some relevant program, it is recorded in an annual report in detail, it scores two.

We use the four RDAP (Reactive, Defensive, Accommodative, Proactive) rating scales introduced by (Carroll 1979; Max 1995; Wartick and Cochran 1985) to determine each bank's social responsiveness levels (Table 1).

Table 1. The reactive-defensive-accommodative-proactive (RDAP) scale.

Achievement	Rating	Posture or Strategy	Performance
0-40%	Reactive (R)	Deny responsibility	Doing less than required
41-50%	Defensive(D)	Admit responsibility but fight it	Doing the least that is required
51-60%	Accommodative (A)	Accept responsibility	Doing all that is required
61-100%	Proactive (P)	Anticipate responsibility	Doing more than is required

3 FINDING AND DISCUSSION

The subsequent table presents the findings of the analysis and assessment of twelve Islamic banks from 2019 to 2023, utilising the RDAP Scale for evaluation.

The output in Table 2 shows that five out of 12 Islamic banks scored above 60%, which receives a Proactive rating. This means they carry out social activities more than needed or their posture and strategy anticipate responsibility. Meanwhile, two Islamic banks still receive a reactive rating with a value below 40%, which means that their social outcomes are below what is required or deny responsibility. Meanwhile, three of the 12 Islamic banks are included in a defensive rating with a score of 41% -50%. They do the least that is required. Finally, the two Islamic banks are included in the accommodative rating, meaning that they do all that is required where they accept responsibility for social aspects.

Table 2. The output of Islamic banking social outcomes.

Bank	Religiosity	Environment	Community	Governance	Employee	Inclusivity	Total	Score
Bank Muamalat	75.0%	100.0%	100.0%	62.5%	75.0%	16.7%	71.5%	P
BSI	75.0%	100.0%	100.0%	75.0%	75.0%	33.3%	76.4%	P
Bank Mega	50.0%	66.7%	90.0%	62.5%	50.0%	0.0%	53.2%	A
BJBS	25.0%	33.3%	60.0%	37.5%	37.5%	0.0%	32.2%	R
Bank Bukopin	25.0%	0.0%	20.0%	50.0%	50.0%	0.0%	24.2%	R

(continued)

Table 2. Continued

Bank	Religiosity	Environment	Community	Governance	Employee	Inclusivity	Total	Score
Syariah								
BCA Syariah	50.0%	33.3%	80.0%	62.5%	87.5%	16.7%	55.0%	A
BTPN Syariah	25.0%	66.7%	90.0%	87.5%	62.5%	66.7%	66.4%	P
Bank Victoria Syariah	0.0%	83.3%	40.0%	62.5%	75.0%	0.0%	43.5%	D
Bank Aladin Syariah	0.0%	66.7%	50.0%	25.0%	37.5%	0.0%	29.9%	R
Bank Panin Dubai Syariah	62.5%	50.0%	10.0%	50.0%	62.5%	16.7%	41.9%	D
Bank Aceh Syariah	75.0%	100.0%	100.0%	87.5%	75.0%	33.3%	78.5%	P
Bank NTB Syariah	100.0%	83.3%	100.0%	75.0%	87.5%	33.3%	79.9%	P

Interestingly, the banks with the best performance in the Social Aspect are those without ties to conventional banks, like NTB Syariah and Aceh Syariah banks, both stemming from conventional bank conversions. Additionally, Muamalat bank, which has no conventional affiliations, also excels. Their lack of conventional bank shareholders allows for greater flexibility in strategy implementation and achieving social goals. BSI, Indonesia's largest Islamic bank, was formed by a merger of three banks, and is a state-owned subsidiary. With significant assets, BSI can prioritise social aspects and assist the government in these efforts due to its merger resulting from government intervention. Despite its ties to conventional banks, BSI's focus on social outcomes enhances its corporate image, reputation, and trust, positively impacting its overall performance. A firm that genuinely addresses the needs of its stakeholders fosters a positive perception of its values, direction, and overall worth. Strong social outcomes reflect the principles inherent within a firm, leading to intangible and unique advantages such as corporate image, reputation, and trust. Ultimately, this contributes to the overall performance of an Islamic bank.

The banks with the lowest social performance are those affiliated with conventional banks, such as Bukopin Syariah, BJBS, Bank Victoria Syariah, and Bank Panin Dubai Syariah. These Islamic banks often prioritise shareholder demands over social aspects, as outlined in shareholder theory, which emphasises maximising profits for shareholders. As a result, their social outcomes are lower than independent Islamic banks. Tackling the challenges faced by Islamic banks necessitates a comprehensive approach that integrates regulatory support, innovative product development, effective governance, and strengthened stakeholder engagement. By adopting these strategies, Islamic banks can enhance their social impact while ensuring financial stability, fulfilling their dual role in society as commercial enterprises and catalysts for positive change.

4 CONCLUSION

Outcomes from the action arena of the Islamic banking industry are economic and social outcomes. This follows the mandate of Law No. 21 of 2008: Islamic banks have commercial and social functions as part of their performance. The results of these outcomes evaluate the effectiveness of policies for policymakers and investors in the Islamic banking industry. Good outcomes can incentivise users, the government, and investors in developing Islamic banking by increasing the investment value and vice versa. From the social outcomes perspective, the results show that Islamic banks with high social outcomes are not affiliated with conventional banks or banks with large capital, such as BSI Bank Aceh Syariah and Bank NTB Syariah. Effective regulation tailored to the distinctive aspects of Islamic banking, paired with strategic incentives, can greatly improve the social outcomes of these institutions while maintaining their commercial viability. By concentrating on these areas, regulators can promote a more inclusive financial system that aligns with economic development objectives and the ethical standards fundamental to Islamic banking. Future studies could investigate the dynamics between primary and secondary stakeholders within the context of stakeholder management strategies, salience, and social performance.

REFERENCES

Aslam, E., and Haron, R. (2020). Does corporate governance affect the performance of Islamic banks? New insight into Islamic countries. *Corporate Governance: The International Journal of Business in Society*, 20 (6), 1073–1090. https://doi.org/10.1108/cg-11-2019-0350

Asutay, M. K., and Harningtyas, A. F. (2015). Developing Maqasid al-Shari'ah Index to Evaluate Social Performance of Islamic Banks: A Conceptual and Empirical Attempt. *International Journal of Islamic Economics and Finance Studies*, 1, 5–64.

Carroll, A. B. (1979). A Three-Dimensional Conceptual Model of Corporate Performance. *The Academy of Management Review*, 4(4), 497–505. https://doi.org/10.2307/257850

Hamidi, L., and Worthington, A. C. (2020). How social is Islamic banking? *Society and Business Review*, 16 (1), 51–70. https://doi.org/10.1108/sbr-03-2020-0036

Haniffa, R. (2002). Social Responsibility Disclosure: An Islamic Perspective. *Indonesian Management and Accounting Research*, 1, 128–146.

Haniffa, R., and Hudaib, M. (2007). Exploring the Ethical Identity of Islamic Banks via Communication in Annual Reports. *Journal of Business Ethics*, 76(1), 97–116. https://doi.org/10.1007/s10551-006-9272-5

Maali, B., Casson, P., and Napier, C. (2006). Social reporting by islamic banks. *Abacus*, 42(2), 266–289. https://doi.org/10.1111/j.1467-6281.2006.00200.x

Max, B. E. C. (1995). A Stakeholder Framework for Analyzing and Evaluating Corporate Social Performance. *The Academy of Management Review*, 20(1), 92–117. https://doi.org/10.2307/258888

Mirakhor, A., and Askari, H. (2010). *Islam and the Path to Human and Economic Development*. https://doi.org/10.1057/9780230110014

Ur Rehman, Z., Zahid, M., Rahman, H. U., Asif, M., Alharthi, M., Irfan, M., and Glowacz, A. (2020). Do Corporate Social Responsibility Disclosures Improve Financial Performance? A Perspective of the Islamic Banking Industry in Pakistan. *Sustainability*, 12(8). https://doi.org/10.3390/su12083302

Wartick, S. L., and Cochran, P. L. (1985). The Evolution of the Corporate Social Performance Model. *Academy of Management Review*, 10(4), 758–769. https://doi.org/10.5465/amr.1985.4279099

Towards Resilient Societies: The Synergy of Religion, Education, Health, Science, and Technology – Rahiem (Ed.)

Islamic philanthropy and tourism entrepreneurship: A systematic literature review

Zenno Noeralamsyah
Universitas Islam Negeri Syarif Hidayatullah Jakarta, Indonesia & Universitas Islam Sultan Maulana Hasanuddin Banten, Indonesia
ORCID ID: 0000-0002-3197-3118

Muhammad Nur Rianto Al Arif
Universitas Islam Negeri Syarif Hidayatullah Jakarta, Indonesia
ORCID ID: 0000-0002-5731-1411

Desmadi Saharuddin
Universitas Islam Negeri Syarif Hidayatullah Jakarta, Indonesia
ORCID ID: 0000-0003-3663-725X

ABSTRACT: This paper reviews studies on Islamic philanthropy, entrepreneurship, and tourism, examining both their interconnections and research focused on each topic individually. A systematic literature review (SLR) approach was employed to collect and analyse relevant data. Bibliometric analysis was conducted using Harzing's Publish or Perish to filter articles indexed in Scopus from 2015 to 2024. VOSviewer, NVivo 12 Pro, and Atlas.Ti 9 software was utilised for data analysis and visualisation. Following a rigorous selection process, 122 out of 500 documents were chosen for detailed examination. The findings reveal a significant gap in research on how Islamic philanthropy can be effectively utilised to develop and support the tourism industry through entrepreneurship. This study highlights the potential role of Islamic philanthropy in advancing the tourism sector and calls for future research to evaluate this integration's social and economic impacts. Additionally, it suggests the need for frameworks that align philanthropic efforts with Sustainable Development Goals (SDGs), contributing to sustainable economic growth and community development.

Keywords: entrepreneurship, Islamic philanthropy, systematic literature review (SLR), tourism

1 INTRODUCTION

Islamic philanthropy significantly promotes sustainable development goals (SDGs) by contributing to poverty alleviation, economic empowerment, and social inclusion (Adnan *et al.* 2024) (Mohammad Jaber Thalgi 2024). Strong economic growth in the past two decades has forced the transformation of Islamic philanthropy into a more inclusive social movement in facing global challenges (Fauzia 2023). The application of Islamic philanthropy has expanded, particularly when linked to waqf, to encompass a wider and more inclusive philanthropic environment (Susamto *et al.* 2024) and behaviours, such as charitable giving, donations, volunteer work, communal solidarity, and philanthropic foundations, are indicators of a civil society that is functioning properly.

DOI: 10.1201/9781003645542-89

The tourism industry has also undergone a remarkable transformation in recent decades, as the concept of philanthropy has become increasingly linked to the pursuit of profit and commercial gain. As tourism has grown into one of the largest and most influential sectors of the global economy, traditional notions of charitable giving and community support have been subtly transformed to serve the interests of tourism industry players (Sainz and M. Samorin 2021). Tourism research has the potential to aid the achievement of SDGs by economically assisting disadvantaged communities and encouraging sustainable and mutually beneficial global tourism (Thompson 2022).

Philanthropy is an activity driven by social principles founded on religious beliefs. In contrast, the tourism sector is an economic activity that aims to satisfy the needs and desires of individuals. The pattern of secular tourism relations to religion studied by scholars sometimes overlaps with theological and religious study approaches (Kuuliala and Rantala 2020). Successful sustainable tourism through rural entrepreneurship is associated with economic, social and environmental sustainability factors (Utami *et al.* 2023). Due to the problematic relationship between religion and tourism, Islamic philanthropy as a portrayal of faith-based conduct becomes more nuanced when it is linked to tourism-related activities. There is an unclear and somewhat vague relationship between sacred religion and secular tourism. Therefore, the following topics will be mapped out in this article: 1) Which articles from 2015 to 2024 address Islamic philanthropy, entrepreneurship, and tourism? What are the current research trends concerning Islamic philanthropy, tourism, and entrepreneurship?

2 METHODS

The SLR method is being implemented to evaluate previous research results on Islamic philanthropy and tourism entrepreneurship. SLR is frequently used to assess the influence of studies, papers, or subjects within a specific field or body of literature (Manu and Akoita 2021). SLR aims to reduce research bias by showing the existence of previous research purchases. SLR will align the research question and endeavour using a desk-based approach involving secondary analysis of explicit knowledge or a literature review, thus producing explorative results in conceptual abstractions from explicit and tacit knowledge (Carrera-Rivera *et al.* 2022). By collecting and analysing publications on specific subjects, SLR helps find research potential and gaps in philanthropy and tourism.

Through SLR, we provided a rigorous and unbiased approach to understanding the interconnection between Islamic philanthropy, tourism and entrepreneurship and the impact of philanthropy in the tourism industry by identifying trends, key themes on philanthropy's role in the SDGs and the influence on social and economic outcomes in tourism. Establish inclusion and exclusion criteria, focusing on peer-reviewed articles, reports, and relevant studies published in reputable journals over a specified timeframe. In this article, we conduct a bibliometric analysis using VOSviewer, Nvivo 12 Pro and Atlas.ti to extract data on authorship, keywords, research methods, publication trends and findings. This study only contains papers from reputable journals that are relevant to the intended integration issue in the Scopus database from 2015 to 2024 with several key terms, including "Islamic philanthropy", "entrepreneurship", and "tourism". We used Harzing's Publish or Perish (PoP) to reduce the collection of 500 documents from Scopus-indexed journals. The remaining 122 articles were examined to ensure that the relevancy of the articles met the objectives of the study.

3 FINDING & DISCUSSION

3.1 *The relationship between Islamic philanthropy and tourism*

Tourism entrepreneurship is a complex field involving many interrelated interactions at various levels. Due to its complexity, comprehensive research reviewing its structure has yet to be conducted. The complexity can greatly increase due to attempts to integrate tourism entrepreneurship with Islamic philanthropy. Researchers have drawn attention to global charity and entrepreneurship, but they have not yet given proper credit to Islamic philanthropy.

Throughout the VOSviewer evaluation process (Figure 1), it was discovered that there is no significant connection to philanthropy or/and Islamic philanthropy in almost all recent empirical research on tourism, both tourism entrepreneurship and the tourism industry. Philanthropy as a religious and social activity, as well as a form of generosity, is being studied more extensively in the context of social entrepreneurship. The discourse around social entrepreneurship has the potential to become an intersection between tourism entrepreneurship and philanthropy, partly because of the associated contribution of both to the achievement of the SDGs in decent work and economic growth.

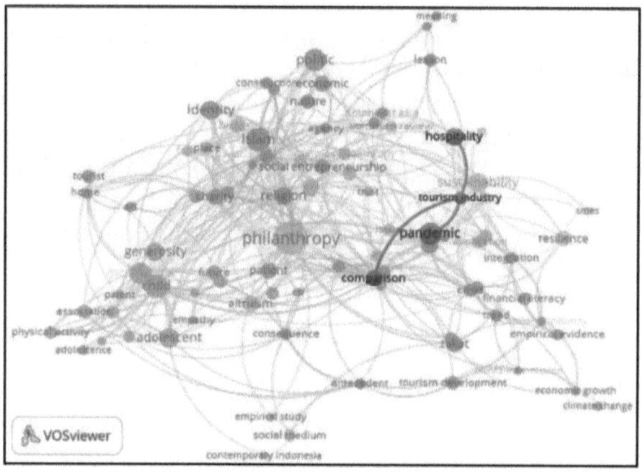

Figure 1. VOSviewer results for keywords "philanthropy" and "tourism".

Comparison studies in the tourism industry and hospitality before COVID-19, which happened after the pandemic, is a dominant study methodology used in tourism. The following visualisation illustrates that most philanthropic research trends are empirical studies conducted in Asian nations. Political and theological viewpoints—Islamic or otherwise—were used throughout the study. Considering most research on philanthropy comes from an Islamic perspective, Islamic social finance instruments such as waqf and zakah are concepts and practices constantly connected to philanthropy.

The tourism and hospitality industry has become the focus of several empirical comparative studies conducted before and after the pandemic. In the meantime, research in tourism up until 2023 is not connected to the philanthropic phenomena. According to the mapping results, the social entrepreneurship theory and religion (Islam) are the two viewpoints' researchers are interested in.

3.2 *Bibliometric analysis of Islamic philanthropy and tourism entrepreneurship*

Based on the reduction of bibliometric searches from 122 articles, 38 journals were generated that were most relevant to the research topic. The keywords "Islamic philanthropy" and "tourism entrepreneurship" yielded 0 results. Meanwhile, the keywords "waqf" and "zakah" were specified as representative instruments of Islamic philanthropy and were reduced to 9 articles. By examining the abstracts and the findings, the selected papers were manually examined to verify that each of the remaining articles meets the aim of the research question. This phase generated 9 items appropriate for the relevant study in Islamic philanthropy and tourism entrepreneurship, as detailed in Table 1.

Table 1. Synthesis matrix on sources by the relevance of "Islamic philanthropy" and "tourism entrepreneurship" themes.

No	Topics	Approaches	Findings
1	CSR; SDGs; social entrepreneurship; social enterprise	Business Canvas Model analysis	The model demonstrates the key components of the existing socio-economic ecosystem and demonstrates how they could cooperate to achieve the SDGs. The role of social enterprises and sustainable businesses can play in accomplishing the SDGs (Rahdari *et al.* 2016).
2	Sustainable entrepreneurship	Exploratory study	Ownership of decision-making in culturally sustainable entrepreneurship fosters long-term empowerment for people and communities that experience discrimination and exclusion (Swanson and DeVereaux 2017).
3	Venture waqf; responsible small businesses	Analytical study	Purpose of the waqf to make impactful small businesses successful and designing a financial contract to loan in favor of responsible businesses (Khan 2019).
4	Moral responsibility; Socio-economic model; system fairness; *zakah*	Survey	Entrepreneurs' *zakah* compliance is heavily impacted by the social fairness of the Islamic system, *zakah* morale, and peer pressure (Bin-Nashwan *et al.* 2020).
5	Entrepreneurial philanthropy; Inequality; neo-liberalism; · philanthrocapitalism	– Interviews – Statistically non-representative matched sampling – Exploratory study	The seemingly successful tool of social justice that is entrepreneurial philanthropy is morally tarnished by a constant pursuit of ideological purity. It plays an ostensibly effective instrument of social justice (Harvey *et al.* 2021).
6	Tourism contribution to the SDGs	– Critical analysis – Thematic analysis	Tourism supports economic growth, locally, regionally and nationally; tourism provides decent work opportunities; diversification through tourism value chains reduces income inequalities (Dwyer 2022).
7	The role of social entrepreneurship in the attainment of the SDGs	– Delphy method; Evaluation, multi-criteria decision-making method (MCDM) – Case study	The absence of specific tools to appraise the impact of social businesses on the fulfilment of the SDGs. The effective implementation of the SDGs in business is highly complicated as reflected in the low number of SDGs and SDG indicators (Diaz-Sarachaga and Ariza-Montes 2022).
8	Integration of zakah and corporate social responsibility (CSR); poverty	– Semi-structured interview – Content analysis	Shariah compliance acts as a guiding force for selecting well-being-oriented projects in zakah-funded CSR resulting in pursuing the priority goals – No Poverty (1), Zero Hunger (2) – of SDGs (Hoque *et al.* 2023).
9	Waqf and sustainability of tourism industry	– The qualitative approach through semi-structured interviews	The waqf model as an alternative mechanism that can assist the government in reducing spending to fulfill the social needs of the community besides it is an innovative way to fund the tourism industry without relying on government funding (Ratnasari *et al.* 2024).

Figure 2 is a tree map generated by NVivo showing the most commonly occurring keywords related to the theme of philanthropy in the 100 articles examined. This indicates a lack of explicit study on philanthropy as a social action in connection with tourism and entrepreneurship.

We map the topic of philanthropy that drew the interest of researchers from a variety of viewpoints, such as the diversity of philanthropic internal motivations (altruism, prosocial

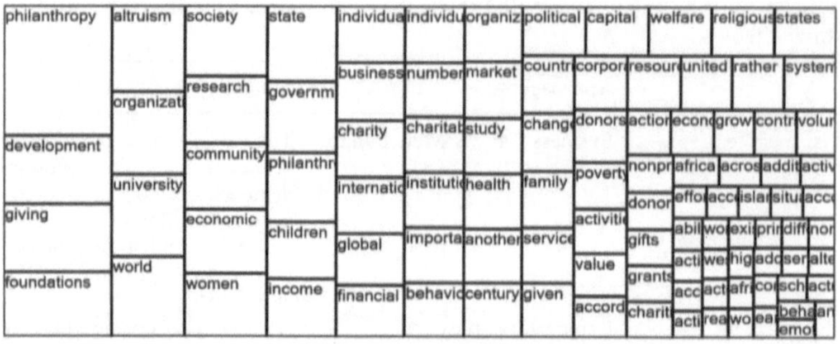

Figure 2. Tree map of codes in the topic of research on philanthropy.

behaviours and religious intention), various philanthropic activities carried out by individuals and organisations (foundations, charitable organisation, university, society, community, corporation and the state government), the purpose of philanthropy (business, political goals, health, social service, charity, social development, to eradicate poverty, family empowerment, donors) and other particular themes.

The development, giving and foundations are more frequently mentioned than the other three aspects of philanthropy. Since development is a critical component that drives economic growth, zakah also plays a role in quality economic growth through economic stimulus programs to encourage increasing household consumption. The term that represents philanthropic internal motivation is altruism a value theory that positively moderates the effect of SDG agenda on pro-sustainable travel behaviour (PSTB). Organised philanthropy is a foundation called the zakah, infaq, sadaqah, and waqf (ZISWAF) institution in Muslim society. A growing number of philanthropic organisations expressed interest in contributing to the SDGs and the international and multi-stakeholder efforts to find solutions to the multiple crises and long-term issues about sustainability. Meanwhile, philanthropy involves community members positioned as reciprocal exchanges built on trust. This relational philanthropy creates social capital that strengthens relationships and solidarity across civil society. The other most frequent terms that appeared are university, followed by state, government, individuals and family, testifying the level of charitable giving and demonstrating the suitable inductive procedure for the analytical framework and structural configuration in a literature review.

The results from selected articles indicate that waqf and zakah are the primary instruments of Islamic philanthropy that are practically linked to accomplishing the SDGs. However, further empirical and normative study is required to prove how waqf and zakat could encourage the 17 SDGs internationally and particularly in Muslim countries. Islamic philanthropy and tourism entrepreneurship can be integrated by paying attention to SDGs. The absence of an Islamic philanthropy-based hospitality and tourism industry in Muslim countries results from the lack of empirical studies in this field.

4 CONCLUSION

The software-based analysis shows that articles published from 2015-2024 are dominated by empirical research regarding the respective roles of Islamic philanthropy and tourism entrepreneurship. The implications of a study demonstrate the range of research approaches used in finding the connection between Islamic philanthropy and tourism entrepreneurship. Various methods are required to examine how Islamic philanthropy contributes to the achievement of the 17 SDGs, regarding the issues of climate action, gender equality, life on land and below water, and the need for strong institutions for justice and peace. In accordance with this, philanthropy cannot be used independently to solve all current challenges that sustainable development faces.

We recommend further studies on the potential Islamic philanthropy from a different view of Islamic social finance instruments such as sukuk, takaful, sadaqah, and infaq, in achieving the SDGs. Including the policies and strategies implemented by Muslim countries and philanthropic organisations to establish Islamic philanthropy in tourism entrepreneurship. Furthermore, there is currently an inadequate number of research on the relationship between Islamic philanthropy and tourism entrepreneurship, so a more in-depth study is required by examining various approaches and multiple theories.

ACKNOWLEDGMENTS

We would like to thank the Ministry of Religious Affairs Republic of Indonesia, in collaboration with the LPDP Ministry of Finance Republic of Indonesia, for funding this research under the BIB (Beasiswa Indonesia Bangkit)-LPDP program for providing the financial support to conduct the research.

REFERENCES

Adnan, N. I. M., Ghani, N. A. R. N. A., Ahyani, H., Syamsudin, Soleh, I., Slamet, M., Mutmainah, N., Lousada, S. A. N., and Sofiani, T. (2024). Leveraging Qardhul Hasan Practices from Zakat Funds for Sustainable Development: A Maqasid Sharia Perspective and Its Contribution to SDGs in Malaysia. *Journal of Lifestyle and SDGs Review*, *5*(2), e04016. https://doi.org/10.47172/2965-730X.SDGsReview.v5.n02.pe04016

Bin-Nashwan, S. A., Abdul-Jabbar, H., Abdul Aziz, S., and Viswanathan, K. K. (2020). A socio-economic model of Zakah compliance. *International Journal of Sociology and Social Policy*, 40(3/4), 304–320. https://doi.org/10.1108/IJSSP-11-2019-0240

Carrera-Rivera, A., Ochoa, W., Larrinaga, F., and Lasa, G. (2022). How-to conduct a systematic literature review: A quick guide for computer science research. *MethodsX*, *9*, 101895. https://doi.org/10.1016/j.mex.2022.101895

Diaz-Sarachaga, J. M., and Ariza-Montes, A. (2022). The role of social entrepreneurship in the attainment of the sustainable development goals. *Journal of Business Research*, *152*, 242–250. https://doi.org/10.1016/j.jbusres.2022.07.061

Dwyer, L. (2022). Tourism contribution to the SDGs: Applying a well-being lens. *European Journal of Tourism Research*, *32*, 3212. https://doi.org/10.54055/ejtr.v32i.2500

Fauzia, A. (2023). Transformation in Muslim Philanthropy Movement. *Journal of Muslim Philanthropy & Civil Society*, *7*(1). https://scholarworks.iu.edu/iupjournals/index.php/muslimphilanthropy/article/view/6211

Harvey, C., Gordon, J., and Maclean, M. (2021). The Ethics of Entrepreneurial Philanthropy. *Journal of Business Ethics*, *171* (1), 33–49. https://doi.org/10.1007/s10551-020-04468-7

Hoque, N., Uddin, M., Islam, M. T., Mamun, A., Nazim Uddin, M., Ahmad, A., and Thowhidul Islam, M. (2023). Pursuing sustainable development goals through integrating the aspirations of zakah and CSR: evidence from the perspective of an emerging economy. *International Journal of Social Economics*, *50*(9), 1232–1247. https://doi.org/10.1108/IJSE-09-2022-0600

Kuuliala, J., and Rantala, J. (eds.). (2020). *Travel, Pilgrimage and Social Interaction from Antiquity to the Middle Ages*. Routledge.

Khan, T. (2019). Venture waqf in a circular economy. *ISRA International Journal of Islamic Finance*, *11*(2), 187–205. https://doi.org/10.1108/IJIF-12-2018-0138

Manu, E., and Akoita, J. (ed.). (2021). *Secondary Research Methods in The Built Environment*. Routledge.

Mohammad Jaber Thalgi. (2024). Altruism and Social Cohesion: An Islamic Perspective of Charitable Acts for Sustainable Development. *Journal of Islamic Thought and Civilization*, *14*(2), 42–62. https://doi.org/10.32350/jitc.142.04

Rahdari, A., Sepasi, S., and Moradi, M. (2016). Achieving sustainability through Schumpeterian social entrepreneurship: The role of social enterprises. *Journal of Cleaner Production*, *137*, 347–360. http://dx.doi.org/10.1016/j.jclepro.2016.06.159

Ratnasari, R. T., Ismail, S., Mahphoth, M. H., Laila, N., and Hidzir, P. A. M. (2024). *Waqf model and sustainability of tourism industry: Malaysian and Indonesian perspectives*. 020059. https://doi.org/10.1063/5.0182331

Sainz, C. C., and M. Samorin, M. C. (2021). Sustainable Tourism in Northern Iloilo: Perceived Impact and Prospects for Development. *Tourism and Sustainable Development Review*, *2*(2), 30–39. https://doi.org/10.31098/tsdr.v2i2.51

Susamto, A. A., Bisri, M., Masyhadi, A. R., and Susamto, B. (2024). Redefining Boundaries: The Case for Enterprise-Structured Waqf Over Corporate Waqf. *ISRA International Journal of Islamic Finance*, *16*(S1), 118–135. https://doi.org/10.55188/ijif.v16iS1.759

Swanson, K. K., and DeVereaux, C. (2017). A theoretical framework for sustaining culture: Culturally sustainable entrepreneurship. *Annals of Tourism Research*, *62*, 78–88. https://doi.org/10.1016/j.annals.2016.12.003

Thompson, J. (2022). Volunteer tourism fields: Spaces of altruism and unsustainability. *Current Issues in Tourism*, *25*(5), 779–791. https://doi.org/10.1080/13683500.2021.1887824

Utami, D. D., Dhewanto, W., and Lestari, Y. D. (2023). Rural tourism entrepreneurship success factors for sustainable tourism village: Evidence from Indonesia. *Cogent Business & Management*, *10*(1), 2180845. https://doi.org/10.1080/23311975.2023.2180845

Towards Resilient Societies: The Synergy of Religion, Education, Health, Science, and Technology – Rahiem (Ed.)

Islamic ethical perspectives on AI development and use

Hamidi Ilhami, Aulia Khafizah & Hidayah Nor
State Islamic University Antasari Banjarmasin, Indonesia

ABSTRACT: In today's era of technological advancement, the emergence of Artificial Intelligence (AI) has become a prominent topic of discussion. Islam encourages the use of human intellect to develop knowledge and contribute positively to life on earth. The presence of AI reflects humanity's ability to utilise intellect for technological progress. However, given the transparency of information enabled by AI, this study addresses the key question of how AI technology aligns with the ethical principles of the Qur'an and Hadith as sources of Islamic law. Using a qualitative method with a descriptive-analytical approach, this research contributes to the understanding of AI ethics within Islamic contexts, enriching ethical discourse, incorporating diverse perspectives, and supporting the responsible adoption of AI across Islamic countries. This framework balances local values with universal relevance. The findings indicate that the Qur'an and Hadith provide clear guidance for the ethical use of technology, promoting progress within Islamic civilisation. References such as QS. Al-Jasiyah verse 13, Hadith narrated by Muslim, QS. Al-Nur verses 27–28, and QS. Al-Zumar verse 18 emphasise the importance of protecting privacy and avoiding misuse driven by personal desires. Instead, AI should be harnessed to advance knowledge, particularly in supporting Islamic *dakwah* (propagation).

Keywords: artificial intelligence, development, Islamic ethics, perspectives

1 INTRODUCTION

Technological sophistication will continue to be a hot topic of conversation with the innovations released, one of which is Artificial Intelligence (artificial intelligence), or can be abbreviated as AI. Artificial Intelligence means intelligence, adapted from the Latin intelligo, meaning I understand. In terms, intelligence is a skill or ability to understand and carry out actions (Karim *et al.* 2020). So, this type of technology can be understood as a form of simulation of intelligence that is almost the same as how humans think in the form of machines.

Artificial Intelligence is a much-needed component in the production of intelligent computer applications. Artificial Intelligence was developed in the 1940s; at that time, only paper was used as a research framework. In the 20th century, this field was developed along with the contribution of new human innovations, especially in the industrial and household fields (Al-Khowarizmi and Lubis 2023). As a technology that is touted as taking over human activities, Artificial Intelligence is formed by utilising algorithms combined with software (Kusnaini 2024). AI technology can compete with the thinking ability of the human brain in general, both in processing incoming information and learning existing movements.

The presence of AI in the world has almost taken over most human jobs. Various countries worldwide, especially developed countries, are now slowly utilising AI as a helper for their work. For example, in the United States, the White House has issued many policies that emphasise the strategic importance of AI. This policy is the United States' strategy to move towards a future that fully utilises AI. Russian President Vladimir Putin has stated that countries that comprehensively utilise AI will become world rulers. Meanwhile, China

 DOI: 10.1201/9781003645542-90

ambitiously predicts that it will win the seat of world power in AI by 2030. Similarly, in 2018, the European Commission launched its AI strategy to improve all aspects of life with the development of AI technology (Marr and Ward 2021). This is proof that many countries are starting to make AI the front line in the fight for the ruling seat of the world.

Recent AI applications or examples in nations with a majority of Muslims give arguments on AI ethics in Islamic contexts a great deal of current significance. These instances offer a basis for incorporating Islamic principles into AI development and show how AI is applied practically to solve societal, economic, and religious issues. Some important areas where AI is being used in nations with a majority of Muslims are listed below.

For example, Saudi Arabia has used AI to enhance healthcare delivery by utilising AI-powered diagnostic tools for disorders like diabetes and heart disease (Albinsaad *et al.* 2024). AI is employed in Malaysia and Turkey to create assistive technology for individuals with disabilities (Dange *et al.* 2023), bolstering the teachings of the Quran on inclusion and caring for those in need. AI has been incorporated into Dubai's urban planning and governance through crime prevention technology, intelligent traffic management systems, and AI-powered public services (Abdulrahim 2024). Malaysia uses big data analysis using AI to improve governance and policy making. Through increasing the efficiency and transparency of decision-making, this application upholds the values of Shura (consultation). AI-powered applications are popular in Pakistan and Indonesia for learning Tajweed, or the recitation and memorisation of the Quran.

AI is being applied in Egypt and Morocco to forecast crop production, enhance irrigation systems, and stop desertification. Regions dealing with water scarcity and climate change need AI-driven solutions (Dewedar *et al.* 2023). To maintain the integrity of the Halal certification procedures, Malaysia and Indonesia are utilising AI. Supply chain analysis, non-Halal element detection, and Shariah compliance are all done using AI tools. Malaysian and Turkish governments utilise AI to track and combat hate speech, extremist content, and false information on the internet.

Saudi Arabia uses AI to handle logistics during the Hajj and Umrah, including crowd management, transportation, and safety surveillance. Robots using artificial intelligence (AI) help pilgrims by offering advice. These instances show how AI can be used in Muslim-majority nations and how ethical discussions can incorporate current relevance.

Looking at some previous studies, the researchers found several studies that have examined the issue of AI technology from an Islamic perspective. Hidayat *et al.*, (2020) examined the implementation of AI in Islamic studies, the Quran, and Hadith. Khoirunnisa *et al.*, (2023) examined the opportunities and threats accompanying AI technology so that its use aligns with Islam. Then Murtopo (2023) examines artificial intelligence being able to present itself in the world of Islamic education and the influence it brings.

Based on several studies related to useful AI technology according to Islamic perspectives that have been studied, the researchers have a gap with three problem formulations, including (1) The views of the Quran and Hadith on the development of technology; (2) Ethics and morality in useful AI according to the Quran and Hadith; and (3) Integration of Islamic values in the development of AI.

This research's distinctive contribution is its capacity to offer a contextualised Islamic viewpoint on AI ethics that directly interacts with sources, suggests workable frameworks for AI development, and connects advanced technology issues with conventional religious teachings. Its analysis grounded in the timeless wisdom of the Quran and Hadith provides a unique voice in the expanding field of AI ethics while tackling the moral challenges of contemporary AI systems.

2 METHODS

Based on the formulation of the existing problems, the author uses a qualitative method with a descriptive-analytical approach to produce relevant discussion results. The techniques used vary, including in-depth interviews, observation, and text analysis to collect data as research material (Rachmad *et al.* 2024).

The methodology presented in this study ensures that innovation advances industry, infrastructure, and institutions while advancing justice and peace, allowing technology to benefit humanity within moral bounds. By connecting Islamic ethics to these objectives, the study provides a means for AI to advance technology while improving society. The alignment of religious principles with the goals of world development is strengthened by this strategy (Kosasih *et al.* 2024).

Specifically, the author uses data analysis techniques, namely text with library research methods sourced from the Quran and Hadith as primary sources, then supported by secondary sources in various related discussion literature. The data analysis techniques needed in this research include collecting data from several literatures, data reduction, presenting data, and finally, making conclusions.

3 FINDINGS AND DISCUSSION

3.1 *The view of the Quran and Hadith on the development of technology*

Islam strongly observes and supports all aspects of life, including the development of science and human knowledge on earth. Islam and technology support each other. Islam needs technology in its development, while technology needs Islam as a balance (cosmic) (Desi *et al.* 2021).

The Quran guides human life and encourages humans to develop their minds. This is evidenced by the Quran repeatedly mentioning questions to encourage humans to always use their minds, with the closing verse: afalā tatafakkarūn (do you not think?), afalā takqilūn (do you not use your minds?) and afalā tatadabbarūn (do you not think?) (Iyubenu 2020). This statement is proof that Islam supports humans in developing all forms of science, including technology.

Particularly in domains like artificial intelligence (AI), where developments might have substantial advantages and moral hazards, striking a balance between technological innovation and ethical constraints is crucial (Hastuti 2023). This method guarantees that technology advances mankind while upholding its moral underpinnings by encouraging responsible innovation that respects individual liberties, cultural norms, and the benefit of society (Zafar 2024).

Technology is evidence of the development of science amid human civilisation as well as evidence of the greatness of Allah. It must be that even if it is born from His creation, it still returns to Him. Everything in the universe happens by the will of Allah, through anything and anyone. As in QS. al-Jasiyah verse 13 below.

"He has subjected for you all that is in the heavens and all that is on the earth (as a mercy) from Him. Surely in such there are signs (greatness of Allah) for those who think."

Al-Maraghi (1946) reveals that Allah has provided everything created by Him on this earth, which is dedicated to helping human interests and livelihoods. This interpretation aligns with the rise of technology in the current era, which facilitates human affairs. One Hadith is relevant to the development of technology as evidence of the development of the human mind on earth. The following Hadith is mentioned.

"When the Prophet (peace and blessings of Allah be upon him) came to Madinah, the people of Madinah were pollinating date palms so that they would bear fruit, which they called mating, so he asked them: what are you doing? They replied: We have always done this. He said: If you had not been doing it, it would have been better. So they left it, only to find that the dates had fallen off. He said: Then they complained to him, and he said: "I am only a human being, so if I order something from your religious affairs, then do it, and if I order something to you based on my own opinion, then know that I am only a human being." (HR. Muslim)

The above Hadith illustrates that the Prophet initially discouraged the people of Medina from making innovations in their gardening activities, namely in pollinating date flowers so that they would bear fruit quickly. However, when they obeyed not to pollinate anymore, several dates miscarried and fell off. In response to this, the Prophet ordered them to follow the orders of the Prophet in matters of religion, the rest was the opinion alone of the Prophet, which Allah did not inspire.

3.2 *Ethical and moral use of AI according to the Quran and Hadith*

The development of technology has given rise to artificial intelligence, which is popularly called Artificial Intelligence. This man-made intelligence facilitates all human activities in various aspects of life. It should be noted, however, that this artificial intelligence has absolutely no ethics and morality. AI cannot know human values and only follows the values set for it (Akbar 2023).

The terms ethics and morals are interchangeable. Their connotations are almost the same, namely understanding wrong and right or good and bad (Sagala 2013). Although the two are virtually the same, they are slightly different when deeply understood.

Ethics is the study of what is considered right or wrong in the context of human behaviour or conduct. In this case, it involves consideration of the moral principles that direct and guide the behaviour of each individual. Morality, on the other hand, is the practical application of ethics in daily life. This scope concerns how humans make decisions and produce actions (Anandari 2024).

Some cases that can be used as evidence that AI technology does not heed ethics and morality are maintaining security and privacy. Therefore, humans as the subject can determine the ethics and morality of AI in society. The point of contention is the misuse and proper use of the technology to create benefits.

Islam maintains the privacy of oneself, and others is a noble character that needs to be promoted. While indulging in disgrace is something Allah hates, it is not a praiseworthy act that should be avoided. As explained by Allah in surah al-Nur verses 27-28 below.

"O you who believe, do not enter a house that is not your own before asking permission and greeting the occupants. That is better for you so that you may learn a lesson. If you find no one there, do not enter until you have received permission. If it is said to you, "Return," (let) you return. That is purer for you. Allah knows best what you do."

The verse explains the prohibition of entering someone's residence before getting permission from the host. Al-Maraghi (1946) suggests that the house is a place that stores privacy, which is very sensitive for everyone. Therefore, one should ask the owner's permission before entering not to disturb the host's right to privacy. This illustrates that in addition to recognising the right to privacy, Islam also protects the privacy of each individual from any circle.

Ethically, the noble dignity of human beings should always be considered in accompanying the advancement of AI technology. This is because technological advancement is solely intended to advance human life on earth (Pabubung 2023), not to set back civilisation with sheer lust. As explained in surah al-Zumar verse 18 below.

"(Those) who listen to speech and follow what is best among it. They are the ones whom Allah has guided, and they are the ulul albab (people of sound mind)."

Al-Saadi (2000), in his commentary, says that if a person puts his passions ahead of his intellect, he is not intelligent. Although the level of quality of his intellect is dhahiriya competent. Thus, humans should balance their intellect with faith and remove lust from the first position. The balance of reason and lust will bring good to mankind. Open disgrace bluntly without caring that their privacy is exposed. This disgraceful disclosure is also done to humiliate others. Allah and His Messenger highly hate this behaviour. As the following sahih Hadith states.

"Whoever covers a disgrace of person, Allah will cover his disgrace in this world and the next."

Covering disgrace is part of the tips for maintaining privacy. For example, the profession of prosecutor requires disgrace for legal purposes. Thus, scrutinising other disgraced people is permissible within certain tolerable limits in the context of self-improvement (Muzadi 2017). This rule of ethics and morality in useful AI needs to be put forward.

3.3 *Integration of Islamic values in AI development*

The challenge of integrating traditional values and technological advances in Islamic civilisation provides a good opportunity to produce innovative and ethical solutions (Maulana 2024). Technological progress is in the hands of people who can integrate themselves with Islamic values, and it will undoubtedly give birth to new thinkers who can make a significant contribution to the world of science and technology. Muslims will not be oppressed just because they are rigid to the progress of the times. Religion and the progress of the times must work together to produce a brilliant generation of people.

The misuse of AI technology will be minimised if humans apply Islamic principles to avoid exposing shame and maintain privacy. So, then the main focus is to advance Islamic civilisation with science, especially Islamic dakwa. As most often echoed in the Quran, humans are highly enthusiastic about science. Using something in its place will give birth to the benefit of the people, which is very important for advancing civilisation.

Islamic ethical principles highlight values like accountability, dignity, and trustworthiness, which offer a strong framework for addressing privacy concerns in AI applications. These principles complement the larger ethical discourse on privacy while providing a distinctive viewpoint based on Islamic teachings (Firnando and Wahyudi 2024). By establishing privacy protections based on these Islamic principles, AI developers and policymakers can create systems that uphold individual rights while promoting ethical accountability (Elmahjub 2023).

4 CONCLUSION

Responding to technological advances with the presence of artificial intelligence or Artificial Intelligence, the Quran and Hadith, which are guidelines for life, certainly direct health so that their use can bring progress to Islamic civilisation. Maintaining privacy and not indulging in disgrace to pay for the interests of lust alone but directed so that the ease of AI technology is used as a medium to advance science, especially Islamic dakwa. That way, artificial intelligence will positively impact Islamic civilisation, especially for future generations.

It is necessary to match technology breakthroughs with Islamic ethical concepts, such as fairness (Adl), benefit to mankind (Maslahah), trustworthiness (Amanah), and avoiding damage (La Darar), to deploy AI ethically in Muslim contexts in fields like healthcare and education. AI has revolutionary potential in healthcare and education, especially in Muslim situations where morality is crucial. Islamic ideals such as justice, reliability, and compassion can be incorporated into modern technologies to benefit society while honouring cultural and religious sensitivities. This strategy promotes diversity and trust in these vital areas while guaranteeing responsible innovation.

REFERENCES

Abdulrahim, A. S. (2024). *Reduction of Crimes in Dubai with the use of Artificial Intelligence* (Master's thesis, Rochester Institute of Technology).

Akbar, J. (2023). *Artificial Intelligence Teman atau Musuh, sih?* Brimedia Global.
Albinsaad, L. S., Alkhawajah, A. A., Abuageelah, B. M., Alkhalaf, R. A., Alfaifi, M. H., Oberi, I. A., and Khan, A. (2024). The Saudi Community View of the Use of Artificial Intelligence in Health Care. *Annals of African Medicine*, *23*(3), 343–351.
Al-Saadi, A. (2000). *Tafsīr al-Saadi.* Muassisah al-Risalah.
Al-Khowarizmi, and Lubis, A. R. (2023). *Artificial Intelligence.* Umsu Press.
Al-Maraghi, A. (1946). *Tafsīr al-Marāghī.* Muṣṭafa al-Bābī al-Ḥalabi.
Anandari, A. A. (2024). *Bijak Beragama di Dunia Maya: Pendidikan Karakter Era Digital.* Jejak Publisher.
Dange, P., Mistry, T., and Mann, S. (2023). AI and Assistive Technologies for Persons with Disabilities-Worldwide Trends in the Scientific Production Using Bibliometrix R Tool. In *International Conference on Artificial Intelligence: Towards Sustainable Intelligence* (pp. 24–43). Cham: Springer Nature Switzerland.
Desi, N, *et al.* (2021). *Masyarakat Cita: Konsepsi & Praktik.* Liyan Pustaka.
Dewedar, O. M., El-Shafie, A. F., Marwa, M. A., and Abdelraouf, R. E. (2023). Improvement of Irrigation Water Management Using Simulation Models And Artificial Intelligence Under Dry Environment Conditions In Egypt: A Review. *Egyptian Journal of Chemistry*, *66*(11), 93–106.
Elmahjub, E. (2023). Artificial Intelligence (AI) in Islamic Ethics: Towards Pluralist Ethical Benchmarking for AI. *Philosophy & Technology*, *36*(4), 73.
Firnando, H. G., and Wahyudi, M. (2024). The Role of Artificial Intelligence in Shaping the Islamic Worldview of the Digital Economy. *Journal of Islamic Economics and Philanthropy*, *6*(3), 231–249.
Hastuti, R. (2023). Ethical considerations in the age of artificial intelligence: balancing innovation and social values. *West Science Social and Humanities Studies*, *1*(02), 76–87.
Hidayat, R., *et al.* (2020). Survey Paper: Tantangan dan Peluang Kecerdasan Buatan dalam Bidang Islam, Quran dan Hadits. *Prosiding Konferensi Integrasi Interkoneksi Islam dan Sains*, *2*.
Iyubenu, A. AH. (2020). *Beginilah Islamku.* Diva Press.
Karim, A., *et al.* (2020). *Pengantar Teknologi Informasi.* Yayasan Labuhanbatu Berbagi Gemilang.
Khairunnisa, A., *et al.* (2023). Islam in the Midst of AI (Artificial Intelligence) Struggles: Between Opportunities and Threats. *International Journal of Islamic Studies*, *35*(1).
Kosasih, E., Islamy, M. R. F., and Wiwaha, R. S. (2024). Artificial Intelligence in the Era of Society 5.0: Compromising Technological Innovation Through theWasathiyyah Approach within the Framework of Islamic Law. *Al-Istinbath: Jurnal Hukum Islam*, *9*(2), 551–572.
Kusnaini, R. P. (2024). *Dampak Penggunaan AI.* Brimedia Global.
Marr, B. and Ward, M. (2021). *Artificial Intelligence in Practice terj. Irene Christin.* Elex Media Komputindo.
Maulana, Y. (2024). *Perkembangan Teknologi dalam Peradaban Islam.* Oyusep.
Murtopo. (2023). Pendidikan Islam dalam Bayangan Artificial Intelligence (Kecerdasan Buatan): Sebuah Kajian Pustaka Mengenai Dampak AI dalam Pendidikan Islam. *PENDALAS: Jurnal Penelitian Tindakan Kelas dan Pengabdian Masyarakat*, *3*(1).
Muzadi, A. H. (2017). *Al-Hikam: Butiran Hikmah Abah Hasyim Muzadi.* Tira Smart.
Pabubung, M. R. (2023). Era Kecerdasan Buatan dan Dampak terhadap Martabat Manusia dalam Kajian Etis. *Jurnal Filsafat Indonesia*, *6*(1).
Rachmad, Y. E., *et al.* (2024). *Integrasi Metode Kuantitatif dan Kualitatif: Panduan Praktis Penelitian Campuran.* Green Pustaka Indonesia.
Sagala, S. (2013). *Etika dan Moralitas Pendidikan: Peluang dan Tantangan.* Kencana Prenadamedia Group.
Zafar, A. (2024). Balancing the scale: navigating ethical and practical challenges of artificial intelligence (AI) integration in legal practices. *Discover Artificial Intelligence*, *4*(1), 27.

Towards Resilient Societies: The Synergy of Religion, Education, Health, Science, and Technology – Rahiem (Ed.)

Copyright in marriage: Legal status and economic rights as joint property

Nurlaili Rahmawati & Fathudin Fathudin
Universitas Islam Negeri Syarif Hidayatullah Jakarta, Indonesia

Sigit Nurhadi Nugraha
Universitas Nahdlatul Ulama Indonesia (UNUSIA), Jakarta

ABSTRACT: This study is normative legal research that employs both conceptual and case study approaches. This research aims to determine the legal status, offer solutions, and address legal issues related to intellectual property rights, specifically, copyright (Hak Kekayaan Intelektual or HKI) acquired during the marriage. The findings conclude that moral rights to a copyrighted work remain exclusively with the creator, even if the work was produced during the marriage. However, the economic rights to the copyrighted work are considered joint property if created during the marriage, regardless of whether the marriage has ended. Consequently, both spouses are entitled to royalties from copyrighted works produced during marriage. This conclusion aligns with the legal ruling in the division of joint property in the case between Virgoun, a musician, and Inara Rusli, a social media influencer.

Keywords: copyrights, divorce, intellectual property rights, joint property, marriage

1 INTRODUCTION

One of the most important arrangements in Indonesian marriage law is the property of property in marriage. Marriage law explicitly divides property in marriage in the form of: (a) inheritance; and (b) joint property. These provisions are regulated in various provisions, among others: Article 35 of the Law No. 1 of 1974 on Marriage, Article 119 of the Islamic Code and Articles 85 to 95 of the Compilation of Islamic Law. Article 35 (1) of the Act No. 1 of 1974 defines the entire property acquired before the marriage, whereas the joint property (in Article 35 (2) of the Law No. 1, 1974) refers to the whole property acquired during marriage or may also be understood as the property produced/acquired by a spouse at the time of marriage (Rochaeti 2013). This joint property in other literature is also called the "Gono-Gini Treasure".

The provisions of the law concerning the mixing of such common property above are a general law (lex generalis), which regulates the common property as a whole. On the other hand, it is necessary to re-question the implementation of the provision with respect to the assets generated by intellectual property rights (IPRs) because the IPRs are regulated by a rule of a special nature. (lex spesialis). Intellectual property is the right to enjoy materially or economically the results of intellectual creativity.(Alfons 2017) The conflict of norms in this matter arises because the ownership of intellectual property rights is personal, that is, based on the principle of personality. The existence of this personality principle means that the right to intellectual property rights (hereinafter referred to as HKI) can only be owned by the creator of the HKI personally (individual), even if the right was acquired during the marriage.

The complexity of the problem in this study is the existence of differences in concepts and legal regulations between: (1) Copyright regulated in the Family Law (Gono-Gini Charter); contrary to (2) Copyright governed in the Law on Intellectual Property Rights (hereinafter referred to as HKI). Marriage law considers that HKI acquired during marriage should be part of the common

DOI: 10.1201/9781003645542-91

property owned by the spouses. However, it is different from the perspective of the HKI Law, which holds that ownership of HKI is personal, so it cannot be a joint property in marriage.

This study focuses on one form of Intellectual Property Rights (Copyright). According to article 4 of Act No. 28 of 2014, Copyright grants two (two) exclusive rights to the creator or copyright holder, namely moral and economic rights. The reason for the selection of the focus of this study is that the copyright was made the object of a joint property lawsuit and has been examined in a trial in the Court. One example is the joint gono-gini lawsuit filed by Inara Rusli and Virgoun artists, which has been investigated and sent to the West Jakarta Religious Court. The lawsuit has been rescinded by a judge of the Western Jakarta Religion Court who declared the work of Virgoon as a song entitled (a) Letter of Love for Starla; (b) Evidence; (c) Goodbye (Selgal Tinggamat) and (d) Same person; (dono) who is a property in a marriage between Virgoun and Inara Rusli (*Menyoal Royalti Lagu sebagai Harta Gono-Gini Perceraian Virgoun Tambunan dan Inara Rusli*, t.t.)

The above facts indicate that the court considers the copyright an object of joint property that can be the object of it in the joint property claim (gono-gini). However, referring to the above, the other legal question is why the establishment as a joint property is only four (four) songs while there are many other works of Virgoon's creation? When the reasoning for the creation of the song is based on the argument that the inspiration for the song comes from the wife and child, it must be less satisfactory. For that, looking for arguments corresponding to the applicable legal provisions is necessary.

The conflict of norms and the implementation of the law in establishing gono-gini property, whose object is HKI (Copyright) as described above, has no legal certainty, so there is ambiguity in the setting of a legal norm and its implementation. At this point, researchers must investigate the above-mentioned legal issues as a matter of urgency.

The planned research is crucial in enriching the scientific knowledge of the field of law, particularly in intellectual property law (copyright) and private law (marriage law as well as property law). Doctrines taught in private law, especially in family law (matrimonial law) and property law, usually refer to the provisions of the Book of Private Law (burgerlijk wetbok). These concepts are classical doctrines that are still in use to this day. Still, on the other hand, the research is confronted with the concept of copyright, one of the forms of intellectual property rights that is a modern, new and modern legal concept.

From this description, we can see the importance of the provisions of civil law in addressing the problem of intellectual property rights, as well as how the law of intellectual rights can synergise with the national civil law that has been applied in Indonesia. So, this research is important to provide the resolution of legal issues and ensure that legal provisions are harmonious and following the interests of national law.

2 METHODS

This type of research is normative legal research, which emphasises research on certain legal principles, rules, and concepts and analyses the relationship between these things to resolve the ambiguity of norms or conflicts of legal norms. The rules and legal concepts in this study include: 1) Provisions for Marriage and Marital Property (Law Number 1 of 1974; Presidential Instruction No. 1 of 1991 concerning the Compilation of Islamic Law (KHI)); and 2) Provisions on Copyright (Law Number 28 of 2014 concerning Copyright).

3 FINDINGS & DISCUSSION

3.1 *Intellectual property copyright ownership concept*

In intellectual property copyright, there are several elements, namely the creation, the creator, the copyright holder, licenses and royalties. Creation is a right arising from works created by any copyrighted works in science, art, and literature produced by inspiration, ability, thought, imagination, dexterity, skill, or expertise expressed in real form. A creator is a person or several people who individually or jointly produce a unique and personal creation. The Copyright Holder is the Creator as the Copyright owner, the party who legally receives the right from the Creator or another party who further receives the rights from the party who legally receives the right. License is a written permission granted by the Copyright Holder or Related Rights Owner to another party to exercise economic rights over his/her Work or Related Rights products under certain conditions. Royalty is a reward for using the Economic Rights of a Creation or Related Rights Product received by the creator or owner of the related rights.

Intellectual property rights are rights arising from works created using human intellectual abilities, whether in the field of technology, art, music, knowledge, written works, etc., that are beneficial to the life of society, nation or country. In addition, HKI is a moral, economic and practical right (Anhar 2024). The existence of the registration of intellectual property rights is a form of state care and appreciation of a person's intellectual work. The legal protection of copyright by the State must be consistent with the objectives of the State of Indonesia as set out in the fourth paragraph of the opening of the 1945 NRI UUD.

In Law no. 28 of 2014 concerning copyright, copyright can give rise to two rights: moral rights (article 5 of Law no. 28 of 2014) and economic rights (article 8 of Law no. 28 of 2014). Moral rights are the right to be named, altered or defended if there is a distortion in the work. Moral rights cannot be transferred as long as they are alive. However, it can be inherited after death. Economic rights are the right to use the creation economically, transferable and valid for 70 (seventy) years after the death of the Creator. If by a legal entity 50 years.

According to labour theory, the creator owns the moral and economic rights of the work he has created. The owner acts as the absolute ownership or absolute owner of the work of his creation (Alfons 2017). For this, this theory requires legal protection of the intellectual work it has created so that the creator is motivated to develop new ideas and innovations to the work of its creators. The government plays an important role in providing legal protection either through regulatory instruments that support the acquisition of copyright creators or copyright holders, as well as support through government instruments and institutions empowered to provide legal protection.

The Labour Theory is a theory put forward by John Locke which states that everyone has the right to the result of his work, its consideration is every process/stage, thought/creativity carried out to create something (Simatupang 2021). This theory is further submitted in three (three) forms of principle, namely:

(1) The creator must receive a reward of ownership of the effort/process to create something.
(2) Assets already produced for the other party can create some rights for the workers, but it depends on the "work agreement" between the workforces.
(3) Claims of a collective property right must be subject to the same provisions of law, all of which are designed to reconcile the rights of individual creators with the claims of the wider community in general.

Robert M. Sherwood argues the importance of copyright ownership to be given to the creator and to be protected based on several theories, namely:

(1) *Reward theory* rewards the creator in exchange for his creative efforts.
(2) *Recovery* theory conforms to the principle that the inventor/creator/designer who has spent time, costs and energy producing his intellectual work must recover what he has produced.
(3) *Incentive theory* is the theory that is in line with reward, which connects the development of creativity by giving incentives to the inventors/creators or designers.
    Based on this theory, incentives must be given to pursue the most intriguing and useful research activities.
(4) *Risk theory* acknowledges that HKI is a product of work that contains risk; for example, research to discover a vaccine against the disease virus can risk the life of the researcher/inventor if not careful. Besides, he has spent little cost, time and energy.
(5) *Economic growth stimulus theory* recognises that the protection of HKI is a tool of economic development, and what is meant by economic development is the whole purpose of building an effective system of intellectual property protection.

3.2 *Joint property in marriage*

The term joint property or commonly referred to as gono gini is the wealth acquired during marriage. Article 35 of Act No. 1 of 1974 states that property acquired during marriage becomes joint property. Further, in article 97 of the Compilation of Islamic Law, it is mentioned that the division of each husband and wife gets half of the share. However, the inheritance of property acquired before the marriage is the property or possession of each spouse as long as there is no other agreement.

The Marriage Law expressly distinguishes property in marriage in the form of: (a) innate property; and (b) joint property. The definition of Innate property in Article 35 paragraph 1 of Law No. 1 of 1974 is all property obtained before the marriage occurs, while joint property (in Article 35 paragraph 2 of Law Number 1 of 1974) is all property obtained during the marriage period or can also be interpreted as property produced/obtained by a married couple during a direct marriage.

The property acquired by law during the marriage is divided into joint property as long as it is not promised in a marriage agreement (Alexander 2019). The provision concerning the mixing of Joint Property acquired during this marriage occurs automatically, regardless of who acquires the wealth. However, if there is a marriage agreement, then the agreement can determine the separation of property. The separation of marital property causes the husband or wife's property to become their respective property, even if it was acquired at the time of marriage.

The concept of joint property in Islam and associating it with the concept of syirkah follows what is presented by KH Ma'ruf Amin, who gives wealth together with syrekah because the same is a partnership between husbands and wives (Sugiswati 2014).

Is it a form of cooperation between the two parties that do business and the third party as a funder, while the abdan is the form of cooperation between two or more people that only do business without any capital (Hertasmaldi and Hafizh 2021). Thus, all property acquired during marriage is joint property, regardless of the husband or wife who produces or on whose behalf the property is registered (Djuniarti 2017).

3.3 *The status of copyright as joint property in marriage*

Marriages carried out by Muslims are recorded at the Office of Religious Affairs (KUA), while marriages carried out by non-Muslims are recorded by the Population and Civil Records Office (Dispenduk Capil). It also refers to which court handles divorces: Religious Courts handle Muslim divorces, while District Courts handle non-Muslim divorces (Kharlie *et al.* 2021). The Compilation of Islamic Law (KHI) article 114 mentions that "the breakup of a marriage caused by divorce can occur due to divorce repudiation/*talak* (by husband) or based on divorce lawsuit (by a wife)" (Nasution 2018). As a consequence of a marriage that has ended either due to death or divorce, the laws and regulations in Indonesia stipulate that any property obtained during the marriage is used as the common property of husband and wife, regardless of who works and whose name the property is registered in. then the property must be divided by two equal amounts (Mesraini 2012).

Copyright is an intangible object with an economic value that can generate economic/material gain for its owners over Intellectual Property Rights (Poetri 2020). Copyright is usually acquired by musicians from creating a work of art, either music or song. A song or music with or without text is one of the inventions that can be protected (Anhar 2024). Article 58 (1) of the Act No. 28 of 2014 on Copyright states that the copyright protection of the creation of a song or music with or without text is valid for the lifetime of the author and continues for 70 (seventy) years after the creator's death, counting from 1 January of the following year.

In Article 1, paragraph 1 of law No. 28 of 2014 concerning copyright, the definition of copyright is the creator's exclusive right, which arises automatically based on declarative principles after a work is realised in real form without reducing restrictions following the provisions of statutory regulations. The copyright contains 2 (two) rights, namely moral and economic rights. Moral rights are the right of a creator to prevent other people from carrying out actions that are detrimental to the creator. In contrast, economic rights are the rights owned by the creator to obtain economic benefits from his creation by obtaining payment from the party who uses his creation based on a contract/agreement (Hapsari 2012).

Furthermore, moral rights will always be attached to the creator (even if there is a transaction or transfer of rights). This right does not change even if a legal marriage event occurs. So, concerning this, moral rights remain with the creator of the copyrighted work. Meanwhile, economic rights to a copyrighted work, which have implications for licenses and royalties for a copyrighted work, are common property. This follows the decision to divide common property between Virgoun and Inara Rusli. Royalties or economic rights to intellectual property rights can be categorised as joint property between husband and wife as long as they are registered during the marriage (Islam, t.t.).

4 CONCLUSION

Moral rights to a copyrighted work remain the creator's rights, even though the copyrighted work was created during the marriage period. The Economic Right to a copyrighted work becomes a joint right (joint property) if the copyrighted work is created during the marriage period, even if the marriage has been broken. So that the husband/wife is still entitled to royalties for copyrighted works created during the marriage. This follows the decision to divide joint property between Virgoun and Inara Rusli.

REFERENCES

Alexander, O. (2019). Efektivitas Pembagian Harta Gono—Gini Pasca Perceraian Dalam Persfektif Yuridis Sosiologis. *El-Ghiroh: Jurnal Studi Keislaman*, 16(01), Article 01. Https://Doi.Org/10.37092/El-Ghiroh.V16i01.70

Alfons, M. (2017). Implementasi Hak Kekayaan Intelektual Dalam Perspektif Negara Hukum. 14(03).

Anhar, H. I. (2024). Analisis Royalti Hak Cipta Lagu Sebagai Harta Bersama Dalam Gugatan Cerai. *Jurnal Res Justitia: Jurnal Ilmu Hukum*, 4(1), Article 1. Https://Doi.Org/10.46306/Rj.V4i1.101

Djuniarti, E. (2017). Hukum Harta Bersama Ditinjau Dari Perspektif Undang-Undang Perkawinan Dan Kuh Perdata. *Jurnal Penelitian Hukum De Jure*, 17(4), 445. Https://Doi.Org/10.30641/Dejure.2017.V17.445-461

Hapsari, F. T. (2012). Eksistensi Hak Moral Dalam Hak Cipta Di Indonesia. *Masalah-Masalah Hukum*, 41(3), Article 3. Https://Doi.Org/10.14710/Mmh.41.3.2012.460-464

Hertasmaldi, H., and Hafizh, A. (2021). Penolakan Pembagian Harta Bersama Alasan Nafkah Anak Menurut Kompilasi Hukum Islam. *Sakena: Jurnal Hukum Keluarga*, 6(1), Article 1.

Nasution, H. (2018). Implementation Of The Principle Of Ultra Petitum Partium In Deciding Children Livelihood In Divorce Lawsuit In Religious Courts. Diambil 1 Januari 2025, Dari Https://Journal.Uinjkt.Ac.Id/Index.Php/Ahkam/Article/View/7488/5249

Islam, H. (T.T.). Pascasarjana Universitas Islam Negeri Sumatera Utara.

Kharlie, A. T., Fathudin, F., and Triana, W. (2021). Reforming Islamic Marriage Bureaucracy In Indonesia: Approaches And Impacts. *Al-Jami'ah: Journal Of Islamic Studies*, 59(2), 255–286. Https://Doi.Org/10.14421/Ajis.2021.592.255-286

Menyoal RoyaltiLagu Sebagai Harta Gono-Gini Perceraian Virgoun Tambunan Dan Inara Rusli. (T.T.). Voi - Waktunya Merevolusi Pemberitaan. Diambil 15 Januari 2024, Dari Https://Voi.Id/Bernas/329716/Menyoal-Royalti-Lagu-Sebagai-Harta-Gono-Gini-Perceraian-Virgoun-Tambunan-Dan-Inara-Rusli

Mesraini. (2012). *Konsep Harta Bersama Dan Implementasinya Di Pengadilan Agama*. Diambil 1 Januari 2025, Dari Https://Journal.Uinjkt.Ac.Id/Index.Php/Ahkam/Article/View/980

Poetri, T. R. (2020). Penyelesaian Pembagian Hak Cipta Dan Hak Atas Merek Sebagai Harta Bersama Dalam Perceraian Islam. *Lex Renaissance*, 5(2), Article 2. Https://Doi.Org/10.20885/Jlr.Vol5.Iss2.Art6

Rochaeti, E. (2013). Analisis Yuridis Tentang Harta Bersama (Gono Gini) Dalam Perkawinan Menurut Pandangan Hukum Islam Dan Hukum Positif. *Jurnal Wawasan Yuridika*, 28(1), Article 1. Https://Doi.Org/10.25072/Jwy.V28i1.61

Simatupang, K. M. (2021). Tinjauan Yuridis Perlindungan Hak Cipta Dalam Ranah Digital. *Jurnal Ilmiah Kebijakan Hukum*, 15(1), Article 1. Https://Doi.Org/10.30641/Kebijakan.2021.V15.67-80

Sugiswati, B. (2014). Konsepsi Harta Bersama Dari Perspektif Hukum Islam, Kitab Undang-Undang Hukum Perdata Dan Hukum Adat. *Perspektif*, 19(3), Article 3. https://doi.org/10.30742/perspektif.v19i3.22

Towards Resilient Societies: The Synergy of Religion, Education, Health, Science, and Technology – Rahiem (Ed.)

Legal uncertainty in cross-border data transfers in Indonesia: A call for reform

Indra Rahmatullah
Universitas Islam Negeri Syarif Hidayatullah Jakarta, Indonesia
ORCID ID: 0000-0002-6160-4225

Pujiyono Suwadi
Universitas Sebelas Maret, Indonesia
ORCID ID: 0000-0002-5971-2446

Hari Purwadi
Universitas Sebelas Maret, Indonesia
ORCID ID: 0000-0002-2226-624

Rosee Laehseng
The Patani Center for Organizational and Public Outreach, Thailand

ABSTRACT: This study examines the legal framework governing cross-border transfers of personal data in Indonesia, focusing on the lack of legal certainty and protection for data subjects. The current regulations are weak and liberal, lacking restrictions on the types of data that may be transferred abroad, clear parameters for data transfer requirements, and a dedicated state institution responsible for personal data protection. These shortcomings complicate law enforcement, particularly in personal data breaches involving Indonesian citizens abroad. This article proposes reconstructing the regulatory framework for cross-border data transfers by implementing data localisation requirements, limiting the transfer of sensitive data, and mandating international agreements between countries. Such restructuring aims to protect individuals' right to privacy, recognised as a fundamental human right, from public and private sectors' surveillance while facilitating law enforcement. Strengthening these conditions is essential for enhancing digital rights and ensuring robust personal data protection in Indonesia's rapidly evolving digital economy.

Keywords: cross-border, digital economic, Indonesia, and personal data

1 INTRODUCTION

In the contemporary era, the world has evolved through digital globalisation, which can be conceptualised as the transfer of information and data flows that have more excellent value than the trade in goods (Qadri and Bhat 2018). Data transfer flow is an essential commodity in free trade, distinct from trade in goods and services (Yakovleva 2020). Data has emerged as a unique form of economic asset. Consequently, the necessity arises for data collection, analysis, and use, as well as cross-border data transfer (Aaltonen *et al.* 2021; United Nations Conference on Trade and Development 2021b). Data is considered an engine or fuel that drives economic innovation and creativity (Casarosa 2020). It can be sold or licensed, used in product development or licensing of new products, and employed to increase the quantity and efficiency of production (Giakoumopoulos *et al.* 2018).

DOI: 10.1201/9781003645542-92

The cross-border transfer of personal data has become a standard practice in the contemporary global economy, particularly among major corporations that have amassed significant data assets. The contestation of cross-border personal data flow in the international context cannot be separated from the dominance of the United States of America and the People's Republic of China. These two countries have a significant capacity for central data from approximately half of the world's population and nearly 90% of market capitalisation on digital platforms worldwide (United Nations Conference on Trade and Development 2021a). Both countries make the legitimacy of global digital companies with a wide range of financial, market, and technological power. These forces give large digital platform companies a dominant position due to their size, profitability, and market value, which is very high, so they control the data users (United Nations Conference on Trade and Development 2021a). In addition to the commitments made by ASEAN countries, the 2016 ASEAN Framework on Personal Data Protection and the subsequent ASEAN Model Contractual Clauses as a Legal Basis for Data Transfers have been established. However, it should be noted that these provisions are not legally binding due to their nature as recommendations rather than mandatory directives.

The free flow of data across borders becomes a significant challenge. The concept of state sovereignty is often confronted with the need to establish clear boundaries between data localisation and the traditional concept of state territoriality (Kong 2010; Mitchellt and Hepburn 2018; United Nations Conference on Trade and Development 2021a). It has led to a situation where countries are free to protect the personal data of their citizens when faced with the need to transfer personal data outside their country (Asian Business Law Institute 2020). In light of this phenomenon, four approaches to regulating the transfer of personal data by countries worldwide have emerged. The first approach is the absence of regulation, such as in Chad, Mali, Syria, Sudan, Brunei Darussalam, and Timor-Leste. Second, the country does not prohibit data transfer and does not require the implementation of specific provisions—third, conditional flows on safeguards such as Indonesia, Malaysia, Singapore, and Australia. Fourth, government authorities (Khumon 2018, 2020; OECD 2020) such as India, China, and Vietnam permit only specific data for transfer. Australia, Brazil, Canada, and Denmark.

The four models illustrate that the international community has placed considerable emphasis on establishing a comprehensive set of international regulations governing cross-border personal data transfer (Table 1). These regulations encompass principles, conditions, and procedures and include stringent provisions (Mattoo and Meltzer 2019). However, when Indonesia is compared to some countries in ASEAN, such as the Philippines, Malaysia, and Singapore, which have previously had Personal Data Protection Laws, it can be observed that the provisions for transferring cross-border personal data in Indonesia are more permissive or liberal because there are no restrictions in Indonesia on the types of data that may not be transferred cross-border. Following Article 56 of Law No. 27 of 2022 on Personal Data Protection, all personal data may be transferred across borders. This stipulation is more permissive than that of the Philippines (Khumon 2020) , Singapore (Khumon 2020) and Malaysia (Khumon 2018; Personal Data Protection Act 2010; Santhi Kandiyah 2019) , prohibiting the transfer of sensitive personal data across borders and requiring additional conditions.

Table 1. Comparison on personal data transfer some countries in ASEAN.

Philippines	Prohibiting sensitive personal data from being transferred abroad.
Singapore	Singapore adds obligations of contract clauses, personal data transfer agreements, and corporate binding rules.
Malaysia	Adds the conditions of the data transfer agreement, due diligence to identify and ensure all actions that can result in personal data breaches, and risk mitigation.
Indonesia	there are no restrictions in Indonesia on the types of data that may not be transferred cross-border.

Secondly, the issue of data localisation (the storage, copying, or backup of data and data centres) requires further examination. Due to the enactment of Law No. 27 of 2022 concerning the Protection of Personal Data, there is currently no obligation for data localisation in the country. The law allows data controllers, including both public and private entities, the autonomy to transfer personal data across national borders without being compelled to store it domestically. Therefore, this situation presents a significant risk to national security and law enforcement efforts to address personal data violations, which could erode the sovereignty of the Unitary State of the Republic of Indonesia (NKRI). Notwithstanding a specific law regulating the cross-border transfer of personal data in Indonesia, the rights of personal data subjects remain unprotected by legislation, rendering them vulnerable to personal data breaches in Indonesia. It contradicts the 1945 Constitution, which vests the state with the authority to protect human rights, particularly in guaranteeing the security of citizens' privacy.

2 METHODS

The research uses normative legal research methods or doctrinal legal research with statute, theoretical, and comparative approaches (Irwansyah and Ahsan Yunus 2021). The research uses secondary data that consists of primary and secondary legal material. Primary legal material used in the research is the Universal Declaration of Human Rights, the International Covenant on Civil and Political Rights), the Convention for the Protection if Individuals concerning the automatic processing of personal data (Convention 108), Guidelines on the Protection of Privacy and Transborder Data Flows of Personal Data by OECD 1980 amended in 2013 and Law Number 27 of 2022 concerning the Protection of Personal Data. Meanwhile, the secondary legal materials consist of several documents such as scientific journals, books, seminar papers, trusted internet sites, and other related documents.

3 FINDING & DISCUSSION

3.1 *Review of cross-border personal data transfer in Indonesia*

Legal provisions governing the transfer of personal data in Indonesia need more certainty and protection for data subjects. The current regulations are flexible and permissive, potentially placing national interests at risk. The following matters are considered: Firstly, there is no obligation to localise and limit the data types that can be transferred across borders. Before Law No. 27 of 2022 was enacted concerning the Protection of Personal Data, Government Regulation 82 of 2012 concerning the Implementation of Electronic Systems and Transactions was in force. Article 17 stipulates that operators of electronic systems for public services are obliged to establish data centres and data recovery centres within Indonesian territory for law enforcement, data protection, and the maintenance of state sovereignty over the data of its citizens. This regulation was reinforced again by the Minister of Communication and Informatics Regulation No. 20 of 2016 concerning the Protection of Personal Data in Electronic Systems. It should be noted, however, that these regulations are not intended to regulate the cross-border transfer of personal data.

The following regulation is Government Regulation No. 71 of 2019 concerning implementing electronic systems and transactions. Article 99 stipulates that strategic electronic data backup must be recorded for the government administration sector, energy and mineral resources, transportation, finance, health, information and communication technology, food, defence, and other sectors as the President determines. It should be noted that the provisions outlined in this Article are not intended for the cross-border transfer of personal data. The legal uncertainty persists despite Law No. 27 of 2022 concerning Personal Data Protection in Indonesia. Article 56, which addresses the transfer of personal data outside Indonesia, fails to stipulate the obligation to localise data and the types of personal data that may be transferred. Consequently, it fails to provide legal protection for the digital rights of Indonesian citizens.

Since the inception of the legislative debate surrounding this legislation in the DPR, the issue of data localisation and the types of personal data that can be transferred across borders has not yet been subjected to rigorous examination by legislators. This is evidenced by the 2020 Personal Data Protection Draft Academic Paper, which the Government produced (Nurbaningsih 2015). Similarly, the Problem Inventory List (DIM) represents the primary discussion material in the Personal Data Protection Bill discussion meeting. Notably, 10 political parties, including Demokrasi Indonesia Perjuangan Party (PDIP), Golongan Karya Party (Golkar), Gerakan Indonesia Raya Party (Gerindra), Nasional Democrat Party (Nasdem), Kebangkitan Bangsa Party (PKB), Democratic Party (PD), Keadilan Sejahtera Party (PKS), Amanat Nasional Party (PAN) and Persatuan Pembangunan Party (PPP), did not provide any commentary on the cross-border transfer of personal data (RI 2020).

To ensure citizens' digital rights are protected, it is necessary to undertake a legal reconstruction of the provisions on the transfer of personal data outside Indonesia. This reconstruction will be based on the following principles: First. It is strengthening the commitment to human rights. The right to privacy is enshrined in Article 12 of the Universal Declaration of Human Rights (UDHR), Article 17 of the International Covenant on Civil and Political Rights (ICCPR), which has been incorporated into Indonesian legislation as Law Number 12 of 2005 concerning the International Covenant on Civil and Political Rights (ICCPR) and Article 28G (1) the 1945 Indonesian Constitution.

Personal data is privacy, and the state must prepare legal instruments that protect privacy rights when personal data is transferred abroad. It requires data localisation in Indonesia and a list of sensitive personal data that cannot be transferred abroad, fulfilling the data protection principle. 1) Fair, lawful, and transparent; 2) Purposes limitation; 3) Minimisation; 4) Accurate; 5) Storage limitation; 6) Integrity and confidentiality; and 7) Accountability. Indonesia must also update its cross-border data transfer rules to protect people's rights. These include a. the right to be informed, b. the right to restrict processing, and c. the right to data portability (Convention for the Protection of Individuals with Regard to Automatic Processing of Personal Data 1981; OECD Priv. Framew. 2013; Giakoumopoulos *et al.* 2018; International, n.d.; Riza Roidila Mufti 2021).

Second, protection from surveillance threats. The reconstruction's objective is to safeguard Indonesian citizens' data from the potential risks associated with surveillance by external parties, including governmental entities, in collaboration with foreign entities. There have been numerous surveillance threats against citizens' data, including the data theft incident perpetrated by Edward Snowden. At some point, Snowden accessed nearly 35,000 computer networks and the personal data of more than 35 world leaders.

Another notable example is the collusion between Australian secret agents working with the United States, Britain, Canada, and New Zealand to obtain the personal data of Australian citizens (Chander and Lê 2015). Historically, the government misused citizens' data around the 1990s, resulting in espionage and surveillance. Bureaucratic problems are evident due to the state intervention required to secure citizens' data. Paradoxically, there is a public distrust of the bureaucracy that uses citizens' data, consequently becoming an issue of politics and power. This phenomenon has historical precedents in the Nazi and Fascist events of the 1930s and 1940s, which were the main reason for the emergence of personal data protection (Bennett 1991).

Third, to simplify law enforcement. Online media can provide avenues for illicit access to personal data, which is susceptible to being divulged into the public domain and exploited by certain parties for illegitimate purposes (Maple 2017; Weber 2015b, 2015a). Personal data collected, stored, used, and transferred digitally can potentially be a personal data breach if used without legal provisions (Waldo *et al.* 2007). Therefore, the reconstruction of the transfer of personal data and the data localisation in Indonesia aims to simplify the process for law enforcement to secure evidence in case of a personal data breach, as data copies are available domestically (John Selby 2017; Siroli 2018; Wong 2020).

The view is because personal data breaches can occur in various ways: a. Social media platforms. Through this platform, data collected by certain accounts is shared with other parties illegally so that it is sold, intercepted, and accessed; b. Geospatial technology. This technology is vulnerable to illegal access to someone's location, which can disturb privacy, c. Biometrics. It works by collecting a person's physical and behavioural data. This technology is vulnerable to cyber crimes such as data theft due to weak data owner control, d. Web tracking: This search engine is susceptible because it can trace someone's track record when accessing the internet. The Internet of Things (IoT) is a system that connects many devices to facilitate digital transactions. This system is also vulnerable because it collects people's sensitive data, f. Big data that profiles a particular subject with the help of algorithms. Big data is at risk of illegal access and data leakage (Quach *et al.* 2022).

The open and accessible nature of the internet makes it susceptible to law violations. It creates an environment conducive to cybercrimes such as cyberattacks, potentially disrupting the established order of life for humans, companies, and governments alike (Kremling and Parker 2018). A data breach is a cybercrime that discloses confidential data or information distributed to the public. It is conducted by hackers, which causes damage and loss. Sabillon defines a data breach as a disclosure of data or information that breaks confidentiality, leading to the distribution in the public domain. The leakage can occur by insider agents or hacker attacks. The resulting damages can affect or trigger corporate reputation, financial losses, lawsuits, share prices, fraud, and physical assets (Sabillon *et al.* 2016).

In light of the above, it is imperative that the provisions for cross-border transfer of personal data, as set out in Article 56 of Law No. 27 of 2022 concerning the Protection of Personal Data, be amended as follows: 1) Personal Data Controllers or electronic system operators, both in public and private institutions, are required to localise data, especially strategic electronic data, within the territory of the Unitary State of the Republic of Indonesia. 2) Personal data of a sensitive nature can only be transferred across borders if there is a bilateral agreement between the countries in question. 3) The provisions of international agreements between countries must be considered in the article on cross-border transfer of personal data.

4 CONCLUSION

The legal issues surrounding transferring certain types of data across borders and localising data are a concern for countries worldwide. There is no consensus on this, as each country has a different stance. Some view data localisation and data type classification as potential barriers to international trade, while others do not. In the meantime, certain countries have taken the initiative to safeguard their national interests by establishing specific terms and conditions. Indonesia has aligned itself with the globalisation trend by establishing terms and conditions for the cross-border transfer of personal data that prioritise legality, fairness, and transparency.

The three principles are implemented by reconstructing Law No. 27 of 2022 concerning the Protection of Personal Data: first, adding conditions of obligation for Personal Data Controllers or electronic system operators in public and private institutions. Second, data localisation, especially strategic electronic data, is required within the territory of the Unitary State of the Republic of Indonesia. Third, limits on the transfer of personal data cross-border.

The proposed amendments aim to reinforce individuals' privacy rights, ensuring that any party does not misuse their data and ensure a. Enhance human rights commitments, b. protect against domestic and foreign surveillance threats, and facilitate law enforcement in the event of a personal data breach. Therefore, Law No. 27 of 2022 concerning the Protection of Personal Data must be amended to guarantee the digital rights of the Indonesian people.

ACKNOWLEDGEMENTS

The authors would like to thank all those who have helped, especially the Indonesian Education Scholarship (BPI) of the Ministry of Education, Culture, Research and Technology of the Republic of Indonesia.

REFERENCES

Aaltonen, A., Alaimo, C., and Kallinikos, J. (2021). The Making of Data Commodities: Data Analytics as an Embedded Process. *Journal of Management Information Systems*, 38(2), 406–407. https://doi.org/10.1080/07421222.2021.1912928

Asian Business Law Institute. (2020). Transferring Personal Data in Asia: a Path to Legal Certainty and Regional Convergence (Issue May 2020).

Bennett, C. J. (1991). Computers, Personal Data, and Theories of Technology: Comparative Approaches to Privacy Protection in the 1990s. *Science, Technology and Human Values*, 16(1), 51–69. https://doi.org/10.1177/016224399101600103

Casarosa, F. (2020). Transnational collective actions for cross-border data protection violations. *Internet Policy Review*, 9(3), 2. https://doi.org/10.14763/2020.3.1498

Chander, A., and Lê, U. P. (2015). Data Nationalism. *Emory Law Journal*, 64(3), 715–716.

Convention for the Protection of Individuals with Regard to Automatic Processing of Personal Data, Pub. L. No. 108 (1981).

Giakoumopoulos, C., Buttarelli, G., and O'Flaherty, M. (2018). Handbook on European data protection law 2018 edition (2018th ed.). European Union Agency for Fundamental (FRA), the EU Agency for Fundamental Rights Europe, the Council of Supervisor, the Registry of the European Court of Protection, Human Rights) and the European Data. https://doi.org/10.2811/58814

International, P. (n.d.). Rights of Data Subjects Rights of Data Subjects. In A Guide for Policy Engagement on Data Protection (p. 2018). Privacy International.

Irwansyah, and Ahsan Yunus. (2021). Penelitian Hukum Pilihan Metode dan Praktik Penulisan Artikel (Revisi). Mirra Buana Media.

John Selby. (2017). Data Localization Laws: Trade Barriers or Legitimate Responses to Cybersecurity Risks, or Both? *International Journal of Law and Information Technology*, 25(3), 213–232.

Khumon, P. (2018). Regulation for Cross-Border Privacy in Southeast Asia: An Institutional Perspective. 29th European Regional Conference of the International Telecommunications Society (ITS): "Towards a Digital Future: Turning Technology into Markets?" http://www2.itif.org/2015-crossborder-data-flows.pdf

Khumon, P. (2020). Cross-Border Data Privacy Regulation in ASEAN: Overcoming Institutional Challenges (Issue June).

Kong, L. (2010). Data protection and transborder data flow in the European and global context. *European Journal of International Law*, 21(2), 441–456. https://doi.org/10.1093/ejil/chq025

Kremling, J., and Parker, A. M. S. (2018). *Cyberspace, Cybersecurity, Cybercrime*. SAGE Publications, Inc. All. https://www.ptonline.com/articles/how-to-get-better-mfi-results

Maple, C. (2017). Security and Privacy in The Internet of Things. *Journal of Cyber Policy* ISSN:, 2(2), 165. https://doi.org/10.1080/23738871.2017.1366536

Mattoo, A., and Meltzer, J. P. (2019). International data flows and privacy: The conflict and its resolution. *Journal of International Economic Law*, 21(4), 769–789. https://doi.org/10.1093/jiel/jgy044

Mitchellt, A. D., and Hepburn, J. (2018). Don't Fence Me In: Reforming Trade and Investment Law to Better Facilitate Cross-Border Data Transfer. The Yale Journal of Law and Technology, 19(128 March), 1–56.

Nurbaningsih, E. (2015). Naskah Akademik RUU Perlindungan Data Pribadi. In Naskah Akademik Rancangan Undang-Undang tentang Perlindungan Data Pribadi.

OECD. (2020). Mapping Approaches to Data and Data Flows: Report for the G20 Digital Economy Task Force. *In OECD*. http://www.oecd.org/termsandconditions.

Personal Data Protection Act 2010, Pub. L. No. 709, 31. (2010). https://doi.org/10.1007/978-0-387-68254-9_5
Qadri, B., and Bhat, M. A. (2018). Interface between Globalization and Technology. *Asian Journal of Managerial Science*, 7(3), 1–6. https://doi.org/10.51983/ajms-2018.7.3.1351
Quach, S., Thaichon, P., Martin, K. D., Weaven, S., and Palmatier, R. W. (2022). Digital Technologies: Tensions in Privacy and Data. *Journal of the Academy of Marketing Science*, 50(6), 1302–1305. https://doi.org/10.1007/s11747-022-00845-y
Riza Roidila Mufti. (2021). A Policy Brief EU General Data Protection Regulation (GDPR) (D. Yusilawati, M. M. Arkan, N. Hindratmo, and M. Chandra(Eds.); 2021st ed.). *Kedutaan Besar Republik Indonesia di Brussel.* https://kemlu.go.id/download/L1NoYXJlZCUyMERvY3VtZW50cy9icnVzc2VsL3Jlc2VhcmNoJTIwc2VyaWVzL0dEUFIlMjAtJTIwdXBkYXRlZC5wZGY=
Sabillon, R., Cano, J., Cavaller, V., and Serra, J. (2016). Cybercrime and Cybercriminals: A Comprehensive Study. *International Journal of Computer Networks and Communications Security*, 4(6), 172–174.
Santhi Kandiyah. (2019). Malaysia. In A. C. Raul (Ed.), *Privacy, Data Protection and Cybersecurity* (6th ed.). Tom Barnes Law Business Research. www.TheLawReviews.co.uk
Siroli, G. P. (2018). Considerations on the Cyber Domain as the New Worldwide Battlefield. *International Spectator*, 53(2), 111–123. https://doi.org/10.1080/03932729.2018.1453583
The OECD Privacy Framework, The OECD Privacy Framework 1 (2013).
United Nations Conference on Trade and Development. (2021a). Digital Economy Report 2021 Cross-border data flows and development: For whom the data flow. *In the United Nations Conference on Trade and Development (Issue September).*
United Nations Conference on Trade and Development. (2021b). TECHNOLOGY AND INNOVATION REPORT 2021 Catching technological waves Innovation with equity. https://unctad.org/system/files/official-document/tir2020_en.pdf
Waldo, J., Lin, H. S., and Millett, L. I. (Eds.). (2007). *Engaging Privacy and Information Technology in a Digital Age*. The National Academies Press. https://doi.org/10.17226/11896
Weber, R. H. (2015a). Internet of Things: Privacy Issues Revisited. *Computer Law and Security Review: The International Journal of Technology Law and Practice*, 31(5), 619. https://doi.org/10.1016/j.clsr.2015.07.002
Weber, R. H. (2015b). The Digital Future - a Challenge for Privacy? *Computer Law and Security Review*, 31 (2), 236–238. https://doi.org/10.1016/j.clsr.2015.01.003
Wong, B. (2020). Data Localization and ASEAN Economic Community. *Asian Journal of International Law*, 10(1), 158–180. https://doi.org/10.1017/S2044251319000250
Yakovleva, S. (2020). Personal data transfers in international trade and EU law: A tale of two "necessities." *Journal of World Investment and Trade*, 21(6), 881–919. https://doi.org/10.1163/22119000-12340189

Towards Resilient Societies: The Synergy of Religion, Education, Health, Science, and Technology – Rahiem (Ed.)

Legal challenges in watershed management addressing criminal aspects in local governance

Nahrowi Nahrowi
Universitas Islam Negeri Syarif Hidayatullah Jakarta, Indonesia
ORCID ID: 0000-0001-5771-3667

Masyrofah Masyrofah
Universitas Islam Negeri Syarif Hidayatullah Jakarta, Indonesia
ORCID ID: 0000-0001-9657-0443

ABSTRACT: This article examines law enforcement concerning the misuse of watershed land in the Situ Cipondoh area of Tangerang, highlighting that such misuse involving immovable property can lead to criminal charges against the perpetrators. This study utilises a literature review approach, analysing the misuse of watershed areas in Situ Cipondoh based on legal provisions, including Article 385 of the Criminal Code, Article 167 of the Criminal Code, Article 6 paragraph (1) of Government Regulation in Lieu of Law No. 51 of 1960 on the Prohibition of Land Use Without Permission from the Rightful Owner or Their Proxy, and the Law on Environmental Protection and Management. Data were also gathered through interviews with local residents. The findings reveal those legal violations in civil land transactions that harm others—particularly, in this case, damage to local government interests or breaches of legal regulations—can escalate into criminal offences. This study further demonstrates how civil disputes involving land use can give rise to criminal consequences when they infringe upon statutory laws.

Keywords: agrarian law, Situ Cipondoh Cidurian-Cisadane watershed, local governance, criminal offences

1 INTRODUCTION

The misuse of watershed land by individuals misusing resources in the Situ Cipondoh Tangerang area became a public spotlight because unscrupulous residents, without the permission of the local government, leased the land to residents who set up business places such as coffee shops and clothing stores. At the same time, rent was taken by individuals who misused resources for personal gain. Land leasing by residents for business activities occurs without legal documents, so utilisation becomes illegal. Therefore, the misuse of watershed land includes elements of the alleged criminal offence of misappropriation of rights to immovable property and/or entering a yard without permission or prohibition of use of land without rights as referred to in Article 385 of the Criminal Code or Article 167 of the Criminal Code or Article 6 paragraph (1) of Government Regulation in Lieu of Law No. 51 of 1960 concerning the Prohibition of Use of Land without the Permission of the Eligible Person or his Proxy.

This study analysed the criminal aspects of watershed management conducted by the community. However, individuals misusing resources actually abuse the function of the area and turn it into a business area by taking rent for the land used as a business place. This case can become a criminal case if there is an element of illegal use of land owned by the local government by collecting several payments from the people who rent the land, so the question arises whether it is permissible for someone to occupy or manage a plot of land owned by the local government under the UPTD Management of the Cidurian-Cisadane River Basin, Banten Province Public Works, and Spatial Planning Office.

 DOI: 10.1201/9781003645542-93

Sanctioning the perpetrators of Situ Cipondoh watershed land abuse is an effort for law enforcement to maintain watershed land in line with Sustainable Development Goals (SDGs), particularly SDG 15 (Life on Land). By 2030, SDG 15 target is to combat desertification, restore degraded land and soil, including land affected by desertification, drought, and flooding, and strive to achieve a world that is neutral to land degradation. The diversion of land functions impacts watersheds that are not functioning correctly, resulting in floods and landslides.

2 METHODS

This paper is a literature study that explains the case of misuse of watershed areas in Situ Cipondoh Tangerang with the data source obtained through Article 385 of the Criminal Code and/or Article 167 of the Criminal Code and/or Article 6 paragraph (1) of the Government Regulation in Lieu of Law No. 51 of 1960 on Prohibition of Land Use without the permission of the Eligible Person or his Proxy, Law on Environmental Protection and Management) and interviews with the local people.

3 FINDINGS & DISCUSSION

3.1 *Situ Cipondoh Cidurian-Cisadane Watershed Tangerang*

Situ Cipondoh is one of the sites located in Tangerang City, Banten Province, and is part of a watershed that is used by the community and needs to be adequately managed. One of the existing problems is the use of the situ boundary, which is not following its designation. The boundary is used for commercial functions (traders, restaurants, etc.) that interfere with the function and use of the boundary as a means of green open space and irrigation/reservoir. These conditions illustrate inappropriate conditions regarding the arrangement of the site boundaries. It is necessary to supervise and control the utilisation of the Cipondoh in-situ border through a series of processes that are carried out in an integrated and collaborative manner to utilise the spatial functions in the area in Situ Cipondoh; steps and efforts are made to restore the function of land use through various instruments of spatial control aimed at restoring the function of land physically, economically, socially, and culturally following its designation (Tranggono 2021) (Tranggono 2021).

In comparison, watershed changes have also occurred in Brazilian Uberada, where the invasion of grasslands by crop agriculture has resulted in a drastic decrease in soil organic matter and potassium content, threatening soil fertility and triggering erosion (Valera 2016) (Valera 2016). Similarly, in China's Shiyang River Watershed, expanding agricultural land and building areas has led to shrinking lake areas and degrading grassland despite several land rehabilitation efforts (Ngabire 2022) (Ngabire 2022). In the Qinghai Lake watershed, converting land into agricultural and residential areas has contributed to declining lake levels and grassland degradation (Li 2008) (Li 2008) (Wang 2019). The limited availability of hydrological data records often constrains efforts to build watershed management models from upstream to downstream. In Asia, including Indonesia, many watersheds still do not have data recording tools to help the government as decision-makers in watershed management. The area around Lake Singkarak, West Sumatra, is a catchment area that has been categorised as critical because of land use change in the upstream area from forest to dry land agriculture. This development affects water availability for downstream use, particularly for wetland agriculture and fisheries (Febriamansyah 2021) (Febriamansyah 2021).

Likewise, the misuse of watershed utilisation in Situ Cipondoh Tangerang will undoubtedly impact the diversion of the actual watershed function. When the Cidurian-Cisadane Watershed Management UPTD of the Banten Provincial Public Works and Spatial Planning Office, which is the manager of the sites in Tangerang City, checked the management of Situ Cipondoh and other sites to inventory the assets owned by the Banten Provincial Government (Pemprov) for the Banten Provincial Government's Original Revenue (PAD), however, it is known that currently several traders are still trading and erecting buildings, such as cafes, stalls, kiosks in Situ Cipondoh without permission. If there are no documents or joint use rights with the Banten Provincial Public Works and Spatial Planning Office, a manager named Br is currently known.

MMH asked for rent to traders who were in the location without the knowledge of the Banten Provincial Public Works and Spatial Planning Office; for this incident, the Banten Provincial Public Works and Spatial Planning Office finally made a complaint report to the Tangerang City Metro Police Resort (Polres) for further investigation.

3.2 *Law enforcement against watershed abuse*

Criminal law enforcement involves taking action against perpetrators of serious violations with sanctions following applicable legal provisions. Equally important is educating the community by raising awareness of the importance of protecting watersheds through socialisation and empowerment programs. In addition, residents were asked to restore the environment by implementing rehabilitation and conservation programs to restore the function of the affected watershed ecosystem, and residents were asked to implement rehabilitation and conservation programs to restore watershed ecosystem functions. Finally, there is the issue of sustainable supervision, which means awareness of future violations.

In applying the principle of *strict liability* in the Law on Environmental Protection and Management (UUPPLH) for violations of environmental crimes by corporations, not by citizens or communities, criminal liability based on this principle can be imposed. UUPPLH only regulates *strict liability* as civil liability and does not provide provisions on whether this principle can be applied to criminal liability. However, *strict liability* is recognised as a fundamental principle of criminal liability. In Indonesia, criminal liability is still based on the principle of fault, meaning there is no punishment without a fault. Although the principle of *strict liability* has been recognised in civil liability, its use in criminal law is still limited and rarely used. In environmental criminal law enforcement, several court decisions show minimal use of the principle of *strict liability*. Therefore, it is necessary to analyse further the application of the principle of *strict liability* in criminal liability for corporations in environmental crimes (M. Fahri Ikhsan 2023) (M. Fahri Ikhsan 2023).

A study conducted by Bochet (Bochet 1983) (Bochet 1983) investigated the environmental conditions in watersheds across multiple countries, including Ethiopia, (Teshome 2018) (Teshome 2018) Indonesia, Korea, India, (Wadhawan 2020) (Wadhawan 2020) Iran, Morocco, Nepal, Pakistan, Philippines, Thailand, (Waiyasusri 2023) (Waiyasusri 2023) and Tunisia. Bochet and fellow researchers determined that excessive human and animal population pressures had significantly damaged the ecosystem. Poor resource management reduced soil productivity, increased soil and water erosion, and sedimentation in riverbeds and dams. Multiple factors, such as shifting cultivation, overgrazing, forest fires, mining activities, deforestation, and infrastructure construction, contributed to this situation. In certain nations, environmental degradation has almost reached a point of no return (Bochet 1983) (Bochet 1983).

Some watersheds in Indonesia are damaged due to land use change, population growth, and lack of public awareness of watershed environmental conservation. Symptoms of watershed ecological damage can be seen from the shrinkage of forest area and land damage, especially in protected areas around the watershed. (Badung 2018) (Badung 2018) The Citarum and Barito watersheds have Indonesia's highest hydrometeorological disaster hazard risk index scores, with 0.74 and 0.62, respectively. The extensive and massive growth of built-up land in both watersheds does not all follow the Regional Spatial Plan (RTRW) document. According to Kompas analysis, 25,132 hectares or 22.3 percent of built-up land in the Citarum watershed does not follow its designation. Meanwhile, in the Barito watershed, 9,916 hectares of built-up land, or 21.9 percent, are not following the designation (Lifidyah 2023) (Lifidyah 2023).

3.3 *Elements in the alleged offence of embezzlement of rights on immovable property*

There are several articles related to the misuse of land that does not belong to the person; the Explanation of Article 385 of the Criminal Code is that the crimes mentioned in this article are commonly called "embezzlement of rights to immovable property," such as land, rice fields, buildings, and others. To be prosecuted under this article, the defendant must have committed the following actions: intent to benefit himself or others unlawfully (illegally); has sold, exchanged, or made subject to a credit bond the title to state-owned land or to land belonging to a particle or buildings, structures, crops, or seeds on land with a title; knows that another person has the right or shares the right, has not informed the other party that there is a loan agreement in place, has not informed the other party that the land has been pledged. Pledged or leased the land of another person. Has sold or exchanged land being pledged to another person without giving notice to the interested party. He has leased out land for a while, knowing it has been leased out to another person.

Some land in Indonesia belongs to the nation, state, indigenous peoples, and individuals (private). Still, all land belongs to the state; the people who own the land are only users, and the rights available to them are only rights of use. Even if in a loan agreement (credit), where someone pledges his land (Hak Tanggungan), what is meant by "credit bond" here is a loan of money from the Bank to the resident who owns such land, with the land as collateral, (Sugandhi 1981) (Sugandhi 1981) does not indicate that the land belongs to the individual in the philosophical sense of land rights.

The elucidation of Article 167 of the Criminal Code is that the acts punishable by this article are unlawfully entering a house, enclosed room, etc., by force, unlawfully remaining in a house, enclosed space, etc., and not immediately leaving the place at the request of a person entitled to the house or room. Such entry alone does not constitute "forcible entry"; "forcible entry" means entry in a manner contrary to the expressed will of the person entitled to it, for example, by word, by deed, by writing "entry prohibited" or other signs of similar meaning which people in the surrounding area understand. The gate or door of a house closed without a key does not mean that others cannot enter. If the gate or door of the house is locked with a locking device or has a sign saying, "entry is prohibited," it fulfils the juridical formality that others are not allowed to enter the place. A bill collector, vegetable seller, or beggar who enters a room or house where the door is not locked or where there is no sign saying "no entry" is not yet a "forced entry" and cannot be punished. However, if the rightful person evicts them, they must leave the premises immediately. They are liable to punishment if the demand is repeated three times in succession and is not heeded. A person who wants to enter another person's house and the owner of the house forbids him from doing so or prevents him from doing so but insists on entering shall be deemed to have entered by force and shall be liable to punishment. What shall also be considered as "forcible entry", according to the second paragraph, is a person who enters using breaking, climbing, false keys, a false order, a false costume, or a person who enters the premises and a person who is in the premises during the night.

Concerning breaking in, climbing in, using a false key, false order, or false costume, see the explanation in Article 363. A person who breaks into a house or an enclosed space during the day and is found there at night falls under this prohibition. However, a person who breaks into a house or enclosed space during the night and is found there the next day does not fall under the prohibition of this section. Thus, the person who can be prosecuted under this article and paragraph is the person who is in place during the night. A house is a boat or other vehicle where people live; in short, all dwellings are used for living (Sugandhi 1981) (Sugandhi 1981).

The Explanation of Article 6, (Sugandhi 1981) (Sugandhi 1981) paragraph (1) of Government Regulation in Lieu of Law No. 51 of 1960 on the Prohibition of the Use of Land without the Permission of the Rightful Owner or His Proxy states that the elements are: (1) whoever; (2) using land; and (3) without the permission of the rightful owner or their authorised representative. What is meant by " in the law is (a) land controlled by the state. (b) Land not included in (a) is held under the right by an individual or legal entity. What is meant by "using land" in this law is to occupy, work on, and control a piece of land or to have a building planted on it, regardless of whether the building is used alone. The utilisation of state land is a series of actions carried out by a person or group of people in terms of land use, for example, in river borders or lake borders without permission from the proper or legal power of attorney or contrary to applicable laws, which can cause damage to the place. For example, along the Krueng Aceh River in Gampong Lamreung Meunasah Baktrieng, river borders are used by residents without permission. The criminal aspect of unauthorised state land use becomes the benchmark for this problem (Handayani 2019).

3.4 *Criminal aspects of residents' occupation and management of watershed land owned by the local government*

A person who occupies a parcel of land must have proof of land rights, either a certificate or letters of evidence of rights (Article 19 of Law No. 5/1960 on the Basic Regulation of Basic Principles of Agrarian. Principles are abbreviated as UUPA). This provision is also in line with

Article 1 (20) and Article 4 of Government Regulation Number 24 of 1997 on Land Registration; Article 1 point (3) of Government Regulation Number 40 of 1996 on Cultivation Rights, Building Rights, and Land Rights; and Article 1 point (4) of Government Regulation Number 18 of 2021 on Management Rights, Land Rights, Flat Units, and Land Registration. Suppose you do not have proof of land rights. In this case, someone cannot use a plot of land, according to Article 1 point (3) of the Government Regulation in Lieu of Law Number 51 of 1960 concerning the Prohibition of Use of Land without the Permission of the Rightful Person or his Proxy (in the future abbreviated as Perppu 51/1960), using land is occupying, working on, and about a plot of land or having plants or buildings on it, regardless of whether the building is used alone or not. The word "occupy" is synonymous with "occupy." According to Article 6-point (1) and letter (a) of Perppu 51/1960: "Without prejudice to the applicability of the provisions in Articles 3, 4, and 5, shall be punished with imprisonment for 3 (three) months and a fine of up to Rp. 5,000 (five thousand rupiahs); a person who uses land without the rightful owner's permission or his legal representative, provided that in the case of plantation and forest lands, those who will be settled according to Article 5, paragraph (1) shall be excluded.

A person or individual who manages a parcel of land owned by the Regional Government must be with a cooperation scheme or granting of Management Rights; in the UUPA Management Rights are not explicitly mentioned, the UUPA only mentions management in General Elucidation Part II number (2) in the last paragraph, which is as follows: "The State may grant such land to a person or legal entity with a right according to its designation and needs, such as ownership rights, business use rights, building use rights, use rights or grant it under management to an authority (Department, Jawatan, or Daerah Swatantra) to be used for the implementation of their respective duties (Article 2 paragraph 4)". Hak Pengelolaan is also mentioned in Article 1 point (2) of Government Regulation No. 40/1996 on HGU, HGB, and Land Rights. However, Management Rights cannot currently be granted to private individuals; management rights regulations include the Regulation of The Minister of Agrarian Affairs No. 9/1999 on procedures for granting and cancelling state land and management rights. Management rights can only be granted to (a) government agencies, including local governments; (b) state-owned enterprises; (c) regionally owned enterprises; (d) Persero Limited Liability Companies ("PT"); (e) authorised entities; and (f) other government legal entities appointed by the government, which is in line with Government Regulation No. 18 of 2021 concerning Management Rights, Land Rights, Flat Units and Land Registration.

According to Articles 61, 62, and 63 of PP No. 37/2012 on the watershed, community empowerment in watershed management activities can be carried out with cooperatives or community organisations instead of individual individuals and with permits and cooperation schemes that do not violate regulations, such as permits with local governments (provincial, district/city). A person or individual who manages a plot of land owned by the state or local government does not exist and cannot; an example of managing is renting to another party, renting according to article 1548 of the Civil Code (KUHPerdata), "renting is an agreement by which one party binds himself to give the other party the enjoyment of something goods, during a certain time and with the payment of a price, which the latter party is willing to pay." The word "goods" here can be movable or immovable objects, such as land or buildings, and the "goods" must belong to the lessee, otherwise it is a forbidden cause according to Article 1335 of the Civil Code so that the actions of the lessee are unlawful. A person who uses a parcel of land who does not have a right to the land according to Article 24 of the UUPA, "The use of land owned by non-owners is limited and regulated by legislation," according to Article 52 points (2) of the UUPA: "Government regulations and regulations referred to in Articles 19, 22, 24, 26 paragraph 1, 46, 47, 48, 49 paragraphs 3 and 50 paragraph 2 may provide criminal penalties for violations of their regulations with imprisonment for up to three months and a fine of up to Rp. 10.000.

Problems or cases related to land objects can involve several fields of law and justice, cases of disputes over ownership and mortgage rights to land enter the realm of civil law, parties who feel aggrieved file a civil lawsuit to the District Court, cases of cancellation of certificates enter the realm of state administration, submit an application to the State Administrative Court, cases of land inheritance for Muslims enter the realm of Islamic civil law, file a lawsuit to the Religious Court, cases of land sales of state assets or local government, land fraud (Article 378 of the Criminal Code), illegal land grabbing (Article 385 of the Criminal Code), use of land without the permission of the rightful owner or his proxy (Perppu 51/1961), falsification of land documents (263, 264, 266 of the

Criminal Code), entering someone else's house or yard without permission or unlawfully (Article 167 (1) of the Criminal Code or Article 257 paragraph (1) of Law Number 1 of 2023), so that the government, in this case the UPTD of Cidurian-Cisadane River Watershed Management, PUPR Office of Banten Province, can follow up according to the chronology of the case.

4 CONCLUSION

There are three findings from this study. First, legal efforts have implications for regulations governing the existence of watersheds in which individuals misuse resources and are often exploited by occupying or managing land plots owned by the Banten Provincial Government. Second, the view of Agrarian law on the existence of individuals misusing resources who use land plots under the management of the Technical Implementation Unit of the Cidurian-Cisadane Watershed Management Office (UPTD DAS) of the Banten Provincial Public Works and Spatial Planning Office. Third, legal events in civil traffic, if they cause harm to others in the context of this case, harm local governments, or violate statutory provisions, can become criminal offenses, also showing civil events in which criminal aspects can arise.

REFERENCES

Badung, D. L. (2018, June 11). Kerusakan Sungai dan Daerah Aliran Sungai di Indonesia. Retrieved from Dinas Lingkungan Hidup dan Kebersihan Kabupaten Badung: https://dislhk.badungkab.go.id/artikel/17937-kerusakan-sungai-dan-daerah-aliran-sungai-di-indonesia

Bochet, J. J. (1983). *Management of upland watersheds. Participation of the mountain communities of FAO. Rome: Conservation Guide 8.* FAO.

Febriamansyah, R. Y. (2021). Land use change, climate change, and river basin management: A preliminary study in the small river basin of Batang Paninggahan, West Sumatra, Indonesia. *Natural Resource Governance in Asia* (pp. 325–342). Elsevier. https://doi.org/10.1016/b978-0-323-85729-1.00021-9, 325-342.

Handayani, S. R. (2019). Aspek Pidana Dalam Pemanfaatan Tanah Negara Tanpa Izin Perspektif Fiqh Jinayah, (Studi Kasus Di Gampong Lamreung Kecamatan Krueng Barona Jaya Aceh Besar). *LEGITIMASI*, Vol. 8 No. 1, Januari – Juni 2019.

Li, X. -Y.-J.-H.-S.-Y. (2008). Impact of land use and land cover change on environmental degradation in lakLakenghai watershed, northeast Qinghai-Tibet Plateau. *Land Degradation & Development*, 20(1), 69–83. https://doi.org/10.1002/ldr.885, 69-83.

Lifidyah. (2023, July 10). Pelanggaran Tata Ruang di DAS Beresiko Tinggi. Retrieved from https://konservasidas.fkt.ugm.ac.id/2023/07/10/pelanggaran-tata-ruang-di-das-berisiko-tinggi/,

M. Fahri Ikhsan, H. H. (2023). Opportunities for Implementing The Principle of Strict Liability as A Basis for for Corporate Criminal Responsibility in Environmental Offenses. Bengkoelen Justice : Jurnal Ilmu HukumP-ISSN 2088-3412 Available at https://ejournal.unib.ac.id/index.php/j_bengkoelenjustice E-ISSN 2686-2867, DOI: 10.33369/j_bengkoelenjust.v12i1.25033,Volume Issue November 2023, 1.

Ngabire, M. W. (2022). Synergic effects of land-use management systems towards the reclamation of Aeolian Desertified Land in the Shiyang River Basin. *Ecological Indicators*, 139, 108882. https://doi.org/10.1016/j.ecolind.2022.108882, 139.

Sugandhi, R. (1981). *KUHP dan Penjelasannya.* Surabaya: Usaha Bersama.

Teshome, Y. B. (2018). Evaluating performance of traditional surface irrigation techniques in Cheleleka watershed in Central Rift Valley, Ethiopia. *Applied Water Science*, 8(8). https://doi.org/10.1007/s13201-018-0862-z.

Tranggono, M. (2021). Pengendalian Ruang Pada Kawasan Situ Cipondoh di Kota Tangerang, Provinsi Banten: Kajian Kebijakan. dinamika Teknik Sipil/Vol. 14/No. 2/Desember 2021, 41-46.

Valera, C. A. (2016). The role of environmental land use conflicts in soil fertility: A study on the Uberaba River basin, Brazil. Science of The Total Environment,. *Science of The Total Environment*, 562, 463–473. https://doi.org/10.1016/j.scitotenv.2016.04.046, 463-473.

Wadhawan, S. K. (2020). *Causative factors of landslides 2019: a case study in Malappuram and Wayanad districts of Kerala*, India. *Landslides*, 17(11), 2689–2697. https://doi.org/10.1007/s10346-020-01520-5.

Waiyasusri, K. S. (2023). Flood Susceptibility Mapping Using Logistic Regression Analysis In Lam Khan Chu Watershed, Chaiyaphum Province, Thailand. *GEOGRAPHY, ENVIRONMENT, SUSTAINABILITY*, 16(2), 41–56. https://doi.org/10.24057/2071-9388-2022-159, 41-56.

Wang, C. J. (2019). Ecological environment assessment based on land use simulation: A case study in the Heihe River Basin. *Science of The Total Environment*, 697, 133928. https://doi.org/10.1016/j.scitotenv.2019.133928, 697.

Towards Resilient Societies: The Synergy of Religion, Education, Health, Science, and Technology – Rahiem (Ed.)

SWOT analysis of Desa Tani: Zakat-based economic empowerment in Indonesia

Fadia Laila Syafitri, Ali Rama & Mohammad Lutfi
Syarif Hidayatullah State Islamic University Jakarta

ABSTRACT: The management of zakat funds presents a promising solution for poverty alleviation, particularly by raising awareness among affluent Muslims to contribute through zakat. This study examines the Desa Tani Program initiated by Dompet Dhuafa, which aims to strengthen community economic development through empowerment initiatives. Using a qualitative approach with SWOT analysis, this study evaluates the program's strengths, weaknesses, opportunities, and threats. The findings indicate that Dompet Dhuafa provides participants with capital, training, land facilities, production resources, and continuous guidance. The analysis yields an internal factor score of 0.63 and an external factor score of 0.86, positioning the program in quadrant I of the SWOT matrix, highlighting strong strengths and opportunities. Based on these results, the recommended strategy is Strengths-Opportunities (SO), leveraging internal strengths to maximise available opportunities through an aggressive growth approach. This program directly supports Sustainable Development Goals (SDGs), particularly SDG 1 (No Poverty) and SDG 2 (Zero Hunger), by fostering zakat-driven economic empowerment and food security.

Keywords: Desa Tani, economic empowerment, poverty alleviation, SWOT analysis, zakat

1 INTRODUCTION

Poverty is a complex issue that extends beyond mere statistics, reflecting society's challenging living conditions (Gweshengwe and Hassan 2020). It requires a decisive and comprehensive approach to address effectively. One aspect of poverty, known as the dimension of poverty, arises from factors limiting societal opportunities. These factors can be internal, such as low levels of education, or external, originating from outside the individual (Paraschiv 2017). Both internal and external factors significantly impact the opportunities available to those who are considered impoverished.

In contemporary Indonesia, research on Islamic economics has expanded beyond the scope of Islamic banking to investigate the concepts and practical applications of zakat deeply (Choiriyah *et al.* 2020; Rama and Yusuf 2019). Zakat, a mandatory charitable contribution in Islam, serves dual roles: it is both a personal obligation that embodies the duty of Muslims to support one another and a critical component of public finance, overseen and regulated by relevant authorities (Herianingrum *et al.* 2024).

Research highlights the multifaced impact of zakat in addressing poverty and fostering economic empowerment. Anis and Kassim (2016) analyse the effectiveness of zakat management for poverty alleviation in Bangladesh and identify several key factors influencing the success of women entrepreneurs, including the age of entrepreneurs, family size, type of business, and the amount of zakat received. Similarly, Zulkifli *et al.* (2021) assess the role of zakat in combating poverty among recipients in Malaysia. Their findings confirm the significant contribution of zakat in eradicating poverty and improving the quality of life for the *asnaf* (zakat-eligible

 DOI: 10.1201/9781003645542-94

individuals). In Pakistan, Ahmad and Ghiasul Haq (2021) demonstrate the vital role of corporate zakat in achieving the SDGs, further emphasising its indispensability in poverty alleviation. Aziz *et al.* (2020) develop a comprehensive mechanism for evaluating zakat's impact on poverty reduction and its alignment with SDGs. Their findings indicate that zakat significantly reduces multidimensional poverty and contributes meaningfully to achieving several SDGs in Pakistan.

The growing body of research on zakat and its role in poverty reduction underscores the strategic importance of zakat as a tool for fostering economic justice and promoting sustainable development globally. This study aims to investigate the Desa Tani Program, initiated by Dompet Dhuafa, focusing on enhancing community economic development through empowerment initiatives. The study uses SWOT analysis to evaluate the program's strengths, weaknesses, opportunities, and threats to provide a comprehensive assessment.

2 METHODOLOGY

This research uses a descriptive study method, employing a SWOT analysis to comprehensively explore the Desa Tani Program in Bandung, West Java. The goal is to provide a detailed narrative of the program's strengths, weaknesses, opportunities, and threats (Puyt *et al.* 2023; van Wijngaarden *et al.* 2012).

To gather data, interviews and observations were conducted to identify strategic factors within the SWOT framework. These factors then integrated into a questionnaire distributed to mustahik (zakat recipients) and program staff. The responses from 23 key respondents, including mustahik and staff, will undergo statistical analysis to determine the program's position within the SWOT matrix. This method ensures that real-world observations and stakeholder input inform strategy development.

3 FINDINGS AND DISCUSSIONS

The Desa Tani program, launched in December 2018 by Dompet Dhuafa, is a significant initiative focused on agricultural empowerment through horticulture. Covering 10 hectares and benefiting 52 participants by September 2022, the program aims to enhance harvest quality, increase agricultural production, and improve farmers' incomes. Comprehensive management support and resources has become a cornerstone of Dompet Dhuafa's economic empowerment efforts.

The program's success was recognised with the Best SDGs Action Award 2023, underscoring its contribution to sustainable development and poverty alleviation. This accolade highlights the program's role in improving agricultural productivity and fostering broader economic and social well-being.

3.1 *The role of Dompet Dhuafa in enhancing mustahik's economic conditions through the Desa Tani program*

The Desa Tani Program, initiated by Mang Ade and supported by Dompet Dhuafa, utilises a creative productive zakat model to address economic hardships in underprivileged communities. Zakat funds are strategically allocated to provide agricultural labourers with capital and land, ensuring long-term empowerment rather than short-term relief. Effective planning has been crucial in turning this vision into reality, with strategic partnerships amplifying the impact of zakat funds. The program also provides hands-on training in planting, fertilisation, and harvesting, equipping mustahik (zakat recipients) with essential agricultural skills. Regular sessions reinforce discipline and positive behavioural change, benefiting farmers with limited educational backgrounds. "We recently held a joint meeting to refresh our understanding of the Desa Tani module. Continuous learning is crucial," said Mang Ade (Interview 2024).

Dompet Dhuafa provides initial capital through a revolving fund system, ensuring continuous reinvestment and sustainable growth. Beneficiaries receive land and essential agricultural resources, maximizing the utility of zakat funds over time. This model allows the program to expand its reach while maintaining financial sustainability. "For agricultural inputs, we use a sustainable funding model. Farmers' invoices include deductions that are reinvested to purchase more inputs," explained Ustadz Jajang (Interview 2024). Through this system, the program ensures that farmers receive short-term support and develop long-term financial stability, fostering a self-sustaining agricultural ecosystem.

Supervision from Dompet Dhuafa ensures the program adheres to Sharia principles while allowing financial independence in daily management. Mang Ade frequently consults with Dompet Dhuafa to align program rules with Islamic guidelines, ensuring ethical and effective governance. "Mang Ade discusses regulations with us to ensure compliance with Sharia, but financial management is handled independently," he stated (Interview 2024). The Desa Tani Program exemplifies how zakat funds can drive long-term economic empowerment and community resilience by combining strategic planning, hands-on training, sustainable funding, and continuous oversight.

3.2 *Factors of strengths, weaknesses, opportunities, and threats in the Desa Tani program*

The Desa Tani Program, initiated and managed by Dompet Dhuafa's Zakat Management Organization (LAZ), strives to enhance community welfare and economic stability through various strategic efforts. This analysis identifies key strategic factors based on research findings:

3.2.1 *Internal factors*

3.2.1.1 *Strengths*

Sustainable empowerment initiatives
The Desa Tani Program exemplifies a long-term empowerment effort that has evolved significantly since its inception. Initially limited in scope and resources, the program has grown through its founders' and supporters' unwavering commitment and enthusiasm. This sustained effort has led to the program's expansion and increased community engagement, reflecting its resilience and effectiveness in fostering lasting change.

Provision of agricultural land and facilities
Beneficiaries in the Desa Tani Program benefit from access to essential agricultural resources, including land and infrastructure. The program provides farmers with the necessary facilities, such as greenhouse setups and land, to enhance their productivity. This support facilitates better crop yields and product quality, as beneficiaries can focus on optimising their agricultural practices without the burden of acquiring land or facilities.

Effective fund management
Dompet Dhuafa's efficient management of program funds ensures that resources are allocated effectively and align with the program's objectives. By precisely targeting the use of donated funds, the program maximises its impact on beneficiaries, enhances its credibility, and builds public trust. This responsible financial management contributes to the program's success and reputation.

Stable market access
The program benefits from establishing buyers for its produce, providing a reliable market for farmers' harvests. This stability reduces the risk of unsold produce and allows the program to focus on maintaining and improving quality standards. The presence of consistent buyers supports long-term planning and sustainability for the program.

Local community development
The Desa Tani Program is committed to the broader development of the local community. Regular evaluations and meetings help beneficiaries and administrators refine their skills and improve their work. This ongoing support contributes to enhancing beneficiaries' capabilities, increasing their income, and improving their overall quality of life.

3.2.1.2 *Weaknesses*

Limited agricultural input availability
The program faces challenges in providing sufficient agricultural inputs, which are crucial for effective farming. Despite these limitations, beneficiaries demonstrate resilience by seeking additional resources independently. This proactive approach helps them maintain their agricultural activities, although it highlights a need for a more reliable input supply.

Inadequate training programs
Regular training is essential for improving beneficiaries' skills and productivity. However, there are gaps in the frequency and quality of training provided within the program. Some beneficiaries have not yet received the necessary training, which can hinder their development and the program's overall effectiveness.

Low engagement with new technologies
Beneficiaries come from diverse backgrounds with varying levels of education and technological familiarity. This diversity can affect their willingness and ability to embrace new technologies. The lack of interest or skill in new agricultural technologies can impact the program's effectiveness and the achievement of its goals.

Educational deficiencies
Many beneficiaries have low educational levels, which limits their ability to engage effectively with program requirements and communicate effectively. This educational gap necessitates additional training and support to improve their performance and integration into the program.

3.2.2 *External factors*

3.2.2.1 *Opportunities*

Enhanced collaboration with other organizations
Strengthening partnerships with other organisations and zakat management entities can significantly benefit the Desa Tani Program. For instance, collaborations with PT. Audy Mandiri Indonesia has proven successful. Expanding and deepening these partnerships can enhance the program's impact and resource availability.

Economic development potential
The Desa Tani Program has substantial potential for further growth and economic impact. Initially starting with 1.2 hectares and 12 beneficiaries, the program now covers approximately 10 hectares and supports 53 beneficiaries. This expansion demonstrates the program's capacity to drive significant economic improvements in the community.

Favourable natural resources
The program benefits from the region's ideal conditions for horticultural crops. The favourable climate and soil quality contribute to high-quality produce. Leveraging these natural advantages supports the program's goals and enhances its success in agricultural production.

Clear policy framework
The Desa Tani Program operates with well-defined policies and procedures, including clear harvest collection processes and fixed pricing. This structured approach ensures smooth operations and emphasises the importance of discipline among all stakeholders.

Potential for Agro-Tourism Development
The scenic beauty and strategic location of Desa Tani present opportunities for agro-tourism. Developing this potential can boost the program's revenue, support local economic development, and increase the program's visibility and attractiveness to tourists.

3.2.2.2 *Threats*

Payment delays from buyers
Delays in buyer payments pose a significant risk to the program and its beneficiaries. These delays can affect the program's financial stability and hinder the ability to cover inputs and beneficiary payment costs. Addressing these payment issues is crucial for maintaining the program's sustainability.

Human resource challenges
The quality of human resources involved in the program impacts its success. Untrained or inadequately skilled personnel can impede the program's progress and effectiveness. Addressing these challenges is essential for ensuring the program's continued success.

Limited administrative capacity
The program's effectiveness can be compromised by having fewer administrators relative to beneficiaries. This limitation can lead to inadequate oversight and management, affecting the program's overall performance and beneficiary outcomes.

Pest and disease issues
Pest and disease outbreaks can significantly impact crop production and the program's sustainability. Beneficiaries in certain areas report frequent pest and disease problems, leading to economic losses and affecting the program's stability.

Analyzing these factors, including the IFAS and EFAS matrices obtained through questionnaires, provides a comprehensive understanding of the Desa Tani Program's strategic position and informs future planning and improvements.

Based on the IFAS Score (the table is skipped), the total score for Strengths is 2.70, and the total score for Weaknesses is 2.07. The overall IFAS score is 0.63. This score is obtained by subtracting the total Weaknesses score from the total Strengths score. The positive IFAS score indicates that the internal strengths of the program outweigh its weaknesses, suggesting a generally favourable internal environment for the program's success. This analysis helps identify the program's capacity to leverage its strengths and address its weaknesses effectively.

Based on the EFAS Score (the table is skipped), the total score for Opportunities is 2.74, while the total score for Threats is 1.88. The overall EFAS (External Factor Analysis Summary) score is 0.86, derived by subtracting the total Threats score from the total Opportunities score. This positive EFAS score indicates that the program's external opportunities outweigh its threats, suggesting a generally favourable external environment for its success. This analysis helps to assess the program's potential for growth and the challenges it needs to address.

The straight-line results indicate that the Desa Tani Program is positioned in Quadrant I, highlighting its strong strengths and opportunities. This favourable positioning suggests the program is well-suited for a Strengths-Opportunities (SO) strategy. By leveraging its internal strengths to capitalise on external opportunities, the program can pursue aggressive growth and maximise its impact. Program managers should focus on competitive advantage strategies to expand the program's reach and effectiveness, ensuring broader community benefits. This approach will drive significant improvements and support the program's long-term success.

Based on the analysis conducted through various stages of the SWOT evaluation, we have formulated Strength-Opportunity (SO) strategies designed to capitalise on the existing opportunities within the Desa Tani Program and leverage its inherent strengths. The proposed strategies are as follows:

3.3 *Fostering collaboration between zakat organisations and other entities to maximise the potential of the Desa Tani program*

Effective collaboration is a strategic management approach emphasising mutual benefits by exchanging information and resources. By working together, all parties involved can achieve their goals more efficiently, as collaboration helps address challenges while respecting each participant's interests, thus leading to shared benefits. For zakat organisations, collaborating with various stakeholders—such as governmental bodies, private sector companies, and other relevant institutions—can significantly enhance the program's reach and impact. This partnership allows for the pooling of resources, sharing of expertise, and coordination of efforts, improving the program's effectiveness in meeting its economic objectives and delivering more substantial benefits to the community.

Expanding collaboration beyond traditional zakat organisations to include diverse partners can also open new avenues for resource acquisition, increase the program's capacity, and drive innovation. For instance, partnerships with private companies might lead to additional funding or technological support, while collaborations with government agencies could offer regulatory advantages and broader outreach.

3.4 *Enhancing information dissemination and expanding agricultural networks*

Providing beneficiaries with extensive and relevant information is crucial for improving their knowledge and skills. A well-structured information dissemination strategy can include comprehensive training programs designed to boost farmers' capabilities in various areas, such as land management, modern agricultural techniques, and business aspects of farming.

Beneficiaries can stay informed about the latest advancements and best agriculture practices by offering tailored training sessions and creating platforms for knowledge exchange. Additionally, expanding agricultural networks can facilitate connections between farmers and industry experts, fostering a community of practice that supports continuous learning and development. These initiatives not only enhance the technical skills of farmers but also empower them to apply innovative solutions, thereby improving their productivity and the overall success of the Desa Tani Program.

Such efforts can be complemented by creating online resources, organising workshops, and establishing mentorship programs to provide ongoing support and guidance. This holistic approach ensures that farmers can manage their resources effectively and adapt to evolving agricultural trends.

4 CONCLUSION

The research reveals Dompet Dhuafa's vital role in empowering communities through the Desa Tani Program, significantly enhancing the livelihoods of local farmers in Bandung, West Java. Key findings indicate that the program has successfully provided participants with essential resources, including land, agricultural inputs, and comprehensive training while ensuring continuous monitoring and guidance. These efforts have led to improved agricultural productivity and economic conditions for beneficiaries. The program's strengths, weaknesses, opportunities, and threats were identified through a SWOT analysis, yielding an Internal Factor Analysis Summary (IFAS) score of 0.63 and an External Factor Analysis Summary (EFAS) score of 0.85. this positioning in Quadrant I of the SWOT matrix highlights a strong foundation for leveraging strengths and opportunities to achieve sustainable growth.

Specific strategies are recommended for Dompet Dhuafa and the Desa Tani Program Staff to capitalise on these strengths and opportunities. The Proposed Strength-Opportunity (SO) strategies include enhancing collaboration with other zakat institutions and relevant organisations to share resources and expertise. Policy recommendations include fostering strategic partnerships with local governments, private sector stakeholders, and other non-governmental organisations to expand the reach and impact of such programs. Expanding information dissemination and agricultural networks is also crucial for supporting farmers through comprehensive training and robust support systems.

REFERENCES

Ahmad, S., and Ghiasul Haq, S. (2021). How far corporate zakat may contribute to sustainable development goals? *Islamic Wealth and the SDGs: Global Strategies for Socio-Economic Impact*, 421–437.

Anis, F. M., and Kassim, S. H. (2016). Effectiveness of zakat-based programs on poverty alleviation and economic empowerment of poor women: a case study of Bangladesh. *Journal of Islamic Monetary Economics and Finance*, 1(2), 229–258.

Aziz, Y., Mansor, F., Waqar, S., and Haji Abdullah, L. (2020). The nexus between zakat and poverty reduction, is the effective utilization of zakat necessary for achieving SDGs: A multidimensional poverty index approach. *Asian Social Work and Policy Review*, 14(3), 235–247.

Choiriyah, E. A. N., Kafi, A., Hikmah, I. F., and Indrawan, I. W. (2020). Zakat and Poverty Alleviation in Indonesia: a Panel Analysis At Provincial Level. *Journal of Islamic Monetary Economics and Finance*, 6(4), 811–832. https://doi.org/10.21098/jimf.v6i4.1122

Gweshengwe, B., and Hassan, N. H. (2020). Defining the characteristics of poverty and their implications for poverty analysis. *Cogent Social Sciences*, 6(1), 1768669.

Herianingrum, S., Supriani, I., Sukmana, R., Effendie, E., Widiastuti, T., Fauzi, Q., and Shofawati, A. (2024). Zakat as an instrument of poverty reduction in Indonesia. *Journal of Islamic Accounting and Business Research*, 15(4), 643–660.

Paraschiv, C. I. (2017). The role of education in poverty alleviation. *Theoretical and Applied Economics*, 24.

Puyt, R. W., Lie, F. B., and Wilderom, C. P. M. (2023). The origins of SWOT analysis. *Long Range Planning*, 56(3), 102304.

Rama, A., and Yusuf, B. (2019). Construction of Islamic human development index. *Journal of King Abdulaziz University: Islamic Economics*, 32(1), 43–64. https://doi.org/10.4197/Islec.32-1.3

van Wijngaarden, J. D. H., Scholten, G. R. M., and van Wijk, K. P. (2012). Strategic analysis for health care organizations: the suitability of the SWOT-analysis. *The International Journal of Health Planning and Management*, 27(1), 34–49.

Zulkifli, M. F., Taha, R., MOHD NOR, M. N., and Ali, A. (2021). Combating poverty in Malaysia: The role of zakat. *The Journal of Asian Finance, Economics and Business*, 8(5), 505–513.

Towards Resilient Societies: The Synergy of Religion, Education, Health, Science, and Technology – Rahiem (Ed.)
© 2026 The Author(s), ISBN: 978-1-041-08471-6

Comparative analysis of personal data protection in Indonesia and Singapore for a sustainable digital future

Ahmad Tholabi Kharlie
Universitas Islam Negeri Syarif Hidayatullah Jakarta, Indonesia

Fathudin Fathudin
Universitas Islam Negeri Syarif Hidayatullah Jakarta, Indonesia
ORCID ID: 0000-0002-3692-0154

Firman El Amny Azra
Universitas Islam Negeri Syarif Hidayatullah Jakarta, Indonesia
ORCID ID: 0000-0003-3333-4739

Solahudeen Cheduerame
Institute Nahdatul Syuban School, Naratiwat, Thailand

ABSTRACT: Law No. 27 of 2022 on Personal Data Protection is considered insufficient for safeguarding personal data, particularly against potential misuse by data controllers outside Indonesian jurisdiction. This article adopts a qualitative and comparative approach, examining case studies and policy aspects to identify differences and challenges in Indonesia and Singapore's cybersecurity and personal data protection frameworks. Comparative analysis demonstrates how both countries address cross-border data protection issues. The findings show that Singapore's framework offers a more transparent and secure model than Indonesia's. Consequently, this study highlights the need to strengthen Indonesia's legal framework to foster a sustainable digital environment that aligns with Sustainable Development Goals (SDG) 9 and 16.

Keywords: cybersecurity, data protection, consent principle, cyber law

1 INTRODUCTION

The development of information technology has impacted changes in human life in many sectors, such as the sociocultural sector, the economy, and even the legal and policy framework. Despite its positive impact, it also presents some challenges, particularly regarding protecting personal data and privacy, which has become a worldwide concern with frequent data breaches and cybercrime. In the context of sustainable development, the protection of personal data is heavily interwoven with SDG 9 and SDG 16. The personal data protection issue is critical for SDG 16 due to the right to privacy, which is considered one of the fundamental rights (EU 2012). protecting personal data is not merely a legal requirement; it is quintessential for the advancement of these SDGs since safe and equitable processing frameworks will be key to sustainable development, protection of individual rights, and building trust in digital infrastructures globally. This is reflected in the Charter of Fundamental Rights of the European Union; Article 8 states that 1) everyone has the right to protect personal data relating to him/her. 2). Such data shall be processed fairly for a specific purpose and based on the consent of the person concerned or other lawful basis provided for by law. Everyone has the right to access data that has been collected about him or her and the right to have it rectified (Fuster 2014).

Personal data issue also relates to SDG 9 due to its role as a key economic resources that drive innovation and ensure digital infrastructure resilience. As Clive Humby once said, "Data is the

DOI: 10.1201/9781003645542-95

new oil". This statement implies that data has now become a new economic commodity along with the pace of development of information technology (Rianarizkiwati 2022). The importance of data in the digital age cannot be downplayed as it is considered as the fuel of digital technology that allows information technology to function and provide value-added services to society(Lock and Seele 2017). As a consequence, data protection is crucial in fostering a safe and productive cyberspace. Without an appropriate and sufficient data protection system, data breaches will occur more frequently and destroy the trust of society in cyberspace, which will hinder the development of the digital industry (Malatras *et al.* 2017).

1.1 *Bottom of form*

The conception of the right to personal data protection developed from the concept of the right to respect for private life. The concept of private life relates to humans as living beings. Thus, the individual owns the right to personal data protection (Macke 2007). Explicitly, privacy is not included in the 1945 Constitution of Indonesia. However, implicitly, the right to privacy is contained in Article 28G paragraph (1) of the 1945 Constitution as follows: "Every person shall have the right to the protection of his or her person, family, honour, dignity and property under his or her control, and shall have the right to security and protection from threats of fear to do or not to do something which is a human right". In addition to being recognised and guaranteed by the Constitution, the guarantee of the right to privacy is also contained in Article 29 paragraph (1) and Article 30 of Law Number 39 of 1999 concerning Human Rights.

Indonesia currently has in Law Number 27 of 2022 concerning Personal Data Protection, which set out how the state guarantees and protects the personal data of its citizens. The policy contained the presence of the law, which provides great hope for the Indonesian people regarding the protection of personal data in the midst of rampant hacking of personal data. However, this law cannot be considered a perfect law because there are still shortcomings and weaknesses in it. One of the weaknesses in implementing this law is the guarantee of legal protection for personal data owners who are faced with the potential misuse of personal data management by personal data controllers in other countries. This problem is based on the absence of the principle of consent in the law when there is a transfer of personal data to data controllers outside the jurisdiction of Indonesia. The absence of this principle will certainly have the potential for the misuse of personal data. This article aims to present a comparative analysis of the policies in Indonesia and Singapore, particularly in the context of applying the principle of consent during collecting, using or disclosing personal data.

2 METHODOLOGY

This research used a normative legal comparative approach to analyse the personal data protection legal framework of Indonesia and Singapore. Indonesia and Singapore are chosen for case studies due to their geographical proximity, economic interdependence and contrasting stages of legal development in personal data protection, which make them particularly appropriate for comparative analysis. This study mainly relies on secondary data sources, including statutory text, a policy document which includes Law No. 27 of 2022 on Personal Data Protection, Singapore's 2012 Personal Data Protection Act (PDPA) as well as a relevant academic papers. The comparative approach is utilised to contrast Indonesia's and Singapore's legal frameworks to identify their strengths, weaknesses and potential for improvement.

3 FINDINGS & DISCUSSION

3.1 *Role of personal data protection in cybersecurity*

Cybersecurity involves a series of activities to protect the integrity of data and computer networks in terms of their physical hardware and software (Aji 2023). These protective measures are geared toward mitigating the diverse threats and interferences that can compromise the security and functionality of these networks. Cyber threats come in various forms, including malware, ransomware, and denial-of-service attacks, and pose substantial risks to the integrity of information systems. Cybersecurity is a barrier against unauthorised surveillance, ensuring that only authorised parties can access and utilise the information contained within these networks. Data confidentiality is critical to cybersecurity, as it ensures data can only be accessed through authorised parties (Septiandy and Nasution 2023). Only individuals or entities with proper authorisation can open and use the data. Any attempt to gain unauthorised access, often through nefarious means such as information theft, jeopardizes data confidentiality. Cybersecurity measures are deployed to fortify data confidentiality by implementing encryption, access controls, and user authentication protocols (Nugroho *et al.* 2021).

Personal data breaches are serious issues in cybersecurity as they can be used for cyber fraud, and the relevant data can be used as essential resources to hack into a secured system. The predominant approach to personal data protection, which relies on the state providing security for its citizens, has often overlooked the residents' capabilities within their own country to safeguard their data in the cyber realm (Aji 2023). This capacity development and individual capability align with Amartya Sen's concept of capability. An approach rooted in the state's role in the political economy, especially concerning personal data protection, becomes more crucial in the context of data sovereignty, both among nations and in relations with large capitalist entities. In the context of data sovereignty laws, a country risks substantial vulnerabilities if it does not control data centres within its borders, particularly those owned by external companies. Therefore, the empowerment and competence of citizens in safeguarding their data assume great significance, ensuring that the state does not exclusively dominate the protection of personal data while simultaneously overseeing cyberspace (Aji 2023).

3.2 *Personal data protection as a fundamental right*

Over the history of human civilisation, the human need for privacy is as old as humanity itself (Westin 1971). Almost all societies, both primitive and modern, have ways to organise distance and avoid contact with others to establish physical boundaries to maintain privacy (DeCew 1997). Therefore, anthropological, biological and sociological studies also show that even in primitive societies, individuals always desire privacy (Schoeman 1984). The importance of the right to privacy is why this is the most basic right needed by humans.

Initially, the right to privacy was only oriented towards physical protection. Right to be, let alone assume that everyone has the right to be alone, not to be disturbed by others (Thomson 1975). The concept of privacy surfaced in the discourse on social and political threats posed by information and communication technology in the modern era. In the United States (US), "privacy" is a concept that is often used in public and academic discourse. The discourse on "privacy" began to be seriously discussed in the 1960s, particularly on the implications of computerised processing of personal data. The concept of "privacy" was then used as a key term to summarise the various concerns raised by the (mis)use of computers (Bygrave 2010). The right to privacy is conceptualised as limited access to the self. This conception emphasises the desire of individuals to hide themselves and be separate from others. However, the conception of limited access is not solitude, seclusion, or withdrawal from others. A broader conception of limited access includes freedom from government interference as well as from intrusion by the press and others (Solove 2002).

Along with the development of information technology, the right to privacy is increasingly related to the many cases of taking personal data, such as photos published on social media, without the owner's permission (Warren and Brandeis 1989). The growing importance of personal data protection has raised public awareness of its significance in safeguarding an individual's private life and legal rights. With more data flowing and being transferred overseas, the need for protection increases. Personal data protection is a fundamental human right in today's information society, requiring individuals to control their information and regulate its collection, processing, and use. The rapid development of information and communication technology has made it even more crucial. The importance of protecting personal data is increasingly reflected in national and international legal documents, such as the Charter of Fundamental Rights of the European Union (Rodotà 2009).

In the Southeast Asian region, there is also a legal document regulating personal data protection, as stated by the Union of Southeast Asian Countries. The ASEAN Human Rights Declaration, published in 2012, especially in Point 21, states that "every person has the right to be free from arbitrary interference with his or her privacy, family, home or correspondence including personal data, or to attacks upon that person's honour and reputation." These two documents confirm the importance of the state's role in providing guarantees and protection of the right to personal data.

3.3 *Comparison between Singapore and Indonesia*

Personal data protection laws in Singapore are governed by the Personal Data Protection Act (PDPA) 2012 (Table 1). In Singapore, consent is one of the crucial principles in the personal data protection policy governed by the Personal Data Protection Act (PDPA) 2012. This principle emphasises the importance of obtaining consent before collecting, using or disclosing their data. This principle also befits SDG 16, ensuring that institutions become more accountable while advancing individuals' rights to privacy in an increasingly digital world. By fostering transparent data handling, the PDPA encourages SDG 9 concerning the development of digital transformation through ethical data processing.

In the context of the PDPA, consent is one of the legal grounds for processing personal data. Some of the articles in the PDPA that relate to the Principle of Consent include: First, there are provisions stipulating that organisations must obtain consent from individuals before collecting their data. Consent must be voluntary, informed, and specific to the purpose of the data collection (Schermer *et al.* 2014). Secondly, an organisation should also only use personal data for purposes that have been informed to the individual at the time consent is given unless there is further permission from the individual or such permission is provided for in law (Schermer *et al.* 2014). Third, if any organisation wishes to disclose personal data to a third party, it must obtain additional consent from the individual unless an exception is permitted by law (Chik 2013). Fourth, individuals have a right to withdraw previously given consent. If the individual withdraws consent, the organisation may have an obligation to stop collecting or using the data. Fifth, there is a right granted to individuals to object to or restrict the collection, use or disclosure of their data for a specific purpose, and organisations must comply with such requests unless there is a legal basis that requires the collection or use of such data (Wong YongQuan 2017). Sixth, in protecting children's data, Singapore's PDPA provides that organisations must obtain the consent of the parent or legal guardian of a child under a certain age before collecting, using, or disclosing that child's personal data (Wong YongQuan 2017). The Consent Principle in Singapore's PDPA underscores the importance of respecting individuals‘ privacy and control over their personal data. Organisations must clearly and transparently explain the purpose of data collection and obtain valid consent before processing individuals' personal data. Without consent or a valid legal basis, the collection and use of personal data is not permitted under the policies in Singapore's PDPA.

The PDPA puts great emphasis on consent not only to protect privacy rights but also in a manner that enables sustainable digital economies. Trust in digital services is key, and consumers are protected from the possible misuse of data. Clear and transparent data governance frameworks and hence SDG 9.5 on sustainable industrialisation and digital transformation through innovation deal with ethical concerns associated with AI, big data, and digital inclusion. The balance of consent-based processing and legitimate interest exceptions in the PDPA will also allow organisations to use personal data responsibly in cases such as employment management, where such practice needs to be in line with workplace regulations while individual rights are upheld. This nuanced approach ensures that businesses can operate efficiently while maintaining ethical standards aligned with global sustainability goals. Singapore's PDPA fosters a regulatory environment of trust in protecting privacy while promoting digital innovation, as envisioned by SDG 16 on account of institutional accountability and SDG 9 on account of resilient infrastructure and innovation. The PDPA gains confidence in digital services through informed consent and responsible data governance, furthering a sustainable and inclusive digital economy.

Specific consent and notification procedures and other regulatory obligations in data management are part of the effort to protect privacy rights (Petters 2020). Data protection and privacy are both considered part of personal data protection. In the preamble of Law No. 27 of 2022 on Personal Data Protection, it is stated that personal data protection is intended to guarantee the rights of citizens to personal protection and foster public awareness and ensure recognition and respect for the importance of personal data protection. The right to privacy is a basic right guaranteed and protected in various international legal instruments and constitutions worldwide. Personal data protection is a special form of respect for the right to privacy (La Rue 2011).

Table 1. Comparison Indonesia-Singapore personal data protection law.

Aspect	Indonesia	Singapore
Legal Framework	Law No. 27 of 2022 on Personal Data Protection.	Personal Data Protection Act (PDPA) 2012.
Principle of Law	Protection, legal certainty, public interest, prudentiality, balance, legal obligation, confidentiality	Explicit consent, specific purpose, limitation, notification, accuracy, protection, access.
Regulatory Enforcement	Basic enforcement mechanism that based on criminal and civil proceeding code.	Well-established enforcement mechanisms, with clear penalties for non-compliance.
Data Subject Right	To have access, have correction, place limitation, objection, compensated for violation, terminate process, destroy and withdraw personal data	To have access, correction, limiting process, raise objection, erasure of data, right for relief in civil proceeding, withdraw personal data
Sanction and Fines	Up to 6 years imprisonment and Rp6 billion fines.	Administrative fine up to SDG1 million or 10% of annual turnover in Singapore

Compared with Singapore PDPA, which leans more toward a balancing act between digital economic growth, innovation, and protection of personal data, the Indonesian Law seeks to ensure accountability of state responsibility and digital risks to the citizens. The PDPA in Singapore remains business-centric; it supports the digital economy in ensuring that companies maintain standards for data privacy. Meanwhile, Indonesia's PDP Law is modelled on the state-centred approach wherein data protection intersects with national sovereignty, individual security, and public interest regulation.

From a philosophical perspective, the Personal Data Protection Law is based on the legal ideals in the Preamble of the 1945 Constitution, namely the purpose of "The State protects the entire Indonesian nation and the entire Indonesian homeland....". Secondly, sociologically, Indonesian society is also part of the information society in a global interconnected space. Therefore, protecting personal data privacy is necessary to provide security for every individual, both Indonesian citizens and foreign citizens in Indonesia, related to the collection, processing, organisation, and dissemination of personal data. In addition, from a legal certainty viewpoint, the overarching legal framework must ensure various laws and regulations in Indonesia are aligned with the principle of data protection. Therefore, a data protection law is needed to guarantee personal data protection and provide legal certainty (Djafar and Santoso 2019).

4 CONCLUSION

Personal data protection is an important issue in today's digital era. The principle of consent in personal data protection is an essential foundation for protecting individual privacy rights. In Singapore, the Personal Data Protection Act (PDPA) 2012 regulates personal data protection with the important principle of consent. This principle emphasises the importance of consent from individuals before collecting, using or disclosing personal data. The PDPA sets out provisions explaining how consent is obtained, the limitations, and the rights of individuals about personal data. However, in Indonesia, Law Number 27 Year 2022 on Personal Data Protection shows weaknesses in personal data protection. Article 56 paragraph (1) does not explicitly mention the requirement of personal data owner's consent before data transfer outside the territory of Indonesia. This ignores the rights of personal data owners and contradicts the main purpose of the law to protect the rights of individuals. While Singapore has a strong legal framework with the PDPA governing the principle of consent, Indonesia should improve its laws' weaknesses to protect personal data owners' rights more effectively.

REFERENCES

Aji, M. P. (2023). Sistem Keamanan Siber dan Kedaulatan Data di Indonesia dalam Perspektif Ekonomi Politik (Studi Kasus Perlindungan Data Pribadi)[Cyber Security System and Data Sovereignty in Indonesia in Political Economic Perspective]. *Jurnal Politica Dinamika Masalah Politik Dalam Negeri Dan Hubungan Internasional*, 13(2), 222–238.

Bygrave, L. A. (2010). Privacy and data protection in an international perspective. *Scandinavian Studies in Law*, 56 (8), 165–200.

Chik, W. B. (2013). The Singapore Personal Data Protection Act and an assessment of future trends in data privacy reform. *Computer Law \& Security Review*, 29(5), 554–575.

DeCew, J. W. (1997). *In Pursuit of Privacy: Law, Ethics, and the Rise of Technology*. Cornell University Press.

Djafar, W., and Santoso, M. J. (2019). *Perlindungan Data Pribadi Konsep, Instrumen, dan Prinsipnya*. Jakarta: ELSAM.

EU. (2012). Charter of fundamental rights of the European Union. *The Review of International Affairs*, 63(1147), 109–123.

Fuster, G. G. (2014). The Emergence of Personal Data Protection as a Fundamental Right of the EU (Vol. 16). Springer Science \& Business.

La Rue, F. (2011). Report of the Special Rapporteur on the promotion and protection of the right to freedom of opinion and expression. *Human Rights Council*, 16.

Lock, I., and Seele, P. (2017). Theorizing stakeholders of sustainability in the digital age. *Sustainability Science*, 12, 235–245.

Macke, J. (2007). European union agency for fundamental rights. Eucrim-the European Criminal Law Associations' Forum, 01–02, 18–19.

Malatras, A., Sanchez, I., Beslay, L., Coisel, I., Vakalis, I., D'Acquisto, G., Sanchez, M. G., Grall, M., Hansen, M., and Zorkadis, V. (2017). Pan-European personal data breaches: Mapping of current practices and recommendations to facilitate cooperation among Data Protection Authorities. *Computer Law \& Security Review*, 33(4), 458–469.

Nugroho, I. I., Pratiwi, R., and Zahro, S. R. A. (2021). Optimalisasi Penanggulangan Kebocoran Data Melalui Regulatory Blockchain Guna Mewujudkan Keamanan Siber di Indonesia. *Ikatan Penulis Mahasiswa Hukum Indonesia Law Journal*, 1(2), 115–129.

Petters, J. (2020). Data privacy guide: definitions, explanations and legislation. Varonis.

Rianarizkiwati, N. (2022). Ius Constituendum Hak Atas Pelindungan Data Pribadi: Suatu Perspektif Hak Asasi Manusia. *Jurnal Hukum Sasana*, 8(2), 324–341.

Rodotà, S. (2009). Data protection as a fundamental right. *In Reinventing Data Protection?* (pp. 77–82). Springer.

Schermer, B. W., Custers, B., and der Hof, S. (2014). The crisis of consent: How stronger legal protection may lead to weaker consent in data protection. *Ethics and Information Technology*, 16(2), 171–182.

Schoeman, F. D. (1984). *Philosophical Dimensions of Privacy: An Anthology*. Cambridge University Press.

Septiandy, M. D., and Nasution, M. I. P. (2023). Urgensi Perlindungan Data Pribadi Guna Menjamin Keamanan Data. *IJM: Indonesian Journal of Multidisciplinary*, 1(4), 1252–1259.

Solove, D. J. (2002). Conceptualizing privacy. California Law Review, 1087–1155.

Thomson, J. J. (1975). The right to privacy. Philosophy \& Public Affairs, 295–314.

Warren, S., and Brandeis, L. (1989). The right to privacy. In *Killing the Messenger: 100 Years of Media Criticism* (pp. 1–21). Columbia University Press.

Westin, A. F. (1971). *Information Technology in a Democracy*. Harvard University Press.

Wong YongQuan, B. (2017). Data privacy law in Singapore: the personal data protection act 2012. *International Data Privacy Law*, 7(4), 287–302.

Towards Resilient Societies: The Synergy of Religion, Education, Health, Science, and Technology – Rahiem (Ed.)
© 2026 The Author(s), ISBN: 978-1-041-08471-6
Open Access: www.taylorfrancis.com, CC BY-NC-ND 4.0 license

Maqashid sharia: A human resource management practice for social sustainability in workplaces

Diamantin Rohadatul Aisy
Universitas Islam Negeri Syarif Hidayatullah Jakarta, Indonesia
ORCID ID: 0000-0001-7994-4351

Prilla Kurnia Ningsih & Bilqis Khoiriyyah
Universitas Islam Negeri Syarif Hidayatullah Jakarta, Indonesia

ABSTRACT: In Indonesia, the assessment of corporate compliance with Islamic principles is typically limited to financial ratios. This research explores alternative perspectives on evaluating adherence to Islamic values, specifically through applying the *maqashid* sharia concept in practical studies at the Islamic philanthropy organisation PPPA Daarul Quran, representing the halal industry. Interviews were conducted with middle-level and first-line management employees to gather information on management policies and practices, particularly in human resource management (HRM). Using the *maqashid* sharia framework to evaluate HRM policies and practices, the study found that this concept is implemented in various regulatory aspects, including committed rules and facilities supporting personal worship, physical and mental health protection, self-development programmes, family-based benefits such as health insurance, scholarships, and employment opportunities, as well as regulations on *zakah* and *shadaqah* payments. Additionally, HRM policies and practices at PPPA Daarul Quran contribute to achieving several Sustainable Development Goals (SDGs), particularly in (1) reducing poverty, (2) reducing hunger, (3) ensuring good health and well-being, (4) providing quality education, (8) promoting decent work and economic growth, (10) reducing inequality within countries, and (16) fostering peaceful and inclusive societies.

Keywords: human resource management, Islamic corporate culture, Islamic philanthropy, maqashid sharia, sustainable development goals

1 INTRODUCTION

The development of Islamic economics today covers more comprehensive aspects, not only related to the Islamic financial sector, such as Islamic banking and capital markets but also include the halal food sector, halal cosmetics and medicines, fashion, Muslim-friendly tourism, education, Islamic media and recreation that has main products and services by Islamic law and values, known as a halal industry (DinarStandard 2022).

The compatibility of the halal industry's products and services with Islamic values can be seen through the legality of halal certification, which market players must have. In addition to the Islamic financial products and services sector, the Financial Services Authority (OJK) adds more criteria related to the interest-based debt ratio and other non-halal income (Financial Services Authority 2017). As the evaluation, the application of Islamic values and principles in corporate is also assessed using the aspect on the financial ratios, namely Islamicity Performance Index (Azwirman *et al.* 2019; Sudirman *et al.* 2023) and Sharia Maqashid Index (SMI) (Ascarya and Masrifah 2023; Asutay and Harningtyas 2015; Tarique *et al.* 2021).

In addition to fulfilling halal certification and the composition of financial ratios, several previous studies have also tried to internalise Islamic values into an Islamic organisational culture in halal industry companies, known as Islamic Work Ethics (IWE) (Aldulaimi 2016; Hakim 2016; Karia and Asaari 2016). However, in this case, the IWE concept internalises Islamic values based on the normative concept of morality in the Al-Quran and Hadith.

On the other hand, the Islamic economic concept in the economic system has main goals in the form of prosperity, or what is often referred to as achieving *maqashid* sharia. It had five principles initiated by Al-Ghazali (2000) that carried out five basic needs fulfilment including caring for someone's religion, soul, intelligence, generation, and wealth. This concept of *maqashid* sharia also has a pivotal role in supporting the Sustainable Development Goals (SDGs) achievement hand in hand through effective *zakah* management that can reduce inequality in wealth distribution (Hasan 2020; Hoque *et al.* 2023; Isman *et al.* 2024; Kahf 2013).

Hence, Asutay and Harningtyas (2015) disclosed several quantitative and qualitative indicators to evaluate *maqashid* sharia performance in Islamic banks, which covers practice in safeguarding the value of human life, the human self, society, and the physical environment. However, there are no specific measurements for human resource management performance. Moreover, due to the lack of applicable measurements, *maqashid* sharia has also not been applied as an assessment in non-financial institutions (Rahman *et al.* 2017).

Therefore, this research examined applying the *maqashid* sharia concept in practical studies, specifically in human resource management regulations. Moreover, *maqashid* sharia is easier to represent as a practical goal that aligns with the SDG targets. This concept should be applied as a framework for creating Islamic values and principles in the workplace rather than applying only the normative concept of morality. This research was conducted in the Islamic philanthropy (*Zakah*) company, Daarul Quran, as a representative of non-financial corporations in the halal industry.

2 METHOD

This research uses qualitative research methods to find phenomena that are happening, in this case, Islamic value creation in Islamic philanthropy companies. The data used in this study is primary data taken through in-depth interview techniques from the employees in middle-level and first-line management specialised in the human resource management division of PPPA Daarul Quran. This method was used to obtain information on management policies and practices in human resource management (HRM) aspects. Secondary data, such as company profiles, financial reports, and program reports, are also used as confirmatory documents to ensure the validity and reliability of the interview process.

Furthermore, data was analysed qualitatively using five *maqashid* sharia elements introduced by Al-Ghazali (2000) and 17 indicators of the SDGs released by the United Nations (2015) that were taken through literature studies. In addition, several interpretations of the Quran and Hadith related to *maqashid* sharia concepts were carried out to deepen the analysis.

3 FINDING AND DISCUSSION

3.1 *Regulation to preserve one's religion (Hifdz ad-Dien)*

To encourage its employees to maintain the quality of their religion, PPPA Daarul Quran implements several policies. First, apart from taking morning attendance through recording on the attendance machine. PPPA Daarul Quran implements morning attendance by reciting the Quran with different daily readings. PPPA Daarul Quran leaders believe this practice can open the door to good fortune. After reciting the Quran together, PPPA Daarul Quran employees and their leaders will pray for each other, not only for the members and internal employees of PPPA Daarul Quran, but also for the people who partner with PPPA Daarul Quran.

Furthermore, the morning attendance will continue with the Dhuha Prayer, addressing each person's beliefs and needs. Management and HRD have no time limit for praying Dhuha in this case. This policy is directly supervised by HRD, whether the main morning attendance in the form of recording on the attendance machine, reciting the Quran, or Duha prayers are included in Daarul Quran's KPI assessment.

Hifdz ad-dien or protection of religion is the thought of rights and freedoms protection in Islam, including protection of belief, worship, and protection of the religion and thought (*mahzab*) preferences that should not be forced (Jauhar 2023). In this context, religious protection applied by PPPA Darul Quran has fulfilled the *maqashid* sharia by freeing its employees to carry out worship to maintain the quality of religion so that they have good performance at work.

Besides, reading and practising the Quran is a priority for a Muslim and they need to be given the freedom to do so. One of the verses that states the virtue of reading the Quran is Surah *Al-Baqarah,* verse 2. Ibn 'Asyur (1984) interpreted that Quran is a reminder, guard, and shield for Muslims, to keep their faith increasing by reciting it. Moreover, the Hadith of Prophet Muhammad SAW that was carried out by Hanzhalah RA in the Muslim Book (1955) explained the importance of consistency in worship, to always get closer to Allah (Hadith 2750).

In this case, PPPA Daarul Qur'an runs a routine program of reciting the Quran together as a form of effort to maintain religion (*Hifdz ad-dien*), which provides several benefits for employees. One of them is to maintain the spiritual balance of employees. By listening and internalising the meaning of the recited Quran together, workers can maintain their faith. This activity indirectly forms positive habits, encouraging everyone to always draw closer to Allah and improve their quality of life. Furthermore, reciting the Quran provides the guidance and motivation to remain strict to the principles of goodness as Islamic values are taught. This routine not only contributes to strengthening the spirituality of individuals but also plays a role in developing their moral character, such as obedience, patience, and kindness. Therefore, maintaining religion (*Hifdz ad-dien*) in this context is not only limited to the spiritual aspect but also includes character-building that aligns with noble religious values.

3.2 *Regulation for taking care of one's soul (Hifdz an-Nafs)*

The activity of reciting the Quran together not only functions as a moment to offer prayers together but also to cleanse the heart and as medicine for the soul. As a message in the Quran Surah *ar-Ra'd* verse 28, reciting the Quran leads to peace of mind.

> "Those who believe and whose hearts find tranquility in the remembrance of Allah, for indeed in the remembrance of Allah do hearts find tranquility." (Quran 13:28)

In his commentary, Ibn 'Asyur (1984) explains that the tranquility here means calmness, which is equated with certainty and without doubt, because doubt is usually equated with anxiety. It is limited to not only physical tranquility but also soul tranquility.

PPPA Daarul Quran gives much attention to this program. Therefore, employees are always encouraged to remember Allah through a routine program of reading the Quran. This teaches all employees to continuously renew their spiritual relationship with Allah, bringing peace in facing all life's challenges. The Quran, as a revelation that comes directly from Allah, is a guide for life, a source of peace, and a reminder for a restless heart. When someone reads and reflects on the verses of the Quran, he is building a relationship with Allah that can bring peace and wisdom in facing problems.

Apart from that, PPPA Daarul Quran also implements leave rights, which provide uninterrupted rest time, besides family and employee gatherings, which are also very important to maintain employees' psychological condition and work spirit. Meanwhile, to resolve problems, PPPA Daarul Quran also has a problem-solving committee that involves several employees from all levels (from manager to director). This internal committee also discusses professional issues and personal and external factors that can influence work performance.

In line with Jauhar (2023), who states that *Hifdz an-Nafs,* or protection of the soul, is related to comprehensive human rights regulation. Islam forms its society by strengthening human rights and the first and most important right of Islam is the right to life. This is stated in the Quran Surah as-Syam verses 7-10, which emphasise the soul with what was made of it, its talents. Hence, recitation activities are provided to maintain the health and protect the souls of its employees so that prosperity is maintained in their lives, so that PPPA Darul Qur'an has fulfilled the *maqashid* of sharia by providing leave rights to its employees.

3.3 *Regulation to preserve one's intelligence (Hifdz al-'Aql)*

To maintain employee intelligence, PPPA Daarul Quran organises mandatory capacity-building activities. Every year, employee participation in capacity development activities is also one of the aspects audited and assessed in the KPI to ensure this is carried out well. This activity is carried out monthly for low and middle management levels.

The virtue of improving individual knowledge and skills to maintain intelligence is also stated in several verses of the Quran. One of them is Surah *al-Mujadila* verse 11.

> " ... Allah will raise in ranks those who believed from among you and those who are given knowledge. And Allah is All-Aware of what you do." (Quran 58:11)

This verse was interpreted by al-Zamakhsyari (1964) that Allah will elevate the status of those who believe by carrying out His commands and the commands of His Messenger, especially those who know. Abdullah bin Mas'ud RA also narrated another Hadith in the al-Zamakhsyari Book (1964), which the Prophet Muhammad SAW said related to the importance of gaining knowledge.

> "On the Day of Judgment, three groups will intercede: the prophets, then the scholars, and then the martyrs." (Hadith 1143)

Based on the interpretation above, there are several benefits from the training program owned by PPPA Darul Qur'an in the context of preserving one's intelligence (*Hifdz al-'Aql*). First, employees will gain an increase in their status by taking training and gaining knowledge. Knowledgeable employees will be appreciated and elevated in status by Allah and the people around them. Second, employees will be able to protect their minds from incorrect ideologies. Third, a person's mind will be honed and sharpened by training. Knowledgeable employees can think more critically, rationally, and objectively, which is part of protecting the mind itself.

This aligns with Busyro (2019), who stated that *Hifdz al-'Aql,* or protection of reason, protects humans' common sense. By guiding Muslims through the Quran and Hadith, Muslims are encouraged to use their minds to obtain guidance through creativity and work hard to make life more meaningful. In conclusion, PPPA Darul Quran's attempts have fulfilled the *maqashid* of sharia as it gives its employees the right to self-development to work better and more optimally in unleashing the potential of their abilities and synergising with the corporation for the better.

3.4 *Regulation to maintain the generation (Hifdz An-Nasl)*

PPPA Daarul Quran provides various facilities for families, such as health and education benefits, to maintain the continuity of the families of PPPA Daarul Quran employees. The benefits include wife and child allowances, or other education, health, and marriage benefits. Educational benefits can also be provided through scholarships for employees' children studying at the Daarul Quran Foundation, whether at Islamic boarding schools, full-day schools, or other educational institutions.

Meanwhile, health benefits provided by PPPA Daarul Quran are offered in several forms, either the compulsory BPJS or other additional health benefits for cases that BPJS does not cover. Second, PPPA Daarul Quran also provides allowance allocations for the families of

PPPA Daarul Quran employees affected by illness or death. Third, maternity benefits are provided for employees or partners of PPPA Daarul Quran employees if the person concerned cannot use BPJS facilities.

Those practices are in line with sharia principles, which place marriage as the only legitimate way to form a family and produce offspring, as the message in Quran Surah *al-Tahrim* verse 6.

> "O you who believe, protect yourselves and your families from a Fire whose fuel is people and stones ... " (Quran 66:6)

Al-Zamakhsyari (1964) interpreted the words 'protect yourselves and your families' in the context of protecting from committing sins and doing good deeds by educating and guiding families to do things for themselves. Another Hadith mentioned that each family member's responsibility is to protect the other.

So, the PPPA Daarul Qur'an program, which focuses on religious and moral education for children, families, and marriages, aligns with the concept of *Hifdz An-Nasl*. This program not only protects families from ignorance by educating and teaching religious knowledge, but also protects descendants from badness. Therefore, this program supports the realisation of a quality generation with good morals and knowledge that can bring happiness in this world and the hereafter.

3.5 *Regulation to protect the wealth (Hifdz al-Mal)*

As a *zakah* distributor (*amil*) institution, PPPA Daarul Quran employees do not receive a fixed income such as a salary, but instead receive *bisyarah* whose value can change. However, PPPA Daarul Qiran remains committed to meeting the UMR, as it is a mandatory regulation. PPPA Daarul Quran also complies with the rights regulated by the government, such as BPJS Employment and Health, and is obedient in paying taxes and *zakah*.

PPPA Daarul Quran employees' *zakah* is deducted directly from their gross salary, while annual *zakah,* such as *zakah* fitrah, is always encouraged to be paid through the PPPA Daarul Quran agency itself. For those whose income does not meet the *nishab*, the PPPA Daarul Quran policy provides concessions to pay *shadaqah.*

Hifdz al-Mal, or protection of assets, is related to protecting wealth that supports human life and achieving happiness in the afterlife. However, assets are not the main component, as they are tools used to meet needs. At a certain level of maintaining assets in Islam, namely *al-tahsiniyah*, it is encouraged for someone to give charity, even though their wealth does not reach the *nishab* and *haul.* Several verses in the Quran also require Muslims to spend their wealth in the way of worship and carry out the commands of Allah SWT and His Messenger as well as *muamalah* among fellow humans, through *zakah*, *infaq,* and *shadaqah.*

> "Believe in Allah and His Messenger and spend [in Allah's cause] out of what He has entrusted you with. For those among you who believe and spend (In the way of Allah), there will be a great reward." (Quran 57:7)

Those verses were interpreted by Ibn 'Asyur (1984) as a motivation to give *shadaqah.* Due to the interpretation, it shows that Allah commands us to give the best possessions for charity rather than deliberately choosing things that are no longer suitable. Based on the verses above, by providing various welfare benefits for its employees, PPPA Darul Qur'an will gain higher income and other benefits in general due to the charity program that has been implemented, which Allah has promised. The assets provided by PPPA Darul Qur'an are expected to be used wisely and within a positive framework, supporting the welfare of the people and employees, and protecting them from financial difficulties. Thus, implementing this policy is in line with Islamic principles, which emphasise the importance of maintaining and using assets correctly and beneficially. In conclusion, PPPA Darul Qur'an strives to integrate Islamic teachings into its operations, providing a living, protecting employee rights, and encouraging them to work in an environment that supports goodness and blessings.

3.6 *Regulation alignment with Sustainable Development Goals*

Based on Table 1, human Resource Management (HRM) regulations and practices in PPPA Daarul Quran address targets that are related to (1) reducing poverty; (2) reducing hunger and achieving food security; (3) ensuring good health and well-being, including providing health-care services and family planning education; (4) ensuring inclusive and equitable quality education, including promoting lifelong learning opportunities; (6) ensuring clean water and sanitation availability; (8) promoting productive employment (reducing the proportion of unemployment, uneducated and untrained youth), decent work and economic growth (level of economic productivity and per capita income), including promote development-oriented policies that support creativity, innovation, safe working environment and protection of labour rights; (10) promoting social inclusion for all religion and reducing inequality within countries; and (16) promoting peaceful and inclusive societies.

Table 1. The SDGs target achieved by HRM policies and activities in PPPA Daarul Quran.

Maqashid Sharia	Policies and Activities	The SDGs Target Achieved
Hifdz ad-Dien	Reciting the Quran together and praying Dhuha	Target 10; Target 16
Hifdz an-Nafs	Leave rights and Family/employee gathering	Target 8
	Internal consultant	Target 3
Hifdz al-'Aql	Capacity development or training	Target 4; Target 8
Hifdz An-Nasl	Family allowance	Target 8
	Health insurance	Target 3
	Marriage benefits	Target 3
	Scholarships	Target 4; Target 8
Hifdz al-Mal	Paying *zakah* and *shadaqah*	Target 1; Target 2; Target 3

Source: Authors (2025); United Nations (2015).

4 CONCLUSION

The concept of *maqashid* sharia could be applied to the Islamic corporation in several aspects of regulation. In order to maintain the employees' religion, corporation might enforce regulations that might encourage personal worship. As the protection of the soul is related to comprehensive human rights regulation, corporations should provide personal space and other facilities related to mental health protection. Islamic corporations also should encourage Muslims to use their minds to obtain guidance through creativity and work hard by providing self-development programs in order to gain better performance and work more optimally. Especially for Islamic philanthropy institutions, protecting wealth and maintaining the next generation could be fulfilled simultaneously. Assets are provisions for worship used to carry out the commands of Allah SWT and His Messenger as well as *muamalah* among fellow humans, through *zakah*, *infaq,* and *shadaqah*. This program could be distributed to support other Muslims' education as a *zakah* receiver (*mustahiq*), hand in hand to benefit employees to access the fund. Besides, other programs might also benefit employees by providing family support through family allowance, educational scholarship, and family-based health insurance, which are allocated from corporate budgeting.

In conclusion, the discussion offered above encourages the *maqashid* sharia framework as a baseline for human resource management regulation on Islamic companies. Several human resource management policies and programs that reflect *maqashi*d sharia principles should be regulated to ensure the Islamic corporate culture in the workplace. Those kind of regulations and programs implemented by PPPA Daarul Quran might be adopted and regulated as the requirements of Islamic corporations in the halal industry. However, based on the discussion above, further evaluation should be involved to measure the effectiveness of *maqashid* sharia practices to gain social welfare, namely *maslahah*, as well as social sustainability. As the corporate performance baseline, *maqashid* sharia practices should also be quantified so that their application can be more measurable.

REFERENCES

Aldulaimi, S. H. (2016). Fundamental Islamic perspective of work ethics. *Journal of Islamic Accounting and Business Research*, 7(1), 59–76. https://doi.org/10.1108/jiabr-02-2014-0006

Al-Ghazali. (2000). *Al-Mustashfa fi 'ilm al-Ushul*. Daar al-Kotob al-Ilmiyah.

Al-Zamakhshari, Abu al-Qasim Mahmud bin Amr bin Ahmad. (1964). *al-Kasyaf 'an Haqaiqi Gawamidhit Tanzil*. Kairo: Darul Ma'arif. Accessed January 1, 2025, from http://www.altafsir.com.

Ascarya, A., and Masrifah, A. R. (2023). Developing maqasid index for Islamic CSR: the case of Ummah's Endowment Fund in Indonesia. *International Journal of Islamic and Middle Eastern Finance and Management*, *16*(4), 835–855. https://doi.org/10.1108/imefm-12-2021-0474

Asutay, M., and Harningtyas, A. F. (2015). Developing Maqasid al-Shari'ah Index to evaluate social performance of Islamic Banks: A conceptual and empirical attempt. *International Journal of Islamic Economics and Finance Studies*, 1(1), 5–64.

Azwirman, A., Zulhelmy, Z., and Suryadi, A. (2019). Analysis of islamicity performance index on sharia banks in Indonesia. *International Journal of Innovation, Creativity and Change*, 10(3), 221–232.

Busyro, M. A. (2019). Maqashid al-syariah: pengetahuan mendasar memahami maslahah. *Prenada Media*.

DinarStandard. (2022). *State of The Global Islamic Report. New York: Dinar Standard*. Accessed June 11, 2023, from https://cdn.salaamgateway.com/reports/pdf/State+of+the+Global+Islamic+Economy+Report+2022_V2_Updated.pdf

Financial Services Authority. (2017). *Financial Services Authority Regulation: Criteria and Issuance of List of Syariah Securities*. Accessed June 11, 2023, from https://ojk.go.id/id/regulasi/Pages/Kriteria-dan-Penerbitan-Daftar-Efek-Syariah-.aspx

Hakim, L. (2016). Islamic organizational culture as an effort to improve performance. *Iqtishadia*, 9 (1). http://dx.doi.org/10.21043/iqtishadia.v9i1.955

Hasan, Z. (2020). Distribution of Zakat Funds To Achieve SDGs Through Poverty Alleviation In Baznas Republic of Indonesia. *ZKA International Journal of Zakat and Social Finance*, 25–43. DOI:10.51377/AZJAF.VOL1NO01.7

Hoque, N., Uddin, M., Islam, M. T., Mamun, A., Nazim Uddin, M., Ahmad, A., and Thowhidul Islam, M. (2023). Pursuing sustainable development goals through integrating the aspirations of zakah and CSR: evidence from the perspective of an emerging economy. *International Journal of Social Economics*, 50(9), 1232–1247. https://doi.org/10.1108/IJSE-09-2022-0600

Ibn 'Asyur, Muhammad Al-Tahrir. (1984). *Tafsir al-Tahrir wa al-Tanwir*. Dar At-Tunisiyyah. Accessed January 1, 2025, from http://www.altafsir.com.

Isman, Zaim, M. A., and Eldeen, A. B. (2024). Maqashid Sharia and Harmonizing Law in Indonesia: Impact for SDGs Global Context. *Studies in Systems, Decision and Control*, 745–759. https://doi.org/10.1007/978-3-031-50939-1_60

Jauhar, A. A. M. H. (2023). *Maqashid Syariah*. Amzah.

Kahf, M. (2013). Potential effects of zakat on government budget. *International Journal of Economics, Management and Accounting*, 5(1). https://doi.org/10.31436/ijema.v5i1.31

Karia, N., and Asaari, M. H. A. H. (2016). Halal value creation: its role in adding value and enabling logistics service. *Production Planning and Control*, 27(9), 677–685. https://doi.org/10.1080/09537287.2016.1166276

Rahman, F. K., Tareq, M. A., Yunanda, R. A., and Mahdzir, A. (2017). Maqashid Al-Shari'ah-based performance measurement for the halal industry. *Humanomics*, *33*(3), 357–370. https://doi.org/10.1108/H-03-2017-0054

Sudirman, S., Blongkod, H., Rasjid, H., Hamin, D. I., and Mahmud, M. (2023). Measuring the Performance of Sharia Commercial Banks in Indonesia Using the Islamicity Performance Index (IPI) and Maqashid Sharia Index (MSI). *Quality-Access to Success*, 24(196). DOI: 10.47750/QAS/24.196.14

Tarique, K. M., Islam, R., and Mohammed, M. O. (2021). Developing and validating the components of Maqasid al-Shari'ah-based performance measurement model for Islamic banks. *International Journal of Islamic and Middle Eastern Finance and Management*, *14*(2), 366–390. https://doi.org/10.1108/imefm-12-2018-0432

United Nations. (2015). United Nations Development - Sustainable Development Goals. Accessed January 2, 2025, from https://www.undp.org/?gad_source=1

Towards Resilient Societies: The Synergy of Religion, Education, Health, Science, and Technology – Rahiem (Ed.)

Integrating halal economy with animal tourism for sustainable development: The case of Ancol Faunaland

Roikhan Mochamad Aziz
Universitas Islam Negeri Syarif Hidayatullah Jakarta

ABSTRACT: This study analyses the relationship between the halal economy and animal tourism, focusing on their relevance to sustainability goals. Using a descriptive-analytical methodology, the research combines desk studies of religious scriptures, empirical data, and financial reports with an empirical survey. Data sources include books, journals, holy texts, and online media. The findings reveal that the halal economy supports animal tourism by promoting ethical and sustainable practices, addressing the sector's vulnerability to external shocks and its impact on sustainable operations. This study contributes to Sustainable Development Goal (SDG) 12 (Responsible Consumption and Production) and SDG 15 (Life on Land) by highlighting the importance of integrating ethical economic principles to ensure long-term resilience and animal welfare in tourism.

Keywords: animals, economy, Faunaland, halal, tourism

1 INTRODUCTION

The practice of animal ecotourism has been identified as a significant attraction for both domestic and international tourists. However, to ensure the continued viability of this tourism sector, it must be aligned with conservation efforts to maintain ecological balance. A notable example is turtle ecotourism at Faunaland Ancol in Indonesia, which integrates conservation and education to promote awareness of local and global biodiversity. However, the sustainability of such initiatives often depends on robust economic frameworks, including principles derived from the halal economy (Vu *et al.* 2022).

Recent studies have indicated a recovery in the tourism sector, with a strong focus on sustainability and resilience. Fátima Brilhante and Rocha (2023) found that Portugal's tourism sector had fully recovered by 2022. The recovery has been attributed to the implementation of strategies such as the promotion of domestic tourism, the adoption of digital technologies, and the integration of health and safety protocols. Sunder and Dixit (2023) recommend implementing destination marketing and domestic tourism promotion as effective recovery strategies. These strategies aim to revitalise the tourism industry and re-establish its preeminent position within the Portuguese economy. Furthermore, sustainable tourism development, which consists of building a safe tourism environment, diversifying and improving high-quality tourism products, human resources, and digital transformation is indispensable.

Despite the well-established role of halal economic principles in fostering ethical and sustainable tourism, their application to animal ecotourism remains underexplored. Moreover, the economic and operational challenges encountered by Faunaland Ancol during the pandemic period highlight the necessity to re-evaluate the resilience of prevailing tourism models. This study aims to address these gaps by examining how halal economy principles can enhance the sustainability and recovery of animal tourism in alignment with global sustainability goals, particularly SDG 12 (Responsible Consumption and Production) and SDG 15 (Life on Land).

 DOI: 10.1201/9781003645542-97

1.1 *Theory*

The environment of the fishery exerts an influence on fishery resources due to their inclusion within the human food chain. It is hypothesised that drivers such as economic expansion, population size and business efficiency will slow down the number of fish caught at sea, resulting in a decline (Rahmizal, n.d.). The unique social and economic relations of production and environmental conditions have historically shaped capture fisheries. Capture fisheries represent a distinct subfield of marine business, encompassing the capture, rearing and development of various fish species and other marine life in both coastal and oceanic environments. Assembly businesses centred on fish and other marine life are also important to this subsector. The Ministry of Maritime Affairs and Fisheries considers the growth of this subsector to be a benchmark for the achievement of national maritime economic development. This incentivises the standard development of fish creations. Frontier production function analysis is utilised to assess the efficiency of these devices in terms of price and technical efficacy.

Faunaland, located in Ancol, North Jakarta, is a 5-hectare zoo comprising terrestrial and aquatic areas. Drawing inspiration from Papua, a crossroads of Asian and Australian influences, Faunaland showcases the region's unique flora, fauna, and cultural arts. The park is home to various unique and rare species, with a particular emphasis on animals indigenous to eastern Indonesia and other regions.

In halal ecotourism, Faunaland plays a significant role in introducing turtles to the general public, particularly younger generations, who are the future custodians of the Earth's ecosystems. One turtle species that is a vehicle for visitors is the 70-year-old Aldabra turtle (Ancol official website). Faunaland Ancol possesses the potential to breed these turtles, which are distinguished by their rarity, uniqueness, and intrigue. This initiative has the potential to serve as a platform for education and tourism, which, as expected, is directly related to the intellectual aspects and morality.

The popularity of reptiles as pets has led to establishing a multi-billion-dollar industry. Their popularity as pets has led to their classification as one of the most traded animals, with evidence suggesting that more than a third of reptiles are actively traded on the Internet. However, in certain instances, previously unclassified species have gained a foothold in the pet trade. To protect vulnerable species from exploitation, the Convention on International Trade in Endangered Species (CITES) was established to manage and monitor the import and export of wild animals internationally. In addition, organisations such as the U.S. Fish and Wildlife Service's Law Enforcement Management Information System (LEMIS) monitor domestic imports. These commonly traded species are often non-endangered, captive-bred, and/or domestically purchased, and as a result, they are largely unregulated. This has resulted in much of the reptile trade being unknown, incoming, untraceable, and difficult to track.

The reptile and amphibian trade has a generally unfavourable reputation, attributable to concerns regarding animal welfare, depletion of local populations, the spread of invasive species and human health concerns. The challenge of assessing the impact of the pet trade on the status of natural populations is exacerbated by the fact that many widely traded species are found in unprotected and/or uncontrolled areas worldwide. Furthermore, in countries that import exotic animals and have hospitable climates, imported animals can become pests and hazards to local fauna.

2 METHODOLOGY

Researchers utilise morphometric data, which is analysed using descriptive anatomy, encompassing quantitative and standardised patterns. Morphology observed includes the shape of the head, tail, limbs, and plastron. The morphometric variables measured included carapace length and width. The qualitative descriptive method was used to describe the species based on morphological characteristics.

This study employed a quantitative paradigm with a comparative research design. The quantitative paradigm is utilised when researchers use a commonly accepted meaning, where a theory possesses certain characteristics, such as a series of variables, definitions, or postulates that are interconnected and form a systematic view of a phenomenon to explain scientific phenomena by determining the relationship between variables. Comparative research is a research design that aims to identify cause-and-effect relationships between two or more variables through correlational studies. On the other hand, explanatory research aims to understand the relationship between two or more concepts or variables of a social phenomenon by considering religious values.

This study uses the documentation method as a data collection method, which refers to literature studies, literature exploration, and financial reports published by PT Pembangunan Jaya Ancol Tbk (PJAA). This study uses secondary data obtained from the financial statements of PT Pembangunan Jaya Ancol Tbk (PJAA) during the 2018-2022 period. The secondary data needed in this study are as follows: Input data in the form of income data, such as revenue and profit, and output or expenditure data, such as operational costs, financing, and taxes.

The Hahslm Reflexivity Dynamics (Hefdyn) method is an application that aims to enhance analytical capabilities in identifying and integrating factors that affect reflexivity in diverse units. Hefdyn facilitates the integration of a more profound understanding of the relationships between units, as well as weight values, thereby enabling the assessment of both internal and external reflexivity dynamics. Hefdyn measures reflectivity by comparing input reflectivity (Appreciation) with depreciation reflectivity (Depreciation). A high Hefdyn score indicates efficient resource utilisation, whereas a low score signifies inefficient and suboptimal resource management. Incorporating religious values or a sense of worship into the assessment process creates a system that combines empirical and non-empirical, material and immaterial, and financial and non-physical aspects. The concept under discussion here is that within the framework of universal religion, the distinction between empirical and non-empirical values is not logical and so there is an attempt to combine the two (Aziz 2020).

The Hefdyn model, initiated by Roikhan in 2023, underscores the significance of acknowledging the intricate interplay among variables as opposed to prioritising the quality of the units under comparison. This approach is distinguished by the integration of religious values within financial evaluation systems, informed by prior experience. Hefdyn is a methodological framework that is utilised to assess the relative reflexivity of analogous units, designated as decision-making units (DMUs). It prioritises technical and allocative efficiency within the context of a distinct industry environment.

3 ANALYSIS

PT Pembangunan Jaya Ancol Tbk (PJAA) is a company engaged in tourism that specifically focuses on the management and development of the Ancol area in Jakarta. The company has a long history, commencing on a certain date with an initial focus on developing the Ancol tourist area. The company's competitive advantage stems from its extensive and varied management of the Ancol tourist area. The company's portfolio encompasses many facilities, including amusement parks, beaches, water sports, wildlife sanctuaries, and other entertainment venues, appealing to diverse visitors with varied interests and recreational activities.

The company is headquartered in Jakarta and operates various facilities in the Ancol area, and it may have branches in strategic locations in Indonesia. Research was conducted to determine the effect of prayer indicators on the company's financial performance. The secondary data utilised in this study were sourced from PJAA's financial statements and obtained from the company's official sources. The present study focuses on analysing the effect of prayer levels on the financial performance of PT Pembangunan Jaya Ancol Tbk, to identify a relationship between religious factors and company performance in the tourism and entertainment sector.

In this study, the Hefdyn method was utilised. The research was obtained from the value of prayer or worship routinely carried out at PT Pembangunan Jaya Ancol Tbk. This was obtained from the financial statements in the company's operating profit and financial cost section from 2018 to 2022. These figures were used as the basis for the analysis reported using the Hefdyn calculation in this study. The following formula can be used to calculate Hefdyn, which uses the value of prayer in evaluating and examining the development of congregational prayer for employees of PT Pembangunan Jaya Ancol Tbk. for the last five years:

$$\text{Reflexivity Value Formula : Result(Change Appreciation)}$$

$$= (x2 - x1)/x1(En1, 2, 3, 4)/4 \text{ Result(Change Depreciation)}$$

$$= ((x1 - x2)/x2)((En1, 2, 3, 4)/4)$$

The data presented herein demonstrates that the appreciation reflectivity value is not constant; it fluctuates. During the 2018-2019 period, the reflectivity value was recorded at 42%, whereas in 2020-2021 it diminished by 5%, representing the smallest recorded value. Thereafter, in 2022, it underwent a substantial increase of 279%. The substantial increase observed in 2022 was attributable to the reopening of Faunaland tourism following the two-year lockdown, which led to a decline in reflectivity. This phenomenon directly impacts the activity value of worship in the Faunaland Ancol area, given the absence of visitors and employees, leading to a decline in reflectivity during the 2020-2021 period.

A number of factors contributed to the observed changes in reflectivity and the effectiveness of depreciation, with 2020-2021 being a year of total lockdown activities. This phenomenon can be attributed to the persistent operational expenses that continued to accrue without generating revenue from visitors, resulting in a substantial budget deficit. The data indicates a decline in the reflectivity and effectiveness of the value of salat at PT Pembangunan Jaya Ancol Tbk (PJAA) during the pandemic. This decline has led to sub-optimal cultivation and maintenance practices for turtle animals at Faunaland, further exacerbated by the budget deficit.

This analysis emphasises the significance of incorporating spiritual values into organisational performance metrics. This finding is supported by several relevant references related to the influence of religious values on productivity. Wijayanti and Nurhayati (2021), Robbie and Sayyaf (2022), Akaba and Abiola Monday (2023) concur that religiosity exerts a favourable influence on work productivity and employee performance. The role of religiosity is to realise quality work, namely the fulfilment of quality objectives at every stage of the production process so that the final product produced is by what the company has set (Widodo *et al.* 2020). Furthermore, the role of religious values in enhancing workers' motivation has been demonstrated in numerous studies (Rahul 2024). This enhances their work ethics and organisational commitment, significantly driving performance and productivity.

4 CONCLUSION

The government must establish close communication with tourism organisations, especially in terms of implementing the regulations issued. Furthermore, tourism industries such as Faunaland Ancol must also pay close attention to the animals they keep and the costs of food and clothing. The financial impact of the pandemic and subsequent lockdown on Faunaland's budget has led to a significant deficit, necessitating the continuation of the operational costs for the animals. However, the absence of visitors has resulted in a decline in ticket sales, leading to concerns regarding the sustainability of the tourism industry, particularly in maintaining the well-being of the existing animals. It is acknowledged that Faunaland is home to various species of animals, including turtles, which are captured in the wild.

A thorough examination of the available data reveals fluctuations in the effectiveness and reflectivity values for depreciation from year to year. From 2020 to 2022, a substantial decline in effectiveness and reflectivity was observed, attributable to the repercussions of the pandemic. However, in 2023, a resurgence in effectiveness and reflectivity was observed, attributable to the reopening of tourism channels that commenced in 2022. The revitalisation of the tourism industry has contributed to the increase in Sunway Berhad's operating expenses, which has resulted in the corporation's financial performance being bolstered, as evidenced by the increase in revenues from the beginning of 2022 onwards. Concurrently, the appreciation exhibited an exponential growth trajectory from 2019, though it slightly declined in 2023. This phenomenon can be attributed, at least in part, to the repercussions of the pandemic, particularly the imposition of stringent lockdown measures that precipitated an ongoing accumulation of operational expenses despite a concomitant dearth of adequate revenue generation across the corporate sector. This had a deleterious effect on the effectiveness and reflectivity of the corporation, as well as on its activities in the hotel and tourism business.

This study contributes significantly to the literature on the nexus between the halal economy and tourism, focusing on animal tourism. In addition, this study underscores the vulnerability of animal tourism to exogenous shocks and its ramifications for sustainable practices. It also contributes to SDG 12 (Responsible Consumption and Production) and SDG 15 (Life on Land) by emphasising the importance of integrating ethical economic principles with sustainable tourism to ensure long-term resilience and animal welfare.

REFERENCES

Akaba, J., and Abiola Monday. (2023). Faith and Workplace Behaviour: The Effect of Religion on Employees' Productivity among Non-Teaching Staff at the University of Benin, Benin City, Nigeria. *International Journal of Research Development*, 14(1), 1–22.

Aziz, R. M. (2020). *Universe Created by Allah Equation of Hahslm 472319 as Big Bang Concept and Eid Pray Symbols*. (Iconist 2019), 1–4.

Fátima Brilhante, M. de, and Rocha, M. L. (2023). COVID-19 Pre-pandemic Tourism Forecasts and Post-pandemic Signs of Recovery Assessment for Portugal. *Research in Globalization*, 7, 1–15.

Rahmizal, M. (n.d.). *Analysis of Indonesia Marine Fisheries with Economic Growth, Population and Effort Effectiveness*. 1(1).

Rahul, N. A. (2024). The Role of Christian Values in Enhancing Efficiency and Effectiveness in Business Administration: A Project Management Perspective. *American Journal of Industrial and Business Management*, 14(04), 603–618.

Robbie, R. I., and Sayyaf, R. T. F. (2022). Impact of Religiosity, Work-related Stress on Job Performance and Workload as Moderating Variable. *International Journal of Social Science and Business*, 6(2), 156–164.

Roikhan, MA. (2018). *Reflexivity of Food and Plants on Islam With Theory H*. Atlantis Press.

Sunder, S., and Dixit, S. (2023). Post-Pandemic Tourism: An Article on the Recovery Strategies. *Integrated Journal for Research in Arts and Humanities*, 3(2), 174–179.

Vu, H. D., Nguyen, A. T. N., Nguyen, N. T. P., and Tran, D. B. (2022). Impacts and restoration strategy of the tourism industry post-COVID-19 pandemic: evidence from Vietnam. *Journal of Tourism Futures*, 1–13.

Widodo, A. A., Fachrurrozi, A., Chotib, M. N., and Hamzah, S. (2020). The Role of Religiosity in Increasing the Productivity of Steel Industry Employees in Gresik-East Java: Study at P.T. Semenanjung Pangeran Agung. *Proceedings of the 1st International Conference on Recent Innovations*, 425–429.

Wijayanti, R., and Nurhayati, M. (2021). The Effect of Employee Religiosity and Organizational Culture on Employee Performance with Intrinsic Motivation as a Mediation Variable in Bank Syariah Indonesia (BSI) Ex BNI Syariah Head Office. *European Journal of Business and Management Research*, 6(5), 100–106.

Towards Resilient Societies: The Synergy of Religion, Education, Health, Science, and Technology – Rahiem (Ed.)

Shari'a bylaws and multiculturalism in Indonesia: Legal framework and political dynamics

Asep Saepudin Jahar
Universitas Islam Negeri Syarif Hidayatullah Jakarta, Tangerang Selatan, Indonesia
ORCID ID: 0000-0002-4957-5273

Kamal Aly
An-Nikmah Al-Islamiyah Phnom Penh, Cambodia

ABSTRACT: In the Indonesian context, Shari'a operates on two levels: as a framework for personal religious observance and as a tool for political identity (Islamism). The former involves individual compliance with religious obligations, such as fasting during Ramadan and performing Hajj, while the latter seeks to regulate all aspects of governance, potentially replacing secular laws, including Pancasila. This paper examines Shari'a-based regional bylaws in Indonesia through two lenses: legal content and the politicisation of Shari'a. By analysing these regional regulations, we demonstrate how they reflect a bureaucratic approach to Shari'a implementation and contribute to the shariatisation of political systems. The paper argues that regulating personal religious practices remains compatible with democracy as long as it aligns with Pancasila's foundational values. Ultimately, this study highlights the intersection between religious law, governance, and Indonesia's commitment to pluralism.

Keywords: democracy, Indonesia, Islamism, legal framework, Pancasila, politicisation, shari'a bylaws

1 INTRODUCTION

The Reformation Era brought greater public participation and political freedom and fuelled religious fundamentalism. The rise of shari'a bylaws, particularly in Muslim-majority regions, is a byproduct of democracy. While shari'a aspirations date back to the 1940s constitutional debates, the term was ultimately omitted due to political disputes, replaced with "the Indonesian state is based on the belief in the One, Supreme God with the obligation of the adherents of Islam to implement the shari'a." (Hooker and Saikal 2004).

Under Soeharto's rule, shari'a aspirations were suppressed due to strict political control. However, after his regime, regional bylaws incorporating shari'a were proposed. These bylaws, drawn from Islamic jurisprudence, conflict with civic values and human rights. Their negative impacts include: (1) a formalistic, symbolic approach that sees shari'a as a solution to social issues based on the Qur'an and Sunna, and (2) the perception that state laws fail to address societal problems, leading to demands for shari'a enforcement. This trend threatens minorities and even some Muslims, especially those whose actions contradict shari'a. Additionally, the push for shari'a bylaws reflects resistance to Western influence, modernisation, and economic domination.

1.1 *Shari'a within secular state*

For most Muslims, life is a dedication to God, with deeds judged in the afterlife. Islamic teachings in the Qur'an and Sunna serve as life's guidance. The interpretation and implementation of shari'a vary among Muslim groups. Khalid Mas'ud identifies two trends:

DOI: 10.1201/9781003645542-98

revivalists, who seek to restore early Islamic laws, and modernists, who advocate reinterpretation in response to contemporary contexts (Masud 2005). A third group, traditionalists, aligns shari'a with fiqh from the Hanafi school. Revivalists take a militant stance, calling for a return to the Qur'an and Sunna while rejecting fiqh (Masud 2005).

Regarding shari'a and politics, Indonesian Muslim intellectuals are divided: some argue shari'a should govern the state, placing sovereignty in divine hands (Effendy 2006), while others believe Islam lacks a defined political system, making state governance a human construct. The former group's militancy has driven efforts to implement shari'a bylaws, despite their questionable relevance.

The push for political Islam dates back to 1945 constitutional debates. The BPUPKI (the Investigating Committee for the Preparation of Independence) called 'Doku-ritsu Zyunbi Tyoosakai' or 'Badan Penyelidik Usaha Persiapan Kemerdekaan' deliberated on whether Islam should be the state ideology or if a secular framework was preferable (Madinier 2007). The resulting Jakarta Charter initially required Muslims to observe shari'a, but Christian objections led to its revision as "Ketuhanan Yang Maha Esa". Some Muslim groups continue advocating its reinstatement.

Under Soeharto (1966–1998), non-Pancasila ideologies were banned, relegating Islam to private life. Muslims faced political distrust and exclusion (Effendy 2006). However, suppression did not eliminate the desire for shari'a, instead fostering latent militancy. In the post-Soeharto era, political Islam resurfaced, partly due to state failures in governance. Shari'a bylaws emerged as a perceived solution to social problems, despite their discriminatory effects in a pluralistic society (Taylor *et al.* 1994).

Beyond religious motives, political opportunism drives the shari'a agenda. Azra notes that some politicians exploit shari'a to gain local power after failing at the national level (Hooker and Saikal 2004). The bylaws, rather than ensuring effective governance, serve as political tools. Human rights activists and scholars criticise their lack of feasibility and incompatibility with multiculturalism. Hardline groups like Laskar Jihad justify violence and moral policing—raiding nightclubs, discotheques, and alleged prostitution houses—in the name of shari'a. These practices highlight the dangers of shari'a bylaws.

1.2 *Debating Shari'a*

When Soeharto rose to power, many Muslims saw hope for Shari'a implementation and the recognition of Muslim political parties like Masyumi. However, like Soekarno, he suppressed political Islam. His depoliticisation strategy included (1) secularising politics under Pancasila and (2) merging political parties into three: Golkar, PDI, and PPP. The PPP, tied to Muslim organisations like NU and Muhammadiyah, was required to adopt Pancasila as its ideological foundation (Jahar 2006). Under the New Order, Islam's role in politics declined, shifting focus to social and educational fields.

Shari'a, meaning "the way" in Arabic, was as vital as water in early Islam (Schimmel and Heiler 1994). In Islamic discourse, it is linked to din (religion) and serves as a guiding legal system based on the Qur'an and Sunna. In Indonesia's pre-reformation era, debates over Shari'a's role in governance remained unresolved. Muslim intellectuals split into two groups: traditionalists, who see Shari'a as fixed fiqh, and modernists, who advocate for reinterpretation. The latter reject classical jurisprudence, arguing it fails to address contemporary challenges (Masud 2005).

During the 1945 constitutional debates, BPUPKI was divided: Muslims sought Shari'a as state law, while secularists opposed it. A compromise led to the Jakarta Charter, which initially mandated Shari'a for Muslims but was later replaced by "Ketuhanan Yang Maha Esa" due to Christian objections (Madinier 2007). Some Muslim groups continue efforts to restore the original wording.

Soeharto's 1966 rise to power, following the communist purge, solidified Pancasila as the sole ideology. Political Islam was repressed, with religious symbols banned from politics (Effendy 2006). However, this strengthened underground Islamic militancy, which resurfaced after Soeharto's fall. The failure of state governance fueled calls for Shari'a bylaws, seen as a solution to social problems (Taylor *et al.* 1994). Yet, these bylaws often violate multicultural values.

Beyond religious motivations, Shari'a is a political tool. Some politicians exploit it after failing to gain power at the national level (Hooker and Saikal 2004). Human rights activists criticise these bylaws as impractical and discriminatory. Hardline groups like Laskar Jihad enforce them through violent raids on clubs and suspected prostitution venues, showcasing Shari'a's potential dangers.

Muslim scholars remain divided on Shari'a's application. Some, like Kasman Singodemidjo, proposed an Islamic Council to regulate aspects like marriage law, zakat, and Islamic education. Others, such as Mohammad Saleh Suaidy, viewed Shari'a implementation as a gradual legal reform (Madinier 2007).

Shari'a interpretations vary across Indonesia. Medan and Padang favour persuasion (da'wah), while Aceh enforces it legally. In Djeunieb, authorities tracked Friday prayer attendance, warning absentees and imposing social sanctions (Al-Banjari 2006). In South Sulawesi, even Christians were required to attend church, while Bima banned non-Islamic religious propagation.

The central government repeatedly intervened to halt regional Shari'a enforcement, notably in Aceh, Riau, Jambi, and West Nusa Tenggara (Muhammad and Azra 2003). Yet, Muslim demands persisted, with Shari'a understood in three ways: (1) as fiqh (ritual practices), (2) as Islamic values (justice, democracy, equality), and (3) as positive law governing society (Mujani 2005b).

Traditionalists emphasise Shari'a as ritual law, aligning with Islamic schools (pesantren) (Bruinessen 1995). Surveys show most Muslims (28.7%) see Shari'a as ritual practice, while 16.3% support its formalisation into bylaws (PPIM-UIN Jakarta 2004). However, no direct correlation exists between Shari'a formalisation and societal improvement.

Conservative groups dominate the push for Shari'a bylaws. Since the 1950s, movements like Darul Islam (DI) and post-New Order Islamic organisations have promoted Shari'a as a solution to social issues (Jamhari 2004). Majlis Mujahidin Indonesia (MMI) supports nationwide Shari'a enforcement, arguing Islam governs all aspects of life and historically ensures social justice. They also view globalisation as a threat to Islamic values (Mujahidin 2001).

For MMI, Shari'a is both a religious duty and a path to an Islamic state. Though they remain politically unaffiliated, they seek to restore the Jakarta Charter in UUD 1945. Laskar Jihad shares similar views but rejects reinterpretation, advocating strict adherence to Qur'anic law, including hudud punishments like hand-chopping for theft (Abuza 2007).

Another key Shari'a proponent is FPI (Front of Defenders of Islam), which adopts a direct-action approach. Unlike MMI and LJ, FPI enforces moral codes through vigilante tactics, targeting nightclubs, liquor sales, and alleged immoral acts.

1.3 *Shari'a as political ideology*

The push for Shari'a bylaws is not new; it predates Indonesia's independence. Their reintroduction post-Soeharto stems from political failures and the ideological rivalry between secularism (Pancasila) and Islam. Under Soeharto, Islam was used to bolster the regime, not to promote political Islam. A key example is Marriage Law No. 1/1974, which sought legal uniformity but faced opposition from Muslim groups like NU and Muhammadiyah for contradicting Islamic doctrines (Lukito 2007). The PPP (United Development Party) contested the law, leading to revisions on interfaith marriage.

With post-New Order political reforms, regional autonomy (Act No. 25/2001) enabled Shari'a bylaws, particularly in Bireun, Tasikmalaya, Bulukumba, Bima, and Tangerang (Abbas 2012). This shift, hailed as democratic progress, led some regions to replace or complement secular laws with Shari'a, often enacted hastily without assessing societal impact. The bylaws, while symbolic of Islamic revival, have undermined multiculturalism, affecting religious minorities, women, and civil liberties (Abbas 2012).

Public opinion on Shari'a bylaws is divided. Economically, their impact is largely negative—27.7% of Muslims felt worsened financial conditions, while only 26.4% saw improvements. A majority (44.5%) reported no change, confirming no direct correlation between Shari'a enforcement and economic benefits.

However, in security matters, Shari'a laws received more support. 55.1% of respondents said religious adherence improved, and 34.3% felt safer due to Shari'a enforcement, often overseen by Shari'a police. Yet, concerns persist over justice and individual rights—6.4% viewed the enforcement as coercive, emphasising its top-down nature rather than persuasion-based governance.

A key issue is the formalistic approach to Shari'a, focusing on symbolism over public welfare. Many Muslim scholars and citizens lack a deep understanding of Shari'a's philosophical foundation (Roy 1994). Consequently, Shari'a bylaws fail to ensure equity, often contradicting Indonesia's constitutional commitment to multiculturalism.

Table 1. Shari'ah bylaws in 5 regions.

No	District	Content	Decree of Issuance
1.	Tasikmalaya	Requirement for students to dress modestly	Circular Letter no. 451/SE/04/ Sos/1
2.	Bulukumba	Muslim dress	Perda no. 5/2003
		Prohibition on sale and distribution of liquor	Perda no. 3/2002
		Zakat, charity and donation	Perda no. 2/2003
		Capable of Reading Qur'an for Students and brides and grooms	Perda no. 6/2005
3.	Tangerang	Prohibition on Prostitution	Perda no. 7/2005
		Prohibition on Sale and distribution of liquor	Perda no. 8/2005
4.	Aceh	Incorporation of aspects of theology, rituals for	1. Qanun 11/2002
		Islamic interest into shari'a bylaws	2. Qanun 12/2003
		Prohibition on liquor, gambling, khalwat	3. Qanun 13/2003
		Management of zakat	4. Qanun 14/2003
			5. Qanun 7/2004
5.	West Nusa Tenggara	Zakat administration for professional workers	Perda no. 9/2002 and Decree of District Head 17/2003

2 RESULT AND DISCUSSION

2.1 *Shari'a bylaws and religious freedom*

The UN's covenant on human rights guarantees civil liberties, including freedom of expression, religion, and protection from state intervention. However, Shari'a's compatibility with these principles remains debatable. A study shows Shari'a laws limit tolerance, with 68.4% rejecting religious propagation outside Islam and 50.4% opposing diverse interpretations. Such beliefs shape social norms, often restricting pluralism. Ahmadiyah communities, for instance, have faced violence in West Java and West Nusa Tenggara due to their theological differences Komisi Nasional Hak Asasi Manusia 2006).

In Aceh, Qanun bylaws recognise only Sunni Islam, excluding other interpretations (Salim 2008). This contradicts Indonesia's 1945 Constitution, Article 29, which guarantees religious freedom. Similarly, Article 28E grants individuals the right to choose their faith, yet Aceh's laws enforce Sunni conformity, violating these rights. 71.7% of residents support adherence to Shari'a-based doctrines, penalising those with differing beliefs. Only 28.3% accept religious diversity. Christian gatherings also face opposition, with 40.4% advocating bans, though 59.6% support their right to worship. These figures indicate Shari'a bylaws undermine multiculturalism and fail as a viable societal solution.

A West Nusa Tenggara case illustrates forced religious taxation under Shari'a bylaws. While zakat benefits the poor, its compulsory enforcement by local government, deducting 2.5% from civil servants' salaries, led to public rejection (Salim 2008). The coercive implementation of Shari'a regulations raises concerns over individual rights and state overreach.

2.2 *Freedom of repression and violence*

Shari'a bylaws are not universally supported, with concerns over their negative impact on rights and security. In some regions, 27.2% reported disruptions of activities during Friday prayers, while 16.6% witnessed penalties for violating Shari'a laws. Women and tourists face particular challenges, with 39.2% of women avoiding public spaces after 10 PM due to fear of being labelled prostitutes and 32.9% worrying about militant sweeps. Restricting women's movement based on subjective morality violates personal freedoms, reinforcing fear and insecurity. This policy contradicts constitutional guarantees of public welfare and civil liberties (Muhtadi 2003). Shari'a bylaws, therefore, fail to protect both men's and women's rights.

2.3 *Women civil rights*

Shari'a bylaws restrict women's rights, particularly in public spaces. While state security issues also contribute, Shari'a enforcement worsens conditions. A survey across five Indonesian cities shows 17.9% of women were forced to wear the jilbab, and 40.1% experienced sweeps, particularly in areas suspected of prostitution. Additionally, 38.1% reported raids on unmarried couples in secluded places, highlighting the complexity of Shari'a implementation.

In Aceh, Shari'a bylaws mandated jilbab for all women, leading to hair-cutting punishments for noncompliance, carried out by religious school students assigned to traffic duty (Muhtadi 2003). Though no official group claimed responsibility, such incidents instilled fear among women.

These cases challenge civil rights in a multicultural democracy, where men and women should be treated equally. In Tangerang, a woman was arrested on suspicion of prostitution for waiting for her husband after 10 PM—a regulation under Shari'a bylaws that prohibits women from being alone in public at night (Abbas 2012). Such laws face widespread criticism, particularly when misuse and discrimination are exposed.

Public opinion on Shari'a and civil liberty is divided. 1.6% of respondents reported experiencing violence for neglecting prayers, a small but telling indicator of coercive enforcement. However, Article 29 of the 1945 Constitution guarantees religious freedom, stating each citizen has the right to faith and worship.

2.4 *From secular to islamic ideology*

The imposition of Shari'a bylaws is linked to state failures and foreign influences, particularly from the West. The U.S., especially post-9/11, played a role, as its attacks on Iraq and Afghanistan fueled Islamist opposition. Protests erupted against U.S. policies, seen as unjust towards Muslims. Western economic and cultural dominance further encouraged Islamists to push Islam as the foundation of governance, viewing it as a counterforce to Western "colonisation" (Turner 2002). Similar to Egypt's Islamism under Sadat, Indonesian Islamists sought to replace secular laws with Shari'a, despite different political conditions.

Though Pancasila is widely accepted, Islamist groups argue only Islam can solve state failures. They reject Western-style modernisation, prioritising capitalism and free markets, leaving peasants vulnerable. As a result, Shari'a is promoted as an alternative political system.

Anti-Americanism also stems from religious factors, not just foreign policy. Indonesian Islamists associate the U.S. with Christianity, believing its policies favour Christians. A study confirms America is largely rejected by Islamists in Indonesia (Mujani 2005a), reflecting a broader rejection of Western values.

3 CONCLUSION

The Shari'a debate in Indonesia involves two perspectives: ritual and ideological (Islamism). The ritual includes fasting, hajj, and prayer, all protected under Pancasila. These personal religious acts are separate from Islamism, which seeks to impose Shari'a bylaws, replacing secular laws. Muslims supporting Shari'a bylaws see them as both a social solution and a religious obligation. However, formalistic enforcement threatens multiculturalism, as criminal laws in Islam are subject to interpretation based on different jurisprudential schools. The viable solution is to promote Islamic values—justice and equity—that align with multicultural principles rather than strict Islamism.

The demand for Shari'a bylaws is driven by two factors: domestic failures (corruption, economic crisis) and foreign influence (Western modernisation, culture, and economy). These pressures have intensified calls for Islamic governance as an alternative system.

REFERENCES

Abbas, A. S. (2012). Shariah in the Era of Regional Autonomy in Indonesia (الشريعة الإسلامية فى عصر الاستقلال الذاتى باندونيسيا). *AHKAM: Jurnal Ilmu Syariah*, 12(2). https://doi.org/10.15408/ajis.v12i2.973

Abuza, Z. (2007). Laskar Jihad: Islam, Militancy, and the Quest for Identity in Post– New Order Indonesia. *Contemporary Southeast Asia*, 29(1), 196–200. https://doi.org/10.1355/CS29-1I

Al-Banjari, M. (2006). *Feeling Threatened: Muslim-Christian Relations in Indonesia*'s New Order. https://www.academia.edu/17201350/Feeling_Threatened_Muslim_Christian_Relations_in_Indonesias_New_Order

Bruinessen, M. V. (1995). *Kitab Kuning, Pesantren Dan Tarekat*. Bandung: Mizan, 198.

Effendy, B. (2006). *Islam in the Contemporary Indonesian Politics*. Ushul Press.

Hooker, V., and Saikal, A. (Eds.). (2004). Political Islam in Post-Soeharto Indonesia. *In Islamic Perspectives on the New Millennium* (pp. 133–149). ISEAS Publishing. https://doi.org/10.1355/9789812305367-010

Jahar, A. S. (2006). The clash of Muslims and the state: Waqf and zakat in post independence Indonesia. *Studia Islamika*, 13(3).

Jamhari, J. (2004). *Gerakan Salafi Radikal di Indonesia*. RajaGrafindo Persada. https://cir.nii.ac.jp/crid/1130282271381291648

Komisi Nasional Hak Asasi Manusia. (2006). *Laporan Tahunan: Komisi Nasional Hak Asasi Manusia* 2006.

Lukito, R. (2007). Sharī'ah and the Politics of Pluralism in Indonesia. *Studia Islamika*, 14(2), Article 2. https://doi.org/10.15408/sdi.v14i2.545

Madinier, R. (2007). Mujiburrahman, Feeling Threatened. *Muslim-Christian Relations in Indonesia's New Order*. Archipel, 73(1), 248–252.

Masud, M. K. (2005). Communicative Action and the Social Construction of Shari'a in Pakistan. In A. Salvatore and M. LeVine (Eds.), *Religion, Social Practice, and Contested Hegemonies* (pp. 155–179). Palgrave Macmillan US. https://doi.org/10.1057/9781403979247_7

Muhammad, H., and Azra, A. (2003). *Revitalisasi syari'at Islam di Aceh: Problem, solusi, dan implementasi*. (No Title). https://cir.nii.ac.jp/crid/1130282271816802944

Muhtadi. (2003). Syariat Islam: Pandangan Muslim Liberal (Jakarta). *Sembrani Aksara Nusantara*. //digilib.uki.ac.id%2Findex.php%3Fp%3Dshow_detail%26id%3D32472%26keywords%3D

Mujahidin, M. P. M. (2001). *Mengenal Majelis Mujahidin*. Markaz tsb.

Mujani, S. (2005a). Anti-Americanism in Contemporary Indonesia. *Studia Islamika*, 12(2), Article 2. https://doi.org/10.15408/sdi.v12i2.588

Mujani, S. (2005b). *Benturan Peradaban: Sikap dan Perilaku Islamis Indonesia terhadap Amerika Serikat*. (No Title). https://cir.nii.ac.jp/crid/1130000796242778624

PPIM-UIN Jakarta. (2004). *Survey on Islam and consolidation of democracy in Indonesia*.

Roy, O. (1994). The failure of political Islam. Harvard University Press. https://books.google.com/books?hl=en&lr=&id=YEGw8S3ovaoC&oi=fnd&pg=PR7&dq=Oliver+Roy,+The+Failure+of+Political+Islam,+trans.+Carol+Volk+(Cambridge:+Harvard+University+Press,+1990).&ots=EDl2wNZ50N&sig=PFh8dWiphVheXUfm5aW8w7wn854

Salim, M. A. (2008). The Sharī'ah Bylaws and Human Rights in Indonesia. *Studia Islamika*, 15(1), Article 1. https://doi.org/10.15408/sdi.v15i1.536

Schimmel, A., and Heiler, F. (1994). Deciphering the signs of God: A phenomenological approach to Islam. *State University of New York Press*. https://ixtheo.de/Record/131645706

Taylor, C., Appiah, K. A., Habermas, J., Rockefeller, S. C., Walzer, M., and Wolf, S. (1994). *Multiculturalism: Expanded Paperback Edition (REV-Revised)*. Princeton University Press. https://doi.org/10.2307/j.ctt7snkj

Turner, P. B. S. (2002). *Orientalism, Postmodernism and Globalism*. Routledge. https://doi.org/10.4324/9780203427255

Towards Resilient Societies: The Synergy of Religion, Education, Health, Science, and Technology – Rahiem (Ed.)

Author index